Foluso Ladeinde

Applications of Complex Variables

Also of interest

Topics in Complex Analysis
Dan Romik, 2023
ISBN 978-3-11-079678-0, e-ISBN (PDF) 978-3-11-079681-0

Integral Transforms and Applications
Nita H. Shah and Monika K. Naik, 2022
ISBN 978-3-11-079282-9; e-ISBN 978-3-11-079285-0

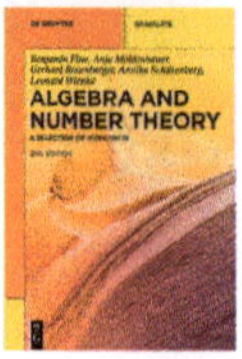

Algebra and Number Theory.
A Selection of Highlights
Benjamin Fine, Anja Moldenhauer, Gerhard Rosenberger, Annika Schürenberg and Leonard Wienke, 2023
ISBN: 978-3-11-078998-0; e-ISBN 978-3-11-079028-3

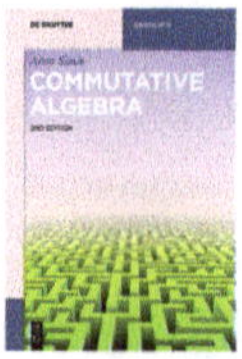

Commutative Algebra
Aron Simis, 2023
ISBN 978-3-11-107845-8, e-ISBN (PDF) 978-3-11-107878-6

Linear Algebra.
A Minimal Polynomial Approach to Eigen Theory
Fernando Barrera-Mora, 2023
ISBN 978-3-11-113589-2, e-ISBN (PDF) 978-3-11-113591-5

Foluso Ladeinde

Applications of Complex Variables

Asymptotics and Integral Transforms

DE GRUYTER

Author
Prof. Foluso Ladeinde
Stony Brook University
100 Nicolls Road
Stony Brook, NY 11794-2300
USA

ISBN 978-3-11-135090-5
e-ISBN (PDF) 978-3-11-135117-9
e-ISBN (EPUB) 978-3-11-135129-2

Library of Congress Control Number: 2023950000

Bibliographic information published by the Deutsche Nationalbibliothek
The Deutsche Nationalbibliothek lists this publication in the Deutsche Nationalbibliografie; detailed bibliographic data are available on the Internet at http://dnb.dnb.de.

Cover image: Anna Bliokh/iStock/Getty Images Plus
Typesetting: Integra Software Services Pvt. Ltd.

www.degruyter.com

About the author

Foluso Ladeinde earned several degrees from Cornell University, including the Ph.D. from The Sibley School of Mechanical and Aerospace Engineering. He is a tenured faculty member in the Department of Mechanical Engineering at Stony Brook University, where he teaches courses in aerospace propulsion, numerical methods, applied mathematics, and theoretical fluid dynamics. He is the founding Chairman and was the Department Head (2015–2019) of the ABET-accredited Department of Mechanical Engineering, State University of New York, Incheon, South Korea. He was a Summer Faculty Fellow for ten years at the United States Air Force at Wright-Patterson AFB, Dayton, Ohio. Dr. Ladeinde's research uses advanced computational science and applied mathematics to investigate theoretical fluid dynamics, fluid flow turbulence theory, two-phase flows and heat transfer, supersonic combustion, and, more recently, atmospheric cloud microphysics. His work has been funded by the United States National Science Foundation (NSF), the United States Air Force (USAF), the United States Navy (USN), the Air Force Office of Scientific Research (AFOSR), the Defense Advanced Research Projects Agency (DARPA), the National Aeronautics and Space Administration (NASA), and commercial industry, including The Boeing Aerospace Corporation. He was two-term Chairman of the External Review Board of the NASA Center for Aerospace Research at North Carolina A&T University, Greensboro, North Carolina. Dr. Ladeinde has served as an Associate Editor (2009–2013) of the *AIAA Journal*, the flagship journal of the American Institute of Aeronautics and Astronautics (AIAA), an Associate Editor (2015–Present) of *The ASCE Journal of Aerospace Engineering*, and a Guest Editor of the *Physics of Fluids* (2020–2021). He is an Associate Fellow of AIAA, a Fellow of the American Society of Mechanical Engineers (ASME), a Life Member of the American Physical Society (APS), and a member of the Society for Industrial and Applied Mathematics (SIAM) and The Combustion Institute. He has published over 250 papers in archival journals and peer-reviewed conference proceedings. Dr. Ladeinde has also won Best Paper and Best Reviewer awards.

https://doi.org/10.1515/9783111351179-202

Preface

Applied complex variable theory is one of the most powerful mathematical tools of engineers, applied mathematicians, and theoretical physicists. The subject is so fundamental that most of the other advanced topics in applied mathematics depend on it. Complex variable theory enables the evaluation of complicated integrals, even in real space, as well as the use in integral transforms. It also allows us to understand the nature of the unexpected restrictions that exist on the validity of certain expansions in real space. The idea of analytic continuation makes possible certain mathematical operations that exploit analyticity beyond points that are formally determined to be singular but are otherwise precluded by convergence criteria.

A driving force for writing this book is the interest of the author in the various ways in which complex numbers are used in engineering analysis. It is also the objective to share the author's passion in the subject matter, hoping that this will encourage graduate students and young professionals. We also want to gather in one source the various applications to which complex variables are directed in advanced engineering analysis.

Why do we need another textbook on complex variables? First, this book has a fairly complete and exhaustive coverage of the subject matter, summarizing the more elementary aspects that you find in most textbooks and delving into the rather more specialized topics that are less commonplace, such as those in Chapters 5 through 7 of the present text. For the latter, the author is aware of only one textbook that is perhaps somewhat competitive. This is the book *Functions of a Complex Variable*, by George F. Carrier, Max Krook, and Carl E. Pearson (Hod Books Publishers, Ithaca, New York, 1983). However, this reference, besides being more than four decades old, is exceedingly difficult to read, it is not structured as a textbook, and doesn't help the student to learn the materials. The present book, which is equally rigorous, presents the materials in a more systematic and relatively easy-to-follow fashion, making it perhaps a better textbook for the subject matter. Thus, the present book is a one-stop reference for complex variables in engineering and theoretical physics, as it represents one of the most complete coverages of the subject for the target audience. Also, it probably has one of the most advanced and complete applications of the Schwarz-Christoffel transformation (Chapter 3). Moreover, the entire Chapter 4 of the book is dedicated to fluid flow and aerodynamics. Also unique is the effort the author has consciously made to render the material easy to read, learn, and access for graduate students, even for otherwise hard topics.

This book is written by an engineer for engineers (and theoretical physicists); therefore, the relevance of the example problems and exercises to engineering analysis, which can be found in Chapters 3 through 7, is another attractive feature of the book. They are carefully drawn from realistic, as opposed to imagined, problems in engineering analysis. While the subject may also be taught by pure or applied mathematicians, it will prove useful for them to collaborate with an experienced engineer-

https://doi.org/10.1515/9783111351179-203

ing professional who must also have the requisite applied mathematics background. To some level, the present book avails the non-engineering instructor the responsibility of content development in the manner thus implied.

The relatively easy presentation of the materials in this book is deliberate and has been informed by the unsuccessful search by the author for a presentation that is kind to students and instructors and facilitates their interest in the subject matter. Whereas the materials in the first two chapters of this book are relatively elementary and can be found in numerous standard textbooks for a first graduate course in Advanced Engineering Mathematics (AEM), the treatment and extensive applications of conformal mapping in Chapters 3 and 4 are more than what is normally available in such books. The application to fluid and aerodynamics, as presented in Chapter 4, as alluded to above, is virtually complete. The treatment of complex Laplace transform in Chapter 5 is unique, as is the asymptotic treatment of transform integrals in Chapter 6. In the same vein, the presentation of complex Fourier transform in Chapter 7 is also extensive. The last chapter of the book, Chapter 8, summarizes the modern developments in complex variables and their application and relevance to engineering analysis.

A case or two can be made for the advanced topics in this book. We know from the undergraduate AEM curriculum that the use of tables previously generated by others is the most expedient procedure for the inversion of a given transform. This approach is obviously limited by the extent of the available tables, and a larger proportion of the model equations in engineering analysis cannot be solved in this manner. The implication is that the advanced engineering and theoretical physics analyst must be equipped with the fundamental knowledge for inverting transforms, which invariably falls within the realm of complex variable theory. This is the subject of Chapter 5 in the present book.

In real Fourier transform, the transform variables in the physical and transform spaces are real. When one or both are allowed to be imaginary, we will have complex Fourier transformation. An advantage of the latter is that a much larger class of functions can be transformed, allowing for an expanded use of the Fourier transform technique in, say, engineering analysis. A case in point is a class of mixed boundary value partial differential equations, or their integral equivalents, for which the complex Fourier-transform-based Wiener–Hopf technique is perhaps the only known method for obtaining closed-form solutions. The procedure could be exceedingly complex and is usually beyond the scope of the standard AEM curriculum for beginning graduate students. The Wiener–Hopf method is presented in Chapter 7 of the present book.

While regular singular differential equations, for example of the Bessel's or Legendre's types, can at least be partially solved with the Frobenius procedure, equations with essential singularities cannot benefit from such an approach, and one is left with approximate procedures such as the method of dominant balance and asymptotic and perturbation analysis. It is often possible to convert a differential equation into an integral one, although the converse is rarely true in general. The asymptotic behavior of integrals provides a very powerful means of analyzing engineering problems with irregular singular points. Incidentally, the fundamental techniques often require working in complex space, hence the interest in the asymptotics of integrals in this book. Although asymptotic series are often not convergent, they are quite invaluable for determining local behaviors at irregular singular points of model equations in engineering analysis.

This book is generous with example problems that we know students would like in their quest for mastery of the subject matter. The section "Miscellaneous Examples," which is contained in most of the chapters of the book, serves to soften the learning curve, and is particularly useful for some of the otherwise difficult topics in the book. At the end of each major chapter, the book gives a fairly large number, more than fifty in some cases, of carefully selected problems (exercises), with varying degrees of difficulty. A solution manual that fully solves all the exercise problems accompanies the textbook for the benefit of the instructors that decide to adopt this book for their courses.

This book has been written primarily for advanced graduate students in engineering, theoretical physics, and applied mathematics disciplines, and for engineering practitioners. All the disciplines of engineering should be able to adopt the book. The book contains enough materials for a single, fifteen-week, three-credit, advanced course in applied complex variables for engineers and applied mathematicians.

Acknowledgments

First, I would like to acknowledge my graduate applied mathematics professors at Cornell University (Herbert Hui, Geoffrey S. S. Ludford (late), Subrata Mukherje, and Richard Herbert Rand), who significantly boosted the love I have always had for mathematics since elementary school, and have facilitated, even if unconsciously, the projection of this love into an all-consuming passion for applying mathematics to solve engineering problems. The materials learned from their respective classes have helped me to consolidate my graduate applied mathematics background, which I still enjoy today. Ayo A. Oyediran, a friend of mine at Cornell, his blessed memory, infectiously propagated his love of applied mathematics, for which I am grateful. In the same token, I wish to appreciate Ernest Eteng, also my contemporary at Cornell.

Some of the materials in this book were taken from lecture notes that I developed for several courses on graduate applied mathematics and theoretical fluid mechanics for engineering students at Stony Brook University (SBU) for a period of several decades. I wish to acknowledge the graduate students at SBU who assisted with the administration of some of these courses.

My collaborators at SBU: Danny Bluestein, James Glimm, Xiaolin Li, and Edward E. O'Brien (late) have been supportive, and I thank them very much. My department and the College of Engineering and Applied Sciences at SBU have provided an environment that supported the development of this book.

Many people outside of SBU, with whom I have interacted professionally at some point in my career, deserve to be acknowledged: Dare Afolabi, Ramesh Agarwal, Hasan Akay, Kehinde (Dr. Ken) Alabi, Louis Albright (late), Charles Aworh, Jose Camberos, Tim Colonius, Douglas Davis, Cesar Dopazo, Frederick Ferguson, Datta Gaitonde, Peyman Givi, Barry Kiel, Heuy-Dong Kim, Harold Kirk, Sidney Leibovich, Wenhai Li, Zhipeng Lou, Kazem Mahdavi, Kirk McDonald, Arun Muley, Ayodele Olorunda, Wole (Wally) Orisamolu, Vishwanath Prasad, Ramons Reba, Syed Rizvi, Bob Schlinker, Simon Senibi, John Spyropoulos, Mike Stoia, Kenneth Torrance (late), Miguel Visbal, Hui Zhang, and Lili Zheng.

I am exceedingly grateful to HyeJin Oh for providing an extensive amount of support. Without her relentless support this project might not have succeeded. HyeJin, I thank you very much indeed.

Finally, I owe this accomplishment to my wife, children, and grandchildren, to whom I express unreserved and special appreciation. Beatrice, my wife, endured extensive periods of absence because of my several intellectual pursuits, including the development of this book. I am sorry for the apparent selfishness. My three smart and lovely daughters – Lola, Tayo, and Nike, provided the real inspiration behind the efforts to write this book. I am also thankful to their respective husbands: Randall Temple, Nathan Black, and Jide Onifade. I wish to thank my grandchildren (Abby, Dara, Bennett,

https://doi.org/10.1515/9783111351179-204

and Elliott) for their love. Abby and Dara, I hope that you now partly understand why Papa was always working. Mine is the best family that anyone could hope for.

Foluso Ladeinde

Stony Brook, Long Island, New York

January 24, 2024

Contents

Abbreviations

APS	asymptotic power series
BC	boundary condition
BV	boundary value
C-R	Cauchy-Riemann
CCW	closed counterclockwise
CPV	Cauchy principal value
FT	Fourier transform
HOT	high-order terms
IV	initial value
IC	initial condition
JL	Jordan's lemma
LD	linearly dependent
LHP	left half plane
LHS	left-hand side
LI	linearly independent
LS	Laurent series
LT	Laplace transform
ODE	ordinary differential equation
PDE	partial differential equation
RHP	right half plane
RHS	right-hand side
RL	Riemann-Lebesgue
SOV	separation of variables
TS	Taylor series
UHP	upper half plane

https://doi.org/10.1515/9783111351179-206

Nomenclature

The following symbols and notations are used in this text. Standard symbols, such as the notation for integrals and sums, are not included.

$W[f,g]$	Wronskian of f and g
$L[f]$	Laplace transform of f
$L[f(x)]$	Laplace transform of f as a function of x
$L^{-1}[F(s)]$	inverse Laplace transform of F
$H(t)$	Heaviside function
$\delta(t)$	Dirac delta function
(a,b,c)	vector with three components
$\\|\mathbf{v}\\|$	norm (magnitude) of a vector $\mathbf{v}$
$\mathbf{F}\cdot\mathbf{G}$	dot product of **F** and **G**
$\mathbf{F}\times\mathbf{G}$	cross product of **F** and **G**
$R^n, \mathbb{R}^n$	n-space; set of all n-vectors
$[a_{ij}]$	matrix whose i,j element is a_{ij}
$\mathbf{T}$, $\mathbf{t}$	a unit tangent vector to a curve
$\boldsymbol{N}, \boldsymbol{n}$	a normal (or unit normal) to a curve
n	magnitude of the unit normal to a curve
∇	del operator
$\nabla\varphi$ or $\text{grad}(\varphi)$	gradient of φ
$D_u\varphi(P)$	directional derivative of φ in the direction of u, evaluated at P
$\int_C fdx+gdy+hdz$	line integral over C
$\int_C f(x,y,z)ds$	line integral of f with respect to arc length ds
$\dfrac{\partial(u,v)}{\partial(x,y)}$	Jacobian of u and v with respect to x and y
$\int\int_S f(x,y,z)d\sigma$	integral of f over a surface S
$f(x_0^-), f(x_0^+)$	left and right limits, respectively, of f at x_0
$f'_L(x_0), f'_R(x_0)$	left and right derivatives (respectively) of f at x_0
$F[f]$, or $\hat{f}$	Fourier transform of f
$F^{-1}[\hat{f}]$, or f	inverse Fourier transform of $\hat{f}$
$L^2(R)$	space of square integrable functions defined on the real line
$P_n(x)$	n^{th} Legendre polynomial
$H_n(x)$	n^{th} Hermite polynomial
$H_v^1(z)$	Hankel function of order v of the first kind
$H_v^2(z)$	Hankel function of order v of the second kind
$\Gamma(x)$	gamma function
$J_n(x)$	Bessel function of the first kind of order n
$Y_n(x)$	Bessel function of the second kind of order n
γ	Euler constant
$I_n(x), K_n(x)$	modified Bessel functions of the first and second kinds, respectively, of order n
$\nabla^2 u$	Laplacian of u
$\text{Re}[z]$	real part of complex variable z
$\text{Im}[z]$	imaginary part of complex variable z

https://doi.org/10.1515/9783111351179-207

$\bar{z}$	complex conjugate of complex variable z
$\|z\|$	magnitude (modulus) of complex variable z
$\arg(z)$	argument of complex variable z
$\int_\Gamma f(z)dz$	integral of a complex function f over a curve Γ
$\text{Res}[(f, z_0)]$	residue of complex function f at z_0
$f: D \rightarrow D^*$	f is a mapping from D into D^*
ρ	density
ν	kinematic viscosity, Poisson ratio
k	thermal conductivity
α	thermal diffusivity
c_p	specific heat at constant pressure

Part I: **Introduction to complex variables**

Introduction to Part I

The evolution of the number system conceivably started with the natural numbers for counting: 1, 2, 3, 4, . . ., which are the integers. The solution of equations such as $5x = 4$ required rational numbers – fractions, basically a/b, where a, b are integers. But these are positive and are therefore inadequate to solve equations such as $3x + 4 = 0$, which required negative numbers, or numbers less than zero. Furthermore, the solution of equations such as $x^2 = 2$ required irrational numbers, or numbers that cannot be written as a/b, where a, b are integers. Last, the solution of equations such as $x^2 + 16 = 0$ or $x^2 + 1 = 0$ required the existence of a number whose square is -1 or $\sqrt{-1} \equiv i$. The latter kind of number is known as an imaginary number or a complex number. Numbers that are not imaginary are referred to as real numbers. Although complex numbers are widely used in applied mathematics today, that achievement did not come easy. There was a strong opposition to their use, not unlike the path of adoption of negative numbers. Famous mathematicians such as Leibnitz and Gauss reportedly registered their skepticism of imaginary numbers before they were widely accepted as necessary for a complete description of the number system.

A complex number is partitioned into two: a real part and an imaginary part. For example, a complex number z is written as $z = a + ib$, where a is the real part and b, which is also real, is the coefficient of the imaginary part. Thus, any complex number can be thought of as the ordered pair (a, b). Fortunately, no new operations are introduced using complex numbers, and the common operations of arithmetic and the rules of calculus are directly applicable to the two parts of a complex number. That is, such operations on complex numbers involve only manipulations of real numbers. For example, adding or multiplying two ordered pairs of complex numbers (a_1, b_1) and (a_2, b_2) could be defined to yield the new ordered pair $(a_1 + a_2,\ b_1 + b_2)$ or $(a_1 a_2 - b_1 b_2,\ a_1 b_2 + a_2 b_1)$. This way, the i merely serves as a label for the order of the pair of real numbers. Evidently, there are numerous subtle manifestations of this decomposition.

The theory of complex variables is one of the most powerful mathematical tools of engineers, applied mathematicians, and physicists. The subject is so fundamental in the sense that all other topics in applied mathematics depend on it.

The treatment of complex variable theory in this part of the book focuses on the fundamentals of the subject. Part II of the book goes into more advanced topics in complex variables, for example, as encountered in complex integral transforms.

In the introductory chapter, Chapter 1, we present preliminary concepts of complex variable theory including writing complex numbers in polar form, finding the roots of equations, and introducing functions of complex variables and their differentiation and integration. We present complex trigonometric and hyperbolic functions and their inverses. Then we briefly discuss harmonic functions, which are functions that satisfy Laplace equation, with their ellipticity and "smoothness." We also discuss branch points, multivaluedness, and the creation of branch cuts, and we identify certain classes of functions that often lead to multivaluedness. Also, in Chapter 1, we review the con-

https://doi.org/10.1515/9783111351179-001

cept of conservative force fields as a prelude to the discussion of analyticity and the Cauchy-Riemann conditions. We present Cauchy-Goursat theorem, singularities, and the Poisson Integral Formula. The latter is based on the Cauchy Integral Formula, and it enables closed-form solution of the Laplace equation inside a circle when a Dirichlet boundary condition is specified on the surface of the circle.

Chapter 2 of the book deals with the Laurent series as a "superset" of the Taylor series, which admits negative exponents of the expansion variable. To this end, we present the development of these two types of series. We also discuss analytic continuation in the complex plane, which is basically saying that if a function is singular at a point in the complex plane, we should be able to use the function at all other points in the plane. We conclude Chapter 2 with the classification of singularities, the residue theorem, and an extensive treatment of contour integration for numerous types of integrand and integral limits, including improper integrals.

The two remaining chapters in Part I of the book go into a very detailed treatment of general conformal mapping and its application (Chapter 3) and an exhaustive presentation of the application to perfect fluid flow (Chapter 4). In Chapter 3, we discuss the preservation of angles and sense of direction by conformal mapping and present a series of useful transformations, and applications to fairly realistic problems involving complex geometries in engineering. Chapter 3 is concluded with an extensive presentation of the Schwarz-Christoffel transformation and a detailed analysis of the application to the free streamline theory.

The topics in the application of complex variable theory and conformal mapping to perfect fluid flow, which is presented in detail in Chapter 4, include the complex potentials for many basic flows, such as uniform parallel flows; sources, sinks, vortices, flow in a sector, flow around a sharp edge, flow due to a doublet, flow around a circular cylinder, and the Blasius integral laws. We also go into the details of the Joukowski transformation and conformal mapping in airfoil design for aircraft aerodynamics, windmills, and general-purpose turbine blades.

1 Introductory concepts

Relatively elementary aspects of the complex variable theory are presented in this chapter in terms of the preliminary operational concepts, the polar form of the complex variable, complex-variable root-finding, and functions of a complex variable. Identities for trigonometric and hyperbolic complex functions and their inverses are given within the framework of a complex variable. We also discuss analyticity and analytic continuation of functions of complex variables, multivaluedness, and the development of branch cuts. Singularities, the Cauchy theorems and the Cauchy-Riemann (C-R) conditions are discussed, as is the Poisson Integral Formula.

1.1 Preliminary concepts

We define $\sqrt{-1} = i$. If the number z is imaginary, we will represent it as $z = a + bi$, where a is the real part of z, $a = \mathrm{Re}[z]$, and b is the imaginary part of z, $b = \mathrm{Im}[z]$. The conjugate of $z = a + bi$, which is denoted as $\bar{z}$, is defined as

$$\bar{z} = a - bi.$$

Example 1.1 $\overline{2 + 3i} = 2 - 3i$

The magnitude of $z = a + bi$, written as $|z|$, is obtained from

$$|z|^2 = z\bar{z} = (a + bi)(a - bi) = a^2 - abi + abi - b^2 i^2 = \left(a^2 + b^2\right).$$

The usual arithmetic operations in real calculus apply to imaginary variables if we allow for the presence of the imaginary component and that this component cannot be directly added to the real component. Let us examine the four common operations.

Sum: If $z_1 = a + bi$ and $z_2 = c + di$, then $z_1 + z_2 \equiv (a + c) + (b + d)i$

Subtraction: If $z_1 = a + bi$ and $z_2 = c + di$, then $z_1 - z_2 = z_1 + (-z_2) = (a - c) + (b - d)i$

Product: If $z_1 = a + bi$ and $z_2 = c + di$, then

$$z_1 z_2 = (a + bi)(c + di) = ac + adi + bci - bd = (ac - bd) + (ad + bc)i.$$

Division:

$$z_1 / z_2 = z_3 \ \text{ if } \ z_1 = z_2 z_3.$$

It is straightforward to prove the following properties of the conjugate operation:

$$\overline{z_1 + z_2} = \overline{z_1} + \overline{z}_2,$$

$$\overline{z_1 z_2} = \overline{z_1}\,\overline{z_2},$$

The original version of this chapter was revised: Figure 1.5 has been corrected. An Erratum is available at DOI: https://doi.org/10.1515/9783111351179-021

https://doi.org/10.1515/9783111351179-002

$$\overline{\left(\frac{z_1}{z_2}\right)} = \frac{\bar{z}_1}{\bar{z}_2}.$$

It should also be noted that

$$z + \bar{z} = a + bi + a - bi = 2a = 2\,\mathrm{Re}[z],$$
$$z - \bar{z} = a + bi - a + bi = 2bi = 2i\,\mathrm{Im}[z].$$

Some aspects of the triangle inequality can be written as follows for imaginary variables z_1 and z_2:

$$|z_1 + z_2| \le |z_1| + |z_2|,$$

$$|z_1 + z_2| \ge ||z_1| - |z_2||,$$
$$|z_1 - z_2| \ge ||z_1| - |z_2||,$$
$$|z_1 + z_2|^2 \le |z_1|^2 + |z_2|^2 + 2|z_1 z_2| = \{|z_1| + |z_2|\}^2,$$
$$|z_1 z_2| = |z_1||z_2|.$$

The inequalities $|z_1 + z_2| \le |z_1| + |z_2|$ can be easily proved. For this purpose, consider the parallelogram in Figure 1.1, where z_1, with $|z_1| = |x_1 + iy_1|$, represents the coordinate of Point A, and z_2, with $|z_2| = |x_2 + iy_2|$, represents the coordinate of Point B. Obviously

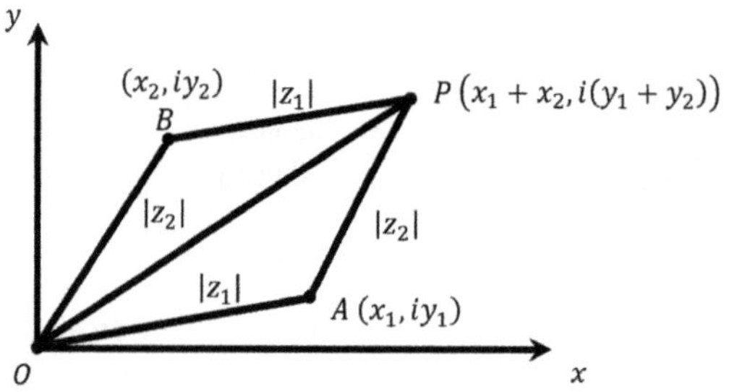

Figure 1.1: The geometry of triangular inequality.

$|z_1| = OA$, $|z_2| = OB$. Since $OB = AP$, the vertical line from $iy = 0$ to point P is $i\,(y_1 + y_2)$, while the horizontal line at P is $(x_1 + x_2)$, giving the coordinate of P as shown in the figure. It is known that the side of any triangle is less than the sum of the other two sides. This means that in the triangle OAP, $OP < OA + AP$, and since $OB = AP$, $OP < OA + OB$, we will have

$$|z_1 + z_2| < |z_1| + |z_2|,$$

where $OP = OA + AP$ if the points O, A, and P are collinear, or

$$|z_1 + z_2| = |z_1| + |z_2|.$$

We could also use algebraic arguments to arrive at the same results. Using ordinary algebraic operations, we see that

$$\begin{aligned}
z_1 + z_2 &= (x_1 + x_2) + i(y_1 + y_2) \\
|z_1 + z_2|^2 &= (x_1 + x_2)^2 + (y_1 + y_2)^2 \\
&= x_1^2 + x_2^2 + 2x_1x_2 + y_1^2 + y_2^2 + 2y_1y_2 \\
&= (x_1^2 + y_1^2) + (x_2^2 + y_2^2) + 2(x_1x_2 + y_1y_2) \\
&= (x_1^2 + y_1^2) + (x_2^2 + y_2^2) + 2\left\{[x_1x_2 + y_1y_2]^2\right\}^{1/2} \\
&= |z_1|^2 + |z_2|^2 + 2\{[x_1^2x_2^2 + y_1^2y_2^2 + 2x_1x_2y_1y_2]\}^{1/2}.
\end{aligned}$$

From the positivity of each term in the last equation and the fact that $[x_1x_2 + y_1y_2]^2 = x_1^2x_2^2 + y_1^2y_2^2 + 2x_1x_2y_1y_2$, we will have $(x_1y_2 - x_2y_1)^2 \geq 0$, or $x_1^2y_2^2 + x_2^2y_1^2 \geq 2x_1x_2y_1y_2$.

We can thus write

$$|z_1 + z_2|^2 \leq |z_1|^2 + |z_2|^2 + 2\ \{[x_1^2x_2^2 + y_1^2y_2^2 + x_1^2y_2^2 + x_2^2y_1^2]\}^{(1/2)}.$$

Factoring the last term on the RHS of this equation we will have

$$\begin{aligned}
|z_1 + z_2|^2 &\leq |z_1|^2 + |z_2|^2 + 2\ [(x_1^2 + y_1^2)(x_2^2 + y_2^2)]^{(1/2)} \\
&\leq |z_1|^2 + |z_2|^2 + 2|z_1||z_2| \\
&\leq [|z_1| + |z_2|]^2,
\end{aligned}$$

which provides the required proof if we take the square root of both sides of this inequality:

$$|z_1 + z_2| \leq |z_1| + |z_2|.$$

For the inequality $|z_1 - z_2| \geq |z_1| - |z_2|$, we can use the fact that $a = b + c$ implies $|a| \leq |b| + |c|$ to write: $|z_1| = |(z_1 - z_2) + z_2| \leq |z_1 - z_2| + |z_2|$. Subtracting $|z_2|$ from both sides provides the required proof.

We usually want to write functions of an imaginary variable as $\lambda_1 + i\lambda_2$, where λ_1, λ_2 are real, even if the function is otherwise expressed as a fraction. We do this by multiplying the numerator and denominator by the conjugate of the denominator. For example,

$$\frac{1}{a+bi}=\frac{1}{a+bi}\times\frac{\overline{a+bi}}{\overline{a+bi}}=\frac{a-bi}{a^2-(bi)^2}=\frac{a-bi}{a^2+b^2}=\frac{a}{a^2+b^2}-\frac{b}{a^2+b^2}i$$

so that $\lambda_1 = a/(a^2+b^2)$, $\lambda_2 = -b/(a^2+b^2)$.

1.2 Polar form of complex variables

It is often convenient to work with the polar form of an imaginary variable. The argand diagram (Figure 1.2(a)) shows the geometric relationship between the Cartesian $(x,\ y)$ coordinate representation of an imaginary number z and its polar $(r,\ \theta)$ coordinate representation. Figures 1.2(b) and 1.2(c), respectively, show positive and negative angles in the azimuth. The transformation between the two coordinate systems can be written as

$$z = x + iy = r\cos\theta + ir\sin\theta = r(\cos\theta + i\sin\theta) = re^{i\theta},$$

where we have used the Euler equation:

$$e^{i\theta} = \cos\theta + i\sin\theta \equiv cis\ \theta, \text{ and } r = \sqrt{x^2+y^2}.$$

The angle $\theta = \tan^{-1}(b/a)$ is the principal argument of the imaginary number. In general, the angle is multivalued, as it takes on the values $\theta + 2n\pi$, $n = 1, 2, \ldots$.

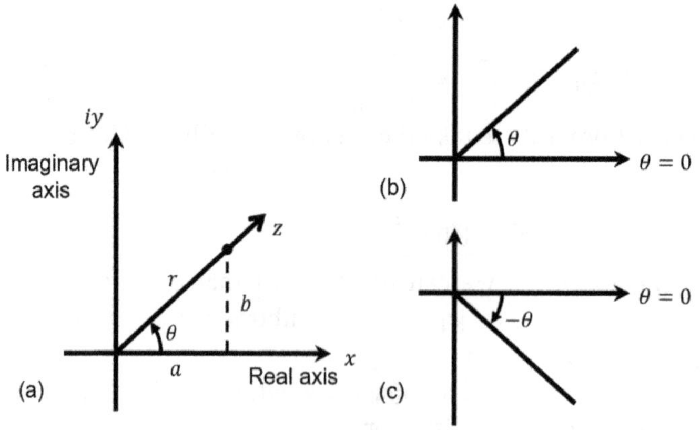

Figure 1.2: Argand diagram and the convention for the sense of angular measure: (a) The Argand diagram, (b) positive angle, and (c) negative angle.

Example 1.2 The polar representation of $z = 3 - 2i$ is shown below (Figure 1.3(a)). We have three units in the positive x-coordinate direction and two units in the negative iy-coordinate direction. In this case, $r\cos\theta = 3$, $r\sin\theta = -2$, and $\theta = \tan^{-1}(-2/3)$. The conjugate of z, or $\bar{z} = 3 + 2i$, shown in Figure 1.3(b), has three units in the positive x-coordinate direction and two units in the positive iy-coordinate direction. Thus, the conjugate operation reflects z about the x-axis. Here, we have $r\cos\theta = 3$, $r\sin\theta = 2$ (Figure 1.3(b)), and $\theta = \tan^{-1}(2/3)$.

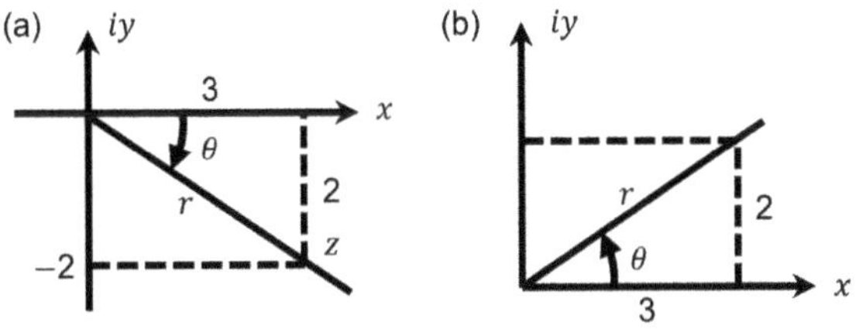

Figure 1.3: (a) Polar representation of $z = 3 - 2i$; (b) conjugate of $z = 3 - 2i$.

1.3 Roots of numbers

Imaginary numbers allow us to extend our knowledge of the roots of numbers. This is best illustrated in the Euler form: $z = re^{i\theta}$, where $r = \sqrt{a^2 + b^2}$ when $z = a + bi$.

Example 1.3 Determine $\sqrt{4}$.

Solution: The number 4 is real and positive and so it is located on $\theta = 0$ and $2n\pi$, where n is an integer. The number -4 on the other hand is real and located on the negative real axis. Therefore, the angle θ for -4 is $\theta = \pi$, or $-\pi$, and $2n\pi$ additions to these. See Figure 1.4.

We can write

$$4 = 4 + 0i \equiv z = re^{i\theta} = 4e^{(0+2n\pi)i}, \qquad n = 0,\ 1,\ 2,\ \ldots;\ r^2 = 4^2 + 0^2,$$

$$z^{\frac{1}{2}} = \left[4e^{(0+2n\pi)i}\right]^{\frac{1}{2}} = 2e^{n\pi i} = 2[\cos n\pi + i\sin n\pi].$$

So $\sqrt{4}$ is not just 2 or -2, but an infinite number of possibilities.

Similarly, $-4 = 4e^{(+\pi+2n\pi)i}$ and

$$\sqrt{-4} = 4^{\frac{1}{2}}e^{(\pi+2n\pi)i/2} = 2\left[\cos\left(\frac{\pi}{2} + n\pi\right) + i\sin\left(\frac{\pi}{2} + n\pi\right)\right].$$

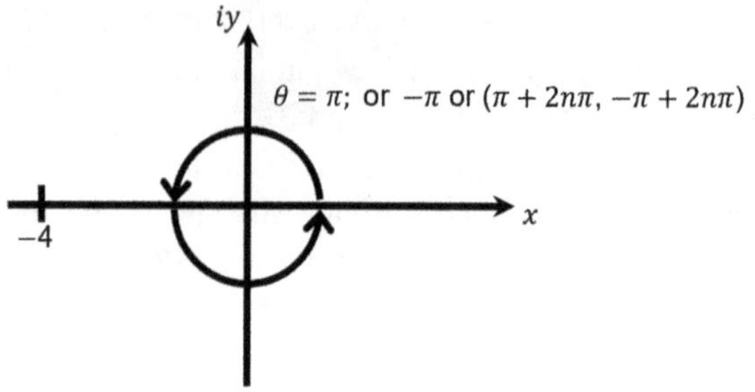

Figure 1.4: Angle of $z = -4$.

For $n = 0$,

$$\sqrt{-4} = 2e^{+\frac{\pi}{2}i} = 2\left[\cos\frac{\pi}{2} + i\sin\frac{\pi}{2}\right] = 2i.$$

For $n = 1$, $\sqrt{-4} = -2i$, etc.

Remark 1.1 (Notation) For convenience, the complex number $z = a + ib$ is sometimes written as $z = a + bi$, where a and b are real and $i = \sqrt{-1}$ in both cases. Also, the ordinate in the geometric representation of a complex number is interchangeably written as "iy" and "y." No confusion need arise because of these preferences.

Example 1.4 What is the locus represented by $|z - 2| = 3$ shown in Figure 1.5?

Solution: The circular domain is centered at $z = 2$ and has a radius of 3 about this center. We can write

$$|z - 2| = 3 \Rightarrow |(x - 2 + iy)| = 3, \Rightarrow \left[(x-2)^2 + y^2\right]^{1/2} = 3,$$

$$\Rightarrow (x-2)^2 + y^2 = 9 \Rightarrow \text{Circle of radius } 3, \text{ center } (2, 0).$$

Note that a circle of radius r_0 located at $z_o = a_o + b_o i$ will be described by $|z - z_o| = r_o$.

Example 1.5 What is the locus represented by $|z - 3| + |z + 3| = 10$?

Solution: We can write this as

$$\sqrt{(x-3)^2 + y^2} + \sqrt{(x+3)^2 + y^2} = 10$$

or

$$\sqrt{(x-3)^2 + y^2} = 10 - \sqrt{(x+3)^2 + y^2}.$$

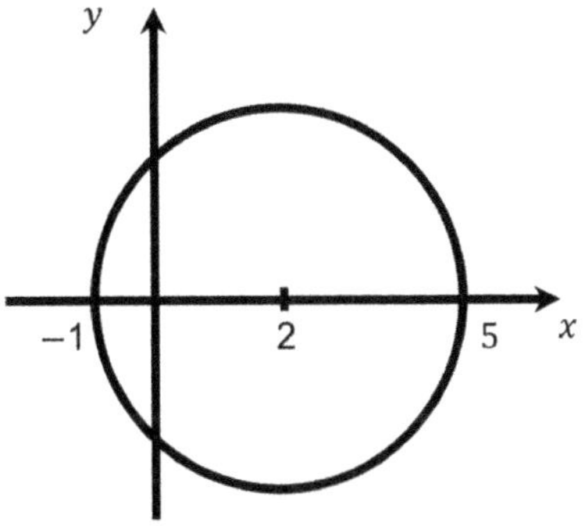

Figure 1.5: The locus of $|z-2|=3$.

Squaring gives

$$25+3x=5\sqrt{(x+3)^2+y^2}.$$

Squaring once more gives

$$\frac{x^2}{25}+\frac{y^2}{16}=1,$$

which is an ellipse shown in Figure 1.6 with center at (0, 0) and semi-major and semi-minor axes of lengths 5 and 4 units, respectively.

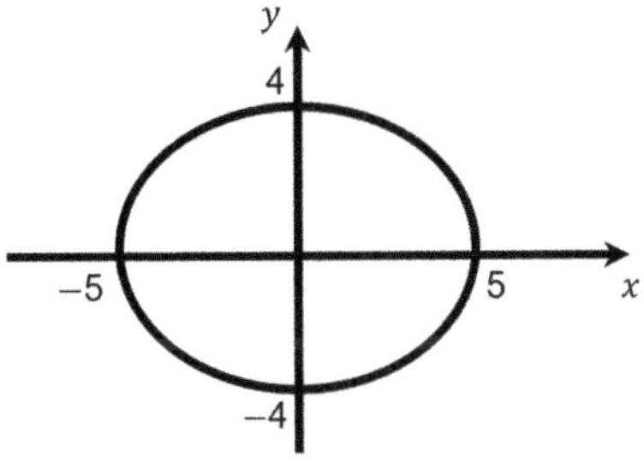

Figure 1.6: The locus of $|z-3|+|z+3|=10$.

1.4 Functions of a complex variable

Any function $f(z)$ of the complex variable z can be written as $u(x,y)+iv(x,y)$, where u and v are both real numbers. This is the rectangular form that we will use in this book.

Example 1.6 Write $f(z)=z^2$ in the form $u+iv$.

Solution: With $z=x+iy$,

$$f(z) = (x+iy)^2 = \underbrace{x^2 - y^2}_{u} + i\,\underbrace{2xy}_{v} \equiv u(x,y) + iv(x,y).$$

Note that the function $f(z) = 1/z$ can be written in polar form as

$$f(z) = \frac{1}{z} = \frac{1}{re^{i\theta}} = \frac{1}{r}e^{-i\theta}.$$

This function reflects $z = re^{i\theta}$ about the x-axis (Figure 1.7). It also changes the modulus from r to $1/r$.

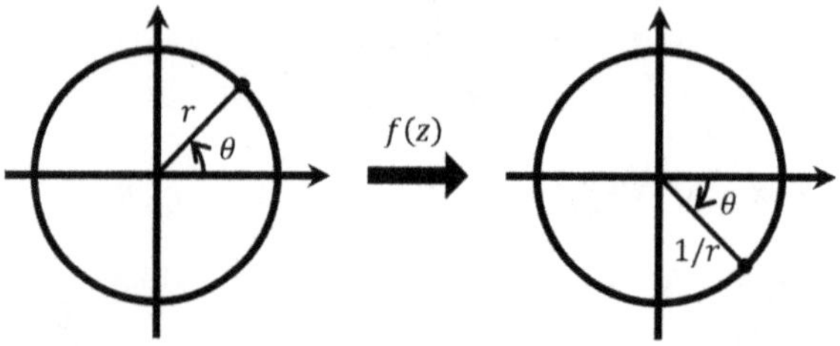

Figure 1.7: Reflection of z by $1/z$.

Theorem 1.1 De Moivre's Theorem
The theorem states that

$$z^n = \{r(\cos\theta + i\sin\theta)\}^n = r^n[\cos n\theta + i\sin n\theta].$$

For example,

$$z^{\frac{1}{n}} = r^{\frac{1}{n}}\left\{\cos\left(\frac{\theta + 2k\pi}{n}\right) + i\sin\left(\frac{\theta + 2k\pi}{n}\right)\right\}, \quad k = 0,\ 1,\ 2, \ldots.$$

The theorem also allows us to write

$$(\cos\theta + i\sin\theta)^n = \cos n\theta + i\sin n\theta.$$

We need to be aware of the following identities:

a) $(\cos\theta_1 + i\sin\theta_1)(\cos\theta_2 + i\sin\theta_2) = \cos(\theta_1 + \theta_2) + i\sin(\theta_1 + \theta_2)$

b) $[r_1(\cos\theta_1 + i\sin\theta_1)][r_2(\cos\theta_2 + i\sin\theta_2)] = r_1r_2[\cos(\theta_1 + \theta_2) + i\sin(\theta_1 + \theta_2)]$

c) $\dfrac{r_1(\cos\theta_1 + i\sin\theta_1)}{r_2(\cos\theta_2 + i\sin\theta_2)} = \dfrac{r_1}{r_2}[\cos(\theta_1 - \theta_2) + i\sin(\theta_1 - \theta_2)]$

Example 1.7 Evaluate $f(z) = \left(-1 + \sqrt{3}\,i\right)^{10}$.

Solution: We need the values of r and θ in order to use De Moivre's theorem:

$$\theta = \frac{2}{3}\pi, \qquad r = \sqrt{(-1)^2 + \left(\sqrt{3}\right)^2} = 2.$$

Therefore,

$$f(z) = \left(-1 + \sqrt{3}i\right)^{10} = 2^{10}\left[\cos\left(\frac{20\pi}{3}\right) + i\sin\left(\frac{20\pi}{3}\right)\right].$$

1.5 Trigonometric and hyperbolic functions

If we use the familiar series expansion for the exponential function, noting that the iy in $z = x + iy$ can be treated as a number for this purpose, then

$$\begin{aligned} e^{iy} &= 1 + (iy) + \frac{1}{2!}(iy)^2 + \frac{1}{3!}(iy)^3 + \ldots + \frac{1}{n!}(iy)^n \\ &= \left(1 - \frac{y^2}{2!} + \frac{y^4}{4!} - \ldots\right) + i\left(y - \frac{y^3}{3!} + \frac{y^5}{5!} - \ldots\right) \\ &= \cos y + i\sin y. \end{aligned} \tag{1.1}$$

Similarly,

$$e^{-iy} = \cos y - i\sin y. \tag{1.2}$$

Addition of eqs. (1.1) and (1.2) gives

$$\cos y = \frac{e^{iy} + e^{-iy}}{2},$$

while subtraction gives

$$\sin y = \frac{e^{iy} - e^{-iy}}{2i}.$$

From

$$e^z = e^{x+iy} = e^x e^{iy} = e^x(\cos y + i\sin y)$$

and adopting the foregoing identities for the series expansion of e^z, we can write the complex cosine and sine functions as

$$\cos z = \frac{e^{iz} + e^{-iz}}{2}, \quad \sin z = \frac{e^{iz} - e^{-iz}}{2i}. \tag{1.3}$$

We can also easily derive the formulas

$$\cosh x = \frac{e^x + e^{-x}}{2}, \quad \sinh x = \frac{e^x - e^{-x}}{2},$$

and

$$\cosh x + \sinh x = e^x, \quad \cosh^2 x - \sinh^2 x = 1$$

so that

$$\cosh z = \frac{e^z + e^{-z}}{2}, \quad \sinh z = \frac{e^z - e^{-z}}{2}. \tag{1.4}$$

Substituting iz for z in eqs. (1.3) and (1.4), we can easily arrive at the following correlations:

$$\begin{aligned} &\cos(iz) = \cosh z, \quad \cosh(iz) = \cos z, \quad \sin(iz) = i\sinh z, \\ \text{and} \quad &\sinh(iz) = i\sin z, \quad \tan(iz) = i\tanh z, \quad \tanh(iz) = i\tan z. \end{aligned}$$

Note that eqs. (1.3) and (1.4) can also be used to show that $\cos^2 z + \sin^2 z = 1$:

$$\begin{aligned} \cos^2 z + \sin^2 z &= \left(\frac{1}{2}\right)^2 (e^{iz} + e^{-iz})^2 - \left(\frac{1}{2}\right)^2 (e^{iz} - e^{-iz})^2 \\ &= \frac{1}{4}(e^{i2z} + 2 + e^{i2z} - e^{-i2z} + 2 - e^{-i2z}) = \frac{4}{4} = 1. \end{aligned}$$

The following familiar identities in real trigonometry also hold for complex numbers:

$$\tan z = \frac{\sin z}{\cos z}, \quad \sec z = \frac{1}{\cos z}, \quad \csc z = \frac{1}{\sin z}, \quad \cot z = \frac{1}{\tan z},$$

and we can define the following:

$$\begin{aligned} \operatorname{sech} z &= \frac{1}{\cosh z} = \frac{2}{e^z + e^{-z}}, \quad \operatorname{csch} z = \frac{1}{\sinh z} = \frac{2}{e^z - e^{-z}}, \\ \tanh z &= \frac{\sinh z}{\cosh z} = \frac{e^z - e^{-z}}{e^z + e^{-z}}, \quad \coth z = \frac{\cosh z}{\sinh z} = \frac{e^z + e^{-z}}{e^z - e^{-z}}. \end{aligned}$$

The exponential identities above can be used to show that

$$\begin{aligned} &\tfrac{d}{dz}\cos z = -\sin z, \quad \tfrac{d}{dz}\sin z = \cos z, \quad \tfrac{d}{dz}\tan z = \sec^2 z, \quad \tfrac{d}{dz}\cot z = -\csc^2 z, \\ &\tfrac{d}{dz}\sec z = \sec z \tan z, \quad \tfrac{d}{dz}\csc z = -\csc z \cot z, \end{aligned}$$

which are identical to the corresponding identities for real variables.

1.6 Inverse trigonometric and hyperbolic functions

Starting with the inverse sine function, $\sin^{-1}z$, we know that it satisfies $\omega = \sin^{-1}z$ provided $z = \sin\omega$ or

$$z = \frac{e^{i\omega} - e^{-i\omega}}{2i} \Rightarrow e^{2i\omega} - 2ize^{i\omega} - 1 = 0,$$

which can be written as

$$\lambda^2 - 2iz\lambda - 1 = 0, \quad \left(\lambda \equiv e^{i\omega}\right)$$

and solved using the quadratic formula to obtain

$$\lambda = e^{i\omega} = iz \pm \left(1 - z^2\right)^{\frac{1}{2}}.$$

Thus,

$$\sin^{-1}z = -i\ln\left[iz \pm \left(1 - z^2\right)^{\frac{1}{2}}\right], \quad \left(\frac{1}{i} = -i\right).$$

Note that because $\left(1 - z^2\right)^{1/2}$ has two values, we can conveniently replace the "$\pm$" symbol with "$+$."

Similarly,

$$\cos^{-1}z = -i\ln\left[z + i\left(1 - z^2\right)^{1/2}\right], \quad \tan^{-1}z = \frac{i}{2}\ln\frac{i+z}{i-z},$$

$$\csc^{-1}z = \frac{1}{i}\ln\left(\frac{i + \sqrt{z^2 - 1}}{z}\right), \quad \sec^{-1}z = \frac{1}{i}\ln\left(\frac{1 + \sqrt{1 - z^2}}{z}\right),$$

$$\text{and } \cot^{-1}z = \frac{1}{2i}\ln\left(\frac{z+i}{z-i}\right).$$

As an example, $f(z) = \sin^{-1}\sqrt{5}$, with $z = \sqrt{5}$, can be found as

$$f(z) = \sin^{-1}\sqrt{5} = -i\ln\left[\sqrt{5}i + \left(1 - \left(\sqrt{5}\right)^2\right)^{1/2}\right]$$

$$-i\left[\ln\left(\sqrt{5} \pm 2\right) + \left(\frac{\pi}{2} + 2n\pi\right)i\right], \quad n = 0, \pm 1, \pm 2, \ldots.$$

where we have used $(-4)^{1/2} = \pm 2i$ and $\ln i = \left(\frac{\pi}{2} + 2n\pi\right)i$.

In terms of the derivatives:

$$z = \sin\omega \Rightarrow \frac{dz}{dz} = \frac{d(\sin\omega)}{dz} = \cos\omega\frac{d\omega}{dz}$$

$$\Rightarrow \frac{d\omega}{dz} = \frac{1}{\cos\omega} = \frac{1}{\left(1-\sin^2\omega\right)^{1/2}} = \frac{1}{(1-z^2)^{1/2}} = \frac{d}{dz}\sin^{-1}z.$$

Similarly,

$$\frac{d}{dz}\cos^{-1}z = \frac{1}{(1-z^2)^{1/2}}; \quad \frac{d}{dz}\tan^{-1}z = \frac{1}{1+z^2}.$$

The following inverse hyperbolic functions can also be derived:

$$\sinh^{-1}z = \ln\left(z+\sqrt{z^2+1}\right), \quad \operatorname{csch}^{-1}z = \ln\left(\frac{1+\sqrt{z^2+1}}{z}\right),$$

$$\cosh^{-1}z = \ln\left(z+\sqrt{z^2-1}\right), \quad \operatorname{sech}^{-1}z = \ln\left(\frac{1+\sqrt{1-z^2}}{z}\right),$$

$$\tanh^{-1}z = \frac{1}{2}\ln\left(\frac{1+z}{1-z}\right), \quad \coth^{-1}z = \frac{1}{2}\ln\left(\frac{z+1}{z-1}\right).$$

1.7 Differentiation of a function of a complex variable

From elementary calculus the differentiation of $f(z) = u + iv$ with respect to z can be expressed as

$$\frac{df(z)}{dz} \equiv f'(z) = \lim_{\Delta z \to 0} \frac{f(z+\Delta z) - f(z)}{\Delta z}.$$

However, since z depends on both x and y, it becomes necessary to determine from which axis (Im or Re) Δz must approach zero. This is solved by the Cauchy-Riemann (C-R) conditions which essentially imply that Δz must approach zero at the same rate from both the x- and y-directions. If we move in the x-direction, we will have the derivative as

$$f'(z) = \lim_{\Delta x \to 0} \frac{f(z+\Delta x) - f(z)}{\Delta x} = \lim_{\Delta x \to 0} \frac{u(x+\Delta x, y) + iv(x+\Delta x, y) - u(x,y) - iv(x,y)}{\Delta x}$$

$$\Rightarrow f'(z) = \frac{\partial u}{\partial x} + i\frac{\partial v}{\partial x} \equiv u_x + iv_x.$$

Also, for perturbation in the y-direction, we obtain

$$f'(z) = v_y - iu_y.$$

Note that in y, the denominator is $i\Delta y$ (not just Δy). Also note that $\frac{1}{i} = \frac{\sqrt{-1}}{\sqrt{-1}} \cdot \frac{1}{\sqrt{-1}} = -i.$

The necessary conditions for complex differentiability are that the C-R conditions are satisfied or that the derivatives of $f(z)$ from the x- and y-directions are equal, or $u_x + iv_x = v_y - iu_y$. This requires that

$$u_x = v_y, \quad u_y = -v_x,$$

which are the formal statements of the C-R conditions.

1.8 A note on the adequacy of the C-R conditions

The C-R conditions as defined above give necessary conditions for a function $f(z)$ to be complex-differentiable. However, the validity of the C-R conditions does not establish that $f(z)$ is in fact differentiable at $z = z_0$. Consider

$$f(z) = \begin{cases} 0, & xy = 0 \\ 1, & \text{otherwise.} \end{cases}$$

Although the C-R conditions are satisfied for this function, the function is not continuous at $z_0 = 0$. We recall the requirements for complex differentiability as stated by Irving and Mullineux [3] and Wegert [4]. The former are stated in a theorem:

Theorem 1.2

(i) The necessary conditions for $f(z) \equiv u + iv$ to be regular at a point z_0 of a domain $\mathcal{D}$ are that u_x, u_y, v_x, v_y should exist and satisfy the C-R conditions.

(ii) The sufficient conditions for the continuous and single-valued function $f(z)$ to be regular in $\mathcal{D}$ are: u_x, u_y, v_x, and v_y exist, are continuous and satisfy the C-R conditions. The proofs are given in the referenced work.

The conditions for establishing the sufficiency of the C-R conditions as stated by Wegert are a bit more expository. The concept of real differentiability ($\mathbb{R}$-differentiability) is first defined:

Definition 1.1 A complex function $f: \mathcal{D} \to \mathbb{C}$ is said to be $\mathbb{R}$-differentiable at a point z_0 in the open domain $\mathcal{D}$ if there are constants $c_x, c_y \in \mathbb{C}$ such that $f(z)$ can be represented as

$$f(z) = f(z_0) + c_x(x - x_0) + c_y(y - y_0) + R(z), \tag{1.5}$$

where $z_0 = x_0 + iy_0$ and the remainder term $R(z)$ satisfies $R(z)/|z - z_0| \to 0$ as $z \to z_0$. An $\mathbb{R}$-differentiable function $f(z)$ satisfies the following equations:

$$c_x = \left.\frac{\partial f}{\partial x}\right|_{z_0} = \left.\frac{\partial u}{\partial x}\right|_{(x_0,y_0)} + i\left.\frac{\partial v}{\partial x}\right|_{(x_0,y_0)},$$

$$c_y = \left.\frac{\partial f}{\partial y}\right|_{z_0} = \left.\frac{\partial u}{\partial y}\right|_{(x_0,y_0)} + i\left.\frac{\partial v}{\partial y}\right|_{(x_0,y_0)}.$$

The Wirtinger derivatives are

$$\frac{\partial f}{\partial z} = \frac{1}{2}\left(\frac{\partial f}{\partial x} - i\frac{\partial f}{\partial y}\right),$$

$$\frac{\partial f}{\partial \bar{z}} = \frac{1}{2}\left(\frac{\partial f}{\partial x} + i\frac{\partial f}{\partial y}\right),$$

where $\partial f/\partial \bar{z}$ is also referred to as the C-R operator. Employing these relations in eq. (1.5) yields

$$\frac{f(z)-f(z_0)}{z-z_0} = \left.\frac{\partial f}{\partial z}\right|_{z_0} + \left.\frac{\overline{(z-z_0)}}{(z-z_0)}\frac{\partial f}{\partial \bar{z}}\right|_{z_0} + R'(z), \tag{1.6}$$

where $R'(z)$ is small compared to unity. Equation (1.6) shows that $f(z)$ is complex-differentiable at z_0 if and only if $\partial f/\partial \bar{z} = 0$, and we can write $f'(z)_0 = (\partial f/\partial z)|_{z_0}$. Thus, in the terminologies of Wegert, a function $f\colon \mathcal{D} \to \mathbb{C}$ which is $\mathbb{R}$-differentiable at $z_0 \in \mathcal{D}$ is complex-differentiable at $z = z_0$ if and only if it satisfies the C-R equations at z_0. With $dz \equiv z - z_0$, $d\bar{z} \equiv \overline{z - z_0}$, and $df = (\partial f/\partial z)|_{z_0} dz + (\partial f/\partial \bar{z})|_{z_0} d\bar{z}$, we see that $f(z)$ is complex-differentiable if and only if its total differential (df) does not depend on $d\bar{z}$.

1.9 Regular functions

A function $f(z)$ is regular (analytic, holomorphic) if it is both single-valued and complex-differentiable.

Definition 1.2 (Complex differentiability). Let $\mathcal{D}$ be an open subset of the complex plane. A function $f\colon \mathcal{D} \to \mathbb{C}$ is said to be complex-differentiable at the point z_0 if the limit

$$\lim_{z\to z_0} \frac{f(z)-f(z_0)}{z-z_0}$$

exists and unique, that is, it is independent of the manner in which $z \to z_0$. If these conditions are satisfied, the limit is called the (complex) derivative of f at z_0, which we denote by $f'(z)$. Using a linear approximation of $f(z)$ about z_0 to first-order, with a remainder term of $R(z)$, we can say that a complex function is differentiable at z_0 if and only if the function can be approximated by a linear function of z, with an error $R(z)$ that is small compared to $|(z-z_0)|$, or $R(z)/(z-z_0) \to 0$ as $z \to z_0$. Note that with $|f(z) - f(z_0)| \sim |f'(z_0)||z - z_0|$, we can write the geometric relation

$$\arg[f(z)-f(z_0)]\sim \arg[z-z_0]+\arg\left[f'(z_0)\right].$$

Example 1.8 Show that the C-R conditions are satisfied for $f(z)=z^2$.

Solution:

$$f(z)=z^2=x^2-y^2+i2xy;\ u=x^2-y^2,\ v=2xy,$$

$$\frac{\partial u}{\partial x}=2x=\frac{\partial v}{\partial y},\ \frac{\partial u}{\partial y}=-2y=-\frac{\partial v}{\partial x}.$$

Hence, the C-R conditions are satisfied and $f(z)$ is complex-differentiable.

Example 1.9 Show that the C-R conditions are not satisfied for $f(z)=\bar{z}$.

Solution:

$$f(z)=\bar{z};\ u=x,\ v=-y,\ u_x=1,\ v_y=-1,\ u_y=0=v_x.$$

The C-R conditions are not satisfied here and $f(z)$ is not complex-differentiable.

Example 1.10 Determine if the C-R conditions are satisfied for $f(z)=-1/z^2$.

Solution:

$$\begin{aligned}f(z)&=-\frac{1}{z^2},\ z\neq 0\\ &=-\frac{(x-iy)^2}{(x+iy)^2(x-iy)^2}=-\frac{(x^2-y^2)-2xyi}{(x^2-i^2y^2)^2}\\ &\equiv u(x,y)+iv(x,y)\end{aligned}$$

where

$$u(x,y)=\frac{-x^2+y^2}{(x^2+y^2)^2},\ v(x,y)=\frac{2xy}{(x^2+y^2)^2}$$

and

$$u_x=u_y=\frac{2x^3-6xy^2}{(x^2+y^2)^3}$$

is differentiable so that $-1/z^2$ is complex-differentiable because the C-R conditions are satisfied:

$$f'(z)=\left(-z^{-2}\right)'=2z^{-3}=\frac{2}{z^3}.$$

Note that the C-R conditions, assuming $x = r\cos\theta$, $y = r\sin\theta$, $f(z) = u + iv$, can be written in cylindrical polar coordinates as $\partial u/\partial r = (1/r)(\partial v/\partial\theta)$, $\partial v/\partial r = -(1/r)(\partial u/\partial\theta)$, while the derivative of $f(z)$ can be written as $f'(z) = (\overline{z}/|z|)\left[\frac{\partial u}{\partial r}(z) + i\frac{\partial v}{\partial r}(z)\right]$.

1.9.1 Analytic functions

Suppose $f(z)$ is differentiable at $z = z_0$ (i.e., C-R conditions are satisfied and u_x, u_y, v_x, v_y are continuous at $z = z_0$). If, in addition, $f(z)$ is differentiable (in the region or small neighborhood of z_0 (Figure 1.8)), $|z - z_0| < \epsilon, \epsilon > 0$, we say that $f(z)$ is analytic (regular, holomorphic) at z_0.

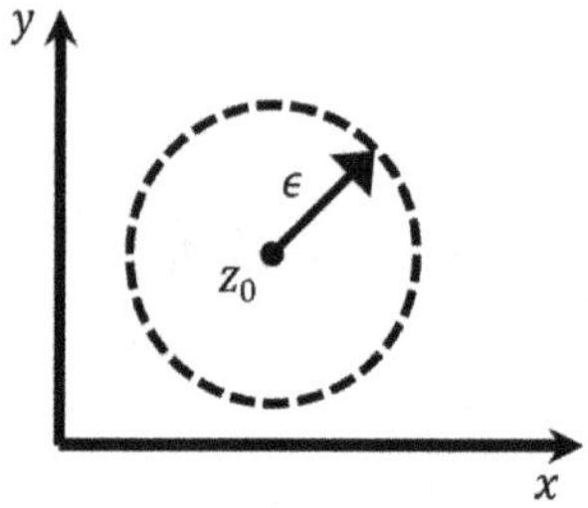

Figure 1.8: Depiction of the neighborhood of analyticity.

Thus, $f(z)$ is analytic at point z_0 if:

(a) The derivatives exist at the point, or the C-R conditions are satisfied (Necessary Condition)

(b) The derivatives exist in a neighborhood of the point, in addition to satisfying the C-R conditions (Sufficient Condition)

The significance of analyticity is that the partial derivatives of all orders of u, v exist and are continuous at z_0, and we can write $u, v \in \mathbb{C}^1$, where $\mathbb{C}^1$ denotes the set of all once-complex differentiable functions. An entire function is one that is analytic at every point of the entire z-plane. Examples include $f(z) = e^z$, $f(z) = \sin z$, and $f(z) = \cos z$. After an entire function, the next level of complication is a function which has only poles in the finite z-plane. Such a function is called a meromorphic function. A singular point z_0 of a function $f(z)$ is a point where the function fails to be analytic.

1.9.2 Harmonic functions

A function g is harmonic if it satisfies Laplace equation in some region $\mathcal{D}$, or

$$\Delta g \equiv \nabla^2 g = 0 \text{ in } \mathcal{D}.$$

This makes g to be analytic everywhere in the complex plane. This property is important in potential theory and in the solution of the Laplace equation in engineering problems with fairly complex geometries. In this case, an advantage can be taken of the properties of conformal transformation, which is the topic of Chapters 3 and 4. Applications of Laplace equation in engineering include temperature distribution in a solid, potential flow, structural mechanics, and electrostatics.

Remark 1.2 If $f(z)$ is analytic throughout D, the C-R conditions imply that

$$u_x = v_y \Rightarrow u_{xx} = v_{xy}, \qquad u_y = -v_x \Rightarrow u_{yy} = -v_{xy}. \tag{1.7}$$

Or, by adding in eq. (1.7):

$$u_{xx} + u_{yy} = 0 \equiv \nabla^2 u.$$

Similarly,

$$\nabla^2 v = 0.$$

Thus, both the real and imaginary parts of the function f satisfy the Laplace equation. The real (φ) and imaginary (ψ) components of the complex potential $F = \varphi + i\psi$ are harmonic mechanics and have widespread applications in several branches of engineering and physics. Some of these applications are shown in Tab. 1.1. Chapter 4 of this book is devoted to fluid mechanics and aerodynamic applications.

Table 1.1: A few harmonic functions in engineering and physics.

Application	$\varphi(x, y)$	$\psi(x, y)$
Fluid mechanics	Velocity potential	Stream function
Heat transfer	Isothermals	Heat flow lines
Elasticity	Strain function	Stress lines
Gravitational field	Gravitational potential	Lines of force
Electrostatics	Potential	Lines of force
Magnetism	Potential	Lines of force
Current flow	Potential	Lines of flow

1.10 Multivalued functions

It is the case that certain functions assume different values at the same point in a complex domain. To illustrate this, define z_1 as the first point in a complex plane and z_2 as the point after encircling the "origin" (branch point) and coming back to the first point. Single-valuedness implies

$$f(z_1) = f(z_2).$$

We cover 2π degrees by encircling the origin in this illustration (Figure 1.9):

$$z = z(r, \theta) = re^{i\theta} = re^{i(\theta+2\pi)} = \ldots = re^{i(\theta+2n\pi)}.$$

Multivaluedness is usually an undesirable property in analytical procedures and is more commonly found in the following kinds of functions:
(a) $f(z) = z^m$, where m is a fraction. That is, m is not an integer.
(b) log functions
(c) Inverse trigonometric functions

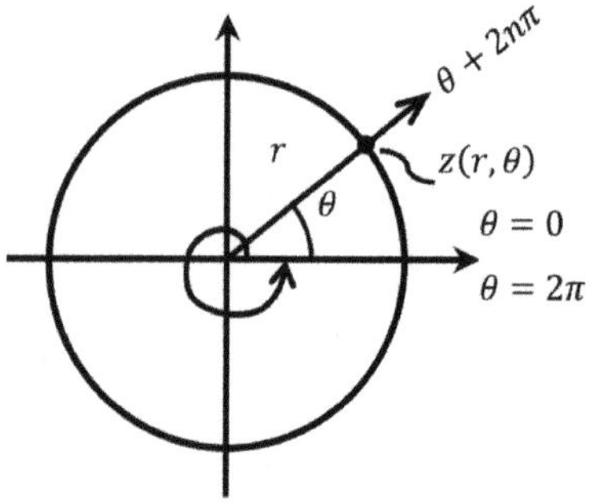

Figure 1.9: Encircling the branch point ((0, 0) in this case) could bring about multivaluedness: $f(r, \theta) \neq f(r, \theta + 2n\pi)$.

Example 1.11
(a) Single-valued functions. As an example of a single-valued function, consider

$$f(z_1) = z = re^{i\theta} = r(\cos\theta + i\sin\theta),$$

$$f(z_2) = re^{i(2\pi+\theta)} = r[\cos(2\pi + \theta) + i\sin(2\pi + \theta)] = r(\cos\theta + i\sin\theta).$$

We see that $f(z_1) = f(z_2)$ for this special case, where single-valuedness is due to cosine and sine being cyclic with period 2π.
(b) Multivalued functions. As examples of multivalued functions, we will consider the three cases below:
   (i) Fractional powers of complex functions. Consider $f(z) = z^{1/2}$:

$$f(z) = z^{1/2} = r^{1/2}e^{i\theta/2}$$

$$f(z_1) = r^{1/2}[\cos(\theta/2) + i\sin(\theta/2)]$$

$$f(z_2) = r^{1/2}e^{i(2\pi+\theta)/2}$$

$$= r^{1/2}[\cos(\pi + \theta/2) + i\sin(\pi + \theta/2)] \neq f(z_1)$$

(ii) Log of complex functions. Consider $\log z$:

$$f(z) = \log z = \log re^{i\theta} = \log r + i\theta$$
$$= \log\, r \,+\, i[2n\pi + \theta] \neq \log\, r \,+\, i\theta,\quad n \neq 0,$$

where $n =$ number of times the branch point (origin in this case) is encircled. Note that $\log e^x \equiv x\log_e e = x$, and that $\log_{10} e^x = \dfrac{\log_e e^x}{\log_e 10} = \dfrac{x}{\log_e 10}$ and more generally $\log_a b = \dfrac{\log_x b}{\log_x a}$.

(iii) Inverse trigonometric functions. Consider $f(z) = \sin^{-1}(z)$:

$$f(z) = \sin^{-1}(z) = \omega,\ \text{or}\ \sin\omega = z = \frac{e^{i\omega} - e^{-i\omega}}{2i}.$$

We can write this as

$$e^{i\omega} - 2iz - e^{-i\omega} = 0.$$

Multiplying by $e^{i\omega}$ yields

$$e^{2i\omega} - 2ize^{i\omega} - 1 = 0,\ \left(\lambda^2 - 2iz\lambda - 1 = 0\right), \lambda \equiv \underbrace{e^{i\omega}}_{(a)} = iz \pm \sqrt{1 - z^2},$$

which, from $\lambda = e^{\log\lambda}$, gives $\lambda = \underbrace{e^{\log\left[iz \pm \sqrt{1-z^2}\right]}}_{(b)}$.

Equating the exponents of (a) and (b) we have

$$f(z) = \omega = \frac{1}{i}\log\left[iz \pm \sqrt{1 - z^2}\right] = \sin^{-1}z.$$

We again have a log function, which is multivalued.

Multivalued functions always occur but are useless. To solve the multivaluedness problem, we introduce branch cuts. An example of a branch cut is shown in Figure 1.10.

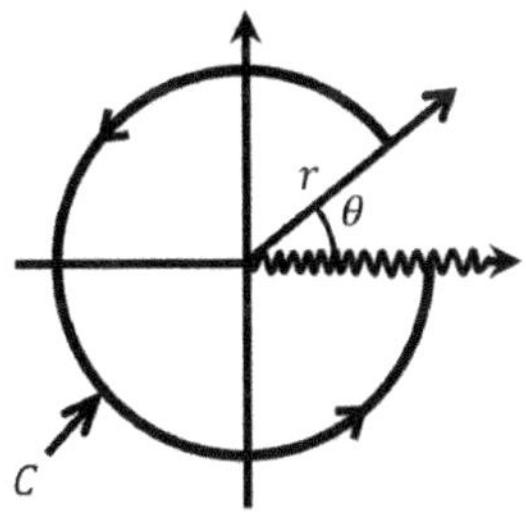

Figure 1.10: A branch cut along $0 \le x < \infty, y = 0$.

The domain of a function $f(z)$ must not include the branch point. Therefore, we modify the domain accordingly, by excluding the line $(0 \le x < \infty, y = 0)$ from the contour.

Because of the cut, $\theta = 0 \neq \theta = 2\pi$, as a discontinuity (Figure 1.11) of the function is now implied along $\theta = 0$ after each circuit around the branch point, we no longer have a problem of multivaluedness. After introducing the cut, contour C is now formally depicted as the "key-hole contour" shown in Figure 1.12. The presence of the new contour is taken into consideration during analysis.

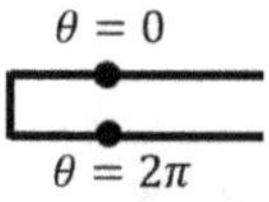

Figure 1.11: An enlarged view of the discontinuity at $\theta = 0$ when you have a branch cut along $x \in [0, \infty)$.

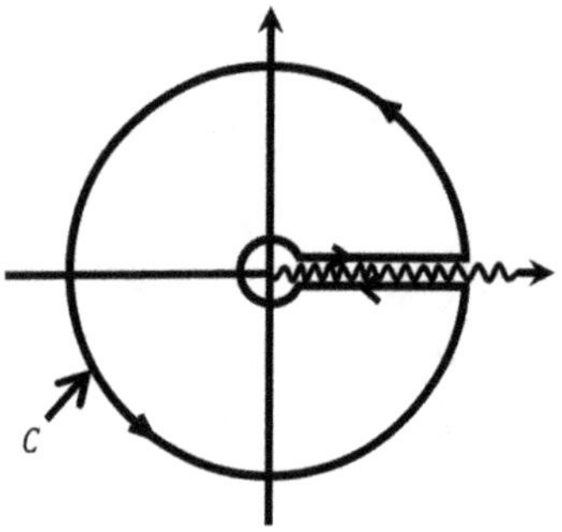

Figure 1.12: The key-hole contour C with a branch cut.

1.10.1 Further remarks on branch cuts

(a) For some functions, we need to encircle a branch point that is not necessarily the origin in order for the function to be multivalued. A redefinition (limitation) of the domain for $f(z)$ removes the multivaluedness. That is, the issue is about the domain. For example, the domain for $\log z$ (Figure 1.13) is such that the function is single-valued since the origin (branch point) is not encircled. It is the encircling of the branch point that causes the multivaluedness. The illustration in Figure 1.14 also shows single-valuedness for $\log z$, as the relevant domain merely touches, and does not encircle, the origin (the branch point).

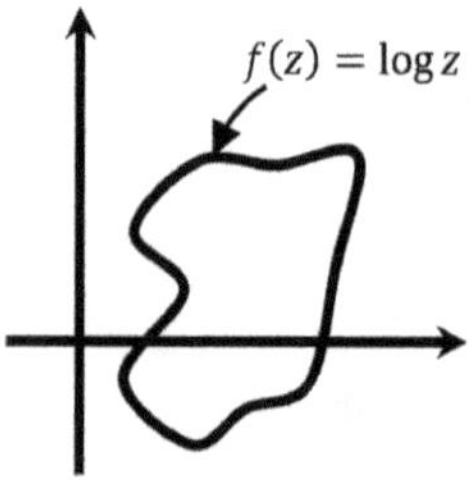

Figure 1.13: $\text{Log} z$ is single-valued when a point other than $(0, 0)$ is encircled.

(b) There is no unique branch cut. Any cut along any direction will be acceptable as long as a range of θ equal to 2π is traversed and the branch point is included in the cut. For example, Figure 1.15 shows another valid branch cut for a branch point at the origin. The cut in this case is in $x \in (-\infty, 0]$.

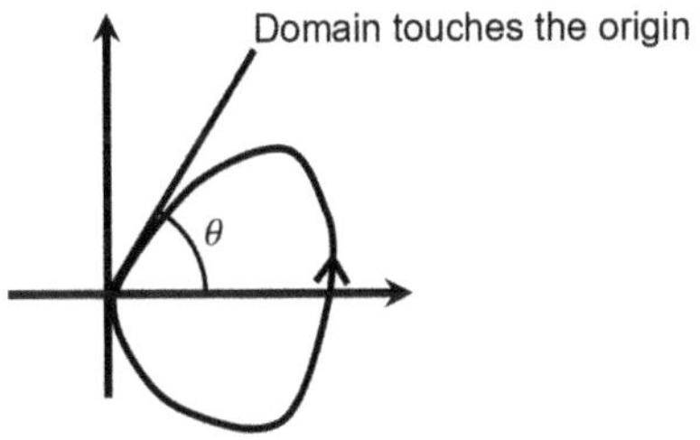

Figure 1.14: A domain merely touching the origin is single-valued even if the origin is otherwise a branch point.

(c) It is emphasized that the branch point of $f(z)$ is the point which, if encircled by $f(z)$, causes the function to be multivalued, and that the point need not be the origin of the complex plane.

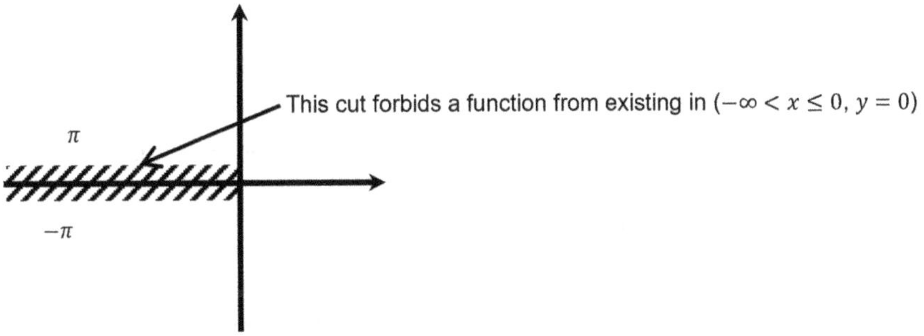

Figure 1.15: A branch cut between $-\pi < \theta < \pi$.

Example 1.12 The function

$$f(z) = \sqrt{z+1} \equiv f_1(z') = \sqrt{z'};\ z' = z + 1,$$

with origin at $z' = 0$ or $(-1, 0)$ is multivalued if it encircles $z' = 0$ or $z = -1$ based on the example for $f(z) = z^{1/2}$ presented previously (Figure 1.16).

Three branch cuts labeled $\mathcal{L}_1$, $\mathcal{L}_2$, and $\mathcal{L}_3$ are shown in Figure 1.17 for the function $f(z) = \sqrt{(z-1)(z-2)}$, which is obviously multivalued because of the square root operation. The branch points for the function are located at $z = 1,\ 2$. The branch cuts, all

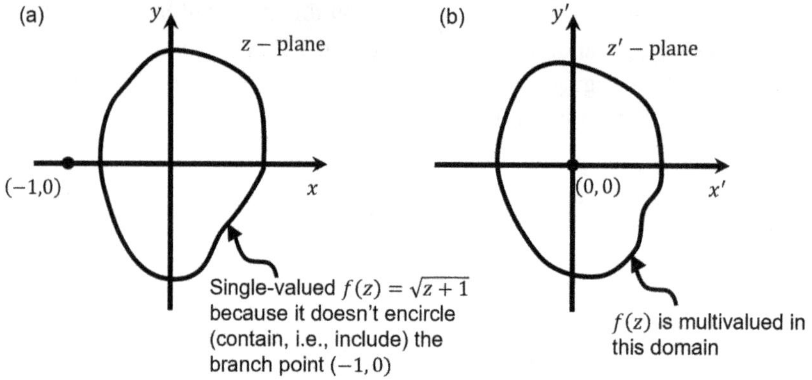

Figure 1.16: A branch point away from the origin (a), transformed to a new origin (b).

on $y = 0$, are, respectively, $-\infty < x \leq 1$, $2 \leq x < \infty$, and $1 \leq x \leq 2$. Note that cuts $\mathcal{L}_1$ and $\mathcal{L}_2$ must be applied together to simultaneously exclude the two branch points.

1.10.2 Implication of branch cuts

Many useful analyses with complex variables involve contour integration. Creating branch cuts leads to different contours, hence different results of contour integration relative to integrals without branch cuts. Appropriate contours and branch cuts must be chosen to ensure that the desired domain of integration is properly represented. A sample contour for integrating the function $f(z) = \sqrt{(z-1)(z-2)}$ is shown in Figure 1.18.

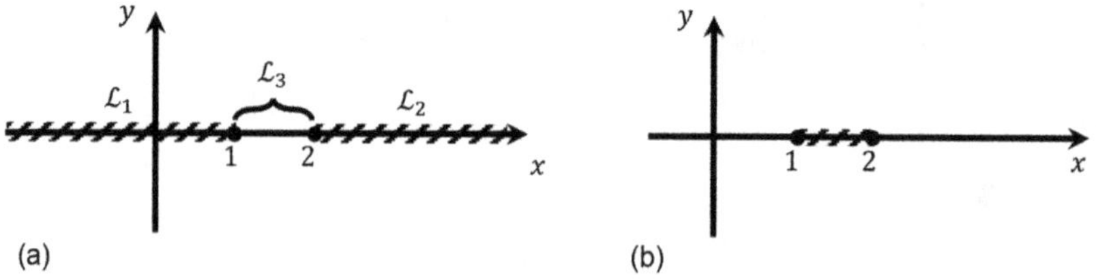

Figure 1.17: Three branch cuts for $f(z) = \sqrt{(z-1)(z-2)}$ are shown in (a). The third branch cut is shown in (b). Note that $\mathcal{L}_1$ and $\mathcal{L}_2$ must be taken together.

The details of the branch cuts $\mathcal{L}_1 \cup \mathcal{L}_2$ and $\mathcal{L}_3$ are shown in Figures 1.18(a) and 1.18(b). Note that $z = \infty$ is not a branch point of $f(z)$. To see this, set $z = 1/t$ so that $f(z) = [(z-1)(z-2)]^{1/2} = [(1-t)(1-2t)]/t$ and limit $t \to 0$ gives $f(t) \approx 1/t$, which is not a multivalued function. With $z - 1 = \rho_1 e^{i\theta_1}$ and $z - 2 = \rho_2 e^{i\theta_2}$, where $\rho_1 = |z-1|$, $\rho_2 = |z-2|$, and $f(z) = (\rho_1\rho_2)^{1/2} e^{i(\theta_1+\theta_2)/2}$, the magnitudes ρ_1 and ρ_2 are of course fixed by the location of the point z, while the angles θ_1 and θ_2 can be chosen freely to span 2π, without passing through a branch cut. Figures 1.18(a) and 1.18(b) show a sample choice of these angles.

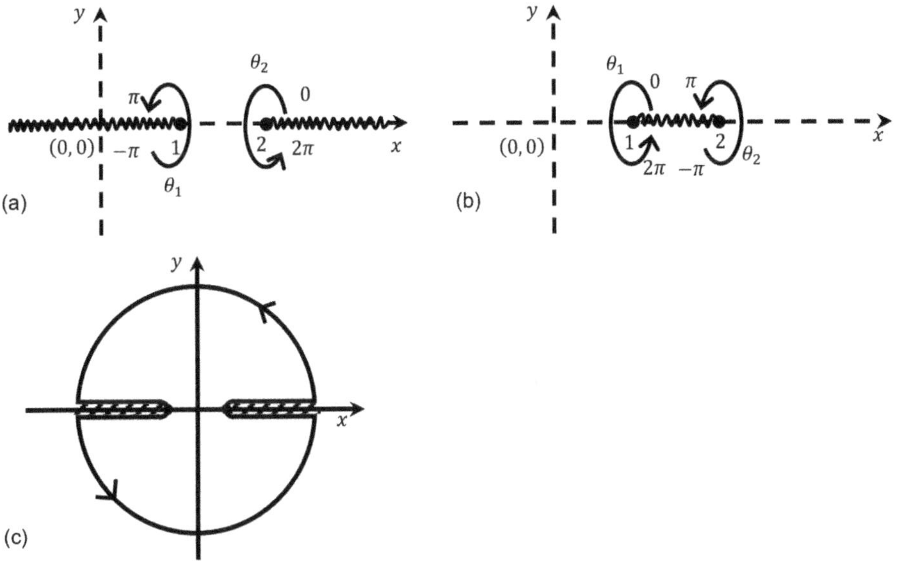

Figure 1.18: Details of the branch cuts for $\mathcal{L}_1 \cup \mathcal{L}_2$ (a), $\mathcal{L}_3$ (b) and a contour with a specific branch cut for integrating $f(z) = \sqrt{(z-1)(z-2)}$ (c).

1.11 Riemann surface

A Riemann surface allows us to generalize the z-plane into a surface composed of sheets, such that each branch of a multivalued function (such as $\log z$; $z^{1/n}$, $n > 1$; and $\sin^{-1} z$) can be imagined to be defined on a sheet of the surface. All sheets are also imagined to be fastened together along a branch cut. The multivalued function will thus become a single-valued function on the surface, allowing the theory of single-valued functions to be applied for the function. The number of sheets in a surface depends on the multiplicity of the function. For example, $f(z) = z^{1/2}$ has two roots and hence two sheets, $f(z) = z^{1/n}$ has n sheets, while $f(z) = \log z$ will have an infinite number of sheets since the function has an infinite number of roots $\left(\log z \equiv \log\left(re^{i\theta}\right) = \log r + i(\theta + 2n\pi),\ n = 1, 2, \ldots\right)$. The function $(z-a)^{1/3}(z-b)^{1/2}$ requires six sheets.

Let us consider the function $f(z) \equiv w(z) = z^{1/2}$. This function has two values for each value of z except at the origin, where w is unique. Hence, the function will be represented by two sheets on a Riemann surface. Let us designate the two sheets as S_1 $\left(A - D - A' - B\right)$ and S_2 $\left(B - E - B' - C\right)$ in Figure 1.19 and assume that they are both cut along the positive x-axis ($0 \le x < \infty$ in $z = x + iy$). We will also place S_2 in front of S_1 so that S_2 is further out of the paper relative to S_1. In the projection of the two surfaces depicted in Figure 1.19, S_1 is shown in a dash line (background) and S_2 in a solid line (foreground).

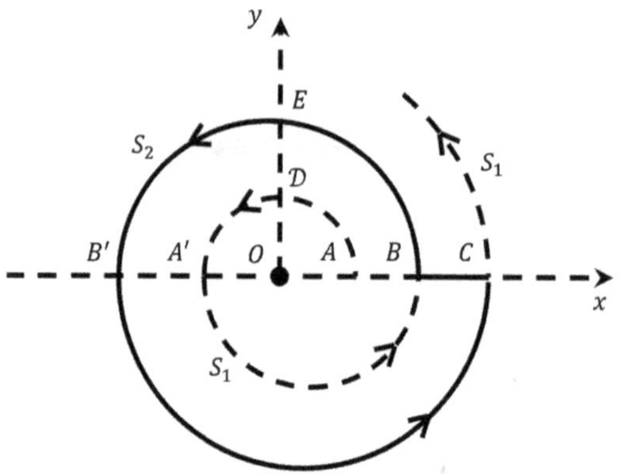

Figure 1.19: A projection of surfaces S_1 and S_2 where S_2 (solid line) is located further out of the paper relative to S_1 (dashed line).

Note that S_1 goes from $\theta = 0$ to $\theta = 2\pi$ (in the z-plane), at which point it is then stitched to S_2, which takes θ from 2π to 4π in the z-plane. Because $z^{1/2}$ has only two roots, S_2 is stitched back to S_1 at $\theta = 4\pi$, from where θ varies either from 4π to 6π, or from 0 to 2π, the choice of which does not affect the value of $w(z)$ because of the cyclic nature of θ. That is, the 2π periodicity in $w(z) = r^{1/2}[\cos(\theta/2 + n\pi) + i\sin(\theta/2 + n\pi)]$ is equivalent to 4π periodicity in z.

The sheets S_1 and S_2 cross each other at the branch cut plane where they are stitched together in some rather complicated manner. At any rate z describes a continuous circuit about the origin, going from $\theta = 0$ to $\theta = 4\pi$, without a multiplicity in $w(z)$. The region $0 \le \theta \le 2\pi$ in the z-plane is mapped to $0 \le \theta < \pi$ or the upper half plane in $w(z)$, while $2\pi \le \theta \le 4\pi$ in the z-plane is mapped to $\pi \le \theta < 2\pi$ (lower half plane) in $w(z)$. Note that the surfaces (sheets) are not parallel to each other but contoured in three dimensions in the vicinity of the cut plane. Moreover, the two sheets have the origin as a common point. Figure 1.20 shows the Riemann surface for $w(z) = z^{1/2}$.

For the function $z^{1/n}$, $n > 1$, there will be n roots and hence n sheets in the Riemann surface. In this case, the sheets S_1 and S_2 in S_1,S_2, . . ., S_n will cover $0 \le \theta \le 2\pi$, $2\pi \le \theta \le 2n\pi$, respectively, while the nth sheet, S_n, will ultimately be stitched to S_1. Figure 1.21 shows the Riemann surface for $w(z) = z^{1/3}$.

The function $f(z) \equiv w(z) = \log z = \log r + i(\theta + 2n\pi)$ has infinitely many values and hence an infinite number of sheets in its Riemann space representation. Figure 1.22 shows a few sheets of the Riemann surface for $\log z$. Specifically, we replace the z-plane by a Riemann surface wherein a new sheet is created whenever the argument of a point in the z-plane is increased or decreased by 2π. Thus, as for the square root function, the argument in Sheet 1 will range from 0 to 2π, while in sheets 2, 3, . . ., n, it

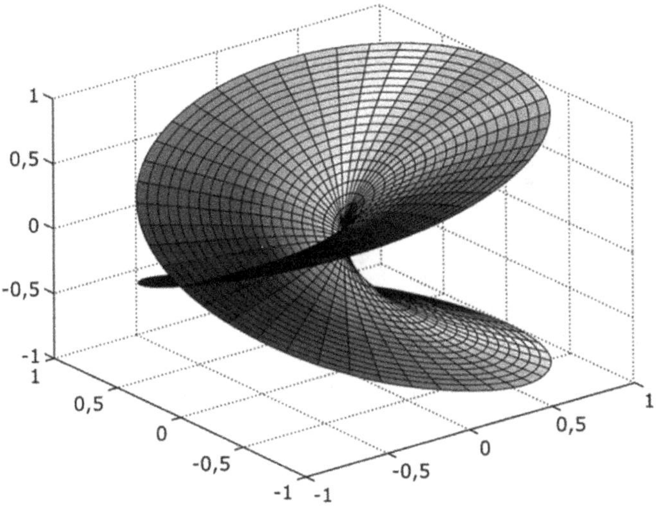

Figure 1.20: The Riemann surface for $w(z) = z^{1/2}$.

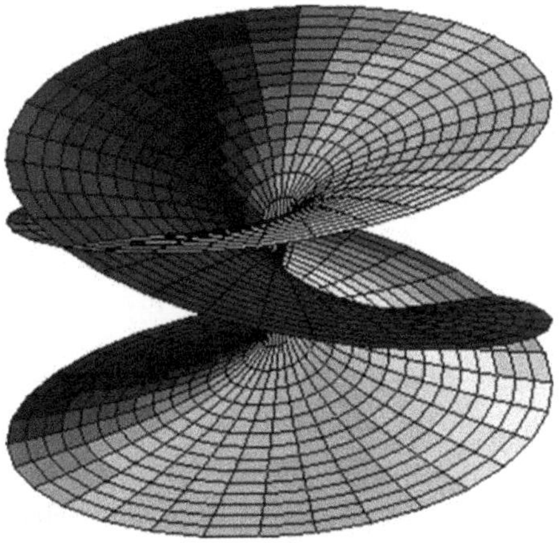

Figure 1.21: The Riemann surface for $w(z) = z^{1/3}$.

will range from 2π to 4π, 4π to 6π, . . ., and $2(n-1)\pi$ to $2n\pi$, respectively. Also sheet S_n is placed in front of S_{n-1}, which is placed in front of S_{n-2}, and so on, as in the case for the square root function, and all sheets are cut along $0 \leq x < \infty$. The projection of the sheets is as shown in Figure 1.19 for the square root function, except that now we have an infinite number of sheets. Note that a sheet S_0 for which the argument varies from zero to -2π could be cut and placed behind S_1, the two being joined at plane A in Figure 1.19. Note that the origin is a point that is common to all sheets, and that the (r, θ) coordinates of a point on any sheet can be represented as the (r, θ) coordinates of the projection of the point on the original z-plane, with the realization that the θ values are unique and span 2π radians on different sheets, as described above.

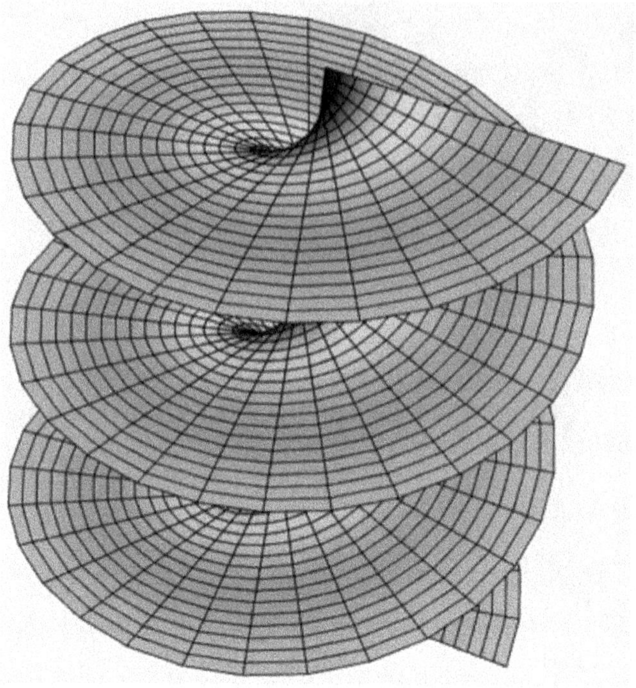

Figure 1.22: The Riemann surface for $w(z) = \log z$.

With the new single-valued function $\log z = \log r + i(\theta + 2n\pi)$ defined on the Riemann surface, we see that $-\pi < \theta \leq \pi$ in z is mapped, for the principal value of $w(z)$, into $-i\pi \leq w \leq i\pi$, or $-\pi < v \leq \pi$, if we write $w = u + iv$. We see that the image of Sheet S_i is the strip $2(i-1)\pi \leq v \leq 2i\pi$ on the imaginary axis of $w(z)$. As noted in [2,3], the strip $2\pi \leq v \leq 4\pi$ represents the analytic continuation of the single-valued analytic function $w(z) = \log r + i\theta$, where $0 < \theta < 2\pi$, upward across the positive x-axis. This implies that $w(z)$ is not only a single-valued function of all points z on the Riemann surface but is also an anlytic function at all points except the origin in the z-plane.

1.12 Two-dimensional integrals

As a prelude to complex integrals, consider the integration of a function $f(z)$ along a curve C, which could be open or close but is piecewise smooth (no gaps) and directed (counterclockwise in our case). It is important to note that this integration is not over the area under a curve since the ordinate in Figure 1.23 is not $f(z)$! (It is iy.) In fact, $f(z)$ is defined *along* the curve. The integration can be defined as follows:

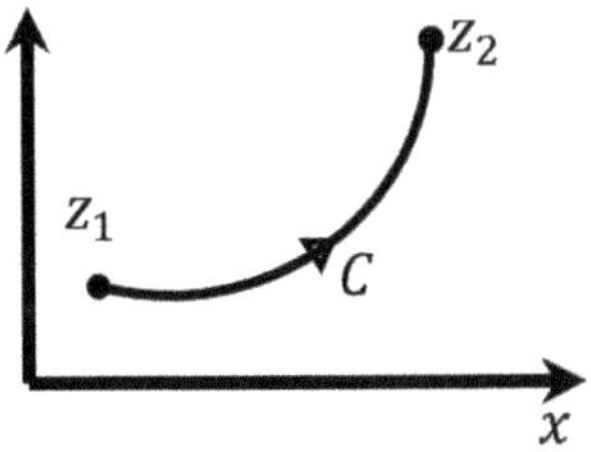

Figure 1.23: Line integral in complex plane. This does not involve the area under the curve C, but integration along C.

Definition 1.3

$$\int_C f(z)dz = \lim_{|p|\to 0} \sum_{k=1}^{n} f(z_{k-1})(z_k - z_{k-1}),$$

where the norm of the partition, $|p| = \max_{1\le k\le n} |z_k - z_{k-1}|$.

1.12.1 Conservative field, F

The idea of a conservative force field that we learned in high school or the freshman class is relevant to contour integration in the complex plane. Let us explore the relationship and define a vector force field **F** and a displacement vector **R** as

$$\mathbf{F} = M\mathbf{i} + N\mathbf{j} + P\mathbf{k}, \qquad \mathbf{R} = x\mathbf{i} + y\mathbf{j} + z\mathbf{k},$$

so that

$$d\mathbf{R} = dx\mathbf{i} + dy\mathbf{j} + dz\mathbf{k}.$$

If **F** is conservative, the integrand in the work (force × distance) integral $\int_A^B \mathbf{F}\cdot d\mathbf{R}$ or

$$\mathbf{F}\cdot d\mathbf{R} = Mdx + Ndy + Pdz$$

is an exact differential. That is, there is a scalar function φ whose *total differential* is equal to the given integrand or

$$d\varphi = \mathbf{F}\cdot d\mathbf{R} \ (= Mdx + Ndy + Pdz).$$

That is, $\int_A^B \mathbf{F}\cdot d\mathbf{R} = \int_A^B d\varphi = \varphi_B - \varphi_A = 0$ if points A and B coincide. This holds if and only if $\mathbf{F} = \nabla\varphi$, such that (since $d\varphi = (\partial\varphi/\partial x)dx + (\partial\varphi/\partial y)dy + (\partial\varphi/\partial z)dz$):

$$M = \frac{\partial \varphi}{\partial x}, \qquad N = \frac{\partial \varphi}{\partial y}, \qquad \text{and} \qquad P = \frac{\partial \varphi}{\partial z}. \tag{1.8}$$

Note that this idea is used in mechanics when we require the gravitational force field to be expressible as the gradient of a scalar in order for it to be conservative.

Theorem 1.3 Let $M(x,y,z)$, $N(x,y,z)$, and $P(x,y,z)$ be continuous, as are their first-order partial derivatives. Then a necessary condition for the expression

$$Mdx + Ndy + Pdz \equiv d\varphi$$

to be an exact differential is that the following equations hold:

$$\frac{\partial M}{\partial y} = \frac{\partial N}{\partial x}, \qquad \frac{\partial M}{\partial z} = \frac{\partial P}{\partial x}, \qquad \frac{\partial N}{\partial z} = \frac{\partial P}{\partial y}.$$

The validity of this equation and the existence of a function φ that meets the criteria above (eq. (1.8)) are the necessary and sufficient conditions for the integral

$$\int_A^B \mathbf{F} \cdot d\mathbf{R} = \int_A^B d\varphi = \int_A^B (M\ dx + N\ dy + P\ dz)$$

to be independent of the path taken from A to B so that the integral depends only on the end points A and B (as per the definition of total differential). Applied to imaginary functions, with $z \equiv x + iy$, the integral

$$I = \int_C f(z)dz$$

can be written as

$$I = \int_C [u(x,y) + iv(x,y)](dx + idy)$$

or

$$I = \int_C [u(x,y)dx - v(x,y)dy] + i\int_C [v(x,y)dx + u(x,y)dy], \tag{1.9}$$

where each of the two integrals is a line integral of the form

$$I' = \int_C Mdx + Ndy,$$

where $M = u$, $N = -v$ for the first integral, and $M = v$, $N = u$ for the second.

Thus, path independence for the first integral requires

$$\frac{\partial M}{\partial y} = \frac{\partial N}{\partial x} \quad \text{or} \quad \frac{\partial u}{\partial y} = -\frac{\partial v}{\partial x},$$

and for the second integral requires

$$\frac{\partial M}{\partial y} = \frac{\partial N}{\partial x} \quad \text{or} \quad \frac{\partial v}{\partial y} = \frac{\partial u}{\partial x}.$$

These conditions: $u_y = -v_x$ and $v_y = u_x$ are simply those of C-R on $f \equiv u + iv$. Thus, if these conditions are satisfied, we can use the idea of a conservative (path-independent) field to obtain the integral $\int_C f(z)dz$ directly using only the end points. For the first integral in eq. (1.9), as an example,

$$\mathbf{F} = u\mathbf{i} - v\mathbf{j}, \qquad \mathbf{R} = x\mathbf{i} + y\mathbf{j}, \qquad d\mathbf{R} = dx\mathbf{i} + dy\mathbf{j}$$

and

$$\mathbf{F} \cdot d\mathbf{R} = (u\mathbf{i} - v\mathbf{j}) \cdot (dx\mathbf{i} + dy\mathbf{j}) = udx - vdy.$$

Theorem 1.4 If $f'(z)$ exists and it is continuous in a simply connected region D (Figure 1.24) and if C, a path from z_0 to z, lies entirely within D then

$$\int_C f(z)dz = \int_{z_0}^{z} f(z)dz = F(z)$$

is analytic and $F'(z) = f(z)\ \forall z \in D$. Also, if $F'(z) = f(z)$ is continuous in D then

$$\int_C f(z)dz = F(z).$$

This is the Fundamental Theorem of Complex Integral Calculus. This theorem is also sometimes stated as follows:

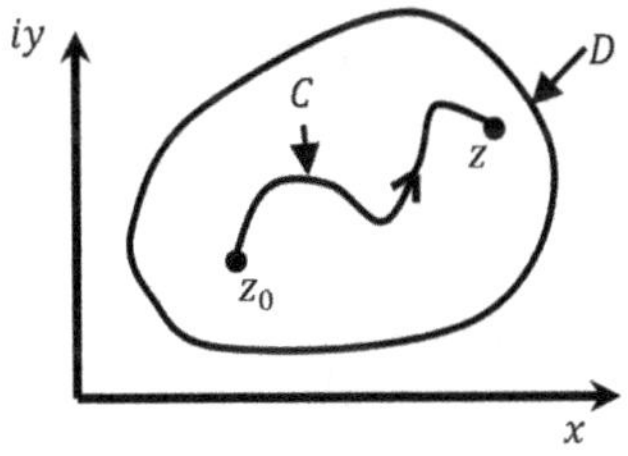

Figure 1.24: A contour in a simply connected domain.

Theorem 1.5 (Fundamental Theorem of Complex Integral Calculus). If F is a primitive of a continuous function $f: \mathcal{D} \to \mathbb{C}$, then, for any piecewise smooth curve C from z_0 to z in $\mathcal{D}$,

$$\int_C f(z)dz = F(z) - F(z_0).$$

Here, $\mathbb{C}$ is the set of all complex numbers.

Example 1.13 Evaluate $\int_C z^2 dz$ where C is the line from $(0,0)$ to $(1,i)$ as shown in Figure 1.25.

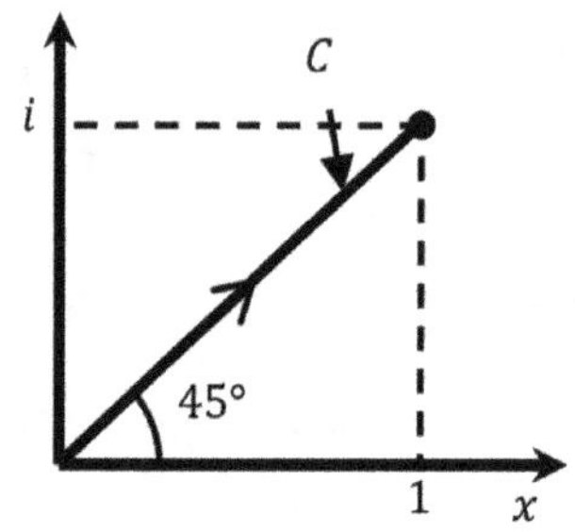

Figure 1.25: The curve from $(0,0)$ to $(1,i)$.

Solution: Two approaches could be used to carry out the required integration:

(a) $f(z) = z^2$ satisfies the C-R conditions. Therefore, we have path independence and can complex integrate directly using the end points:

$$\int_C f(z)dz = \int_{z_1}^{z_2} z^2 dz = \left[\frac{z^3}{3}\right]_0^{1+i} = \frac{2}{3}(-1+i).$$

(b) We could also carry out real integrations using a first-principle approach. Note that $x = y$ along C and in this relationship between x and y, we do not need to concern ourselves with the fact that y axis is imaginary, as this is allowed for in the definition of z in terms of x and y, or $z = x + iy$. Using the equation of the contour, we convert to real integration along the fixed curve (line here) $x = y$, as an integral in one direction. Thus, with $z = x + iy$, $dz = dx + idy$, and $x = y$, we have

$$\int_C z^2 dz = \int_C (x+iy)^2(dx+idy) = \int_0^1 x^2(1+i)^2[1+i]dx = (1+i)^3\frac{1}{3} = \frac{2}{3}(-1+i).$$

Example 1.14 Evaluate $\int_C \sqrt{z}dz$, where C is the circle $z = 2e^{i\theta}$ from 0 to 2π and $\sqrt{z}$ is defined by the branch cut shown in Figure 1.26.

Solution:

(a) From Theorem 1.4 (since $f(z)=\sqrt{z}$ is analytic and $f'(z)$ is continuous in D and C lies entirely in D):

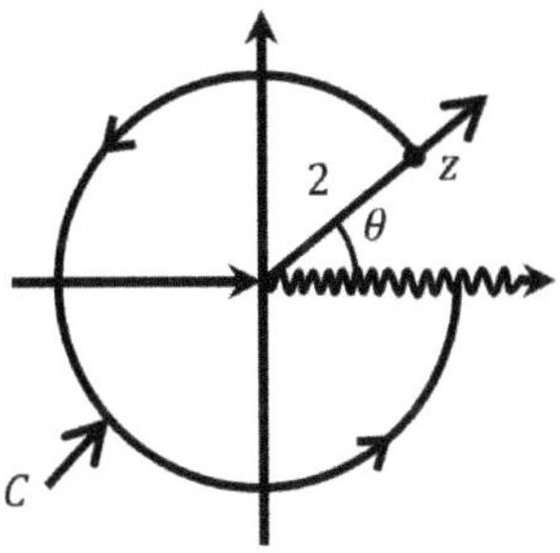

Figure 1.26: Contour with a branch cut when the integrand is multivalued.

$$\int_C \sqrt{z}dz = \frac{2}{3}z^{3/2}\Big|_{2e^{i0}}^{2e^{i2\pi}} = -\frac{8\sqrt{2}}{3}$$

Note that $\theta = 0$ and $\theta = 2\pi$ are distinct on account of the branch cut.

(b) From first principle: That is, integrating along a curve or line as you would in a problem with one independent variable. We do this by using the equation of the curve to eliminate one of the two variables (r,θ). The procedure is completely analogous for (x,y) and (r,θ). We will integrate along θ here by eliminating r:

$$z = re^{i\theta}, \qquad dz = \frac{\partial}{\partial r}\left(re^{i\theta}\right)dr + \frac{\partial}{\partial \theta}\left(re^{i\theta}\right)d\theta = e^{i\theta}dr + ire^{i\theta}d\theta = ire^{i\theta}d\theta = 2ie^{i\theta}d\theta$$

$(r=2$ and $dr=0$ on $C)$. Therefore,

$$\int_C \sqrt{z}dz = \int_C \sqrt{2}e^{i\theta/2}2ie^{i\theta}d\theta = -2\sqrt{2}\int_0^{2\pi}\sin 3\theta/2d\theta + i2\sqrt{2}\int_0^{2\pi}\cos 3\theta/2d\theta = -\frac{8\sqrt{2}}{3}.$$

1.13 Cauchy-Goursat theorem

The Cauchy-Goursat theorem, also sometimes simply referred to as the Cauchy theorem, could be regarded as a formal statement of the complex variable interpretation of a conservative (path-independent) field. The theorem states that if $f(z)$ is single-valued and analytic inside and on a closed counter-clockwise curve C, then $\int_C f(z)dz = 0$. That is, since we have path independence, the two end points coincide (Figure 1.27) so that

$\int_C f(z)dz = 0$. The implication of this theorem is significant in the sense that to have a nontrivial value for contour integration around C, we need to have some singularities inside the contour! This is perhaps one of the instances where having a singularity is desirable. The functions $f(z) = 3$, $f(z) = z$, and $f(z) = z^3$ are analytic with continuous derivatives in $\mathcal{D} \subset \mathbb{C}$. Therefore, integrating them around a closed loop will give zero. Practical applications of complex integrations in engineering and physics must involve integrands that have singularities. Also, if you integrate the pressure field around a closed loop containing an asymmetric aircraft wing at a fixed chord location, you will get a nonzero pressure force, the lift force on the wing. On the other hand, if such a loop contains only atmospheric air, you will get a zero force. Chapter 4 goes into more details on the use of complex variable theory in aerodynamics.

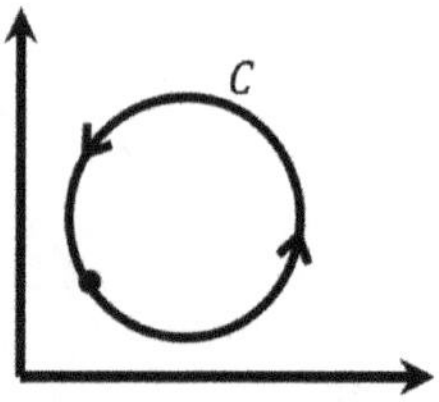

Figure 1.27: Coincident points in a closed curve C.

Theorem 1.6 Corollary (Deformation Theorem)
If $f(z)$ is analytic between and on C_1 and C_2 (Figure 1.28), then

$$\int_{C_1} f(z)dz = \int_{C_2} f(z)dz.$$

This means that if there are no singularities of $f(z)$ between and on contours C_1 and C_2, the contour integration of $f(z)$ over the two curves gives the same result.

Figure 1.29 illustrates contour deformation. If $f(z)$ is analytic inside C_1 in addition, then

$$\int_{C_1} = \int_{C_2} = 0.$$

To see this, heuristically, consider that

$$-\int_A^B = \int_B^D + \int_D^C + \int_C^A.$$

Since

$$\int_B^D = -\int_C^A,$$

we have

$$-\int_A^B = \int_D^C.$$

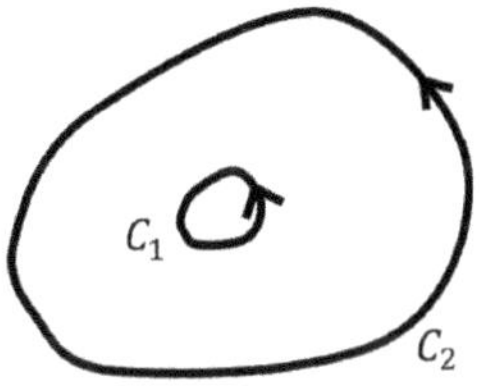

Figure 1.28: Deformation of contour C_2 to contour C_1 if there is no singularity between the two contours.

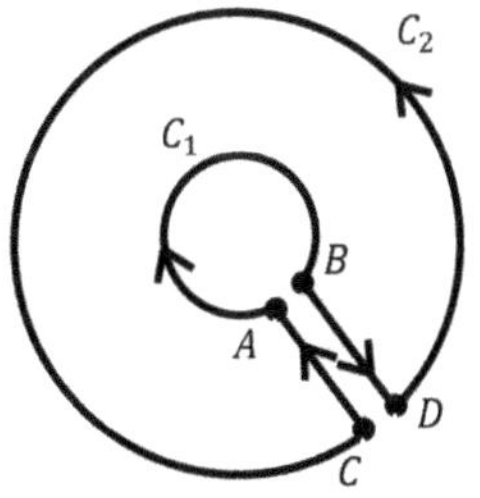

Figure 1.29: Demonstration of contour deformation.

1.14 Singularities and Cauchy's theorems

The significance of having singularities inside a closed curve was alluded to in the previous subsection. In this subsection, we will discuss an approach to carry out contour integration in a closed curve that harbors one or more singularities. In real calculus, when we have

$$G(x) = \frac{f(x)}{g(x)},$$

and $g(x_0) \to 0$ at a point x_0, we say that $G(x) \to \infty$ at $x = x_0$ and use this to mean that $G(x)$ has a singularity at x_0. The same concept applies to complex variables, and we could merely replace x by z, of course with the full awareness that z approaches zero from both the x and iy coordinate directions. As pointed out above, having a singularity is not necessarily a bad thing, as physical objects could represent singularities that enable integrals to yield nontrivial results in engineering and physics applications.

To illustrate, examine $I = \oint_C (z-a)^n dz$, where n is an integer, C is a closed counterclockwise (CCW), piecewise smooth curve, and a lies inside C (Figure 1.30). The following values of the integral exist for different n:

(a) For $n = 0, 1, 2, \ldots$

$$I = \int_C (z-a)^n dC = 0$$

since $f(z) \equiv (z-a)^n$ is analytic at all points inside the closed curve C, as per the Cauchy-Goursat theorem.

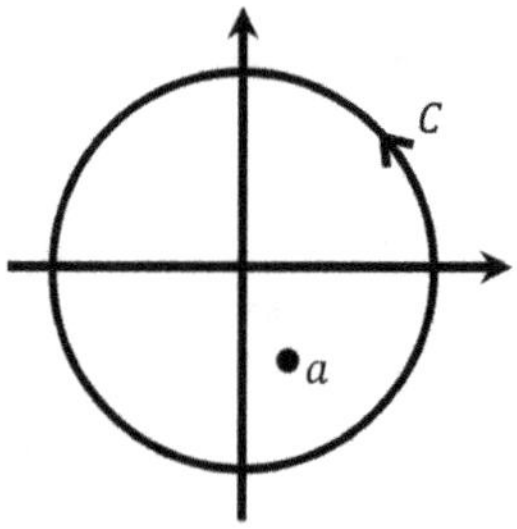

Figure 1.30: A singular point inside a CCW curve C.

(b) For $n = -2, -3, \ldots$

$$I = \int_C (z-a)^n dz = \left.\frac{[z-a]^{n+1}}{n+1}\right|_{z_1}^{z_2}, \qquad n \neq -1,$$

where z_1 and z_2 are the initial and final points on the curve C. Because C is closed, the points z_1, z_2 coincide and since $(z-a)^{n+1}$ is single-valued (n is an integer), it follows that $I = 0$. Note that the integrand is not analytic inside C, and hence the Cauchy's theorem does not apply.

(c) For $n = -1$, let $z - a = \rho e^{i\theta}$. We can write

$$\int_C \frac{dz}{z-a} = [\ln(z-a)]_{z_1}^{z_2} = (\ln \rho + i\theta)|_{z_1}^{z_2}.$$

Multivaluedness is implied because of the appearance of the log function. However independent of the branch cut chosen for $\ln(z-a)$, we will have $\rho_1 = \rho_2$ so that $\ln \rho|_{z_1}^{z_2} = 0$ and $i\theta|_{z_1}^{z_2} = 2\pi i$. Alternatively, we could deform C to a small circle C' (Figure 1.31) with origin at a since there is no singularity in between C and C'. Using a first principle approach, translate $z - a = \rho e^{i\theta} = 0$ when $z = a$ (Figure 1.32). Thus, $dz = i\rho e^{i\theta} d\theta$ and

$$\int_C \frac{dz}{z-a} = \int_{C'} \frac{dz}{z-a} = \int_0^{2\pi} \frac{i\rho e^{i\theta} d\theta}{\rho e^{i\theta}} = 2\pi i.$$

Therefore, the integral of the function $f(z) = 1/(z-a)$ around a CCW curve enclosing a singularity at $z = a$ is $2\pi i$. This is an important result.

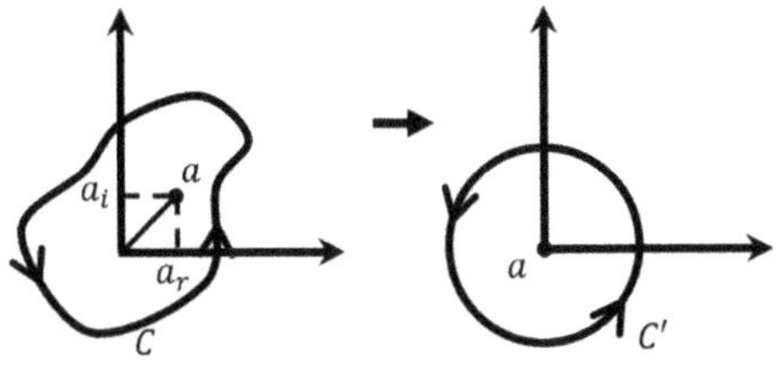

Figure 1.31: Deformation of contour C to a circular contour C'.

Theorem 1.7 (Cauchy Integral Formula)
If $f(z)$ (not $f(z)/(z-a)$) is analytic inside (including at a) and on a simple closed CCW curve C and if the point a (Figure 1.30) is inside (but not on) C, then

$$I \equiv \frac{1}{2\pi i}\int_C \frac{f(z)}{z-a}\,dz = f(a)$$

or,

$$I' \equiv \int_C \frac{f(z)}{z-a}\,dz = 2\pi i f(a).$$

We emphasize that $f(z)$ is the function that is required to be analytic inside C, not $f(z)/(z-a)$, and that an integrand such as $f(z)/(z+2)$, is singular inside a circle such as $|z|=3$ even as $f(z)$ could be analytic inside this domain.

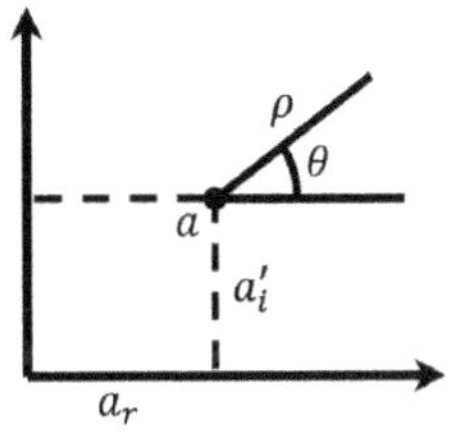

Figure 1.32: Depiction of point a in the complex plane.

1.14.1 The proof of the Cauchy integral theorem

The theorem has the significance that if a regular function is defined on C, then its value at every point inside C is known. To prove this theorem, let Γ be a small circle with center at $z=a$ and radius ρ. Because $z=a$ is an interior point of C, the circle Γ is entirely enclosed by C if ρ is sufficiently small. Then $f(z)/(z-a)$ is regular between and on C and Γ so that by the deformation theorem we have

$$\int_C \frac{f(z)}{(z-a)}dz = \int_\Gamma \frac{f(z)}{(z-a)}dz = \int_\Gamma \frac{f(z)-f(a)}{z-a}dz + \int_\Gamma \frac{f(a)}{z-a}dz \equiv I_1 + I_2.$$

On Γ, $z-a=\rho e^{i\theta}$, $0\le\theta\le 2\pi$. Since $f(z)$ is regular, it is continuous so that for an arbitrary $\varepsilon>0$, $|f(z)-f(a)|<\varepsilon$ if $|z-a|=\rho<\delta$. We also see that

$$|I_1| = \left|\int_\Gamma \frac{f(z)-f(a)}{(z-a)}dz\right| \le \int_\Gamma \frac{|f(z)-f(a)|}{|z-a|}dz < \frac{\varepsilon}{\rho}\cdot 2\pi\rho = 2\pi\varepsilon,$$

leading to $I_1\to 0$ as $\rho\to 0$. (Above we have used the fact that $e^{i\theta}$ goes like unity and $\int_\Gamma dz = 2\pi\rho$.)

We can write

$$I_2 = \int_\Gamma \frac{f(a)}{(z-a)}dz = f(a)\int_0^{2\pi} \frac{\rho i e^{i\theta}d\theta}{\rho e^{i\theta}} = 2\pi i f(a)$$

so that

$$I_1 + I_2 = \int_C \frac{f(z)}{(z-a)}dz = 2\pi i f(a),$$

and

$$f(a) = \frac{1}{2\pi i}\int_C \frac{f(z)}{(z-a)}dz.$$

1.14.2 The formulas for the first and higher derivatives

If $f(z)$ is regular inside and on a closed contour C then it possesses a unique derivative at all points $z=a$ inside C given by

$$f'(a) = \frac{1}{2\pi i}\int_C \frac{f(z)}{(z-a)^2}dz.$$

To prove this, let δ_a be such that $a+\delta_a$ is located in the interior of the domain $\mathcal{D}$ bounded an enclosed by C. We have from Cauchy's integral theorem

$$f(a+\delta_a) = \frac{1}{2\pi i}\int_C \frac{f(z)}{(z-a-\delta_a)}dz,$$

so that

$$f(a+\delta_a)-f(a)=\frac{1}{2\pi i}\int_C\left[\frac{1}{(z-a-\delta_a)}-\frac{1}{(z-a)}\right]f(z)dz=\frac{\delta_a}{2\pi i}\int_C\frac{f(z)dz}{(z-a)(z-a-\delta_a)}.$$

We can divide this equation by δ_a and write it as

$$\frac{f(a+\delta_a)-f(a)}{\delta_a}=\frac{1}{2\pi i}\int_C\frac{f(z)}{(z-a)^2}dz+\frac{\delta_a}{2\pi i}\int_C\frac{f(z)\,dz}{(z-a)^2(z-a-\delta_a)}.$$

Using familiar arguments, the last term in this equation can be shown to approach zero as $\delta_a \to 0$. Hence the proof.

In a similar manner, we can show that

$$f^{(2)}(a)=\lim_{\delta_a\to 0}\frac{f'(a+\delta_a)-f'(a)}{\delta_a}=\frac{2!}{2\pi i}\int_C\frac{f(z)}{(z-a)^3}dz,$$

and that, in general,

$$f^{(n)}(a)=\frac{n!}{2\pi i}\int_C\frac{f(z)}{(z-a)^{n+1}}dz.$$

It is worth noting from the foregoing, that is, as a consequence of the Cauchy Integral Formula, that if $f(z)$ is analytic within a simple closed contour C, it can be differentiated any number of times. This means that the nth derivative $f^n(z)$ is also analytic. It is not true in real variable theory that if $f(x)$ is differentiable once then it is differentiable any number of times.

As an example, for $C=$ the circle $|z-1|=0.5$ as shown in Figure 1.33, there is no singularity of $f(z)/(z+2)$; therefore,

$$I\equiv\int_C\frac{f(z)}{z+2}dz=0.$$

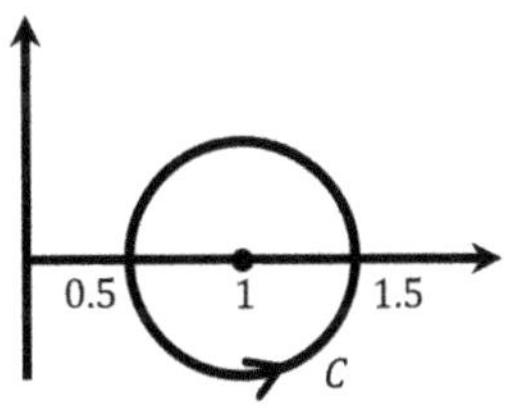

Figure 1.33: Curve C defined by $|z-1|=0.5$.

Example 1.15 Evaluate $I = \int_C e^{-z}/(z+2)dz$ where C is the circle $|z| = 3$ taken CCW.

Solution: $f(z) = e^{-z}$ is analytic inside and on C, and $a = -2$ (which is in C). Thus,

$$I = 2\pi i f(-2) = 2\pi i e^2$$

Note that any C is good as long as a lies inside it, by virtue of the deformation theorem. Also note that the integrand $e^{-z}/(z+2)$ is not analytic everywhere in $|z| = 3$; otherwise, the integral will be zero.

Note that the integral is zero when the nth derivative of $f(z)$ is zero. For example, when the denominator is raised to the power of $(n+1)$ as above and $f(z)$ is the polynomial $\sum_{k=1}^{n-1} a_k \xi^k$, the integral will be zero.

Example 1.16 Evaluate

$$I = \int_C \frac{1}{(\xi - z)^2} d\xi,$$

where z is inside C.

Solution: $n = 1$ and $f(\xi) = 1$ are certainly analytic inside any C. Thus, for an arbitrary C enclosing z, we have

$$\int_C \frac{f(\xi)}{(\xi - z)^2} d\xi = \frac{2\pi i \frac{d^{2-1}}{d\xi^{2-1}} f(\xi)|_{\xi = z}}{(2-1)!} = \frac{f'(z) 2\pi i}{1!} = 0,$$

since $d(1)/d\xi = 0$.

1.14.3 A few more examples on Cauchy's theorem and Cauchy's formula

Example 1.17 Evaluate

$$I = \oint \frac{dz}{z^2 - 5z + 6} = \oint \frac{dz}{(z-2)(z-3)},$$

where C is the unit circle contour shown in Figure 1.34.

There are at least three ways to solve this problem. We could define

$$f(z) \equiv \frac{1}{(z-2)(z-3)},$$

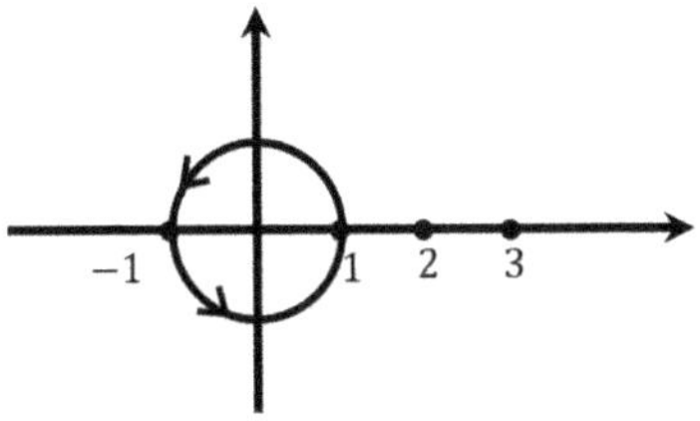

Figure 1.34: The contour $|z|=1$.

which has poles at $z=2,\ 3$. But these poles are not inside C. So $f(z)$ is analytic inside C and

$$\int_C f(z)dz = 0,$$

by the Cauchy-Goursat theorem. We could also set $f(z)=1/(z-3)$ and $a=2$ and use the Cauchy integral theorem. Last, we could set $f(z)=1/(z-2)$, $a=3$ and use the same theorem. In all cases, the Integral will be zero. For this latter case,

$$I = \oint_C \frac{f(z)}{z-3}dz, \qquad f(z) = \frac{1}{z-2},$$

where $a=3$ is not inside C. Therefore, $\oint_C = 0$.

It will indeed be useful to keep the following result in mind:

$$\oint_C (z-a)^n dz = \begin{cases} 2\pi i, & n=-1 \\ 0, & n \neq -1. \end{cases}$$

Example 1.18 Evaluate

$$I = \oint_C \frac{dz}{z^2(z-2)(z-4)},$$

where C is the contour shown in Figure 1.35.

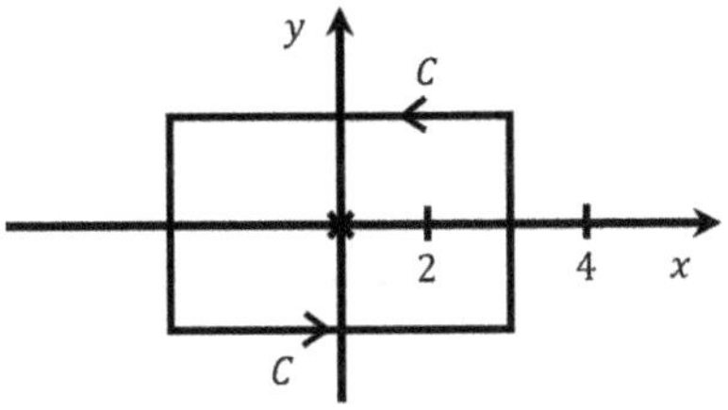

Figure 1.35: A box contour.

Solution: Break the integral up into partial fractions:

$$I = \frac{3}{32}\oint_C \frac{dz}{z} + \frac{1}{8}\oint_C \frac{dz}{z^2} - \frac{1}{8}\oint_C \frac{dz}{z-2} + \frac{1}{32}\oint_C \frac{dz}{z-4}$$
$$= \frac{3}{32} \times 2\pi i(1_{z=0}) + \frac{1}{8} \times \frac{2\pi i[d(1)/dz]}{1!} - \frac{1}{8} \times 2\pi i(1_{z=2}) + \frac{1}{32} \times 0.$$

The second term on the RHS contributes zero because $f(z)$ is a constant so that $f^n(z) = 0$ $(n = 1)$. The last term is a consequence of $z = 4$ being located outside of C, coupled with the fact that there are no poles in C (Cauchy-Goursat).

Thus

$$\mathrm{I} = \frac{3}{32}(2\pi i) + \frac{1}{8}(0) - \frac{1}{8}(2\pi i) + 0 = -\frac{\pi i}{16}.$$

Note that because $z = 0$ and $z = 2$ are located inside the contour in Figure 1.35, it is not easy to factor out a $g(z)$ that is analytic inside the contour that allows us to use the Cauchy Integral theorem.

Example 1.19 Evaluate

$$I = \oint \frac{e^z}{z^3} dz$$

where C is $|z| = 1$, the unit circle (Figure 1.36),

$$I = \oint \frac{e^z}{(z-0)^3} dz.$$

Solution: Compared to

$$I \equiv \oint \frac{f(z)}{(z-a)^{n+1}} dz = \frac{2\pi i}{n!} f^{(n)}(a),$$

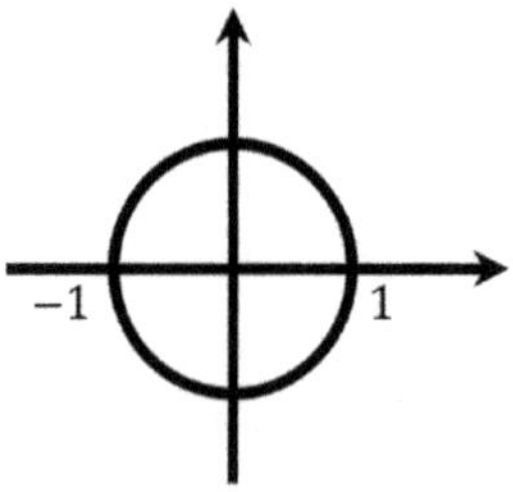

Figure 1.36: The unit circle $|z| = 1$.

we have $n = 2$, $a = 0$, and $f(z) = e^z$. Therefore,

$$I = \frac{2\pi i}{2!}\frac{d^2}{dz^2}e^z|_{z=0} = \frac{2\pi i}{2!}e^z|_{z=0} = \pi i e^0 = \pi i.$$

Example 1.20 Evaluate

$$I = \oint_C \frac{z+1}{z(z-2)(z-4)^3}dz$$

where C is the counterclockwise circle $|z-3| = 2$ (Figure 1.37).

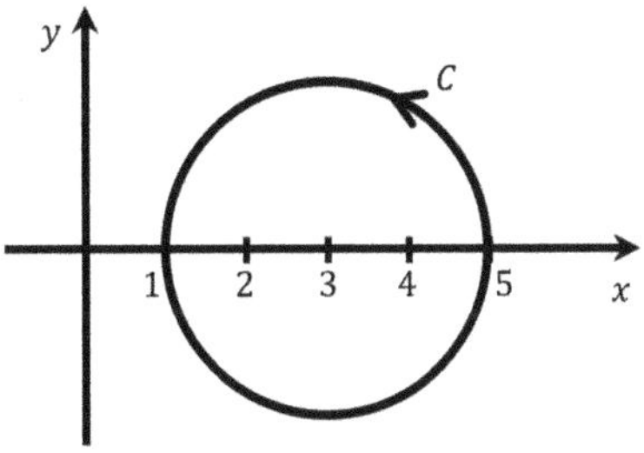

Figure 1.37: The contour $|z-3| = 2$.

Solution: Singularities exist at $z = 0$ (outside of C) and $z = 2$, 4 (within C). We could break up the integrand into partial fractions and integrate, or deform C into contours, respectively, centered around the two poles within the original contour, as in Figure 1.38.

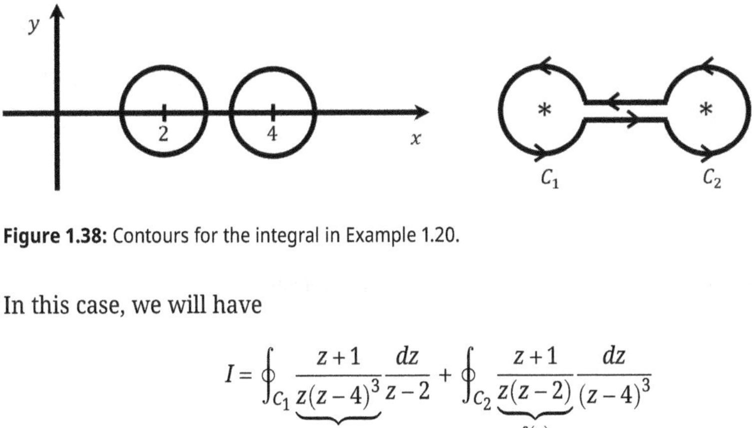

Figure 1.38: Contours for the integral in Example 1.20.

In this case, we will have

$$I = \oint_{C_1} \underbrace{\frac{z+1}{z(z-4)^3}}_{f(z)}\frac{dz}{z-2} + \oint_{C_2} \underbrace{\frac{z+1}{z(z-2)}}_{f(z)}\frac{dz}{(z-4)^3} \tag{1.10}$$

$$= 2\pi i\left[\frac{z+1}{z(z-4)^3}\right]\bigg|_{z=2} + \frac{2\pi i}{2!}\frac{d^2}{dz^2}\left[\frac{z+1}{z(z-2)}\right]\bigg|_{z=4}$$

$$= -\frac{3\pi i}{8} + \frac{23\pi i}{64} = -\frac{\pi i}{64}.$$

Note in eq. (1.10) that $f(z)$ in the first term on the RHS is analytic inside C_1 and $f(z)$ in the second term is analytic inside C_2.

1.15 Poisson's integral formula

One rather simple analysis based on the Cauchy Integral Formula, and due to Poisson, enables a closed-form solution of the Laplace equation inside a unit circle when a Dirichlet condition is specified on the surface of the circle. That is, if we know the value $\phi(\theta)$ of a harmonic function ϕ on the boundary of the unit circle $\zeta = e^{i\theta}$, and $z = re^{i\vartheta}$ is located in the interior of the unit circle ($r < 1$), we can easily obtain $\phi(r, \vartheta)$ at all points in the interior of the circle. This analysis can directly be used in engineering to determine the temperature distributions in a disc or the cooling load in thermal management with a device of similar geometry. We start the analysis with the Cauchy Integral Formula:

$$f(z) = \frac{1}{2\pi i}\int\limits_{C}\frac{f(\zeta)}{\zeta - z}d\zeta. \tag{1.11}$$

With $d\zeta = ie^{i\theta}\ d\theta = i\zeta\ d\theta$, eq. (1.11) can be written as

$$f(z) = \frac{1}{2\pi}\int\limits_{0}^{2\pi}\frac{f(\zeta)\zeta}{\zeta - z}d\theta. \tag{1.12}$$

With z inside the unit circle, $1/\bar{z}$, where $\bar{z}$ is the conjugate of z, must be located outside the unit circle and hence it is not a point of singularity, we can write, by virtue of the Cauchy-Goursat theorem:

$$0 = \frac{1}{2\pi}\int\limits_{0}^{2\pi}\frac{f(\zeta)\zeta}{(\zeta - 1/\bar{z})}d\theta. \tag{1.13}$$

Multiplying the numerator and denominator of the integrand of eq. (1.13) by $\bar{z}/\zeta$ and adding the resulting equation to eq. (1.12), we will have

$$f(z) = \frac{1}{2\pi}\int\limits_{0}^{2\pi}f(\zeta)\left[\frac{\zeta}{\zeta - z} \pm \frac{\bar{z}}{\bar{\zeta} - \bar{z}}\right]d\theta.$$

The positive sign is the one of interest so that

$$f(z) = \frac{1}{2\pi}\int\limits_{0}^{2\pi}f(\zeta)\frac{1 - |z|^2}{|\zeta - z|^2}d\theta \equiv \frac{1}{2\pi}\int\limits_{0}^{2\pi}f(\zeta)Q(\zeta, z)d\theta. \tag{1.14}$$

We note that $Q(\zeta, z)$ is a real quantity. $f(z)$ can be decomposed in the usual manner: $f(z) = \phi + i\psi$, where both ϕ and ψ are real. Taking the real part of eq. (1.14) yields

$$\phi(r, \vartheta) = \frac{1}{2\pi} \int_0^{2\pi} \phi(\theta) \frac{1 - r^2}{|e^{i\theta} - re^{i\vartheta}|^2} d\theta$$

or

$$\phi(r, \vartheta) = \frac{1}{2\pi} \int_0^{2\pi} \phi(\theta) \frac{(1 - r^2)}{[1 - 2r\cos(\vartheta - \theta) + r^2]} d\theta.$$

The factor $\phi(\theta)$ is the distribution of the harmonic function ϕ on the boundary. This result is known as the Poisson's integral formula. Utilizing a procedure similar to the one just given, we can obtain the analogous solution for a circle of arbitrary radius, R:

$$u(r, \vartheta) = \frac{1}{2\pi} \int_0^{2\pi} \phi(\theta) \frac{(R^2 - r^2)}{(R^2 - 2Rr\cos(\vartheta - \theta) + r^2)} d\theta.$$

Note that the second factor in the integrand for $\phi(r, \vartheta)$ or $u(r, \vartheta)$ is a mere real number once the coordinates of the desired point inside the circle have been specified in terms of (r, ϑ).

1.16 Miscellaneous examples

Example 1.21 Obtain the values of i^i.

Solution:

$$i^i = \left[e^{i\left(2k\pi + \frac{\pi}{2}\right)}\right]^i = e^{-\left(2k\pi + \frac{\pi}{2}\right)} \quad (k = 0, \pm 1, \pm 2, \ldots)$$

Example 1.22 Determine the principal value of

$$f(z) = (1 + i)^{2-i}$$

in the form $u + iv$.

Solution:

$$\begin{aligned} f(z) &= (1 + i)^{2-i} = \exp[(2 - i)\mathrm{In}(1 + i)] \\ &= \exp\left\{(2 - i)\left[In\sqrt{2} + i\left(\frac{\pi}{4} + 2n\pi\right)\right]\right\} \equiv f(z). \end{aligned}$$

Now, $n = 0$ gives the principal value or

$$f(z) = \exp\left\{(2-i)\left(\text{In}\sqrt{2} + i\frac{\pi}{4}\right)\right\}$$
$$= \exp\left\{2\left(\text{In}\sqrt{2} + \frac{\pi}{4}\right) + i\left(-\text{In}\sqrt{2} + \frac{\pi}{2}\right)\right\}$$
$$= \exp\left(\text{In}\,2 + \frac{\pi}{4}\right)\left[\cos\left(\frac{\pi}{2} - \text{In}\sqrt{2}\right) + i\sin\left(\frac{\pi}{2} - \text{In}\sqrt{2}\right)\right]$$
$$\approx e^{1.4785}(\sin 0.3466 + i\cos 0.3466)$$
$$= 1.490 + 4.126i$$

Example 1.23 Suppose $f(z) = (1+z^2)^{1/2}$: (a) Show that $z = \pm i$ are branch points of $f(z)$. (b) Show that a complete circuit around the two branch points produces no change in the branches. (c) Suggest a few branch cuts for the function.

Solution:

$$(1+z^2) = (z-i)(z+i) \Rightarrow (1+z^2)^{1/2} = [(z-i)(z+i)]^{1/2}.$$

With the square root operation, the function is multivalued, and the branch points are $z = \pm i$.

(a) Let $z - i = r_1 e^{i\theta_1}$, $z + i = r_2 e^{i\theta_2}$.
Thus,

$$f(z) = \left[r_1 r_2 e^{i(\theta_1+\theta_2)}\right]^{1/2} = \sqrt{r_1 r_2}\, e^{i\theta_1/2} e^{i\theta_2/2} \equiv f_1(z).$$

Let us encircle only $z = i$ by 2π so that

$$f_2(z) \equiv \sqrt{r_1 r_2}\, e^{i(\theta_1+2\pi)/2} e^{i\theta_2/2} = -\sqrt{r_1 r_2}\, e^{i\theta_1/2} e^{i\theta_2/2} \neq f_1(z).$$

Hence, we have multivaluedness and $z = i$ is a branch point. We can establish the same result for $z \equiv -i$.

(b) Here $\theta_1 \to \theta_1 + 2\pi$, $\theta_2 = \theta_2 + 2\pi$ so that

$$f_3 \equiv \sqrt{r_1 r_2}\, e^{i(\theta_1+2\pi)/2} e^{i(\theta_2+2\pi)/2} = \sqrt{r_1 r_2}\, e^{i\theta_1/2} e^{i\theta_2/2} = f_1.$$

Therefore, no change in the function occurs if we make a complete circuit around the two branch points.

(c) The branch cuts below are plausible. Obviously, there are numerous ways to create branch cuts that avoid the two branch points.

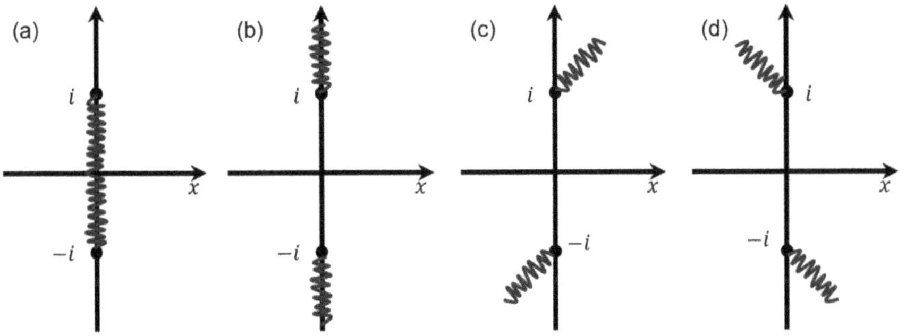

Figure 1.39: A sketch for Example 1.23.

Problems

1.1 Evaluate 1^{i+i}.

1.2 Determine the 4th root of 2.

1.3 Determine the 6th root of -3.

1.4 Express the function e^{z^2} in the form $u(x,y)+iv(x,y)$.

1.5 Express the function $e^{\ln z}$ in the form $u(x,y)+iv(x,y)$.

1.6 If $z=x+iy$, find the equation of the locus $\left|\frac{z+1}{z-1}\right|=2$.

1.7 Determine the curve in the complex plane that is described by $|z-1|=2$.

1.8 Express $z=-1+i$ in polar form.

1.9 For what values of $z\equiv z_0$ does $f'(z)$ exist if $f(z)=2x+3iy$?

1.10 Integrate the function $f(z)=1/z$ once around the circle $|z|=\rho$.

1.11 Integrate

$$\frac{1-z^2}{z^2-1}$$

in the CCW sense around a circle of radius 1 with center at the point

(a) $z=1$, (b) $z=\frac{1}{2}$, (c) $z=-1$, (d) $z=i$.

1.12 Determine the harmonic conjugate of

$$u=\sin x\sinh y.$$

1.13 If the imaginary part of an analytic function is $2x(1-y)$, determine (a) the real part of the function in terms of x, y, (b) the function itself, in terms of z.

1.14 Describe geometrically the region in the z plane determined by the following inequality:

$$|z-1|+|z+1|\le 4.$$

1.15 Show that the function

$$u(z) = y^3 - 3x^2y$$

is harmonic and determine its harmonic conjugate.

1.16 Does the following limit exist?

$$\lim_{z\to 0}\left(\frac{1}{1+e^{1/x}} + iy^3\right).$$

1.17 Integrate $f(z) = \operatorname{Re} z = x$ along the segment from $z = 0$ to $z = 1 + i$.

1.18 Evaluate the function

$$\left(1 - \sqrt{2}i\right)^{50}$$

in the form $u + iv$.

1.19 Integrate $f(z) = 1/z$ around a unit circle in the CCW sense.

1.20 Integrate $f(z) = (z - z_0)^m$, where m is an integer and z_0 is a constant, in a CCW sense around the circle C of radius ρ with center at z_0.

1.21 Integrate

$$\frac{z^2+1}{z^2-1}$$

in the CCW sense around a circle of radius 1 with center at the point.

$$\text{(a) } z = 1, \quad \text{(b) } z = \frac{1}{2}, \quad \text{(c) } z = -1, \quad \text{(d) } z = i.$$

1.22 What is the principal value of $(-1+i)^{3-i}$?

1.23 Determine the curve in the complex plane that is described by $|z-3| + |z+3| = |10-3|$.

1.24 Express the function $\frac{1}{(1-z)^2}$ in the form $u(x,y) + iv(x,y)$.

1.25 Evaluate the integral

$$\int_C z^3$$

where the curve C is the straight line joining $1+i$ and $2+4i$.

a) From first principles, that is, using real integration.
b) Using the fact that the line integrals are independent of the path from $1+i$ to $2+4i$, that is, using complex integration.

1.26 Express $f(z) = \tan^{-1}(a+ib)$ in the form $f(z) = u + iv$.

1.27 Find the residue of

$$f(z) = \frac{z^{(1-p)}(z-1)^p}{z^2+1}.$$

1.28 Show that $\sin^3\theta = \frac{3}{4}\sin\theta - \frac{1}{4}\sin 3\theta$.

1.29 Show that $\cos^4\theta = \frac{1}{8}\cos 4\theta + \frac{1}{2}\cos 2\theta + \frac{3}{8}$.

1.30 Show that $\lim\limits_{z\to 0}\left(\frac{\bar{z}}{z}\right)$ does not exist.

1.31 Determine whether $u = e^{-x}(x\sin y - y\cos y)$ is harmonic and, if so, determine the harmonic conjugate.

1.32 Write the function $f(z) = \tanh(z)$ in the form $f(z) = u + iv$.

1.33 Evaluate the integral

$$I = \int_C \frac{\sin z}{z-1}\,dz,$$

where C is the contour $z(t) = 2e^{it},\ -\pi \le t \le \pi$.

1.34 Evaluate the integral

$$I = \int_C \frac{e^z}{2z^2 - 5z + 2}\,dz,$$

where C is the contour $z(t) = e^{it},\ -\pi \le t \le \pi$.

1.35 Write the equation $|z|^2 + 3Re\left[z^2\right] = 4$ in the form $y = y(x)$.

1.36 Suppose that

$$f(z) = \begin{cases} z^5/|z|^4, & z \neq 0 \\ 0, & z = 0. \end{cases}$$

Show that the C-R conditions are satisfied but that $f(z)$ is not differentiable at z if $z \neq 0$.

1.37 Investigate the analyticity of

$$f(z) = \frac{x}{x^2+y^2} - \frac{iy}{x^2+y^2}.$$

1.38 Investigate the analyticity of $f(z) = x^2y^2 + 2x^2y^2i$.

1.39 Prove the Cauchy-Goursat's theorem that if $f(z)$ is analytic inside and on a simple closed curve C, then $\int_C f(z)dz = 0$.

1.40 Write the function $f(z) = \cos(1+2i)$ in the form $u + iv$ and give numerical values for u and v.

1.41 Show that $f(z) = z + 1/2$ satisfies the polar form of the C-R conditions.

1.42 Evaluate

$$\int_C \frac{\sinh z}{z^2 + (\pi/2)^2}\,dz,$$

where C contains the point $i\pi/2$ but excludes the point $-i\pi/2$.

1.43 Evaluate

$$I = \int_C \frac{\cos z}{(z - \pi/4)^3}\, dz,$$

where C is $|z - \pi/2| = 1$.

1.44 Suppose $f(z)$ is analytic in a simply connected domain containing the circle $|z - z_o| = \rho$, show that

$$f(z_o) = \frac{1}{2\pi} \int_0^{2\pi} f(z_o + \rho e^{i\theta})\, d\theta.$$

(This is the Gauss mean value theorem.)

1.45 Find all the values of $\sin^{-1}\sqrt{2}$.

1.46 Determine df/dz at $z = \sqrt{2}$ if $f = \sin^{-1}\sqrt{z}$.

1.47 Express $f(z) = (\cos\theta + i \sin\theta)^8/(\sin\theta + i\cos\theta)^4$ in the form $u + iv$, where u and v are real.

1.48 Show that

$$f(z) = \frac{(\cos\theta + i\sin\theta)^n}{(\sin\theta + i\cos\theta)^m} = (-i)^m\{\cos(n+m)\theta + i\sin(n-m)\theta\}.$$

1.49 Evaluate

$$I = \int_C \frac{\log z}{(z-1)^3}\, dz,$$

where C is $|z - 1| = 1/2$, using Cauchy integral formula.

1.50 Integrate $f(z) = e^{iz}$ along the three directed curves $(\Gamma_1, \Gamma_2, \Gamma_3)$ shown in Figure P1.50 and find the sum of the three integrals.

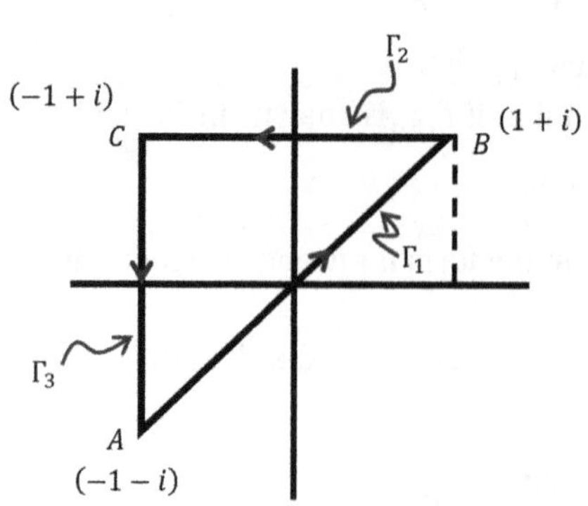

1.51 Evaluate

$$I = \int_{|z|=3} \frac{\sin \pi z^2 + \cos \pi z^2}{(z-1)(z-2)} dz.$$

1.52 Evaluate

$$I = \int_{|z|=10} \frac{e^z}{(z-1)(z+3)} dz.$$

1.53 Evaluate

$$I = \int_{|z|=8} \frac{e^z}{(z-1)(z+3)^2} dz.$$

1.54 Find the Laurent series of $f(z) = z/[(z+1)(z+2)]$ about $z = -1$.

1.55 Evaluate

$$I = \int_{|z-1|=1} \frac{(3z^2+z)}{(z^2-1)} dz$$

by using the Cauchy integral theorem.

1.56 Classify all the singularities of

$$f(z) = \frac{\pi + z + \sin z}{(1 + \cos z)^2}.$$

1.57 Evaluate $I = \int_{1+i}^{4+i} z^2 dz$ along the curve $x = t, y = t^2$, where $1 \leq t \leq 2$.

1.58 Express

$$f(z) = \frac{z - 2 + \sqrt{3}}{(2-\sqrt{3})z - 1}$$

in the form $f(z) = \phi + i\psi$.

Suggested reading

[1] Brown, J.W. and Churchill, R.V. Complex Variables and Applications, 6th Edition, New York, NY, USA, McGraw-Hill, 1996.

[2] Churchill, R.V. Introduction to Complex Variables and Applications. New York, NY, US, McGraw-Hill Book Company, 1948.

[3] Irving, J. and Mullineaux, N. Mathematics in Physics and Engineering, Cambridge, MA, USA, Academic Press, 1959.

[4] Wegert, E. Visual Complex Functions, Birkhauser, Freiberg, Germany (2012).

2 Laurent series, residue theorem, and contour integration

This chapter focusses on the required elements for complex variable contour integration. Thus, we start with the Taylor series and its negative-power extension, the Laurent series, with demonstrations of the approaches that could be taken to generate the series. We go into singularities (isolated and essential) and discuss the residue theorem and contour integration. Many examples of the latter are given, including the evaluation of several types of improper integrals, integrals with poles on the real axis and indented contours, Cauchy principal value (CPV) integrals, integrands that contain trigonometric functions, and the inverse Laplace transform (LT) integral involving direct employment of the Bromwich contour. A list of suggested reading materials is provided in [1–59].

2.1 Taylor series

2.1.1 Taylor series for real functions

From the calculus of real variables, we usually write the Taylor series for a function $f(x)$ about a point x_0 as a way of predicting the function at a neighboring point $(x_0+\xi)$ to x_0 in the following manner:

$$f(x_0+\xi)=f(x_0)+\left.\frac{df}{dx}\right|_{x_0}\xi+\left.\frac{d^2f}{dx^2}\right|\frac{\xi^2}{2!}+\ldots+\left.\frac{d^{n-1}f}{dx^{n-1}}\right|_{x_0}\frac{\xi^{(n-1)}}{(n-1)!}+\left.\frac{d^nf}{dx^n}\right|_{x_0}\frac{\xi^n}{n!}+R_n, \tag{2.1}$$

where $R_n\equiv\frac{f^n(\lambda)}{n!}\xi^n$ is the remainder, which has to be evaluated in $x_0<\lambda<x_0+\xi$. Obviously, ξ has to be infinitely small relative to x_0, and the derivatives of all orders must exist. A simple way to understand the Taylor series is that if we know $f(a)$, $f'(a)$, $f''(a), \ldots$ at a point "a," we can estimate the value of f at a neighboring point to a by simple extrapolation. The convergence and the interval of convergence of the series are usually of interest when employing the Taylor series.

Example 2.1 Obtain the Taylor series for $f(x)=e^x$ about $x=0$.

Solution: $x_0=0$; $x_0+x=x$; $\Delta x=x\equiv\xi$; $e^0=1, f'(x_0)=f''(x_0)=\ldots=f^n(x_0)=e^{x_0}$ so that

$$f'(0)=f''(0)=\ldots=f^n(0)=e^0=1.$$

https://doi.org/10.1515/9783111351179-003

Substituting into eq. (2.1) we have

$$e^x = 1 + x + \frac{x^2}{2!} + \ldots + \frac{x^{n-1}}{(n-1)!} + \frac{x^n}{n!} + R^n.$$

Similarly,

$$\sin x = x - \frac{x^3}{3!} + \frac{x^5}{5!} - \frac{x^7}{7!} \pm \ldots, \qquad \cos x = 1 - \frac{x^2}{2!} + \frac{x^4}{4!} - \frac{x^6}{6!} \pm \ldots,$$

$$\tan^{-1} x = x + \frac{x^3}{3!} + \frac{x^5}{5!} + \frac{x^7}{7!} + \ldots \; (x^2 < 1).$$

Series summation is defined as

$$\sum_{k=1}^{N} a_k = a_1 + a_2 + \ldots + a_N.$$

No problems are encountered in summing a finite series. For an infinite series, summation is usually understood as follows:

$$\sum_{k=1}^{\infty} a_k = \lim_{n\to\infty} \sum_{k=1}^{n} a_k.$$

The partial sums S_i are obtained as follows:

$$S_1 = a_1,\; S_2 = a_1 + a_2,\; S_3 = a_1 + a_2 + a_3, \ldots,\; S_n = a_1 + a_2 + a_3 + \ldots + a_n = \sum_{i=1}^{n} a_i.$$

2.1.2 Convergence

The convergence of a series may be desirable in some applications. There are at least three ways in which convergence can be examined:

(1) $a_n \to 0$ as $n \to \infty$: Successively smaller terms of a series suggest convergence.
Cauchy sequence: We say that the partial sum S_n is Cauchy sequence if, to each $\epsilon > 0$ (no matter how small), there corresponds an $N(\epsilon)$, such that $|S_m - S_n| < \epsilon \forall\; m$ and $n > N$. That is, we can make $|S_m - S_n|$ as small as we want. (S_n above may be the partial sum of a certain series or simply some sequence of an ordered set of numbers without a reference to any series.) Consider the two partial sums as shown in the expression (2.2):

$$\underbrace{\underbrace{a_1 + a_2 + a_3 + \ldots + a_n}_{S_n} + \ldots + a_m}_{S_m} + \ldots \tag{2.2}$$

The idea is that we can make the difference $|S_m - S_n|$ (in the expression (2.2)) to be as small as we want by choosing $n < m$ to be large as a_n becomes progressively smaller as n increases. Any sequence that meets this condition is convergent. Thus, as a corollary, a necessary (though not sufficient) condition for a series $\sum_{k=1}^{\infty} a_k$ to converge is that $a_n \to 0$ as $n \to \infty$. This is the first thing you look for when you are trying to determine if a series is convergent. Note that the Harmonic series $1 + 1/2 + 1/3^2 + \ldots$ defies this test as $a_n \to 0$ as $n \to \infty$ but the series diverges.

(2) Ratio test: We can find an interval of convergence of the power series in x (i.e., a series whose terms are the powers of x) by using the ratio test. The test is good for a sequence of numbers and functions.

Let

$$\lim_{n\to\infty} \left| \frac{a_{n+1}}{a_n} \right| = r.$$

If

$$\begin{aligned} &r < 1 \text{ convergence} \\ &r > 1 \text{ divergence} \\ &r = 1 \text{ test fails} \end{aligned}$$

As an example, for the exponential function above,

$$\lim_{n\to\infty} \left| \frac{a_{n+1}}{a_n} \right| = \lim_{n\to\infty} \left| \frac{x}{n+1} \right| = 0,$$

which is always less than 1, confirming the **entire nature** of e^x. This is also the case for $\sin x$ and $\cos x$.

(3) Comparison test: Consider two series $\sum a_n$, $\sum b_n$ of positive terms. If we can find some $k > 0$ such that every term of the two series satisfies $a_n < kb_n$, then $\sum a_n$ converges if $\sum b_n$ converges. The ratio test will be used exclusively to test for convergence and determine its radius in this book.

2.1.2.1 Analytic functions

We have discussed analyticity of a function in Chapter 1. It can also be defined in terms of a Taylor series, if the function is real. That is, a (real) function is analytic at a point a if its Taylor series exists at the point. For real functions, if $R_n(x) \to 0$ as $n \to \infty$, the Taylor series converges if the equality (in the series expression) holds for some $|x - a| < \delta$, $\delta > 0$. In this case, $f(x)$ is analytic at $x = a$ (i.e., there is an interval $|x - a|$ where the series exists).

Example 2.2 Convergence of power series
Applying the ratio test to the power series:

$$a_0 + a_1(x-a) + a_2(x-a)^2 + \ldots + a_n(x-a)^n + \ldots,$$

we have

$$\lim_{n\to\infty}\left|\frac{a_{n+1}(x-a)^{n+1}}{a_n(x-a)^n}\right| = \lim_{n\to\infty}\left|\frac{a_{n+1}}{a_n}\right||x-a| \equiv L|x-a|,$$

which converges if

$$L|x-a| < 1 \text{ or } |x-a| < \frac{1}{L};$$

diverges if

$$L|x-a| > 1 \text{ or } |x-a| > \frac{1}{L},$$

and we cannot tell if

$$|x-a| = \frac{1}{L}.$$

Therefore, the radius of convergence is $-1/L < (x-a) < 1/L$. Figure 2.1 shows the interval of convergence for real series, which is located on a line. As we show later, for complex series, the boundary of convergence is no longer on a line but in a disk.

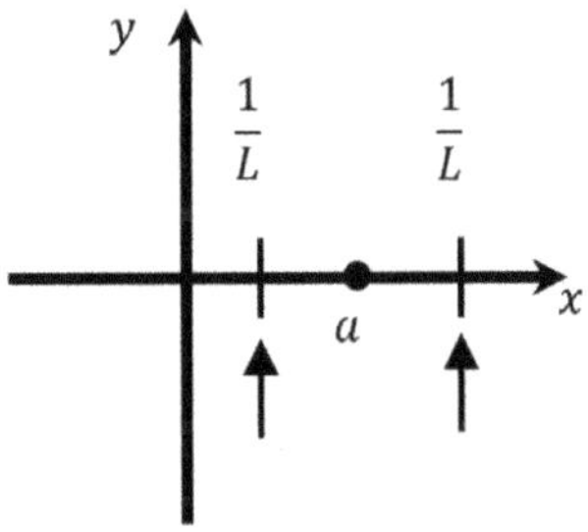

Figure 2.1: Radius of convergence for real series. Open interval of convergence is 1-D for a real series but 2-D for a complex series.

2.1.2.2 Uniform convergence

The discussions above pertain to point convergence. Uniform convergence represents a stronger condition of convergence. The sum $\sum_{n=1}^{\infty} a_n(x)$ converges uniformly to some $S(x)$ over $\alpha \le x \le \beta$ if for each ϵ, there is an $N(\epsilon)$ such that $|S_n(x) - S(x)| < \epsilon\ \forall\ n > N$ between $\alpha \le x \le \beta$ (Figure 2.2). That is, the partial sum $S_n(x)$ lies in the band shown in the figure for all $n > N$, and the error is contained within the band.

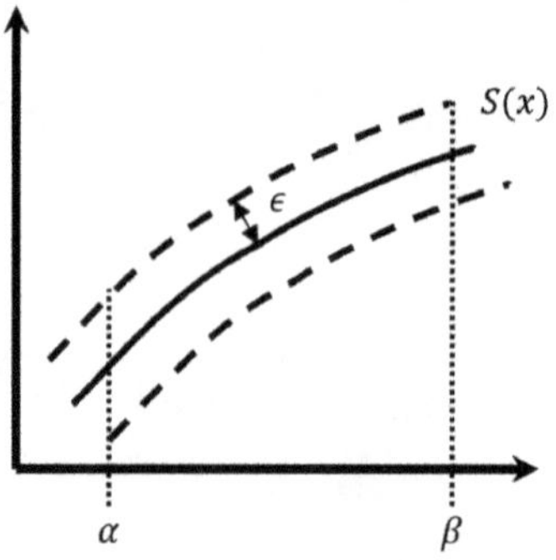

Figure 2.2: Illustration of uniform convergence, showing the band of convergence.

Example 2.3 Determine the convergence of the Taylor series for $f(x) = \ln x$ about $x \equiv a = 3$ so that $\Delta x = x - 3$.

Solution: $a \equiv x_0 = 3, \Delta x = x - 3, x_0 + \Delta x = x, f(x) = \ln x, f(3) = \ln 3,\ f'(3) = \frac{1}{3},\ f''(3) = -\frac{1}{9}$. Thus, from

$$f(x_0 + \Delta x) = f(x_0) + \left.\frac{df}{dx}\right|_{x_0} \Delta x + \left.\frac{d^2 f}{dx^2}\right| \frac{\Delta x^2}{2!} + \dots + \left.\frac{d^{n-1} f}{dx^{n-1}}\right|_{x_0} \frac{\Delta x^{(n-1)}}{(n-1)!} + \left.\frac{d^n f}{dx^n}\right|_{x_0} \frac{\Delta x^n}{n!} + R_n$$

$$= f(x_0) + \sum_{n=1}^{\infty} \left.\frac{d^n f}{dx^n}\right|_{x_0} \frac{\Delta x^n}{n!} + R_n$$

we will have

$$\ln x = \ln 3 + \frac{x-3}{3} - \frac{(x-3)^2}{2 \cdot 3^2} + \dots + \frac{(-1)^n (x-3)^{n-1}}{3^{n-1}(n-1)} + \dots \tag{2.3}$$

To obtain the interval of convergence, apply the ratio test:

$$\left|\frac{(n+1)^{\text{th}}}{n^{\text{th}}}\right| = \left|\frac{(x-3)^n}{3^n n} \cdot \frac{3^{n-1}(n-1)}{(x-3)^{n-1}}\right| = \frac{(n-1)|x-3|}{3n}.$$

Take $\lim n \to \infty$ and divide top and bottom by n (to remove n at the bottom). (We could also use L'Hopital's rule.) The result is

$$\lim_{n \to \infty} \left|\frac{(1-1/n)(x-3)}{3}\right| = \left|\frac{x-3}{3}\right|.$$

Therefore, the series converges for $|x-3| < 3$ or $-3 < x - 3 < 3$. (Actually $-3 < x - 3 < 3$ or $0 < x < 6$ since $\ln 0$ is not defined.) Note, however, that we cannot test for convergence at $x = 6$ by the ratio test. The series (eq. (2.3)) can be used for a neighborhood x of 3 defined by $0 < x < 6$. Divergence occurs outside of this range. For example, we cannot obtain $\ln 8$ with this series.

Example 2.4 Expand $f(x)=\sqrt{1+x}$ about $x=0$ (i.e., Maclaurin series) and determine the interval of convergence.

Solution: We can use the binomial expansion to generate the series:

$$f(x)=(1+x)^{\frac{1}{2}}=1+\frac{x}{2}-\frac{x^2}{8}+\frac{x^3}{16}+\ldots+\frac{\{(1/2)(-1/2)\ldots[(3-2n)/2]x^n\}}{n!}+\ldots$$

$$\left|\frac{a_{n+1}}{a_n}\right|=\left|\frac{(1-2n)x}{2(n+1)}\right|\Rightarrow\lim_{n\to\infty}\left|\frac{(1/n-2)x}{2(1+1/n)}\right|=|x|.$$

Thus, the series for $f(x)=\sqrt{1+x}$ converges for $|x|<1$ or $-1<x<1$.

Remark Take note of the following related expansions:

(a) If n is a positive integer, x is any value:

$$(1+x)^n=\sum_{r=0}^{n}\binom{n}{r}x^r;\ \binom{n}{r}=\frac{n!}{r!(n-r)!}$$

(b) If n is not a positive integer, but $|x|<1$:

$$(1+x)^n=1+nx+\frac{n(n-1)}{1\times2}x^2+\frac{n(n-1)(n-2)}{1\times2\times3}x^3+\ldots.$$

Note that while the series in (a) is finite, that in (b) is infinite.

(c) The series

$$(s+t)^n=s^n+ns^{n-1}t+\frac{n(n-1)}{2!}s^{n-2}t^2+\frac{n(n-1)(n-2)}{3!}s^{n-3}t^3+\ldots$$

is valid for all values of n if $|s|>|t|$. If $|s|<|t|$, expansion is valid only for non-negative integers. If $|s|=|t|$, the expansion is valid if n is a non-negative integer but may not be valid for other values of n.

Example 2.5 Evaluate $(1+2)^3$ by the use of a series expansion.

Solution: Using the expansion in (c), with $s=1$ and $t=2$, we have

$$(1+2)^3=3^3=1^3+3\times1^{3-1}\times2^1+\frac{3(3-1)}{2!}1^{3-2}2^2+\frac{3(3-1)(3-2)}{3!}1^{3-3}2^3$$

$$=1+6+12+8=27,$$

which is obviously the value of 3^3.

Example 2.6 Determine the interval of convergence of the real function $f(x) = 1/(1+x^2)$ about $x = 0$. Note that this function appears to be a well-behaved real function where x could be 0,−1, or any (real) value. Therefore, based on the real variable theory, we do not anticipate any restrictions on its convergence.

However, the series for this function,

$$f(x) = 1 - x^2 + x^4 - x^6 + \dots$$

converges for $|x| < 1$. So, contrary to our intuition, there is indeed a bound on x. The problem occurs in the complex plane, where $1 + x^2$ can be zero if the restriction $|x| < 1$ is violated. That is, when $x = \pm i$. This result underscores why the knowledge of real theory may not be adequate for general analysis.

2.1.2.3 Complex series

Complex series are those defined for variables which are complex. Complex power series is the series of complex variables that can be expressed as

$$f(z) = \sum a_n (z - a)^n.$$

In principle, many aspects of the theory of complex series are similar to those of real series if we allow for the two-dimensionality in the former. Convergence and the test for it are identical.

Example 2.7 Determine the radius of convergence of the complex power series

$$f(z) = \sum_{n=1}^{\infty} \frac{(z-2)^n}{n}.$$

Solution: Apply the ratio test and L'Hopital's Rule, as you would to a real series:

$$\lim_{n\to\infty} \left| \frac{(z-2)^{n+1} n}{(n+1)(z-2)^n} \right| = |z-2|.$$

Thus, the given series converges if $|z-2| < 1$, which is a circle of radius 1 with center at $z = 2 + 0i$ (Figure 2.3). Note that we don't know what happens on the circle, but the series diverges outside and converges inside.

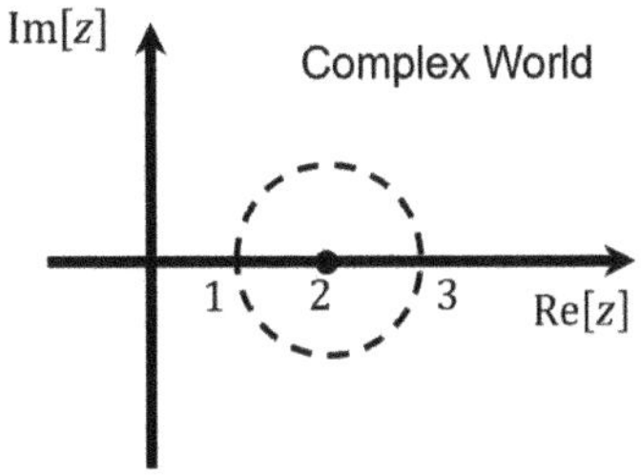

Figure 2.3: A circle of radius 1 centered at $z = 2 + 0i$.

2.1.3 Taylor series for complex functions

Let $f(z)$ be analytic inside and on a closed circle C, center a (point z is inside C). Then with the help of Cauchy Integral Formula, we will show that the complex Taylor series has a similar form to that for real functions:

$$f(z) = f(a) + (z-a)f'(a) + \frac{(z-a)^2}{2!}f''(a) + R_n,$$

where the remainder $R_n \to 0$ as $n \to \infty$. The contour C for developing the Taylor series in the complex plane is shown in Figure 2.4. Note that there is no singularity of the function inside the contour.

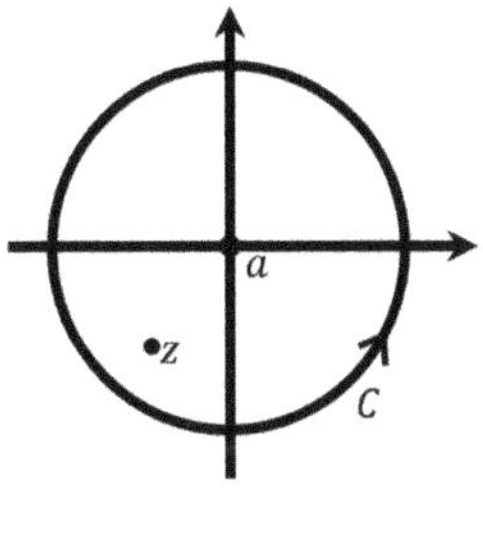

Figure 2.4: Contour for Taylor series development in the complex plane.

Theorem 2.1 If $f(z) = \sum a_n(z-a)^n$ (the sum function), then $f(z)$ is analytic everywhere inside the circle of convergence. Furthermore, the derivatives of f (of all orders) can be found by repeated termwise differentiation of the power series, and each of the resulting series has the same circle of convergence as $\sum a_n(z-a)^n$.

2.1.4 Development of complex Taylor series

In order to develop the complex Taylor series, we suppose that $f(z)$ is analytic inside and on the circle C with center a (Figure 2.5). For any point z inside C, we have, according to the Cauchy Integral Formula:

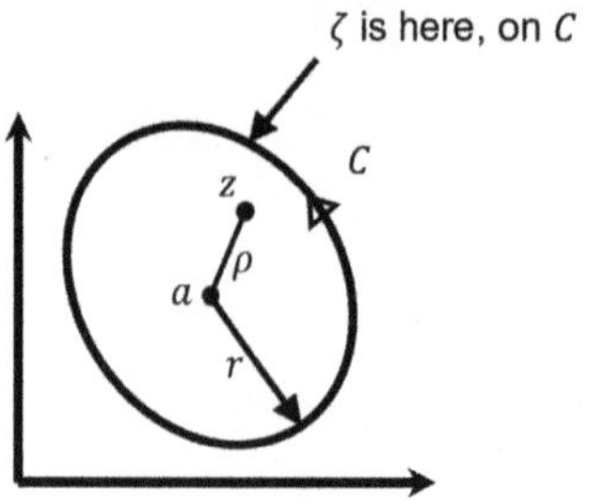

Figure 2.5: The contour for the development of the complex Taylor series. Note that $\zeta > z$ so that $t \equiv \left|\frac{z-a}{\zeta-a}\right| < 1$.

$$f(z) = \frac{1}{2\pi i}\int_C \frac{f(\zeta)}{(\zeta - z)}d\zeta = \frac{1}{2\pi i}\int_C \frac{f(\zeta)}{(\zeta - a)}\underbrace{\left[\frac{1}{1-\frac{z-a}{\zeta-a}}\right]}_{\text{the source of the series}}d\zeta. \tag{2.4}$$

Noting that $|z| < |\zeta|$ and using the geometric series, with $t \equiv (z-a)/(\zeta-a)$; $|t| < 1$:

$$\frac{1}{1-t} = 1 + t + t^2 + \ldots + t^{n-1} + \frac{t^n}{1-t}, (t \neq 1) \tag{2.5}$$

on the square bracket in eq. (2.4), we have

$$f(z) = \frac{1}{2\pi i}\left\{\int_C \frac{f(\zeta)}{(\zeta-a)}d\zeta + (z-a)\int_C \frac{f(\zeta)}{(\zeta-a)^2}d\zeta + \ldots + (z-a)^{n-1}\int_C \frac{f(\zeta)}{(\zeta-a)^n}d\zeta\right\} + R_n$$

where the remainder term is

$$R_n = \frac{(z-a)^n}{2\pi i}\int_C \frac{f(\zeta)}{(\zeta-a)^n(\zeta-z)}d\zeta.$$

(Note that the geometric series in eq. (2.5) is a Maclaurin series or an expansion about $t = 0$.)

Using the Cauchy Integral Formula,

$$f^n(a) = \frac{n!}{2\pi i}\int_C \frac{f(\zeta)}{(\zeta-a)^{n+1}}d\zeta,$$

we will have

$$f(z) = f(a) + f'(a)(z-a) + \ldots + \frac{f^{n-1}(a)}{(n-1)!}(z-a)^{n-1} + R_n. \tag{2.6}$$

This is the complex version of the Taylor formula, with remainder. Letting $n \to \infty$, we write this as the Taylor series representation

$$f(z) = f(a) + f'(a)(z-a) + \frac{f''(a)}{2!}(z-a)^2 + \ldots$$

provided we can show that $R_n \to 0$ as $n \to \infty$. In summary, we were able to develop the complex form of the Taylor series by introducing the geometric series and employing the Cauchy Integral Formula. Note that eq. (2.6) has exactly the same form as the Taylor series in real-variable theory.

2.1.4.1 Examples of complex series

The familiar real-valued series for e^x, $\sin x$, and $\cos x$ retain their form in the complex plane:

$$e^z = 1 + z + \frac{z^2}{2!} + \frac{z^3}{3!} + \ldots, \quad \sin z = z - \frac{z^3}{3!} + \frac{z^5}{5!} - \ldots, \quad \cos z = 1 - \frac{z^2}{2!} + \frac{z^4}{4!} - \ldots$$

The fact that the geometric series in eq. (2.5) is being used to develop the Taylor series for $f(z)$ can be criticized. This is true, but the function $f(z)$ is intended to be more general than $1/(1-t)$. Moreover, t is a complex function in this case.

Example 2.8 Determine the Taylor series about $z = 0$ of the function

$$f(z) = \frac{1}{1+z^2}.$$

Solution: To do this, we merely need to determine the derivatives of the function of all orders and evaluate them at $z = 0$. The result is $1/(1+z^2) = 1 - z^2 + z^4 - \ldots$, which is similar to the expression for the analogous real series presented previously. The circle of convergence of this series (by the ratio test) is in the unit disk $|z| < 1$ because of the poles at $z = \pm i$. This makes more sense than the restriction on $f(x) = 1/(1+x^2)$ in the real case alluded to earlier. The restriction $|x| < 1$ in the real case occurs because of singularities off the real axis at $z = \pm i$, and the ratio test is able to catch this.

2.2 Analytic continuation

The function $f(z) = 1/(1+z^2)$ is analytic everywhere in the complex plane except at the two points $z = \pm i$. However, as discussed above, the series

$$g(z) = 1 - z^2 + z^4 - \ldots$$

converges and is analytic in the disk $|z| < 1$ and is therefore only a partial representation of the useful part of $f(z)$. The function $g(z)$ actually converges to $1/(1+z^2)$ within the circle $|z| < 1$ and hence coincides with $f(z)$ in this region. Since $f(z)$ extends beyond $|z| < 1$, we say that $f(z)$ is the analytic continuation of $g(z)$. We can extend $g(z)$ outside of the unit disk by employing the method known as analytic continuation, as depicted in Figure 2.6(a).

For another example, the geometric series $g_0(z) = \sum_n z^n$ converges and is analytic for $|z| < 1$ but diverges for $|z| > 1$. It converges to $1/(1-z)$ within $|z| < 1$. Similarly, the series $g_1(z) = \sum_n [(z-i)^{n-1}/(n-i)^n]$ converges to $1/(1-z)$ but with a disk of convergence of $|z-i| < \sqrt{2}$. The domains of these two functions are shown in Figure 2.6(a) as $\mathcal{D}_0$ and $\mathcal{D}_1$, respectively, with $\mathcal{D}_0 \cap \mathcal{D}_1$ shown as the hatched region in the figure. We can define a new region $g(z)$ as follows:

$$g(z) = \begin{cases} g_0(z), & z \in \mathcal{D}_0 \\ g_1(z), & z \in \mathcal{D}_1 \end{cases}$$

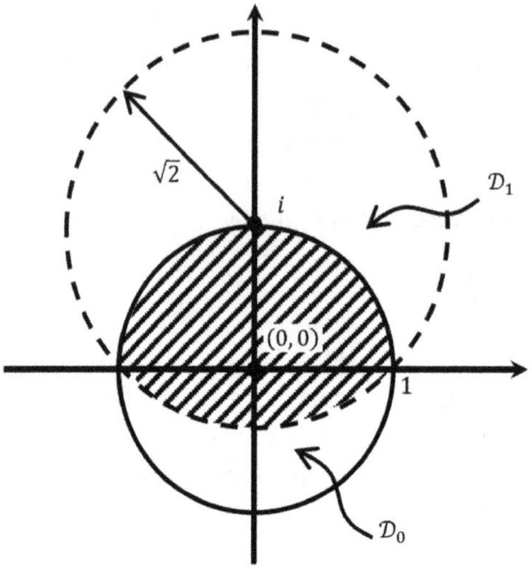

Figure 2.6(a): Circles of convergence of $g_0(z)$ and $g_1(z)$. The hatched region is $\mathcal{D}_0 \cup \mathcal{D}_1$.

and think of $g(z)$ as extending $g_0(z)$ into a larger domain $\mathcal{D}_0 \cup \mathcal{D}_1$. In this manner, $g(z)$ is called the analytic continuation of $g_0(z)$ into $\mathcal{D}_0 \cup \mathcal{D}_1$. We can generalize this idea in a definition.

Definition Suppose $g_0(z)$ is analytic in a region $\mathcal{D}_0$ bounded by a simple closed contour C_0 and $g_1(z)$ is analytic in the region $\mathcal{D}_1$, such that $\mathcal{D}_0$ and $\mathcal{D}_1$ have a common region $\mathcal{D}_0 \cap \mathcal{D}_1$ in which $g_0(z) = g_1(z)$, then $g_1(z)$ is called the analytic continuation of $g_0(z)$ into $\mathcal{D}_0 \cup \mathcal{D}_1$. In the first example above, $f(z)$ analytically continues $g(z)$ into the union of the two domains containing the two functions.

Another pair of functions of $z \equiv x + iy$ that we use to illustrate analytic continuation is $g_1(z) = 1/z$ and $g_0(z) = \int_0^\infty e^{-zt} dt$. The latter can be written as

$$\int_0^\infty e^{-xt}\cos yt\,dt - i\int_0^\infty e^{-xt}\sin yt\,dt,$$

which are real integrals that give

$$g_0(z) = -\frac{1}{z}e^{-zt}\Big|_0^\infty = \frac{1}{z},$$

provided $x > 0$ since otherwise we get a blow up from the exponential term. Thus, whereas $g_1(z)$ is analytic everywhere in the z-plane with the exception of the origin, the domain of $g_0(z)$ precludes the right-hand plane of the complex plane. Thus, $g_1(z)$ analytically continues $g_0(z)$ into the union of the domains for the two functions.

Some functions cannot be continued. This is the case for functions that have singularities that are so closely packed, as to form what is known as a "natural boundary," across which the function cannot be continued. Consider the function (Titchmarsk [55])

$$g_0(z) = 1 + \sum_{n=0}^{\infty} z^{2n},$$

which is analytic when $|z| < 1$ and is the disk inside which the power series converges. However, if q and p are any two positive integers, it can be proved that the limit of $g_0(z)$ does not exist as z approaches any one of the points $e^{(2\pi q/2^p)}$ from the interior of the circle. A function that continues $g_0(z)$ must be analytic throughout a neighborhood of a point on the disk, and it must be equal to $g_0(z)$ at all interior points of the unit disk that belongs to that neighborhood. In this example, the limit of a function that continues $g_0(z)$ as z approaches a point $e^{(2\pi i q/2^p)}$ on the unit disk in that neighborhood does not exist.

The basic steps for the analytic continuation method depicted in Figure 2.6(b) are as follows for the sample series $f(z) = 1 - z^2 + z^4 - \ldots$.

(1) Sum the series on the RHS of eqs. (2.7–2.9):

$$f(z) = 1 - z^2 + z^4 - \ldots \tag{2.7}$$

$$f'(z) = -2z + 4z^3 - 6z^5 + \ldots \tag{2.8}$$

$$f''(z) = -2 + 12z^2 - 30z^4 + \ldots \tag{2.9}$$

at a point $P_0(z = P_0)$ inside the unit disk C_0 to obtain the values of $f(P_0), f'(P_0), f''(P_0), \ldots$ for the Taylor series

$$f(z) = f(P_0) + f'(P_0)(z - P_0) + \frac{f''(P_0)}{2!}(z - P_0)^2 + \ldots \tag{2.10}$$

about P_0.

(2) Applying the ratio test to the new series (eq. (2.10)) we find that it is valid (convergent) inside circle C_1 about P_0.

(3) With the coefficients in the series of eq. (2.10) known, we take the new series (eq. (2.10)), generate and sum the series for f, f', f'', etc. at some point P_1 inside the new circle of convergence C_1. This way, we have the coefficients for the Taylor series

$$f(z) = f(P_1) + f'(P_1)(z - P_1) + \frac{f''(P_1)}{2!}(z - P_1)^2 + \ldots$$

about P_1, which will be found to be valid inside C_2, after a ratio test.

(4) Repeating the process indefinitely, we can develop a set of expansions that collectively represents $f(z)$ throughout the entire z-plane, except for the two singular points $z = \pm i$. We may define e^z as the analytical continuation of e^x off the real axis and into the complex plane. Similarly $\sin z$ and $\cos z$ could be considered as the analytic continuation of $\sin x$ and $\cos x$. The concept of analytic continuation is used extensively in the theory of advanced complex variables in Chapters 5 through 7.

The procedure described above for exploring analytic functions is called the Weierstrass disk chain method.

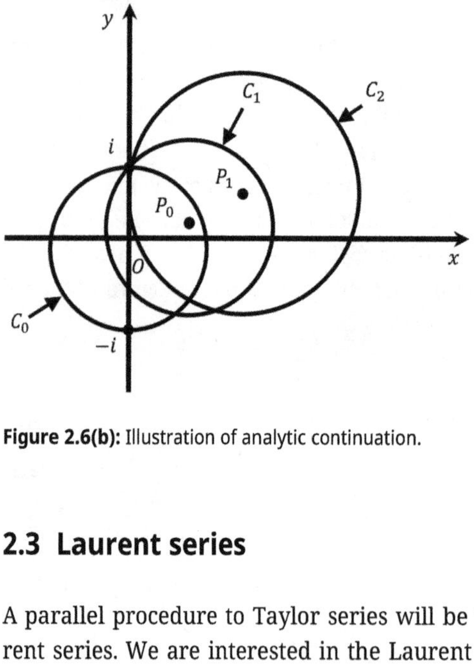

Figure 2.6(b): Illustration of analytic continuation.

2.3 Laurent series

A parallel procedure to Taylor series will be used in this section to develop the Laurent series. We are interested in the Laurent expansion of $f(z)$ about "a," where z is in the neighborhood engulfing a (Figure 2.7). We require that $f(z)$ be analytic at least in an annulus, that is, between and on concentric circles C_i and C_0 with their center at a. (Note that the scheme used for developing the Taylor series doesn't have C_i.) This, or the annulus, is the source of the negative powers that you will see in the Laurent series. (The required sense of the curve for the development is CCW.)

Note that for C_0

$$\frac{|z-a|}{|\zeta-a|}<1$$

requires that $|\zeta|>|z|$, where z is a point inside the annulus and ζ runs around the contour C_0. Conversely for C_i: $|\zeta|<|z|$. The slit contour in Figure 2.7(b) is required in cases where we have to introduce branch cuts because there exists no annulus in which certain functions, such as $\ln z$, are analytic because of multivaluedness.

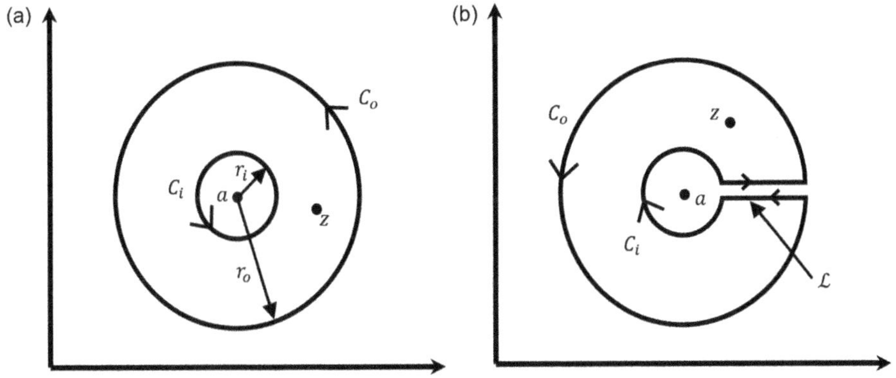

Figure 2.7: Contour for developing the Laurent series.

If C is the slit contour in Figure 2.7(b), we have, from Cauchy Integral Formula,

$$\begin{aligned} f(z) &= \frac{1}{2\pi i}\int_C \frac{f(\zeta)}{\zeta-z}d\zeta \\ &= \underbrace{\frac{1}{2\pi i}\int_{C_0+(C_i)+\rightarrow+\leftarrow} \frac{f(\zeta)}{\zeta-z}d\zeta}_{\text{Directly gives us the Laurent series}} \\ &= \frac{1}{2\pi i}\int_{C_0} \frac{f(\zeta)}{\zeta-z}d\zeta+\frac{1}{2\pi i}\int_{C_i} \frac{f(\zeta)}{\zeta-z}d\zeta, \end{aligned}$$

as the contributions from the two straight line segments cancel exactly. Following the same procedure as in our development of the Taylor series, the C_o integral is treated as follows:

$$\frac{1}{2\pi i}\int_{C_o}\frac{f(\zeta)}{\zeta-z}d\zeta=\frac{1}{2\pi i}\int_{C_o}\frac{f(\zeta)}{\zeta-a}\left[\frac{1}{1-\underbrace{(z-a)/(\zeta-a)}_{\text{the } t \text{ trick again = source of series}}}\right]d\zeta;\quad (\zeta>z \text{ for } C_o)$$

$$=\text{etc.}=\sum_{n=0}^{\infty}a_n(z-a)^n.$$

That is, using Cauchy Integral Formula, etc., where

$$a_n=\frac{1}{2\pi i}\int_{C_o}\frac{f(\zeta)}{(\zeta-a)^{n+1}}d\zeta$$

and $n=0,1,\ldots,\infty$. Note that C_o can be deformed to any path between C_o and C_i since the integrand in a_n is analytic in this annulus. Unlike the Taylor series case, f is not necessarily analytic inside C_o.

For C_i, note that

$$\frac{1}{\zeta-z}=-\frac{1}{z-a}\left[\frac{1}{1-\underbrace{(\zeta-a)/(z-a)}_{\equiv t \text{ again, where } z>\zeta \text{ on } C_i}}\right]=-\frac{1}{z-a}\sum_{n=0}^{\infty}\left(\frac{\zeta-a}{z-a}\right)^n$$

This is done to get $t=(\zeta-a)/(z-a)$ so that we can use the standard t expansions. We can proceed as follows:

$$\frac{1}{2\pi i}\int_{C_i}\frac{f(\zeta)}{\zeta-z}d\zeta=\frac{-1}{2\pi i(z-a)}\int_{C_i}f(\zeta)\left[\frac{1}{1-(\zeta-a)/(z-a)}\right]d\zeta$$

$$=-\frac{1}{2\pi i}\int_{C_i}\frac{f(\zeta)}{z-a}\sum_{n=0}^{\infty}\left(\frac{\zeta-a}{z-a}\right)^n d\zeta$$

$$=\sum_{n=0}^{\infty}\left[-\frac{1}{2\pi i}\int_{C_i}f(\zeta)(\zeta-a)^n d\zeta\right](z-a)^{-(n+1)}$$

$$=\sum_{n=-1}^{-\infty}\left[\frac{1}{2\pi i}\int_{-C_i}\frac{f(\zeta)}{(\zeta-a)^{n+1}}d\zeta\right](z-a)^n$$

Deform to C (since $-C_i$ is now CCW) in the manner that C_o was deformed to C so that

$$\frac{1}{2\pi i}\int\limits_{C_i}\frac{f(\zeta)}{\zeta - z}d\zeta = \sum_{n=-1}^{-\infty}\left[\frac{1}{2\pi i}\int\limits_{C}\frac{f(\zeta)}{(\zeta - a)^{n+1}}d\zeta\right](z-a)^n$$

$$= \sum_{n=-1}^{-\infty} b_n(z-a)^n$$

where

$$\underbrace{b_n}_{\text{limit: } -1 \text{ to } -\infty} = \frac{1}{2\pi i}\int\limits_{C}\underbrace{\frac{f(\zeta)}{(\zeta - a)^{n+1}}d\zeta}_{\neq 0 \text{ since integrand could have poles inside } C}.$$

Note that the summation index has been shifted to obtain the powers of $(z-a)$:

$$\sum_{n=0}^{\infty}\left[\int\limits_{C_i} f(\zeta)(\zeta - a)^n d\zeta\right](z-a)^{-(n+1)} \equiv \sum_{m=0}^{\infty}\left[\int\limits_{C_i} f(\zeta)(\zeta - a)^m d\zeta\right](\zeta - a)^{-(m+1)}$$

$$\underline{\underline{-(m+1) = n}} \quad \sum_{n=-1}^{-\infty}\int\limits_{C_i} f(\zeta)\left[(\zeta - a)^{-(n+1)}d\zeta\right](z-a)^n.$$

(Be mindful of the equal sign over which "$-(m+1) = n$" is written.) We similarly have

$$\sum_{k=1}^{\infty} a_{-k}(z - z_0)^{-k} = \sum_{k=-1}^{-\infty} a_k(z - z_0)^k.$$

Combining the two series, we have

$$f(z) = \sum_{n=-\infty}^{\infty} C_n(z-a)^n, \tag{2.11}$$

which is the Laurent series, where

$$C_n = \frac{1}{2\pi i}\int\limits_{C}\frac{f(\zeta)}{(\zeta - a)^{n+1}}d\zeta. \tag{2.12}$$

A few differences between the Laurent series and Taylor series can be observed:
(1) The Laurent series has negative powers of $(z-a)$, but the Taylor series does not.
(2) C_n is constant in the Taylor series; not so in the Laurent series.

Comments

(1) If $f(\zeta)$ (not $f(\zeta)/(\zeta - a)$) has no singularities inside C_i so that it is analytic everywhere inside and on C_o, then Cauchy's theorem tells us that

$$C_n \equiv \frac{1}{2\pi i}\int\limits_{C_0} \frac{f(\zeta)}{(\zeta - a)^{n+1}} d\zeta$$

$$\equiv \frac{1}{2\pi i}\int G(\zeta) d\zeta$$

$$= 0 \text{ for } n = -1, \ -2, \ -3, \ \ldots : \text{(Cauchy – Goursat)}$$

$$= \frac{f^n(a)}{n!} \text{ for } n = 0, \ 1, \ 2, \ \ldots : \text{(Generalized Cauchy Integral Formula)}$$

and the Laurent expansion reduces to the Taylor expansion of $f(z)$ about a, as it should.

(2) If $f(z)$ is singular at points within C_i then the Laurent series will contain at least one and perhaps an infinite number of negative powers.

2.3.1 Computing the C_n's in the Laurent series

We need the C_n's (eq. 2.12) in order to obtain the Laurent series in eq. (2.11). These can be obtained in one of two ways:

(1) By evaluating the integral for C_n
(2) It's generally possible to bypass the evaluation of the integral for C_n and compute the C_n's rather easily. The examples below illustrate this approach.

2.3.2 Evaluation of C_n without integration

Note that the goal is to obtain a series in positive and negative powers of z, so you should use any method that allows you to do this. The following examples illustrate sample procedures.

Example 2.9 Obtain the Laurent series for $f(z) = 1/(z-1)$ (Figure 2.8).

Solution: Expand about $z = 0$. (If the point about which expansion is carried out is not zero, translate it to zero). The singularity is at $z = 1$. Two expansions are possible: in $|z| < 1$ and in the "annulus" $1 < |z| < \infty$ or $|z| > 1$.

(i) For $|z| < 1$, the function is analytic, and so the Laurent series is simply the Taylor series:

$$f(z) = \frac{1}{z-1} = -\frac{1}{1-z} = -\left(1 + z + z^2 + \ldots\right) = -1 - z - z^2 - \ldots$$

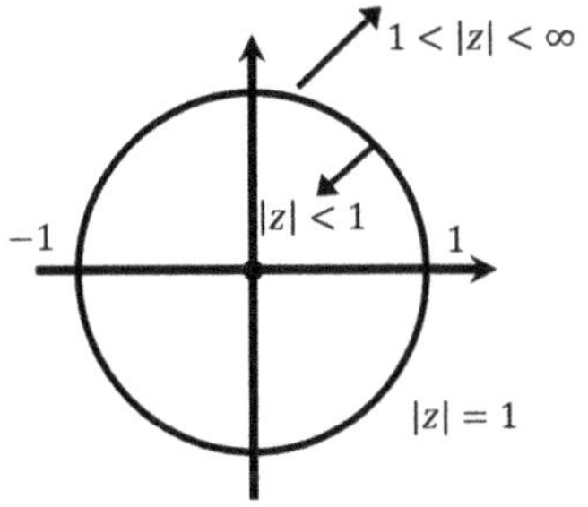

Figure 2.8: Inside and outside of $|z| = 1$.

(ii) For $|z| > 1$ the trick is to rewrite $f(z)$ as

$$f(z) = \frac{1}{z-1} = \frac{1}{z} \cdot \frac{1}{1-1/z}$$

so that
$t \equiv \left|\frac{1}{z}\right|$ is now < 1, and we can use the Taylor (geometric) series expansion for $\frac{1}{1-1/z}$:

$$f(z) = \tfrac{1}{z}\left[1 + \left(\tfrac{1}{z}\right) + \left(\tfrac{1}{z}\right)^2 + \ldots\right] = \tfrac{1}{z} + \tfrac{1}{z^2} + \tfrac{1}{z^3} + \ldots$$

Example 2.10 Expand $f(z) = 1/z(1-z)$:

(i) About $z = 0$ in $0 < |z| < 1$ (Note that we can expand about a singular point.)

$$f(z) = \frac{1}{z} \cdot \frac{1}{1-z} = \frac{1}{z}\left[1 + z + z^2 + \ldots\right]$$

since $t \equiv |z| < 1$. However, the radius of convergence of the expansion will exclude the singularity. That is, the expansion is not valid at singularity even though we can expand about the point; it will be excluded for convergence purposes. In terms of the validity of the expansion, we can factor $f(z)$ as

$$f(z) = \underbrace{\underbrace{\frac{1}{z}}_{\text{valid for } |z|>0} \left[\underbrace{1 + z + z^2 + \ldots}_{\text{valid for } |z|<1}\right]}_{\text{valid for } (|z|>0) \cup (|z|<1)},$$

where ∪ is the union symbol.

(ii) About $z = 0$ in $|z| > 1$ (annulus)

$$\begin{aligned} f(z) &= -\frac{1}{z^2} \cdot \frac{1}{1-1/z} = -\frac{1}{z^2}\left[1 + \left(\frac{1}{z}\right) + \left(\frac{1}{z}\right)^2 + \ldots\right] \\ &= -\frac{1}{z^2} - \frac{1}{z^3} - \frac{1}{z^4} - \ldots \end{aligned} \tag{2.13}$$

Note that we could guess the factor $(1-1/z)$ in the denominator of eq. (2.13) first and find the x in $(1-1/z)x = z(1-z) \Rightarrow x = z^2$. The base expansion associated with $1/(1-t)$ is about $t = 0$ (Maclaurin), also a singularity point. You will need to translate the point about which expansion is sought to some origin in order to use the Maclaurin series. This is done in an example below.

(iii) Expand about $z = i$ in the annulus $|z-i| > \sqrt{2}$ (Figure 2.9). We seek expansion in powers of $z-i$, which suggests a change of variables $t = z-i$ so that

$$|z-i| > \sqrt{2} \Rightarrow |t| > \sqrt{2}.$$

We can then write $f(z)$ as

$$f = \frac{1}{z(1-z)} = \frac{1}{(t+i)(1-i-t)} = \frac{1}{-t\Big(1+\underbrace{i/t}_{(a)}\Big)} \frac{1}{t\Big[1-\underbrace{(1-i)/t}_{(b)}\Big]}$$

(a) $|i/t| = 1/|t| < 1$ in $|t| > 1$ or $|t| > \sqrt{2}$ and hence in the expansion annulus, and
(b) $|(1-i)/t| = \sqrt{2}/|t| < 1$ in $|t| > \sqrt{2}$.

On $|1-i| = \sqrt{2}$, recollect that

$$|z| = (z\bar{z})^{\frac{1}{2}} = [(x+iy)(x-iy)]^{\frac{1}{2}} = (x^2 - i^2y^2)^{\frac{1}{2}} = (x^2+y^2)^{\frac{1}{2}}.$$

Applying the conditions in (a) and (b) above, we have

$$\begin{aligned} f &= -\frac{1}{t^2}\left[1 - \frac{i}{t} + \left(\frac{i}{t}\right)^2 + \dots\right]\left[1 + \left(\frac{1-i}{t}\right) + \left(\frac{1-i}{t}\right)^2 + \dots\right] \\ &= -\frac{1}{t^2}\left[1 + (1-2i)\frac{1}{t} - (2+3i)\frac{1}{t^2} + \dots\right] \\ &= -\frac{1}{(z-i)^2} - \frac{(1-2i)}{(z-i)^3} + \frac{2+3i}{(z-i)^4} - \dots . \end{aligned}$$

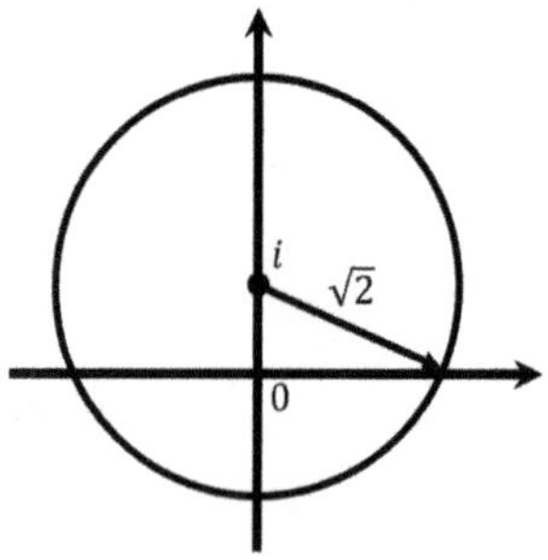

Figure 2.9: The domain $|z-i| > \sqrt{2}$.

Example 2.11 Expand $f(z) = \cos z/z$ about $z = 0$ in $|z| > 0$.

Solution:

$$f(z) = \frac{1}{z}\left[1 - \frac{z^2}{2!} + \frac{z^4}{4!} - \dots\right] = \frac{1}{z} - \frac{z}{2!} + \frac{z^3}{4!}\dots$$

Example 2.12 Expand $f(z) = \sec z = 1/\cos z$ about $z = \pi/2$ in the annulus $0 < |z - \pi/2| < \pi$.

Solution: $f(z)$ is singular at the zeroes of $\cos z$ or $z = \pm\pi/2, \pm 3\pi/2, \dots$Now the Taylor series of $\cos z$ about $\pi/2$ is

$$\begin{aligned}\cos z &= -\left(z - \frac{\pi}{2}\right) + \frac{1}{3!}\left(z - \frac{\pi}{2}\right)^3 - \frac{1}{5!}\left(z - \frac{\pi}{2}\right)^5 + \dots \\ &= \left(z - \frac{\pi}{2}\right)\left\{1 + \frac{1}{3!}\left(z - \frac{\pi}{2}\right)^2 - \frac{1}{5!}\left(z - \frac{\pi}{2}\right)^4 + \dots\right\}\end{aligned} \tag{2.14}$$

We can write $f(z)$ as

$$f(z) = \frac{1}{z - \pi/2} \cdot \frac{z - \pi/2}{\cos z},$$

where $((z - \pi/2)/\cos z)$ should admit a Taylor series in $|z - \pi/2| < \pi$ since $\cos z \sim -(z - \pi/2)$ as $z \to \pi/2$ and the function $(z - \pi/2)/\cos z$ is well-behaved.

Set

$$t = z - \frac{\pi}{2}$$

so that

$$f(z) = \frac{1}{t} \cdot \frac{t}{1} \cdot \underbrace{\frac{1}{t}\frac{1}{\left[-1 + \frac{1}{3!}t^2 - \frac{1}{120}t^4 + \dots\right]}}_{1/\cos z \text{ from Eqn.}(2.14)}. \tag{2.15}$$

We can then carry out a long division of the last factor in eq. (2.15). The Laurent series from this exercise can be written as

$$f(z) = \frac{1}{t}\left[-1 - \frac{1}{6}t^2 - \frac{7}{360}t^4 - \dots\right] = -\frac{1}{z - \pi/2} - \frac{1}{6}\left(z - \frac{\pi}{2}\right) - \frac{7}{360}\left(z - \frac{\pi}{2}\right)^3 - \dots.$$

Example 2.13 Expand $f(z) = e^{-1/z^2}$ in Laurent series.

Solution: Since $e^{\lambda} = 1 + \lambda + \lambda^2/2! + \ldots$ for all $|\lambda| < \infty$, it follows that

$$f(z) = e^{\lambda}\left(\lambda = -\frac{1}{z^2}\right) = 1 - \frac{1}{z^2} + \frac{1}{2!z^4} - \ldots$$

for all $|z| > 0$. Note that the series diverges for $z = 0$.

2.4 Classification of isolated singularities

Generally speaking, we can classify singularities into two types: isolated and nonisolated singularities. While the presence of singularities inside a closed curve may be desirable to avoid trivial integrals in engineering applications, we cannot use the complex variable theory per se to directly deal with nonisolated singularities. The residue calculation, to be presented shortly, will not apply to nonisolated singularities. Procedures to handle nonisolated singularities are discussed in Chapter 6.

2.4.1 An example of nonisolated singularity

Consider the harmonic series $\{1/n\}$; $n = 1, 2, 3, \cdots$. As $n \to \infty$ we have a crowding of points near the origin (Figure 2.10), which is the limit point. Thus, the series has an essential singularity. In order to classify isolated singularities, let us suppose that z_0 is such a singularity of $f(z)$. That is, whereas f is singular at z_0, there is an annulus $0 < |z - z_0| < r$ for some $r > 0$ inside which f is analytic (Figure 2.11).

In this case, $f(z)$ admits a Laurent expansion

$$f(z) = \ldots + \frac{C_{-2}}{(z - z_0)^2} + \frac{C_{-1}}{(z - z_0)} + C_0 + C_1(z - z_0) + \ldots$$

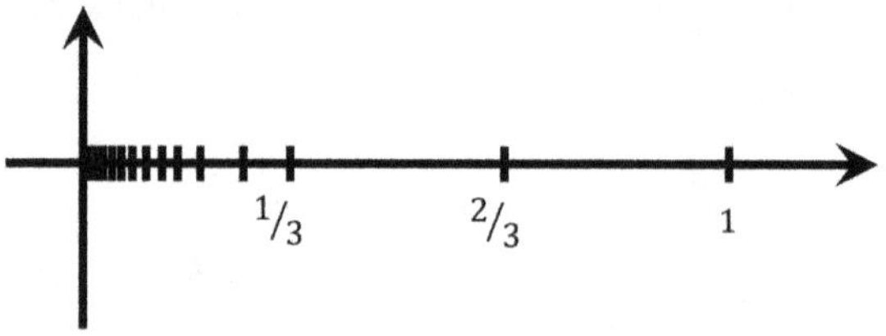

Figure 2.10: Illustration of essential singularity with the crowding of points as $n \to \infty$.

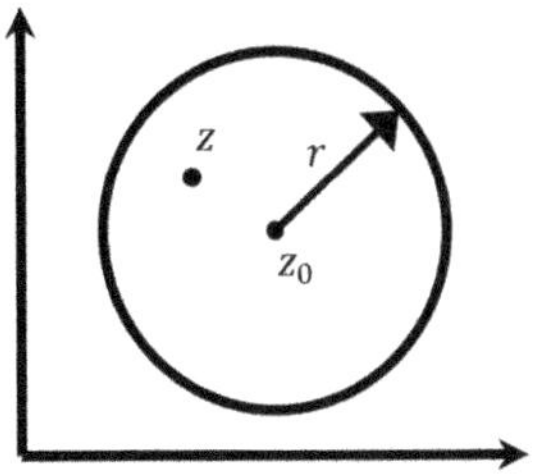

Figure 2.11: Domain for expansion, with an isolated singularity at $z = z_0$.

in that annulus. If the series terminates on the left, that is,

$$f(z) = \frac{C_{-m}}{(z - z_0)^m} + \frac{C_{-m+1}}{(z - z_0)^{m-1}} + \ldots + C_0 + C_1(z - z_0) + \ldots$$

where $C_{-m} \neq 0$, then we classify f's singularity at z_0 as an mth-order pole. If there are infinite number of negative powers present, then f has essential singularity at z_0.

Example 2.14 Given $f(z) = 1/z(1 - z)$; classify the singularity at $z = 0$ (or $z_0 = 0$).

Solution: To classify the singularity (at z_0), we first need to have the series. Consider expansion in (a) $0 < |z| < 1$ and (b) $|z| > 1$. The expansion in (b) is irrelevant to the classification of the singularity at $z \equiv z_0 = 0$. Thus, we need to consider the expansion

$$f(z) = \frac{1}{z} + 1 + z + z^2 + \ldots$$

(using the geometric series and $|z| < 1$).

Thus, $m = 1$, given the term $1/z$, and we have a first-order pole.

Example 2.15 Classify the singularity of $f(z) = e^{-1/z^2}$.

Solution:

$$e^{-1/z^2} = 1 - \frac{1}{z^2} + \frac{1}{2!}\frac{1}{z^4} - \frac{1}{3!}\frac{1}{z^8} + \ldots$$

in $0 < |z| < \infty$. Therefore, we have an essential singularity or an infinite number of negative powers of z.

2.5 Residue theorem and contour integration

We are interested in the evaluation of the contour integral

$$I = \int_C f(z)dz,$$

where C is closed and f may have isolated singularities inside C, say at $z_1, z_2, \ldots, z_m$. We show four $(m=4)$ of such singularities in Figure 2.12. (Note that if $f(z)$ does not have singularities inside of C, the integral will be zero. If $f(z)$ has essential singularities inside of C, the technique of complex variable contour integration cannot be used directly.)

We can decompose I as follows:

$$I = \int_{C_1} f(z)dz + \int_{C_2} f(z)dz + \int_{C_3} f(z)dz + \int_{C_4} f(z)dz \equiv I_1 + I_2 + I_3 + I_4.$$

(Note that the C_i's are in the same sense as C.) To evaluate the I_j's, recollect the Laurent series:

$$f(z) = \sum_{n=-\infty}^{\infty} C_n (z-a)^n,$$

where

$$C_n = \frac{1}{2\pi i} \int_C \frac{f(\zeta)}{(\zeta - a)^{n+1}} d\zeta. \tag{2.16}$$

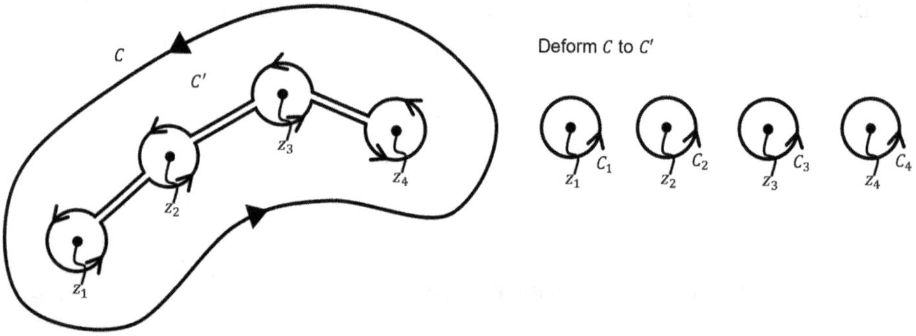

Figure 2.12: Four isolated singularities.

The expression in eq. (2.16) is valid for all n. If in particular we set $n = -1$, we have

$$2\pi i C_{-1} = \int_C f(z)dz.$$

Thus, we see that the integral of a function $f(z)$ around a contour C is given by $2\pi i$ times the C_{-1} term in the Laurent series for the function. Because only the C_{-1} term "survives" in the Laurent series for $f(z)$, we call the C_{-1} term the **residue**. The integral of $f(z)$ over contour C_j can thus be written as

$$I_j = \int_{C_j} f(z)dz = 2\pi i\, C_{-1}^{(j)},$$

where $C_{-1}^{(j)}$ comes from the Laurent series for $f(z)$ for the singularity at z_j. An annulus of analyticity surrounding each pole in Figure 2.12 is depicted in Figure 2.13. The overall integral can thus be written as the sum of the individual integrals:

$$I = \sum_j \int_{C_j} f(z)dz = \sum_j^{\#\text{singularities}} I_j = 2\pi i \underbrace{\sum_j C_{-1}^{(j)}}_{\text{Residues}}.$$

This is the Residue Theorem.

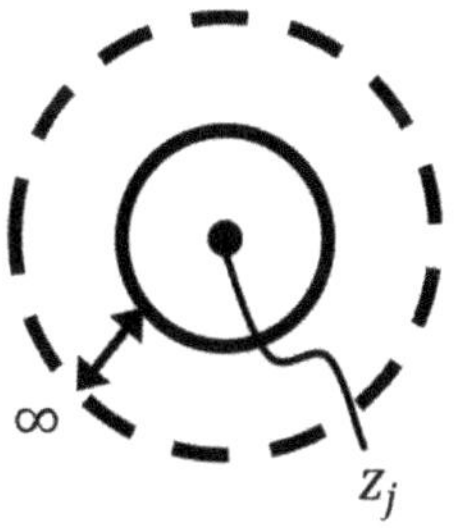

Figure 2.13: An annulus of analyticity around the singularity z_j.

2.5.1 Calculating the residues

We may be able to pick up the residues directly from the Laurent series for the function that we want to integrate. That is, we might not need the whole series and could just pick up the coefficient of the $1/z$ term. There are formulas for that.

Case I: First-order poles
If f(z) has a simple pole (i.e., a first-order pole) at z = a, then the Laurent expansion in an annulus with an arbitrarily small inner circle radius is of the form

$$f(z) = \frac{C_{-1}}{z-a} + C_0 + C_1(z-a) + \dots$$

$$\Rightarrow \quad \underbrace{(z-a)f(z)}_{\text{Not necessarily 0 for } z\to a} = C_{-1} + \underbrace{C_0(z-a) + C_1(z-a)^2 + \dots}_{\to 0 \text{ as } z\to a} \tag{2.17}$$

If $f(z) \to \infty$ faster than $(z-a) \to 0$, then $(z-a)f(z)$ does not go to zero as $z \to a$, unlike the $(z-a)$ terms on the right-hand side of eq. (2.17). Hence the need for this procedure, which is to let $z \to a$ so that

$$\lim_{z\to a}[(z-a)f(z)] = C_{-1}.$$

Example 2.16 Evaluate $\int_C f(z)dz$ where

$$f(z) = \frac{2z+3}{z^2-4z}$$

and C contains all the singularities of $f(z)$, which is the integrand of the contour integration problem.

Solution: $f(z)$ has simple poles at $z = 0, 4$. The residues can be obtained as follows:

$$z=0:\ \ C_{-1} = \lim_{z\to 0} \overbrace{z}^{(z-a)} \underbrace{\frac{(2z+3)}{z^2-4z}}_{f(z)} = \lim_{z\to 0}\frac{2z+3}{z-4} = -\frac{3}{4}$$

$$z=4:\ C_{-1} = \lim_{z\to 4} \overbrace{(z-4)}^{(z-a)} \underbrace{\frac{(2z+3)}{(z-4)(z)}}_{f(z)} = \lim_{z\to 4}\frac{2z+3}{z} = \frac{11}{4}$$

Hence we can calculate

$$I = \int_C f(z)dz \equiv \sum_{j=1}^{2} I_j = 2\pi i \sum_j C_{-1}^{(j)} = 2\pi i \left[-\frac{3}{4} + \frac{11}{4}\right] = 4\pi i.$$

Note that the points of singularity of $f(z)$ may not be as obvious as the $z = 0$ and $z = 4$ in this example. The following example illustrates this.

Example 2.17 $f(z) = 1/\sin z$ has poles at $z = n\pi$ of $\sin z$. They are simple poles since

$$\sin z = (-1)^n[z-n\pi] - \frac{(-1)^n}{3!}(z-n\pi)^3 + \ldots \sim (-1)^n(z-n\pi)$$

The residue at $z = n\pi$ is

$$C_{-1} = \lim_{z\to n\pi}\left(\frac{z-n\pi}{\sin z}\right) = \lim_{z\to n\pi}\frac{1}{\cos z} = (-1)^n,$$

where we have used the L'Hopital's rule. We could also obtain the residue as follows:

$$C_{-1} = \lim_{z\to n\pi}\frac{z-n\pi}{\sin z} = \lim_{z\to n\pi}\frac{z-n\pi}{\underbrace{(-1)^n(z-n\pi)\left[1+O(z-n\pi)^2\right]}_{\text{1st 2 terms in the series for sinz about } n\pi}} = (-1)^n,$$

using the expansion

$$\sin z = z - \frac{z^3}{3!} + \frac{z^5}{5!} + \ldots$$

about $z = n\pi$.

Case II: Second-order poles
Suppose $f(z)$ has a second-order pole at $z = a$, then

$$f(z) = \frac{C_{-2}}{(z-a)^2} + \frac{C_{-1}}{(z-a)} + C_0 + C_1(z-a) + \ldots,$$

which implies that

$$(z-a)^2 f(z) = C_{-2} + C_{-1}(z-a) + C_0(z-a)^2 + \ldots$$

so that

$$\frac{d}{dz}\left[(z-a)^2 f(z)\right] = C_{-1} + 2C_0[z-a] + \ldots$$

and

$$C_{-1} = \lim_{z\to a} \frac{d}{dz}\left\{[z-a]^2 f(z)\right\},$$

as the other terms $\to 0$ as $z \to a$.

Case III: Generalization
Generalizing to a pole of order k, we have

$$C_{-1} = \frac{1}{(k-1)!} \lim_{z\to a} \frac{d^{k-1}}{dz^{k-1}}\left[(z-a)^k f(z)\right].$$

Example 2.18 Find the residues of

$$f(z) = \frac{z^2}{(z^2+4)^2}.$$

Solution: This function has second-order poles at $z = \pm 2i$.
At $z = 2i$,

$$C_{-1} = \lim_{z\to 2i} \frac{d}{dz} \frac{(z-2i)^2 z^2}{(z+2i)^2(z-2i)^2} = \text{etc.} = -\frac{i}{8}.$$

Similarly, $C_{-1} = i/8$ at $z = -2i$.

2.5.2 A remark on the calculation of the residue

The procedure described above for calculating the residues for first-order poles, $\text{Res}[f(z), z_0] = \lim_{z \to z_0}(z - z_0)f(z)$, works well when $f(z)$ is a rational function. Suppose $f(z)$ is not a rational function but can be written as the quotient $f(z) = g(z)/h(z)$, where both $g(z)$ and $h(z)$ are analytic at z_0. If $g(z) \neq 0$ and h has a zero of order one at z_0, then f has a simple pole at $z = z_0$ and

$$\text{Res}[f(z), z_0] = \frac{g(z_0)}{h'(z_0)}.$$

(Note that the analyticity of $h(z)$ at z_0 does not conflict with $h(z)$ being zero at z_0.)

To see this, first note that $h(z)$ having a zero of order one at $z = z_0$ means that $h(z_0) = 0$ so that

$$\frac{z - z_0}{h(z)} = \frac{z - z_0}{h(z) - h(z_0)} = \frac{1}{h'(z)},$$

and

$$\text{Res}[f(z), z_0] = \lim_{z \to z_0}(z - z_0)\frac{g(z)}{h(z)} = \frac{g(z_0)}{h'(z_0)}.$$

The procedure can be extended to higher-order poles. As an example for the case of a first-order pole, consider $f(z) = \sin z/\cos z$ and $z_0 = -\pi/2$, where $g \equiv \sin z$ and $h \equiv \cos z$. Then

$$2\pi i\{\text{Res}[f(z), z_0]\} = \frac{\sin(-\pi/2)}{-\sin(-\pi/2)} = -1.$$

2.6 Contour integration

In advanced mathematics, the complex variable theory typically finds applications in contour integration and not necessarily of complex integrals. This section illustrates the process of carrying out contour integration using several examples. The details of the techniques depend on the type of integrand, as the examples and miscellaneous problems below show.

Example 2.19 Evaluate

$$I = \int_0^\infty \frac{x^2}{(x^2 + 4)^2}\,dx. \tag{2.18}$$

Solution: Although x is real and I is a real integral, we could pose the problem as one for complex variable contour integration where the contour is the positive real axis or $-R$ to R, $R \to \infty$ (Figure 2.14). However, we need a closed contour in order to apply the residue theorem, which says that

$$I_j = \int_{C_j} f(z)dz = 2\pi i\ c_{-1}^{(j)}$$

for all the j poles contained in a given closed contour. Note that we can write

$$I = \frac{1}{2}\int_{-\infty}^{\infty} \frac{x^2}{(x^2+4)^2}dx$$

since $f(x)$ is an even function. $(f(x) = f(-x))$. We will therefore consider the complex integral

$$J = \int_C \frac{z^2}{(z^2+4)^2}dz, \tag{2.19}$$

where C (Figure 2.14) is made up of the line $-R < x < R$ on the real axis and the semicircle C^1. If we can show that the contribution of the semicircle to the integral in eq. (2.19) is zero and we let $R \to \infty$, we would have found the real-valued integral I in eq. (2.18).

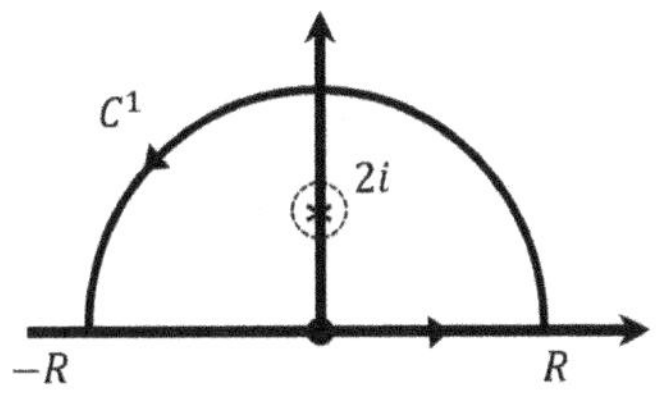

Figure 2.14: An upper half plane contour with a simple pole at $z = 2i$.

To this end, we note that the function $f(z)$ is singular only at $z = \pm 2i$, which are second-order poles. Only the pole at $+2i$ is relevant to the upper half plane (UHP) domain shown in Figure 2.14. Since any closed contour containing $-R < x < R$, $R \to \infty$ will be appropriate for our purpose, we could have chosen the closed contour in the lower half plane (LHP) and obtain the same result. We have arbitrarily chosen the UHP.

For $z = 2i, C_{-1} = -i/8$ (from a previous example), and

$$J = 2\pi i \frac{(-i)}{8} = \int_C f(z)dz = \int_C \frac{z^2}{(z^2+4)^2} dz$$

$$= \underbrace{\int_{-R}^{R} \frac{x^2}{(x^2+4)^2} dx}_{(A)} + \underbrace{\int_{\frown} \frac{z^2}{(z^2+4)^2} dz}_{(B)}.$$

On the integral in (B): For large R, $z^2 + 4 \sim z^2$ on the semicircle (with length $\int dz \sim \pi R$, the perimeter of the semicircle) so that

$$\left| \int_{\frown} \cdot \right| = O\left(\frac{R^2}{R^4} \cdot \pi R \right) = O\left(\frac{1}{R} \right) \to 0 \text{ as } R \to \infty.$$

Thus, letting $R \to \infty$

$$J = 2\pi i \left(-\frac{i}{8} \right) = 2I + O \Rightarrow I = \frac{\pi}{8}.$$

Example 2.20 Evaluate

$$I = \int_0^\infty \frac{x^2 dx}{x^6+1} = \frac{1}{2} \int_{-\infty}^{\infty} \frac{x^2}{x^6+1} dx. \tag{2.23}$$

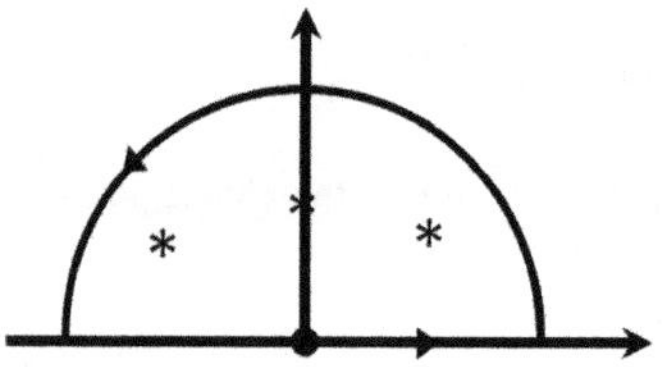

Figure 2.15: Upper half plane contour for I in eq. (2.23), with three poles.

Solution: Using the contour $\curvearrowleft$ as shown in Figure 2.15:

$$I = \frac{1}{2} \cdot 2\pi i \sum_{j=1}^{3} \underbrace{\text{Res}\left[\frac{z^2}{z^6+1}, a_j \right]}_{\text{Residue of } \frac{z^2}{z^6+1} \text{ at } a_j},$$

where the a_j's are the roots of $z^6 + 1 = 0 \Rightarrow z = (-1)^{1/6}$; and the principal value of -1 has arg $\pm\pi$, and the arguments of z are given by $\pm(\pi + 2n\pi)/6$ or $\pm\pi/6$, $\pm\pi/3$, and $\pm 5\pi/6$ for the first three values. The magnitude of $z \equiv (-1)^{1/6}$ is unity. Thus,

$$I = \pi i \sum_{j=1}^{3} \lim_{z \to a_j} \left[\frac{z^2}{z^6 + 1} (z - a_j) \right].$$

Using L'Hopital's Rule, we see that

$$\frac{z^3 - z^2 a_j}{z^6 + 1} \Rightarrow \frac{3z^2 - 2za_j}{6z^5} = \frac{a_j^2}{6a_j^5} \text{ as } z \to a_j.$$

Thus,

$$I = \pi i \sum_{j=1}^{3} \frac{a_j^2}{6a_j^5} = \frac{\pi i}{6} \sum_{j}^{3} a_j^{-3}$$

and

$$I = \frac{\pi i}{6} \left[\left(e^{\frac{i\pi}{6}} \right)^3 + \left(e^{\frac{i\pi}{2}} \right)^3 + \left(e^{\frac{i5\pi}{6}} \right)^3 \right] = \frac{\pi}{6}.$$

Example 2.21 Evaluate the integral

$$I = \int_0^\infty \frac{dx}{x^3 + \lambda^3}, \qquad \lambda > 0.$$

Solution: We cannot use the domain $-\infty < x < \infty$ to evaluate this integral because $f(x) \equiv 1/\left(x^3 + \lambda^3\right)$ is odd so that we do not have a symmetry. However, there is a radial line $\theta = 2\pi/3$ (Figure 2.16) where $\lambda^3 + z^3$ is a constant multiple of $a^3 + r^3$. That is, the denominator of the integrand is the same. To see this, consider the integral $\int_0^\infty dz/(z^3 + \lambda^3)$. On C_I, $z = re^{i\theta} = r$ so that the integral is $\int_{C_I} dr/(r^3 + \lambda^3)$. On C_{III}, $z = re^{i2\pi/3}$, $dz = e^{i2\pi/3} dr$ and the integral is $\int_{C_{III}} (e^{i2\pi/3} dr)/[(re^{i2\pi/3})^3 + \lambda^3] = \int_{C_{III}} [(e^{i2\pi/3} dr)/(r^3 + \lambda^3)]$, which has the same denominator as the integral over C_I, allowing factorization. Note that instead of the cube power in the denominator if we have the nth power, then the sector angle that allows this type of factorization is $2\pi/n$ so that $x^n + \lambda^n$ becomes $(re^{i2\pi/n})^n + \lambda^n = r^n + \lambda^n$.

The integral over $C \equiv C_{\text{I}} \cup C_{\text{II}} \cup C_{\text{III}}$ is given by the residue theorem as

$$I' = \int_C \frac{dz}{z^3 + \lambda^3} = \int_{C_{\text{I}}} \frac{dz}{z^3 + \lambda^3} + \int_{C_{\text{II}}} \frac{dz}{z^3 + \lambda^3} + \int_{C_{\text{III}}} \frac{dz}{z^3 + \lambda^3} = 2\pi i \sum_j \text{Res} \left[\frac{1}{z^3 + \lambda^3}, z_j \right],$$

where z_j satisfies $z_j^3 + \lambda^3 = 0$ or $z_j^3 = -\lambda^3 = \lambda^3 e^{i(\pi + 2n\pi)}$ or $z_j = \{\lambda^3 e^{i(\pi + 2n\pi)}\}^{1/3}$. This value $\lambda e^{i\pi/3} \equiv z_0$ lies inside C and is used, yielding $I' = (2\pi i/3\lambda^2) e^{-2\pi i/3}$. On C_{II}, the integral goes like $R/R^3 = 1/R^2 \to 0$ as $R \to \infty$.

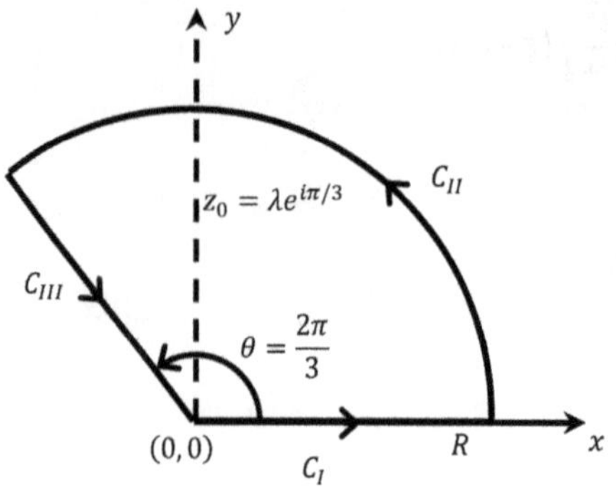

Figure 2.16: A sector of angle $2\pi/3$ for evaluating the integral in Example 2.21.

Hence

$$\int_{C_I} \frac{dr}{r^3+\lambda^3} + \int_{C_{\mathrm{III}}} \frac{e^{i2\pi/3}}{r^3+\lambda^3} dr = \frac{2\pi i}{3\lambda^2} e^{-2\pi i/3}.$$

We can write this equation as

$$\lim_{R\to\infty} \left(\int_0^R \frac{dr}{r^3+\lambda^3} + \int_R^0 \frac{e^{i2\pi/3}}{r^3+\lambda^3} dr \right) = \frac{2\pi i}{3\lambda^2} e^{-2\pi i/3}$$

or,

$$\lim_{R\to\infty} \left[1 - e^{\frac{2\pi i}{3}}\right] \int_0^R \frac{dr}{r^3+\lambda^3} = \frac{2\pi i}{3\lambda^2} e^{-2\pi i/3}$$

$$\Rightarrow \int_0^R \frac{dr}{r^3+\lambda^3} = \frac{2\pi i}{3\lambda^2} e^{-\frac{2\pi i}{3}} \Big/ \left[1 - e^{\frac{2\pi i}{3}}\right].$$

2.6.1 Evaluation of improper real integrals and real Fourier integrals

In this section, we discuss the evaluation of improper real integrals such as $\int_{-\infty}^{\infty} f(x)dx$ or of real integrals such as

$$\int_{-\infty}^{\infty} f(x) \sin ax dx \text{ and } \int_{-\infty}^{\infty} f(x) \cos ax dx, \tag{2.24}$$

which can be found in problems on Fourier integrals. Suppose $f(x)$ is a rational function whose denominator is different from zero for all real x and is of degree at least two units higher than the degree of the numerator. The Fourier integrals in eq. (2.24) may be evaluated by considering the corresponding integral $\int_C f(z)e^{iaz}dz$, where a is real and positive. Note that above we have also assumed that the improper integrals

$$\int_{-\infty}^{\infty} f(x)\sin ax dx \equiv \int_{-\infty}^{\infty} g(x)dx$$

and

$$\int_{-\infty}^{\infty} f(x)\cos ax dx \equiv \int_{-\infty}^{\infty} h(x)dx$$

are such that the two limits on the RHS of the following split integrals exist:

$$\int_{-\infty}^{\infty} g(x)dx = \lim_{p\to -\infty}\int_{p}^{0} g(x)dx + \lim_{q\to\infty}\int_{0}^{q} g(x)dx$$

$$\int_{-\infty}^{\infty} h(x)dx = \lim_{p\to -\infty}\int_{p}^{0} h(x)dx + \lim_{q\to\infty}\int_{0}^{q} h(x)dx.$$

If both limits on the RHS exist, then the integrals on the left are said to be convergent, but divergent otherwise. The significance of this condition is that we know ahead of time that if an integral $\int_{-\infty}^{\infty} g(x)dx$ converges, we can evaluate it by means of a single limiting process or

$$\int_{-\infty}^{\infty} g(x)dx = \lim_{R\to\infty}\int_{-R}^{R} g(x)dx, \tag{2.25}$$

rather than splitting it into the two subdomains $-R$ to 0 and 0 to R, $R \to \infty$. The limit in eq. (2.25) is called the CPV of the integral (see the next section) and is expressed as

$$P.V.\int_{-\infty}^{\infty} g(x)dx = \lim_{R\to\infty}\int_{-R}^{R} g(x)dx. \tag{2.26}$$

It should be noted that the limit on the RHS of eq. (2.26) may exist even when the integral is divergent because we avoid integrating over the singular points of the integrand, integrating around them instead. That is, if $z_o = a$ is a singular point, we will evaluate the integral as

$$\int_{-\infty}^{\infty} g(x)dx = \lim_{R\to\infty}\left\{\int_{-R}^{a^-} g(x)dx + \int_{a^+}^{R} g(x)dx\right\} \equiv P.V.\int_{-\infty}^{\infty} g(x)dx.$$

To see this consider $\int_{-\infty}^{\infty} xdx$, whose CPV is zero even when $\lim_{R\to\infty}\int_0^R x\,dx \to \infty$. From the foregoing, we see that if the integral $\int_{-\infty}^{\infty} g(x)dx$ converges, its CPV will be the same as the value of the integral. On the other hand, if the integral diverges, in which case the value of the integral is infinite, the integral may still possess a CPV. The obvious question then is how useful the CPV is in such cases.

Under the foregoing conditions, we will have

$$\lim_{r\to\infty}\int_{-r}^{r} g(x)dx = 2\pi i \sum \operatorname{Res}\left[g(z), z_j\right] - \int_{\frown} g(z)dz.$$

Since we require that the degree of the denominator of $g(z)$ be at least two units higher than that of the numerator, we will have $\int_{\frown} = 0$, if we use familiar arguments as $R \to \infty$:

$$|g(z)| < \frac{k}{|z|^2}, \qquad \left|\int g(z)dz\right| < \frac{k}{R^2}\pi R = \frac{k\pi}{R} \to 0 \text{ as } R \to \infty.$$

Above, πR is the perimeter of the semicircular contour $\frown$. Thus,

$$\int_{-\infty}^{\infty} g(x)dx = 2\pi i \sum_{j=1}^{N_p} \operatorname{Res}\left[f(z), z_j\right],$$

where N_p is the number of poles contained inside the semicircular domain in the UHP.

Similarly,

$$\int_{-\infty}^{\infty} f(z)e^{iaz}dz = 2\pi i \sum_j \operatorname{Res}\left[f(z)e^{iaz}, z_j\right] \quad (a > 0 \text{ and real}), \tag{2.27}$$

where the summation is carried out over the residues of $f(z)e^{iaz}$ at its poles in the UHP. Equating the real and the imaginary parts on both sides of eq. (2.27), we have

$$\int_{-\infty}^{\infty} f(x)e^{iax}dx = \int_{-\infty}^{\infty} f(x)\cos axdx + i\int_{-\infty}^{\infty} f(x)\sin axdx,$$

from which we have

$$\int_{-\infty}^{\infty} f(x)\cos axdx = -2\pi \sum_j \operatorname{Im}\left[\operatorname{Res}\left[f(z)e^{iaz}, z_j\right]\right]$$

$$\int_{-\infty}^{\infty} f(x)\sin axdx = -2\pi \sum_j \operatorname{Re}\left[\operatorname{Res}\left[f(z)e^{iaz}, z_j\right]\right].$$

In the foregoing we have used the fact that

$$\left|e^{iaz}\right| = \left|e^{iax}\right|\left|e^{-ay}\right| = e^{-ay} \le 1 \text{ as } y \to +\infty$$

so that

$$\left|f(z)e^{iaz}\right| = |f(z)|\left|e^{iaz}\right| \le |f(z)|.$$

Example 2.22 Evaluate

$$I = \int_0^\infty \frac{\cos ax}{x^2+4}\,dx, \quad a \ge 0.$$

Solution: We can attempt to use the same contour as in the example above and evaluate $J = \int_C \cos az/(z^2+4)dz$. (Note that $\cos az = \left(e^{iaz}+e^{-iaz}\right)/2$.) However, the contribution of $\int_{\curvearrowleft}$ can no longer be neglected as $R \to \infty$. For example, when $z = iR$, at the top of the semicircle $\cos az = \left(e^{aR}+e^{-aR}\right)/2 \sim e^{aR}/2$ as $R \to \infty$, which is very large. Instead of using $\cos az$ in the numerator, let us use the part of $\cos az$ that is bounded as $R \to \infty$ or e^{iaz}. That is, we will consider the integral

$$J = \int_C \frac{e^{iaz}}{(z^2+4)}\,dz,$$

where the contribution from the contour $\curvearrowleft$ is negligible since

$$\left|e^{iaz}\right| = \left|e^{iax-ay}\right| = \left|e^{iax}\right|\left|e^{-ay}\right| = \left|e^{-ay}\right| \le 1$$

as $y \to \infty$ so that

$$O\underbrace{\left(\frac{1}{R^2}\right)}_{(z^2+4)^{-1}} \cdot \underbrace{\pi R}_{dz} = O\left(\frac{1}{R}\right) \to 0 \text{ as } R \to \infty$$

and

$$J = 2\pi i \operatorname{Res} @ 2i = \underbrace{\int_{-\infty}^{\infty} \frac{\cos ax}{x^2+4}\,dx}_{\text{Real}} + i\underbrace{\int_{-\infty}^{\infty} \frac{\sin ax}{x^2+4}\,dx}_{\text{Real}}, \text{ with } \int_{\curvearrowleft} (\cdot)dz = 0.$$

For the residue at $2i$:

$$\operatorname{Res}[f(z), 2i] = \lim_{z\to 2i} \frac{(z-2i)e^{iaz}}{z^2+4} = \lim_{z\to 2i} \frac{e^{iaz}}{z+2i} = \frac{e^{-2a}}{4i}$$

$$\Rightarrow J = 2\pi i \frac{e^{-2a}}{4i} = +\frac{\pi e^{-2a}}{2} \text{ (Real)} \tag{2.22}$$

Equating real and imaginary parts of J_i

$$\int_{-\infty}^{\infty} \frac{\cos ax}{x^2+4} dx = 2\int_{0}^{\infty} \frac{\cos ax}{x^2+4} dx \Rightarrow \int_{0}^{\infty} \frac{\cos ax}{x^2+4} dx = \frac{J}{2} = \frac{\pi e^{-2a}}{2} = \left(\frac{\pi}{4}\right) e^{-2a}$$

and

$$\int_{-\infty}^{\infty} \frac{\sin ax}{x^2+4} dx = 0,$$

The latter is obviously predictable from the antisymmetry of $\sin ax/(x^2+4)$ about $x = 0$.

2.6.2 Cauchy principal value

Let us discuss the concept of the CPV in detail in this subsection. It is understandable that if the integrand $f(x)$ in the integral

$$I = \int_{a}^{b} f(x)dx \tag{2.28}$$

is unbounded (undefined) at the lower and upper limits of the integral, we could avoid these limits and approximate I, respectively, as follows:

$$I = \lim_{\varepsilon_1 \to 0} \int_{a+\varepsilon_1}^{b} f(x)dx, \qquad I = \lim_{\varepsilon_2 \to 0} \int_{a}^{b-\varepsilon_2} f(x)dx,$$

where ε_1 and ε_2 can be made infinitesimally small, but need not be equal. Moreover, if $f(x)$ is unbounded at both a and b in eq. (2.28), we could approximate I as

$$I = \lim_{\varepsilon_1, \varepsilon_2 \to 0} \int_{a+\varepsilon_1}^{b-\varepsilon_2} f(x)dx,$$

where ε_1 and ε_2 could go to zero independently. However, if the singularity is located at an interior point x_o in the domain $x \in [a, b]$, we will approximate I as

$$I = \lim_{\varepsilon_1 \to 0} \int_{a}^{x_o-\varepsilon_1} f(x)dx + \lim_{\varepsilon_2 \to 0} \int_{x_o+\varepsilon_2}^{b} f(x)dx. \tag{2.29}$$

If we set $\varepsilon_1 = \varepsilon_2 \equiv \varepsilon$ and take the limit $\varepsilon \to 0$ of the sum of the integrals in eq. (2.29), the value obtained for I in this case if the limit exists is called the CPV, and several notations have been used to denote it. We will use the symbol '$P.V.\int$' in this book:

$$I = P.V.\int_a^b f(x)dx = \lim_{\varepsilon\to 0}\left\{\int_a^{x_0-\varepsilon} f(x)dx + \int_{x_0+\varepsilon}^b f(x)dx\right\}.$$

In a similar fashion, for the improper integral, we can write

$$P.V.\int_{-\infty}^{\infty} f(x)dx = \lim_{\varepsilon\to 0}\left\{\lim_{a\to\infty}\int_{-a}^{x_0-\varepsilon} f(x)dx + \lim_{b\to\infty}\int_{x_0+\varepsilon}^{b} f(x)dx\right\}.$$

So, basically, with the CPV, we essentially "ignore" the presence of point x_0 and effectively evaluate

$$P.V.\int_{-\infty}^{\infty} f(x)dx = \lim_{R\to\infty}\int_{-R}^{R} f(x)dx$$

so that the point x_0 does not contribute to the original integral I. Needless to say that this could be a dangerous thing to do, especially if the physical problem involves a contour that must pass through the point x_0. This is the case when integration is along the real axis and there is a singularity at x_0. In this case, we will do the analysis in the complex plane, using the CPV value as described above, but combine the integral obtained with that from integrating over a small semicircular region surrounding z_0. Using Figure 2.17, what we are saying is that the CPV evaluates the integral over the contour $C' \equiv C_1 \cup C_2 \cup C_4$, assumed closed so that

$$P.V.\int_{C'\cup C_3} f(z)dz = \int_{C'} f(z)dz = 2\pi i\sum_j \operatorname{Res}[f(z), z_j],$$

where the z_j's are the singular points inside the closed contour $C' \cup C_3$. This can be done when the contour does not have to pass through z_0. If it does, we will need to augment the CPV value with the integral over C_3, which, as we have shown elsewhere in this chapter, is equal to $\pi i \operatorname{Res}[f(z), z_0]$.

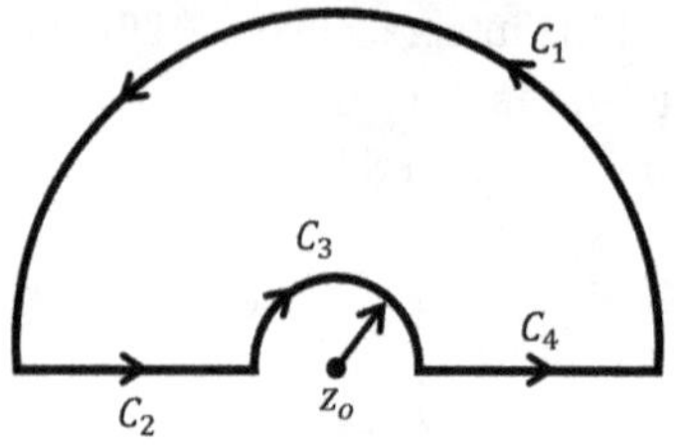

Figure 2.17: Depiction of a pole on the real axis.

To appreciate the nuances in the foregoing discussions, consider $I = \int_{-\infty}^{\infty} f(x)dx$, where $f(x) = x$. The CPV for this can be obtained as

$$P.V. \int_{-\infty}^{\infty} x\,dx = \lim_{R\to\infty} \int_{-R}^{R} x\,dx = \lim_{R\to\infty} \left(\frac{R^2}{2} - \frac{R^2}{2}\right) = \lim_{R\to\infty}(0) = 0.$$

On the other hand,

$$\begin{aligned}\int_{-\infty}^{\infty} x\,dx &\equiv \lim_{R_1\to\infty} \int_{-R_1}^{0} x\,dx + \lim_{R_2\to\infty} \int_{0}^{R_2} x\,dx \\ &= \lim_{R_2\to\infty} \frac{R_2^2}{2} - \lim_{R_1\to\infty} \frac{R_1^2}{2}.\end{aligned} \tag{2.30}$$

We see divergence of the integral because neither of the two terms on the RHS of eq. (2.30) can be defined. The examples below further illustrate the technique of calculating the CPV.

Example 2.23 Evaluate the integral $\int_a^b \frac{dx}{(x-1)^3}$, where $a = -1$, $b = 5$: (a) in the usual sense and (b) in the CPV sense. Note the singularity at $x = 1$.

Solution:

(a)

$$\int_a^b \frac{dx}{(x-1)^3} = \int_{-1}^{5} \frac{dx}{(x-1)^3} = \lim_{\varepsilon_1\to 0^+} \int_{-1}^{1-\varepsilon_1} \frac{dx}{(x-1)^3} + \lim_{\varepsilon_2\to 0^+} \int_{1+\varepsilon_2}^{5} \frac{dx}{(x-1)^3}$$

$$= \lim_{\varepsilon_1\to 0^+} \left[\frac{1}{8} - \frac{1}{2\varepsilon_1^2}\right] + \lim_{\varepsilon_2\to 0^+} \left[\frac{1}{2\varepsilon_2^2} - \frac{1}{32}\right].$$

The limits do not exist for either term on the RHS. Therefore, the integral does not converge in the standard sense.

(b)

$$P.V. \int_{-1}^{5} \frac{dx}{(x-1)^3} = \lim_{\varepsilon\to 0^+} \left\{ \int_{-1}^{1-\varepsilon} \frac{dx}{(x-1)^3} + \int_{1+\varepsilon}^{5} \frac{dx}{(x-1)^3} \right\} = \lim_{\varepsilon\to 0^+} \left\{ \frac{1}{8} - \frac{1}{2\varepsilon^2} + \frac{1}{2\varepsilon^2} - \frac{1}{32} \right\} = \frac{3}{32}.$$

Thus, the integral converges to 3/32 in the CPV sense. Also note that $I = \int_{-\infty}^{\infty} \sin x dx$ does not exist when integrated as

$$\lim_{R_1\to\infty, R_2\to\infty} \int_{-R_1}^{R_2} \sin x \, dy = \lim_{R_1\to\infty, R_2\to\infty} [\cos R_2 - \cos R_1],$$

whereas

$$P.V. \int_{-\infty}^{\infty} \sin x \, dx = \lim_{R\to\infty} \int_{-R}^{R} \sin x \, dx = \lim_{R\to\infty} [\cos R - \cos(-R)] = 0,$$

which is convergent. Similarly,

$$\int_{-\infty}^{\infty} \frac{x}{1+x^2} dx = \lim_{R_1\to\infty, R_2\to\infty} \int_{-R_1}^{R_2} \frac{x}{(1+x^2)} dx = \lim_{R_1\to\infty, R_2\to\infty} \frac{1}{2} \{ \ln(1+R_2^2) - \ln(1+R_1^2) \},$$

which is not defined, but

$$P.V. \int_{-\infty}^{\infty} \frac{x}{(1+x^2)} dx = 0.$$

Example 2.24 Determine the Cauchy principal value of

$$\int_{-\infty}^{\infty} \frac{\sin x}{x(x^2 - 2x + 2)} dx.$$

Solution: The integrand is of the form $f(x) \sin ax$ or $f(x) \cos ax$. Therefore, the integral to consider is

$$I = \int_{-\infty}^{\infty} \frac{e^{iz}}{z(z^2 - 2z + 2)} dz \equiv \int_{-\infty}^{\infty} f(z) e^{iz} \, dz.$$

Now $f(z)$ has simple poles at $z = 0, 1 \pm i$, of which $z = 1 + i$ is located in the UHP and $z = 0$ is on the real axis. The indented contour is shown in Figure 2.18. Inside the indented contour the only singularity is the pole at $z = 1 + i$. Therefore,

$$\int_C (\cdot)\,dz = \int_{C_R} (\cdot)dz + \int_{-R}^{-r} (\cdot)dx + \int_{-C_r} (\cdot)dz + \int_r^R (\cdot)dx$$

$$= 2\pi i \operatorname{Res}\left[f(z)e^{iz}, 1+i\right] = -\frac{e^{-1+i}}{4}(1+i)$$

Note that $\int_{-C_r} = -\int_{C_r}$ and that $\int_{C_R}(\cdot)dz \to 0$ as $R \to \infty$, consistent with previous discussions in this subsection. Furthermore,

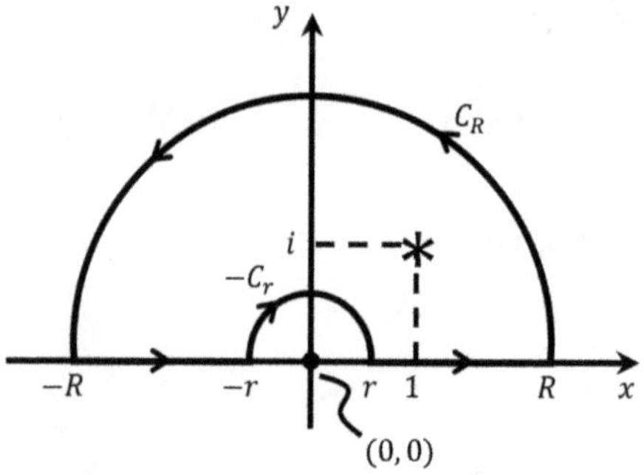

Figure 2.18: A sketch for Example 2.24.

$$\int_{-C_r} (\cdot)\,dz = -\int_{C_r} (\cdot)\,dz = \pi i\left\{\operatorname{Res}\left[f(z)e^{iz}, 0\right]\right\}$$

$$= \pi i \operatorname{Res}\left[f(z)e^{iz}, 0\right] = \frac{\pi i}{2}.$$

Thus, from

$$\text{P.V.} \int_{-\infty}^{\infty} \frac{e^{ix}}{x(x^2 - 2x + 2)} dx - \int_{C_r} f(z)e^{iz} dz = \int_C f(z)e^{iz} dz,$$

we have

$$\text{P.V.} \int_{-\infty}^{\infty} \frac{e^{ix}}{x(x^2 - 2x + 2)} dx = 2\pi i\left[-\frac{e^{-1+i}}{4}(1+i)\right] + \frac{\pi i}{2}.$$

Since

$$e^{-1+i} = e^{-1}(\cos 1 + i \sin 1),$$

we have

$$P.V.\int_{-\infty}^{\infty}\frac{e^{ix}}{x(x^2-2x+2)}dx=\frac{\pi}{2}\left[1+e^{-1}(\sin 1-\cos 1)\right]. \tag{2.31}$$

Obviously, the procedure used to obtain eq. (2.31) can be extended to include n poles inside the UHP and m poles on the real axis. In that case, the theorem can be stated as follows:

Theorem 2.2 Let $f(z)$ be an analytic function except for n poles at the points z_1, z_2, . . ., z_n in the UHP and m poles at the points x_1, x_2, . . ., x_m on the real axis. If $\lim_{R\to\infty}\int_{\frown} f(z)dz=0$ where $\frown$ is the semicircle $|z|=R$ in the UHP, then

$$P.V.\int_{-\infty}^{\infty} f(x)dx=\pi i\sum_{k=1}^{m}\text{Res}[f(z),x_k]+2\pi i\sum_{k=1}^{n}\text{Res}[f(z),z_k].$$

2.6.3 Poles on the real axis and indented contours

Let us formally discuss the evaluation of the real integral

$$I=\int_{-\infty}^{\infty} f(x)dx,$$

when $f(x)$ has poles on the real axis. The result of the analysis in this section has been used in the previous section, so we are just formalizing the procedure here. We could use an indented contour if there is indeed a pole on the real axis. Say the pole is located at $z=z_o$ on the real axis, the indented contour, assumed to be on the UHP, could be depicted as shown in Figure 2.19. C_r in the figure is a "small" semicircular contour with center at $z=z_o$, and whose orientation relative to our counterclockwise convention is opposite to that of C_R. Hence the negative sign on C_r (Figure 2.19). A theorem would help with the evaluation of $\int_{C_r} f(z)dz$.

Theorem 2.3 (Behavior of integral as $r\to 0$) Suppose f has a simple pole at $z=z_o$ on the real axis, and C_r is the contour defined by $z=z_o+re^{i\theta}$, where $0\le\theta\le\pi$, then

$$\lim_{r\to 0}\int_{C_r} f(z)dz=\pi i\,\text{Res}[f(z),z_o]. \tag{2.32}$$

Note that the factor "2" is absent on the RHS of eq. (2.32) relative to the usual statement of the residue theorem regarding the evaluation of the integral around a closed-contour inside which $f(z)$ has a simple pole.

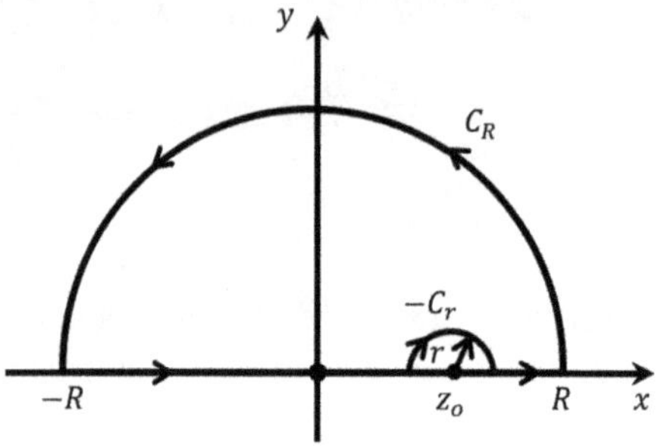

Figure 2.19: Depiction of a pole on the real axis.

Proof 2.1 Because $f(z)$ has a simple pole at $z = z_0$, its Laurent series can be written as

$$f(z) = \frac{C_{-1}}{(z - z_0)} + h(z),$$

where C_{-1} is the residue of $f(z)$ at $z = z_0$ (or $C_{-1} = \text{Res}[f(z), z_0]$). The function $h(z)$ is analytic. Let $z - z_0 = re^{i\theta}$ on C_r so that $dz = ire^{i\theta} d\theta$ and

$$\int_{C_r} f(z)dz = \int_{C_r} \frac{C_{-1}}{(z - z_0)} + \int_{C_r} h(z)dz$$

$$= \int_0^\pi \frac{(C_{-1})ire^{i\theta} d\theta}{re^{i\theta}} + ir \int_0^\pi h\left(z_0 + re^{i\theta}\right) e^{i\theta} d\theta \equiv I_1 + I_2.$$

I_1 can be expressed as

$$I_1 = C_{-1} \int_0^\pi i d\theta = \pi i C_{-1} = \pi i \ \text{Res}[f(z), z_0].$$

Since $h(z)$ is analytic at z_0, it is continuous at the point and also bounded in a neighborhood of the point. This means that there exists an $M > 0$ such that $|h(z)| \equiv \left|h\left(z_0 + re^{i\theta}\right)\right| \le M$ and

$$|I_2| \le r \int_0^\pi M \, d\theta = \pi r M \to 0 \text{ as } r \to 0.$$

Example 2.25 Evaluate the integral

$$I = \int_{-\infty}^{\infty} \frac{\cos x}{(x^2 + 1)(x^2 - 3x + 2)} dx.$$

Solution: Following previous treatments, we consider the integrand $f(z) = e^{iz}/\left[(z^2+1)(z^2-3z+2)\right]$, which has simple poles at $z = \pm i$, $z = 1$, and $z = 2$. The first two poles are inside $\curvearrowleft$, whereas the other two are located on the real axis.

Noting that

$$\mathrm{Res}[f(z), i] = \frac{(3-i)}{20e}, \quad \mathrm{Res}[f(z), 1] = -\frac{1}{2}[\cos(1) + i\sin(1)],$$

$$\mathrm{Res}[f(z), 2] = \frac{1}{5}[\cos(2) + i\sin(2)],$$

and that $z = -i$ is not contained in the UHP, we can invoke the above theorem to evaluate I as follows:

$$\begin{aligned} I &= P.V. \int_{-\infty}^{\infty} \frac{e^{ix}}{(x^2+1)(x^2-3x+2)} dx \\ &= 2\pi i \,\mathrm{Res}[f(x), i] + \pi i \,\mathrm{Res}[f(z), 1] + \pi i \,\mathrm{Res}[f(z), 2] \\ &= \frac{2\pi i(3-i)}{20e} + \pi i\left\{-\frac{[\cos(1) + i\sin(1)]}{2}\right\} + \pi i\left\{\frac{[\cos(2) + i\sin(2)]}{5}\right\}. \end{aligned} \tag{2.33}$$

Equating the real parts on the LHS and RHS of this equation gives

$$P.V. \int_{-\infty}^{\infty} \frac{\cos x}{(x^2+1)(x^2-3x+2)} dx = \frac{\pi}{10}\left[5\sin(1) - 2\sin(2) + \frac{1}{e}\right],$$

which is the required result. However, as a bonus, we can obtain the following result by balancing the imaginary part of eq. (2.33):

$$P.V. \int_{-\infty}^{\infty} \frac{\sin x}{(x^2+1)(x^2-3x+2)} dx = \frac{\pi}{10}\left[2\cos(2) - 5\cos(1) + \frac{3}{e}\right].$$

2.6.4 Integrals of the form $I = \int_0^{2\pi} F(\cos\theta, \sin\theta)\, d\theta$

The idea here is to convert I to a complex integral, where the contour C is the unit circle, or $|z| = 1$, which has a radius of one and is centered at the origin [i.e., $(0, 0)$]. The contour can be parameterized for the unit circle as follows. Employing the Euler formula:

$$z = \cos\theta + i\sin\theta = e^{i\theta}, \qquad 0 \le \theta \le 2\pi,$$

we have

$$dz = ie^{i\theta} d\theta, \qquad \cos\theta = \frac{e^{i\theta} + e^{-i\theta}}{2}, \qquad \sin\theta = \frac{e^{i\theta} - e^{-i\theta}}{2i},$$

$$d\theta = \frac{dz}{iz}, \qquad \cos\theta = \frac{1}{2}\left(z+z^{-1}\right), \qquad \sin\theta = \frac{1}{2i}\left(z-z^{-1}\right).$$

Therefore,

$$I = \oint_C F\left[\frac{1}{2}\left(z+z^{-1}\right), \frac{1}{2i}\left(z-z^{-1}\right)\right]\frac{dz}{iz}$$

where C is $|z| = 1$.

Example 2.26 Evaluate

$$I = \int_0^{2\pi} \frac{1}{(2+\cos\theta)^2}\,d\theta$$

Solution: This yields

$$I = \frac{4}{i}\int_C f(z)dz,$$

where C is $|z| = 1$ and

$$\frac{1}{(2+\cos\theta)^2} = \frac{1}{\left[2+\frac{1}{2}\left(z+z^{-1}\right)\right]^2} = \frac{1}{\frac{4z+z^2+1}{4}\frac{1}{z^2}},$$

so that

$$f(z) = \frac{z}{(z^2+4z+1)^2} = \frac{z}{(z-z_0)^2(z-z_1)^2}$$

and

$$z_0 = -2-\sqrt{3}, \qquad z_1 = -2+\sqrt{3}.$$

We see that only z_1 is located inside the unit circle, and it is a pole of order 2:

$$\operatorname{Res}[f(z), z_1] = \lim_{z\to z_1}\frac{d}{dz}(z-z_1)^2 f(z) = \lim_{z\to z_1}\frac{d}{dz}\left(\frac{1}{(z-z_0)^2}\right) = -\lim_{z\to z_1}\frac{z+z_0}{(z-z_0)^3} = \frac{1}{6\sqrt{3}}.$$

Therefore,

$$I = \frac{4}{i}\oint_C \frac{z}{(z^2+4z+1)^2}dz = \frac{4}{i}2\pi i \operatorname{Res}[f(z), z_1] = \frac{4}{i}2\pi i\frac{1}{6\sqrt{3}} = \frac{4}{3}\pi\frac{1}{\sqrt{3}}$$

and

$$\int_0^{2\pi} \frac{1}{(2+\cos\theta)^2} d\theta = \frac{4\pi}{3\sqrt{3}}.$$

Example 2.27 Show that if $|f(z)| \le M/R^k$ for $z = Re^{i\theta}$ and $k > 0$ and M are constants, then

$$\lim_{R\to\infty} \int_\Gamma e^{imz} f(z)dz = 0,$$

where Γ is the semicircular arc of the contour shown in Figure 2.20 and m is a positive constant.

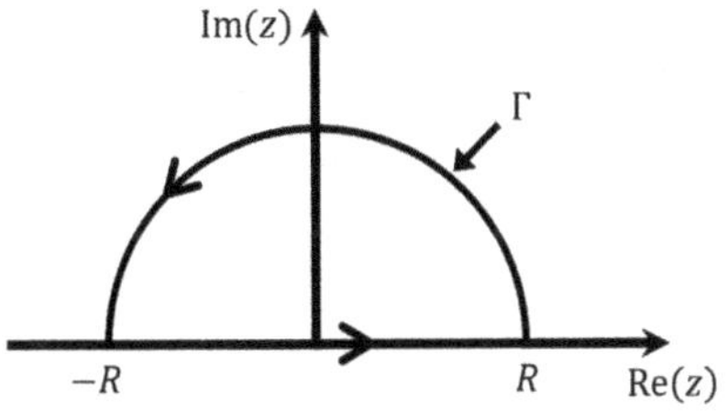

Figure 2.20: Semicircular arc on the upper half plane of domain.

Solution: Let $z = Re^{i\theta} \Rightarrow dz = iRe^{i\theta}d\theta$, $\int_\Gamma e^{imz} f(z)dz = \int_0^{2\pi} e^{imRe^{i\theta}} f(Re^{i\theta}) iRe^{i\theta} d\theta$. Then, noting that $e^{i\phi}$ is bounded for any real ϕ, we can write

$$\left|\int_0^\pi e^{imRe^{i\theta}} f\left(Re^{i\theta}\right) iRe^{i\theta} d\theta\right| \le \int_0^\pi \left|e^{imRe^{i\theta}} f\left(Re^{i\theta}\right) iRe^{i\theta}\right| d\theta$$

$$= \int_0^\pi \left|e^{i(mR\cos\theta)-mR\sin\theta} f\left(Re^{i\theta}\right) iRe^{i\theta}\right| d\theta = \int_0^\pi e^{-mR\sin\theta} \left|f\left(Re^{i\theta}\right)\right| R d\theta$$

$$\le \frac{M}{R^{k-1}} \int_0^\pi e^{-mR\sin\theta} d\theta = \frac{2M}{R^{k-1}} \int_0^{\frac{\pi}{2}} \underbrace{e^{-mR\sin\theta}}_{\substack{\Rightarrow 0 \text{ as } R\to\infty \\ \text{faster than } R^{k-1}}} d\theta$$

using symmetry of $\sin\theta$ about $\theta = \frac{\pi}{2}$ in $\theta \in [0, \pi]$.

But $\sin\theta \ge 2\theta/\pi$ for $0 \le \theta \le \pi/2$, therefore

$$\frac{2M}{R^{k-1}} \int_0^{\frac{\pi}{2}} e^{-mR\sin\theta} d\theta \le \frac{2M}{R^{k-1}} \int_0^{\frac{\pi}{2}} e^{-\frac{2mR\theta}{\pi}} d\theta$$

$$= \frac{\pi M}{mR^k} \left(1 - e^{-mR}\right) \to 0 \text{ as } R \to \infty.$$

2.6.5 Improper integrals having integrands of the form $e^{imz}Q(z)$

This section could be considered a generalization and formalization of the result in Example 2.27. We have seen that the success with the evaluation of many real but improper integrals of rational functions without poles on the real axis by using the residue theorem rests on the property that the integral around the semicircular part of the contour vanishes in the limit as the radius of the semicircle $R \to \infty$. That is, the residue theorem approach works if we can set

$$\lim_{R\to\infty} \int_{\curvearrowleft} f(z)dz = 0. \tag{2.34}$$

It will be useful to establish the conditions that will ensure the validity of eq. (2.34) if $f(z)$ is of the form $e^{imz}Q(z)$ and the method of residue is to be used. To this end, we cite Jordan Inequality and Jordan Integral Inequality. The Jordan Inequality states that

$$\frac{2\theta}{\pi} \le \sin\theta \le \theta \text{ for } 0 \le \theta \le \frac{\pi}{2}, \tag{2.35}$$

while the Jordan Integral Inequality states that

$$\int_0^{\pi/2} e^{-k\sin\theta} d\theta \le \frac{\pi}{2k}\left(1 - e^{-k}\right), k > 0.$$

To prove the inequality, assume that it is true and divide eq. (2.35) by θ to obtain

$$1 \ge \frac{\sin\theta}{\theta} \ge \frac{2}{\pi}, 0 \le \theta \le \frac{\pi}{2}.$$

Let $H(\theta) = \sin\theta/\theta$ so that $H(\pi/2) = \sin(\pi/2)/(\pi/2) = 2/\pi$. Using L'Hopital's rule, we can show that

$$H(0) = \lim_{\theta\to 0} H(\theta) = \lim_{\theta\to 0} \frac{\sin\theta}{\theta} = 1.$$

We need to show that $H(\theta) < 0$ for $0 \le \theta \le \pi/2$, which will establish $H(\theta)$ as a decreasing function of θ in $\theta \in [0, \pi/2]$. If $H(\theta) = \sin\theta/\theta$ is differentiated, we have

$$H'(\theta) = \frac{\theta\cos\theta - \sin\theta}{\theta^2}.$$

Consider the numerator, $g(\theta) \equiv \theta\cos\theta - \sin\theta$. We see that $g(0) = 0$ and $g'(\theta) = \theta(-\sin\theta) + \cos\theta - \cos\theta = -\theta\sin\theta$. We also see that $g'(\theta) = -\theta\sin\theta \le 0$ for $0 \le \theta \le \pi/2$. This shows that $g(\theta)$ and hence $H(\theta)$ are decreasing functions of θ in $\theta \in [0, \pi/2]$ since the denominator of $H'(\theta)$ is positive. The Jordan's lemma, which is based on the Jordan Integral Inequality, will now be presented.

Jordan's lemma: Let m be a constant, with $m > 0$, and let $Q(z)$ be a continuous function in the upper half of the complex plane (UHP), such that $|z| \geq R_o$, and let

$$Q_m \equiv \max_{z \in \curvearrowleft} |Q(z)| \to 0 \text{ as } R \to \infty,$$

where $\curvearrowleft$ is the semicircle $|z| = R$ in the UHP (Figure 2.21). Then

$$\lim_{R \to \infty} \int_{\curvearrowleft} e^{imz} Q(z) dz = 0.$$

Proof 2.2 Let z lie on $\curvearrowleft$ with $R > R_0$ so that $z = Re^{i\theta}$ and $dz = iRe^{i\theta} d\theta$. We can write

$$\left| e^{imz} \right| = \left| e^{imR(\cos\theta + i \sin\theta)} \right| = \left| e^{-mR\sin\theta} \right|$$

so that on $\curvearrowleft$ we shall have

$$\left| \int_{\curvearrowleft} e^{imz} Q(z) dz \right| \leq \max_{z \in \curvearrowleft} |Q(z)| \int_0^{\pi} e^{-mR\sin\theta} R \, d\theta$$
$$= RQ_m \int_0^{\pi} e^{-mR\sin\theta} d\theta.$$

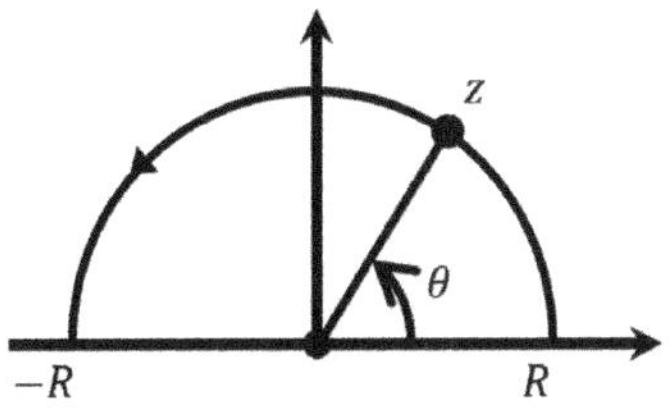

Figure 2.21: Depiction of the angular measure θ in relation to the coordinate point z.

We use the symmetry of $\sin\theta$ about $\theta = \pi/2$ to write

$$RQ_m \int_0^{\pi} e^{-mR\sin\theta} d\theta = 2RQ_m \int_0^{\pi/2} e^{-mR\sin\theta} d\theta,$$

which now has the same integral limits and interval $(0 \leq \theta \leq \pi/2)$ as in our statement of the Jordan Integral Inequality, allowing us to use the inequality:

$$2RQ_m \int_0^{\pi/2} e^{-mR\sin\theta} d\theta \leq \frac{\pi Q_m}{m} \left(1 - e^{-mR}\right)$$

so that

$$\left|\int_{\frown} e^{imz}Q(z)dz\right| \le \frac{\pi Q_m}{m}\left(1-e^{-mR}\right). \tag{2.36}$$

Employing the hypothesis that $Q_m \to 0$ as $R \to \infty$, the RHS of eq. (2.36) becomes zero. The Jordan lemma is then proved since the "less than" sign does not apply to the absolute sign on the LHS of eq. (2.36). It must be pointed out that with the hypothesis imposed on Q_m and $m > 0$, Jordan's lemma is in fact an expected result, given that e^{imz} is bounded.

Together, Jordan's lemma and the residue theorem allow us to arrive at the following theorem for evaluating improper integrals where the integrand is of the form $f(z) = e^{imz}Q(z)$ with $m > 0$ and $Q(z)$ is a rational function. Note that the invocation of $\lim_{R\to\infty} \int_{\frown} f(z)dz = 0$ is required, but this has been provided for by the Jordan's lemma.

Theorem 2.4 (Estimation of $\int_{\frown} f(z)dz$ when $f(z)$ decays rapidly for large $|z|$ and $\frown$ is the part of the circle $|z| = R$ that lies in the UHP): Let $f(z)$ be an analytic function in the UHP with the exception of a finite number of poles at the points $z_1, z_2, \ldots, z_N$. Then if for $|z| > R$ the function $f(z)$ is such that $|f(z)| < K/|z|^{(1+\delta)}$, where K and δ are positive constants, then

$$\lim_{R\to\infty} \int_{\frown} f(z)dz = 0.$$

Writing $z = Re^{i\theta}$ and noticing that

$$\begin{aligned}\left|\int_{\frown} f(z)dz\right| &= \left|\int_0^{\pi} f\left(Re^{i\theta}\right)Rie^{i\theta}d\theta\right| \le \int_0^{\pi}\left|f\left(Re^{i\theta}\right)Rie^{i\theta}\right|d\theta \\ &< \int_0^{\pi}\left(\frac{K}{R^{1+\delta}}\right)R\,d\theta = \frac{K\pi}{R^{\delta}} \to 0 \text{ as } R \to \infty,\ \delta > 0,\end{aligned}$$

and the theorem is proved.

2.7 A few other theorems on improper integrals

We have found the following theorems and results on certain improper integrals in the following three subsections to be of interest.

2.7.1 $\int_0^\infty x^{a-1}P(x)dx$

Theorem 2.5 If $P(x)$ is analytic everywhere in the z-plane except at a finite number of poles $z_1, z_2, \ldots, z_n$, none of which lies on the positive half of the real axis, and if $z^a P(z)$ converges uniformly to zero as $z \to 0$ and as $z \to \infty$, then

$$\int_0^\infty x^{a-1}P(x)dx = \left(\frac{\pi}{\sin a\pi}\right)\left(\Sigma \text{ Residues of } (-z)^{a-1}P(z) \text{ at all its poles}\right)$$

provided that $\arg z$ is taken in the interval $(-\pi, \pi)$.) The reader can consult Whittaker and Watson [58] for a proof.

Example 2.28 A typical example of this theorem is the evaluation of $\int_0^\infty x^{a-1}P(x)dx$, where $P(x) = 1/(1+x^2)$ so that $I = \int_0^\infty \left[x^{a-1}/(1+x^2)\right]dx$. Determine I.

Solution: $P(x) = 1/(1+x^2)$ is analytic everywhere except at $z = \pm i$. We need to compute the residues of $(-z)^{a-1}P(z) \equiv f(z) = (-z)^{a-1}/[(z-i)(z+i)]$ at all of its poles:

$$\text{Res}[f(z), i] = \lim_{z\to i}(z-i)\frac{(-z)^{a-1}}{(z-i)(z+i)} = \frac{e^{-i\pi(a-1)/2}}{2i}$$

$$\text{Res}[f(z), -i] = \lim_{z\to -i}(z+i)\frac{(-z)^{a-1}}{(z+i)(z-i)} = \frac{e^{i\pi(a-1)/2}}{-2i}.$$

Therefore

$$I = -\frac{\pi}{\sin a\pi}\left\{\frac{e^{i\pi(a-1)/2} - e^{-i\pi(a-1)/2}}{2i}\right\} = -\frac{\pi}{\sin a\pi}\sin\left[\frac{(a-1)\pi}{2}\right] = \frac{\pi}{2\sin(a\pi/2)}.$$

2.7.2 $\int_0^\infty x^{a-1}P(x)dx, P(z) = \frac{\sum_{i=0}^m a_i z^{m-i}}{\sum_{j=0}^n b_j z^{n-j}}$

Theorem 2.6 Let $f(z) = z^{a-1}P(z)$, where a is not an integer, and let

$$P(z) = \frac{a_0 z^m + a_1 z^{m-1} + \ldots + a_m = \sum_{i=0}^m a_i z^{m-i}}{b_0 z^n + b_1 z^{n-1} + \ldots + b_n = \sum_{j=0}^n b_j z^{n-j}},$$

where $a_0, a_1, \ldots, a_m$ and $b_0, b_1, \ldots, b_n$ are real, and $0 < a < n - m$, and $P(z)$ has neither a pole nor a zero at the origin. Let the poles of $P(z)$, which are located at $z_1, z_2, \ldots, z_n$ be such that none of them lies on the positive half of the real axis. Then

$$\int_0^\infty x^{a-1}P(x)dx = \frac{2\pi i}{1 - e^{2\pi i a}}\sum_j \text{Res}\left[f(z), z_j\right].$$

At the heart of this integral is the fractional power to which x (or z) is raised, which indicates the occurrence of multivaluedness if no restrictions are imposed on $\arg z$. You will recollect that a previous section deals with the case where $\arg z$ is restricted to $-\pi < \arg z < \pi$, and so there was no need to consider multivaluedness and the complications that it introduces into contour integration. Moreover, the poles in that case are restricted to the negative half of the real axis. Of course, the price we pay for allowing multivaluedness is the need to consider the distortion of an otherwise normal contour to avoid the branch point. We can no longer restrict our contour to the UHP.

Other than these factors, we can have a well-behaved integral that allows us to neglect the effect of the circular component C_R in the contour of Figure 2.22 if we recognize that as $z \to \infty$, $\int_0^\infty z^{a-1}P(z)dz$ goes like $G \equiv \frac{z^{a-1} \times z^m}{z^n} \times z$, where the last factor, or z, comes from the circumference of C_R, or $2\pi z$, of course with the 2π factor in this expression suppressed. Since we require G to go to zero as $R \to \infty$, the inequality $R^{a-1}R^m R < R^n$, or $R^{m+a} < R^n$ must be satisfied, or $m + a < n$.

Other than these considerations, the analysis becomes ordinary. As Figure 2.22 shows, we have made a cut along the positive real axis which includes the origin, and $\arg z = \theta + 2k\pi$, $k = 0, \pm 1, \pm 2, \ldots$, with $0 \le \theta \le 2\pi$. The contour C_R has the equation $|z| = R$. The circular contour C_ε has the equation $|z| = \varepsilon$. We consider the branch with $k = 0$ and any point within C_R can be described as $z = re^{i\theta}$ so that $z^{a-1} = r^{a-1}e^{(a-1)\theta i}$.

The modulus-argument form of z is convenient to use for the various contours in Figure 2.22, which are $A - B$, C_R, C_ε, and $C - D$. On AB, $z = re^{\theta i} = re^{0i} = r$ and $dz = r\, de^{0i}d\theta + e^{0i}dr = dr$ so that on this segment $z^{a-1}P(z) = r^{a-1}P(r)$. On CD, $z = re^{i2\pi}$, $dz = e^{2\pi i}\, dr$, giving $z^{a-1}P(z) = r^{a-1}r^{(a-1)2\pi i}P(r)$.

Let $f(z) \equiv z^a P(z)$ have poles at $z_1, z_2, \ldots, z_n$ between AB, C_R, CD, and C_ε, such that none of these poles lies on the positive real axis where we have made a cut. The integral over $(AB) \cup (C_R) \cup (CD) \cup (C_\varepsilon)$ can be written as

$$\int_\varepsilon^R r^{a-1}P(r)dr + \int_{C_R} z^{a-1}P(z)dz + \int_R^\varepsilon r^{a-1}\exp[(a-1)2\pi i]P(r)e^{2\pi i}dr + \int_{C_\varepsilon} z^{a-1}P(z)dz$$
$$= 2\pi i \sum_{j=1}^n \operatorname{Res}[f(z), z_j].$$

Now, the integral around C_ε goes to zero as $\varepsilon \to 0$, and we have restricted m, n, and a so that the integral around C_R goes to zero as $R \to \infty$. This means that if $\varepsilon \to 0$ and $R \to \infty$, we will have

$$\int_0^\infty r^{a-1}P(r)dr + e^{2\pi i a}\int_\infty^0 r^{a-1}P(r)dr = 2\pi i \sum_{j=1}^n \operatorname{Res}[f(z), z_j].$$

If we replace the integration variable r by x, we will have

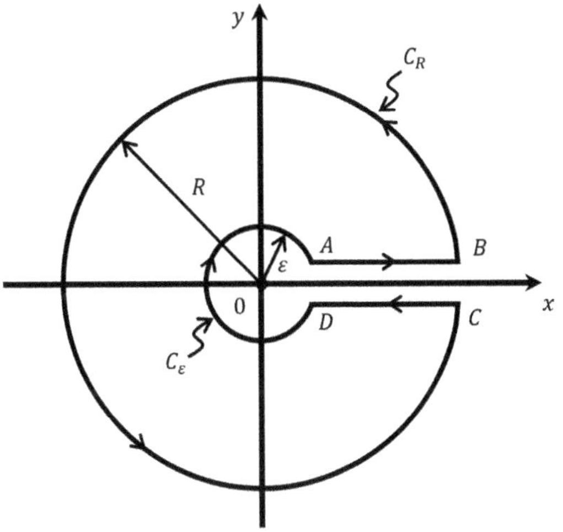

Figure 2.22: The key-hole contour.

$$\int_0^\infty x^{a-1}P(x)dx = \frac{2\pi i}{1-e^{2\pi i a}}\sum_{j=1}^{n}\text{Res}\left[f(z), z_j\right],$$

which proves the theorem.

Example 2.29 Evaluate the real integral

$$I = \int_0^\infty \frac{x^{1/3}}{x(x^2+1)}dx.$$

Solution: Let us check for the validity of the condition $m+a<n$ that allows us to neglect the contribution of C_R:

$$a-1=\frac{1}{3} \Rightarrow a=\frac{4}{3}, m=0, n=3;\; m+a=\frac{4}{3}<n=3,$$

so the condition $m+a<n$ is satisfied. Here,

$$P(z)=\frac{1}{z(z^2+1)}, \qquad f(z)=z^a\frac{1}{z(z^2+1)}=z^{a-1}\frac{1}{(z^2+1)}.$$

With $f(z)=z^aP(z)$, the poles are $z=\pm i$, neither of which lies on the positive real axis. Hence the theorem applies and

$$\text{Res}[f(z), i] = \lim_{z\to i}(z-i)\frac{z^{4/3}}{z(z-i)(z+i)} = \lim_{z\to i}\frac{-i^{4/3}}{2i^2} = -\frac{e^{2\pi/3}}{2},$$

$$\operatorname{Res}[f(z), -i] = \lim_{z\to -i}(z+i)\frac{z^{4/3}}{z(z+i)(z-i)} = \frac{-i^{4/3}}{2i^2} = -\frac{e^{2\pi}}{2}.$$

Thus,

$$\int_0^\infty x^{1/3}\frac{1}{z(z^2+1)}dx = \frac{-\pi i}{\left(1-e^{8\pi i/3}\right)}\left[e^{2\pi/3}+e^{2\pi}\right].$$

Example 2.30 Given the real integral

$$I = \int_0^\infty \frac{x^{a-1}}{x^2+1}dx,$$

determine the range of the parameter a that ensures the existence of I.

Solution: $P(z) = 1/\left(z^2+1\right)$, $m = 0$, $n = 2$. The relationship between m, n, and a is $m + a < n$, or, here, $0 + a < 2$ or $a < 2$. Moreover, $a > 0$ so that $0 < a < 2$.

We can evaluate the integral in terms of a by finding the residues of $f(z) \equiv z^{a-1}/\left(z^2+1\right)$, which has simple poles at $z = \pm i$:

$$\operatorname{Res}[f(z), i] = \lim_{z\to i}\left[(z-i)\frac{z^{a-1}}{(z-i)(z+i)}\right] = \lim_{z\to i}\left[\frac{z^{a-1}}{z+i}\right] = \frac{i^{a-2}}{2} = -\frac{e^{a\pi i/2}}{2}$$

$$\operatorname{Res}[f(z), -i] = \lim_{z\to -i}\left[(z+i)\frac{z^{a-1}}{(z+i)(z-i)}\right] = \lim_{z\to -i}\left[\frac{z^{a-1}}{z-i}\right] = \frac{(-i)^{a-2}}{2} = -\frac{1}{2}e^{3a\pi i/2}$$

Thus,

$$\int_0^\infty \frac{x^{a-1}}{1+x^2}dx = \frac{2\pi i}{\left(1-e^{2\pi a i}\right)}\left[-\frac{e^{a\pi i/2}}{2} - \frac{e^{3a\pi i/2}}{2}\right] = \pi i\left[\frac{e^{a\pi i/2}+e^{-a\pi i/2}}{e^{a\pi i}-e^{-a\pi i}}\right] = \frac{\pi\cos(a\pi/2)}{\sin(a\pi)}.$$

2.7.3 Inversion integral for the Laplace transform

If $\varphi(s)$ is the LT of $f(t)$, then $f(t)$ can be obtained by evaluating the integral

$$f(t) = \frac{1}{2\pi i}\int_{\gamma-i\infty}^{\gamma+i\infty}\varphi(z)e^{tz}dz,$$

along the line $\eta \in (\gamma - i\infty, \gamma + i\infty)$, where s is the LT variable, written as z in this integral. As exhaustively discussed in the later chapters of this book, the Bromwich contour (Figure 2.23) is employed for the inversion. In this section, we will merely present a theorem for carrying out the inversion and prove it. Even then, this treatment falls under the real space LT for the most part. Complex LT is treated in Chapters 5 and 6.

Theorem 2.7 Let $\varphi(s)$ be the LT of $f(t)$ and an analytic function of s except at a finite number of poles each of which lies to the left of the vertical line $Re[s]=\gamma$. Let $\varphi(s)$ be bounded as s becomes infinite through the half plane $Re[s]\le\gamma$, then $f(t)=\mathcal{L}^{-1}\{\varphi(s)\}=\sum$ residues of $\varphi(s)e^{st}$ at each of the poles.

Figure 2.23 depicts the contour used for the proof. The residue theorem can be stated for this problem as follows:

$$\lim_{\eta\to\infty}\left\{\frac{1}{2\pi i}\int_{\gamma-i\eta}^{\gamma+i\eta}\varphi(s)e^{st}ds+\frac{1}{2\pi i}\int_{C_R}\varphi(s)e^{st}ds\right\}=\sum\text{Residues of }\varphi(s)e^{st},$$

where η is a measure along $s=\gamma$. Examining the absolute values, we can write

$$\left|\frac{1}{2\pi i}\int_{\gamma-i\eta}^{\gamma+i\eta}\varphi(s)e^{st}ds-\sum\text{ Residues of }\varphi(s)e^{st}\right|=\left|-\frac{1}{2\pi i}\int_{C_R}\varphi(s)e^{st}ds\right|.$$

On C_R: $s=\gamma+\eta e^{i\theta}=\gamma+\eta(\cos\theta+i\sin\theta)$, with a restriction to the left of the vertical line $s=\gamma$ or $\pi/2\le\theta\le3\pi/2$. If s is sufficiently large, we can say that $|s-\gamma|\le|s|+|\gamma|<2|s|$, where $|s\varphi(s)|<M$. We can also write

$$\left|-\frac{1}{2\pi i}\int_{C_R}\varphi(s)e^{st}ds\right|\le\frac{1}{2\pi}\int_{C_R}|\varphi(s)|\left|e^{st}\right|ds\equiv I.$$

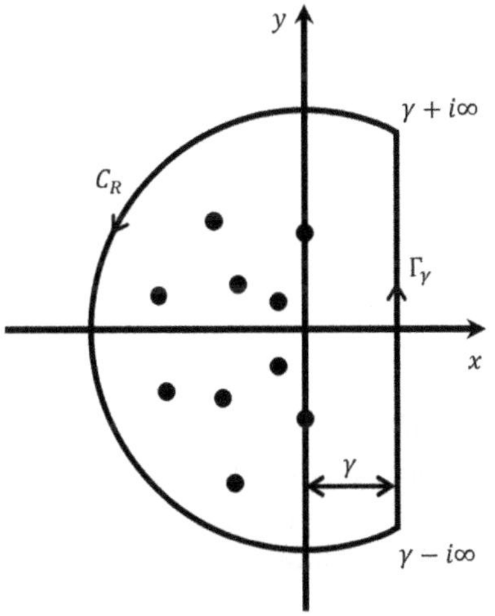

Figure 2.23: The Bromwich contour with singularities in the left half plane.

Using $s = \gamma + \eta e^{i\theta}$, $ds = i\eta e^{i\theta} d\theta = i(s-\gamma)d\theta$, we can further write I as

$$
\begin{aligned}
I &= \frac{1}{2\pi} \int_{\pi/2}^{3\pi/2} |\varphi(s)||\exp\{t[\gamma + \eta(\cos\theta + i\sin\theta)\}||i\eta e^{i\theta} d\theta| \\
&= \frac{1}{2\pi} \int_{\pi/2}^{3\pi/2} |\varphi(s)||s-\gamma| \exp[t(\gamma + \eta\cos\theta)]d\theta; \ |e^{t\eta i \sin\theta}| = 1 \\
&\le \frac{1}{2\pi} \int_{\pi/2}^{3\pi/2} 2|s\varphi(s)|e^{\gamma t} \exp[\eta t \cos\theta]d\theta; \ |s-\gamma| < 2s \\
&\le \frac{1}{\pi} M e^{\gamma t} \int_{\pi/2}^{3\pi/2} \exp(\eta t \cos\theta)d\theta; \ |s\varphi(s)| < M
\end{aligned}
$$

Since the region of interest is $\pi/2 \le \theta \le 3\pi/2$, we can define a new angle $\theta' = \theta - \pi/2$, use the relation $\cos(\theta' + \pi/2) = -\sin\theta'$, and the fact that the integrand that we obtain with this substitution is symmetric (about $\theta = \pi/2$). The result is that

$$
I \le \frac{M}{\pi} e^{\gamma t} \int_0^{\pi} \exp[-\eta t \sin\theta']d\theta' = \frac{2M}{\pi} \int_0^{\pi/2} e^{\gamma t} \exp[-\eta t \sin\theta']d\theta'.
$$

By graphing the functions $y = \sin\theta'$ and $y = (2/\pi)\theta'$ (Figure 2.24), we can see that $\sin\theta' \ge 2\theta'/\pi$ if $0 \le \theta' \le \pi/2$. This means that when we replace $\sin\theta'$ by $2\theta'/\pi$ in the exponent of the integrand above, we would be overestimating the integral. Therefore, we can write

$$
\left| \frac{1}{2\pi i} \int_{C_R} \varphi(s)e^{st}ds \right| \le \frac{2M}{\pi} e^{\gamma t} \left[\frac{e^{-2\eta + \theta'/\pi}}{-2\eta t/\pi} \right]_0^{\pi/2} = \frac{2M}{\pi} e^{\gamma t} \left[-\frac{\pi}{2\eta t}(e^{-\eta t} - 1) \right].
$$

Since $e^{-\eta t} \to 0$ as $\eta \to \infty$, we have

$$
\begin{aligned}
\lim_{\eta\to\infty} \frac{1}{2\pi i} \int_{\gamma - i\eta}^{\gamma + i\eta} \varphi(s)e^{st}ds &\equiv \frac{1}{2\pi i} \int_{\gamma - i\infty}^{\gamma + i\infty} \varphi(s)e^{st}ds = \mathcal{L}^{-1}\{\varphi(s)\} \\
&= \Sigma \text{ Residues of } \varphi(s)e^{st}.
\end{aligned}
$$

Note that if $\varphi(s)$ has an infinite number of poles, the proof provided above will not work because we will always be able to obtain semicircles C_R on which $|s\varphi(s)|$ is not bounded.

It should be noted that the inversion formula stated and proved above does not consider the case of multivalued $\varphi(s)$, for which branch cuts must be introduced.

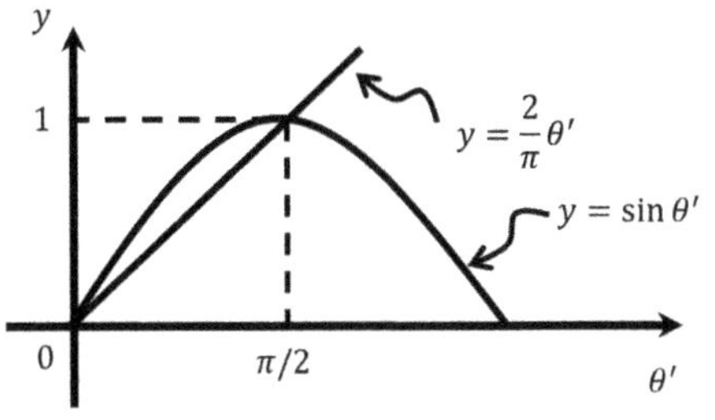

Figure 2.24: Graph of $y = \sin\theta'$ and $y = \frac{2}{\pi}\theta'$.

Example 2.31 Determine $\mathcal{L}^{-1}\left\{(s^2 - a^2)/(s^2 + a^2)^2\right\}$, $a > 0$.

Solution:

$$g(z) = \frac{e^{zt}(z^2 - a^2)}{(z^2 + a^2)^2};$$

The singularity points of $g(z)$ are located at $z = \pm ia$. The residues are

$$\mathrm{Res}[g(z), ia] = \frac{t}{2}\exp(iat),\ \mathrm{Res}[g(z), -ia] = \frac{t}{2}\exp(-iat).$$

Thus

$$f(t) = \mathcal{L}^{-1}\left\{\frac{s^2 - a^2}{(s^2 + a^2)^2}\right\} = \frac{t}{2}\left[e^{iat} + e^{-iat}\right] = t\cos at.$$

Example 2.32 Evaluate $\mathcal{L}^{-1}\left\{1/\left[(s + a)^2 + b^2\right]\right\}$.

Solution:

$$g(z) = \frac{e^{zt}}{\left[(z + a)^2 + b^2\right]},$$

which has two poles at $z = -a \pm ib$. The residues are

$$\mathrm{Res}[g(z),\ -a + ib] = \lim_{z\to(-a+ib)} \frac{[z - (-a + ib)]e^{zt}}{[z - (-a + ib)][z - (-a - ib)]} = \frac{e^{(-a+ib)t}}{2ib}$$

Similarly

$$\mathrm{Res}[g(z),\ -a - ib] = \lim_{z\to(-a-ib)} \frac{[z - (-a - ib)e^{zt}]}{[z - a(-a - ib)][z - a(-a + ib)]} = \frac{e^{(-a-ib)t}}{-2ib}.$$

Thus,

$$f(t)=\mathcal{L}^{-1}\{\varphi(s)\}=\frac{e^{(-a+ib)t}}{2ib}+\frac{e^{(-a-ib)t}}{-2ib}=e^{-at}\frac{\left(e^{ibt}-e^{-ibt}\right)}{2ib}=\frac{e^{-at}\sin bt}{b}.$$

Example 2.33 Determine $\mathcal{L}^{-1}\{e^{-s}/(s^2+1)\}$.

Solution:

$$g(z)=\frac{e^{-z}e^{zt}}{(z^2+1)},$$

with simple poles at $z=\pm i$. We want to calculate the residues:

$$\mathrm{Res}[g(z),i]=-\frac{i}{2}\exp[i(t-1)],\quad \mathrm{Res}[g(z),-i]=\frac{i}{2}\exp[-i(t-1)]$$

$$f(t)=\mathcal{L}^{-1}\left\{\frac{e^{-s}}{s^2+1}\right\}=\left\{-\frac{i}{2}\exp[i(t-1)]+\frac{i}{2}\exp[-i(t-1)]\right\}=\sin(t-1).$$

Since the transform of $f(t)$ is not defined for $t<0$, we need to require that $\mathcal{L}^{-1}\{e^{-s}/(s^2+1)\}$ be zero when $t<1$, which we can do with the Heaviside function H, which is defined as

$$H(t-a)=\begin{cases}0, & t<a\\ 1, & t>a\end{cases}.$$

Thus, we can write $f(t)$ as $f(t)=H(t-1)\sin(t-1)$, for $t>0$, where $H(t-1)$ is zero for t less than 1 and unity otherwise.

Example 2.34 Determine $\mathcal{L}^{-1}\{1/\sqrt{s}\}$.

Solution: The Bromwich contour and the LT inversion theorem given earlier in this section need to be modified for this problem because of the presence of the square root sign which tells us that we have multivaluedness. The modified contour used is shown in Figure 2.25, where we have taken the liberty of denoting s, which is of course complex, by $z=x+iy$.

With the exclusion of the branch point the function $\varphi(z)=1/\sqrt{z}$ is analytic everywhere inside the modified contour $C\equiv\Gamma_\gamma\cup AB\cup BC\cup\Gamma_\varepsilon\cup DE\cup EF$ in Figure 2.25. Let us restrict analysis to the principal branch for which $-\pi\le\theta\le\pi$. Let us change variables: $\theta=\pi/2+\theta'$ in anticipation of the argument of the domain being $\pi/2\le\theta\le3\pi/2$ or $0\le\theta'\le\pi$. On contour C_R (the big semicircle), $z=\gamma+Re^{i\theta}$ $(\pi/2\le\theta\le3\pi/2)$ or $z=\gamma+iRe^{i\theta'}$ $(0\le\theta'\le\pi)$.

With the change of variables, we have $dz=-Re^{i\theta'}d\theta'\Rightarrow|dz|=Rd\theta'$. When $R\to\infty$ we can assume that $|\gamma+iRe^{i\theta'}|\ge||Re^{i\theta'}|-|\gamma||=R-\gamma$. Note that we also have $|e^{zt}|=|\exp\{t(\gamma-R\sin\theta')+iR\cos\theta'\}|=e^{\gamma t}\exp[-Rt\sin\theta']$.

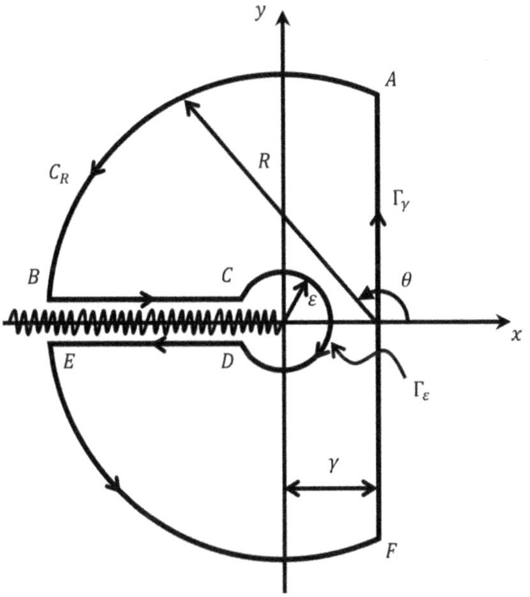

Figure 2.25: The complex plane of the Laplace variable s here denoted as $z = x + iy$.

We can evaluate the integral around $C_R \equiv ABEF$ as follows:

$$I_{C_R} = \left| \int_{C_R} \frac{e^{zt}}{\sqrt{z}} dz \right| \le \int_{C_R} \frac{|e^{zt}|}{|z|^{1/2}} |dz| \le \frac{e^{\gamma t} R}{(R-\gamma)^{1/2}} \int_0^{\pi} \exp[-Rt \sin \theta'] d\theta'.$$

This can be written as

$$I_{C_R} \le \frac{2e^{\gamma t} R}{(R-\gamma)^{1/2}} \int_0^{\pi/2} \exp[-Rt \sin \theta'] d\theta',$$

if we note that $\sin \theta'$ is symmetric about $\theta' = 0$. Furthermore, if we invoke the integral form of Jordan Inequality, we can write

$$I_{C_R} \le \frac{\pi e^{\gamma t}}{(R-\gamma)^{1/2} t} \left(1 - e^{-Rt}\right).$$

With γ constant, $t > 0$, and $R \to \infty$, we can approximate I_{C_R} as zero, meaning that it does not contribute to the integration around the composite contour C.

For the small circle $z = \varepsilon e^{i\phi}$, where ϕ is $\arg z$ for any z inside $|z| = \varepsilon$, we have $dz = i\varepsilon e^{i\phi} d\phi$ so that $z^{1/2} = e^{i\phi/2}\sqrt{\varepsilon}$ and

$$I_\varepsilon \equiv \int_{\Gamma_\varepsilon} \frac{1}{\sqrt{z}} dz = \int_{-\pi}^{\pi} \frac{1}{e^{i\phi/2}\sqrt{\varepsilon}} e^{[\varepsilon t(\cos\phi + i\sin\phi)]} i\varepsilon e^{i\phi}\, d\phi,$$

which tends to zero as $\varepsilon \to 0$. On the line segment BC, $z = re^{i\pi} = -r \Rightarrow \sqrt{z} = e^{i\pi/2}\sqrt{r} = i\sqrt{r}$. Also, $dz = -dr$. On the line segment DE, $z = re^{-i\pi} = -r \Rightarrow \sqrt{z} = e^{-i\pi/2}\sqrt{r} = -i\sqrt{r}$ and $dz = -dr$.

Applying the Cauchy theorem, and noting that there is no singularity inside the composite curve C, we will have

$$0 = \frac{1}{2\pi i} \lim_{R\to\infty, \varepsilon\to 0} \left\{ \int_{\Gamma_\gamma} \frac{e^{zt}}{\sqrt{z}} dz + \int_{R}^{\varepsilon} \frac{1}{i\sqrt{r}} e^{-\gamma t}(-dr) + \int_{\Gamma_\varepsilon} \frac{e^{zt}}{\sqrt{z}} dz + \int_{\varepsilon}^{R} \frac{1}{(-i)\sqrt{r}} e^{-rt}(-dr) + \int_{C_R} \frac{e^{zt}}{\sqrt{z}} dz \right\},$$

or, with the previous results for $\int_{\Gamma_\varepsilon}$ and $\int_{C_R}$,

$$\frac{1}{2\pi i} \int_{\Gamma_\gamma} \frac{e^{zt}}{\sqrt{z}} dz \equiv \frac{1}{2\pi i} \int_{\gamma - i\infty}^{\gamma + i\infty} \frac{e^{zt}}{\sqrt{z}} dz = \frac{1}{2\pi i} \left\{ -\int_{\infty}^{0} \frac{ie^{-rt}}{\sqrt{r}} dr + \int_{0}^{\infty} \frac{ie^{-rt}}{\sqrt{r}} dr \right\}$$

$$= \frac{1}{\pi} \int_0^\infty \frac{e^{-rt}}{\sqrt{r}} dr = \frac{2}{\pi\sqrt{t}} \int_0^\infty e^{-u^2} du = \frac{1}{\sqrt{\pi t}}, \quad Re[z] > 0.$$

2.8 Miscellaneous examples

Example 2.35 Expand the following function in a Laurent Series and determine the region of convergence

$$f(z) = \frac{2}{z(z-1)}$$

valid for $1 < |z-2| < 2$.

Solution:

$$f(z) = \frac{2}{z(z-1)};$$

$$z - 2 = z' \Rightarrow z = 2 + z';\ z - 1 = z' + 1$$

$$f(z')=\frac{2}{(z'+2)(z'+1)}=\frac{2}{2\left(\frac{z'}{2}+1\right)(z')\left(1+\frac{1}{z'}\right)},\quad 1<|z'|<2$$

$$=\frac{1}{z'\left(\frac{z'}{2}+1\right)\left(1+\frac{1}{z'}\right)}$$

$$=\frac{1}{z'}\frac{1}{\left\{1+\frac{z'}{2}+\left(\frac{z'}{2}\right)^2+\left(\frac{z'}{2}\right)^3+\dots\right\}\left\{1+\frac{1}{z'}+\left(\frac{1}{z'}\right)^2+\left(\frac{1}{z'}\right)^3+\dots\right\}}$$

$$=\frac{1}{z'}\frac{1}{\left\{1+\frac{1}{z'}+\left(\frac{1}{z'}\right)^2+\left(\frac{1}{z'}\right)^3+\frac{z'}{2}+\left(\frac{z'}{2}\right)\left(\frac{1}{z'}\right)+\frac{z'}{2}\left(\frac{1}{z'}\right)^2+\dots\right\}}$$

Substitute $z'=z-2$, etc.

Example 2.36 Evaluate by the residue theorem the integral

$$I=\oint_C\frac{e^z dz}{z^4+5z^3}$$

where C is the contour $|z|=2$.

Solution:

$$I=\oint_C\frac{e^z dz}{z^4+5z^3}$$

$$z^4+5z^3=0\Rightarrow\begin{cases}z=0 & \text{is a third-order pole}\\ z=-5 & \text{is a first-order pole (outside of C)}\end{cases}$$

$$\operatorname*{Res}_{z=0}f(z)=\frac{1}{2!}\lim_{z\to0}\frac{d^2}{dz^2}\left[z^3\frac{e^z}{z^4+5z^3}\right]=\frac{1}{2}\lim_{z\to0}\frac{d^2}{dz^2}\left(\frac{e^z}{z+5}\right)$$

$$=\frac{1}{2}\lim_{z\to0}\frac{d}{dz}\left(\frac{e^z}{z+5}-\frac{e^z}{(z+5)^2}\right)$$

$$=\frac{1}{2}\lim_{z\to0}\left(\frac{e^z}{z+5}-\frac{e^z}{(z+5)^2}-\frac{e^z}{(z+5)^2}+\frac{2e^z}{(z+5)^3}\right)$$

$$=\frac{1}{2}\left(\frac{1}{5}-\frac{2}{25}+\frac{2}{124}\right)$$

$$=\frac{17}{250}$$

$$\therefore I=2\pi i\operatorname*{Res}_{z=0}f(z)=\frac{17}{125}\pi i.$$

Example 2.37 Answer the following questions:

a) Characterize the poles of the function

$$g(z) = \frac{z^2}{(z^2+1)^2(z^2+2z+2)}.$$

b) Determine the various residues associated with the function $g(z)$.

c) Hence evaluate the real integral

$$I = \int_0^\infty \frac{x^2}{(x^2+1)^2(x^2+2x+2)}dx.$$

Solution: Poles are $z = i$ (order 2), $z = -1+i$ (order 1) on the upper contour

$$\text{Res}[f(z), i] = \frac{9i-12}{100},$$

$$\text{Res}[f(z), -1+i] = \frac{3-4i}{25},$$

$$\sum_i^2 C_{-1}^i = \left(\frac{9i-12}{100} + \frac{3-4i}{25}\right)2\pi i = \frac{7\pi}{50},$$

$$\int_C = \int_{-R}^{R} \frac{x^2 dx}{(x^2+1)^2(x^2+2x+2)} + \int \frac{z^2 dz}{(z^2+1)^2(z^2+2z+2)} = \frac{7\pi}{50},$$

where $\int_{\curvearrowleft}$=0 using familiar arguments.

Example 2.38 Evaluate the integral

$$I = \int_0^\infty \frac{\cos x}{a^2+x^2}dx,$$

where $a > 0$ is a real number.

Solution: Consider

$$f(z) = \frac{e^{iz}}{z^2+a^2}$$

and the contour shown in Figure 2.26, when $R > a$, the only singular point is at $z = ai$.

$$\underset{z\to ai}{\text{Res}} f(z) = \lim_{z\to ai}(z-ai)f(z) = \lim_{z\to ai}\frac{e^{iz}}{z+ai} = \frac{e^{-a}}{2ai},$$

Therefore,

$$\oint_C f(z)dz = \int_{-R}^{R} \frac{\cos x}{a^2+x^2}dx + \int_{C_R} \frac{e^{iz}}{z^2+a^2}dz = 2\pi i \operatorname*{Res}_{ai} f(z) = 2\pi i \cdot \frac{e^{-a}}{2ai} = \frac{\pi}{a}e^{-a}. \tag{2.37}$$

Let $z = Re^{i\theta}$

$$\therefore \left| \int_{C_R} \frac{e^{iz}}{z^2+a^2}dz \right| = \left| \int_0^{\pi} \frac{e^{iR(\cos\theta - i\sin\theta)}}{R^2e^{i2\theta}+a^2} d\left(Re^{i\theta}\right) \right| \le \frac{\pi Re^{-R\sin\theta}}{R^2-a^2}.$$

Note that $\lim_{R\to\infty} \frac{e^{-R}}{R} = \lim_{R\to\infty} \frac{(e^{-R})'}{(R)'} = \lim_{R\to\infty}\left(-e^{-R}\right) = 0$. Thus $\lim_{R\to\infty} \int_{C_R} \frac{e^{iz}}{z^2+a^2}dz = 0$.

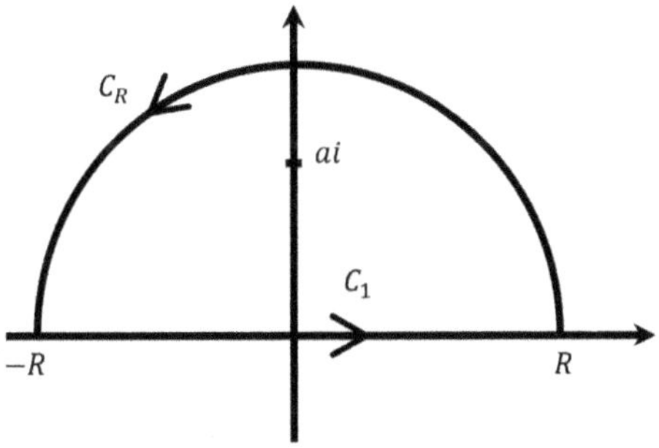

Figure 2.26: Contour for Example 2.38.

Take $\lim_{R\to\infty}$ in eq. (2.37)

$$\Rightarrow \int_{-\infty}^{\infty} \frac{\cos x}{a^2+x^2}dx = \frac{\pi e^{-a}}{a}$$

$$\therefore I = \int_0^{\infty} \frac{\cos x}{a^2+x^2}dx = \frac{\pi e^{-a}}{2a}.$$

Example 2.39 Expand the following in a Laurent Series and determine the regions of convergence

a) $f(z) = \dfrac{3z+1}{z(1-z)}$
valid for $0 < |z| < 1$.

b) $f(z) = \dfrac{2}{z(z-1)}$
valid for $1 < |z-1| < 2$.

c) $f(z) = e^{5/z}$
valid for $0 < |z|$.

Solution:

a) $f(z) = \dfrac{3z+1}{z(1-z)} = \dfrac{4}{1-z} + \dfrac{1}{z}$

$$\frac{1}{1-z} = 1 + z + z^2 + \ldots \text{ for } |z| < 1,$$

$$f(z) = \frac{1}{z} + 4 + 4z + 4z^2 + \ldots \text{ for } |z| < 1.$$

b) $f(z) = \frac{2}{z(z-1)}$,

Since $|z-1| < 2$ implies $|z-2| < 1$, we could set $t = z-2$ so that $|t| < 1$. The condition $|z-1| > 1$ becomes $|t| > 0$. We can proceed with

$$f(z) = f(t) = -2\left(\frac{1}{t+2} - \frac{1}{t+1}\right),$$

where $|t| > 1 \Rightarrow |1/t| < 1$

$$\Rightarrow \frac{1}{t+2} = \frac{1}{2}\frac{1}{1+\frac{t}{2}} = \frac{1}{2}\left[1 + \left(-\frac{t}{2}\right) + \left(-\frac{t}{2}\right)^2 + \left(-\frac{t}{2}\right)^3 + \ldots\right]$$

$$\therefore f(t) = -\left[1 + \left(\frac{t}{2}\right) + \left(\frac{t}{2}\right)^2 + \left(\frac{t}{2}\right)^3 + \ldots\right] + \frac{2}{t}\left[1 + \left(-\frac{1}{t}\right) + \left(-\frac{1}{t}\right)^2 + \left(-\frac{1}{t}\right)^3 + \ldots\right]$$

$$\therefore f(z) = \ldots - \frac{2}{(z-2)^4} + \frac{2}{(z-2)^3} - \frac{2}{(z-2)^2} + \frac{2}{z-2} - 1 + \frac{z-2}{2} - \frac{(z-2)^2}{4} + \frac{(z-2)^3}{8} \ldots$$

when $1 < |z-2| < 2$.

c) $f(z) = e^{\frac{5}{z}}$

Note

$$e^t = 1 + t + \frac{t^2}{2!} + \frac{t^3}{3!} + \ldots \text{ for all } t.$$

Let $t = 5/z$ for $z \neq 0$ so that

$$e^{\frac{5}{z}} = 1 + \frac{5}{z} + \frac{5^2}{z^2 \cdot 2!} + \frac{5^3}{z^3 \cdot 3!} + \ldots.$$

Example 2.40 Evaluate

$$\int_C \frac{1}{(z-1)^2(z-3)} dz,$$

where C is the domain $(x, y) \in [0, 4] \times [-1, 1]$.

Solution: C can be sketched as shown in Figure 2.27.

$$I \equiv \int_C \frac{1}{(z-1)^2(z-3)}\,dz.$$

The poles are $z_o = 1$ (second order) and $z_1 = 3$ (first order), both of which lie inside C. Thus,

$$I = 2\pi i\{\text{Res}[f(z), z_o] + \text{Res}[f(z), z_1]\} = 2\pi i\left(-\frac{1}{4} + \frac{1}{4}\right) = 0.$$

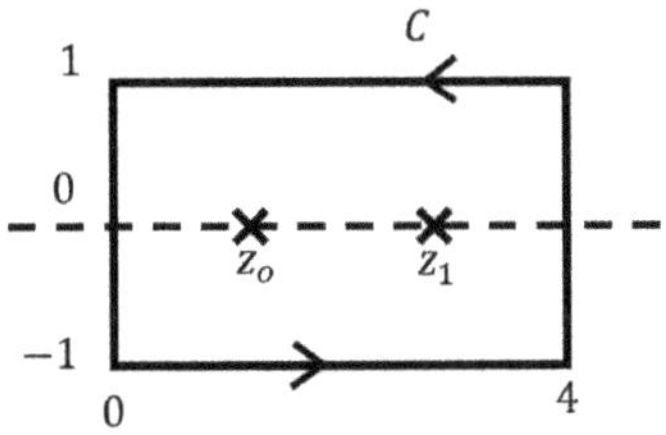

Figure 2.27: Contour for Example 2.40.

Example 2.41 Evaluate

$$\int_C \frac{1}{(z-1)^2(z-3)}\,dz$$

where C is defined by $|z| = 2$.

Solution: The poles are $z_o = 1$ (second order) and $z_1 = 3$ (simple pole). C can be sketched as shown in Figure 2.28.

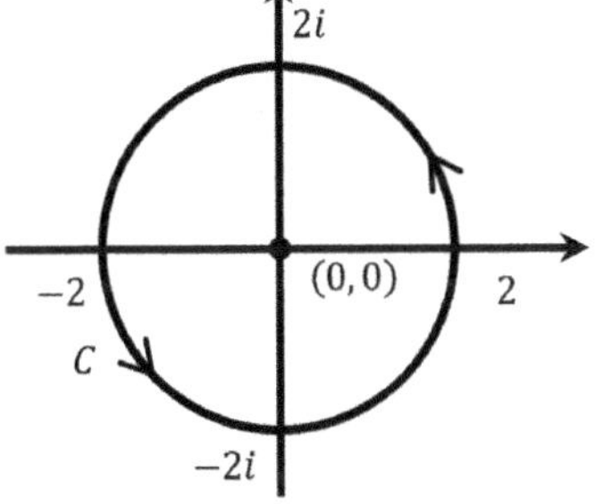

Figure 2.28: Contour for Examples 2.41.

We see that only z_o lies inside C. Therefore,

$$I \equiv \int_C \frac{1}{(z-1)^2(z-3)} dz = 2\pi i\left(-\frac{1}{4}\right) = -\frac{\pi}{2}i.$$

Example 2.42 Evaluate

$$\int_C \tan z \, dz,$$

where C is the circle $|z| = 2$.

Solution:

$$\tan z = \frac{\sin z}{\cos z},$$

which has simple poles at the points, where $\cos z = 0$ or $z = (2n+1)\pi/2$, $n = 0, \pm 1, \pm 2, \ldots$, or $z = \pi/2, 3\pi/2, 5\pi/2$, etc. for $n = -1, -2, -3, \ldots$.

Only the poles at $z = -\pi/2$ and $\pi/2$ are located inside the contour. Thus,

$$I \equiv \int_C \tan z \, dz = 2\pi i\left\{\text{Res}\left[f(z), -\frac{\pi}{2}\right] + \text{Res}\left[f(z), \frac{\pi}{2}\right]\right\} = 2\pi i(-1-1) = -4\pi i.$$

Example 2.43 Evaluate

$$I = \int_C e^{\frac{3}{z}} dz,$$

where C is the unit circle.

Solution: The Laurent series for $f(z) = e^{3/z}$ shows an essential singularity at $z_o = 0$. This means that the formula given for finding residues (for simple poles) is not applicable. However, from

$$e^{3/z} = 1 + \frac{3}{z} + \left(\frac{3}{z}\right)^2 \frac{1}{2!} + \left(\frac{3}{z}\right)^3 \frac{1}{3!} + \ldots,$$

we see that $C_{-1} = \text{Res}[f(z), z_o] = 3$. Thus,

$$I = 2\pi i C_{-1} = 6\pi i.$$

Example 2.44 Evaluate

$$I=\int_0^\infty \frac{dx}{1+x^4}$$

using (1) UHP and (2) LHP.

Solution: $f(x)=1/(1+x^4)$ is even, so

$$I=\frac{1}{2}\int_{-\infty}^{\infty}\frac{dx}{1+x^4}dx.$$

The function $f(z)=1/(1+z^4)$ has four simple poles, at the points

$$z_o=e^{\pi i/4},\ z_1=e^{3\pi i/4},\ z_2=e^{-3\pi i/4},\ z_3=e^{-\pi i/4},$$

which are shown in the contour of Figure 2.29.

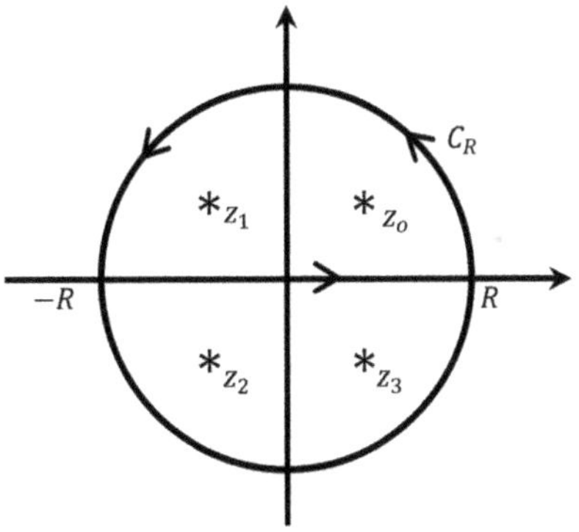

Figure 2.29: Contour for Example 2.44.

We see that the first two poles lie in UHP, whereas the other two are in the LHP. Now

$$\mathrm{Res}[f(z),z_o]=\left[\frac{1}{\frac{d}{dz}(1+z^4)}\right]_{z=z_o}=\left[\frac{1}{4z^3}\right]_{z=z_o}=\frac{1}{4}e^{-\frac{3\pi i}{4}}=-\frac{1}{4}e^{\frac{\pi i}{4}},$$

$$\mathrm{Res}[f(z),z_1]=\left[\frac{1}{\frac{d}{dz}(1+z^4)}\right]_{z=z_1}=\left[\frac{1}{4z^3}\right]_{z=z_1}=\frac{1}{4}e^{-\frac{9\pi i}{4}}=\frac{1}{4}e^{-\frac{\pi i}{4}},$$

$$I'\equiv\int_{-\infty}^{\infty}\frac{dx}{1+x^4}dx+\int_{C_R}\frac{dz}{1+z^4}dz=\int_C dz=\frac{2\pi i}{4}\left(-e^{-\frac{\pi i}{4}}+e^{-\frac{\pi i}{4}}\right)=\pi\sin\frac{\pi}{4}=\frac{\pi}{\sqrt{2}}.$$

The semicircular contour C_R in the UHP contributes zero to I'.

Therefore,

$$I' = \int_{-\infty}^{\infty} \frac{dx}{1+x^4} = 2\int_{0}^{\infty} \frac{dx}{1+x^4} = \frac{\pi}{\sqrt{2}}$$

$$\Rightarrow I = \frac{1}{2} I' = \frac{\pi}{2\sqrt{2}}.$$

Using the LHP gives the same result as the UHP.

Example 2.45 Show that

$$\sin z = \sin x \cosh y + i \cos x \sinh y.$$

Solution:

$$\sin z = \sin(x+iy) = \frac{e^{i(x+iy)} - e^{-i(x+iy)}}{2i} = \sin x\left(\frac{e^{y}+e^{-y}}{2}\right) + i\cos x\left(\frac{e^{y}-e^{-y}}{2}\right)$$
$$= \sin x \cosh y + i \cos x \sinh y$$

since

$$\sinh y = \frac{e^{y}-e^{-y}}{2}, \qquad \cosh y = \frac{e^{y}+e^{-y}}{2}.$$

Similarly,

$$\cos z = \cos(x+iy) = \cos x \cosh y - i \sin x \sinh y.$$

Note that if we write $\omega = \cos z \equiv u + iv$, we can eliminate x to obtain the ellipse

$$\frac{u^2}{\cosh^2 y} + \frac{v^2}{\sinh^2 y} = 1$$

for a prescribed value of y and the hyperbola

$$\frac{u^2}{\cos^2 x} - \frac{v^2}{\sin^2 x} = 1$$

for a prescribed value of x.

Problems

2.1 Answer the following questions:

a) Characterize the poles of the function

$$g(z) = \frac{z^2}{(z^2+1)^2(z^2+2z+2)}.$$

b) Determine the various residues associated with the function $g(z)$.

c) Hence evaluate the integral

$$I = \int_0^\infty \frac{x^2}{(x^2+1)^2(x^2+2x+2)} dx.$$

2.2 Evaluate

$$\int_{C_1} \frac{z^2+2z-3}{z^3+2z^2-11z-12} dz,$$

where C_1 is the circle $z(t) = 5e^{it}$, $-\pi \le t \le \pi$.

2.3 Evaluate the integral

$$I = \int_0^\infty \frac{\sin x}{x} dx.$$

2.4 Evaluate the integral

$$I = \int_{-\infty}^{+\infty} \frac{e^{i\omega t}}{(\omega - \omega_0)^m} d\omega,$$

where ω_0 is real.

2.5 Evaluate the integral

$$I = \int_0^\infty \frac{x^{\alpha-1}\sin(\pi\alpha/2 - x)}{x^2+\beta^2} dx = \frac{\pi}{2} e^{-\beta}\beta^{\alpha-2},$$

where α is real, $0 < \alpha < 3$, β is real, $\beta > 0$.

2.6 Determine the principal value of

$$f(z) = (1+i)^{2-i}$$

in the form $u + iv$.

2.7 Find the residue of the function

$$f(z) = \frac{1+z}{1-\cos z}.$$

2.8 Find the Laurent series about $z = 1$ for the function

$$f(z) = \frac{e^z}{(z-1)^2}$$

and determine the region of convergence.

2.9 Find the Inverse Laplace Transform of

$$f(s) = \frac{se^{-x\sqrt{s}}}{s^2 + \omega^2},$$

where s is the LT variable.

2.10 Evaluate the integral

$$f(\theta) = \int_{-\infty}^{\infty} \frac{\theta}{\sinh\theta - a} d\theta,$$

where a is an imaginary number ($\operatorname{Im} a \neq 0$.)

2.11 Evaluate the integral

$$I(a) = \int_{-\infty}^{\infty} \frac{e^{ia\xi}}{\sqrt{\xi + i} + \sqrt{\xi + 3i}} d\xi,$$

for the cases: (a) $a > 0$, (b) $a < 0$, where a is a real number.

2.12 Evaluate the integral

$$I(x, y, a, k) = \int_{-\infty}^{\infty} \frac{|y|}{y} \frac{e^{i\xi x - |y|[(\xi + ia)(\xi - ik)]^{1/2}}}{(\xi - ik)^{1/2}} d\xi,$$

where a, k, x, and y are real numbers.

2.13 Find the Taylor series of

$$f(z) = \frac{1}{1+z^4}$$

about $z = 0$ (by carrying out the required (repeated) differentiation).

2.14 Test the convergence or otherwise of the following series:

(a) $\sum_{n=1}^{\infty} \frac{n^2}{2^n}$ (b) $\sum_{n=0}^{\infty} \frac{(3-4i)^n}{n!}$ (c) $\sum_{m=0}^{\infty} \frac{z^m}{m!}$ (d) $\sum_{n=0}^{\infty} n! z^n$.

2.15 Find the region of convergence of the series

$$1+\frac{1}{2^2}\frac{z+1}{z-1}+\frac{1}{3^2}\left(\frac{z+1}{z-1}\right)^2+\frac{1}{4^2}\left(\frac{z+1}{z-1}\right)^3+\ldots.$$

2.16 Use the geometric series (a Maclaurin's series)

$$\frac{1}{1-z}=\sum_{n=0}^{\infty}z^n=1+z+z^2+\ldots$$

to find the Maclaurin's series of

(a) $f(z)=1/(1+z^2)$ and (b) $f(z)=\tan^{-1}z$.

2.17 Use the contour integration method to evaluate the integral

$$I=\int_0^{2\pi}\frac{1}{(2+\cos\theta)^2}d\theta.$$

2.18 Use contour integration techniques to evaluate $f(t)$ where

$$f(t)=\frac{1}{2\pi i}\int_{\gamma-i\infty}^{\gamma+i\infty}\bar{f}(s)e^{st}ds,$$

$\bar{f}(s)=1/\sqrt{2}$, and the symbols used are those appropriate for the Laplace transformation method.

2.19 Evaluate the integral

$$I=\int_0^{\infty}\frac{\cos x}{a^2+x^2}dx,$$

where $a>0$ is a real number.

2.20 Use the residue theorem to evaluate the integral

$$I=\oint_C\tan z\,dz,$$

where C is the contour $|z|=2$.

2.21 Expand the following in a Laurent Series and determine the regions of convergence

a)

$$f(z)=\frac{3z+1}{z(1-z)}$$

valid for $0 < |z| < 1$;

b)

$$f(z) = \frac{2}{z(z-1)}$$

valid for $1 < |z-1| < 2$;

c)

$$f(z) = e^{5/z}$$

valid for $0 < |z|$.

2.22 Evaluate the integral

$$I = \oint_{-\infty}^{+\infty} \frac{dx}{x^3 - x^2 + x - 1}.$$

The decoration on the integral symbol suggests that you take the CPV.

2.23 Solve the algebraic equation: $x^3 - 2x - 4 = 0$. That is, find the three values of x that satisfy this equation.

2.24 Express the function $e^{\ln z}$ in the form $u(x,y) + iv(x,y)$.

2.25 Calculate the integral

$$I = \int_0^{\infty} \frac{x^2 dx}{(x^2+9)(x^2+4)^2}.$$

2.26 Calculate the integral

$$I = \int_0^{\infty} \frac{x^{-k} dx}{x+1},$$

where k is a real constant and $0 < k < 1$.

2.27 Calculate the integral

$$I = \int_0^{\pi} \frac{\cos 2\theta}{1 - 2a\cos\theta + a^2} d\theta \quad (a^2 < 1).$$

2.28 Calculate the integral

$$I = \int_0^{2\pi} \frac{\cos 3\theta}{5 - 4\cos\theta} d\theta.$$

2.29 Find the inverse Laplace transform of the function

$$h(s) = \frac{e^{-a\sqrt{s}}}{s}$$

by evaluating the complex integral

$$I = \frac{1}{2\pi i} \int_{\gamma - i\infty}^{\gamma + i\infty} \frac{e^{st - a\sqrt{s}}}{s}\, ds.$$

Note that in the expression $\gamma \pm iT$, $\lim T \to \infty$, γ represents a point on the real axis. Also, $\gamma > 0$. The parameter a is a positive, real constant.

2.30 Calculate the integral

$$I = \int_0^{\infty} \left(1 + x^2\right)^{-1} \log x\, dx.$$

2.31 Evaluate the real integral

$$\int_{-\infty}^{\infty} \frac{x \sin \pi x}{x^2 + 2x + 2}\, dx$$

using contour integration procedures.

2.32 Evaluate the integral

$$I = \int_C \sin\left(\frac{z}{z+1}\right) dz,$$

where C is the circle $|z+1| = 1$.

2.33 Evaluate the integral

$$I = \int_{-\infty}^{\infty} \frac{x^2}{\left(1 + x^2\right)^4}\, dx.$$

Must we treat the answer as a CPV? Explain your answer.

2.34 Evaluate the integral

$$I = \int_{-\infty}^{\infty} \frac{\cos x}{(x^2 + 1)(x^2 - 3x + 2)}\, dx.$$

2.35 Evaluate the real integral

$$I = \int_{-\infty}^{\infty} \frac{x^2}{(x^2 + c^2)(x^2 + d^2)} dx,$$

where c and d are positive, real constants.

2.36 Evaluate

$$I = \int_{|z|=1} \frac{dz}{z}.$$

2.37 Evaluate the real integral

$$I = \int_{-\infty}^{\infty} \frac{\cos kx}{s^2 + x^2} dx,$$

where k and s are each positive constants.

2.38 Evaluate the integral

$$\mathrm{I} = \int_{|z|=3} \frac{dz}{(z^2 + 1)}.$$

2.39 Evaluate the integral

$$I = \int_{|z|=2} \frac{e^z \, dz}{(z-1)(z+3)^2}.$$

2.40 Evaluate the integral

$$I = \int_{|z|=4} \frac{e^z \, dz}{(z-1)(z+3)^2}.$$

2.41 Evaluate the integral

$$I = \int_{0}^{2\pi} \frac{1}{2 - \sin\theta} d\theta.$$

2.42 Determine the residue of the function

$$f(z) = \frac{e^z}{z^2}.$$

2.43 Evaluate the integral

$$f(t) = \frac{1}{2\pi i}\int_{-i\infty}^{i\infty} \frac{e^{st}}{(a+s)^2}\,ds.$$

2.44 Evaluate the real integral

$$I = \int_{-\infty}^{\infty} \frac{e^{3ix}}{x^3 - 8i}\,dx.$$

2.45 Evaluate the real integral

$$I = \int_{0}^{\infty} \frac{2\cos^2 x}{(4+x^2)^2}\,dx.$$

2.46 Evaluate the real integral

$$I = \int_{-\infty}^{\infty} \frac{\cos(ax)}{(x^2+b^2)(x^2+c^2)}\,dx,$$

where a, b, and c are positive numbers and $b \neq c$.

2.47 Evaluate the real integral

$$I = \int_{0}^{\infty} \frac{x^{1/3}}{x(x^2+1)}\,dx.$$

2.48 Use complex variable theory techniques to find the inverse LT of

$$\tilde{f}(s) = \frac{1}{(s+1)(s^2+1)}.$$

2.49 Determine the inverse LT of

$$\tilde{f}(s) = \frac{s^2 - s}{(s^2+1)^2}$$

using complex variable theory.

2.50 Use the complex variable contour integration method to evaluate

$$I = \int_{0}^{2\pi} \frac{\cos 2\theta}{1 - 2a\cos\theta + a^2}\,d\theta, a^2 < 1.$$

2.51 Use the complex variable integration procedure to evaluate the integral

$$I = \int_0^{2\pi} e^{\cos\theta} \cos(\sin\theta - n\theta) d\theta.$$

2.52 Use complex variable contour integration techniques to evaluate the integral

$$I = \int_0^{2\pi} \frac{d\theta}{(a + b\cos\theta)^2};\ a > 0, b > 0, a > b.$$

2.53 Evaluate

$$\int_{-\infty}^{\infty} \frac{x^2}{(x^2 + a^2)(x^2 + b^2)} dx;\ a > 0, b > 0.$$

Suggested reading

[1] Ablowitz, M.J. and Fokas, A.S. Complex Variables: Introduction and Applications, 2nd Edition, New York, NY, USA, Cambridge University Press, 2003.

[2] Ahlfors, L.V. Complex Analysis, 3rd Edition, Burr Ridge, IL, USA, McGraw-Hill Higher Education, 1979.

[3] Antimirov, M.Y., Kolyshkin, A.A., and Vaillancourt, R. Complex Variables, San Diego, CA, USA, Academic Press, 1998.

[4] Bak, J. and Newman, D.J. Complex Analysis, 2nd Edition, New York, NY, USA, Springer-Verlag, 1997.

[5] Bieberbach, L. Conformal Mapping, Providence, RI, USA, American Mathematical Society, 2000.

[6] Boas, R.P. (1987, 1964) Invitation to Complex Analysis, New York: The McGraw-Hill Companies, Another Proof of the Fundamental Theorem of Algebra, Amer Math Monthly USA 1987 and 1964, 71(2), 180.

[7] Bowman, F. Introduction to Elliptic Functions, with Applications, London, UK, English Universities Press, 1953.

[8] Brown, J.W. and Churchill, R.V. Complex Variables and Applications, 6th Edition, New York, NY, USA, McGraw-Hill, 1996.

[9] Brown, J.W. and Churchill R.V. Fourier Series and Boundary Value Problems, 6th Edition, Burr Ridge, RI, USA, McGraw-Hill Higher Education, 2001.

[10] Brown, J.W. and Churchill R.V. Complex Variables and Applications, 8th Edition, Burr Ridge, RI, USA, Mcgraw-Hill Higher Education, 2009.

[11] Caratheodory, C. Conformal Representation, Mineola, NY, USA, Dover Publications, 1952.

[12] Caratheodory, C. Theory of Functions of a Complex Variable, Providence, RI, USA, American Mathematical Society, 1954.

[13] Carrier, G.F., Krook, M., Pearson, C.E. Functions of Complex Variables, Ithaca, NY, USA, Hod Books, 1983.

[14] Churchill, R.V. Operational Mathematics, 3rd Edition, Burr Ridge, RI, USA, McGraw-Hill Higher Education, 1972.

[15] Churchill, R.V. Introduction to Complex Variables and Applications. New York, NY, USA, McGraw-Hill Book Company, 1948.

[16] Conway, J.B. Functions of One Complex Variable, 2nd Edition, 6th Printing, New York, NY, USA, Springer-Verlag, 1997.
[17] Copson, E.T. Theory of Functions of a Complex Variable. London, UK, Oxford University Press, 1962.
[18] Dettman, J.W. Applied Complex Variables. Mineola, NY, USA, Dover Publications, Inc., 1955.
[19] Evans, G.C. The Logarithmic Potential, Discontinuous Dirichlet and Neumann Problems, Providence, RI, USA, American Mathematical Society, 1927.
[20] Flanigan, F.J. Complex Variables: Harmonic and Analytic Functions, Mineola, NY, USA, Dover Publications, 1983.
[21] Fisher, S.D. Complex Variables, 2nd Edition, Mineola, NY, USA, Dover Publications, 1990.
[22] Fourier, J. The Analytical Theory of Heat. (A. Freeman, Trans.), Mineola, NY, USA, Dover Publications, 1955.
[23] Hayt, W.H. Jr. and Buck, J.A. Engineering Electromagnetics, 6th Edition, Burr Ridge, RI, USA, McGraw-Hill Higher Education, 2000.
[24] Henrici, P. Applied and Computational Complex Analysis, Volumes 1, 2, and 3, New York, NY, USA, John Wiley & Sons, 1988, 1991, and 1993.
[25] Hille, E. Analytic Function Theory, Volumes. 1 and 2, 2nd Edition, New York, NY, USA, Chelsea Publishing Co., 1973.
[26] Jeffrey, A. Complex Analysis and Applications. Boca Raton, FL, USA, CRC Press, Boca Raton, 1992.
[27] Kober, H. Dictionary of Conformal Representations, New York, NY, USA, Dover Publications, 1952.
[28] Krantz, S.G. Handbook of Complex Variables, Berlin, Germany, Springer Science+Business Media, 1999.
[29] Krantz, S.G. Complex Analysis: The Geometric Viewpoint, Carus Mathematical Monograph Series, Washington, WA, USA, The Mathematical Association of America, 1990.
[30] Krantz, S.G. Handbook of Complex Variables, Cambridge, MA, USA, Birkhauser Boston, 2000.
[31] Krzyz, J.G. Problems in Complex Variable Theory, New York, NY, USA, Elsevier Science, 1972.
[32] Lang, S. Complex Analysis, 3rd Edition, New York, NY, USA, Springer-Verlag, 1993.
[33] Lebedev, N.N. Special Functions and Their Applications. (R. Silverman, Trans.), Rev. Edition, Mineola, NY, USA, Dover Publications, 1972.
[34] Levinson, N. and Redheffer, R.M. Complex Variables. New York, NY, USA, The McGraw-Hill Companies, 1988.
[35] Love, A.E. Treatise on the Mathematical Theory of Elasticity, 4th Edition, Mineola, NY, USA, Dover Publications, 1944.
[36] Markushevich, A.I. Theory of Functions of a Complex Variable, 2nd Edition, Providence, RI, USA, American Mathematical Society, 1977.
[37] Marsden, J.E. and Hoffman, M.J. Basic Complex Analysis, 2nd Edition, New York, NY, USA, W. H. Freeman & Company, 1987.
[38] Mathews, J.H. and Howell, R.W. Complex Analysis for Mathematics and Engineering, 4th Edition, Sudbury, ON, Canada, Jones and Bartlett Publishers, 2001.
[39] Milne-Thomson, L.M. Theoretical Hydrodynamics, 5th Edition, Mineola, NY, USA, Dover Publications, 1996.
[40] Mitrinovic, D.S. Calculus of Residues. Groningen, Netherlands, P. Noordhoff, Ltd., 1966.
[41] Nahin, P.J. (1998) An Imaginary Tale: The Story of $\sqrt{-1}$. Princeton, NJ, USA, Princeton University Press.
[42] Nehari, Z. Conformal Mapping. Mineola, NY, USA, Dover Publications, 1975.
[43] Newman, M.H.A. Elements of the Toplogy of Plane Sets of Points, 2nd Edition, Mineola, NY, USA, Dover Publications, 1999.
[44] Oppenheim, A.V., Schafer, R.W., and Buck, J.R. Discrete-Time Signal Processing, 2nd Edition, Paramus, NJ, USA, Prentice-Hall PTR, 1999.

[45] Osler, T.J. and Waterpeace, S.P. Intuitive Introduction to Complex Analysis: Volume I, Draft version 1, 2019 Publ. (TBD).

[46] Pennisi, L.L. Elements of Complex Variables, 2nd Edition, Austin, TX, USA, Holt, Rinehart & Winston, Inc., 1976.

[47] Rubenfeld, L.A. A First Course in Applied Complex Variables, New York, NY, USA, John Wiley & Sons, Inc., 1985.

[48] Saff, E.B. and Snider, A.D. Fundamentals of Complex Analysis, 3rd Edition, Paramus, NJ, USA, Prentice-Hall PTR, 2001.

[49] Silverman, R.A. Complex Analysis with Applications. Mineola, NY, USA, Dover Publications, 1984.

[50] Sokolnikoff, I.S. Mathematical Theory of Elasticity, 2nd Edition, Malaba, FL, USA, Krieger Publishing Company, 1983.

[51] Springer, G. Introduction to Riemann Surfaces, 2nd Edition, Providence, NJ, USA, American Mathematical Society, 1981.

[52] Streeter, V.L., Wylie, E.B., and Bedford, K.W. Fluid Mechanics, 9th Edition, Burr Ridge, RI, USA, McGraw-Hill Higher Education, 1997.

[53] Taylor, A.E. and Mann, W.R. Advanced Calculus, 3rd Edition, New York, NY, USA, John Wiley & Sons, Inc, 1983.

[54] Thron, W.J. Introduction to the Theory of Functions of a Complex Variable, New York, NY, USA, John Wiley & Sons, Inc., 1953.

[55] Titchmarsh, E.C. Theory of Functions, 2nd Edition, New York, NY, USA, Oxford University Press, Inc., 1976.

[56] Volkovyskii, L.I., Lunts, G.L., and Aramanovich, I.G. A Collection of Problems on Complex Analysis, Mineola, NY, USA, Dover Publications, 1992.

[57] Wen, G.-C. Conformal Mappings and Boundary Value Problems; Translations of Mathematical Monographs, Volume 106, Providence, NJ, USA, American Mathematical Society, 1992.

[58] Whittaker, E.T. and Watson, G.N. A Course of Modem Analysis, 3rd Edition, New York, NY, USA, Cambridge University Press, 1943.

[59] Whittaker, E.T. and Watson, G.N. A Course of Modem Analysis, 4th Edition, New York, NY, USA, Cambridge University Press, 1996.

3 Conformal mapping and its applications

Mappings between two complex planes that preserve the angles of intersection between smooth curves and the sense of orientation of the curves are known as conformal mappings. The subject of conformal mapping is rich in engineering applications. We start the chapter by discussing the preservation of angles and sense of direction, as required for conformal mapping, and the invariance of Laplace equation under this transformation. We examine several transformations and their usefulness, including the linear transformation, inversion transformation, log transformation, z^2 transformation, and linear fractional transformation (LFT). We then discuss the application of eccentric circle conformal mapping in engineering analysis and the transformation of Dirichlet and Neumann types of boundary conditions in harmonic boundary value problems. We also present the techniques of multiple transformations and the calculation of a harmonic function in the upper half of the complex plane. An extensive discussion of the Schwarz-Christoffel transformation is given, with applications to the free streamline flow theory, where we introduce several transformations including those in the hodograph plane, log-hodograph (Kirchhoff) plane, complex flow velocity plane, and a couple of Schwarz-Christoffel transformations; all of which are combined in order to analyze the complex problem of free streamline in potential flows.

3.1 Preservation of angles and sense of direction

A curve Γ_1 in the z-plane with points A and B on it, or Γ_2 with points A and C on it (Figure 3.1(a)) are said to be directed if a sense of direction along the curves can be defined. This is perhaps more obvious if we define the curve parametrically [1]. In the parametric form, the position vector of a point on a curve in the complex z-plane can be represented as

$$P(t) = [x(t)\ \ iy(t)]. \tag{3.1}$$

In this form, the points on a curve are not axis-dependent but are defined by a single value of the parameter, which is t in our illustration (eq. (3.1)). As an example, a parametric representation of a straight line could be written as

$$P(t) = P_1 + (P_2 - P_1)t, \qquad 0 \le t \le 1,$$

where each of the components of $P(t)$ has the parametric representation:

$$x(t) = x_1 + (x_2 - x_1)t, \qquad iy(t) = iy_1 + i(y_2 - y_1)t, \qquad 0 \le t \le 1.$$

Note that the points (x_1, iy_1) and (x_2, iy_2) will be represented as

$$P(t) = [x_1\ \ iy_1] + ([x_2\ \ iy_2] - [x_1\ \ iy_1])t, \qquad 0 \le t \le 1.$$

https://doi.org/10.1515/9783111351179-004

A point θ in the azimuth of a unit circle can be parametrized as

$$P(\theta) = [x\ y] = [\cos\theta\ \ \sin\theta], \qquad 0 \le \theta \le 2\pi,$$

while a point t on a unit circular arc in the first quadrant can conveniently be written as

$$P(t) = \left[\frac{(1-t^2)}{(1+t^2)} \quad i\frac{2t}{(1+t^2)}\right], \qquad 0 \le t \le 1,$$

with t being zero and unity, respectively, at $(x, iy) = (1, 0)$ and $(x, iy) = (0, 1)$. With the normalized parameter t used in this illustration, we travel on the curve starting from $t = 0$ and reach $t = 1$ at the end of the curve. In this context, a reference to the left and right of a curve makes sense. This sense of direction in the z-plane is also mapped onto a similar sense of direction in the image of the curve in the w-plane under the transformation $f(z)$. By way of notation in the w-plane, we use a prime for the corresponding variables in the z-plane. That is, the curves Γ_1, Γ_2 in the z-plane are referenced as Γ_1', Γ_2' in the w-plane. Similarly for the points A, B, and C. The point of intersection of the two curves Γ_1 and Γ_2 is referred to as z_0, whereas the corresponding image in the w-plane is w_0. The preservation of the sense of direction by conformal mapping implies that regions located on the left or right of Γ_1 will be similarly located on Γ_1'; and similarly for Γ_2 and Γ_2'. Conformal mapping also preserves the angle formed by the intersection of any two curves after the transformation so that the angle between Γ_1 and Γ_2 is equal to that between Γ_1' and Γ_2'.

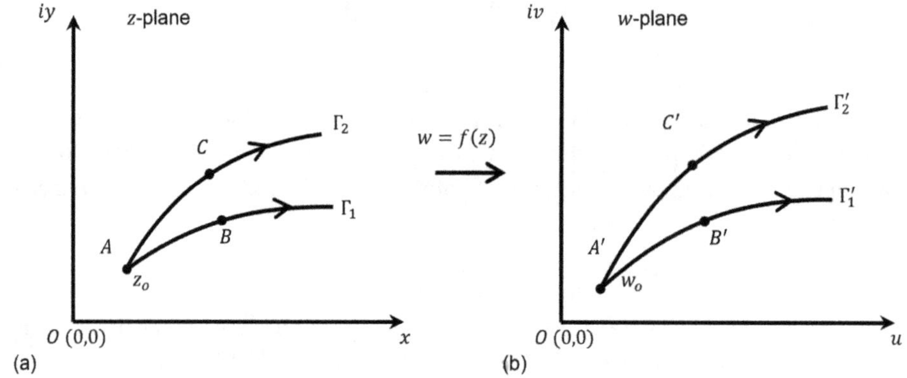

Figure 3.1: Transformation of points and lines.

There are requirements on $f(z)$ for it to be suitable for the purposes identified above. The function must be analytic and single-valued. This allows us to define $z = f^{-1}(w)$ unambiguously. We will also show that the derivative of $f(z)$ with respect to z must not be zero, or $f'(z) \neq 0$, if the magnitudes of angles are to be preserved after transformation. The sense of rotation is also preserved under these conditions. That is, if the tangent to Γ_2 at A (Figure 3.1) is obtained by rotating the tangent to Γ_1 at A counterclockwise through an angle $\beta°$, then the tangent to Γ_2' at A' will be obtained by rotating the tangent to Γ_1' at A' through the same angle $\beta°$. These last two properties (the preservation of intersecting angles and the sense of rotation) formally define a conformal mapping, and allow us to solve boundary value problems for the two-dimensional Laplace equation. The theorems below summarize the results.

Theorem 3.1 Let $f(z)$ be single-valued and analytic in a region of the complex z-plane. Then, at every point z where $f'(z) \neq 0$, the conformal mapping $w = f(z)$ preserves the sense of rotation between intersecting directed curves.

Theorem 3.2 Let $f(z)$ be analytic. Then the function $w = f(z)$ will have a single-valued inverse in the neighborhood of points where $f'(z) \neq 0$. Points where $f'(z) = 0$ are referred to as critical points. Although angles are not preserved at critical points, this property finds application in aerofoil design. For instance, the critical point of $f(z) = z^2$ is zero, which means that angles with a vertex at the origin will not be preserved, but, in this case, doubled.

We can derive the relationship between elemental lengths in the preimage (z) and image $(w(z))$ planes. If $f(z)$ is analytic, and we let $\Delta z = |\Delta z| e^{i\theta}$, $\Delta w = |\Delta w| e^{i\varphi}$, we can define

$$f'(z) = \lim_{\Delta z \to 0} \frac{\Delta w}{\Delta z} = \lim_{\Delta z \to 0} \frac{|\Delta w| e^{i\varphi}}{|\Delta z| e^{i\theta}} = \lim_{\Delta z \to 0} \left| \frac{\Delta w}{\Delta z} \right| e^{i(\varphi - \theta)},$$

which means that $\lim_{\Delta z \to 0} |\Delta w / \Delta z| = |f'(z)|$ and $\lim_{\Delta z \to 0} (\varphi - \theta) = \arg f'(z)$. (We see from here that the argument of a quotient is the difference between the arguments of the numerator and the denominator. So, if $c = a/b$, $\arg c = \arg a - \arg b$. It is also obvious that if $c = a^n / b^m$, then $\arg c = n \arg a - m \arg b$ while $\arg c = n \arg a + m \arg b$ if $c = a^n b^m$.

Thus,

$$|\Delta w| = |f'(z)| \cdot |\Delta z|; \ \varphi = \theta + \arg f'(z).$$

This equation shows that $|f'(z)| \neq 0$ is required for $|\Delta w|$ to be different from zero, and we see that $|f'(z)|$ is in fact a scale factor for elemental lengths between the z and w planes. Correspondingly, the area factor is $|f'(z)|^2$. These results can be stated in a theorem.

Theorem 3.3 In the mapping $w = f(z)$, where $f(z)$ is analytic, the lengths of elemental segments, regardless of their direction, are changed by a factor $|f'(z)|$, and this depends only on the point from which the segments are drawn. Similarly, infinitesimal areas are changed by a factor of $|f'(z)|^2$, which is also the Jacobian $J(u, v/x, y)$ where, employing the C-R conditions $u_x = v_y$ and $v_x = -u_y$,

$$J\left(\frac{u,v}{x,y}\right)=\begin{vmatrix}\frac{\partial u}{\partial x} & -\frac{\partial v}{\partial x}\\ \frac{\partial v}{\partial x} & \frac{\partial u}{\partial x}\end{vmatrix}=\left(\frac{\partial u}{\partial x}\right)^2+\left(\frac{\partial v}{\partial x}\right)^2=\left|\frac{\partial u}{\partial x}+i\frac{\partial v}{\partial x}\right|^2=|f'(z)|^2.$$

The magnification or reduction in the angle between intersecting curves after going through a conformal transformation is of interest. Of course, this is relevant only to locations where $f'(z)=0$. We suppose that $f'(z)$ has an n-fold zero at the critical point z_0. In this case, a series for $f'(z)$ must contain the factor $(z-z_0)^n$ and we can write $f'(z)$ as follows:

$$f'(z)=(n+1)a(z-z_0)^n+(n+2)b(z-z_0)^{n+1}+\ldots, \tag{3.2}$$

where the coefficients a, b, . . . are in general complex but are independent of z. The premultiplying factors $(n+1)$, $(n+2)$, . . ., have been added for convenience so that after integrating eq. (3.2), we have

$$f(z)-f(z_0)\equiv\Delta w=a(z-z_0)^{n+1}+b(z-z_0)^{n+2}+\ldots$$

or

$$\Delta w=a(\Delta z)^{n+1}+b(\Delta z)^{n+2}+\ldots$$

Division by the first term on the RHS gives

$$\frac{\Delta w}{a(\Delta z)^{n+1}}=1+\frac{b}{a(\Delta z)^{n+1}}(\Delta z)^{n+2}+\ldots=1+\frac{b\Delta z}{a}+\ldots \tag{3.3}$$

The RHS of eq. (3.3) approaches unity in the limit that Δz approaches zero. Considering only the angles of the complex variables in eq. (3.3), we can write

$$\lim_{\Delta z\to 0}(\arg\Delta w)-\lim_{\Delta z\to 0}\left[\arg a(\Delta z)^{n+1}\right]=\arg 1=0,$$

remembering that $1=1e^0$. To an arbitrary degree of approximation,

$$\arg\Delta w\equiv\arg a+(n+1)\arg\Delta z. \tag{3.4}$$

Let Δz_1 and Δz_2 be two elemental segments in the preimage and Δw_1 and Δw_2 be their respective images in the range. We can use eq. (3.4) to write

$$\arg\Delta w_1=\arg a+(n+1)\arg\Delta z_1, \tag{3.5}$$

$$\arg\Delta w_2=\arg a+(n+1)\arg\Delta z_2. \tag{3.6}$$

If we substitute eq. (3.5) into eq. (3.6), we will have

$$\arg \Delta w_2 - \arg \Delta w_1 = (n+1)[\arg \Delta z_2 - \arg \Delta z_1] = (n+1)\theta.$$

The following theorem summarizes this result.

Theorem 3.4 Angles are in general preserved in magnitude and sense in the mapping defined by an analytic function $f(z)$. An exception can be found when the vertex of the angle is an n-fold zero of $f'(z)$, in which case the measure of the angle is changed by the factor $n+1$. This result explains why $w = f(z) = z^2$, which has $f'(z) = 2z = 0$, with a one-fold multiplicity at $z = 0$, will have the angle at $z = 0$ in the preimage doubled in the image.

3.2 Invariance of Laplace equation under conformal mapping

One important motivation for studying conformal transformation is the invariance of the Laplace equation $\Delta\phi = 0$ under this transformation. It is well known that this equation governs the distributions of a variety of quantities of engineering interest, including steady-state heat conduction, ideal fluid flow, and electrostatics. The domains of interest in those applications are usually very complex, with the implication that the (Laplace) equation cannot be solved directly in them. However, the equation can often be solved in closed-form in the standard simple geometries (disks and rectangles). With conformal mapping from the complex geometries to the simple ones, we are able to obtain the solution of the Laplace equation in the complex domains by inverse-transforming the solution in the simple domain. The invariance of the Laplace equation under a conformal transformation is stated in a theorem.

Theorem 3.5 If $\phi(x,y)$ is a solution of the Laplace equation $\phi_{xx} + \phi_{yy} = 0$ in the complex z-plane, then when $\phi(x,y)$ is mapped into the function $\Phi(u,v)$ in the w-plane, by a conformal transformation, $\Phi(u,v)$ will satisfy

$$\Phi_{uu} + \Phi_{vv} = 0$$

everywhere except possibly at the images of points where the derivative of the mapping function is equal to zero. Note that $\phi(x(u,v), y(u,v))$ is written here as $\Phi(u,v)$.

Proof: Let $w = u(x,y) + iv(x,y)$ define a conformal transformation in which $\phi(x,y)$ is transformed into a function of u and v.

Then, with partial differentiation,

$$\frac{\partial\phi}{\partial x} = \frac{\partial\Phi}{\partial u}\frac{\partial u}{\partial x} + \frac{\partial\Phi}{\partial v}\frac{\partial v}{\partial x}, \qquad \frac{\partial\phi}{\partial y} = \frac{\partial\Phi}{\partial u}\frac{\partial u}{\partial y} + \frac{\partial\Phi}{\partial v}\frac{\partial v}{\partial y}$$

and

$$\frac{\partial}{\partial x}\left(\frac{\partial\phi}{\partial x}\right)=\frac{\partial\Phi}{\partial u}\left(\frac{\partial^2 u}{\partial x^2}\right)+\frac{\partial u}{\partial x}\left[\frac{\partial}{\partial x}\left(\frac{\partial\Phi}{\partial u}\right)=\frac{\partial}{\partial u}\left(\frac{\partial\Phi}{\partial u}\right)\frac{\partial u}{\partial x}+\frac{\partial}{\partial v}\left(\frac{\partial\Phi}{\partial u}\right)\frac{\partial v}{\partial x}\right]+\frac{\partial\Phi}{\partial v}\left(\frac{\partial^2 v}{\partial x^2}\right)$$
$$+\frac{\partial v}{\partial x}\left[\frac{\partial}{\partial x}\left(\frac{\partial\Phi}{\partial v}\right)=\frac{\partial}{\partial u}\left(\frac{\partial\Phi}{\partial v}\right)\frac{\partial u}{\partial x}+\frac{\partial}{\partial v}\left(\frac{\partial\Phi}{\partial v}\right)\frac{\partial v}{\partial x}\right].$$

Similarly,

$$\frac{\partial}{\partial y}\left(\frac{\partial\phi}{\partial y}\right)=\frac{\partial^2 u}{\partial y^2}\frac{\partial\Phi}{\partial u}+\frac{\partial u}{\partial y}\left[\frac{\partial}{\partial y}\left(\frac{\partial\Phi}{\partial u}\right)=\frac{\partial}{\partial u}\left(\frac{\partial\Phi}{\partial u}\right)\frac{\partial u}{\partial y}+\frac{\partial}{\partial v}\left(\frac{\partial\Phi}{\partial u}\right)\frac{\partial v}{\partial y}\right]+\frac{\partial^2 v}{\partial y^2}\frac{\partial\Phi}{\partial v}$$
$$+\frac{\partial v}{\partial y}\left[\frac{\partial}{\partial y}\left(\frac{\partial\Phi}{\partial v}\right)=\frac{\partial}{\partial u}\left(\frac{\partial\Phi}{\partial v}\right)\frac{\partial u}{\partial y}+\frac{\partial}{\partial v}\left(\frac{\partial\Phi}{\partial v}\right)\frac{\partial v}{\partial y}\right].$$

Adding these two equations, we will obtain

$$\frac{\partial^2\phi}{\partial x^2}+\frac{\partial^2\phi}{\partial y^2}=\underbrace{\frac{\partial\Phi}{\partial u}\left(\frac{\partial^2 u}{\partial x^2}+\frac{\partial^2 u}{\partial y^2}\right)}_{A}+\underbrace{\frac{\partial^2\Phi}{\partial u^2}\left[\left(\frac{\partial u}{\partial x}\right)^2+\left(\frac{\partial u}{\partial y}\right)^2\right]}_{B}+\underbrace{2\frac{\partial^2\Phi}{\partial u\partial v}\left(\frac{\partial u}{\partial x}\frac{\partial v}{\partial x}+\frac{\partial u}{\partial y}\frac{\partial v}{\partial y}\right)}_{C}$$
$$+\underbrace{\frac{\partial\Phi}{\partial v}\left(\frac{\partial^2 v}{\partial x^2}+\frac{\partial^2 v}{\partial y^2}\right)}_{D}+\underbrace{\frac{\partial^2\Phi}{\partial v^2}\left[\left(\frac{\partial v}{\partial x}\right)^2+\left(\frac{\partial v}{\partial y}\right)^2\right]}_{E}.$$

As $w \equiv u+iv$ is analytic, u, v themselves will satisfy Laplace equation (i.e., $u_{xx}+u_{yy}=v_{xx}+v_{yy}=0$). This implies that the terms A, $D=0$. Also, u, v satisfy the C-R conditions $(u_x=v_y,\ u_y=-v_x)$, implying that the term $C=0$.

Therefore,

$$\phi_{xx}+\phi_{yy}=B+E=\Phi_{uu}\left(u_x^2+u_y^2\right)+\Phi_{vv}\left(v_x^2+v_y^2\right).$$

Using the C-R conditions again, we have

$$\phi_{xx}+\phi_{yy}=\Phi_{uu}[u_x^2+(-v_x)^2]+\Phi_{vv}[v_x^2+u_x^2]=(u_x^2+v_x^2)(\Phi_{uu}+\Phi_{vv})$$
$$=|f'(z)|^2(\Phi_{uu}+\Phi_{vv}).$$

Now, $f'(z)\neq 0$ for a conformal transformation, therefore $\Phi_{uu}+\Phi_{vv}=0$.

3.3 Common transformations

3.3.1 The linear transformation $w(z) = az + b$

The linear transformation

$$f(z) \equiv w(z) = az + b,$$

where $a \neq 0$ and b are constants, which could be complex, is the simplest conformal transformation. The transformation basically scales z by a and translates the result by b, where "scaling" could mean magnification or contraction, depending on the magnitude of a. The inverse transformation $z(w)$ is

$$z(w) = \left(\frac{1}{a}\right)w - \frac{b}{a},$$

which is defined and unique (one-to-one) because $a \neq 0$. Also, the transformation is conformal because $w(z) = az + b$ is analytic for all z. Moreover, $w'(z) \equiv f'(z) = a \neq 0$; so there are no critical points for this transformation.

Let us assume that a is an imaginary number, $|a|e^{i\theta_a}$, where θ_a is arg a. We can further write $a = |a| \exp[i \arg a]$ so that

$$w = |a|z \exp[i \arg a] + b \equiv w_2 + b,$$

where $w_2 \equiv \exp[i \arg a]w_1$ and $w_1 = |a|z$. Note that if we write $z = re^{i\theta_z}$, then $w_2 = r|a|e^{i(\theta_z + \theta_a)}$. We see that the original complex number z is uniformly magnified if $|a| > 1$ and uniformly contracted if $|a| < 1$. Moreover, from the expression for w_2, we see that the transformation rotates the given imaginary number by the angle θ_a, if a is a complex number. A depiction of the linear transformation $w = az + b$ is shown in Figure 3.2, with four panes, respectively showing an original shape in the z-plane, the scaling of the original shape by $|a|$ assuming magnification ($|a| > 1$), rotation of the scaled shape by arg a, and the translation of the rotated scaled shape by b. Note that the shapes of the boundaries of the domain are preserved by the linear transformation.

3.3.1.1 Determining the region of the w-plane

For many transformations, determining the region in the w-plane that corresponds to that in the z-plane can be challenging, depending on the domain (pre-image) and the complexity of the transformation. For example, the interior of a closed contour in the z-plane may be transformed into the exterior of its image in the w-plane. Moreover, the boundaries of a general two-dimensional region in the z-plane could be transformed into a straight line in the w-plane. However, shape preservation in conformal mapping comes in handy for this purpose. What we do is to determine the location of the region in the z-plane relative to a few specific points on the boundary of the region in that plane. For example, for the shape in Figure 3.2, we see that the region is located to the left of the points A, B, C, D, and

E, consistent with the context of directed curves as discussed earlier in this section. Shape preservation implies that the transformed shape in the w-plane will also be located to the left of the boundary formed by the transformed points A', B', C', D', and E'.

3.3.2 The inversion transformation $w(z) = 1/z$

This simple-looking transformation, $f(z) = 1/z$, is powerful, particularly in its ability to map circles in the z-plane to circles and straight lines in the w-plane, and lines in the z-plane to circles and lines in the w-plane. Except for the point $z = 0$ where $f(z)$ is undefined, the function is otherwise single-valued and analytic. Also, $f'(z) = -1/z^2$ is not defined at $z = 0$, making this point a critical point of the transformation.

If we write $z = re^{i\theta}$, we see that $w = 1/z = (1/r)e^{-i\theta}$, which gives a magnitude of $|w| = (u^2 + v^2)^{1/2} \equiv \rho = 1/r$ and $\arg w = -\theta$. This shows that the region $|z| < 1$ is mapped to $|w| > 1$, $|z| > 1$ to $|w| < 1$, and $|z| = 1$ to $|w| = 1$. Moreover, the opposite signs of $\arg z$ and $\arg w$ show a reflection about the real axis, so that, for example, the upper half plane z is mapped onto the lower half plane in w and vice-versa. Figure 3.3 shows a hypothetical mapping between the z-plane and w-plane. Going in the direction of $A - B - C - D$ in the z-plane, we see that the interior of $|z| = 1$ is located to the left, whereas the left of $A' - B' - C' - D'$ is located in the exterior of $|w| = 1$. Conversely for the exterior of the z-plane and interior of the w-plane. Note the locations of the points B' and D' in the w-plane, relative to those of B and D in the z-plane. Consequent to this is the reversed directions of traversing the circles in the z- and w-planes. These are shown in Figure 3.3. Some of the treatments in Sections 3.3 through 3.8 in this book are similar to those available in the open engineering mathematics literature, such as [5], but of course the presentation style used here is slightly different.

The idea of inversion that is associated with this mapping derives from the fact $|z||w| = r \times 1/r = 1$. That is, the point w_0 lies on the radius drawn through the point z_0, and its distance from the center of the circle is such that $|w_0||z_0| = 1$. The inversion operation here noted is obviously followed by the reflection of z_0 in order to obtain $w(z)$. That is, $z = re^{i\theta}$, $w_1 = (1/r)e^{i\theta}$, $w = \bar{w}_1 = (1/r)e^{-i\theta}$, remembering that the conjugate operation represents a reflection with the real axis as the mirror line.

The geometric inverse property of $w(z) = 1/z$ discussed above pertains to the unit circle. Let us examine inversion for a circle of arbitrary radius, R. Consider Figure 3.4 for this purpose. In this figure, point P is located outside of the circle with center O at $z = c$, and the point B at w_0 lies inside the circle on the radial line OP at the point of intersection with the chord AC drawn from the points A and C. The lines from P are tangent to the circle at these points. We see that the triangles OBC and OCP are similar so that

$$\frac{OB}{OC} = \frac{OC}{OP},$$

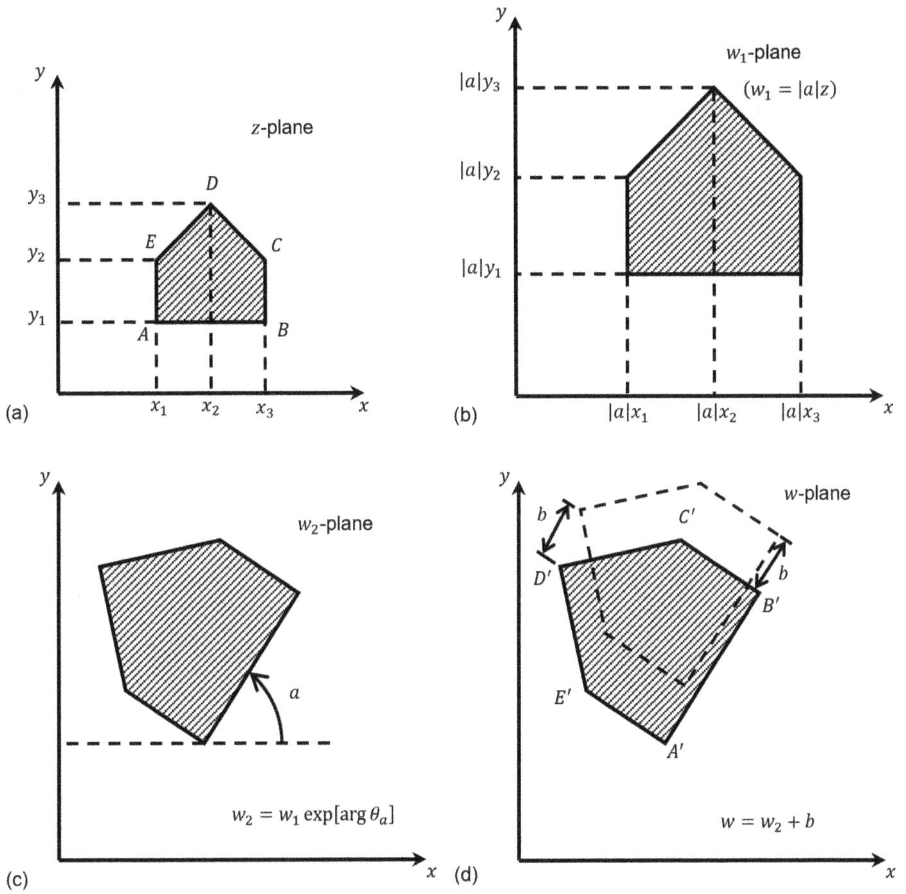

Figure 3.2: Depiction of the linear transformation $w = az + b$: (a) Original shape in the z-plane, (b) scaling by $|a|$, assuming $|a| > 1$, (c) rotation of scaled shape by arg a, and (d) translation of rotated, scaled shape by b.

which implies that $|OB| \times |OP| = |OC|^2$, or $|z_0 - c||w_0 - c| = R^2$. The points P and B in Figure 3.4 are said to be symmetric with respect to the circle with its center at point O, while B is said to be inverse to point P, and P is inverse to point B. For a unit circle, $c = 0$ and $R = 1$, giving $|z_0||w_0| = 1$, as stated earlier. Obviously z_0 and w_0 are in different complex planes, but we have depicted them in the same plane in Figure 3.4 to simplify the discussion of the inversion phenomenon. If $c = 0$, that is the circle in Figure 3.4 is centered on the origin, then $w_0 = R^2/z_0$. Note that the inversion transformation has the fixed points of $z \equiv z_f = \pm 1$ since $f(z_f) = z_f$ or $f(1) = 1/1 = 1$ and $f(-1) = 1/-1 = -1$.

The way that constant x or y lines in the z-plane are transformed into the w-plane is of interest. With $w = u + iv = 1/z = 1/(x + iy)$, we have $u = x/(x^2 + y^2)$, $v = -y/(x^2 + y^2)$, and $x = u/(u^2 + v^2)$, $y = -v/(u^2 + v^2)$. With these equations, we see that the line $x =$ con-

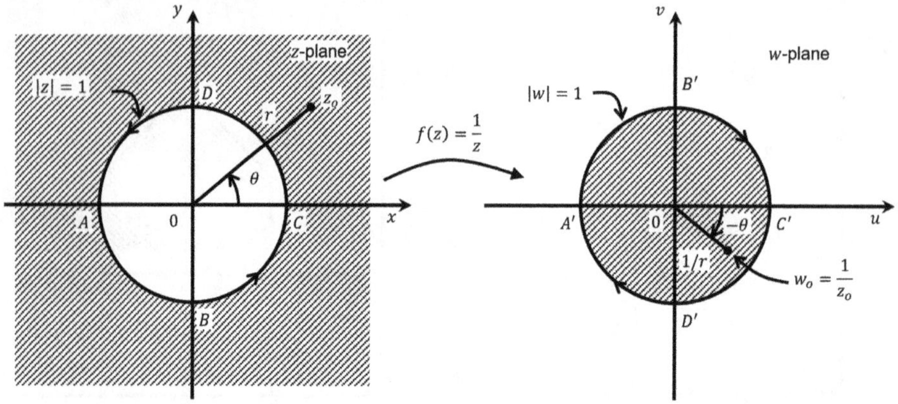

Figure 3.3: The inverse transformation $w \equiv f(z) = 1/z$.

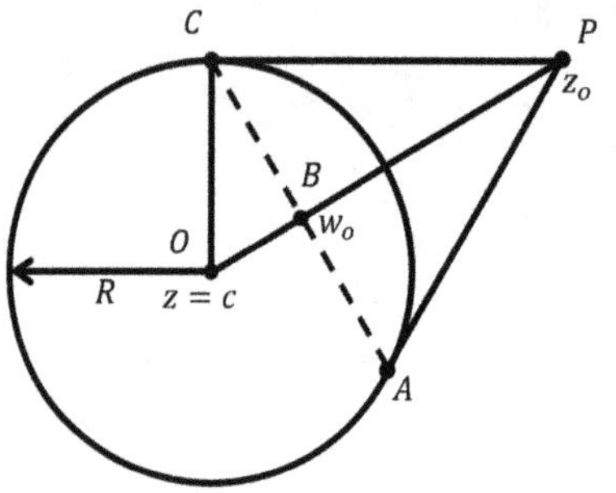

Figure 3.4: Inversion in a circle of arbitrary radius.

stant $\equiv c_1$ in the z-plane gives the circle $u^2 + v^2 - u/c_1 = 0$ in the w-plane, which can be easily drawn for various values of c_1. Similarly, the line $y =$ constant $= c_2$ in the z-plane is represented by $u^2 + v^2 + v/c_2 = 0$ in the w-plane. Figure 3.5 shows this mapping for various values of c_1 and c_2.

The half plane $x > c_1$ has the image $u/(u^2 + v^2) > c_1$. When $c_1 > 0$, we see that

$$\left(u - \frac{1}{2c_1}\right)^2 + v^2 < \left(\frac{1}{2c_1}\right)^2,$$

meaning that points on this z-plane are mapped to a circle in the w-plane that is tangent to the v axis at the origin. From the foregoing, we see that every point inside the circle is the image of some point in the half plane, and the image of the half plane is the entire circular region.

More information on how the inversion transformation maps circles and straight lines can be garnered by considering the general equation of a circle, which is $x^2 + y^2 + 2gx + 2fy + c = 0$. The center of this circle is located at $(-g, -f)$ and the radius

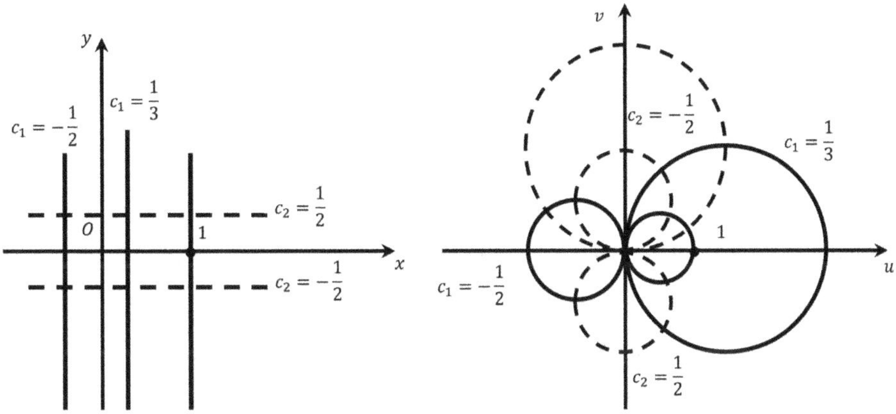

Figure 3.5: The transformation $w(z) = 1/z$ for various $x = \text{constant} \equiv c_1$, $y = \text{constant} \equiv c_2$ lines in the z-plane.

is $\sqrt{(g^2+f^2)-c}$. For the convenience of the reader, the following bits of information on the tangent to a circle are provided:

(a) The tangents of gradient m to the circle $x^2+y^2=r^2$ are $y=mx\pm r\sqrt{(1+m^2)}$.
(b) The equation of the tangent at P_1 with coordinates (x_1,y_1) to the circle $x^2+y^2+2gx+2fy+c=0$ is $xx_1+yy_1+g(x+x_1)+f(y+y_1)+c=0$.
(c) The square of the length of the tangent from point P_1 with coordinates (x_1,y_1) to $ax^2+ay^2+2gx+2fy+c=0$ is $(x_1^2+y_1^2+2gx_1+2fy_1+c)$.

We will consider the following generalized form of the equation of a circle to illustrate other features of the inversion transformation:

$$a(x^2+y^2)+bx+cy+d=0, \tag{3.7}$$

where the coefficients a, b, c, and d are real, and eq. (3.7) describes a circle with radius $R=\left\{(b^2+c^2-4ad)^{1/2}\right\}/2|a|$ with center located at $(-b/2a, -c/2a)$ provided $b^2+c^2>4ad$ and $a\neq 0$. Evidently when $a=0$, we have the straight line $bx+cy+d=0$. The distance of the center of the circle from the origin is $(b^2+c^2)^{1/2}/2|a|$. We see that the circle will not pass through the origin if $d\neq 0$ (since from eq. (3.7) and $x=0, y=0$), $y=-(b/c)x-(d/c)$.

If we utilize the mapping between x and (u,v) and y and (u,v), as given earlier in this section, and substitute into eq. (3.7) we obtain the equation:

$$d(u^2+v^2)+bu-cv+a=0 \tag{3.8}$$

as the general equation of the circle in the w-plane that is the image of the circle represented by eq. (3.7) in the z-plane. Note how the roles of a and d are switched between the z-plane and the w-plane (eqs. 3.7 and 3.8). Equation (3.8) describes a circle in the w-plane with radius $\rho=(b^2+c^2-4ad)^{1/2}/2|d|$, and its center located at $(-b/2d, c/2d)$. The circle will not pass through the origin (in the w-plane) if $a\neq 0$ (since from eq. (3.8) and $u=v=0$), $v=\frac{b}{c}u+\frac{a}{c}$. We see that $f(z)=1/z$ transforms a circle in the z-plane that does not pass through the origin into a circle in the w-plane that does not pass through the origin.

Note, however, that when $a = 0$, $d \neq 0$, a straight line in the z-plane $(bx + cy + d = 0)$ maps to the circle $d(u^2 + v^2) + bu - cv = 0$ with radius $\rho = (b^2 + c^2)^{1/2}/2|d|$ and center $(-b/2d, c/2d)$ in the w-plane. This circle passes through the origin of the w-plane since the radius of the circle and the distance of its center from the origin are equal. Moreover, $a = 0$ in $v = (b/c)u + (a/c)$. In the same token, if $d = 0$ and $a \neq 0$, a straight line in the w-plane will be mapped onto a circle that passes through the origin in the z-plane. If $a = d = 0$, the straight line $y = -\frac{b}{c}x$ in the z-plane, which passes through the origin, will map to the straight line $v = \frac{b}{c}u$, which also passes through the origin, in the w-plane. Note the opposite signs of the slopes of the respective lines in the two planes, which obviously imply some reflection about the real axis.

Finally, note that $f(0)$ is located in the extended complex plane, which is formed by including the point at infinity in the ordinary complex plane. The set of points at infinity is defined as the limit as $R \to \infty$ of all points in the z-plane that lie outside the circle $|z| = R$. Therefore, the inversion transformation maps the origin of the z-plane to the point at infinity in the w-plane, while the point at infinity in the z-plane is mapped to the origin of the w-plane. That is $f(0) = \infty$ and $f(\infty) = 0$.

We see that the extended complex plane unifies the transformation of circles and straight lines by $f(z) = 1/z$, wherein a straight line is regarded as a circle in the limit of an infinite radius.

3.3.3 The linear fractional transformation

The transformation

$$w(z) = \frac{az + b}{cz + d}$$

is called the LFT or the bilinear transformation. It is sometimes called the Möbius transformation. Note that when $c = 0$, the LFT becomes the linear transformation $w(z) = a'z + b'$, where $a' = a/d$ and $b' = b/d$. Also note that $ad - bc \neq 0$, otherwise we will have the condition $w(z) = (az + b)/(cz + d) = a/c$, which is a constant.

The derivative $f'(z) \equiv w'(z)$ of the LFT is $f'(z) = (ad - bc)/(cz + d)^2$, which is one-to-one and conformal everywhere except at the point $z = -d/c$, where the denominator of $f'(z)$ is zero and $f'(z)$ is therefore undefined. Note that we can write the transformation as

$$w(z) = K\left(\frac{z - \alpha}{z - \beta}\right); \quad \text{where } \alpha = -\frac{b}{a}, \quad \beta = -\frac{d}{c}, \quad K = \frac{a}{c}.$$

We can also write the transformation as

$$w(z) = \left(\frac{a}{c}\right) + \frac{(bc - ad)}{c} \cdot \frac{1}{(cz + d)} \equiv a_o + a_1 w_2; \; w_2 = \frac{1}{cz + d} \equiv \frac{1}{w_1},$$

where a_0, a_1 are constants, w_1 is the linear transformation, and w_2 is the inverse of this linear transformation, which obviously has properties similar to the inversion transformation discussed in the previous subsection of this chapter. As expected, the LFT transforms straight lines and circles onto straight lines and circles. From the symmetry properties of the inversion transformation, it is not a surprise that the LFT maps a pair of points that are symmetric with respect to a circle in the z-plane into a pair of points that are symmetric with respect to the image of the circle in the w-plane.

Written as

$$w(z) = \frac{(a/c)z + (b/c)}{z + d/c},$$

we see that the transformation can be completely determined once the values of the three quantities a/c, b/c, and d/c are specified. An interesting exercise is determining the transformation $w(z)$ that maps three specified distinct points z_1, z_2, and z_3 in the z-plane onto three specified distinct points w_1, w_2, and w_3 in the w-plane.

Three distinct points, assumed noncollinear, define a circle, so that the given points will describe a circle in both the z-plane and the w-plane. If the three points are collinear in one plane and noncollinear in the other, we will get a mapping of a straight line described by the collinear points into a circle for the noncollinear points, and obviously this can go both ways. Collinear points in the two planes will involve the mapping of a straight line into another straight line.

To obtain the transformation, let us obtain the expression $w - w^*$, where $w \equiv w(z)$ is the desired transformation and w^* is a specified image for a specified point z^* in the preimage. We can write

$$\begin{aligned} w - w^* &= \frac{az+b}{cz+d} - \frac{az^*+b}{cz^*+b} = \frac{(az+b)(cz^*+d) - (az^*+b)(cz+d)}{(cz+d)(cz^*+d)} \\ &= \frac{ad(z-z^*) - bc(z-z^*)}{(cz+d)(cz^*+d)} = \frac{(ad-bc)}{(cz+d)(cz^*+d)}(z-z^*). \end{aligned}$$

We then form the differences between w and w^* for the three given points to obtain

$$\frac{(w-w_1)(w_3-w_2)}{(w-w_2)(w_3-w_1)} = \frac{(z-z_1)(z_3-z_2)}{(z-z_2)(z_3-z_1)}. \tag{3.9}$$

It should be noted that the specific form of eq. (3.9) depends on the order in which Points 1, 2, and 3 are numbered, as well as the sense (clockwise, counterclockwise, or arbitrary). A clockwise numbering system has been used to obtain eq. (3.9).

The specified points z_1, z_2, and z_3 and w_1, w_2, and w_3 should be inserted into eq. (3.9) to obtain the desired $w(z)$. Note that if one of the three points in any of the two planes is at infinity, we set the factor containing it to 1. For example, if $z_2 = \infty$, then from eq. (3.9), we will have

$$\lim_{z_2\to\infty}\left[\frac{(z-z_1)(z_3-z_2)}{(z-z_2)(z_3-z_1)}\right]=\frac{(z-z_1)}{(z_3-z_1)}\lim_{z_2\to\infty}\frac{(z_3-z_2)}{(z-z_2)}$$
$$=\frac{(z-z_1)}{(z_3-z_1)}\lim_{z_2\to\infty}\frac{(z_3/z_2-1)z_2}{(z/z_2-1)z_2}=\frac{(z-z_1)}{(z_3-z_1)}.$$

Similarly, if $z_3\to\infty$, we wil have

$$\lim_{z_3\to\infty}\left[\frac{(z-z_1)(z_3-z_2)}{(z-z_2)(z_3-z_1)}\right]=\frac{(z-z_1)}{(z-z_2)}. \tag{3.10}$$

Note in passing that $\lim_{|z|\to\infty} f(z)=\lim_{|z|\to\infty}\frac{a+b/z}{c+c/z}=a/c$.

Example 3.1 Determine the LFT that transforms the points $z_1=-1$, $z_2=1$, and $z_3=i$ in the z-plane onto the respective points $w_1=0$, $w_2=1$, and $w_3=-i$ in the w-plane.

Solution: Application of eq. (3.9) gives

$$\left(\frac{w}{w-1}\right)\left[\frac{-(1+i)}{-i}\right]=\frac{z+1}{z-1}\cdot\frac{i-1}{i+1}\Rightarrow w(z)=\frac{z+1}{(2+i)z-i}.$$

As a check, the given points in the z-plane have the following images in the w-plane:

$$z_1=-1\text{: } w=\frac{-1+1}{(2+i)(-1)-i}=0,\qquad z_2=1\text{: } w=\frac{1+1}{(2+i)1-i}=1,$$
$$z_3=i\text{: } w=\frac{i+1}{(2+i)i-i}=-i.$$

The mapping is shown in Figure 3.6 below, where the given points are denoted by A, B, and C in the z-plane, and A', B', and C' in the w-plane, respectively. We see that the

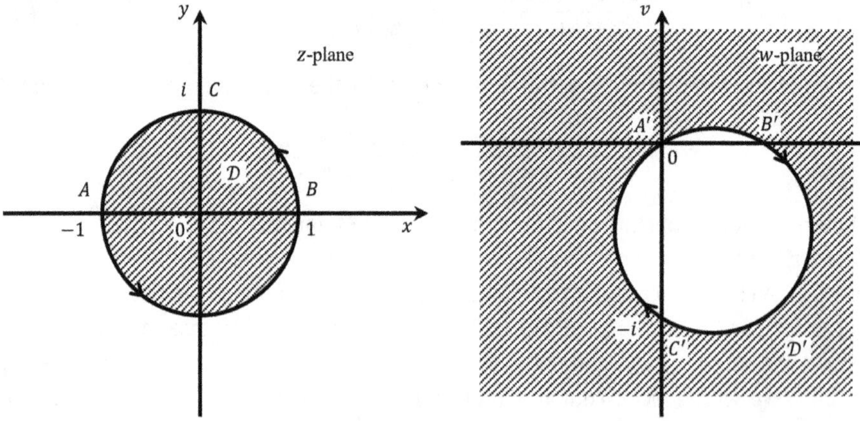

Figure 3.6: The mapping in Example 3.1.

interior of the circle in the z-plane, denoted by $\mathcal{D}$, is mapped to the exterior of the domain in the w-plane which we denote by $\mathcal{D}'$. Note the directions of traversing the circle in the two planes.

Example 3.2 Which LFT transformation maps the points $z_1 = -1$, $z_2 = 0$, and $z_3 = i$ in the z-plane onto the respective points $w_1 = 0$, $w_2 = 1$, and $w_3 = \infty$ in the w-plane?

Solution: Using eqs. (3.9) and (3.10) we obtain

$$\left(\frac{w}{w-1}\right) = \left(\frac{z+1}{z}\right) \times \left(\frac{i}{i+1}\right), \text{ or } w = \frac{z+1}{(iz+1)}.$$

The given points transform as follows:

$$z_1 = -1\text{: } w = \frac{z+1}{iz+1} = \frac{-1+1}{i+1} = 0, \qquad z_2 = 0\text{: } w = \frac{z+1}{iz+1} = \frac{1}{1} = 1,$$

$$z_3 = i\text{: } w = \frac{z+1}{iz+1} = \frac{i+1}{-1+1} = \frac{i+1}{0} = \infty.$$

The mapping is shown in Figure 3.7. The circle in the z-plane is mapped into a straight line in the w-plane. Since $\mathcal{D}$ is to the left of the directed curve $A - B - C$, the region $\mathcal{D}'$ in the w-plane has to be located to the left of the directed curve $A' - B' - C'$, which is in the UHP.

A few other properties of the LFT are of interest. One of these is that successive bilinear transformations are equivalent to a single LFT. For example, if

$$w = \frac{\alpha z + \beta}{\gamma z + \delta} \text{ and } z = \frac{\alpha' z' + \beta'}{\gamma' z' + \delta'},$$

we see by direct substitution that $w = (az' + b)/(cz' + d)$, where a, b, c, and d are complex constants.

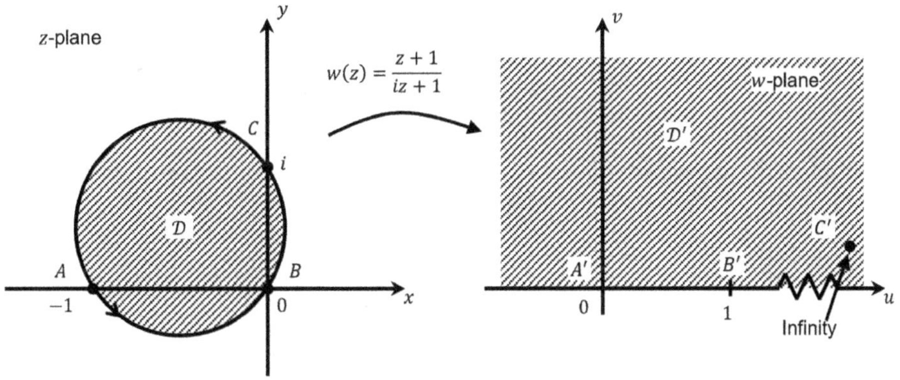

Figure 3.7: The transformation for Example 3.2.

3.3.3.1 Matrix methods

It turns out that 2×2 matrices can be used to describe the LFT transformation and its inverse. To this end, we can associate the constants in the LFT $T(z)$:

$$T(z)=\frac{az+b}{cz+d}$$

with the matrix $\mathbf{A}\equiv\begin{bmatrix} a & b\\ c & d\end{bmatrix}$, and the inverse of $T(z)$, $T^{-1}(w)$, with the inverse of $\mathbf{A}$ save the determinant $(ad-bc)$. That is $T^{-1}(w)=\operatorname{adj}\mathbf{A}$. Since the inverse of any nonsingular matrix $\mathbf{C}$ is given by

$$\mathbf{C}^{-1}=\frac{\operatorname{adj}\mathbf{C}}{\det\mathbf{C}},$$

we recognize that our definition of $\mathbf{A}$ above is not unique since the numerator and denominator in $T(z)$ can be multiplied by a nonzero constant, such as the determinant $(ad-bc)$, which has been left out of the definition of $\mathbf{A}$ in the computation of $T^{-1}(w)$. With this background, we can associate $T^{-1}(w)$ with the adjoint of $\mathbf{A}$:

$$\operatorname{adj}\mathbf{A}=\begin{bmatrix} d & -b\\ -c & a\end{bmatrix},$$

so that $z\equiv T^{-1}(w)=(dw-b)/(-cw+a)$.

Composition of LFTs can be accomplished with this matrix representation approach. For example, if $T(z)=(az+b)/(cz+d)$, $T_1(z)=(a_1z+b_1)/(c_1z+d_1)$, and $T_2(z)=(a_2z+b_2)/(c_2z+d_2)$, then $T(z)=T_2(T_1(z))$ can be written as

$$\begin{pmatrix} a & b\\ c & d\end{pmatrix}=\begin{pmatrix} a_2 & b_2\\ c_2 & d_2\end{pmatrix}\begin{pmatrix} a_1 & b_1\\ c_1 & d_1\end{pmatrix}=\begin{pmatrix} a_2a_1+b_2c_1 & a_2b_1+b_2d_1\\ c_2a_1+d_2c_1 & c_2b_1+d_2d_1\end{pmatrix}$$

so that

$$T(z)=\frac{(a_2a_1+b_2c_1)z+(a_2b_1+b_2d_1)}{(c_2a_1+d_2c_1)z+(c_2b_1+d_2d_1)}.$$

Example 3.3 Suppose we have two LFT transformations: $T(z)=(2z-1)/(z+2)$ and $H=(z-i)/(iz-1)$. Determine $H^{-1}(T(z))$.

Solution: Let

$$H^{-1}(T(z))=\frac{az+b}{cz+d},$$

where

$$\begin{bmatrix} a & b \\ c & d \end{bmatrix} = \text{adj}\begin{bmatrix} 1 & -i \\ i & -1 \end{bmatrix}\begin{bmatrix} 2 & -1 \\ 1 & 2 \end{bmatrix} = \begin{bmatrix} -1 & i \\ -i & 1 \end{bmatrix}\begin{bmatrix} 2 & -1 \\ 1 & 2 \end{bmatrix} = \begin{bmatrix} -2+i & 1+2i \\ 1-2i & 2+i \end{bmatrix}$$

so that

$$H^{-1}(T(z)) \equiv \begin{bmatrix} a & b \\ c & d \end{bmatrix} = \frac{(-2+i)z+(1+2i)}{(1-2i)z+(2+i)}.$$

Example 3.4 Given the domain in the z-plane with the normalized temperature boundary conditions (Figure 3.8), obtain an LFT that transforms the region into the UHP of the w-plane and solve for the temperature.

Solution: We recall the two theorems below:

Theorem: Circle-Preserving Property: An LFT maps a circle in the z-plane to either a line or a circle in the w-plane. The image is a line if and only if the original circle passes through (touches) a pole of the LFT.

Theorem: Let $f(z)$ be an analytic function that maps a domain $\mathcal{D}$ into a domain $\mathcal{D}'$. If T is harmonic in $\mathcal{D}'$, then the real-valued function $u(x,y) = T(f(z))$ is harmonic in $\mathcal{D}$.

The two circles $|z| = 1$ and $|z - 1/2| = 1/2$ pass through $z = 1$, as shown in the figure. We can map the boundaries of each of the two circles onto a line by selecting an LFT that has $z = 1$ as a pole. Furthermore, if we use $w(i) = 0$ and $w(-1) = 1$, then, using the points A, B, C, E in z-plane, we can obtain

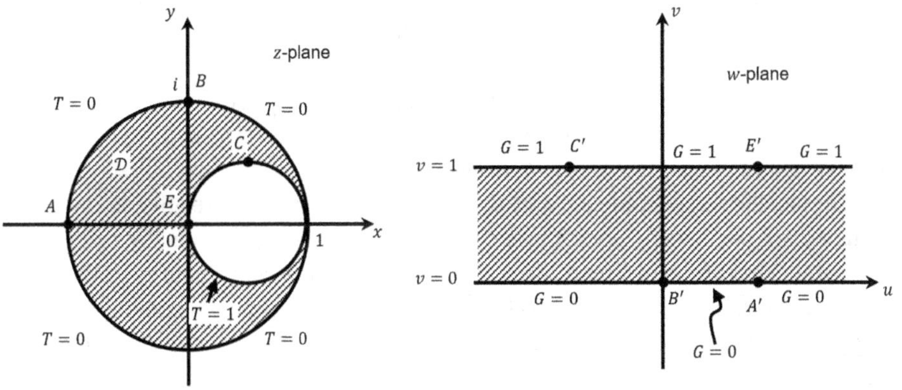

Figure 3.8: Physical and transform domain for temperature calculation in eccentric circles.

$$w(z) = \frac{(z-i)}{(z-1)}\frac{(-1-1)}{(-1-i)} = (1-i)\frac{(z-i)}{(z-1)}.$$

With $w(0) = 1+i$ and $w(\frac{1}{2}+\frac{1}{2}i) = -1+i$, we see that $w(z)$ maps the interior of the circle $|z| = 1$ onto the UHP and the circle $|z-1/2| = 1/2$ onto the line $v = 1$. The region between the eccentric circles in the z-plane is mapped onto the semi-infinite region $-\infty \leq u \leq \infty$, $0 \leq v \leq 1$ in the w-plane.

Now, with the boundary conditions in the w-plane, the transformed temperature G does not depend on u, and the harmonic equation $d^2G/du^2 + d^2G/dv^2 = 0$ becomes $d^2G/dv^2 = 0$. Using the boundary conditions $G(v=0) = 0$ and $G(v=1) = 1$, we have $G = v$ as the temperature solution in the w-plane. Thus, by the second theorem above, $T(x,y) = G(w(z))$. The imaginary part of $w(z) = (1-i)\frac{(z-i)}{(z-1)}$ is $v = \frac{1-x^2-y^2}{(x-1)^2+y^2}$.

Therefore,

$$T(x,y) = \frac{1-x^2-y^2}{(x-1)^2+y^2}.$$

What this really means is that if the solution in the w-plane is $G(u,v)$, then the solution in the z-plane will be $T(x,y) = G(u(x,y), v(x,y))$. That is, we directly insert the relationship between u and (x,y) and v and (x,y) into the expression for G in order to obtain $T(x,y)$.

3.3.3.2 Mapping eccentric circles onto concentric circles

Physical models represented by eccentric circles are abundant in engineering, where the governing equations are sometimes of the Laplace or Poisson types. This geometry is not amenable to easy analysis because the domain cannot be represented by the standard orthogonal coordinate systems, such as the rectangular, cylindrical polar, or the spherical polar types. However, a conformal transformation can be employed to transform the eccentric circular geometry to concentric circles which are readily described in the cylindrical polar coordinate system. We will use Figure 3.9 to discuss the derivation of the required transformation. The larger circles in the z- and w-planes are unit circles, $|z| = 1$, $|w| = 1$, respectively.

The smaller circle in the z-plane, $|z - z_o| = \rho$, is centered at z_o relative to the origin of the z-plane, so that any point z inside this circle could be represented by $z = z_o + \rho e^{i\theta}$. The point C has the coordinate $z_1 \equiv z_o + \rho$, while B has the coordinate $z_2 = z_o - \rho$. It is assumed that z_o is located on the real axis of the z-plane. The point z_o in the z-plane is mapped to the origin of the w-plane. The images of C and B (or C' and B') have the coordinates $w = -\delta$ and $w = \delta$ in the w-plane.

The derivation of the required $w(z)$ starts with the form

$$w(z) \equiv T(z) = K\left(\frac{z-\alpha}{z-\beta}\right)$$

of the LFT, where K could be complex. If z_0 is to map to the origin of the w-plane, we must set $\alpha = z_0$ so that

$$T(z) = K\left(\frac{z - z_0}{z - \beta}\right).$$

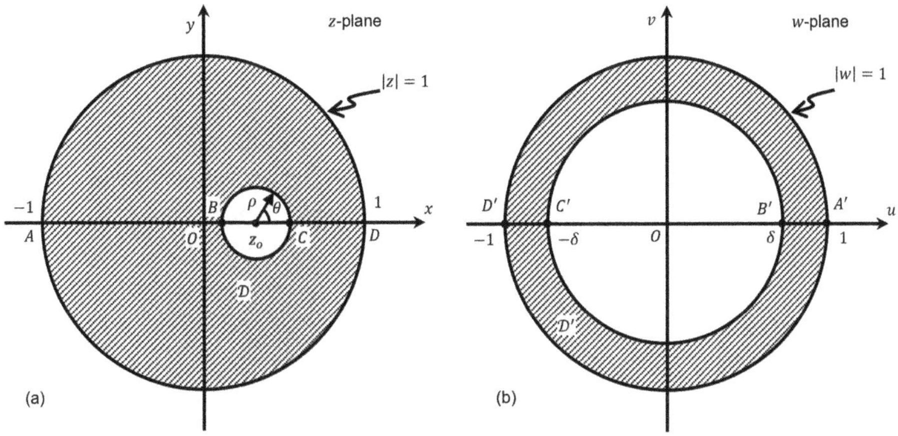

Figure 3.9: Mapping of eccentric circles to concentric circles via the LFT.

The symmetry of the two circles in the z-plane will be preserved when they are mapped to the w-plane. Also, a point z^* that is symmetric relative to $z = z_0$ with respect to the unit circle in the z-plane will be mapped onto a point in the w-plane that is symmetric relative to the origin $w = 0$ with respect to the circle $|w| = 1$ since z_0 is mapped to $w = 0$. This means that z^* will be mapped to infinity in the w-plane, or "$0 \times z^* = 1$," which requires that $\beta = z^*$, and

$$T(z) = K\left(\frac{z - z_0}{z - z^*}\right). \tag{3.11}$$

We will have $z_0 z^* = 1$ since z_0 and z^* are symmetric with respect to the circle $|z| = 1$. Since z_0 is located on the real axis, $z^* \equiv 1/z_0$ must also be a real number. Multiplying and dividing eq. (3.11) by z_0 and using $z_0 z^* = 1$, the equation for $T(z)$ become

$$T(z) = z_0 K\left(\frac{z - z_0}{z_0 z - 1}\right) = w(z).$$

In order to obtain an order of magnitude for the product $z_0 K$, remember that $|z| = 1$ is mapped to $|w| = 1$, so that $|w| = |w\bar{w}|^{1/2} = 1$. With z_0 being real, $\bar{w} = z_0 \bar{K}\left(\frac{\bar{z} - z_0}{z_0 \bar{z} - 1}\right)$, and we have

$$|w| = |w\bar{w}|^{1/2} = 1 = z_0^2 K\bar{K}\left(\frac{z - z_0}{z_0 z - 1}\right)\left(\frac{\bar{z} - z_0}{z_0\bar{z} - 1}\right) \sim z_0^2 K\bar{K}$$

$$\Rightarrow |z_0^2 K\bar{K}| \equiv |z_0 K||z_0\bar{K}| = 1 = |z_0 K|^2 \equiv |k|^2 \Rightarrow |k| = 1.$$

Thus,

$$T(z) = k\left(\frac{z - z_0}{z_0 z - 1}\right), \qquad |k| = 1.$$

The two circles in the w-plane must be concentric. That is, the distance from C' and B' (and also D' and A') to the origin $w = 0$ must be equal so that the images ($-\delta$ and δ) of z_1 $(= z_0 + \rho)$ and $z_2(= z_0 - \rho)$ must be symmetric with respect to $w = 0$, respectively, at the points $w = \pm\delta$, which are the images of z_1 and z_2 (or C and B). With C' and B' being equally located from $w = 0$, though in opposite directions of $w = 0$, we see that

$$T(z_1) \equiv k\left(\frac{z_1 - z_0}{z_0 z_1 - 1}\right) = -T(z_2) \equiv -k\left(\frac{z_2 - z_0}{z_0 z_2 - 1}\right), \text{ or}$$

$$\frac{z_0 - \rho - \delta}{\delta(z_0 - \rho) - 1} = -\left[\frac{z_0 + \rho - \delta}{\delta(z_0 + \rho) - 1}\right], \text{ or}$$

$$z_0\delta^2 - (1 + z_0^2 - \rho^2)\delta + z_0 = 0.$$

This equation needs to be solved for δ to obtain the final form of the transformation from eccentric circles to concentric circles:

$$w(z) \equiv T(z) = k\left(\frac{z - \delta}{\delta z - 1}\right), \qquad |k| = 1,$$

where δ is the radius of the inner circle in the concentric-circle configuration in the w-plane.

3.4 Application of eccentric circle conformal mapping in engineering analysis

Let us apply the previous analysis to the problem of steady-state temperature distribution between eccentric circles with uniform surface temperatures of T_1 and T_2 for the inner and outer circles, respectively. Let us consider the specific location of the inner circle as in Figure 3.10. Note that the outer circle in the z-plane is $|z| = 1$, and that in the w-plane is $|w| = 1$.

For this case, $z_o = 1/4$, $\rho = 1/4$ so that

$$z_0\delta^2 - (1 + z_0^2 - \rho^2)\delta + z_0 = 0$$

becomes

$$\delta^2 - 4\delta + 1 = 0 \Rightarrow \delta = \frac{+4 \pm 2\sqrt{3}}{2}.$$

With $|w| = 1$, we choose $\delta = 2 - \sqrt{3}$. That is, B' is located at $+2 - \sqrt{3}$, while C' is at $-(2 - \sqrt{3})$.

Therefore, the transformation is

$$w(z) = \frac{z - 2 + \sqrt{3}}{(2 - \sqrt{3})z - 1}.$$

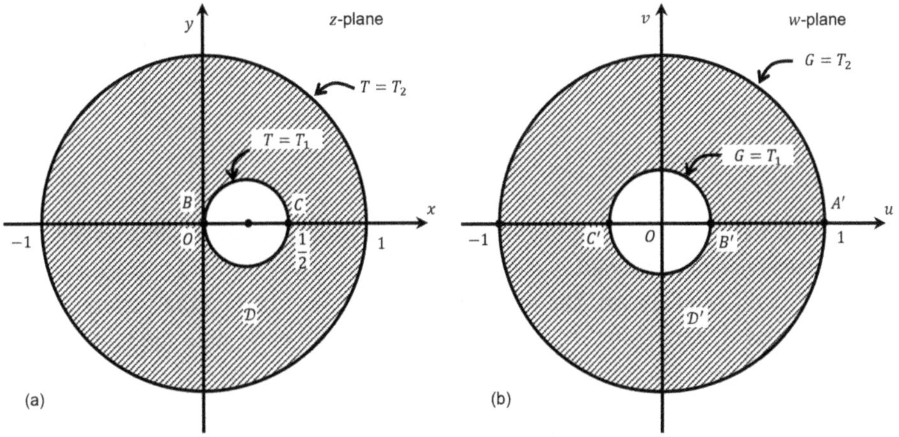

Figure 3.10: Transformation of an eccentric circle.

The governing equation for temperature in the w-plane is

$$\frac{\partial^2 G}{\partial r^2} + \frac{1}{r}\frac{\partial G}{\partial r} + \frac{1}{r^2}\frac{\partial^2 G}{\partial \theta^2} = 0.$$

However, an examination of the boundary conditions on G on the surfaces $r = \delta$ and $r = 1$ shows a uniform distribution of $G = 0$ and 1, respectively. Therefore, there is no temperature variation in the azimuth in the w-plane, and temperature varies only in the radial direction. The governing equation becomes

$$\frac{d^2 G}{dr^2} + \frac{1}{r}\frac{dG}{dr} = 0, \qquad G(\delta) = T_1, \qquad G(1) = T_2, \qquad \delta = 2 - \sqrt{3}.$$

This solves as

$$G(r) = T_2 - \left(\frac{T_2 - T_1}{\ln(2 - \sqrt{3})}\right) \ln r.$$

The solution in the z-plane is obtained by writing $r = |w|$. So we set $z = x + iy$ in the expression for the transformation (w), to obtain, using $r = |z|$,

$$T(x,y) = T_2 - \left[\frac{T_2 - T_1}{\ln(2-\sqrt{3})}\right] \ln\left|\frac{x + iy - 2 + \sqrt{3}}{(2-\sqrt{3})(x+iy) - 1}\right|.$$

The argument of the log term, written as $\ln|a_o|$, has the value

$$a_o = \frac{ab + y^2}{(2-\sqrt{3})[b^2 + y^2]} - \frac{y(a-b)}{(2-\sqrt{3})[b^2+y^2]} i \equiv \phi + i\psi$$

and

$$|a_o| = |\phi^2 + \psi^2|^{1/2} = \left\{\left[\frac{ab + y^2}{(2-\sqrt{3})[b^2+y^2]}\right]^2 + \left[\frac{y(a-b)}{(2-\sqrt{3})[b^2+y^2]}\right]^2\right\}^{1/2}.$$

where $a = x - 2 + \sqrt{3}$, $b = x - 1/(2-\sqrt{3})$.

Example 3.5 Consider the physical problem described in Figure 3.11(a), which pertains to steady-state temperature distribution in the hatched region $\mathcal{D}$, which is external to the two circles shown. The temperature T, is prescribed on the surfaces of the two circles, as shown. Using the image in Figure 3.11(b), determine the temperature distribution in the region $\mathcal{D}$ of the physical domain. Use $x_1 = 3$, $x_2 = 2$.

Solution: First of all, it should be appreciated that because of the geometry of the physical problem, the technique of the separation of variables is not directly applicable. However, with conformal mapping, the problem is particularly simple, provided we avail ourselves of the development of the mapping and just adopt the one given in Appendix H. Thus,

$$w = \frac{z - a}{az - 1}, \qquad a = \frac{1 + x_1 x_2 + \sqrt{(x_1^2 - 1)(x_2^2 - 1)}}{(x_1 + x_2)},$$

$$R_o = \frac{x_1 x_2 - 1 - \sqrt{(x_1^2 - 1)(x_2^2 - 1)}}{(x_1 - x_2)} \qquad (x_2 < a < x_1,\ 0 < R_o < 1 \text{ when } 1 < x_2 < x_1).$$

R_o is the radius of the inner circle in the w-plane.

The G in Figure 3.11(b) is the image of T in the w-plane. We can calculate a as $a = (7 + 2\sqrt{6})/5$. In the w-plane, $\nabla^2 G = 0$ can be written as

$$\nabla^2 G = \frac{\partial^2 G}{\partial \rho^2} + \frac{1}{\rho}\frac{\partial G}{\partial \rho} + \frac{1}{\rho^2}\frac{\partial^2 G}{\partial \phi^2} = 0.$$

Given the boundary conditions in the w-plane, we see that the solution G does not depend on the azimuthal coordinate ϕ. Therefore $\partial/\partial\phi$ can be set to zero in the above equation. This leads to the simple one-dimensional ODE

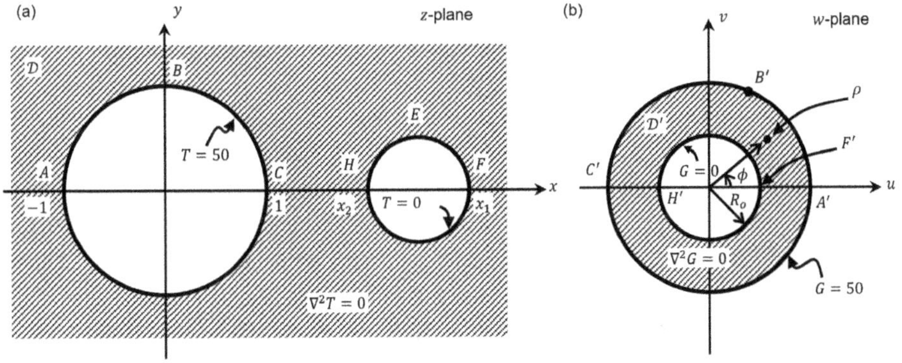

Figure 3.11: Sketch for Example 3.5.

$$\frac{\partial^2 G}{\partial\rho^2}+\frac{1}{\rho}\frac{\partial G}{\partial\rho}=0,$$

which has the general solution

$$G(\rho)=A+B\ln\rho.$$

Applying the boundary conditions $G(R_0)=0$ and $G(1)\equiv 50$, we have

$$G(\rho)=50\left(1-\frac{\ln\rho}{\ln R_0}\right).$$

The next step is to express ρ in the w-plane in terms of x and y in the z-plane so that

$$T(x,y)=G(\rho(x,y))=\frac{50}{\ln R_0}\left[\ln R_0-\frac{1}{2}\ln\frac{(x-a)^2+y^2}{(ax-1)^2+a^2y^2}\right],$$

where we have used

$$|\rho|=|w[u(x,y),v(x,y)]|=\left|\frac{z-a}{az-1}\right|=\left|\frac{(x-a)+iy}{(ax-1)+iay}\right|.$$

3.5 Transformation of boundary conditions on harmonic functions

We have seen that if $w \equiv u + iv = f(z)$ is a single-valued analytic function, with its domain in the z-plane ($z = x + iy$), and $\phi(x,y)$ is a harmonic function in that domain, then $\Phi(u,v)$, which is the image of $\phi(x,y)$ when we change variables from x and y in the z-plane to $u = u(x,y)$ and $v = v(x,y)$ in the w-plane, which is in the range of $f(z)$, will also be harmonic. In fact, $\phi(x,y) = \Phi(u(x,y), v(x,y))$. If $P'(u,v)$ is the image (in the w-plane) of a point $P(x,y)$ in the z-plane, then the Dirichlet condition $\phi(P)$ in the z-plane becomes $\Phi(P')$ in the w-plane. Thus, if $\phi(P) = 5$ units, then $\Phi(P') = 5$ units.

If the Dirichlet boundary conditions are specified as a function of x and y at a point (x,y) in the z-plane, say $\phi = 5x + 2y$, we will need to find the image of $5x + 2y$ and give Φ that value at the point $(u(x,y), v(x,y))$ in the w-plane. Obviously, this requires that we explicitly know $u = u(x,y)$ and $v = v(x,y)$, which we always do. That is, the condition in the w-plane will be $5(x(u,v)) + 2(y(u,v))$. The conformal nature of the transformation means that both $(\partial\phi/\partial n)_P$ and $(\partial\Phi/\partial n)_{P'}$ will be normal to the boundary in their respective planes.

As an example, let the boundary section $A - B$ in the z-plane be represented by the equation $xy = 3$ and the value of a harmonic function T on that boundary be represented by $T = 10e^{-x}$. Let the z-plane be mapped to the w-plane via $w(z) = z^2 \equiv u + iv$, where $u = x^2 - y^2$, $v = 2xy$. We see that the curve $xy = 3$ in z is mapped to

$$u = x^2 - \left(\frac{3}{x}\right)^2 = x^2 - \frac{9}{x^2} \quad \text{and} \quad v = 2x \times \frac{3}{x} = 6.$$

That is, the side $xy = 3$ in z is mapped onto the horizontal line $v = 6$ on which $u = x^2 - 9/x^2$.

From this last equation, we can write

$$x^4 - ux^2 - 9 = 0, \quad \text{or} \quad x^2 = \frac{u \pm \sqrt{u^2 + 36}}{2} \quad \text{and} \quad x = \sqrt{\left(\frac{u + \sqrt{u^2 + 36}}{2}\right)},$$

where we have chosen the positive sign in the last equation so as to keep x real.

We now need to map $T = 10e^{-x}$ on the boundary section $xy = 3$ to the mapped boundary section $v = 6$, where $x = \sqrt{[(u + \sqrt{u^2 + 36})/2]}$. The result, using the symbol G as the temperature in the w-plane, is

$$G(u,6) = 10 \exp\left[-\sqrt{\left(\frac{u + \sqrt{u^2 + 36}}{2}\right)}\right].$$

Similarly, the boundary $x = 0$ in the z-plane, which would be mapped to $u = x^2 - y^2 = -y^2$ and $v = 2xy = 0$, has the condition $T = 100/(3 + y)$ specified. This boundary condition will appear on the line $v = 0$ in the w-plane, with a value of

$$G(u,0) = \frac{100}{3+\sqrt{-u}}.$$

Regarding the Neumann boundary conditions, $\partial\phi/\partial n_z$ at a point P in the z-plane will be mapped to $\partial\Phi/\partial n_w$ at the image point P' in the w-plane, where n_z and n_w are the direction normals to the respective boundary points in the two planes: It will be necessary to determine how $\partial\phi/\partial n_z$ transforms under $f(z)$. Consider Figure 3.12 for this purpose. Let z be a point on the boundary surface $A-B-C$, and $F-E-B$, a line normal to the boundary surface at z. Let A', B', C', E', and F' be the respective images of A, B, C, E, and F under $f(z)$. The symbols $\mathcal{D}$ and $\mathcal{D}'$ in the figure denote domains. The line $F-E-B$ in z is assumed mapped into the curve $F'-E'-B'$ in w, while the boundary point z is mapped into w, and the respective normals to $A-B-C$ $(A'-B'-C')$ at z and w are denoted as n_z and n_w.

Let $\Delta z = z - z_0$ and $\Delta w = w - w_0$. We can approximate the directional derivative of ϕ at the surface point z as

$$\left.\frac{\partial\phi}{\partial n_z}\right|_z = \lim_{\Delta z\to 0}\frac{\phi_z - \phi_{z_0}}{|\Delta z|}.$$

From the preservation of Dirichlet conditions, we have

$$\phi|_z = \Phi|_w, \qquad \phi|_{z_0} = \Phi|_{w_0}.$$

We also have $\lim_{\Delta z\to 0}\Delta w/\Delta z = f'(z)$. Therefore,

$$\left.\frac{\partial\phi}{\partial n_z}\right|_z = \lim_{\Delta z\to 0}\frac{\Phi_w - \Phi_{w_0}}{|\Delta w|}\cdot\frac{|\Delta w|}{|\Delta z|} = \frac{\partial\Phi}{\partial n_w}|f'(z)|,$$

where we have assumed that Δw and Δz approach zero at a comparable rate. This is the case if $w = f(z)$ is continuous, which is the case because we require that $f(z)$ be differentiable for a conformal transformation. Thus, outside of the critical points of $f(z)$, we see that Neumann conditions on a harmonic function ϕ transform as

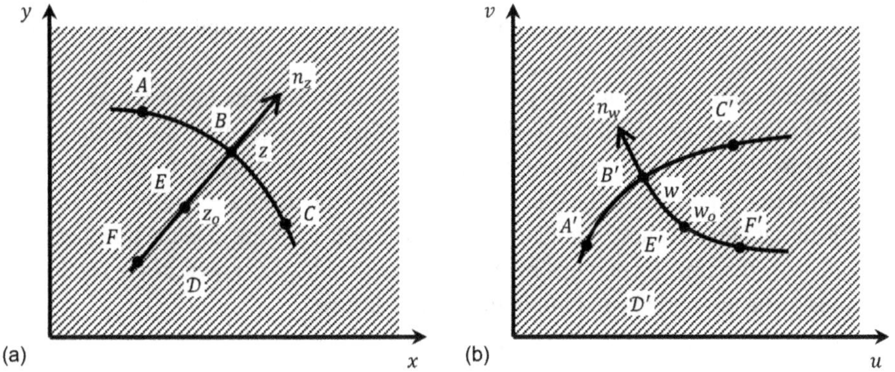

Figure 3.12: Transformation of normals.

$$\frac{\partial \Phi}{\partial n_w} = \frac{1}{|f'(z)|} \frac{\partial \phi}{\partial n_z}.$$

The special case $\partial\phi/\partial n_z = 0$ in the z-plane, which, for the case of heat conduction, implies adiabatic conditions, also transforms as adiabatic conditions in the w-plane or

$$\frac{\partial \Phi}{\partial n_w} = 0.$$

3.6 Multiple transformations

It is often necessary to employ a series of conformal transformations in order to solve some physical problems of interest to the engineer. To illustrate this, consider the physical problem in Figure 3.13(a) involving the solution of the equation

$$\nabla^2 \phi \equiv \frac{\partial^2 \phi}{\partial x^2} + \frac{\partial^2 \phi}{\partial y^2} = 0$$

in the three-sided, semi-infinite rectangular strip or the cavity $A-B-C-E$, assumed located in the z-plane. The Dirichlet conditions $\phi = 0$, 100, and 0 are specified on ϕ, respectively, at the boundaries $A-B$, $B-C$, and $C-E$.

Besides temperature, ϕ could also represent the voltage (electric potential). In this case $\phi = 0$ indicates grounding on $A-B$ and $C-E$, while $B-C$ is maintained at 100 volts. It is desired to calculate ϕ in the domain $\mathcal{D}$. Note that the physical problem is considered to be in the UHP of the z-plane.

The first transformation uses $w(z) \equiv f(z) = -\cos \pi z$, the result of which is shown in Figure 3.13(b). Note that the domain boundary in the z-plane is made up of straight lines, and that the LFT is not able to accomplish the required transformations. Hence the choice of $w(z) = -\cos \pi z$, which has been selected from the table of conformal transformations in the appendix. You can confirm the appropriateness of this transformation by reconciling the coordinates of the points in the z- and w-planes, vis-à-vis the mapping $w(z)$.

Toward solving for $\Phi(u, v)$ in the w-plane, consider a single (generic) point P, coordinate w, in domain $\mathcal{D}'$ of the w-plane (Figure 3.13(c)). Even then, the configuration represented by the w-plane is not convenient for the solution of Φ. Therefore, we transform the w-plane into the ζ-plane, where $\zeta \equiv \xi + i\eta$, using $\zeta = \log[(w-1)/(w+1)]$. The resulting plane is shown in Figure 3.13(d). The $\Phi(u, v)$ in the w-plane now becomes $G(\xi, \eta)$, with the transformations of the boundary values as shown in Figure 3.13(d). Obviously, the governing equation in the ζ-plane is now

$$\frac{\partial^2 G}{\partial \xi^2} + \frac{\partial^2 G}{\partial \eta^2} = 0.$$

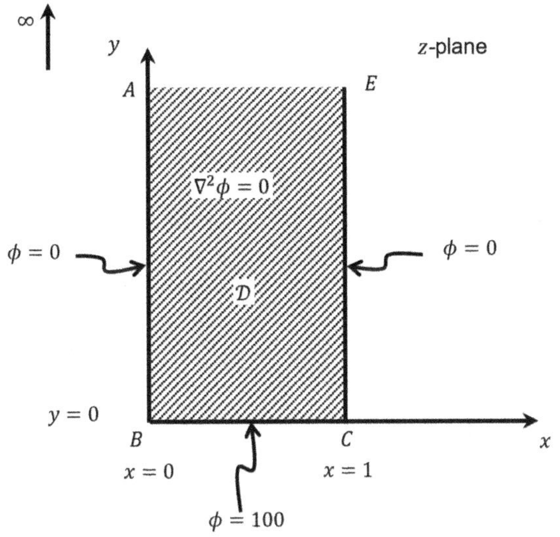

Figure 3.13(a): The physical problem for illustrating multiple conformal transformations.

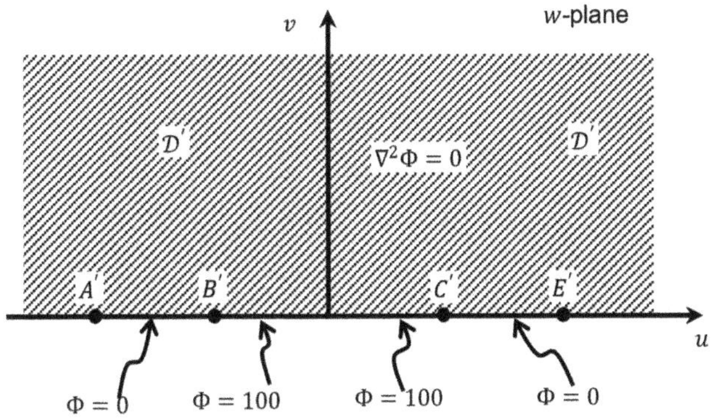

Figure 3.13(b): The image (w-plane) of the problem in Figure 3.13(a).

The boundary conditions in the ζ-plane clearly show that G is independent of the ξ-coordinate direction so that the governing equation simplifies to

$$\frac{d^2G}{d\eta^2} = 0, \qquad G(0) = 0, \qquad G(\pi) = 100,$$

with the solution $G = C_1\eta + C_2$, or $G = (100/\pi)\eta$ when the boundary conditions are applied. Obviously, the solution in the physical (x, y) plane can be obtained as

$$\phi(x, y) = G(\xi(u(x, y)),\ \eta(v(x, y))).$$

To obtain the solution in the w-plane, which is the first step in determining $\phi(x,y)$, we need to express η in terms of u and v. To this end, we can write $w-1=r_1e^{i\theta_1}$, $w+1=r_2e^{i\theta_2}$ in the w-plane so that

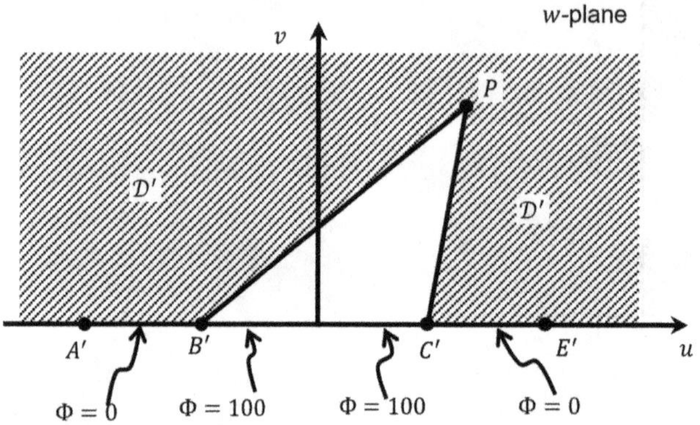

Figure 3.13(c): A generic point P, with coordinate w in the w-plane.

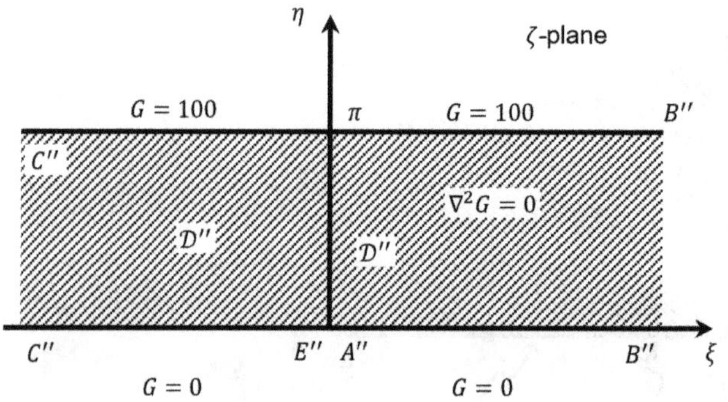

Figure 3.13(d): The ζ-plane into which w is mapped using $f(z) = \log\left(\frac{w-1}{w+1}\right)$.

$$\zeta \equiv \xi + i\eta = \log\left(\frac{w-1}{w+1}\right) = \log\left(\frac{r_1}{r_2}e^{i(\theta_1-\theta_2)}\right) = \ln\frac{r_1}{r_2} + i(\theta_1-\theta_2), \text{ or } \eta = \theta_1 - \theta_2.$$

Thus, in the w-plane, the solution is

$$\Phi(u,v) = \frac{100}{\pi}(\theta_1 - \theta_2),$$

where θ_1, θ_2 can be written respectively as $\tan^{-1}[(1-u)/v]$, $\tan^{-1}[(-1-u)/v]$, yielding

$$\Phi(u,v) = \frac{100}{\pi}\left[\tan^{-1}\left(\frac{1-u}{v}\right) - \tan^{-1}\left(\frac{-1-u}{v}\right)\right],$$

where $-\pi/2 < \tan^{-1}[(1-u)/v] < \pi/2$, $-\pi/2 < \tan^{-1}[(-1-u)/v] < \pi/2$. To obtain $\phi(x,y)$ from $\Phi(u,v)$, we employ the transformation

$$w = f(z) = -\cos \pi z = -\cos \pi(x+iy) = -\cos \pi x \cosh \pi y + i \sin \pi x \sinh \pi y$$

$$\Rightarrow u = -\cos \pi x \cosh \pi y, \qquad v = \sin \pi x \sinh \pi y.$$

Therefore,

$$\phi(x,y) = \frac{100}{\pi}\left[\tan^{-1}\left(\frac{\cos \pi x \cosh \pi y + 1}{\sin \pi x \sinh \pi y}\right) - \tan^{-1}\left(\frac{\cos \pi x \cosh \pi y - 1}{\sin \pi x \sinh \pi y}\right)\right].$$

3.7 Finding a harmonic function in the UHP

The solution for the harmonic function $\Phi(u,0)$ at a point P in the UHP of the w-plane in the previous example (Figure 3.13(c)) can be generalized for partitions of the u-coordinate direction; that is, $u \in [a,b]$ is divided into the N cells $x_0, x_1, \ldots, x_N$, as shown in Figure 3.14. The conditions on $\Phi(u,0)$ are specified as Φ_i for cell i, so we have piecewise-discontinuous values of Φ in the u-coordinate direction.

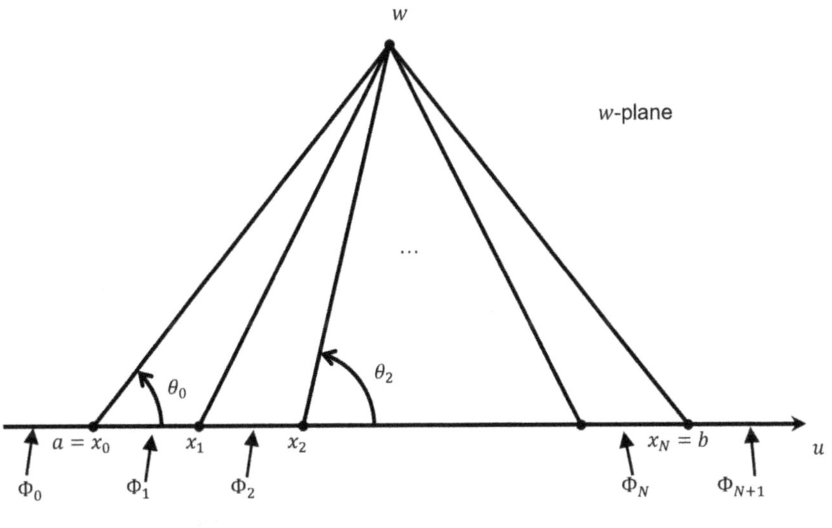

Figure 3.14: Generalization of the solution of a harmonic function.

The solution in the ζ-plane is $G = (100/\pi)\eta$. For the three-segment problem in Figure 3.15, we would find the solution in the w-plane by expressing η in terms of u and v, or r and θ if w is written as $re^{i\theta}$ at the points $w = u_1$ and $w = u_2$, or $w - u_1 = r_1 e^{i\theta_1}$, $w + u_2 = r_2 e^{i\theta_2}$. The variable η in terms of the w variables will be given by

$$\zeta = \xi + i\eta = \log\left(\frac{w-u_1}{w+u_2}\right) = \log\left(\frac{r_1}{r_2}e^{i(\theta_1-\theta_2)}\right) = \ln\frac{r_1}{r_2} + i(\theta_1 - \theta_2)$$

$$\Rightarrow \xi = \ln\frac{r_1}{r_2} \quad \text{and} \quad \eta = \theta_1 - \theta_2.$$

Therefore, the solution in the w-plane is $(100/\pi)(\theta_1 - \theta_2)$, as shown earlier. We note in passing that if we have the problem shown in Figure 3.15,

$$\nabla^2\Phi = 0, \qquad -\infty < u < \infty,\ v > 0$$

$$\Phi(u,0) = \begin{cases} \Phi_1 & u < u_1, \quad v = 0 \\ \Phi_2 & u_1 < u < u_2, \quad v = 0\ . \\ \Phi_3 & u > u_2, \quad v = 0 \end{cases} \tag{3.12a}$$

The solution is

$$\Phi(P) = \Phi_3 + \frac{1}{\pi}[(\Phi_1 - \Phi_2)\theta_1 + (\Phi_2 - \Phi_3)\theta_2].$$

To see this, apply the solution $\Phi(u,v) = (100/\pi)(\theta_1 - \theta_2)$ for the problem in Figure 3.13 (d) and use the linearity of the Laplace equation to superpose solutions (i.e., e.g., $\Phi_i + 0 = \Phi_i$). We will have, for the interior cells in Figure 3.14:

$$\begin{aligned}\Phi(u,v) &= \frac{\Phi_1}{\pi}(\theta_1 - \theta_2) + \frac{\Phi_2}{\pi}(\theta_2 - \theta_1) + \ldots + \frac{\Phi_n}{\pi}(\theta_n - \theta_{n-1}) \\ &= \frac{1}{\pi}(\Phi_1 - \Phi_2)\theta_1 + \frac{1}{\pi}(\Phi_2 - \Phi_3)\theta_2 + \ldots + \frac{1}{\pi}(\Phi_{n-1} - \Phi_n)\theta_n.\end{aligned}$$

Now $\theta_1 = \arg(w - u_1)$ and $\theta_2 = \arg(w - u_2)$ so that in Figure 3.15 we will have

$$\Phi(u,v) = \Phi_3 + \frac{1}{\pi}(\Phi_1 - \Phi_2)\arg(w - u_1) + \frac{1}{\pi}(\Phi_2 - \Phi_3)\arg(w - u_2). \tag{3.12}$$

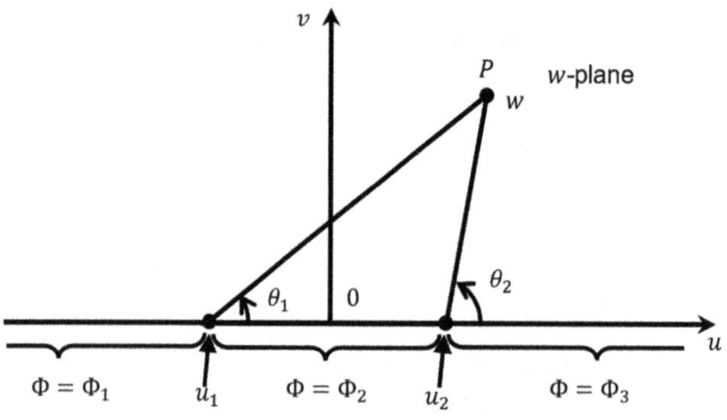

Figure 3.15: A three-segment problem.

Noting that $w = u_2 + r_2e^{i\theta_2}$ and $w = u_1 + r_1e^{i\theta_1}$ so that $w - u_2 = r_2e^{i\theta_2}$, $w - u_1 = r_1e^{i\theta_1}$, and $\log(w - u_2) = \log r_2 + i\theta_2$, $\log(w - u_1) = \log r_1 + i\theta_1$. We can then see that the RHS of eq. (3.12) is the imaginary part of the complex function

$$s = i\Phi_3 + \frac{1}{\pi}(\Phi_1 - \Phi_2)\log(w - u_1) + \frac{1}{\pi}(\Phi_2 - \Phi_3)\log(w - u_2).$$

This function, s, is analytic when $w \neq u_1$, u_2 so that its real and imaginary parts are also harmonic for $w \neq u_1$, u_2. Therefore, Φ must be harmonic for $w \neq u_1$, u_2. The uniqueness of the solutions of the Dirichlet boundary value problems for harmonic functions leads us to the conclusion that the problem described in eq. (3.12a) is

$$\Phi(u, v) = \Phi_3 + \frac{1}{\pi}(\Phi_1 - \Phi_2)\arg(w - u_1) + \frac{1}{\pi}(\Phi_2 - \Phi_3)\arg(w - u_2). \tag{3.12b}$$

Obviously, this can be extended for cases where Φ is required to assume more than three different constant values on the u-axis. That is, if a fourth value, Φ_4, is added, and the Φ values are separated discontinuously on the u-axis by the points u_1, u_2, u_3, then we will have the solution as

$$\Phi(u, v) = \Phi_4 + \frac{1}{\pi}(\Phi_1 - \Phi_2)\arg(w - u_1) + \frac{1}{\pi}(\Phi_2 - \Phi_3)\arg(w - u_2) + \frac{1}{\pi}(\Phi_3 - \Phi_4)\arg(w - u_3)$$

and so on.

The determination of the angles in eq. (3.12b), that is, $\arg(w)$, in terms of the inverse of the tangent function $\tan^{-1}\theta$, must be carefully carried out. This is because $\tan\theta$ is defined only for $-\frac{\pi}{2} < \tan\theta < \pi/2$ and by periodicity elsewhere. The correct values of $\tan^{-1}\theta$ can be obtained by using (a) $\tan^{-1}\theta$ as is when $\theta > 0$, (b) $\tan^{-1}\theta = \pi/2$ when $\theta = \pm\infty$, and (c) $\tan^{-1}\theta \leftarrow \pi\tan^{-1}\theta$ when $\theta < 0$.

Notice that we have developed the preceding solutions in the w-plane, without a reference to a pre-image, or, in our convention, the z-plane. The problem in the w-plane as presented is as much a harmonic problem as that in a z-plane, and no connections need to be made. This means that the given solution in the w-plane could represent a physical problem in its own right, if the conditions of the physical problem are as stated in eq. (3.12a). However, the problem solved in the w-plane could also have come from a prior transformation into this plane. Let us give an example to illustrate the latter.

Example 3.6 Solve the equation $\nabla^2\phi = 0$ in domain $\mathcal{D}$ of the physical problem sketched in Figure 3.16a.

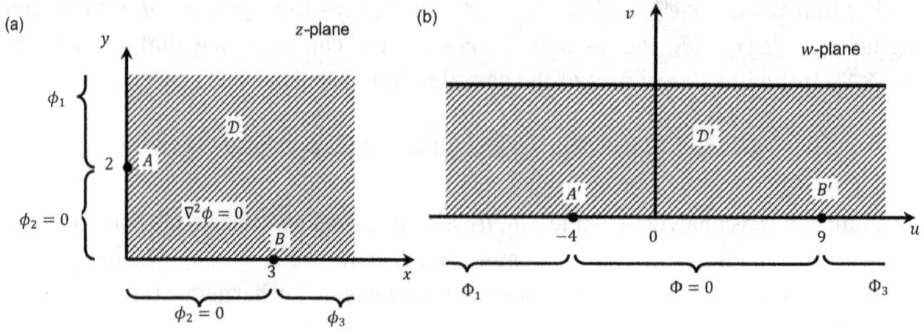

Figure 3.16: The sketch for Example 3.6: (a) Physical problem, (b) transform domain.

Solution: Use $w(z) = z^2$ to map the problem to the w-plane. Note that $w(2i) = -4$, $w(3) = 9$, $w(0) = 0$. Hence the image in Figure 3.16(b). By inspection, the solution in the w-plane is

$$\Phi(u, v) = 20 + \frac{60}{\pi}\arg(w + 4) - \frac{20}{\pi}\arg(w - 9).$$

We now need to take the solution back to the z-plane, using the given transformation: $z = x + iy$, $w \equiv u + iv = z^2$; $u = x^2 - v^2$, $v = 2xy$ so that $w + 4 = x^2 - y^2 + 4 + i2xy$ and

$$\arg(w + 4) = \tan^{-1}\left(\frac{2xy}{x^2 - y^2 + 4}\right), \quad \arg(w - 9) = \tan^{-1}\left(\frac{2xy}{x^2 - y^2 - 9}\right).$$

The solution in the z-plane is therefore

$$\phi(x, y) = 20 + \frac{60}{\pi}\tan^{-1}\left(\frac{2xy}{x^2 - y^2 + 4}\right) - \frac{20}{\pi}\tan^{-1}\left(\frac{2xy}{x^2 - y^2 - 9}\right)$$

for (x, y) in the first quadrant of the z-plane.

3.8 More examples of conformal transformation

Additional examples of useful conformal transformations are presented in this section.

3.8.1 The transformation $w \equiv f(z) = z^2$

With this transformation,

$$w = u + iv = z^2 = (x + iy)^2 = x^2 - y^2 + 2ixy, \quad u = x^2 - y^2, \quad v = 2xy.$$

Thus, lines parallel to the y-axis, that is, $x = c_1$ are mapped into curves (in the w-plane) whose parametric equations are

$$u = c_1^2 - y^2, \qquad v = 2c_1 y.$$

We can eliminate y to get $u = c_1^2 - v^2/4c_1^2$, which is a family of parabolas with the origin of w-plane as the focus and the line $v = 0$ as the axis – all opening to the left, as shown in Figure 3.17. Furthermore, lines parallel to the x-axis or $y = c_2$ are mapped into

$$u = x^2 - c_2^2, \qquad v = 2c_2 x, \qquad u = \frac{v^2}{4c_2^2} - c_2^2,$$

which is a family of parabolas with the origin of the w-plane as focus, also with $v = 0$ as the axis, but opening to the right. The equations $v^2 = 4c_1^2(c_1^2 - u)$ and $v^2 = 4c_2^2(u + c_2^2)$ can be used to construct the graphs in the w-plane that correspond to the lines $x = c_1$ and $y = c_2$ in the z-plane (Figure 3.17).

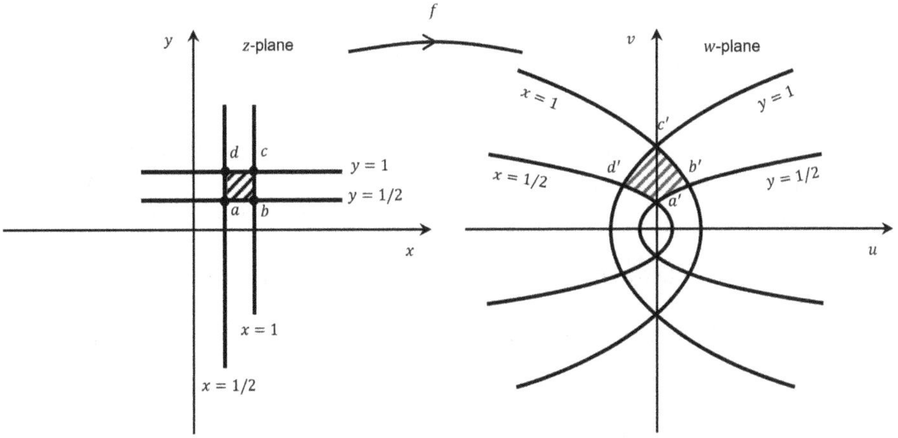

Figure 3.17: The transformation $w(z) = z^2$.

3.8.1.1 Further remarks on $f(z) = z^2$

a) These are relations for two sets of orthogonal hyperbolas. The vertical u lines in the w-plane become hyperbolas with 45° lines through the origin as asymptotes, whereas the horizontal v lines become hyperbolas asymptotic to the coordinate axes in the z-plane.

b) The whole of w-plane is mapped into only the upper half of the z-plane. The transformation $f(z) = z^2$ is not one-to-one since the inverse transformation $z = \pm w^{1/2}$ has two values of z for every value of w or from $z = re^{iv}$, $w = Re^{i\theta}$, $z^2 = r^2 e^{2iv} \Rightarrow R = r^2, \theta = 2v$ so that the magnitude of an angle in the z-plane is doubled when transformed to the w-plane. There is also a squaring of the magnitude $r = |x + iy|$, or $|u + iv| = |x + iy|^2$. Thus, all the four quadrants, 2π, in the z-plane, correspond only to two quadrants, π, in the w-plane. Note the singularity at $w = 0$ in the inverse transformation or $dz/dw = 1/(2w^{1/2}) \to \infty$ for $w = 0$. Thus, a singularity occurs at the origin of the z-plane.

Example 3.7 $w = f(z) = z^2$ is everywhere analytic and angle measurements are in general preserved. However, $f'(z) = 2z$ has a simple zero at $z = 0 \Rightarrow$ angles with vertex at the origin in the z-plane are not preserved but doubled in the w-plane. This knowledge is used in airfoil design (see Chapter 4).

3.8.2 The inverse function $z \equiv f^{-1}(w) \equiv \pm w^{1/2}$

The inverse function $z \equiv f^{-1}(w) \equiv \pm w^{1/2}$ maps $u = k_1$ ($k_1 \neq 0$) into rectangular hyperbolas $x^2 - y^2 = k_1$ and $v = k_2$ ($k_2 \neq 0$) into the rectangular hyperbolas $xy = (1/2)k_2$. Figure 3.18 shows the inverse transformation for $u = 1/2, 1$ and $v = 1/2, 1$.

Example 3.8 Find the curve in the w-plane into which the line $y = 2x + 1$ is transformed by $f(z) = z^2$.

Solution: While transformations of the type discussed in this chapter are point-to-point, it is also possible to transform lines with a function such as z^2. One way to do this is shown here. With $u = x^2 - y^2$, $v = 2xy$, we will substitute for y and eliminate x to obtain

$$u = x^2 - (2x + 1)^2 = -3x^2 - 4x - 1, \qquad v = 2x(2x + 1) = 4x^2 + 2x.$$

Regard this set of two equations as two simultaneous equations for x and x^2, which can be solved as

$$x = \frac{4u + 3v + 4}{-10}, \qquad x^2 = \frac{u + 2v + 1}{5}.$$

Squaring the first equation and setting it equal to the second, we have

$$\frac{u + 2v + 1}{5} = \left(\frac{4u + 3v + 4}{-10}\right)^2$$

or

$$16u^2 + 24uv + 8v^2 + 12u - 16v = 4,$$

which is an equation of a parabola; the curve in the w-plane to which the straight line $y = 2x + 1$ in the z-plane is transformed by $w(z) = z^2$.

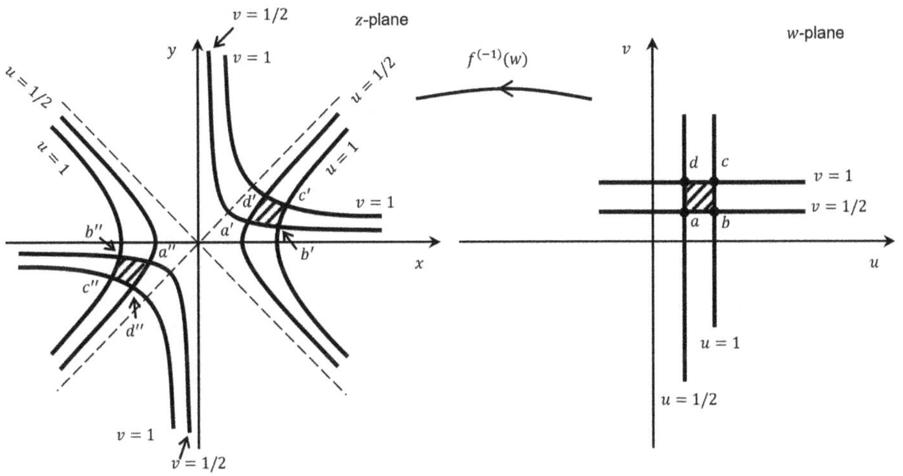

Figure 3.18: Mapping (of $u = k_1$, $v = k_2$ for $k_1, k_2 = 1/2, 1$) from w-plane to hyperbolas in z-plane. Note that the "inverse" mapping is not one-to-one and onto.

3.8.3 The transformation $w \equiv f(z) = \cos z$

Here,

$$w \equiv u + iv = \cos(x + iy) = \cos x \cos iy - \sin x \sin iy = \cos x \cosh y - i \sin x \sinh y$$

$$\Rightarrow u = \cos x \cosh y, \; v = -\sinh y \sin x \tag{3.14}$$

and

$$\left(\frac{u}{\cosh y}\right)^2 - \left(\frac{v}{\sinh y}\right)^2 = \cos^2 x + \sin^2 x = 1.$$

For a given value of x or y, we can eliminate y or x and find the curve in u, v. That is, we express y in terms of u, v in eq. (3.14), and eliminate y to obtain $u = u(v)$. Using this procedure, lines of $y =$ constant become ellipses:

$$\frac{u^2}{\cosh^2 y} + \frac{v^2}{\sinh^2 y} = 1,$$

while lines of $x =$ constant become hyperbolas:

$$\frac{u^2}{\cos^2 x} - \frac{v^2}{\sin^2 x} = 1.$$

Constant x and y lines are transformed as shown in Figure 3.19. Both of these lines have foci at $w = \pm 1$ (or $u = \pm 1$) (corresponding to $z = 0, \pi$ or $x = 0, \pi$), which are singular points that will need to be excluded in a conformal transformation since these are critical points of the transformation as $f'(z) = dw/dz = -\sin z = 0$.

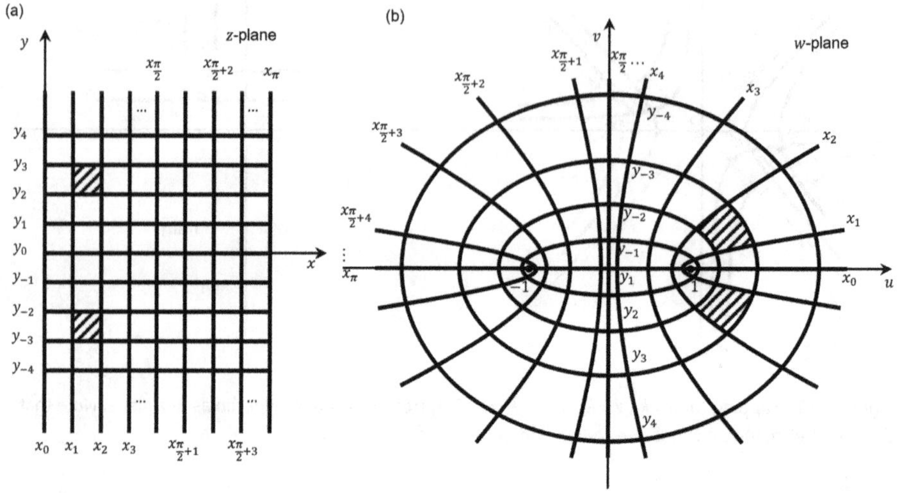

Figure 3.19: The images of constant x and y lines from z in w: (a) z-plane; (b) w-plane.

It should be noted that the strip from x_0 to x_π is mapped into the entire w-plane. The half strip $y \geq 0$ is mapped into the lower half of the w-plane and the complete strip into the whole of the w-plane. As pointed out by Robertson [2], the pattern in the w-plane could be considered to represent the circulating flow about an ellipse of length greater than 2 on a flat plate of length 2, if the w-plane were considered to be a complex potential

$$w = f(z) = \phi(x,y) + i\psi(x,y)$$

with orthogonal lines on the plane representing those of constant ϕ and ψ. The y lines will represent streamlines. Here, ϕ is the scalar velocity potential while ψ is the stream function (see Chapter 4).

3.8.4 The transformation $w = f(z) = \ln z$ or $z = e^w$

This mapping is similar to the inverse of $w = \cos z$, except that the pattern in the z-plane is circular rather than elliptical. With $w = u + iv \equiv Re^{i\theta}$ and $z = x + iy \equiv re^{i\vartheta}$, we see that $u = \ln r$ or $r = e^u$ or $r^2 = e^{2u} = x^2 + y^2$, and $v = \vartheta$ or $y/x = \tan\vartheta = \tan v$. Between these equations, we could prescribe some values for x or y to remove y or x and obtain an equation for v in terms of u or u in terms of v, respectively. With this transformation, vertical u lines ($u^* \equiv u_1, u_2, \ldots$) in the w-plane become circles in the z-plane, whereas the horizontal v lines ($v^* \equiv v_1, v_2, \ldots$) become radial lines in the z-plane (Figure 3.20(a) and (b)). The configuration in Figure 3.20(b) could represent potential fluid flow where the radial lines correspond to the streamlines in a source flow and the azimuthal curves, the equipotential lines. Chapter 4 focuses on the use of the complex variable theory in fluid mechanics and goes into more details.

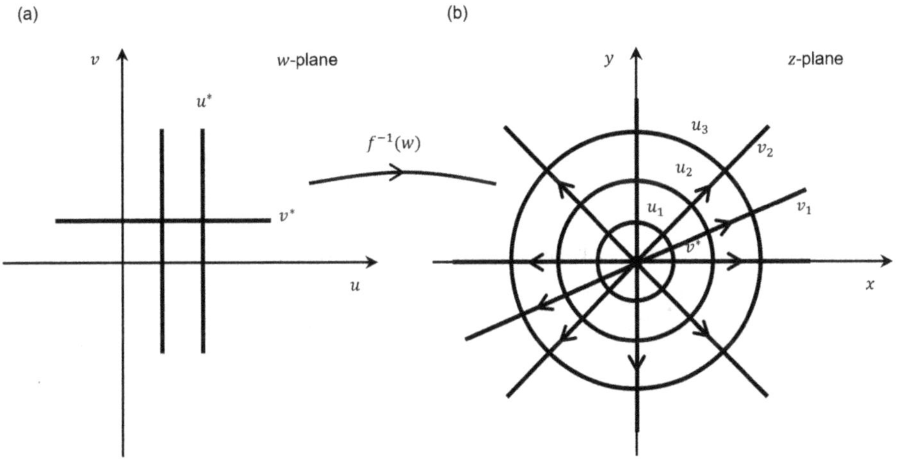

Figure 3.20: Inverse transformation of $w = \ln z$.

Note that the region in the w-plane from $v = 0$ to 2π is mapped into the entire z-plane. Singularity in $w = \ln r$ occurs at the origin $(w'(z) = 1/z)$. So, we will need to stay away from the origin with this transformation.

Finally, this transformation has application in turbomachinery, where a cascade of blades in a circular arrangement could be transformed to a rectangular cascade (Figure 3.21). Although this is essentially a geometric change, it may be one step in the analysis of the inward flow through a guide-vane system such as that which occurs at the inlet to a reaction turbine. An additional transformation or two may be necessary in Figure 3.21 to obtain a domain in which $\nabla^2 G = 0$ can be easily solved to model a blade cascade.

3.9 The Schwarz-Christoffel transformation

To motivate this transformation, consider $f(\zeta) = z(\zeta) = \zeta^{\alpha_1/\pi}$ (α_1 in degrees), or, with translation, $(z - z_1) = (\zeta - \xi_1)^{\alpha_1/\pi}$, which can be written as $z' = (\zeta')^{\alpha_1/\pi}$ with a change of variables. Here, $z = x + iy$ and $\zeta = \xi + i\eta$. If $\zeta' = a' e^{i\pi}$, then $z' = a' e^{i\pi \times \alpha_1/\pi} = a' e^{i\alpha_1}$. This means that an angle of π or 180° in the transformed domain (pre-image) here defined as the ζ-plane, will be transformed to α degrees in the image, or the "physical" domain, which is here taken as the z-plane. That is, $f(\zeta)$ will take a segment of the ξ-axis (in the

ζ-plane) containing ξ_1 in its "interior" – basically a straight angle, π, with "vertex" at ξ_1, and fold it into a subtended angle α_1 degrees with vertex at z_1 in the z-plane.

The reader should be careful with the notations used here, which are different from those in the previous sections on conformal transformation but have been adopted to be consistent with most of the literature and therefore facilitate cross-referencing for the reader. In the previous discussions on conformal mapping, where we had only one transformation, we used $f(z) \equiv w(z)$ or $z = f^{-1}(w) \equiv g(w)$, where $z \equiv x + iy$ is the physical domain (pre-image) and $w(z) \equiv u + iv$ is the image. Compared to what we have here, z remains the physical domain in both cases, but whereas it is the preimage in the previous treatments, $w = w(z)$, it is the image here, $z = z(\zeta)$. In the present treatment ζ is the preimage, and we are transforming from the simple domain that it represents into the physical (complex) one that is the z-plane.

If we could simultaneously do this for a number of points on the ξ-axis of the ζ-plane, or $\xi_1, \xi_2, \ldots, \xi_n$, the ξ-axis would be mapped into a polygon (in the z-plane) whose angles are, respectively, $\alpha_1, \alpha_2, \ldots, \alpha_n$ degrees, and conversely. This is actually possible.

From $z - z_1 = (\zeta - \xi_1)e^{\alpha_1/\pi}$, we have $dz/d\zeta = (\alpha_1/\pi)(\zeta - \xi_1)e^{(\alpha_1/\pi)-1}$. This form suggests the form of the derivative of the transformation that could accomplish the multi-vertex mapping in the z-plane to a straight line (the ξ-axis) in the ζ-plane:

$$\frac{dz}{d\zeta} = K(\zeta - \xi_1)^{\left(\frac{\alpha_1}{\pi}-1\right)}(\zeta - \xi_2)^{\left(\frac{\alpha_2}{\pi}-1\right)} \cdots (\zeta - \xi_n)^{\left(\frac{\alpha_n}{\pi}-1\right)} = K\Pi_{j=1}^{n}(\zeta - \xi_j)^{\left(\frac{\alpha_j}{\pi}-1\right)}. \tag{3.15}$$

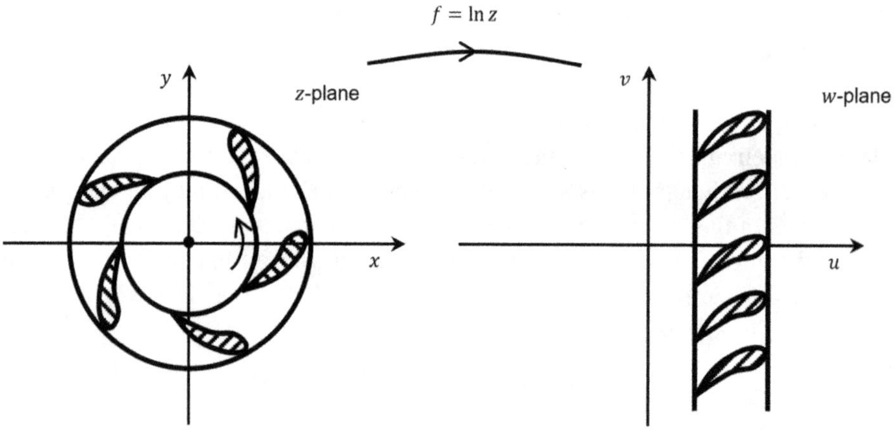

Figure 3.21: Application of $f(z) = \ln z$ in a blade cascade in turbomachinery.

This is the idea behind the Schwarz-Christoffel (SC) transformation. It is emphasized that the α_i's are in degrees in eq. (3.15). In some literature, the exponents $\alpha_i/\pi - 1$ are written as $-\nu_i$, where $\nu_i\pi$ is the complement of α_i, in the sense that $\alpha_i + \nu_i\pi = \pi$ in degrees so that $\alpha_i/\pi - 1 = -\nu_i$ and $\sum \nu_i = 2$. A configuration that illustrates the SC transformation is shown in Figure 3.22, where the interior, subtended angles are denoted by α, β, γ, and λ in place of the α_1, α_2, α_3, and α_4 used above. Moreover, the ξ_i's show up as a, b, c, d in the figures. No confusion is warranted by so doing. A few plausible SC transformations are given in Figure 3.23.

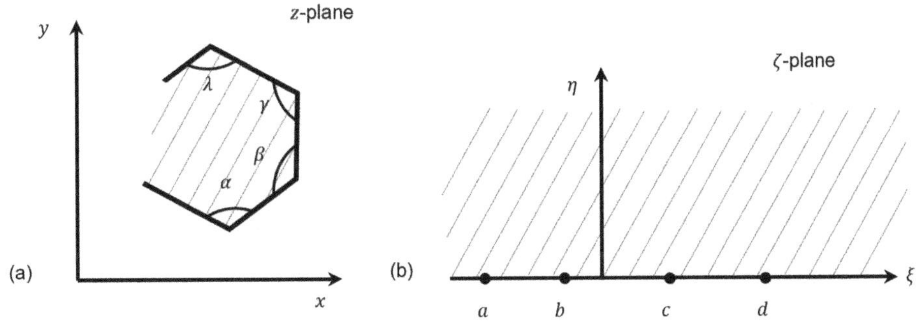

Figure 3.22: Schwarz-Christoffel transformation of interior angles in the z-plane to the UHP in the ζ-plane.

Note that the aspect of the SC transformation that we focus on in this chapter is how to derive the transformation $z(\zeta)$, which obviously requires the integration of eq. (3.15) with respect to ζ. The ξ_i's are points on the real axis of the ζ-plane that correspond to the vertex points a_i in the z-plane. This correspondence is used to determine K and C. K is a constant that determines the scale of the polygon and its orientation, and, again, it is obtained by relating the coordinates of the ξ_i points in the ζ-plane to the subtended angles. The constant of integration, C, determines the location of the origin in the z-plane. For closed polygons, $\sum a_i = (n-2)\pi$ degrees ($\pi = 180°$) where n is the number of polygon vertices, which alternatively can be written as $(2n-4)$ right angles.

A detailed presentation of the necessary and sufficient conditions for the existence of a conformal transformation in general, and the SC transformation, in particular, is beyond the scope of the present text, but the interested reader could consult Carathéodory [3]. The Riemann's theorem provides the theoretical foundation for the existence of conformal mapping.

Riemann Theorem: Any two simply connected domains, each having more than one boundary point, can be mapped conformally one on the other.

In addition to the transformations discussed previously in this chapter, which pertain to relatively simple regions, conformal transformations of regions bounded by polygons that have a finite number of vertices represent another class of regions of practical interest, and the SC transformations are particularly useful. Note that some of the vertices alluded to could lie at infinity. Also, the UHP obtained from SC mapping could further be mapped to any region into which a half plane can be transformed. An example is the use of $\log[(w-1)/(w+1)]$ to map onto the region between two parallel horizontal plates. If we integrate eq. (3.15) with respect to ζ, we obtain

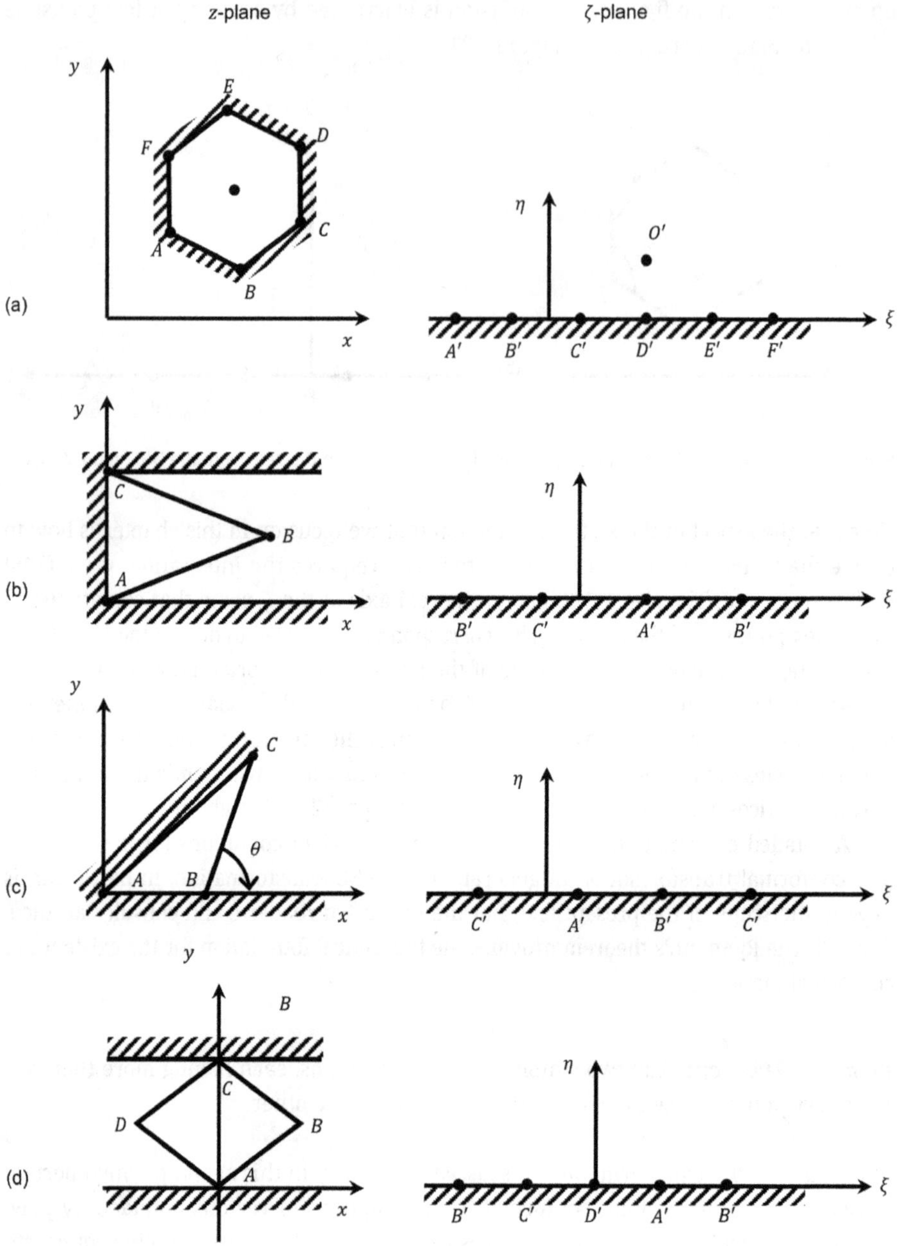

Figure 3.23: Other plausible Schwarz-Christoffel transformations.

$$z = K \int \left[(\zeta - \xi_1)^{\frac{\alpha_1}{\pi} - 1} (\zeta - \xi_2)^{\frac{\alpha_2}{\pi} - 1} \cdots (\zeta - \xi_n)^{\frac{\alpha_n}{\pi} - 1} \right] dz + C.$$

This can be considered to be composed of two transformations:

$$t = \int \left[(\zeta - \xi_1)^{\left(\frac{\alpha_1}{\pi}\right) - 1} (\zeta - \xi_2)^{\left(\frac{\alpha_2}{\pi}\right) - 1} \cdots (\zeta - \xi_n)^{\left(\frac{\alpha_n}{\pi}\right) - 1} \right] dz, \qquad z = Kt + C.$$

The first transformation, maps the ξ_1 axis in the ζ-plane to some polygon in the z-plane, while the second is the same linear transformation we presented earlier in this chapter. It stretches (contracts or magnifies), rotates, and translates the polygon obtained from the first transformation.

The procedure to obtain $z(\zeta)$ via the SC transformation uses the property of similar polygons that the corresponding angles be congruent and the corresponding sides be proportional. These are automatically satisfied for triangles. For quadrilaterals, an additional requirement is that two pairs of corresponding sides must have the same ratio. However, two such conditions are required for pentagons. For a polygon of n sides, $n-3$ additional conditions over and above the requirement that angles be congruent must be imposed to ensure similarity. As a consequence, when a polygon with n sides is mapped onto an UHP of the ζ-plane, three of the points $\xi_1, \xi_2, \ldots, \xi_n$ in the ζ-plane can be assigned arbitrarily, while the remaining $n-3$ points are determined by the conditions of similarity. One vertex of a polygon, such as an infinite one, might correspond to $\zeta = \infty$. This will use up one of the three degrees of freedom in choosing arbitrary points on the ξ-axis. In this case, there will be one less finite image point on the RHS of eq. (3.15) for $dz/d\zeta$ so that only two of the finite image points $\xi_1, \xi_2, \ldots, \xi_{n-1}$ can be specified arbitrarily.

The values normally chosen for the arbitrary points in the ζ-plane are –1, 0, and 1, for convenience. Since not more than three of the vertices can be chosen arbitrarily, when a polygon has more than three sides, some of the images must be determined in order to make the polygon, or any polygon congruent to it, map onto the ξ-axis. The selection of the conditions required to determine those points may require some ingenuity and could pose some challenges. Another potential challenge is carrying out the integration that is involved in determining $z(\zeta)$ from $dz/d\zeta$, as it might not be possible to evaluate the integral in terms of elementary functions. Numerical integration techniques could be explored in such situations. See Driscoll and Trefethen [4], for some details. We will now give a few examples of the procedure to obtain the SC transformation, starting from integrals $z(\zeta)$ that represent elementary functions. A few original applied descriptions of the SC method can be found in [6].

Example 3.9 Find the $z(\zeta)$ that maps the semi-infinite strip $-\pi/2 \le x \le \pi/2$, $y \ge 0$ in Figure 3.24 into the half plane $\eta \ge 0$.

Solution: The strip is considered as the limiting form of a triangle with vertices z_1, z_2, and z_3 where z_3 is located at infinity. Because $|z_2 - z_1| \ll |z_3 - z_2|$, we consider the subtended angle at z_3 or α_3, to be 0°, while those at z_1 (α_1) and z_2 (α_2) are each $\pi/2$. We arbitrarily choose $\xi_1 = -1$ as the image of z_1 and $\xi_2 = +1$ as the image of z_2. For z_3, we have $\xi_3 = \infty$. Noting that $\alpha_1/\pi - 1 = -1/2 = \alpha_2/\pi - 1$, we will have

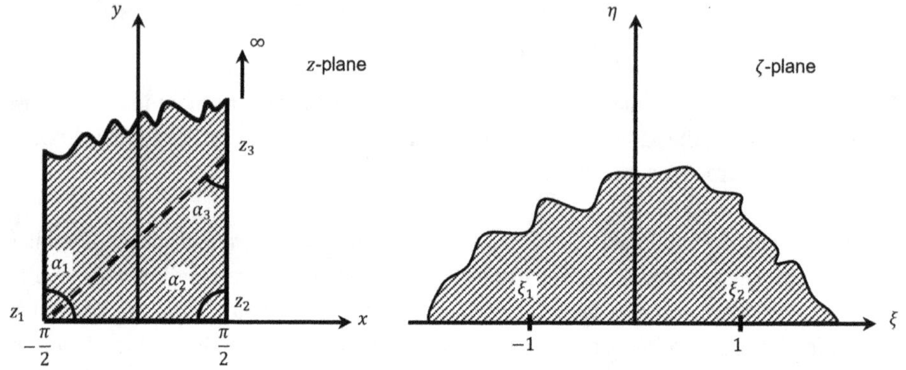

Figure 3.24: Semi-infinite strip in z is mapped to UHP in ζ.

$$z = K\int(\zeta+1)^{-\frac{1}{2}}(\zeta-1)^{-\frac{1}{2}}dz + C = iK\int_0^{\zeta}\frac{dz'}{\sqrt{1-\zeta'^2}} + C,$$

which gives $z = iK\sin^{-1}\zeta + C$. We apply the known conditions to determine K and C: $z = -\pi/2$ when $\zeta = -1$ and $z = \pi/2$ when $\zeta = 1$ so that

$$-\frac{\pi}{2} = -iK\frac{\pi}{2} + C, \qquad \frac{\pi}{2} = iK\frac{\pi}{2} + C,$$

which gives $C = 0$ and $iK = 1$. Therefore $z = \sin^{-1}\zeta$ or $\zeta = \sin z$.

Example 3.10 Determine the mapping $z(\zeta)$ into the UHP of the ζ-plane of the infinite strip $0 < y < \pi$, $-\infty < x < \infty$, with the limiting form of a rhombus with vertices z_1, z_2, z_3, and z_4, where $z_2 = 0$, $z_4 = \pi i$ (Figure 3.25), as the points z_3 and z_4 are moved infinitely far to the right and left, respectively.

Solution: In the limit as z_1 and z_3 become $-\infty$ and ∞, respectively, the "subtended" angles α_1 and α_3 become zero. Meanwhile, the subtended angles α_2 and α_4 are each π. We will arbitrarily select $\xi_1 = 0$, $\xi_2 = 1$, $\xi_3 = \infty$, leaving ξ_4 to be determined. With $\alpha_1 = \alpha_3 = 0$ and $\alpha_2 = \alpha_4 = \pi$, we will have $\alpha_1/\pi - 1 = -1$, $\alpha_2/\pi - 1 = 0$, $\alpha_4/\pi - 1 = 0$, while

$$z = K\int\zeta^{-1}(\zeta-1)^0(\zeta-\xi_4)^0 d\zeta + C = K\int\frac{d\zeta}{\zeta} + C = K\log\zeta + C.$$

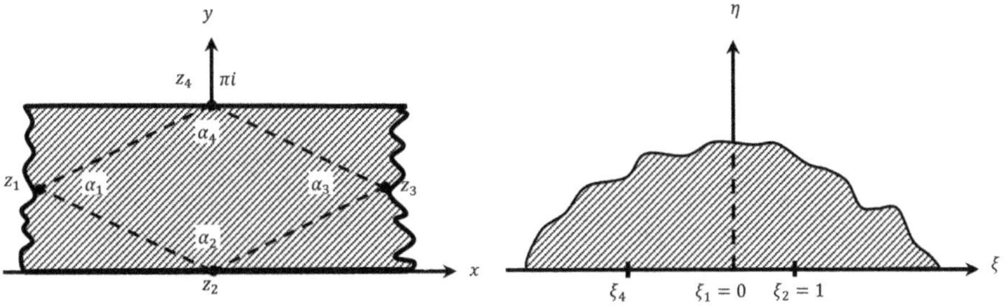

Figure 3.25: Infinite strip in z is mapped into the UHP in ζ.

Using $\zeta = 1$ when $z = 0$ gives $C = 0 \Rightarrow z = K \log \zeta$. K is a constant that must be real because the point z lies on the real axis when $\zeta = \xi$ and $\xi > 0$. The image of the point $z = \pi i$ is $\zeta = \xi_4$ where ξ_4 is a negative number. So, we have $\pi i = K \log \xi_4 = K \log|\xi_4| + K\pi i$. Equating the real and imaginary coefficients, we see that $|\xi_4| = 1 \Rightarrow K = 1$. Thus, $z = \log \zeta$. We also have $\xi_4 = -1$.

Example 3.11 Map the interior of the region shown in Figure 3.26(a) into the UHP of the region in Figure 3.26(b).

Solution: $\alpha_1 = \alpha_4 = 0$, $\alpha_2 = \alpha_3 = \pi/2$. We see that we have enough arbitrary points to select from since, with $\xi_1 = \xi_4 = \infty$, we only need two ξ_i values for which we have selected $\xi_2 = -1$, $\xi_3 = 1$.

Now, $\alpha_2/\pi - 1 = \pi/(2 \times \pi) - 1 = -1/2$ and $\alpha_3/\pi - 1 = \pi/(2 \times \pi) - 1 = -1/2$. Hence, $z = K(\zeta+1)^{-1/2}(\zeta-1)^{-1/2} = K/\left(\zeta^2-1\right)^{2}$, giving $z = K\cos^{-1}\zeta + C$. Consider the point P in the ζ-plane as shown in Figure 3.27 and examine $\theta = \cosh^{-1}(\zeta) \Rightarrow \zeta = \cosh\theta = \frac{1}{2}\left(e^{\theta} + e^{-\theta}\right)$, or $2\zeta e^{\theta} = e^{2\theta} + 1 \Rightarrow e^{\theta} = (1/2)(2\zeta \pm \sqrt{4\zeta^2 - 4}) = \zeta \pm \sqrt{\zeta^2 - 1}$. The limit of $\theta \to \infty$ gives $\zeta \to (1/2)e^{\theta}$. We take the positive root of $e^{\theta} = \zeta \pm \sqrt{\zeta^2-1}$ or $\theta = \ln\left(\zeta + \sqrt{\zeta^2-1}\right)$. We can select the principal value: $\zeta - 1 = r_1 e^{i(\theta_1 + 2n\pi)}$, $\zeta + 1 = r_2 e^{i(\theta_2 + 2m\pi)}$, or $\sqrt{\zeta^2-1} = \sqrt{r_1 r_2}$ $e^{i[(\theta_1+\theta_2)/2 + \pi(n+m)]}$.

The principal value implies $m = n = 0$, which is equivalent to cutting the ζ-plane on the ξ-axis. We can then use the following conditions in order to obtain K and C:

At $z = 0$, $\zeta = 1$: $\quad 0 = K\ln(1) + C \Rightarrow C = 0.$
At $z = ih$, $\zeta = -1$: $\quad ih = K\ln(-1) = Ki\pi \Rightarrow K = h/\pi.$

Thus,

$$z = \frac{h}{\pi}\ln\left(\zeta + \sqrt{\zeta^2-1}\right).$$

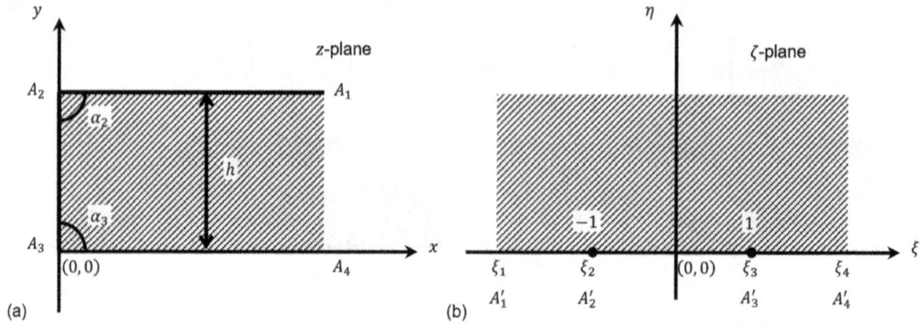

Figure 3.26: Mapping of a semi-infinite region $0 \le x < \infty$, $0 \le y \le h$.

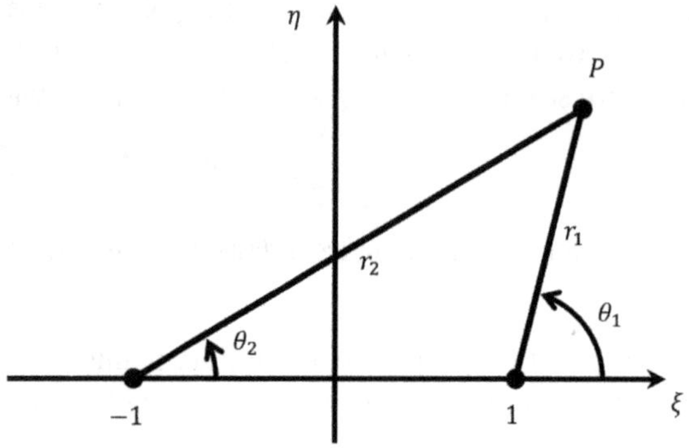

Figure 3.27: A typical point in the "interior" of the ζ-plane.

Example 3.12 Find the $z(\zeta)$ that maps the backward-facing step region in Figure 3.28(a) to the UHP of the ζ-plane in Figure 3.28(b).

Solution: $\alpha_1 = \alpha_4 = 0$, $\alpha_2 = 3\pi/2$, $\alpha_3 = \pi/2$; $\xi_1 = \xi_4 = \infty$, $\xi_2 = -1$, $\xi_3 = 1$. Also, $\alpha_2/\pi - 1 = 3\pi/2\pi - 1 = 1/2$, $\alpha_3/\pi - 1 = \pi/(2 \times \pi) - 1 = -1/2$. So, we have

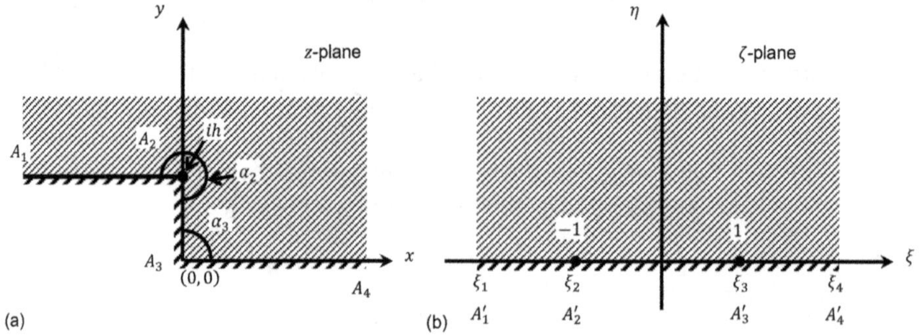

Figure 3.28: Mapping of a backward-facing step.

$$\frac{dz}{d\zeta} = K(\zeta+1)^{\frac{1}{2}}(\zeta-1)^{-\frac{1}{2}} = K\left(\frac{\zeta+1}{\zeta-1}\right)^{1/2}$$

$$\Rightarrow z = K\int \frac{\zeta+1}{\sqrt{\zeta^2-1}}d\zeta + C = K\int \frac{\zeta\, d\zeta}{\sqrt{\zeta^2-1}} + K\int \frac{d\zeta}{\sqrt{\zeta^2-1}} + C.$$

Carrying out the integration we have

$$z = K\left(\sqrt{\zeta^2-1} + \cos^{-1}\zeta\right) + C.$$

If we apply the known conditions at $\zeta = -1, 1$, we have

$$z = 0,\ \zeta = 1{:}\ 0 = K\ln(1) + C \Rightarrow C = 0$$

$$z = ih,\ \zeta = -1{:}\ ih = K\ln(-1) + C \Rightarrow B \equiv h/\pi.$$

The solution is $z = (h/\pi)\left[\sqrt{\zeta^2-1} + \ln(\zeta + \sqrt{\zeta^2-1}\right]$. This result could in principle be employed to solve for the distribution of a variable over the backward-facing step shown in Figure 3.28(a), whose physics is governed by the Laplace equation.

3.9.1 The free streamline flow theory

The free streamline flow problem typically involves flow through an orifice or vena contracta, and a change in the characteristics of the flow at unknown locations. Examples include the Stefan problem (melting, which involves an unknown surface), shock problem (which involves a change in the characteristic of a flow: hyperbolic to elliptic), and the free streamline, in which we are interested in finding the location of the free boundary. We will illustrate the latter with the problem of emptying a reservoir, as sketched in Figure 3.29. The streamlines are shown as $A_1 - A_2 - A_3$, $A_5 - z - A_4$, and $A_1' - A_2' - A_3'$ in the figure, of which $A_2 - A_3$ and $A_2' - A_3'$ are the free streamlines. Some foundation of the application of complex variable theory to fluid flow, which may be required for this section, is provided in Chapter 4.

On $A_2 - A_3$, the pressure p is uniform on a free streamline so that $q \equiv \sqrt{u^2 + v^2}$ is uniform from Bernoulli principle, where u and v are the velocity components. We have

$$\frac{1}{2}\rho q^2 = p_r - p_0,$$

which expresses the balance between the kinetic energy per unit volume of the flow and the difference between the reference pressure, p_r and atmospheric pressure, p_0. (Note that in the absence of gravity, the flow is driven by pressure difference, or $p_r - p_0$.) We

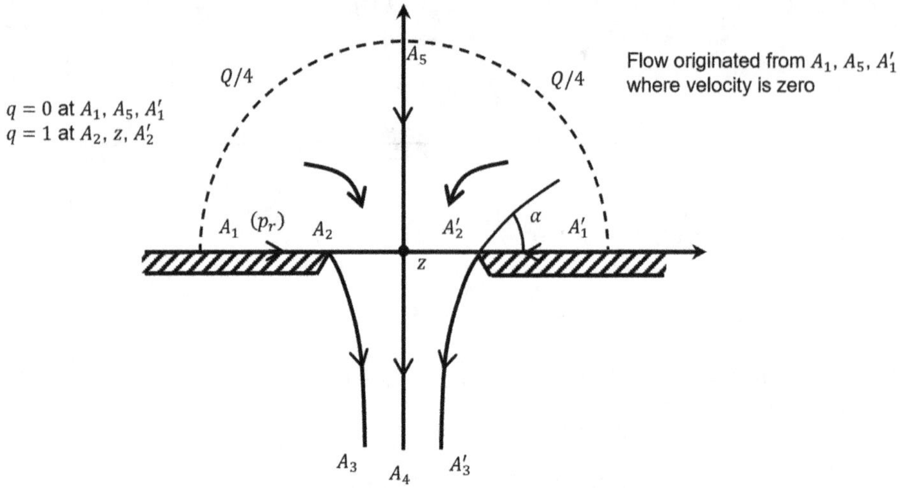

Figure 3.29: Flow through an orifice. This problem requires multiple conformal transformations to solve.

scale by setting $q = 1$. In this problem, the flow originates from the boundary defined by $A_1 - A_5 - A_1{}'$ and leaves through the boundary $A_3 - A_4 - A_3{}'$. The total flow rate through the latter is assumed to be $Q/2$ or $Q/4$ over the left or right of the symmetry streamline $A_5 - A_4$.

Thus, at arc $A_1 - A_5 - A_1{}'$ there is a fluid source $Q/2$ (half plane). Similarly, at $A_3 - A_4 - A_3{}'$, we have a sink of strength $-Q/2$. A series of mappings is required in order to tackle this problem:

I. W-plane (u,v) (also called the hodograph plane): $W = u - iv = qe^{-i\alpha}$. This is a non-geometric mapping.
II. Kirchhoff $(\alpha, \ \ln q)$ (log-hodograph plane): $\Omega = \ln(dz/dF) = \ln(1/W) = -\ln q + i\alpha$, where $W = qe^{-i\alpha}$, α is the angle in the W-plane and q is the radius R in the W-plane, which is the magnitude of the velocity of flow.
III. F-plane: $F = \phi + i\psi$, the complex velocity potential, where ϕ is the scalar velocity potential and ψ is the stream function.
IV. t-plane: (We used ζ previously in our general discussions of the SC mapping. We are going to use t in this section.)

Each of these mappings involves a complex variable $\lambda = a + bi$, and we need to plug in what we know about the flow into these planes in order to find $|\lambda|$ and $\arg \lambda$. Note that all transformations are regular, except for some boundary singularities. For the transformations in I through III, we have explicit real and imaginary coordinates. We need to use Figure 3.29 to locate the origin of each of the complex planes and the ranges in their respective real and imaginary axes. In IV, we work from the physical problem in the sense of the SC transformation. The various complex planes are now discussed in turn.

I. W-plane (u,v): At A_1, A_5, $q = 0$ (where velocities originate from), while along $A_2 - A_3$, $q = 1$, $u^2 + v^2 = 1$. A_1, A_5 is a source and A_3, A_4 is a sink. This is depicted in Figure 3.30. Thus, we see that the flow has an x (horizontal) component u and negative y (vertical) component of $-v$.

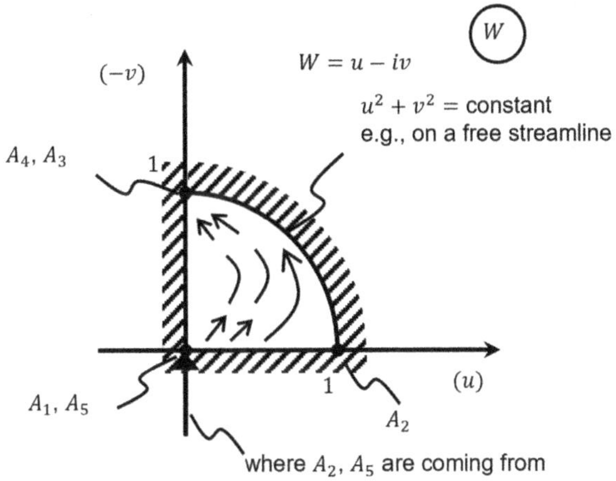

Figure 3.30: Depiction of the hodograph plane.

II. Kirchhoff plane: The Kirchhoff (log-hodograph) plane Ω (Figure 3.31) is defined by

$$\Omega = \ln \frac{dz}{dF} = \ln \frac{1}{W} = -\ln q + i\alpha \ \left(W = qe^{-i\alpha}\right)$$

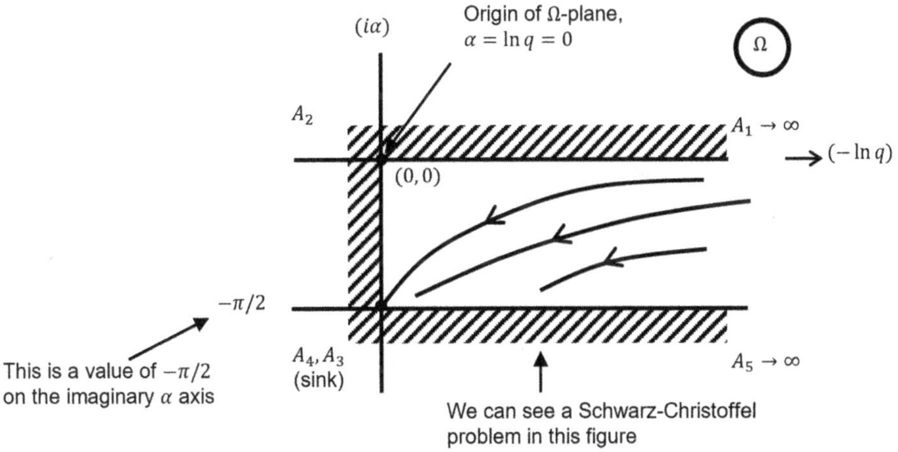

Figure 3.31: The Kirchhoff (log-hodograph) plane.

At A_2: $q=1$, $\ln q=0$, $v=0$, $\alpha=0$, whereas at A_5, $\alpha=-\pi/2$, $\ln q\to-\infty$ $(q=0)$. At A_3, $\alpha=-\pi/2$, $q=1$, and $\alpha=0$, $q=0$, $-\ln q\to\infty$ at A_1. We can obviously see a configuration that can be mapped by the SC method in the Kirchhoff field.

III. F-plane: The complex potential is defined as

$$F=\varphi+i\psi,$$

where φ is the real potential and ψ is the stream function. On the $A_1-A_2-A_3$ streamline: $\psi=0$ (taken as a reference), whereas on the A_5-A_4 streamline: $\psi=Q/4$ (so that the volume flux across $A_3-A_4=Q/4$). The total flow rate is $Q/2$. We define $\phi=0$ at A_2, $\phi\to-\infty$ at A_1 and $\phi\to+\infty$ at A_3. The schematic for this is shown in Figure 3.32(a) and 3.32(b). Note that we can also see a Schwarz-Christoffel problem in the complex potential plane.

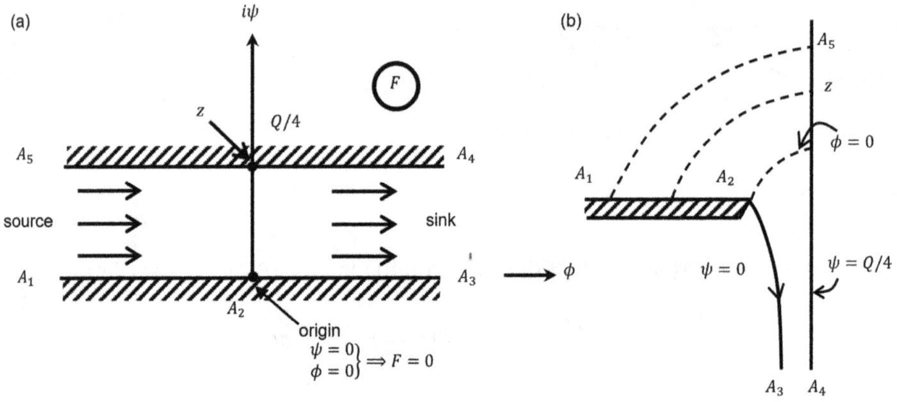

Figure 3.32: Schematic for the real potential and stream function.

IV. The $F-t$ and $\Omega-t$ planes: Here, we are going to map the F (complex potential) and Kirchhoff (log-hodograph, Ω) planes via the SC transformation.

(1) $F-t$ planes: The transformation between F and t can be written as

$$\frac{dF}{dt}=Kt^{-1}\left(\text{or } K(t-0)^{-\frac{1}{2}}(t-0)^{-\frac{1}{2}}\right)$$

so that

$$F=K\ln t+C\equiv\phi+i\psi.$$

We have the following conditions at A_2 and z:
A_2: $0=K\ln(-1)+C$ $(\psi=\phi=0\Rightarrow F=0$ at $A_2)$, z: $iQ/4=K\ln 1+C$ $(F=\phi+i\psi,\ \phi=0$ at $z,\psi=\ Q/4)$. We see from the foregoing that

$$C = \frac{iQ}{4}, \quad K = -\frac{Q}{4\pi}, \quad F = -\frac{Q}{4\pi}\ln t + i\frac{Q}{4} = -\frac{Q}{4\pi}(\ln t - i\pi) = \ldots = -\frac{Q}{4\pi}\ln(-t) = F.$$

The F and t complex planes are shown in Figure 3.33. We set A_1' (or ξ_1) $= -1$, A_2' (or ξ_2) $= 0$, and A_3' (or ξ_3) $= 1$.

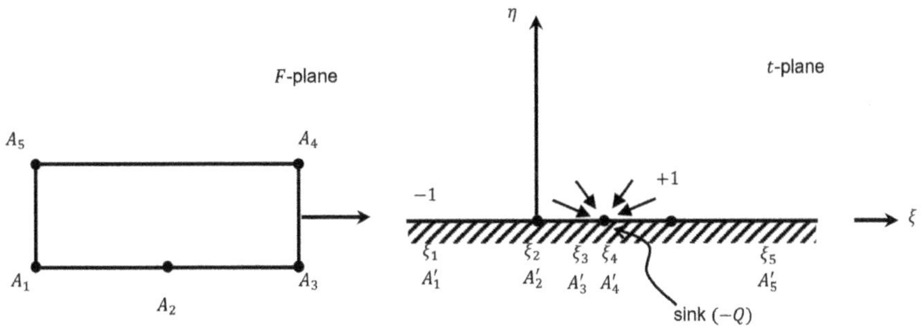

Figure 3.33: The F and t planes.

(2) Ω-t planes: The physical plane for the log-hodograph function is shown in Figure 3.34. Thus, the SC mapping can be written as

$$\frac{d\Omega}{dt} = K(t+1)^{-\frac{1}{2}}(t-0)^{-\frac{1}{2}},$$

from which we can solve for Ω:

$$\Omega = K\int \frac{dt}{\sqrt{t(t+1)}} + C.$$

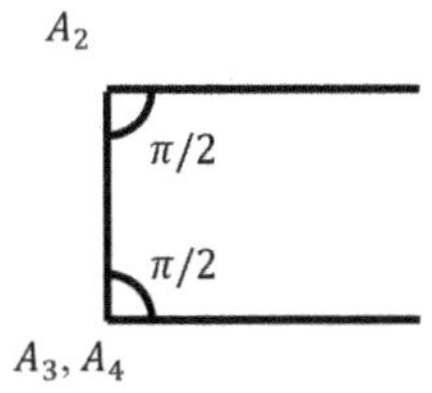

Figure 3.34: Physical plane for the log-hodograph transformation.

If we introduce τ:

$$t = \frac{1}{2}(\tau - 1), \quad t + 1 = \frac{1}{2}(\tau + 1),$$

we can write Ω as

$$\Omega = K\int \frac{d\tau}{\sqrt{\tau^2-1}} + C = K\cosh^{-1}\tau + C, \text{ and}$$

$$\cosh^{-1}\tau = \ln\left(\tau + \sqrt{\tau^2-1}\right) = 0\ (\tau=1) = \ln i\ (\tau=0),\ \left(\ln i = \ln e^{\frac{i\pi}{2}}\right).$$

Introducing the following conditions at A_2, A_4, and A_3:

$$A_2\text{: } 0 = Ki\pi + C,\ \text{At}\, A_4,\ A_3\text{: } -i\frac{\pi}{2} = C,\ K = \frac{1}{2}\frac{i\pi}{i\pi} = \frac{1}{2},$$

we can obtain

$$\Omega = \frac{1}{2}\cosh^{-1}\tau - \frac{i\pi}{2},\ \Omega + \frac{i\pi}{2} = \frac{1}{2}\cosh^{-1}\tau,\ \text{or}\ \Omega = \frac{1}{2}\ln\left(\tau+\sqrt{\tau^2-1}\right) - \frac{i\pi}{2}.$$

We can now obtain some link between Ω, F, and z: Introduce the definition of Ω to eliminate t: $\Omega = \ln(dz/dF)$. We can write

$$\ln\frac{dz}{dF} + \frac{i\pi}{2} = \ln i\frac{dz}{dF} = \ln\left(\tau+\sqrt{\tau^2-1}\right)^{\frac{1}{2}};\quad \frac{dz}{dF} = -i\left(\tau+\sqrt{\tau^2-1}\right)^{\frac{1}{2}},$$

$$\text{and } -\exp(4\pi F/Q) = t, \qquad \tau = 2t+1.$$

On the free (boundary) streamline, $\psi = 0$ and $F = \phi + i(0) = \phi$. We can also write $\underbrace{d\phi/ds}_{\text{speed}} =$ $q = 1 \Rightarrow \phi = s = F$. Note that $\mathbf{u} = \nabla\phi$ and $dF/dz = u - iv$. The zero streamline ($\psi = 0$) and potential ($\phi = 0$) are shown in Figure 3.35.

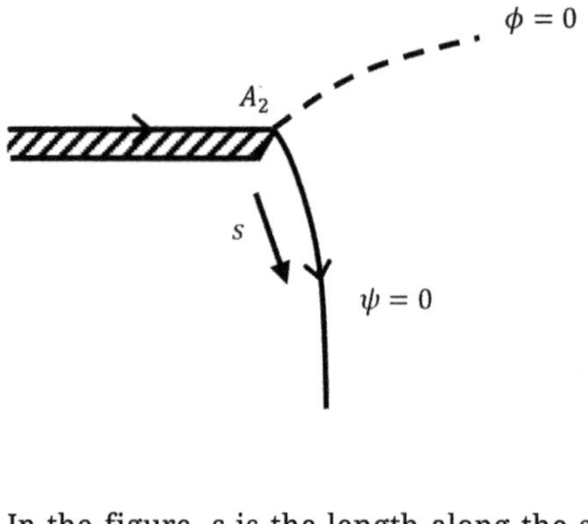

Figure 3.35: Free streamline coordinate direction s; starts from the corner. The zero potential and zero streamline curves are shown.

In the figure, s is the length along the streamline, which starts from the corner at A_2 (Figure 3.35). With $\phi = F = s$, we can write $-\exp(-4\pi s/Q) = t$. Also,

$$2\Omega + i\pi = \cosh^{-1}\tau \Rightarrow \tau = \cosh(2\Omega + i\pi),\quad \tau|_{\psi=0} = \cosh(2\Omega + i\pi),$$

$$\Omega = -\ln q + i\alpha = i\alpha = \cosh i(2\alpha + \pi) = \cos(2\alpha + \pi) = -\cos 2\alpha = 1 + 2t$$

$$t = -\frac{1}{2}(1 + \cos 2\alpha) = -\cos^2\alpha = -\exp\left(-\frac{4\pi s}{Q}\right)$$

$$F = \phi = -\frac{Q}{4\pi}\ln(-t) = s = -\frac{Q}{4\pi}\ln\left(\frac{1-\tau}{2}\right)$$

$$s = -\frac{Q}{2\pi}\ln\frac{1+\cos 2\alpha}{2} = -\frac{Q\ln\cos^2\alpha}{2\pi} = -\frac{Q}{2\pi}\ln\cos\alpha \Rightarrow \cos\alpha = \exp\left(-\frac{2\pi s}{Q}\right).$$

This is the shape of the free boundary. Squaring both sides, we have

$$\cos^2\alpha = \exp\left(-\frac{4\pi s}{Q}\right).$$

Using the elemental lengths in Figure 3.36, we can write $\cos\alpha = dx/ds$, which is equivalent to $dx/ds = \exp(-2\pi s/Q)$.

The distance the fluid moves in the x-direction can be obtained by integration:

$$x = -\frac{Q}{2\pi}\exp\left(-\frac{2\pi s}{Q}\right) + C.$$

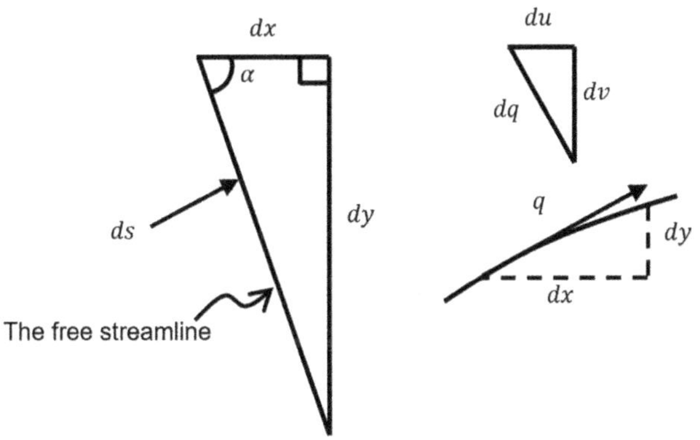

Figure 3.36: Elemental lengths.

At A_2, we have the boundary conditions $s = 0$, $x = -a = -Q/2\pi + C$, giving $x = -a + (Q/2\pi)[1 - \exp(-2\pi s/Q)]$.

As $s \to \infty$, $x \to -a + Q/2\pi$ (a cylindrical exit flow profile because it is independent of s). Let this constant be proportional to a as ka. This is shown in Figure 3.37. We can examine the dimensions of Q:

$$\frac{Q}{4} = \underbrace{ka}_{\text{area}} \cdot \overbrace{1}^{\text{velocity}} \Rightarrow Q = 4ka$$

Thus, at $s \to \infty$, $4ka/2\pi - a = -ka \Rightarrow k = 1/(1 + 2/\pi) \sim .61 + \ldots$.

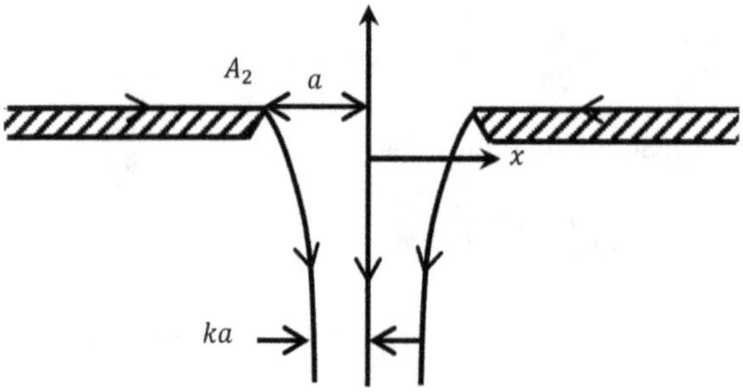

Figure 3.37: Geometry of the free streamline flow showing an equilibrium exit profile.

3.10 Miscellaneous examples

Example 3.13 Blasius Integral Laws. The velocity resolutes at distant points in an infinite liquid which streams past an elliptic cylinder are $-V\cos\beta$, $-V\sin\beta$ parallel to the major and minor axes respectively of the cross-section, and there is a circulation of amount κ about the cylinder. Find the force and couple resultants, per unit thickness, exerted by the fluid on the cylinder [6].

Solution: (The reader may need to consult the materials in Chapter 4 of this book in order to fully understand our solution of this problem.) The given ellipse can be transformed into a circle by using the Joukowski transformation:

$$z = \zeta + \frac{c^2}{4\zeta},$$

where $c^2 = a^2 - b^2$ (Figure 3.38). This preserves the free-stream and circulation conditions at $|z| \to \infty$. The complex potential $\tilde{F}(\zeta)$ and complex velocity $\tilde{W}(\zeta)$ in the ζ-plane are well known:

$$\tilde{F}(\zeta) = -V\zeta e^{-i\beta} - Ve^{i\beta}\frac{(a+b)^2}{4\zeta} - \frac{i\kappa}{2\pi}\ln\zeta,$$

$$\tilde{W}(\zeta) = -Ve^{-i\beta} + Ve^{i\beta}\frac{(a+b)^2}{4\zeta^2} - \frac{i\kappa}{2\pi\zeta}.$$

The components X and Y of the force on the ellipse, by Blasius' formula [6], can be written as

$$X - iY = \frac{i\rho}{2}\oint W^2 dz = \frac{i\rho}{2}\oint \frac{\tilde{W}^2}{dz/d\zeta} d\zeta = \frac{i\rho}{2} 2\pi i \,\mathrm{Res}\left[\frac{\tilde{W}^2}{dz/d\zeta}, \zeta = 0\right],$$

where ρ is the mass density of the liquid. The only singularity is at $\zeta = 0$, so the integration can be done on a circle large enough for $c^2/2\zeta^2 < 1$. Let us expand $\left(\frac{\tilde{W}^2}{dz/d\zeta}\right)$ and retain only relevant terms:

$$\frac{\tilde{W}^2}{dz/d\zeta} = \left(-Ve^{-i\beta} + Ve^{i\beta}\frac{(a+b)^2}{4\zeta^2} - \frac{i\kappa}{2\pi\zeta}\right)^2 \left(1 - \frac{c^2}{4\zeta^2}\right)^{-1} = \left(\frac{2iV\kappa e^{-i\beta}}{2\pi\zeta} + \ldots\right)\left(1 + \frac{c^2}{3\zeta^2}\right).$$

Thus,

$$X - iY = \frac{i\rho}{2} 2\pi i \frac{2iV\kappa e^{-i\beta}}{2\pi} = -i\rho V\kappa e^{-i\beta} = \rho V\kappa e^{-i(\beta + \pi/2)},$$

$$X + iY = \rho V\kappa e^{i(\beta + \pi/2)}.$$

The net force is perpendicular to the free stream, and its magnitude is $(\rho V\kappa)$. The moment is [6]

$$M = -\frac{\rho}{2}\mathrm{Re}\left(\oint W^2 z dz\right) = -\frac{\rho}{2}\mathrm{Re}\left[\oint \frac{\tilde{W}^2}{dz/d\zeta} z(\zeta) d\zeta\right]$$

Carrying out an expansion, we will have

$$\frac{\tilde{W}^2 z}{dz/d\zeta} = \left[V^2 e^{-i2\beta} - \frac{\kappa^2}{4\pi^2\zeta^2} - \frac{2V^2(a+b)^2}{4\zeta^2} + \ldots\right]\left(\zeta + \frac{c^2}{4\zeta}\right)\left(1 + \frac{c^2}{4\zeta^2} + \ldots\right)$$

$$= \frac{1}{4\zeta}\left[2V^2c^2e^{-i2\beta} - \frac{\kappa^2}{\pi^2} - 2V^2(a+b)^2\right] + \ldots,$$

and

$$M = -\frac{\rho}{2}\mathrm{Re}\left[2\pi i \frac{1}{4}\left(2V^2c^2e^{-i2\beta} - \ldots\right)\right] = -\rho\pi V^2c^2 \sin\beta\cos\beta.$$

The moment is in the clockwise direction. It vanishes for $\beta = 0$, $\pi/2$ and also for a circle ($c = 0$).

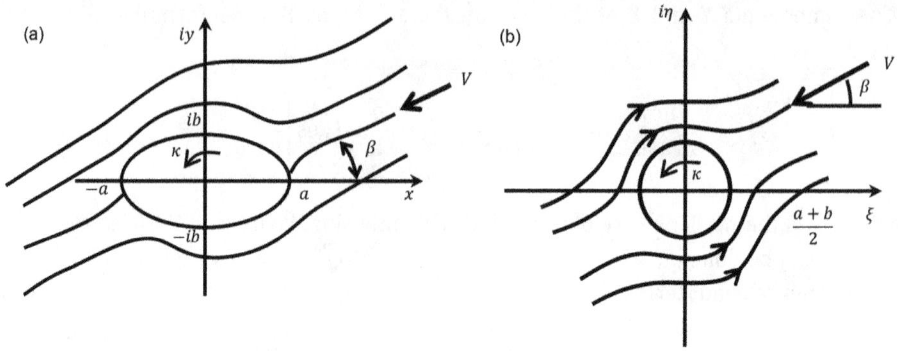

Figure 3.38: Flow over an ellipse (a) and circle (b).

Example 3.14 (Joukowski Transformation): An aerofoil is derived from the circle $\left|\zeta - be^{i\beta}\right| = a$ by the conformal transformation

$$z = \zeta + \sum_{r=1}^{n} \frac{a_r}{\zeta^r},$$

which is such that the zeros of $dz/d\zeta$ all lie within the circle except one which falls on the circumference at $\zeta = -l = be^{i\beta} - ae^{i\alpha}$, where a, b, and l are real and in general the coefficients a_r are imaginary. Show that, if the circulation about the aerofoil is chosen in accordance with Joukowski's hypothesis, then there is a lift at right angles to a steady stream in which an aerofoil is placed, vanishing for a certain angle of incidence, and find the form of a_1 if the moment about the center of the circle vanishes with the lift [6]. Denote the free-stream velocity by U.

Solution: Given the transformation

$$z = \zeta + \sum_{r=1}^{n} \frac{a_r}{\zeta^r},$$

we have

$$\frac{dz}{d\zeta} = 1 - \sum_{r=1}^{n} \frac{ra_r}{\zeta^{r+1}}.$$

The circulation Γ from the Kutta condition:

$$\Gamma = 4\pi a U \sin(\theta + \alpha).$$

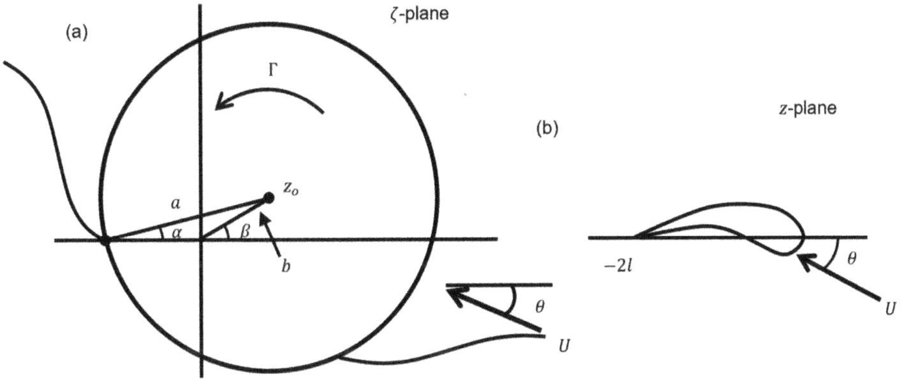

Figure 3.39: A sketch for Example 3.14.

The forces can be found by Blasius' formula [6]:

$$X - iY = \frac{i\rho}{2}\oint W^2 dz = \frac{i\rho}{2}\oint \frac{\tilde{W}^2}{dz/d\zeta}d\zeta.$$

We can integrate over a large circle in the ζ-plane since $dz/d\zeta$ has no zeroes in the region $|\zeta| > a$ and

$$\frac{dz}{d\xi} > 0 \Rightarrow \sum_{r=1}^{n}\frac{ra_r}{\zeta^{r+1}} < 1, \qquad \frac{1}{dz/d\zeta} = 1 + \sum_{r=1}^{n}\frac{ra_r}{\zeta^{r+1}} + \dots.$$

Note also that

$$\frac{1}{\zeta - \zeta_o} = \frac{1}{\zeta}\left(\frac{1}{1 - \zeta_o/\zeta}\right) = \frac{1}{\zeta} + \frac{\zeta_o}{\zeta^2} + \dots$$

on the large circle.

$$\frac{\tilde{W}^2}{dz/d\zeta} = \left[-Ue^{i\theta} + Ue^{-i\theta}\frac{a^2}{(\zeta-\zeta_o)^2} - \frac{i\Gamma}{2\pi(\zeta-\zeta_o)}\right]^2\left(1 + \frac{a_1}{\zeta^2} + \dots\right) = \frac{2iU\Gamma e^{i\theta}}{2\pi\zeta} + \dots,$$

and the x- and y-forces can be obtained from

$$X - iY = \frac{i\rho}{2}2\pi i\frac{2iU\Gamma e^{i\theta}}{2\pi} = -i\rho U\Gamma e^{i\theta} = \rho U\Gamma e^{i(\theta-\pi/2)}.$$

The resultant force is perpendicular to the free stream direction and vanishes when $\Gamma = 0$:

$$\Gamma = 0 \Rightarrow \sin(\theta + \alpha) = 0, \qquad \theta_o = -\alpha,$$

where α is given. We also have the sine formula: $b/\sin\alpha = a/\sin(180 - \beta) = a/\sin\beta$.

Note that in this notation, θ is positive in the clockwise direction (see Figure 3.39). To determine the moment, we need to expand

$$\frac{\tilde{W}^2 z}{dz/d\zeta} = \left[U^2 e^{i2\theta} - \frac{2U^2a^2}{(\zeta-\zeta_o)^2} - \frac{\pi^2}{4\pi^2(\zeta-\zeta_o)^2} + \ldots\right]\left(\zeta + \frac{a_1}{\zeta} + \ldots\right)\left(1 + \frac{a_1}{\zeta^2} + \ldots\right)$$

$$= \frac{2a_1}{\zeta}U^2 e^{i2\theta} - \frac{2U^2a^2}{\zeta} - \frac{\Gamma^2}{4\pi^2\zeta} + \ldots$$

$$M = -\frac{\rho}{2}\mathrm{Re}\left(\oint W^2 z dz\right) = -\frac{\rho}{2}\mathrm{Re}\left[\oint \frac{\tilde{W}^2 z(\zeta)}{dz/d\zeta}d\zeta\right],$$

or

$$-\frac{\rho}{2}\mathrm{Re}\left[2\pi i\left(2a_1U^2e^{i2\theta} - 2U^2a^2 - \frac{\pi^2}{4\pi^2}\right)\right] = -2\pi\rho U^2\mathrm{Re}\left(ia_1e^{i2\theta}\right).$$

If M is to vanish at $\theta = \theta_o$, then: $\mathrm{Re}\left(ia_1e^{i2\theta_o}\right) = 0$.

Setting a_1 as $a_1 = Ae^{i\varphi}$,

$$\mathrm{Re}\left[Ae^{i\left(\frac{\pi}{2}+\varphi+2\theta_o\right)}\right] = A\cos\left(\frac{\pi}{2} + \varphi + 2\theta_o\right) = 0,$$

$$\varphi = -2\theta_o + n\pi, \quad n = 0, 1, \ldots$$

$$a_1 = Ae^{-2i\theta_o + in\pi} = Ae^{+2i\alpha + in\pi}, \quad n = 0, 1.$$

Example 3.15 (Schwarz-Christoffel Transformation): Liquid flows in the negative direction of the axis of y between two planes defined by $x = \pm a$, $y > b$ and meets a barrier defined by $y = 0$, $l > x > -l$. The speed for large positive values of y is V. Show how to determine the ultimate velocity of the two jets and the resultant thrust on the barrier.

Solution: A sketch of the problem is given in Figure 3.40. The magnitude of the jet velocity can be obtained by using the Bernoulli equation, given the pressures P_v, P_o:

$$P_v + \frac{1}{2}\rho V^2 = P_o + \frac{1}{2}\rho U^2,$$

$$U^2 = V^2 + \frac{2}{\rho}(P_v - P_o).$$

The problem is symmetric, therefore we will only consider $x \geq 0$. We will give the center streamline the reference value 0, the external streamline the value Q (since the total flow is $2Q = 2aV$) and the potential at B: $\varphi_B = 0$. Using the same transformations as in the text, we can obtain the complex planes shown in Figure 3.41.

Let us introduce the parameter t_E and use it to define τ:

$$\tau = \frac{2}{1-t_E}t - \frac{1+t_E}{1-t_E}.$$

The following transformations can be derived:

(a) $F \to t$, where $F = \phi + i\psi$:

$$\frac{dF}{dt} = kt^{-1}, \qquad F = k_1 \ln t + k_2.$$

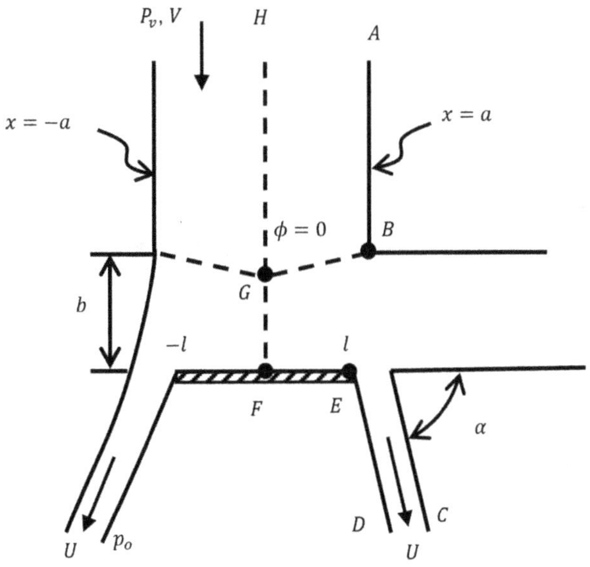

Figure 3.40: A sketch for Example 3.15.

At B: $F = iQ = k_1 \ln(1) + k_2$. At G: $F = 0 = k_1 \ln(-1) + k_2$. Thus,

$$F = -\frac{Q}{\pi}\ln t - iQ = -\frac{Q}{\pi}\ln(-t).$$

(b) $\Omega \to t$, where $\Omega = -\ln q + i\alpha$ and $W = qe^{-i\alpha} \equiv u - iv$:

$$d\Omega/dt = k(t-1)^{-\frac{1}{2}}(t-t_E)^{-1/2} = k\left(\tau^2 - 1\right)^{-1/2};$$

$$\Omega = k_1 \cosh^{-1}(\tau) + k_2.$$

At B: $\Omega = -i(\pi/2) = k_1 \cosh^{-1}(1) + k_2$, at E: $\Omega = 0 = k_1 \cosh^{-1}(-1) + k_2$. That is,

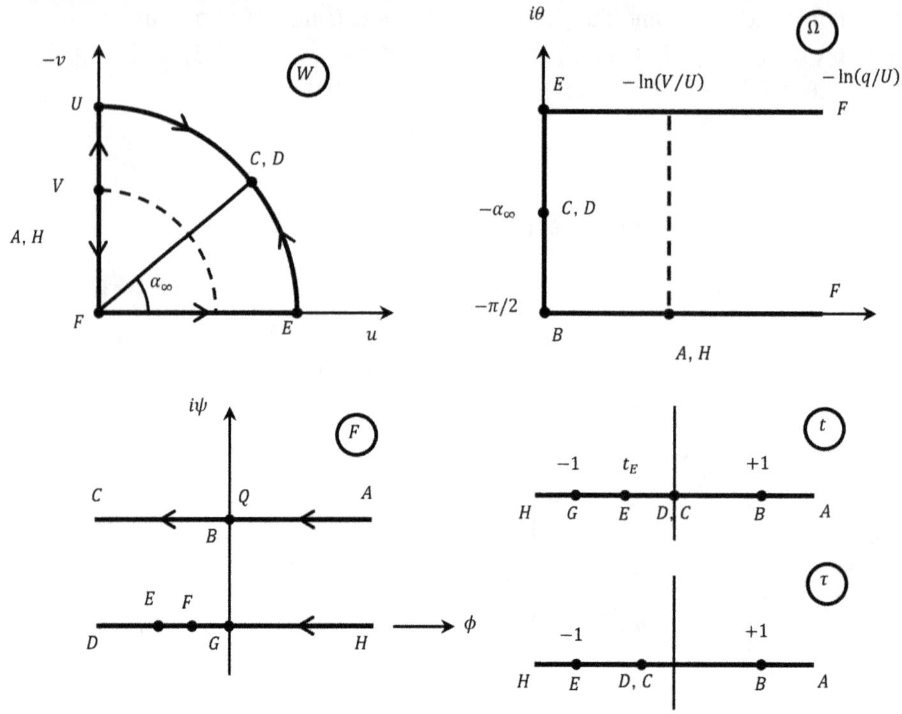

Figure 3.41: The various complex planes for Example 3.15.

$$\Omega = \frac{1}{2}\cosh^{-1}(\tau) - i\frac{\pi}{2}.$$

On the interior streamline (ED): $q = 1$ (scaled by U) $\Rightarrow \ln q = 0$, $\psi = 0$;

$$\Omega = i\alpha, \quad F = \varphi_E + s,$$

$$\tau = \cosh(2\Omega + i\pi) = \cosh i(2\alpha + \pi) = \cos(2\alpha + \pi) = -\cos 2\alpha = 1 - 2\cos^2\alpha,$$

$$t = -\exp\left(-\frac{\pi F}{Q}\right) = -\exp\left(-\frac{\pi\varphi_E}{Q}\right)\exp\left(\frac{\pi s}{Q}\right) = t_E \exp\left(-\frac{\pi s}{Q}\right).$$

We can also write

$$\left(\frac{dx}{ds}\right)^2 = \cos^2\alpha = \frac{1}{2}(1 - \tau) = \frac{1}{2}\left(1 - \frac{2t_E}{1 - t_E}e^{-\frac{\pi s}{Q}} + \frac{1 + t_E}{1 - t_E}\right),$$

which can be used to obtain $x_1(s)$, with the boundary condition $x_1(0) = l$. On the exterior streamline, we have the boundary condition: $q = 1$, $\psi = Q$

$$\Omega = i\alpha, \quad F = s + iQ,$$

$$\tau = \cosh(2i\alpha + i\pi) = -\cos 2\alpha = 1 - 2\cos^2\alpha,$$

$$t = -\exp\left(-\frac{\pi s}{Q} - i\pi\right) = \exp\left(-\frac{\pi s}{Q}\right),$$

$$\left(\frac{dx}{ds}\right)^2 = \frac{1}{2}\left(1 - \frac{2}{1-t_E} e^{-\pi s/Q} + \frac{1+t_E}{1-t_E}\right).$$

We can use this to obtain $x_2(s)$, with the boundary condition $x_2(0) = a$. The profiles $y_1(s)$, $y_2(s)$ can be found from

$$\left(\frac{dx}{ds}\right)^2 + \left(\frac{dy}{ds}\right)^2 = 1,$$

with the boundary conditions $y_1(0) = 0, y_2(0) = b$. The free streamlines are dependent on t_E. As $s \to \infty$, the curves become parallel. The mass balance:

$$aV = Q = U\left\{ [x_2(s_2) - x_1(s_1)] \frac{dy}{ds}\bigg|_\infty \right\},$$

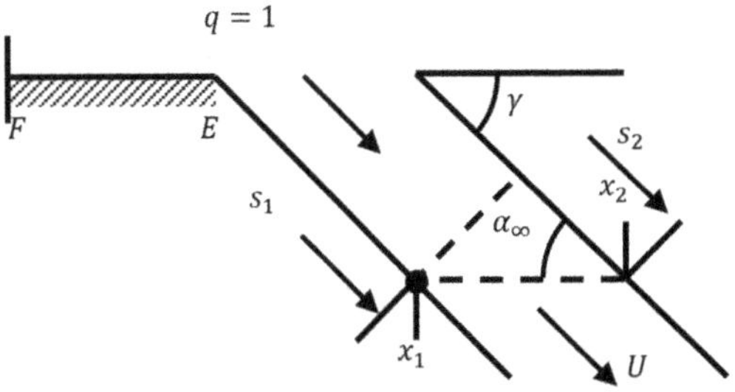

Figure 3.42: Details of the outflow streamline for Example 3.15.

where $y_1(s_1) = y_2(s_2)$ allows us to determine t_E and hence α_∞. The outflow region is shown in Figure 3.42, while Figure 3.43 shows the reaction force F on the plate. The required force on the barrier can be found from an overall momentum balance in the y-direction:

$$F = M_{\text{in}} - 2(M_{\text{out}}) \sin \alpha_\infty = 2QV - 2QU \frac{dy}{ds}\bigg|_\infty.$$

Problems

3.1 What are the precise conditions for a function $f(z)$ to be a conformal mapping?

3.2 Find the image of the point $z = i + 1$ under the transformation $w(z) = z^2$.

3.3 Find the image of the point $z = i - 1$ under the transformation $w(z) = z^2$.

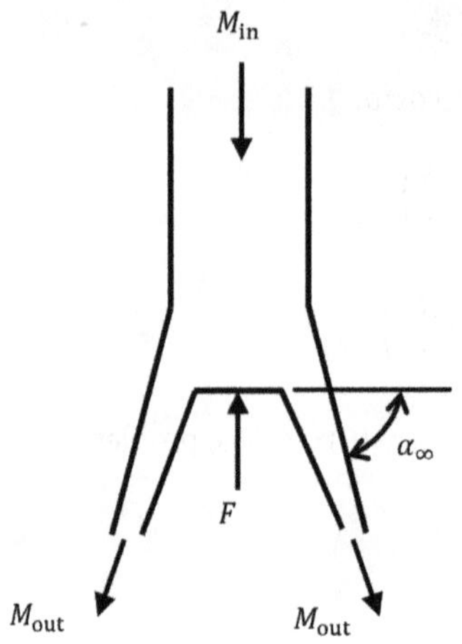

Figure 3.43: Depiction of mass inflow and outflow for Example 3.15.

3.4 Find the image of the point z in the figure below under $w(z) = z^2$.

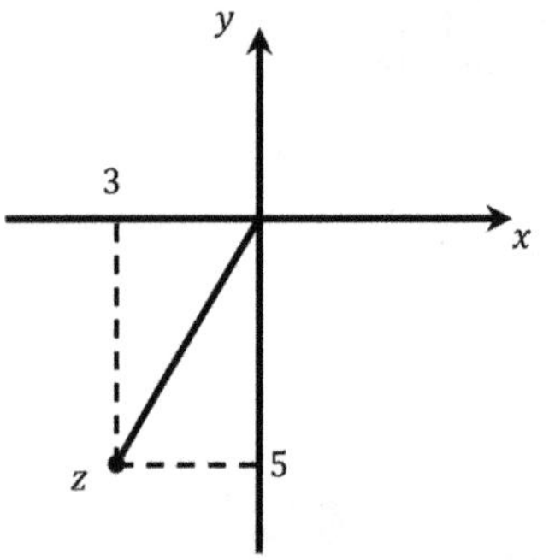

3.5 Find the image of the point z in the figure below under the transformation $w(z) = z^2$.

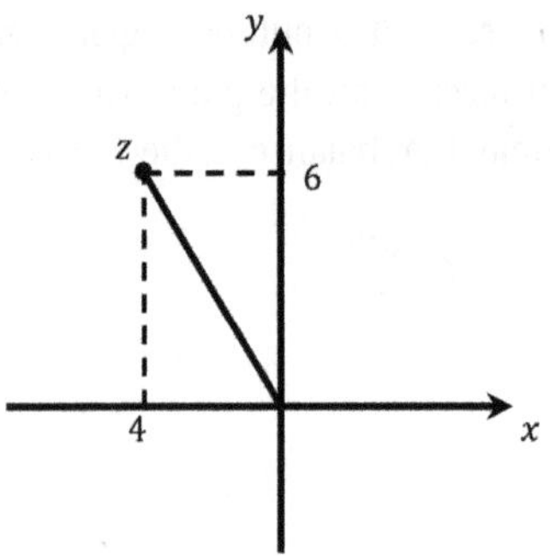

3.6 If $w(z) = 1/z$, determine $u(x,y)$, $v(x,y)$, $x(u,v)$, and $y(u,v)$.

3.7 Determine the image of z in the figure below under the transformation $w = iz$.

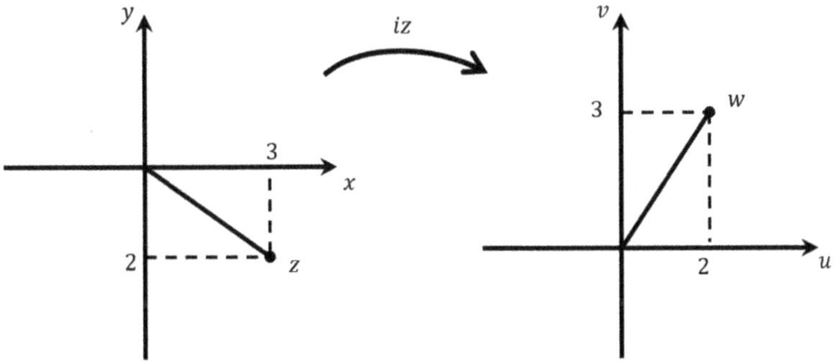

3.8 Show that the function $w(z) = 4z^2/(1-z)^2$ maps the point $z = \infty$ to the point $w = 4$.

3.9 Consider the transformation $f(z) = w(z) = z + 1/z$. If $z = re^{i\theta}$, determine $u(r, \theta)$ and $v(r, \theta)$, where $w = u + iv$.

3.10 Obtain the solution of the Laplace equation governing steady-state temperature distribution in a semi-infinite strip $-\pi/2 \le x \le \pi/2, y \ge 0$, satisfying the conditions below. Normalized temperature values are shown.

$$\frac{\partial^2 T}{\partial x^2} + \frac{\partial^2 T}{\partial y^2} = 0, \quad -\frac{\pi}{2} < x < \frac{\pi}{2}, \; y > 0,$$

$$T\left(-\frac{\pi}{2}, y\right) = T\left(\frac{\pi}{2}, y\right) = 0, \quad y > 0,$$

$$T(x, 0) = 1, \qquad \left(-\frac{\pi}{2} < x < \frac{\pi}{2}\right).$$

Also assume that T is bounded, so that it does not blow up at $y \to \infty$.

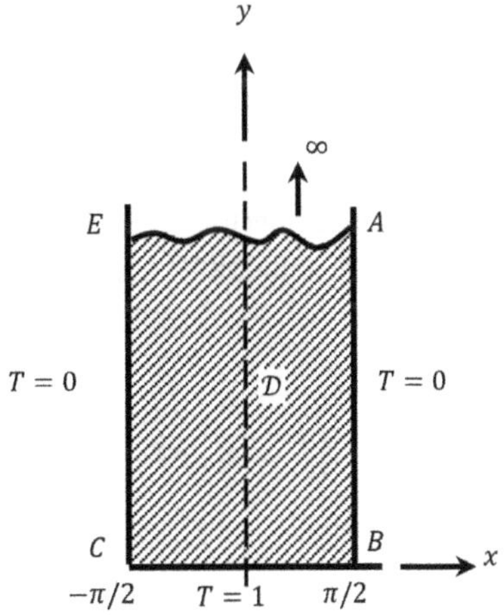

3.11 Consider the first quadrant of the unit disk shown in Figure P3.11, with the conditions $\phi(0,y) = b$, $\phi(x,0) = a$, and $\phi(x^2 + y^2 = 1) = c$. Determine the distribution of ϕ in $\mathcal{D}$, assuming ϕ is harmonic.

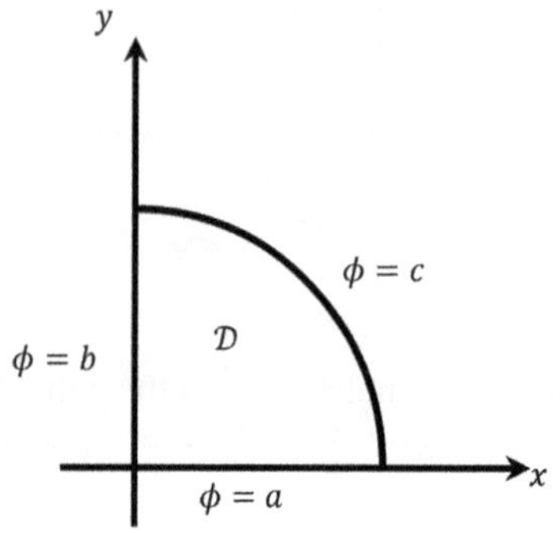

3.12 Determine the steady state temperature distribution exterior to two circles $|z| = 1$ and $|z - (9/2)| = 5/2$ (Figure P3.12), if the boundary of the first circle is maintained at T_o and the boundary of the second at T_1.

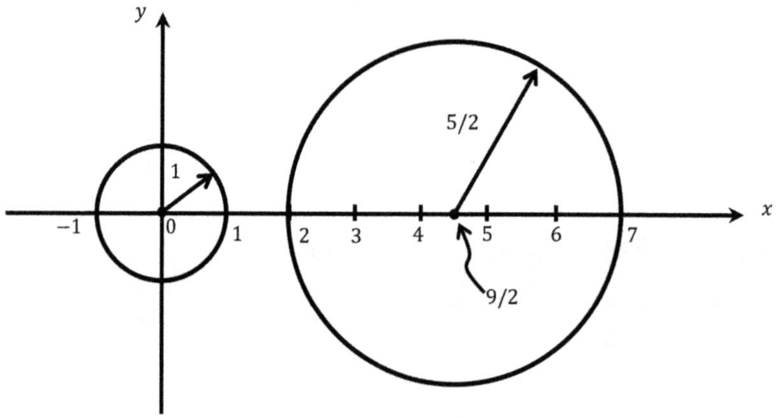

3.13 Determine the composite mapping that transforms the sector $-\pi/6 \leq \arg z \leq \pi/6$ onto the right half plane of the w-plane and then to the unit disk $|\zeta| \leq 1$, as ζ sketched in Figure P3.13.

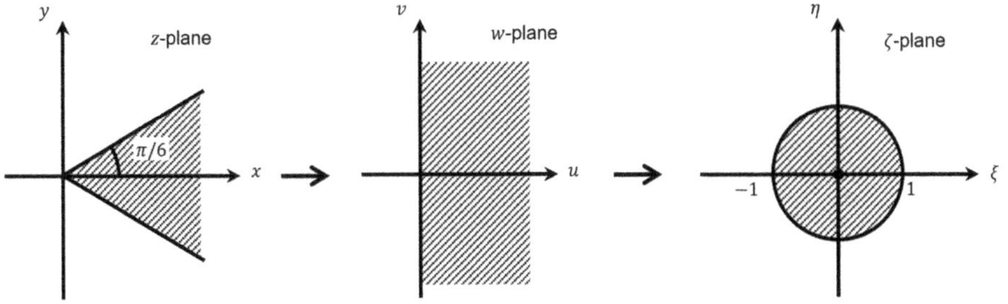

3.14 Find the equation of the locus $\arg(z^2) = -\pi/4$.

3.15 Consider the triangular region shown in Figure P3.15(a), with the mapping $w(z) = z^2$.

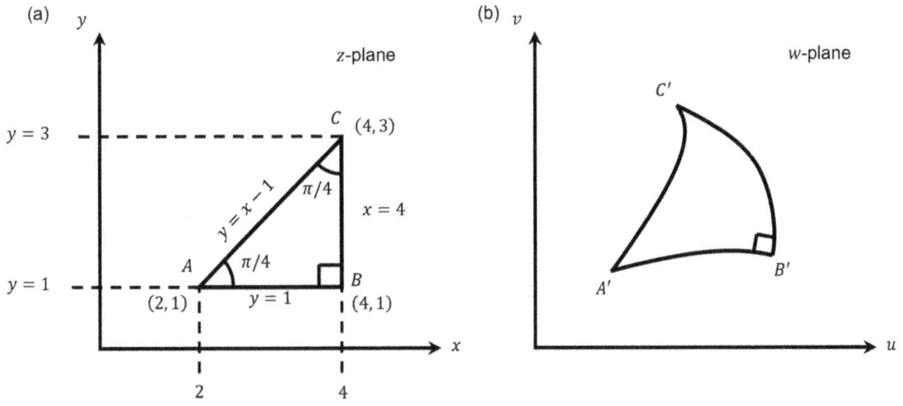

(a) Determine the image of the points A, A, A.

(b) Determine the equation of the curves A, A, and $C' - A'$.

(c) Determine the interior angles of the curvilinear region $A' - B' - C'$.

3.16 Suggest a transformation that maps the disk $|z| < 2$ onto the domain $\mathcal{D}'$: $\xi + \eta > 0$ in the $\zeta \equiv \xi + i\eta$ plane (Figure P3.16).

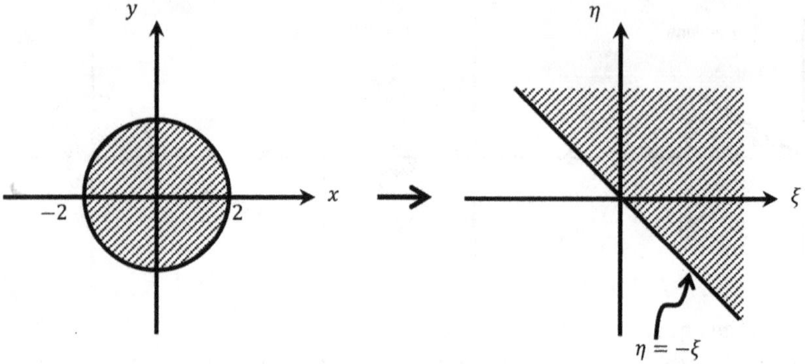

3.17 Determine the equation of the curve in the w-plane that is the image of $x = a$ (or $Re\ z = a$) in the z-plane under the transformation $w(z) = z^2$ (Figure P3.17).

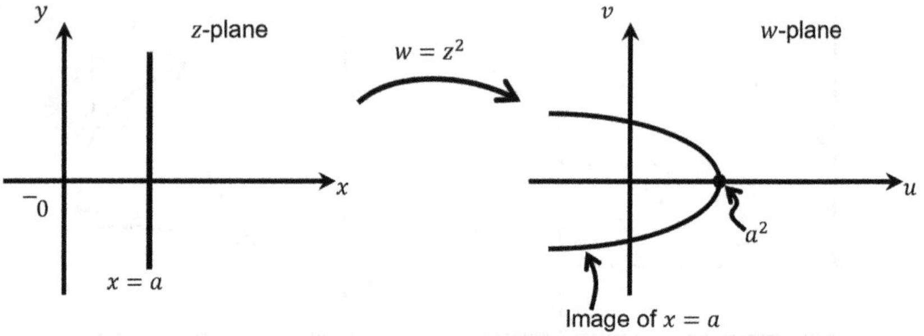

3.18 Find an LFT that maps the points 3, $1 - i$, and $2 - i$ in the z-plane to the corresponding points i, 4, and $6 + 2i$ in the w-plane.

3.19 Find an LFT that maps the points i, 1, $2 + i$ in the z-plane to the corresponding points $4i$, $3 - i$, and ∞ in the w-plane.

3.20 Find a transformation that maps $Re(z) > 0$ conformally onto the disk $|\zeta - i| < 3$ (Figure P3.20).

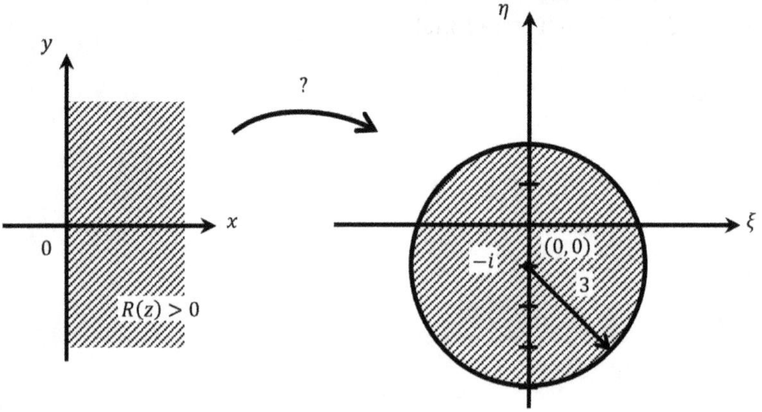

3.21 Determine the LFT that maps the points $z_1 = 1$, $z_2 = 0$, $z_3 = -1$ in the z-plane to the corresponding points $w_1 = i$, $w_2 = 1$, $w_3 = \infty$ in the w-plane.

3.22 Find the images of the circle $|z| = 1$ under the transformation $w(z) = (z+2)/(z-1)$. What are the images of the interiors of this circle?

3.23 Find an LFT that maps the points $z_1 = \infty$, $z_2 = 0$, and $z_3 = 1$ on the real axis to the points 1, i, and -1 on the unit circle $|w| = 1$.

3.24 Determine the steady-state temperature distributions in a quadrant when the boundary planes are kept at fixed temperatures except for strips of equal width at the corner that are insulated (Figure P3.24). Note that insulation implies $\partial T/\partial n = 0$, or an adiabatic condition.

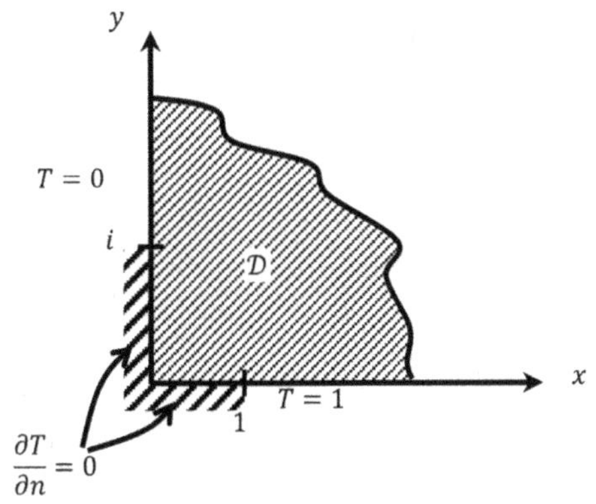

3.25 Determine the steady-state temperature distributions in a solid extending over the infinite region $0 \le \theta \le 3\pi/2$ (Figure P3.25), if strips of the boundary planes are unit wide at the corner are insulated, and if $T = 0$ on the rest of the plane $\theta = 0$ and $T = 1$ on the rest of the plane $\theta = 3\pi/2$.

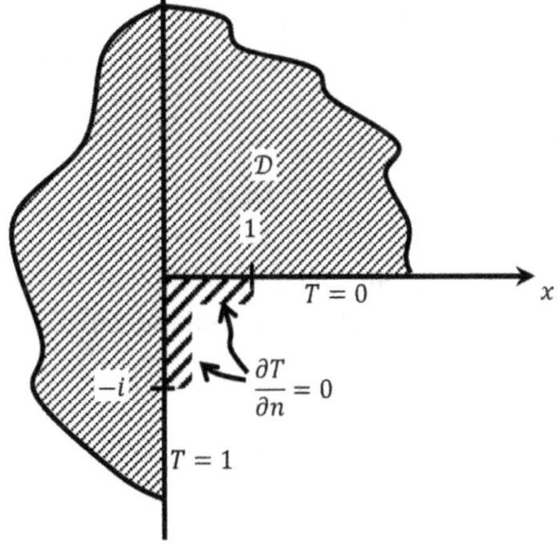

3.26 Determine the distributions of a harmonic function $\phi(r,\theta)$ in the space bounded by the planes $\theta=0$ and $\theta=\pi/4$ and the cylinder $r=1$ if $\phi=0$ on the planes and $\phi=1$ on the cylinder.

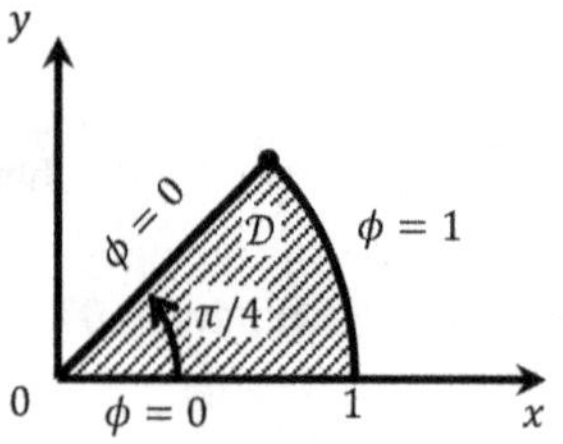

3.27 Determine the solution of the boundary value problem (Figure P3.27):

$$\frac{\partial^2\phi}{\partial x^2}+\frac{\partial^2\phi}{\partial y^2}=0,\ \left(0<x<\frac{\pi}{2},\ y>0\right),$$

$$\phi(x,0)=0,\quad \phi(0,y)=1,\ \phi\left(\frac{\pi}{2},y\right)=0.$$

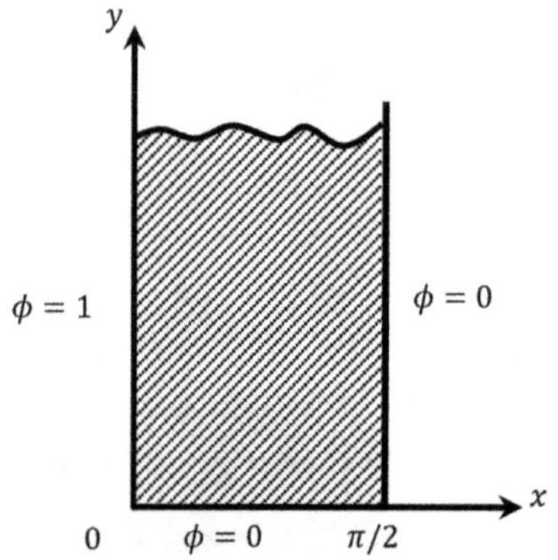

3.28 What is the image of $|z+1|$ under the transformation $w(z)=1/z$?

3.29 How does the hyperbola $x^2-y^2=1$ transform under $w(z)=1/z$?

3.30 (a) Show that $w(z)=i\frac{1-z}{1+z}$ maps the circle $|z|=1$ onto the real axis of w-plane.
(b) Also show that the interior of the circle in the z-plane corresponds to upper half of the w-plane.

3.31 Find the fixed points of the transformation $w(z)=(3z-4)/(z-1)$.

3.32 Find the fixed points of the transformation $w(z)=(z-1)/(z+1)$.

3.33 Find the fixed point of $w(z)=(5z+4)/(z+5)$.

3.34 Find the image of $|z-1|=1$ under the conformal transformation $w(z)=z^2$.

3.35 Determine the temperature distributions in UHP of the z-plane, subject to the boundary conditions

$$T(x,0) = \begin{cases} T_a & x < -1 \\ T_b & x > 1 \end{cases} \quad \text{and} \quad \frac{\partial T}{\partial y}(x,0) = 0 \; -1 < x < 1.$$

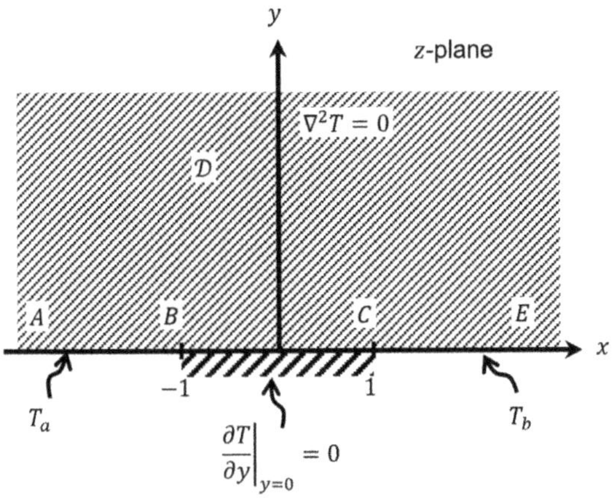

3.36 Show that the composition of the three transformations

$w(z) = \frac{1+z}{1-z}$, $t = w^2$, and $\zeta = \frac{t+1}{t-1}$

is equivalent to the Joukowski transformation $(1/2)[z + (1/z)]$.

3.37 Map the sector in Figure P3.37(a) to the UHP in Figure P3.13.

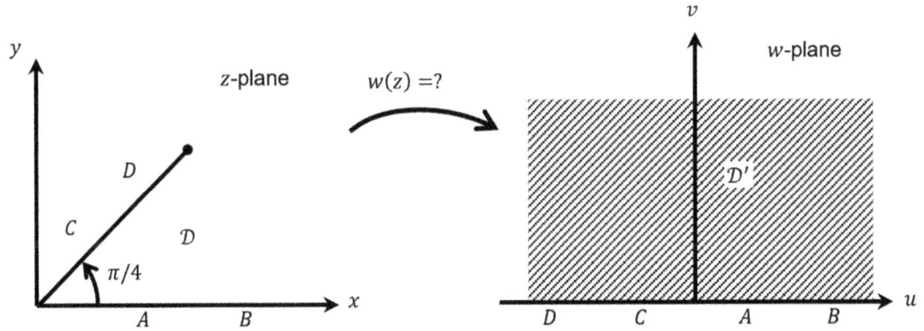

3.38 Determine the image of the unit circle $|z| = 1$, $\arg z = \theta$, under the Joukowski map $w(z) = (1/2a)[z + (1/z)]$. Determine the mapping of circles with radius R_o, $R_o > 1$.

3.39 Describe how you could obtain airfoil shapes with the Joukowski transformation $w(z) \equiv (1/2a)[z + (1/z)]$.

3.40 Show that the LFT that takes z_1, z_2, z_3 to w_1, w_2, w_3 can be expressed as the determinant of a matrix $\mathbf{A}$ equal to zero, where

$$\mathbf{A} = \begin{bmatrix} 1 & z & w & zw \\ 1 & z_1 & w_1 & z_1 w_1 \\ 1 & z_2 & w_2 & z_2 w_2 \\ 1 & z_3 & w_3 & z_3 w_3 \end{bmatrix}.$$

3.41 Map the UHP of the z-plnae onto the domain consisting of the fourth quadrant (of the w-plane) together with the strip $0 < v < 1$ (Figure P3.41). (This has been referred to as a model of the continental shelf.)

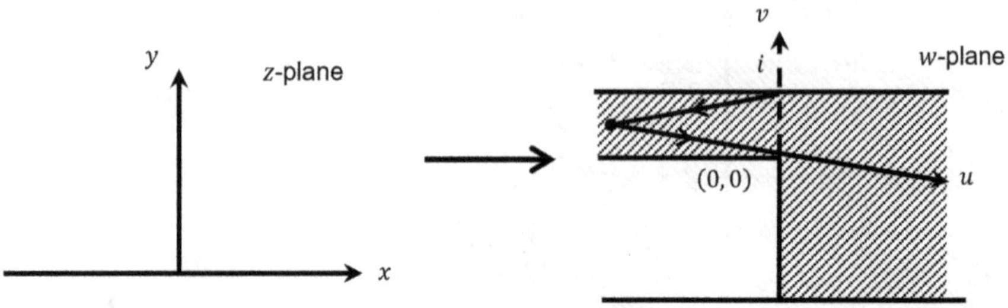

3.42 Determine a mapping that is harmonic in the doubly-split plane (Figure P3.42) and takes the values $\phi = -1$ and $\phi = +1$ on the two slits of the plane.

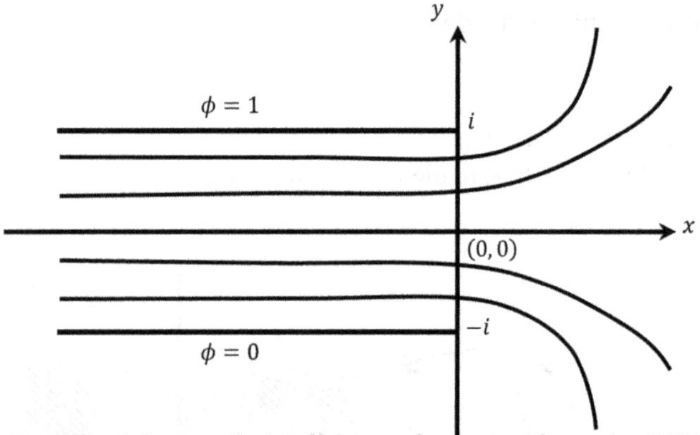

3.43 Suggest a Schwarz-Christoffel transformation from the UHP of the z-plane that could be used to obtain the interior of the triangular region in Figure P3.43.

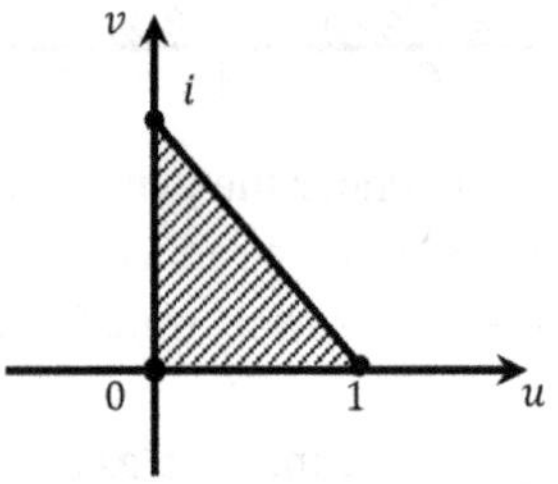

3.44 Show that $w(z) = e^{i\theta_0}\left(\frac{z - z_0}{z - \bar{z}}\right)$ transforms the UHP of the z-plane to the unit circle $|w| = 1$, and show that the transformation is one-to-one.

3.45 Consider the unit circle in Figure P3.45, whose center is in the UHP but circle passes through $z = x = 1$ (as shown) and encloses $z = x = -1$. Determine the image of the unit circle under the transformation $w(z) = (1/2)(z + 1/z)$. Note that C in the figure denotes "Contour."

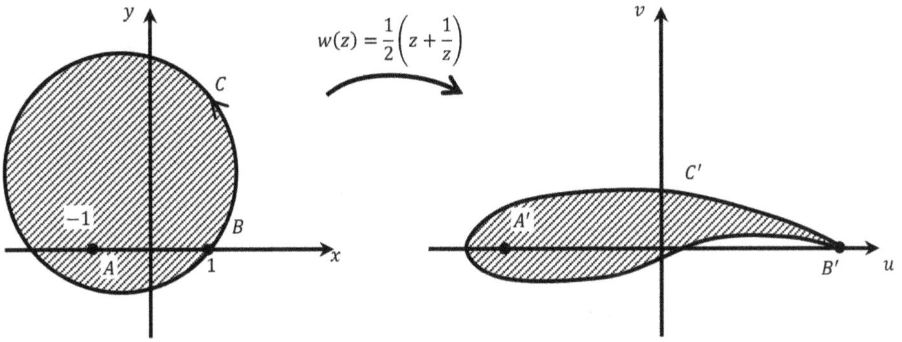

3.46 Consider the unit circle $|z| = 1$ in Figure P3.13, whose center is located on the real axis. Suppose that the circle in addition passes through $z = 1$ (the critical point of the transformation $f(z) = (1/2)[z + (1/z)]$ and the point $z = x = -1$ is located at an interior point on the real axis of the circle. Determine the image of the transformation. Note that C in the figure denotes "Contour."

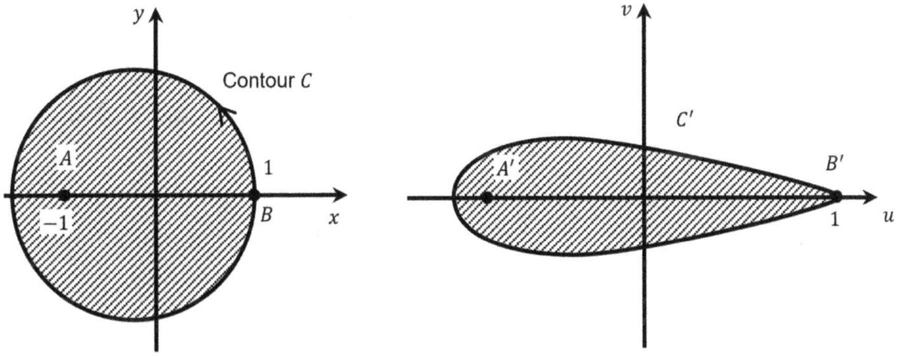

3.47 Suggest a transformation from the infinite section in Figure P3.47(a) and semi-infinite parallel horizontal to lines in Figure P3.47(b). Take

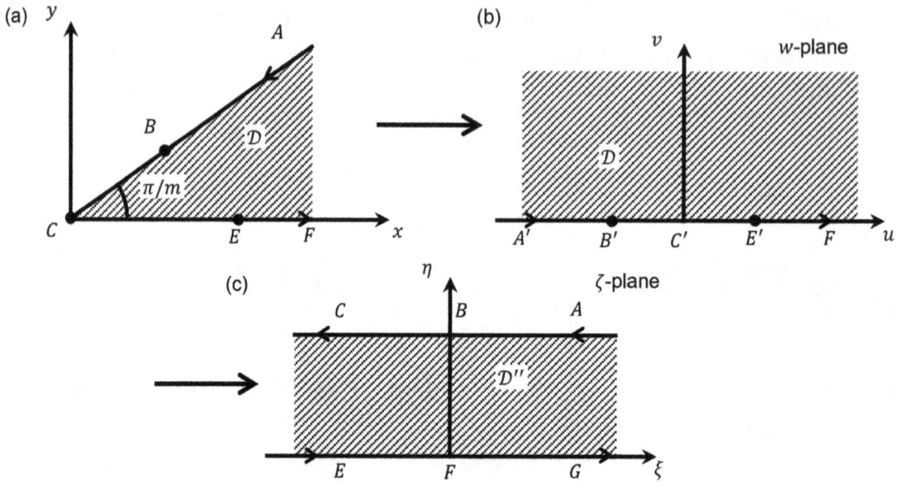

3.48 Suggest a transformation from Figure P3.48(a) onto Figure P3.48(b).

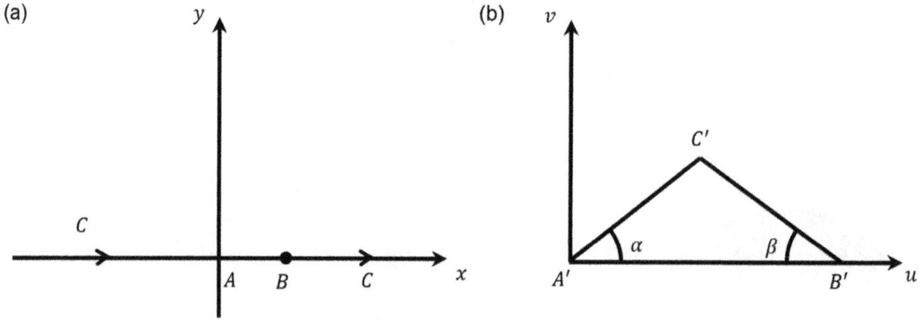

3.49 What is the image of $|z-3i|=3$ under the mapping $w(z)=1/z$?

3.50 Find the LFT whose fixed points are 1 and 2.

Suggested reading

[1] Rogers, D.F. and Adams, J.A. Mathematical Elements for Computer Graphics, 2nd Edition, New York, NY, USA, McGraw-Hill, 1990.

[2] Robertson, J.M. Hydrodynamics in Theory and Application, London, UK, Prentice Hall, 1965.

[3] Carathéodory, C. Conformal Representation, Cambridge, Cambridge, UK, Cambridge University Press, 1932.

[4] Driscoll, T.A. and Trefethen, L.N. Schwarz-Christoffel Mapping, Cambridge, UK, Cambridge University Press, 2002.
[5] Brown, J.W. and Churchill, R.V. Complex Variables and Applications, 6th Edition, New York, NY, USA, McGraw-Hill, 1996.
[6] Milne-Thomson, L.M. Theoretical Hydrodynamics, Fifth Edition, London, UK, Macmillan Press, 1968.

(See the Suggested Reading items of Chapter 2.)

4 Application of complex variable theory and conformal mapping to perfect fluid flow

This chapter is dedicated to the application of complex variable theory and conformal mapping to perfect fluid flow. We start out by giving the complex potential and complex velocity for various canonical and superposed flows, including uniform flows; sources, sinks, vortex flows, flow in a sector, flow around a sharp edge, flow due to a doublet, and flow around a circular cylinder with and without circulations. The Blasius integral laws, which require complex variable contour integration techniques, are presented for calculating the forces and moments exerted on a solid body due to a potential flow around the body. The Joukowski transformation and conformal mapping are presented in the manner in which they are used for generating flows over realistic airfoil geometries, as in aircraft wings, wind turbines, and general turbomachinery blades. The foundational theories for the materials contained in this chapter can be found in [1, 2], but our applied presentation parallels that in [3]. For the most part, the notations used in the latter have been adopted in this chapter.

4.1 Notations

If we denote the velocity vector by $\mathbf{u} = (u, v)$, the scalar velocity potential by ϕ, and the stream function by ψ, we have these definitions from two-dimensional (2D) potential (perfect) flow theory:

$$\mathbf{u} = \nabla\phi, \quad u = \frac{\partial \phi}{\partial x}, \quad v = \frac{\partial \phi}{\partial y}, \quad u = \frac{\partial \psi}{\partial y}, \quad v = -\frac{\partial \psi}{\partial x}.$$

We see that the introduction of the stream function ψ allows us to satisfy the incompressible flow condition identically:

$$\frac{\partial u}{\partial x} + \frac{\partial v}{\partial y} = \frac{\partial}{\partial x}\frac{\partial \psi}{\partial y} - \frac{\partial}{\partial y}\frac{\partial \psi}{\partial x} = \frac{\partial^2 \psi}{\partial x \partial y} - \frac{\partial^2 \psi}{\partial y \partial x} = 0.$$

We also see that $\phi_x = \psi_y$ and $\phi_y = -\psi_x$, which are precisely the Cauchy-Riemann (C-R) conditions. As a consequence, we can define the complex velocity potential $F(z)$ as

$$F(z) = \phi(x, y) + i\psi(x, y),$$

which is analytic since ϕ, ψ satisfy the C-R conditions. For every analytic function $F(z)$ the real part is (automatically) a valid velocity potential and the imaginary part is a valid stream function. From complex variables: $\nabla^2\phi = 0$ and $\nabla^2\psi = 0$, which are precisely the equations for 2D ideal fluid flow.

https://doi.org/10.1515/9783111351179-005

There is a mapping implicit in a function such as $F(z)$. By studying this mapping for $\psi = \text{constant}$ (just as we have done in the geometric treatment of mapping) we can determine the flow field. The velocities may then be obtained from ψ distribution. $W(z) = dF/dz$ is also of interest (F is analytic):

$$W(z) = \text{complex velocity} = \frac{\partial\phi}{\partial x} + i\frac{\partial\psi}{\partial x} \equiv u - vi,$$

where u and v are the x, y components of the velocity field assumed to be in the complex plain $z = x + iy$. $W \equiv f(z)$ can be considered a transformation from z to w both of which are complex planes. The mapping is conformal if $f'(z) \neq 0$.

Note that

$$W\bar{W} = (u - iv)(u + iv) = u^2 + v^2 = \mathbf{u} \cdot \mathbf{u} = \nabla\phi \cdot \nabla\phi,$$

which is numerically equal to twice the kinetic energy per unit mass. In cylindrical coordinates (Figure 4.1) with (u_R, u_θ) as the velocities in (r, θ) coordinate directions, we can write

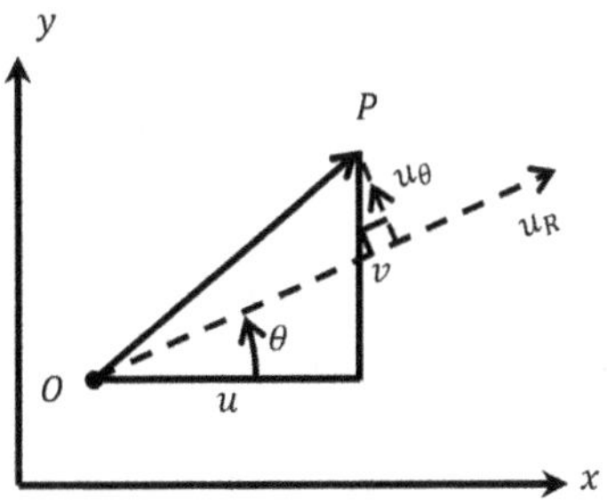

Figure 4.1: The resolution of the velocity components in polar coordinates.

$$u = u_R \cos\theta - u_\theta \sin\theta, \quad v = u_R \sin\theta + u_\theta \cos\theta \Rightarrow W = (u_R - iu_\theta)e^{-i\theta},$$

using $W = u - iv$ and $\cos\theta - i\sin\theta = e^{-i\theta}$.

For a 2D potential flow, knowing the complex potential is tantamount to knowing the velocities, stream function, the forces, and the moments. As we briefly discuss in this subsection, this quantity is known for a variety of simple flows including uniform flows, sources, sinks, vortex flows, flow in a sector, flow around a sharp edge, flow due to a doublet, and flow around a circular cylinder with and without circulation. Some details on the complex potential function for these flows are presented in this chapter.

4.2 The complex potential for familiar flows

4.2.1 Uniform flows

In presenting the complex potential for uniform flows, we distinguish between uniform flows aligned with the x-coordinate direction, y-coordinate direction, and at an angle to the two orthogonal coordinate directions defined by x and y.

(i) Flow in the x-direction: The $F(z)$ for this flow is $F(z) = cz$ where c is a real number. With $z = x + iy$, we have $c(x + iy) = cx + icy$ so that $\phi = cx$, $\psi = cy$, and constant ψ lines are parallel to the x-axis. The velocity is given by $W(z) = dF/dz = c = u - iv = u$ and $v = 0$. This describes a uniform rectilinear flow in the x-direction, with magnitude c, as shown in Figure 4.2(a).

Note that the complex potential for a flow whose magnitude is U in the positive x-direction will be given by $F(z) = Uz$.

(ii) Flow in the y-direction: The $F(z)$ for this flow is $F(z) = -icz$, where again c is a real constant and $i = \sqrt{-1}$ as usual. With this complex potential, $\phi = cy$, $\psi = -cx$, and $W(z) \equiv u - iv = -ic$ so that $u = 0$ and $v = c$. $F(z) = -icz$ $(= -ic(x + iy) = -icx + cy \Rightarrow \phi = cy,\ \psi \equiv -cx)$. This is a uniform rectilinear flow in y-direction, with magnitude c.

Thus, we see that the complex potential for a flow whose magnitude is V in the positive y-direction will be given by $F(z) = -iVz$. Such a flow is shown in Figure 4.2(b).

(iii) Flow inclined to x and y: The $F(z)$ for this flow is $F(z) = ce^{-i\alpha}z$. This gives $W(z) = u - iv = c\cos\alpha - ic\sin\alpha$ so that $u = c\cos\alpha$, $v = c\sin\alpha$, which is a uniform flow inclined at an angle α to the x-axis (Figure 4.2(c)). The complex potential for such a flow with magnitude V is $F(z) = Ve^{-i\alpha}z$.

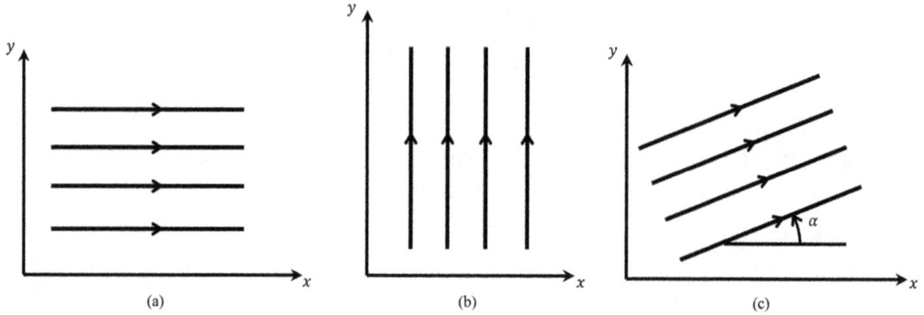

Figure 4.2: The streamlines for uniform flow (a) in the x-direction, (b) in the y-direction, (c) at an angle to the x- and y-coordinate directions.

4.2.2 Sources, sinks, vortex flows

(a) Sources: $F(z)$ for sources is $F(z) = c\log z$. We will assume there is a branch cut that limits the angle θ to $0 < \theta < 2\pi$ for this multivalued function. We can write $F(z) = c\log z = c\log Re^{i\theta} = c\log R + ic\theta \equiv \phi(x,y) + i\psi(x,y) \Rightarrow \phi = c\log R$. Thus, a constant ϕ implies that R is constant, so that the equipotential lines are the circles $R = \text{constant}$. The $\psi = c\theta$, with streamlines that are the radial lines $\theta = \text{constant}$. The complex velocity is

$$W(z) = \frac{c}{z} = \frac{c}{R}e^{-i\theta},$$

or using the cylindrical form of $W(z)$, $W(z) = (u_R - iu_\theta)e^{-i\theta}$, we have $u_R = c/R$, $u_\theta = 0$. The flow field is shown in Figure 4.3. This flow field is called a source: There is no azimuthal component of velocity so that the velocity is purely radial, with a magnitude that decreases with increasing radius. The origin is a singular point as $F'(z) \to \infty$.

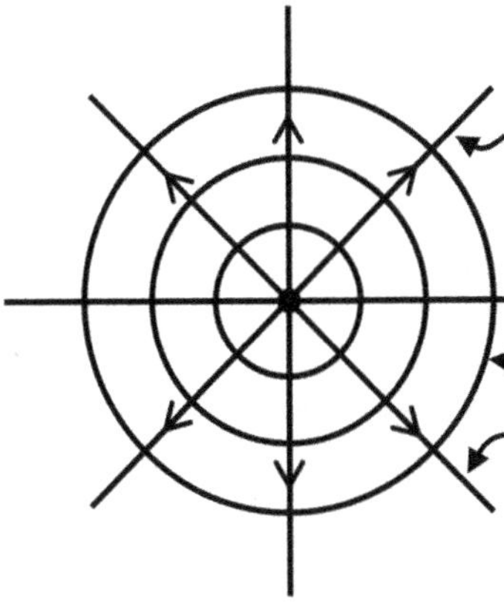

Figure 4.3: The flow field for a source.

Sources are characterized by their strength, m, which is the *volume* of fluid leaving the source (radially and at a fixed radius) per unit time per unit depth of the flow field. Utilizing the elemental sector in Figure 4.4, we can write

$$m = \int_0^{2\pi} u_R R d\theta = \int_0^{2\pi} c d\theta = 2\pi c, \text{ or } c = \frac{m}{2\pi} \text{ and } F(z) = \frac{m}{2\pi}\log z.$$

This $F(z)$ is the complex potential for a source of strength m located at $z = 0$. If the source is located at $z = z_0$, the complex potential will be $F(z) = (m/2\pi)\log(z - z_0)$.

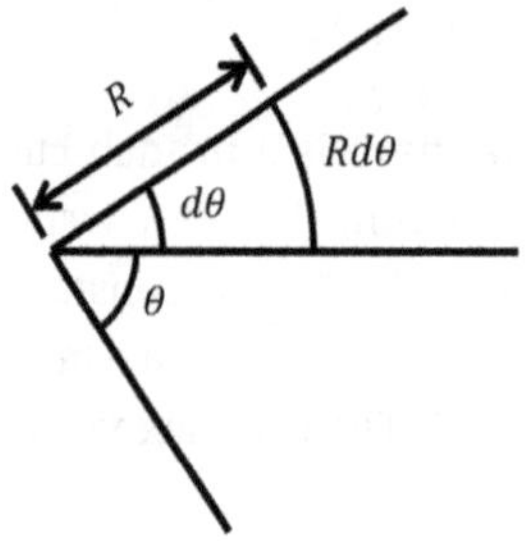

Figure 4.4: Elemental sector for defining the source strength. Note that the strength is basically the radial volume flow rate at a fixed radius per unit distance along the axis of the cylinder.

(b) Sinks: The complex potential for a sink (negative source) is obtained by replacing m by $-m$ in $F(z)=(m/2\pi)\log(z-z_0)$, or $F(z)=-(m/2\pi)\log(z-z_0)$.

(c) Vortex: The complex potential for a vortex is $F(z)=-ic\log z=-ic\log Re^{i\theta}=-ic\log R-ic(i\theta)=c\theta-ic\log R\Rightarrow\phi=c\theta,\ \psi=-c\log R$. Therefore, equipotential lines are the radial lines $\theta=$ constant, while the streamlines are the circles $R=$constant. See Figure 4.5.

The velocities are given by

$$W(z)=\frac{dF}{dz}=-i\frac{c}{z}=-i\frac{c}{R}e^{-i\theta},\quad \text{or } u_R=0,\ u_\theta=\frac{c}{R}.$$

The sign of c determines the direction of flow: clockwise or counterclockwise. A vortex is characterized by its strength, which may be measured by the circulation Γ associated with it. The Γ associated with the singularity at the origin is

$$\text{Strength of vortex}=\Gamma=\oint \boldsymbol{u}\cdot \boldsymbol{dl}=\int_0^{2\pi} u_\theta R\,d\theta=\int_0^{2\pi} c\,d\theta=2\pi c$$

This is the dot product of the azimuthal velocity and an elemental length in the azimuth, integrated along the circumference at a fixed radius.

(The reader should connect this with the Stokes law: $\int_V \boldsymbol{\omega}\cdot d\mathbf{A}=\int_l \mathbf{u}\cdot d\mathbf{l}$, where $\boldsymbol{\omega}$ is the vorticity vector, V is volume of domain, dA is elemental area, and $\boldsymbol{dl}$ is elemental length vector.) Here, $F(z)=-ic\log z=-i(\Gamma/2\pi)\log z$ for a positive (i.e., CCW) vortex, when the center is located at $z=0$, and $F(z)=-(i\Gamma/2\pi)\log(z-z_0)$ for a positive vortex whose center is at z_0.

Remark 4.1 The circulation is zero for any closed contour which does not include a singularity, consistent with the Cauchy-Goursat theorem. The flow will be irrotational in this case.

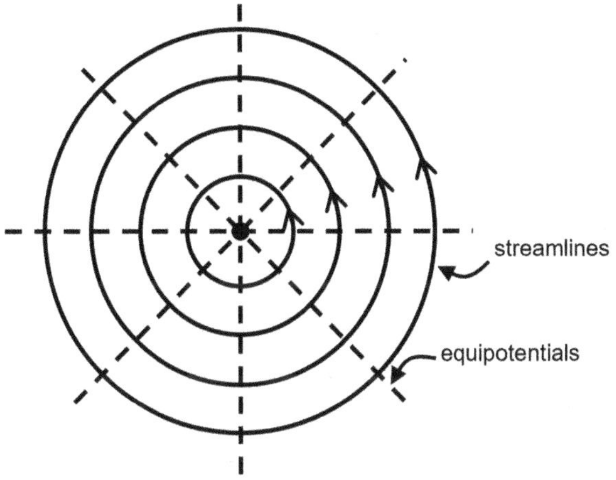

Figure 4.5: The flow field for a vortex.

4.2.3 Flow in a sector

The flows in sharp bends have $F(z) \propto z^n$, $n \geq 1$, where $n = 1$ implies a uniform rectilinear flow, with a sector angle π in Figures 4.6 and 4.7. For a far-field velocity of magnitude U, we will have

$$F(z) = Uz^n = UR^n e^{in\theta} = UR^n(\cos n\theta + i \sin n\theta),$$

which gives $\phi = UR^n \cos n\theta$, $\psi = UR^n \sin n\theta$, with a zero streamline when $\theta = 0$, π/n, or $k\pi/n$, $k = 0, 1, \ldots$, or when $\theta = \pi$, $\pi/2$ for $n = 1, 2$ respectively. The origin and angular measure in a sector are shown in Figure 4.6.

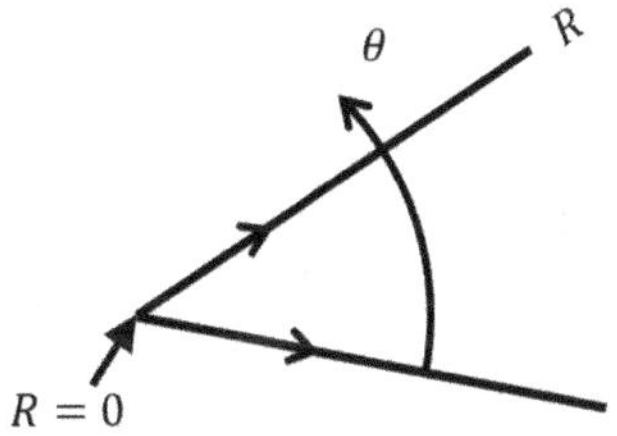

Figure 4.6: The origin and angular measure in a sector.

The streamline $\psi = 0$ corresponds to the radial lines $\theta = 0$ and $\theta = \pi/n$. Other streamlines between these two lines are given by $R^n \sin n\theta =$ constant (see Figure 4.7). With $F(z) = Uz^n$, we will have

$$W(z) = \frac{dF}{dz} = nUz^{n-1} = nUR^{n-1}e^{i(n-1)\theta} = \left(nUR^{n-1}\cos n\theta + inUR^{n-1}\sin n\theta\right)e^{-i\theta}$$

$$\Rightarrow u_R = nUR^{n-1}\cos n\theta,\ u_\theta = -nUR^{n-1}\sin n\theta.$$

In $F(z) = Uz^n$: $n=1$ gives a uniform rectilinear flow, while $n=2$ gives the flow in a right-angled corner.

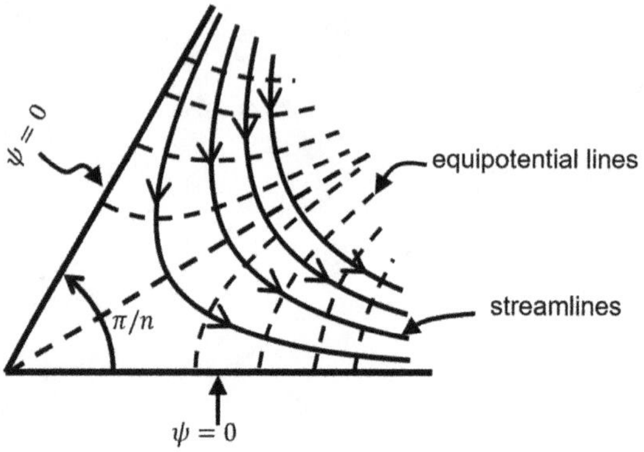

Figure 4.7: Flow in a sector.

The reader is reminded that equipotential lines and streamlines are orthogonal, as is evident from:

$$\nabla\phi\cdot\nabla\psi = (u\mathbf{i} + v\mathbf{j})\cdot\left(\frac{\partial\psi}{\partial x}\mathbf{i} + \frac{\partial\psi}{\partial y}\mathbf{j}\right) = \left(\frac{\partial\psi}{\partial y}\mathbf{i} - \frac{\partial\psi}{\partial x}\mathbf{j}\right)\cdot\left(\frac{\partial\psi}{\partial x}\mathbf{i} + \frac{\partial\psi}{\partial y}\mathbf{j}\right) = 0.$$

4.2.4 Flow around a sharp edge

The complex potential for this case is $F(z) \propto z^{1/2}$ or $F(z) = cz^{1/2}$, where c is real, and θ in z is $0 < \theta < 2\pi$. Setting $z = Re^{i\theta}$, we have $F(z) = cR^{1/2}e^{i\theta/2}$, or $\phi = cR^{1/2}\cos(\theta/2)$, $\psi = cR^{1/2}\sin(\theta/2)$. Lines with $\theta = 0,\ 2\pi$ correspond to $\psi = 0$ and the other streamlines between these are given by $R^{1/2}\sin(\theta/2) =$ constant. The streamlines and equipotential lines are shown in Figure 4.8. The velocities can be obtained from

$$W(z) = \frac{c}{2z^{1/2}} = \frac{c}{2R^{1/2}}e^{-i\theta/2} = \frac{c}{2R^{1/2}}e^{-i\theta}\cdot e^{+\frac{i}{2}\theta} = \frac{c}{2R^{1/2}}\left(\cos\frac{\theta}{2} + i\sin\frac{\theta}{2}\right)e^{-i\theta},$$

yielding

$$u_R = \frac{c}{2R^{1/2}}\cos\frac{\theta}{2},\quad u_\theta = -\frac{c}{2R^{1/2}}\sin\frac{\theta}{2}.$$

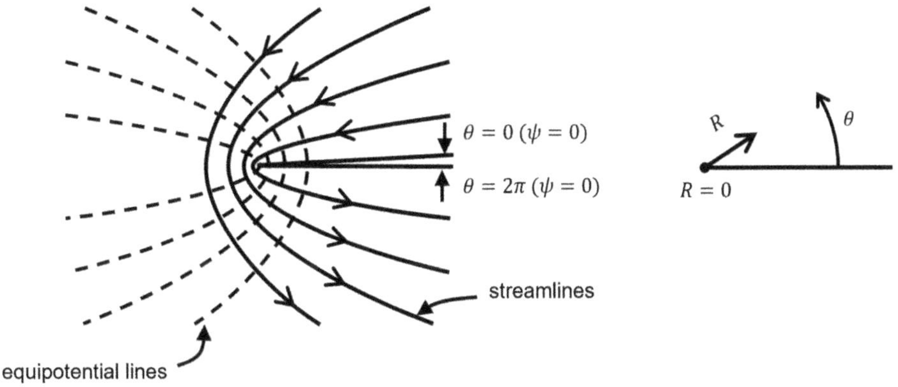

Figure 4.8: The flow field and coordinate system for flow around sharp edge.

Note that the corner is a singular point at which u_R, $u_\theta \to \infty$ since both u_R, u_θ vary as $R^{-1/2}$.

4.2.5 Flow due to a doublet

The singularity at $z = 0$ of the complex potential $F(z) \equiv 1/z$ is called a doublet. We can determine ϕ, ψ, u_R, and u_θ in a straightforward manner as we have for the previous examples. However, here, we will use a limiting procedure to expose some physics [3]. We will consider a source of strength m and a sink of strength m, each of which is located on the real axis a small distance ε from the origin (Figure 4.9). We start by superposing the complex potentials for the source and sink:

$$F(z) = \frac{m}{2\pi}\log(z+\varepsilon) - \frac{m}{2\pi}\log(z-\varepsilon) = \frac{m}{2\pi}\log\left(\frac{z+\varepsilon}{z-\varepsilon}\right) = \frac{m}{2\pi}\log\left(\frac{1+\varepsilon/z}{1-\varepsilon/z}\right).$$

Since $\varepsilon/z \ll 1$, we expand the denominator in the argument of log as

$$F(z) = \frac{m}{2\pi}\log\left[\left(1+\frac{\varepsilon}{z}\right)\left(1+\frac{\varepsilon}{z}+O\left(\frac{\varepsilon^2}{z^2}\right)\right)\right] = \frac{m}{2\pi}\log\left[1+2\frac{\varepsilon}{z}+O\left(\frac{\varepsilon^2}{z^2}\right)\right].$$

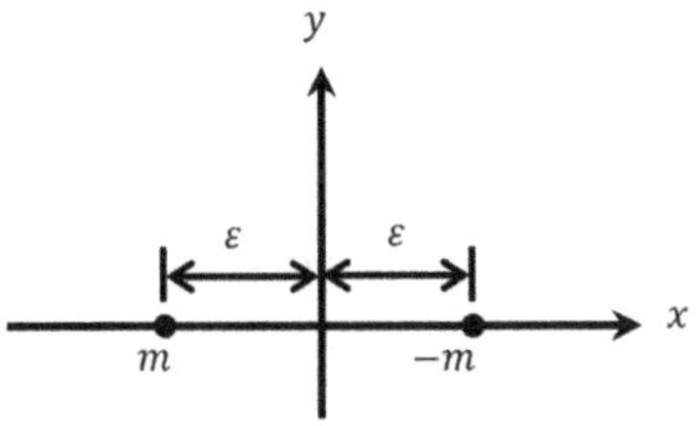

Figure 4.9: Source and sink.

Since $2\varepsilon/z \equiv \gamma \ll 1$ and $\log(1+\gamma) = \gamma - \gamma^2/2 + \gamma^3/3 - \gamma^4/4 + \ldots, \ |\gamma| < 1$, we have $\log(1+2\varepsilon/z) = 2\varepsilon/z + O(2\varepsilon/z)^2$ so that $F(z) = (m/2\pi)\left[2(\varepsilon/z) + O(\varepsilon^2/z^2)\right]$.

We then let $\varepsilon \to 0$ while $m \to \infty$ in such a way that $\lim_{\varepsilon\to 0}(m\varepsilon) = \pi$, giving

$$F(z) = \frac{\mu}{z},$$

where μ is a constant. (Note $\varepsilon \to 0$, $m \to \infty$ ensures $m\varepsilon$ (in $(m/2\pi)\cdot 2(\varepsilon/z)$) is $O(1)$ if the assumption is made that the approach to 0 and ∞ takes place at the same rate.) Obviously, an analogous requirement or restriction must be made for the physical flow for which the resultant $F(z)$ is valid.

Thus, $F(z) = \mu/z$ is the complex potential of the flow produced by superposing a very strong ($m \to \infty$) source and a very strong sink which are very close together ($\varepsilon \to 0$), and the doublet is centered at $z = 0$. We will have $F(z) = \mu/(z - z_0)$ if the doublet is centered at $z = z_0$. We can now use this complex potential to calculate the flow field, using $F(z) = \phi + i\psi$ and $z = x + iy$:

$$F(z) = \frac{\mu}{z} = \frac{\mu}{x+iy} = \frac{\mu\overline{x+iy}}{(x+iy)\overline{(x+iy)}} = \mu\frac{(x-iy)}{x^2+y^2} = \underbrace{\mu\frac{x}{x^2+y^2}}_{\phi} + i\underbrace{\frac{-\mu y}{(x^2+y^2)}}_{\psi}.$$

The streamlines $\psi =$ constant are given by $\psi = -\mu y/(x^2+y^2)$, which is $x^2+y^2 = -\mu y/\psi$ or $x^2 + (y+\mu/2\psi)^2 = (\mu/2\psi)^2$. This is a circle of radius $(\mu/2\psi)$, center $(x=0,\ y=\mu/2\psi)$ (Figure 4.10).

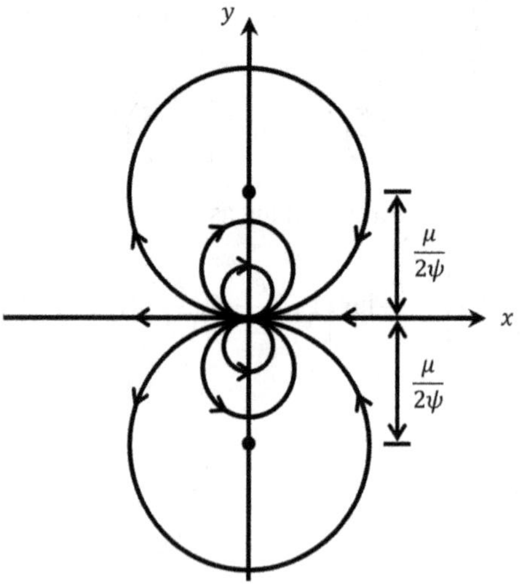

Figure 4.10: The streamline pattern for $\lim \varepsilon \to 0$ with $m\varepsilon =$ constant $= O(1)$.

The velocities can be obtained from

$$W(z) = -\frac{\mu}{z^2} = -\frac{\mu}{R^2}e^{-i2\theta} = -\frac{\mu}{R^2}e^{-i\theta}e^{-i\theta} = -\frac{\mu}{R^2}(\cos\theta - i\sin\theta)e^{-i\theta}$$

or

$$u_R = -\frac{\mu}{R^2}\cos\theta,\ u_\theta = -\frac{\mu}{R^2}\sin\theta.$$

(Remember that the cylindrical form of the complex velocity is $W = (u_R - iu_\theta)e^{-i\theta}$.)

Remark 4.2
1) The singularity at $z = 0$ is very important to the flow in doublets.
2) The complex potential for a doublet of strength μ located at $z = z_0$ is $F(z) = \mu/(z - z_0)$.
3) The principal use of the doublet is in superposition to generate more complex flows such as the flow around a circular cylinder, which is presented in the next subsection.

4.2.6 Flow around a circular cylinder, without circulation

Since the governing (Laplace) equation for ϕ and ψ is linear, we can superpose the complex potentials of simpler flows to obtain those of more complicated flows. For example, the $F(z)$ for the flow around a cylinder without circulation is obtained by adding the $F(z)$ for a uniform rectilinear flow and that of a doublet:

$$F(z) = Uz + \frac{\mu}{z}.$$

If we make the circle of radius $R = a$ a streamline of both the doublet and the parallel flow so that $z = ae^{i\theta}$ at the surface, we will have $F(z) = Uae^{i\theta} + (\mu/a)e^{-i\theta}$ on the surface of a cylinder. Therefore, on the surface using Euler's formula, we will have $F(z) = \underbrace{(Ua + \mu/a)\cos\theta}_{\phi} + i\underbrace{(Ua - \mu/a)\sin\theta}_{\psi|_{R=a}}$. If we make the common streamline the reference streamline $\psi = 0$ or $\psi|_{R=a} = 0$, then $\mu = Ua^2$, $\sin\theta \neq 0$, which is the strength of the doublet that makes the surface of the cylinder to be the $\psi = 0$ streamline.

It should be noted that the observed flow field outside the cylinder is due to the interaction of the doublet and the parallel flow (Figure 4.11). The doublet flow is entirely contained within the cylinder $R = a$, while the uniform flow is deflected by the doublet in such a way that it is entirely outside the region $R = a$. The circle $R = a$ is common to the two flow fields. As it were, for $R \geq a$ the flow field due to a doublet of strength $\mu = Ua^2$ and a uniform flow of magnitude U gives the same flow as that for a uniform flow of magnitude U past a circular cylinder of radius a. This is depicted in Figure 4.12.

The complex potential $F(z) = U(z + a^2/z)$ gives a flow that is symmetric about the x-axis. That is, the pressure is the same at corresponding points in the lower and upper surface $(p_j = p_j')$ (Figure 4.13). Hence, the lift force $= \int pds = 0$, where ds is an elemental surface on the circle. The other quantities u, v, ϕ, and ϕ also have the same symmetry about the x-axis.

We also have symmetry about the y-axis, which means that there is also no drag force. In a real system, viscosity is needed to obtain a hydrodynamic force on the cylinder. This also creates a boundary layer at the wall. However, for the perfect flow field discussed here, there is no viscosity, which means that we have to artificially introduce any forces that we want to model. In the next section, circulation is artificially added to give a lift force, so that the mathematical model for the flow over a cylinder could give results that agree with physical observation.

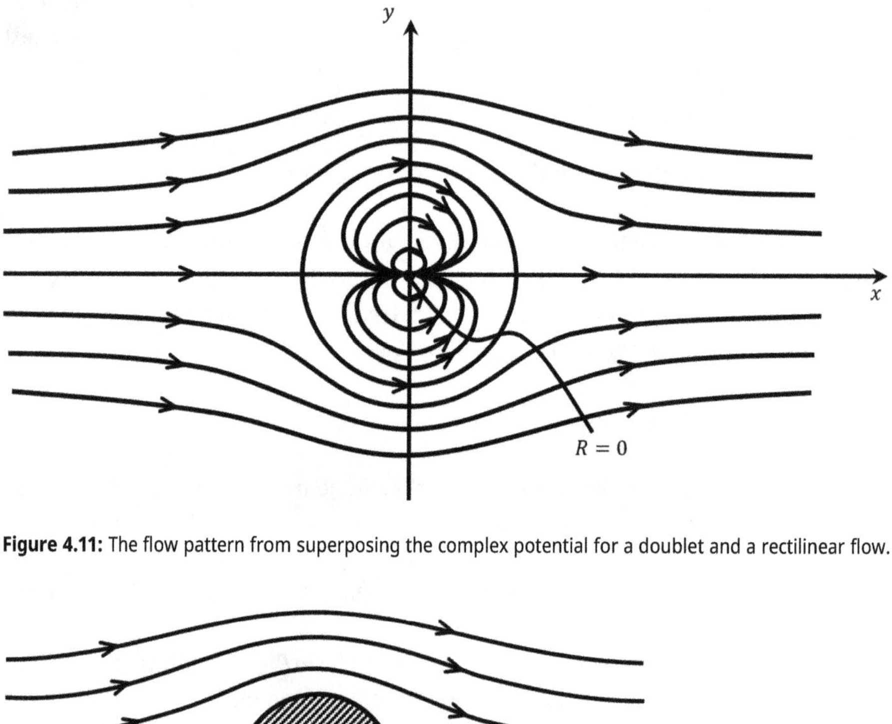

Figure 4.11: The flow pattern from superposing the complex potential for a doublet and a rectilinear flow.

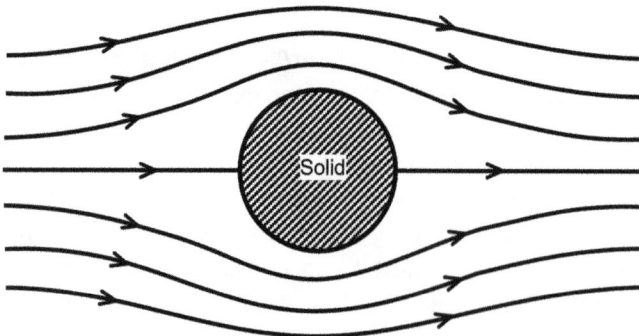

Figure 4.12: The external flow over a circular cylinder.

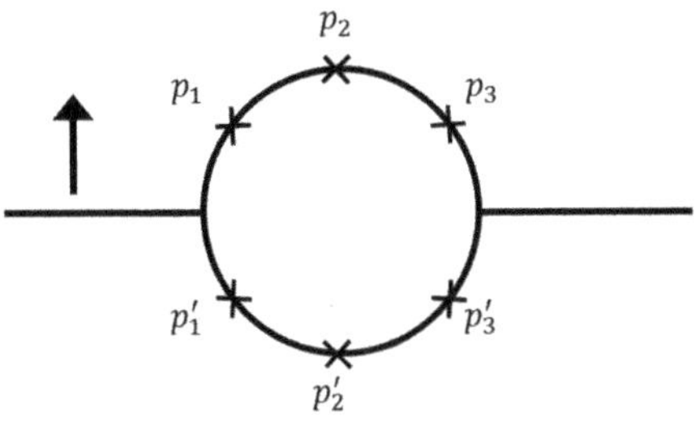

Figure 4.13: Depiction of the symmetry of the pressure field about the x-axis.

4.2.7 Flow around a circular cylinder, with circulation

Since a vortex has circulation, we artificially add one to the flow around a cylinder as a way to add circulation to the flow:

$$F(z) = U\left(z + \frac{a^2}{z}\right) + \frac{i\Gamma}{2\pi}\log z + c; \quad \Gamma = \oint \mathbf{u} \cdot d\mathbf{l} \text{ for the vortex.}$$

Later, c will be used to select the streamline. A negative vortex is added as this leads to a positive lift. Figure 4.14 depicts the sense of a vortex, where clockwise implies a negative vortex, and anticlockwise, a positive vortex.

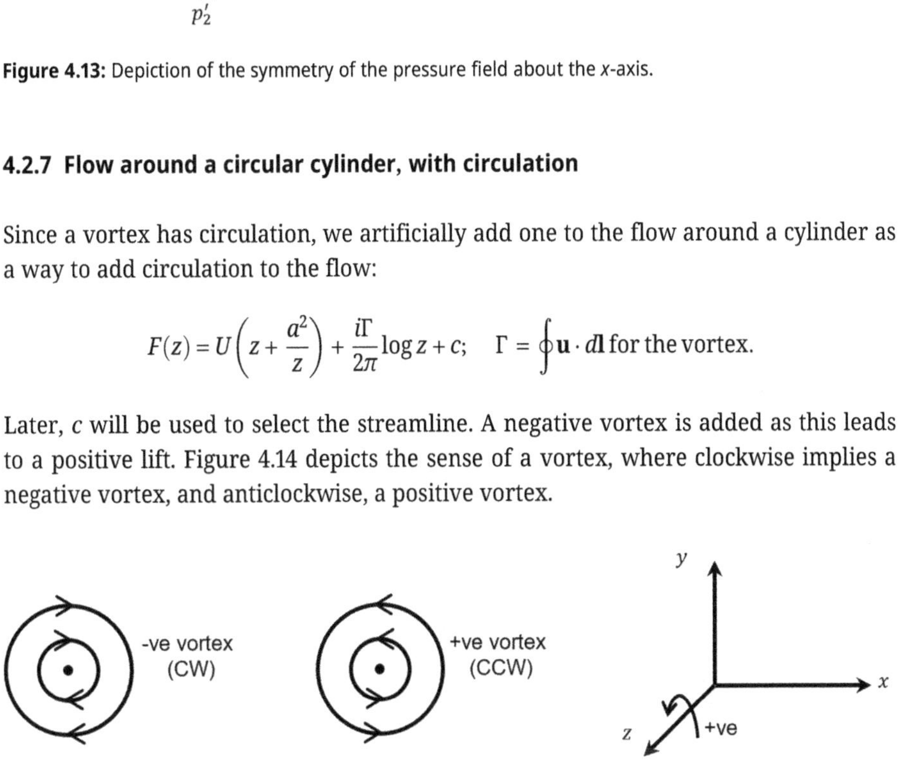

Figure 4.14: Sense of a vortex.

We note that $\psi|_{R=a} \neq 0$ with the addition of $(i\Gamma/2\pi)\log z$ (the vortex). That is, the stream function on the surface of the cylinder, $\psi|_{R=a}$, is not automatically zero when you add a vortex so that some adjustments have to be made if we need the cylinder surface to be a streamline. We can adjust c so that $\psi|_{R=a} = 0$ or $=$ any desired value. Specifically, with $z = ae^{i\theta}$ on $R = a$,

$$F(z) = U\left(ae^{i\theta} + ae^{-i\theta}\right) + \frac{i\Gamma}{2\pi}\log ae^{i\theta} + c = \underbrace{2Ua\cos\theta - \frac{\Gamma}{2\pi}\theta}_{\phi} + i\underbrace{\frac{\Gamma}{2\pi}\log a + c}_{\psi_a,}\quad .$$

assuming $Re(c) = 0$

With $i\psi|_{R=a} = (i\Gamma/2\pi)\log a + c$, we can obtain any value of $\psi|_{R=a}$ by choosing c accordingly. In particular, $\psi|_{R=a} = 0$ if $c = -(i\Gamma/2\pi)\log a$, and

$$F(z) = U\left(z + \frac{a^2}{z}\right) + \frac{i\Gamma}{2\pi}\log z - \frac{i\Gamma}{2\pi}\log a = U\left(z + \frac{a^2}{z}\right) + \frac{i\Gamma}{2\pi}\log\frac{z}{a}.$$

The flow field can be obtained from the complex velocity:

$$W(z) = \frac{dF}{dz} = u - iv = (u_R - iu_\theta)e^{-i\theta}, \text{ etc., yielding}$$

$$u_R = U\left(1 - \frac{a^2}{R^2}\right)\cos\theta, \quad u_\theta = -U\left(1 + \frac{a^2}{R^2}\right)\sin\theta - \frac{\Gamma}{2\pi R}; \ \left(z = Re^{i\theta}\right),$$

$$u_R|_{R=a} = 0, \quad u_\theta|_{R=a} = -2U\sin\theta - \frac{\Gamma}{2\pi a}.$$

The stagnation points are given by $u_R = u_\theta = 0$ so that on the cylinder (Figure 4.15) these are located at $\sin\theta_s = -\Gamma/4\pi Ua \Rightarrow \theta_s = \theta_s(\Gamma)$. Above, Γ is the strength of vortex, U is the magnitude of the incoming velocity in x, and a is the radius of cylinder. Note that $0 < \Gamma/4\pi Ua < 1$.

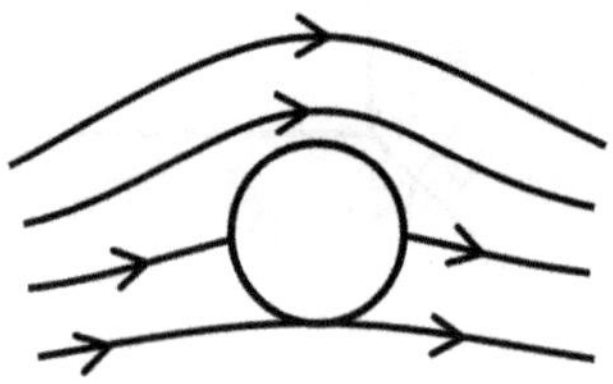

Figure 4.15: The flow field around a circular cylinder with circulation.

4.3 Blasius integral laws

The Blasius integral laws [1] enable us to obtain quantitative expressions for the forces (lift and drag) and the moments exerted on a solid body that is placed inside a potential flow field. The knowledge of complex variable contour integration technique is required to evaluate the integrals.

Blasius Integral Law I: This law states that the forces acting on a solid body due to a potential fluid flow can be expressed as

$$F_x - iF_y = i\frac{\rho}{2}\int_{C_0} W^2 dz, \tag{4.1}$$

where F_x and F_y are, respectively, the x- and y-components of the net hydrodynamic forces on the body. The forces are assumed to act at the center of gravity (COG) of the body. Figure 4.16 depicts the body in a potential flow field, where the C_i and C_o are, respectively, the contours on the outer boundary of the body and an arbitrary surface in the fluid that wholly surrounds the body and detached from it.

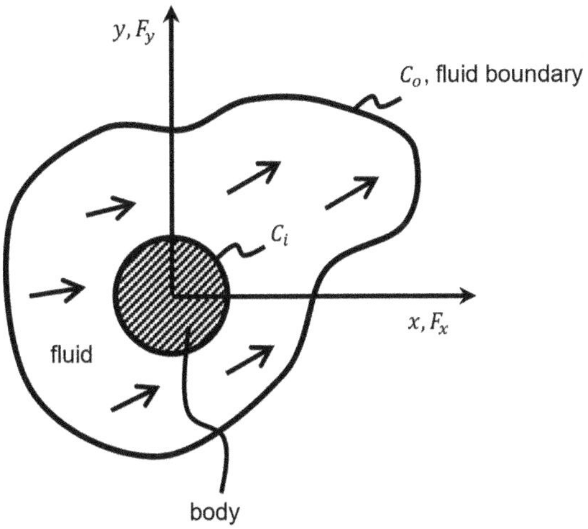

Figure 4.16: Potential flow over a solid body.

The force conservation law can be used to derive eq. (4.1). One statement of this law is that the net external force acting on a solid (body) in a coordinate direction must be equal to the net rate of increase of momentum in that direction. Let us first consider the x-coordinate direction: Let the x-component of total force per unit length acting on the body with surface C_i be F_x so that the reaction of the fluid is $-F_x$. The elemental force per unit depth (in the x-direction on C_o) due to pressure $\Delta F_x/\text{depth} = p\ dy$. The total force on $C_o = \int_{C_0} p\ dy$, where dy is the area per unit depth normal to the x-direction (Figure 4.17).

There is the inertial force through C_o. The volume flow rate per unit depth through element AB (Figure 4.18) on $C_o = d\psi = u\ dy - v\ dx$, where ψ is the stream function. Thus, the mass flow rate per unit depth through element AB on $C_o = \rho\ d\psi = \rho(u\ dy - v\ dx)$. The x-momentum per unit time through $AB = \rho\ d\psi u = \rho(u\ dy - v\ dx)u$, giving a total x-momentum per unit time through C_o of $\int_{C_0} \rho(u\ dy - v\ dx)u$. Note that there is no transfer of momentum through the surface C_i since C_i is a streamline. Moreover, the contribution of $\int_{C_i} d\psi = 0$ since $\psi = \text{constant}$ on C_i. The force balance can therefore be written as

$$-F_x - \int_{C_0} p\,dy = \int_{C_0} \rho\,(u\,dy - v\,dx)\,u. \tag{4.2}$$

It is pointed out that the net balance equation above is written for the fluid. Hence the negative sign in F_x. Similarly, for F_y we have:

$$-F_y + \int_{C_0} p\,dx = \int_{C_0} \rho(u\,dy - v\,dx)v. \tag{4.3}$$

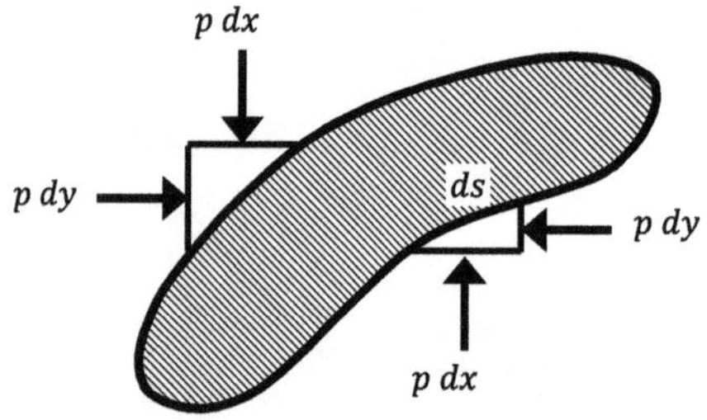

Figure 4.17: Pressure forces on a solid body.

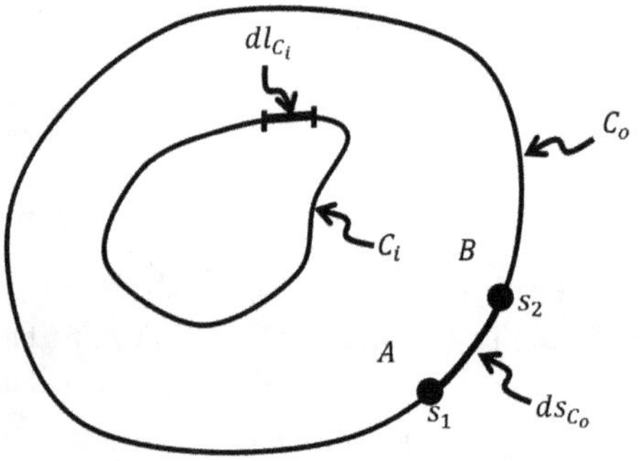

Figure 4.18: Elemental lengths *dl* and *ds* on C_i and C_o, respectively.

We can now solve for F_x, F_y from eqs. (4.2) and (4.3). We eliminate the pressure by using $p + (1/2)\rho(u^2 + v^2) = \text{constant} \equiv B$. We then form the sum $F_y - iF_y$, which can be factored as

$$F_x - iF_y = i\frac{\rho}{2}\int_{C_0} (u - iv)^2(dx + i\,dy) = i\frac{\rho}{2}\int_{C_0} W^2 dz,$$

where $W(z) = u - iv$ and $z = x + iy$. In obtaining this expression we have assumed that B is indeed a conservative field or $\int_{C_0} B\,ds = 0$ since B is a constant, and therefore zero when integrated around the closed contour.

Example 4.1 Suppose the complex potential $F(z)$ for a potential flow field around a body is given by

$$F(z) = U\left(z + \frac{a^2}{z}\right) + \frac{i\Gamma}{2\pi}\log\frac{z}{a},$$

where U is the far stream velocity, a is a constant, and Γ is the circulation around the solid, which could also be treated as a constant. Determine the forces F_x and F_y on the body.

Solution:

$$W(z) = \frac{dF}{dz} = U - \frac{Ua^2}{z^2} + \frac{i\Gamma}{2\pi z},$$

$$[W(z)]^2 = \left(\frac{dF}{dz}\right)^2 = U^2 - \frac{2U^2a^2}{z^2} + \frac{U^2a^4}{z^4} + \frac{iU\Gamma}{\pi z} - \frac{iU\Gamma a^2}{\pi z^2} - \frac{\Gamma^2}{4\pi^2 z^2}.$$

The only singularity of $[W(z)]^2$ is at $z = 0$ (as a result of the doublet and vortex that are located there) and W^2 is in the form of its Laurent Series about $z = 0$. The only term of the form C_{-1}/z is the fourth term on the RHS. Hence the residue C_{-1} is given by $C_{-1} = \text{Res}\left[W^2,\ z = 0\right] = iU\Gamma/\pi$. Thus,

$$F_x - iF_y = i\frac{\rho}{2}\left[2\pi i\left(\frac{iU\Gamma}{\pi}\right)\right] = -i\rho U\Gamma$$

so that $F_x = 0$ and $F_y = \rho U\Gamma$. We see that there is no force (drag) in the x-coordinate direction but that the body experiences a lift force of magnitude $\rho U\Gamma$.

The equation $F_y = \rho U\Gamma$ is a statement of the Kutta-Joukowski law. That is, for flow around a cylinder there will be no lift force on the cylinder if there is no circulation around it, and with circulation, the lift force will be given by the product of the magnitude of this circulation with free-stream velocity of the fluid and the density of the fluid. $F_y > 0$ implies an upward force.

Blasius Integral Law II: Blasius' second integral law pertains to the moment M about the COG of a body due to the force on the body from the potential flow around the body. According to this law [1],

$$M = -\frac{\rho}{2} Re\left[\int_{C_0} zW^2 dz\right],$$

where Re is the real of its argument and the convention (Figure 4.19) is that clockwise moment is positive. The moment is calculated as the force multiplied by the perpendicular distance. This is depicted in Figure 4.20.

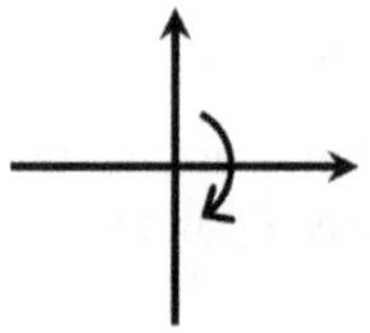

Figure 4.19: Convention for moment: Clockwise is positive.

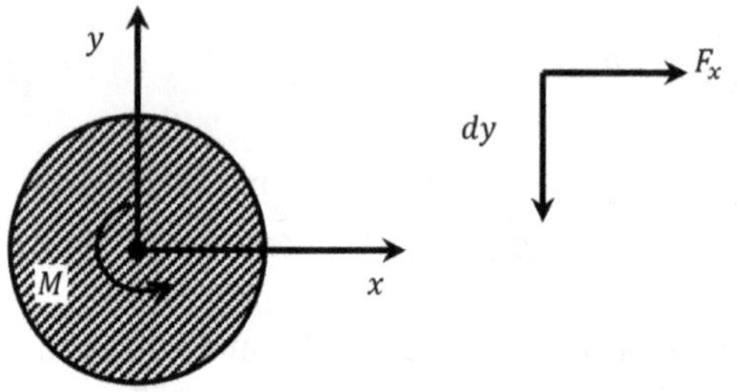

Figure 4.20: Moment of force component F_x due to potential fluid flow on a solid body.

The elemental moment dM_x due to the elemental pressure force in x is equal to $y\Delta F_x$, which is equal to $yp\,dy$, and is perpendicular to ΔF_x. The total moment on the body can be expressed as $\int_{C_0} yp\,dy$. The analogous expression for dM_y is $\int_{C_0} xp\,dx$. The moment due to the fluid inertia force, respectively, in the y- and x-coordinate directions are $\int -\rho(u\,dy - v\,dx)vx$, $\int \rho(u\,dy - v\,dx)uy$. Equating the various components of the moment leads to

$$M = -\frac{\rho}{2}\int_{C_0}\left[(u^2 - v^2)(x\,dx - y\,dy) + 2uv(x\,dy + y\,dx)\right]$$

$$= -Re\left[\frac{\rho}{2}\int_{C_0}(x + iy)(u - iv)^2(dx + idy)\right] = -Re\left(\frac{\rho}{2}\int_{C_0} zW^2 dz\right).$$

Example 4.2 Determine the moment associated with the flow over a body when the complex potential is given by

$$F(z) = U\left(z + \frac{a^2}{z}\right) + \frac{i\Gamma}{2\pi}\log\frac{z}{a}.$$

Solution: The complex velocity is

$$W(z) = \frac{dF(z)}{dz} \Rightarrow zW^2 = U^2 z + \frac{C_{-1}^{(1)}}{z} + \frac{C_{-3}}{z^3} + C_0 - \frac{C_{-2}}{z^2} + \frac{C_{-1}^{(2)}}{z},$$

where $C_{-1}^{(1)}$ and $C_{-1}^{(2)}$ are the two terms consisting of the residues. The point $z = 0$ is a first-order pole. The moment can be calculated as

$$M = -\frac{\rho}{2} Re\left[2\pi i\left[+C_{-1}^{(1)} + C_{-1}^{(2)}\right]\right] \text{ where } C_{-1}^{(1)} = -2U^2 a^2,\ C_{-1}^{(2)} = -\frac{\Gamma^2}{4\pi^2}. \tag{4.4}$$

This gives $M = 0$ since the moment in eq. (4.4) is imaginary. This means that for the given complex potential, there is no hydrodynamic moment acting on the body.

4.4 Joukowski transformation and conformal mapping in airfoil design

In this section, we will discuss the manner in which the complex variable method of conformal transformation is used to aid airfoil design data generation in fields as diverge as aerospace engineering, wind energy, and hydroelectricity. The particular transformation that we shall discuss is Joukowski's, which is perhaps the most important in this class. It enables us to obtain solutions for the flow around ellipses and a family of airfoils. The transformation can be written as

$$z = f(\zeta) = \zeta + \frac{c^2}{\zeta}, \tag{4.5}$$

where c^2 is usually a real constant and $z = z(x,y)$ and $\zeta = \zeta(\xi, \eta)$ are complex planes. The airfoil is in the z-plane, whereas a circular geometry is in the ζ-plane. Thus, z is the relatively complex geometry, while ζ is the relatively simple geometry where analysis can be carried out and the results transformed to the airfoil domain. Note that the choice of z and ζ to represent the two planes has been made in order to be consistent with the literature on this topic and therefore aid the reader in going through the literature. In our general treatment of conformal transformation, we represented the complex geometry in the z-plane as well, but the simple geometry was represented in the w-plane, when only one transformation is involved.

Because $z \to \zeta$ when $|\zeta| \to \infty$ in eq. (4.5), we see that the transformation becomes an identity mapping far from the origin. This means that a uniform flow approaching from infinity in the z system also gives a uniform flow approaching from infinity in system ζ. W is the same far field, so is the angle of attack.

The singularity of $f(\zeta)$ is inconsequential if external flow over a body is of interest, as $\zeta = 0$ will lie inside the body. (Note that it is the inverse of this transformation that we will be dealing with for the most part and mostly circles in ζ-plane.) The critical points of the Joukowski transformation are given by $f'(\zeta) \equiv dz/d\zeta = 0 = 1 - c^2/\zeta^2$.

The function $f(\zeta)$ is undefined at $\zeta = 0$, while the critical points are given by $\zeta = \pm c$ or $z = \pm 2c$. The conformal transformation is not valid at these points. However, the critical points of the transformation do play an important role: allowing us to obtain a sharp trailing or leading edge, which is of significant hydrodynamic relevance. To see this, let us investigate what happens when smooth curves pass through the critical points of a Joukowski transformation. We study the inverse transformation for this purpose (Figure 4.21).

Consider the coordinate system in Figure 4.21 to see what happens. With translation,

$$z = 2c + R_1 e^{i\theta_1} = -2c + R_2 e^{i\theta_2}, \qquad \zeta = c + \rho_1 e^{i\vartheta_1} = -c + \rho_2 e^{i\vartheta_2}. \tag{4.6}$$

Also,

$$z + 2c = \frac{(\zeta + c)^2}{\zeta} \text{ and } z - 2c = \frac{(\zeta - c)^2}{\zeta},$$

which are obtained by using the Joukowski transformation:

$$z = \frac{\zeta^2 + c^2}{\zeta}, \text{ or } z + 2c = \frac{\zeta^2 + c^2 + 2c\zeta}{\zeta}.$$

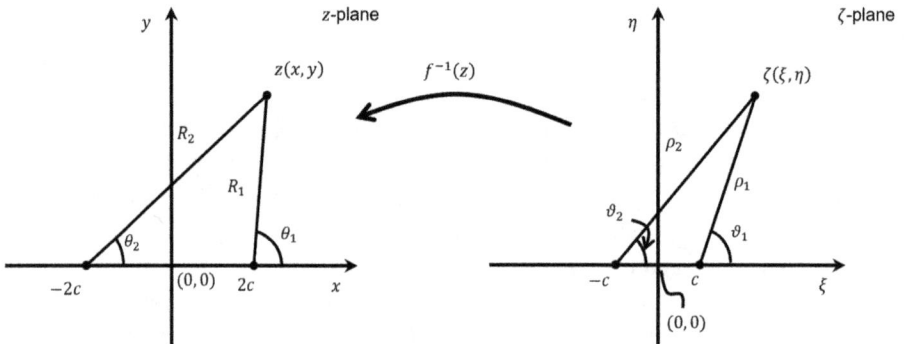

Figure 4.21: Inverse transformation to investigate when smooth curves pass through critical points.

We see that

$$\frac{z-2c}{z+2c} = \left(\frac{\zeta - c}{\zeta + c}\right)^2 = \frac{R_1 e^{i\theta_1}}{R_2 e^{i\theta_2}} = \left(\frac{\rho_1 e^{i\vartheta_1}}{\rho_2 e^{i\vartheta_2}}\right)^2.$$

By using eq. (4.6) we see that

$$\frac{R_1}{R_2} e^{i(\theta_1 - \theta_2)} = \left(\frac{\rho_1}{\rho_2}\right)^2 e^{i2(\vartheta_1 - \vartheta_2)}. \tag{4.7}$$

Equating the modulus and argument of the LHS and RHS of eq. (4.7) gives

$$R_1/R_2 = (\rho_1/\rho_2)^2,$$
$$\underbrace{\theta_1 - \theta_2}_{z} = \underbrace{2(\vartheta_1 - \vartheta_2)}_{\zeta}.$$

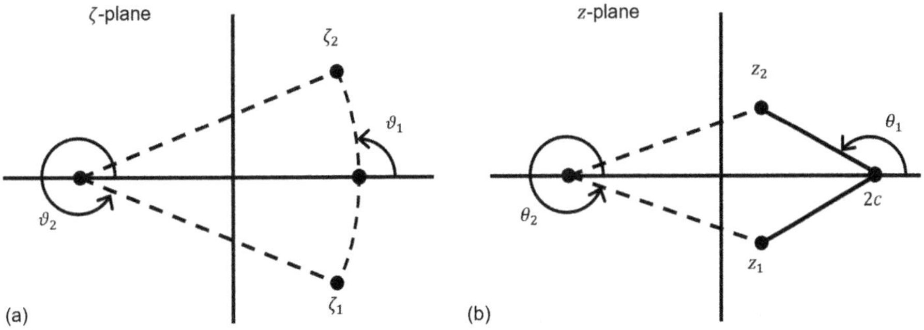

Figure 4.22: The critical points under a Joukowski transformation.

Thus, ζ_1 and ζ_2 transform to z_1 and z_2, respectively, as shown in Figure 4.22. We see that a smooth curve through the point $\zeta = c$ (on the cylinder) produces a knife-edge or cusp in the z-plane (airfoil). Or, a difference of π in ϑ_1 and ϑ_2 produces a difference of 2π in θ_1, θ_2 (Figure 4.23). This explains why we are able to obtain sharp trailing or leading edges or cusps in an airfoil by making those points coincide with the critical points of the transformation, even though we have relatively blunt (flat) surfaces at those locations in the ζ (circle) domain.

That is, we obtain a knife-edge or cusp in the z-plane which allows us to enforce a sharp leading/trailing edge of the airfoil in that plane by making the circle in the ζ-plane pass through the critical points $\zeta = \pm c$.

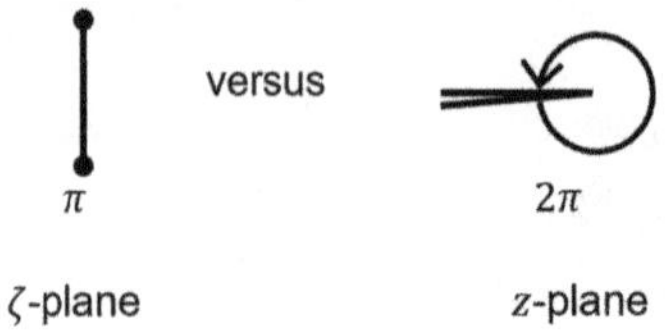

Figure 4.23: Transformation of angular difference in the ζ-plane to angular difference in the z-plane.

4.4.1 Flow around ellipses

If we write $\zeta = ae^{i\vartheta}$ as an imaginary number representation in the circular plane in Euler form, the Joukowski transformation becomes

$$z = ae^{i\vartheta} + \frac{c^2}{a}e^{-i\vartheta} = \underbrace{\left(a + \frac{c^2}{a}\right)\cos\vartheta}_{x} + i\underbrace{\left(a - \frac{c^2}{a}\right)\sin\vartheta}_{y},$$

where $x/\left[a + (c^2/a)\right] \equiv \cos\vartheta$, $y/\left[a - (c^2/a)\right] \equiv \sin\vartheta$. Squaring both sides and adding, we eliminate ϑ (using $\cos^2\vartheta + \sin^2\vartheta = 1$) to obtain

$$\left(\frac{x}{a + c^2/a}\right)^2 + \left(\frac{y}{a - c^2/a}\right)^2 = 1$$

in the z-plane. This is an ellipse with major semiaxis of length $(a + c^2/a)$, aligned with the x-axis, and minor semi-axis of length $(a - c^2/a)$, aligned with the y-axis.

The inverse Joukowski's transformation in this case transforms rectilinear x-flow ($U = 1$) over a cylinder (circle) in the ζ-plane to a flow over an ellipse in the z-plane. Now, in the ζ-plane, and consistent with the previous discussions in this chapter, it can be shown that the complex potential for a uniform flow of magnitude U approaching a circular cylinder of radius a and an angle of attack α to the reference axis is

$$F(\zeta) = U\left(\zeta e^{-i\alpha} + \frac{a^2}{\zeta}e^{i\alpha}\right). \tag{4.8}$$

Of course, in the previous discussions, the z-plane was the physical plane in which we had a circular geometry, and we had the complex potential (with zero angle of attack) as

$$F(z) = U\left(z + \frac{a^2}{z}\right).$$

Here, z is no longer the plane of the circle, but the plane is used to represent the various aerodynamic objects (airfoils, ellipses, etc.). The ζ-plane is now the one used for the circular geometry. With $z = \zeta + c^2/\zeta$, we obtain the inverse transformation as

$$\zeta \equiv g(z) = \frac{z}{2} \pm \sqrt{\left(\frac{z}{2}\right)^2 - c^2} \tag{4.9}$$

for the ellipse in the z-plane. Since a complex potential transforms identically, we will have $F(z) = F(g(z))$. We will substitute eq. (4.9) for ζ into eq. (4.8). Since $\zeta \to z$ as $z \to \infty$, we have to choose the positive root; otherwise, $\zeta \to 0$ as $z \to \infty$. The result is

$$\begin{aligned}
F(z) &= U\left\{\left[\frac{z}{2} + \sqrt{\left(\frac{z}{2}\right)^2 - c^2}\right]e^{-i\alpha} + \frac{a^2 e^{i\alpha}}{z/2 + \sqrt{(z/2)^2 - c^2}}\right\} \\
&= U\left\{\left[z - \frac{z}{2} + \sqrt{\left(\frac{z}{2}\right)^2 - c^2}\right]e^{-i\alpha} + \frac{a^2}{c^2}\left[\frac{z}{2} - \sqrt{\left(\frac{z}{2}\right)^2 - c^2}\right]e^{i\alpha}\right\} \\
&= U\left\{ze^{-i\alpha} + \left[\frac{a^2}{c^2}e^{i\alpha} - e^{-i\alpha}\right]\left[\frac{z}{2} - \sqrt{\left(\frac{z}{2}\right)^2 - c^2}\right]\right\}.
\end{aligned}$$

Hence, we can describe the flow (ϕ, ψ, u_R, u_θ, forces, moments) in the z-plane (the flow over an ellipse). A qualitative depiction of the flow field in the ζ (circular geometry) and z (elliptical geometry) is shown in Figure 4.24.

4.4.2 Stagnation points on the ellipse (z-plane)

In the ζ-plane (circle) stagnation points are located at $\zeta = ae^{i\alpha}$ and $\zeta = ae^{i(\alpha+\pi)}$, where α is the angle of attack (i.e., on the circle at angles α and $\alpha + \pi$). Using $W(\zeta) = 0$ at the stagnation points and

$$F(\zeta) = U\left(\zeta e^{-i\alpha} + \frac{a^2}{\zeta}e^{i\alpha}\right),$$

the corresponding points in the z-plane are

$$z = \pm ae^{i\alpha} \pm \frac{c^2}{a}e^{-i\alpha}\left(\text{using } z = \zeta + \frac{c^2}{\zeta}\right) = \underbrace{\pm\left(a + \frac{c^2}{a}\right)\cos\alpha}_{x} \pm i\underbrace{\left(a - \frac{c^2}{a}\right)\sin\alpha}_{y},$$

so that the coordinates of the stagnation points in z are given as

$$x = \pm\left(a + \frac{c^2}{a}\right)\cos\alpha, \quad y = \pm\left(a - \frac{c^2}{a}\right)\sin\alpha.$$

The location of the stagnation points as a function of the angle of attack α is depicted in Figure 4.25.

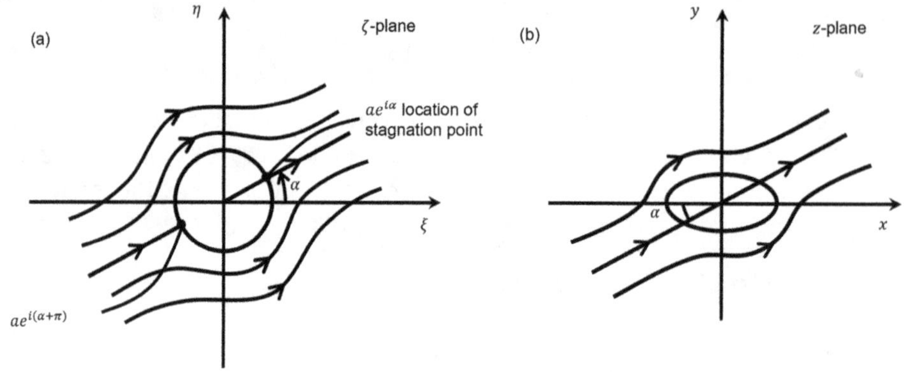

Figure 4.24: A qualitative depiction of the flow over a circular geometry in the ζ-plane and elliptical geometry in the z-plane.

Note that Case d ($\alpha = \pi/2$) in Figure 4.25 can be obtained directly from the Joukowski transformation, using $c = ib$ (b is real, positive). The reader might be wondering that since $u = v = 0$ is the definition of a stagnation point, then it means that all points on the surface of a real (viscous) cylinder are stagnation points. This is not the case because stagnation in the present context is an inviscid concept that applies strictly to potential (perfect) fluid flow in which viscosity is assumed to be zero and fluid particles at the wall are able to slip against the wall and not take on the velocity of the wall.

4.4.3 Flow over a flat plate

If the parameter c that appears in the Joukowski transformation $z = F(\zeta) = \zeta + c^2/\zeta$ is set equal to the radius of the circle in the ζ-plane, we will have a flat plate in the z-plane. These configurations are displayed in Figure 4.26. Writing $\zeta = ae^{i\alpha}$, the Joukowski mapping takes the form

$$z = ae^{i\alpha} + \frac{a^2}{ae^{i\alpha}} = a\left(e^{i\alpha} + e^{-i\alpha}\right).$$

In this case, by examining $W(\zeta) = 0$, the stagnation points do not coincide with the leading/trailing edge because $\alpha \neq 0$. They are located at $x = \pm 2a \cos \alpha$. A few remarks are in order. The velocities at the leading and trailing edges approach infinity mathematically because of the zero thickness of the flat plate at those locations. Real airfoils have a finite thickness at the leading edge, so the velocity is finite there, but the trailing edge is usually sharp and the possibility of an infinitely high velocity at the trailing edge exists. This problem can be overcome if the stagnation point, which is close to the trailing edge, is actually made to move to the trailing edge. This can be accom-

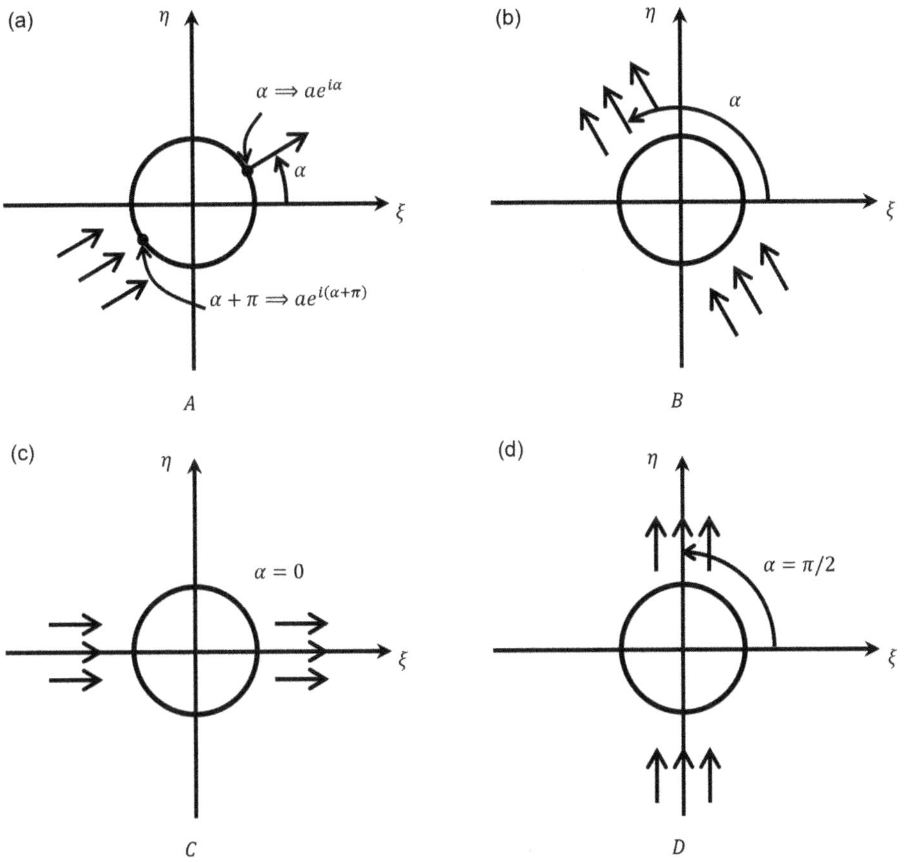

Figure 4.25: The location of the stagnation points as a function of the angle of attack, α.

plished if a circulation existed around the flat plate, with a magnitude that is just sufficient to rotate the rear stagnation point so that its location coincides with the trailing edge. The introduction of this circulation is done to comply with the so-called Kutta condition.

Kutta condition: The Kutta condition states that for bodies with sharp trailing edges which are at small angles of attack to the free stream, the flow will adjust itself in such a way that the rear stagnation point coincides with the trailing edge. Note that the Kutta condition applies to physical systems, that is, to what have been visually observed. Thus, we must modify our mathematical model (i.e., $F(\zeta)$) so as to make the stagnation point coincide with the trailing edge. The issue then is to determine the amount of circulation that is required to do this and comply with the Kutta condition. The trailing edge is located at $z = 2a$ (flat plate) or $\zeta = a$ (circle). In the ζ-plane, the stagnation points on the circle are given by $\sin\theta_s = -\Gamma/4\pi Ua$.

Therefore, a circulation of $\Gamma = 4\pi Ua \sin\alpha$ is required to rotate the stagnation point at the downstream face of the circular cylinder through an angle α, as shown in Figure 4.27. It can be shown that adding the circulation defined here gives the required mathematical result that is consistent with the physical Kutta condition.

With this modification, the complex potential can be written as

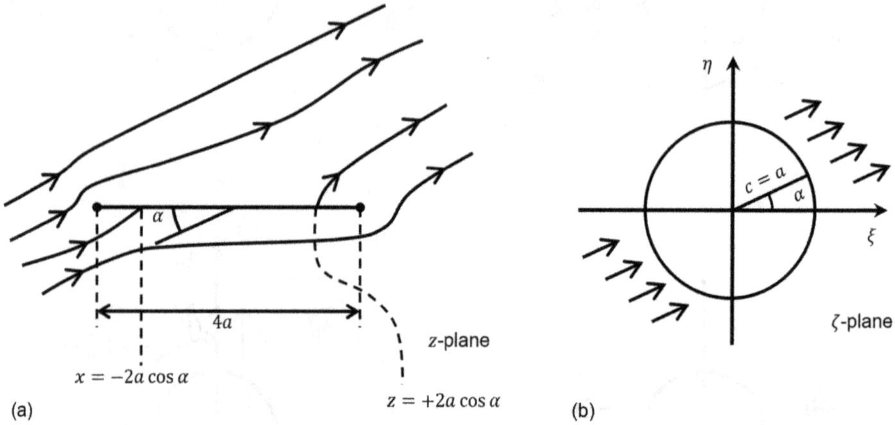

Figure 4.26: Transformation of a flat plate to a circle by setting $c = a$, the radius of the cylinder in the ζ-plane.

$$F(\zeta) = U\left(\zeta e^{-i\alpha} + \frac{a^2}{\zeta} e^{i\alpha}\right) + i2Ua\ \sin\alpha \log\frac{\zeta}{a}$$

where the second term is from

$$\frac{i\Gamma}{2\pi}\log\frac{z}{a}.$$

what's needed to make $R = a$ the $\psi = 0$ streamline

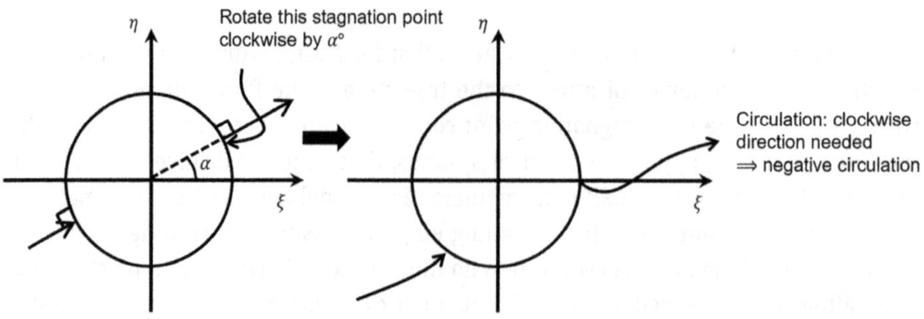

Figure 4.27: Rotating the stagnation point clockwise by $\alpha°$ to make it coincide with the trailing edge.

The mapping is

$$z = \zeta + \frac{a^2}{\zeta} \Rightarrow \zeta = g(z) = \frac{z}{2} + \sqrt{\left(\frac{z}{2}\right)^2 - a^2},$$

where we have chosen the positive sign in the quadratic equation for the inverse transformation to be consistent with $\zeta \to z$ as $z \to \infty$. The negative sign gives $\zeta \to 0$ as $z \to \infty$, which is not appropriate for our purpose. The complex potential in the z-plane becomes

$$F(z) = U\left\{\left[\frac{z}{2} + \sqrt{\left(\frac{z}{2}\right)^2 - a^2}\right] e^{-i\alpha} + \frac{a^2 e^{i\alpha}}{z/2 + \sqrt{(z/2)^2 - a^2}}\right.$$
$$\left. + i2a \sin\alpha \log\left[\frac{1}{a}\left(\frac{z}{2} + \sqrt{\left(\frac{z^2}{2}\right)^2 - a^2}\right)\right]\right\}.$$

With $F(z)$ known, $W(z)$, ϕ, ψ, u_R, u_θ, F_X, F_Y, and M can be determined for the flat plate. A schematic of the flow over a flat plate is shown in Figure 4.28.

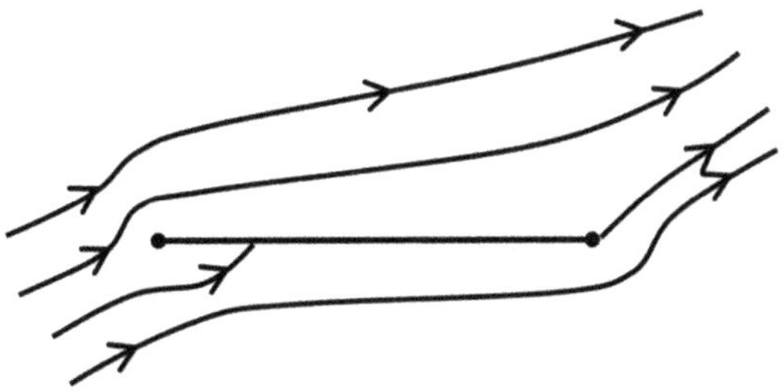

Figure 4.28: The flow pattern over a flat plate obeying the Kutta condition.

In particular, using the Blasius Law I [1], we obtain

$$F_y = \rho U \Gamma = 4\pi\rho U^2 a \sin\alpha,$$

from which we can define a lift coefficient, C_L:

$$C_L = \frac{F_y}{(1/2)\rho U^2 l} = 2\pi \sin\alpha, \text{ where } l = \text{ length of the chord of the airfoil} = 4\text{a for the flat plate.}$$

We see that for the flat plate, the lift force increases with α and is zero when $\alpha = 0$. Of course, if α is sufficiently large, we will encounter massive flow separation, leading to the failure in the ability of the airfoil to be lifted.

4.4.4 Joukowski airfoils

A family of airfoils can be obtained with the Joukowski transformation by using a series of circles (in the ζ-plane) whose centers are slightly displaced from the origin. We will present three of such airfoils: symmetric Joukowski, circular arc, and the non-symmetric Joukowski, which combines the circular arc and the symmetric Joukowski.

4.4.4.1 Symmetrical Joukowski airfoils

The symmetric Joukowski airfoil has a symmetry about the x-axis as shown in Figure 4.29. The airfoil is of course in the z-plane, while the circular geometry is in the ζ-plane. To obtain this airfoil, we need to displace the center from the origin of the circle in the ζ-plane. Questions to answer include what radius should be used, relative to c? In what direction (x-, y-, or both x-and y-) should we displace the center? What will the new radius be relative to c? We know that if the circle passes through any of the critical points ($\zeta = \pm c$), then a sharp edge or cusp is obtained in z-plane. Now, if the leading edge is to have a finite radius of curvature and we don't want singularity in the flow field, $\zeta = -c$ must be placed inside the circle (airfoil). (The flow field is exterior to the airfoil.). Furthermore, if the trailing edge of the airfoil must be sharp, the circumference of the circle must pass through $\zeta = c$, as previously demonstrated. We can satisfy these two conditions by placing the center at $\zeta = -m$, $m > 0$, giving a radius of $c + m \equiv a$. (Note that making the radius to be c will cause the airfoil to be sharp at both the leading and trailing edges.) Figure 4.30 illustrates the configuration for the symmetric Joukowski airfoil, with a focus on the displacements that are required in the ζ-plane in order to produce the correct airfoil shape in the z-plane.

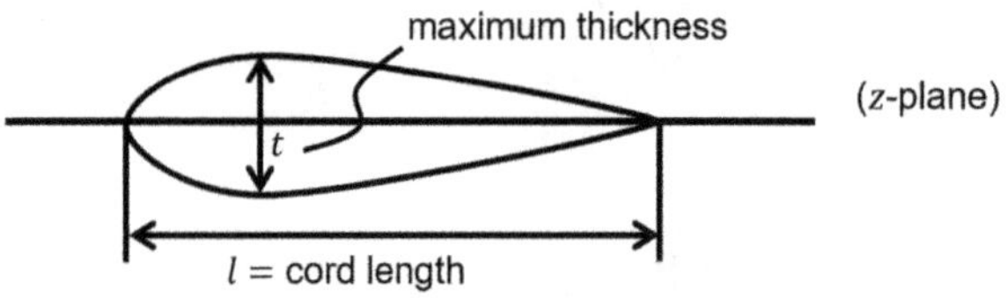

Figure 4.29: A schematic of symmetric Joukowski.

Now we can write

$$a = c + m = c(1 + m/c) \equiv c(1 + \varepsilon);\ \varepsilon \equiv m/c,$$

where ε is the fraction of the old radius by which the leading edge singularity is moved inside. This places the leading edge at the position

$$\zeta = -m - (m + c) = -(c + 2m)$$

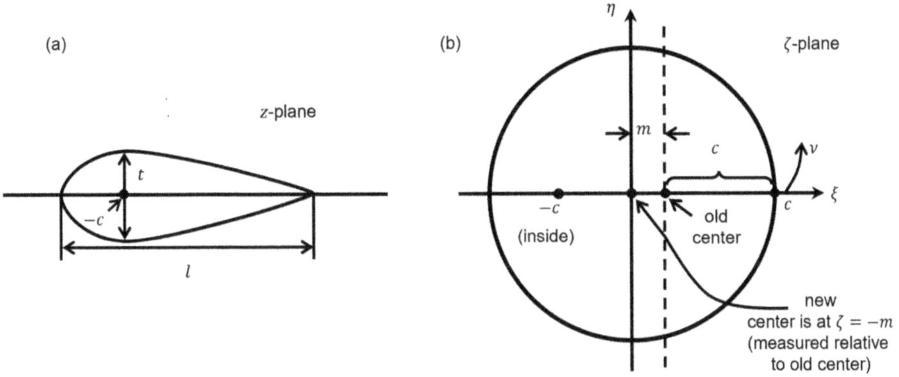

Figure 4.30: Displacements in the ζ-plane required to produce the symmetric Joukowski airfoil in the z-plane.

relative to the old center. Note that $\varepsilon = 0$ gives the flat plate, while $\varepsilon \ll 1$ gives a thin airfoil. These values for ε suggest that ε could be used as a perturbation parameter in some related analytical studies. (Also, we see from the foregoing how $\pm c$, the locations of the critical points, play significant roles in airfoil design.)

Toward determining the equation for the airfoil profile in the z-plane, note that the trailing edge location is $\zeta = c$ (relative to old center), while the leading edge is located at $\zeta = -(c + 2m)$ (relative to old center). The corresponding points in z-plane (with $z = \zeta + a^2/\zeta$) are $z = 2c$ and $z = -c(1+2\varepsilon) - c/(1+2\varepsilon) = -2c + O(\varepsilon^2)$ with linearization ($\varepsilon \ll 1$). This gives $l = 4c$ (in the z-plane) to first order. The maximum thickness, t, of the airfoil can be shown to be equal to

$$t = \frac{3\sqrt{3}}{4} \cdot \varepsilon l \Rightarrow \varepsilon \approx 0.77t/l \ (z-\text{plane}), \tag{4.10}$$

while the equation of the airfoil profile (in the z-plane) can be shown to be

$$y = 2c\varepsilon\left(1 - \frac{x}{2c}\right)\sqrt{1 - \left(\frac{x}{2c}\right)^2}, \quad \frac{y}{t} = \pm 0.385\left(1 - 2\frac{x}{l}\right)\sqrt{1 - \left(2\frac{x}{l}\right)^2},$$

using eq. (4.10). The maximum, minimum value of $y/t = (0.5, -0.5)$, both of which occur at $x = -c$ (Figure 4.31).

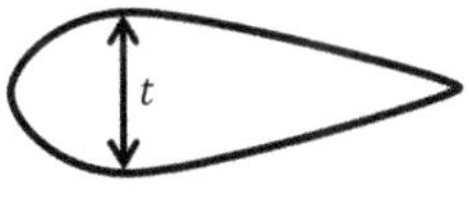

Figure 4.31: The location of the maximum thickness of the symmetric Joukowski airfoil is at $x = -c$.

The circulation that is required to satisfy the Kutta condition is $\Gamma = 4\pi U a \sin\alpha$, where $a = c + m$ (= new radius), $m = c\varepsilon$, $\varepsilon = 0.77(t/l)$, and $c = l/4$ so that $a = c + m = c(1+\varepsilon) = (l/4)\,[1 + 0.77(t/l)]$.

This leads to $\Gamma = \pi U \underbrace{l[1 + 0.77(t/l)]}_{4a} \sin\alpha$. The lift force and lift coefficient are given by $F_y = \rho U\Gamma = \pi\rho U^2 l[1 + 0.77(t/l)]\sin\alpha$ and $C_L = F_y/\left[(1/2)\rho U^2 l\right] = 2\pi[1 + 0.77(t/l)]\sin\alpha$.

Thus, we still need $\alpha \neq 0$ to obtain a lift. The effect of thickness t is to increase the lift coefficient, C_L. However, a large t leads to a bluff body, which encourages separation and destroys lift. Therefore, t cannot be too large. Note that we don't need a big t to encounter separation, as it can occur for slender (nonbluff) bodies if α (angle of attack) is large. Separation is sometimes referred to as stalling, in this context. This is illustrated in Figure 4.32. Note how α is measured, which is shown in Figure 4.33.

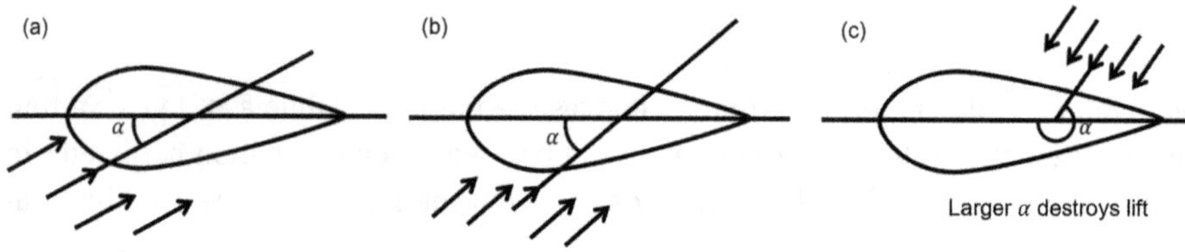

Figure 4.32: Illustration of the variation in the angle of attack, α.

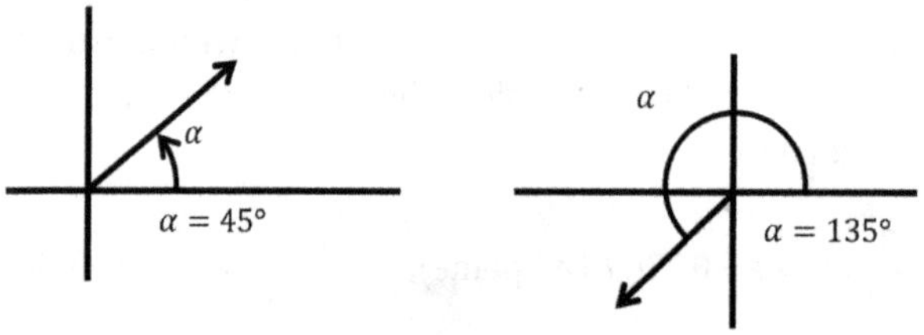

Figure 4.33: The way the angle of attack α is measured.

4.4.4.2 Flow field for the symmetrical Joukowski airfoil

With the center of the disk located at $\zeta = -m$ (relative to old center) so that $\zeta \to \zeta - (-m)$, the complex potential can be written as

$$F(\zeta) = U\left[(\zeta + m)e^{-i\alpha} + \frac{a^2}{\zeta + m}e^{i\alpha}\right] + \frac{i\Gamma}{2\pi}\log\left(\frac{\zeta + m}{a}\right),$$

where

$$a = \frac{l}{4} + 0.77\frac{tc}{l},\ c = \frac{l}{4} \Rightarrow a = \frac{l}{4}\left(1 + 0.77\frac{t}{l}\right),\ m = 0.77\frac{tc}{l},\ \text{and}\ \Gamma = \pi U l\left(1 + 0.77\frac{t}{l}\right)\sin\alpha.$$

4.4.4.3 Flow over a circular-arc air foil

Before we present the circular-arc airfoil of Joukowski, let us summarize the various transformations that we have introduced so far to arrive at the configurations of the ellipse, flat plate, and symmetric Joukowski. This is shown in Tab. 4.1.

Note that all the transformations in the table are based on Joukowski's: $z = \zeta + c^2/\zeta$. The required transformations for the circular arc airfoil are shown in item (f) in the table. These are illustrated more vividly in Figure 4.34. Rather than moving the center toward negative ξ, as in the case of the symmetric Joukowski, we move it in the positive η, keeping the ξ-coordinate of the circle unchanged. The vertical displacement is also denoted by m as in the case of the symmetric Joukowski. The vertical displacement helps create camber in that direction.

Let us examine the equations of the circular arc. If we write $\zeta = Re^{i\vartheta}$ in the ζ-plane, the Joukowski transformation in that plane can be written as

$$z = Re^{i\vartheta} + \frac{c^2}{R}e^{-i\vartheta} = \left(R + \frac{c^2}{R}\right)\cos\vartheta + i\left(R - \frac{c^2}{R}\right)\sin\vartheta$$

or from $z = x + iy$, $x = \left(R + c^2/R\right)\cos\vartheta$, $y = \left(R - c^2/R\right)\sin\vartheta$.

The coordinate along the circular arc or the parametric equation of the circular arc can be obtained in the z-plane by carrying out the steps below:

i) Square both sides of the equations for x and y above.
ii) Multiply the resulting x equation by $\sin^2\vartheta$ and y equation by $\cos^2\vartheta$ and subtract the latter from the former.
iii) Use the cosine rule and eliminate ϑ to obtain the equation of the circular arc in the z-plane as

$$x^2 + \left[y + c\left(\frac{c}{m} - \frac{m}{c}\right)\right]^2 = c^2\left[4 + \left(\frac{c}{m} - \frac{m}{c}\right)^2\right].$$

Note that no approximations have been introduced in arriving at this result. Now, to first order in $\varepsilon = m/c$, we have

$$x^2 + \left(y + \frac{c^2}{m}\right)^2 = c^2\left(4 + \frac{c^2}{m^2}\right),$$

where the center is located at $x = 0, y = -c^2/m$, and the radius is $c\sqrt{4 + c^2/m^2}$.

The ends of the circular arc are located at $\zeta = \pm c$ in the ζ-plane and $l = 4c$ in the z-plane, just as for the flat plate and the symmetrical Joukowski airfoils. The camber height can be obtained as the maximum of y, which occurs at $x = 0$ (center of arc is at $x = 0$). That is $\vartheta = \vartheta_{max} = \pi/2$ (from $x = 0 = \left(R + c^2/R\right)\cos\vartheta$) so that $y_{max} = 2m\sin^2\vartheta_{max} = 2m = h$.

If we eliminate R from $y = \left(R - c^2/R\right)\sin\vartheta$, we will get $y = 2m\sin^2\vartheta$. Note that $m = h/2$ is the amount of translation of the origin in the imaginary axis. Substituting for c and m, the profile of the airfoil can be obtained as

$$x^2 + \left(y + \frac{l^2}{8h}\right)^2 = \frac{l^2}{4}\left(1 + \frac{l^2}{16h^2}\right).$$

Table 4.1: Summary of the transformations so far.

ζ-Plane (c, a are arbitrary)	**z-Plane**
(a) Circular geometry	(a) Other configurations: ellipse, flat plate, symmetric Joukowski, circular arc, Joukowski
(b) Circle, $\zeta = ae^{i\vartheta}$; a is radius. $F(\zeta) = U\left(\zeta e^{i\alpha} + \frac{a^2}{\zeta}e^{i\alpha}\right) \neq F(c)$	(b) Flow over ellipse, arbitrary orientation $F(z) = F(z, c)$ Presence of c is a consequence of transformation.
(c) $\zeta = ae^{i\vartheta}$; make $c = a$ in the transformation $F(\zeta) = U\left(\zeta e^{-i\alpha} + \frac{a^2}{\zeta}e^{i\alpha}\right)$ 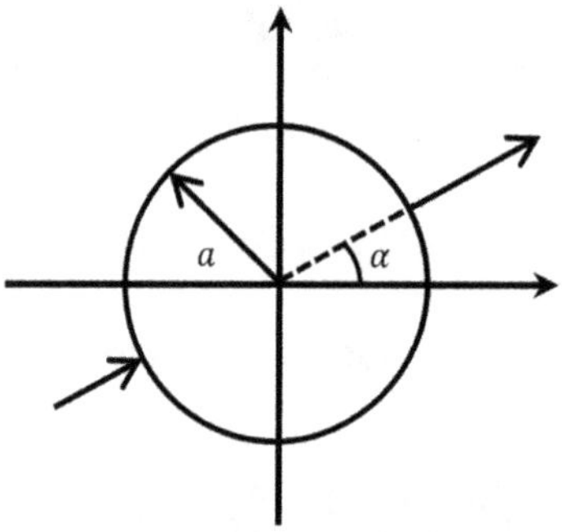	(c) Flat plate of length $4a$ or $4c$ 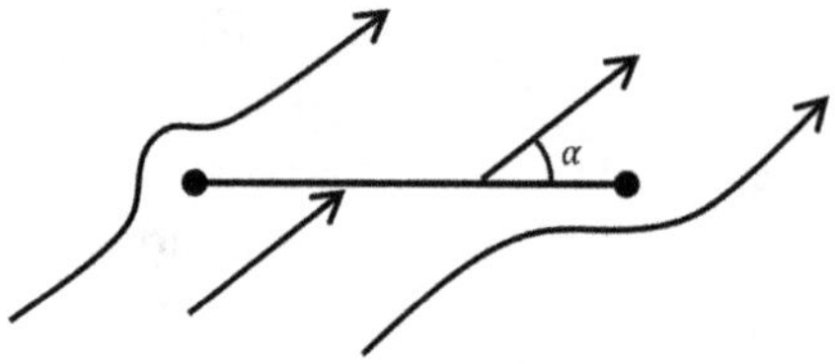
(d) Introduce Γ to move stagnation point to the trailing edge of flat plate in the z-plane $F(\zeta) = U\left(\zeta e^{-i\alpha} + \frac{a^2}{\zeta}e^{i\alpha}\right) + i2Ua\sin\alpha\log\frac{\zeta}{a}$	(d) Satisfy Kutta condition 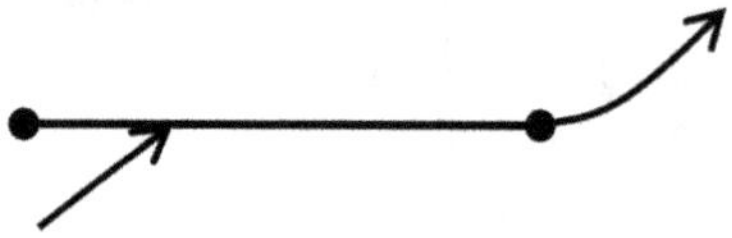

Table 4.1 (continued)

ζ-Plane (c, a are arbitrary)	**z-Plane**
(e) Make the radius of circle to be slightly larger than c to keep leading edge c inside, that is, $$a \to c(1+\varepsilon);\ \varepsilon = \frac{m}{c};\ \varepsilon \ll 1\ \left(a = \frac{l}{4} + 0.77\frac{tc}{l}\right)$$ Move a small distance m $(= 0.77\frac{tc}{l} = \varepsilon c)$ to the left of the real axis so that center is at $\zeta = -m$ (relative to the old origin) and not $\zeta = 0$ 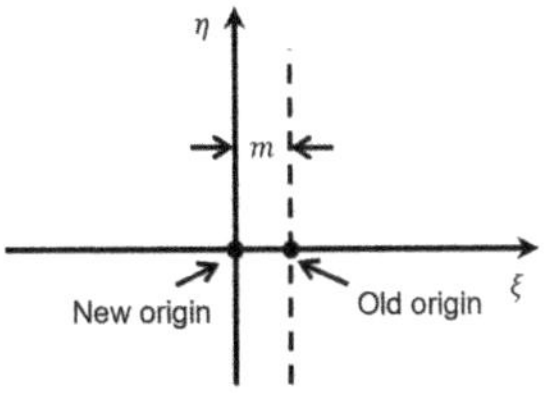	(e) Symmetric Joukowski airfoil: Increase C_L by $2\pi\varepsilon \sin\alpha$ over flat plate's value of $2\pi \sin\alpha$, where $\varepsilon = 0.77\frac{t}{l}$ 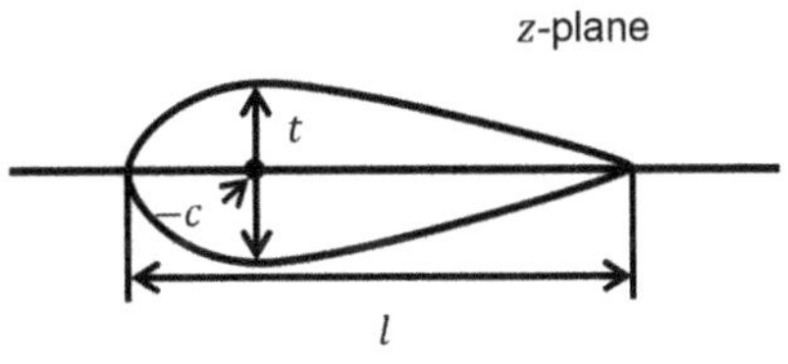
(f) i) Make the radius of the circle to be slightly larger than c. ii) Move a small distance m along the positive imaginary axis $(m = h/2)$ 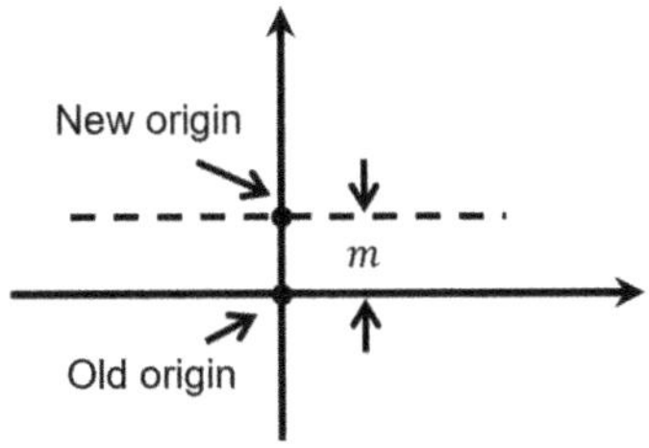 iii) Both leading and trailing edges are sharp, so the circle in ζ must pass through $-c$, $+c$.	(f) Circular air airfoil, no thickness, has curvature or camber (positive camber) of height h (Figure 4.34)

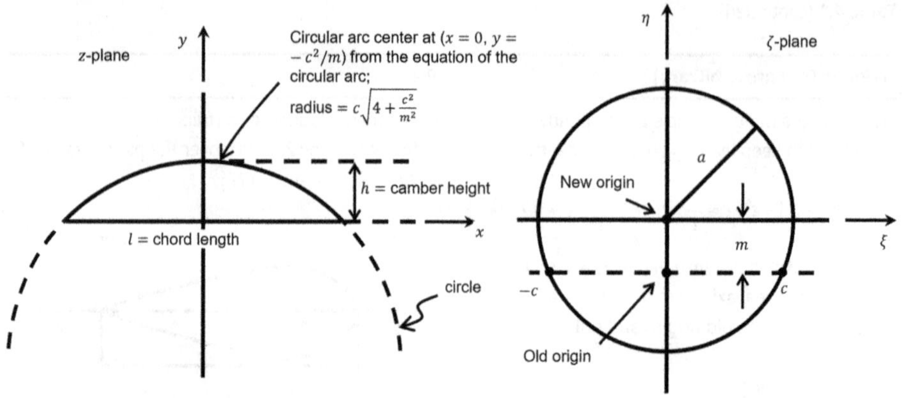

Figure 4.34: The required transformations for the circular arc airfoil.

To move the stagnation point to the trailing edge at the “new” position of the airfoil, use Γ to rotate by $\alpha + \theta$, where

$$\theta = \tan^{-1}\frac{m}{c} \approx \frac{m}{c} \text{ for small} \frac{m}{c}.$$

The extra θ relative to the previous treatments is a result of translating “m” units in the η direction in order to obtain the circular arc. The qualitative flow distributions are shown in Figure 4.35 in the ζ- and z-planes.

To obtain the complex potential $F(\zeta)$ in the ζ-plane, we replace ζ by $\zeta - im$ in

$$F(\zeta) = U\left(\zeta e^{-i\alpha} + \frac{a^2}{\zeta}e^{i\alpha}\right)$$

and add circulation. The result is

$$F(\zeta) = U\left[(\zeta - im)e^{-i\alpha} + \frac{a^2}{\zeta - im}e^{i\alpha}\right] + \frac{i\Gamma}{2\pi}\log\left(\frac{\zeta - im}{a}\right),$$

where $a \approx c = l/4$, $m = h/2$. We can easily obtain the following results from the complex potential distributions:

$$\Gamma = 4\pi U a \sin\left(\alpha + \frac{m}{c}\right), \quad F_Y = \rho U l = 4\pi\rho U^2 c \sin\left(\alpha + \frac{m}{c}\right),$$

$$\text{and } C_L = 8\pi\frac{c}{l}\sin\left(\alpha + \frac{m}{c}\right) = 2\pi\sin\left(\alpha + \frac{2h}{l}\right).$$

We see that the circular arc can develop a lift force even when the angle of attack is zero. This is the case as long as we have a camber ($h \neq 0$).

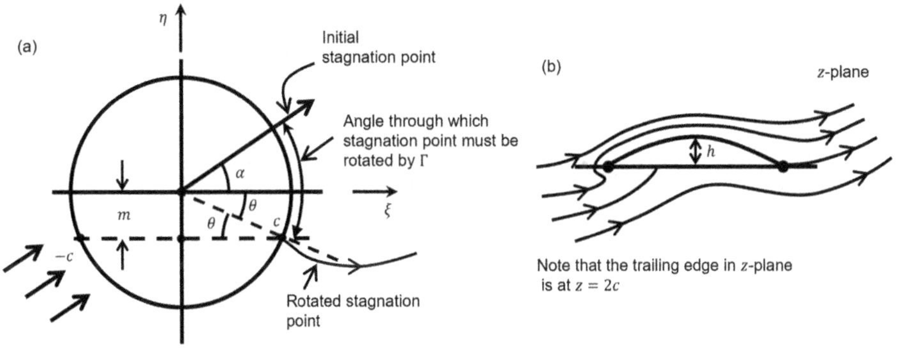

Figure 4.35: Qualitative flow distributions over the circular arc airfoil in the ζ- and z-planes.

4.4.4.4 Flow over Joukowski airfoil

The Joukowski airfoil (Figure 4.36a) is obtained by combining the symmetric Joukowski and the circular arc airfoils. The combining is done in a rather practical sense by moving the center (relative to the ellipse) in both the (negative) real and (positive) imaginary axis by "radius" m. This is shown in Figure 4.36b.

We place $-c$ inside the circle because we do not want a sharp leading edge and a singularity in the flow field. That is the reason for moving the old center to the left. We also move it to the positive η because we want to create a camber. The centerline of the Joukowski airfoil is the circular arc, which is symmetric about $x = 0$. However, both the top and bottom surfaces are from the symmetric Joukowski airfoil. To obtain the profile of the Joukowski airfoil, we directly add those for the symmetric Joukowski and the circular arc airfoils. The result is

$$y = \underbrace{\sqrt{\frac{l^2}{4}\left(1+\frac{l^2}{16h^2}\right) - x^2} - \frac{l^2}{8h}}_{\text{profile from circular arc}} \overset{\text{upper surface}}{\underset{\text{lower surface}}{\pm}} \underbrace{0.385t\left(1-2\frac{x}{l}\right)\sqrt{1-\left(\frac{2x}{l}\right)^2}}_{\text{profile of symmetric Joukowski}}.$$

To obtain $F(\zeta)$, we need to add circulation and replace ζ by $\zeta - me^{i\delta}$, where $me^{i\delta}$ is the new center. The result is

$$F(\zeta) = U\left[\left(\zeta - me^{i\delta}\right)e^{-i\alpha} + \frac{a^2 e^{i\alpha}}{\zeta - me^{i\delta}}\right] + \frac{i\Gamma}{2\pi}\log\left(\frac{\zeta - me^{i\delta}}{a}\right),$$

where

$$m\cos\delta = -0.77\frac{tc}{l},\ \text{and}\ m\sin\delta = \frac{h}{2},\quad a = \frac{l}{4} + 0.77\frac{tc}{l}.$$

This $F(\zeta)$, together with $\zeta = g(z)$, provides the information required to calculate $W(z)$, ϕ, ψ, F_x, F_y, and M. The lift coefficient for the Joukowski airfoil has contributions from the coefficients for symmetric Joukowski and circular arc. The equation for it is $C_L = 2\pi[1 + 0.77(t/l)]\sin(\alpha + 2h/l)$, and the circulation is $\Gamma = \pi Ul[1 + 0.77(t/l)]\sin(\alpha + 2h/l)$. As sketched in Figure 4.37, the new origin relative to the old is located at $\zeta = me^{i\delta}$.

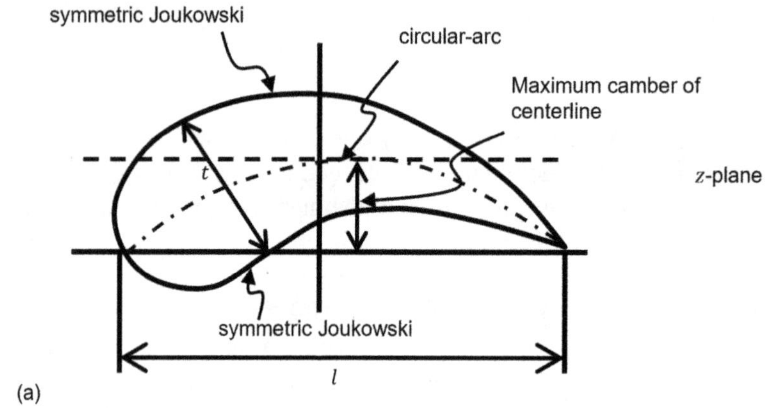

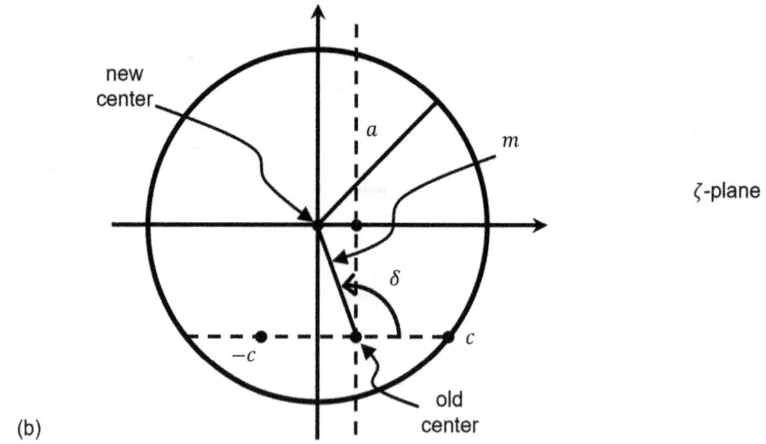

Figure 4.36: Creation of the Joukowski airfoil by combining symmetric Joukowski and the circular arc airfoils. The new origin relative to old is at $\zeta = me^{i\delta}$.

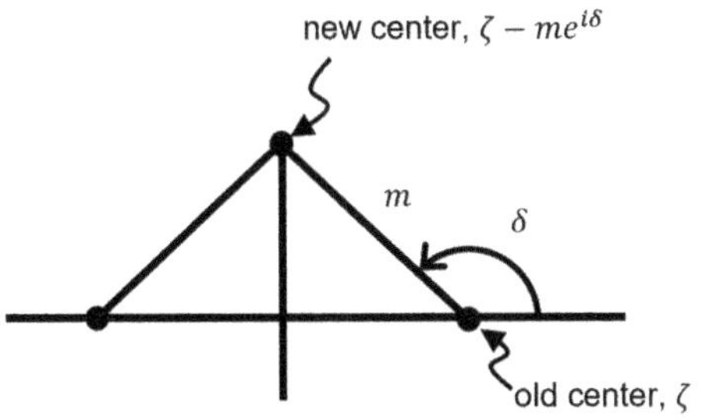

Figure 4.37: The diagonal offset of the new center from the old center of circle.

Suggested reading

[1] Milne-Thomson, L.M. Theoretical Hydrodynamics, Fifth Edition, London, UK, Macmillan Press, 1968.
[2] Robertson, J.M. Hydrodynamics in Theory and Application, London, UK, Prentice Hall, 1965.
[3] Currie, I.G. Fundamental Mechanics of Fluids, 4th Edition, Boca Raton, FL, CRC Press/Taylor and Francis Group, 2013.

Part II: **Complex integral transforms**

Introduction to Part II

Closed-form analytical solutions of many types of engineering problems cannot be obtained with the level of treatment of complex variables presented so far in this book. In fact, the materials in Part I are introductory. Tables of inversion of the common integral transforms, such as the Laplace and Fourier, are expedient for obtaining closed-form solutions of engineering problems. However, such tables exist only for simple and familiar functions, motivating the need for foundational knowledge on the procedure for inverting integral transforms. For example, in elementary treatments, the transform variables are treated as real, but many inversions cannot be carried out with this assumption. Moreover, transform integral asymptotics require series expansion in a complex plane, as opposed to the conventional asymptotic procedure. Also, the analytical solution of certain mixed boundary-value partial differential equations requires techniques that are beyond the scope of a first course in complex variables, while the kernels in certain classes of integral equations must be split to solve the equations, say, by the Fourier transform method. These involve the decomposition of functions into a part that is analytic in the upper half of the complex plane of the transform variable and another that is analytic in the lower half. Advanced concepts of complex variables are required for the foregoing techniques. This part of the book is dedicated to the aforementioned topics.

The author is aware that some graduate engineering programs might optionally choose to use numerical methods – finite difference, finite volume, finite element, or the boundary integral method – to tackle the kinds of problems alluded to above. While these approaches certainly have their place in addressing more realistic engineering problems, they are not a substitute for the fundamental approach presented in this book. Moreover, numerical methods may not be as expedient as closed-form analytical approaches for determining certain asymptotic behaviors of engineering interest.

This part of the book is divided into four chapters: Complex Laplace Transform (Chapter 5), Asymptotic Behavior of Integral Transforms (Chapter 6), Complex Fourier Transform (Chapter 7), and Modern Applications of Complex Variables (Chapter 8). In Chapter 5, we present the theoretical foundations of the techniques of complex Laplace transform in terms of the required conditions for the existence of the transform and its inverse. As there are no general techniques, we illustrate with several ones for a range of PDEs of varying complexities in engineering analysis. These include unsteady diffusion in an infinite rod, unsteady diffusion with a source term in a finite domain, the complexity introduced by periodic boundary conditions, and variable coefficient PDEs. We also include a section on the solution of generalized unsteady temperature or chemical species diffusion problems with variable thermophysical properties and different heat (or species) dissipation mechanisms and sources. The emphasis is placed on the appropriate techniques for complex-variable Laplace inversion for a particular class of problems.

https://doi.org/10.1515/9783111351179-006

The subject of Chapter 6 is the asymptotic behavior of integrals. Engineering problems governed by differential equations that are singular are commonplace. But these are difficult to solve, particularly when the singularity is of the essential type, as no closed-form solutions can easily be found. Asymptotic methods offer an avenue for us to analyze such problems. Since most differential equations of engineering relevance can be transformed to integral equations, and the converse is not generally true, we focus on integrals in this chapter and explore limiting behaviors at small time and longtime dynamics of the systems. The characterization of the types of solutions (exponential decay, exponential growth, algebraic growth or decay, oscillatory solutions, etc.) as functions of the type of singularities involved in the transform space is discussed. The asymptotic methods of Laplace, steepest descent, saddle point, and stationary phase are presented, as are the challenges posed by singularities and multivaluedness in closing the contours for inverting transforms.

In Chapter 7, we focus on the Fourier transform and its inverse from the standpoint of the extension of the transform variable to the complex plane so that certain transforms can be inverted, enabling a broader class of engineering problems to be analyzed by this approach. We give the details of the procedures that could be used to extend the domain of the Fourier transform operation from semi-infinite to the fully infinite domain that is typically required for the transformation. The splitting of functions in the Fourier space is an integral part of this procedure. This decomposition step is illustrated with several examples involving PDEs and integral equations. The Wiener-Hopf technique for solving mixed boundary value problems, which requires the type of decomposition herein alluded to, is presented in detail. Multidimensional Fourier transforms are presented, with several illustrative applications, including the acoustic speaker problem.

In the last chapter of Part II (Chapter 8), we summarize modern developments and applications of complex variables. To this end, we cover recent advances in prime functions, ordinary differential equations in the complex plane, Schwarz-Christoffel transformations, and numerical complex variable simulation.

5 The techniques of complex Laplace transform

The theoretical foundations of the techniques of complex Laplace transform (LT) are presented in this chapter in terms of the required conditions for the existence of the transform and its inverse. There is no general technique for the complex LT solution of partial differential equations; however, we illustrate several techniques for a range of PDEs of varying complexities in engineering analysis. These include unsteady diffusion in an infinite rod, unsteady diffusion with a source term in a finite domain, the complexity introduced by periodic boundary conditions (BCs), and variable coefficient PDEs. We also include a section on the solution of generalized unsteady diffusion problems with variable thermophysical properties and different heat (or species) dissipation mechanisms and sources. The emphasis in all the treatments is the technique for complex-variable Laplace inversion, which tend to be problem-dependent.

5.1 Theoretical background

From elementary treatments of LT, we know that a function $f(t)$ must be piecewise continuous on $[0, \infty)$ and also be of exponential order, for its transform to exist, and for us to be able to recover the function from its transform. References [1–15] contain some level of treatment of the LT method and could be of interest to the reader. The phrases piecewise continuous, piecewise regular, or piecewise smooth are used in this chapter to mean the same thing. By piecewise continuous, we mean that every interval of the form $0 \le t_1 \le t \le t_2$ (Figure 5.1) can be divided into a finite number of subintervals such that $f(t)$ is continuous in the interior of each subinterval and approaches finite limits as t approaches either endpoint of the interval from the interior.

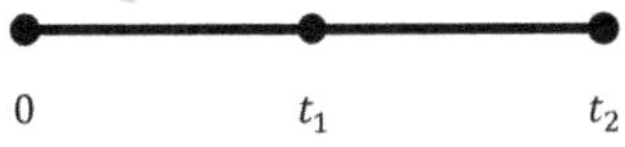

Figure 5.1: An interval of the independent variable t of $f(t)$.

If $f(t), 0 < t < \infty$ is an integrable and smooth function, then we can define its LT as

$$\tilde{f}(s) \equiv \int_0^\infty e^{-st} f(t)dt, \tag{5.1}$$

where s is the transform variable and the operator is $\int e^{-st}dt$. In elementary texts, s is treated as real, but in this chapter, s is a complex variable, with real and imaginary parts, and with properties similar to those complex variables that we have seen in previous chapters of this book. This chapter is principally intended for ab initio Laplace inversion using the complex variable technique, with limited use of tabulated inversion formulas.

https://doi.org/10.1515/9783111351179-007

Is the above LT in eq. (5.1) a well-defined problem? Take as an example, $f(t) = t^n, n \geq 0, n$ is an integer. For any complex s, this function is not integrable from $0 \to \infty$. But if we, for the sake of illustration, assume that the function is integrable, then by repeated application of integration by parts

$$\tilde{f}(s) = n! s^{-n-1}.$$

If $f(t)$ blows up at ∞, for example,

$$f(t) = e^{+t^2},$$

then there is no complex s that can make it integrable because t^2 blows up too fast. Therefore, we need to restrict our functions to those of exponential order, that is, there is an $a > 0$ such that $|f(t)|e^{-at}$ is integrable. Thus, if $f(t)$ is of exponential order, then we can pick some constant "a" large enough such that $\text{Re}[a] > \text{Re}[a_0]$. That is

$$\int_0^\infty |f(t)| e^{-at} dt$$

exists so that $\tilde{f}(s)$ also exists for all s such that $\text{Re}[s] > \text{Re}[s_0]$ (Figure 5.2).

From the foregoing, therefore, we see that $f(t)$ has to satisfy the following conditions for feasible LT and inverse operations:

a) Piecewise smooth over every finite interval
b) Be of exponential order; or there exists real constants K, c, T such that

$$|f(t)| < Ke^{ct} \Rightarrow |f(t)|e^{-ct} < K$$

for all $t > T$. For example, $f(t) = 1/(t+2), t^6, e^{27t}$ are of exponential order but $f(t) = e^{t^2}$ is not.

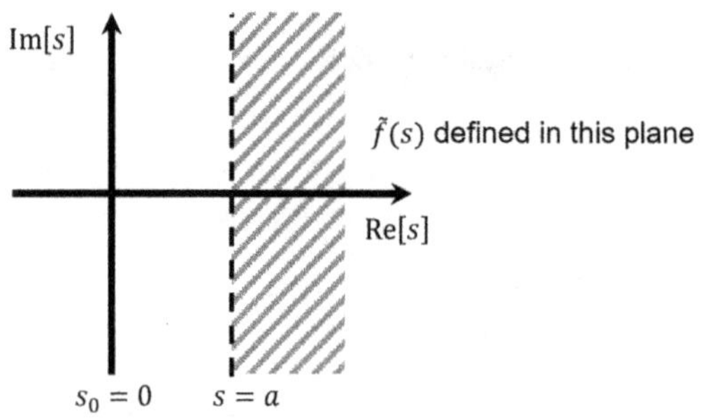

Figure 5.2: Region of definition of s in the complex plane.

Thus,

$$F(t)=\begin{cases}0, & t<0\\ e^{-\gamma t}f(t), & t>0\end{cases},$$

is integrable (although $f(t)$ itself may not be) if we choose $\gamma > c$.

Basically, s_0 is some number such that $I \equiv \int_0^{\infty} e^{-s_0 t} f(t)dt$ exists. For example, $\mathrm{Re}[s]=0$ or $\mathrm{Re}[s]<0$ are not integrable, whereas $s=10^5$ is. Simply put, $f(t)$ cannot grow faster than the rate at which the exponential function $e^{-\gamma t}$ decays to zero since we need a net exponential decay of the integrand.

If $\tilde{f}(s)$ is an analytic function of s that is convergent for $s=s_0$, then it is also convergent in the half plane $\mathrm{Re}[s]>\mathrm{Re}[s_0]$. To see this, we can write

$$\tilde{f}(s)=\int_0^{\infty} e^{-st}f(t)dt$$

as

$$\tilde{f}(s)=\lim_{T\to\infty}\int_0^T e^{[-(s-s_0)t]}e^{(-s_0 t)}f(t)dt,$$

and define $p(t)=\int_0^t e^{-s_0 t}f(t)dt$, which is bounded since it is assumed that $\tilde{f}(s)$ is analytic for $s=s_0$. Integrating the penultimate equation gives

$$\tilde{f}(s)=\lim_{T\to\infty}\int_0^T (s-s_0)e^{[-(s-s_0)t]}p(t)dt,$$

which converges for $\mathrm{Re}[s]>\mathrm{Re}[s_0]$.

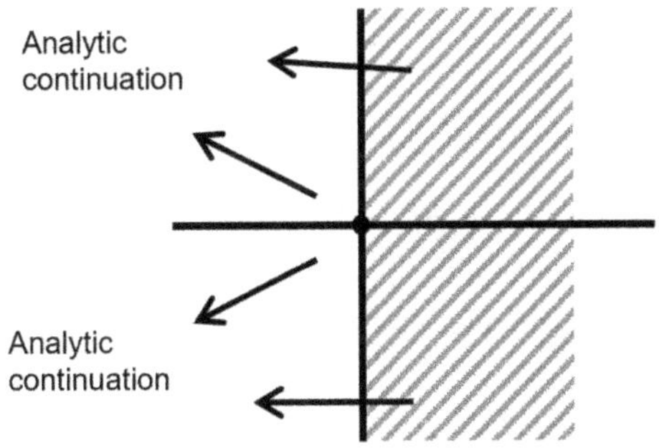

Figure 5.3: Analytic continuation allows us to define $\tilde{f}(s)=1/s=\int_0^{\infty} e^{-st}dt$ everywhere in the complex s plane except at $s=0$, in defiance of the convergence criterion $\mathrm{Re}[s]>0$.

Note that $\tilde{f}(s)$ may exist outside $\mathrm{Re}[s]>\mathrm{Re}[s_0]$ since it might be extensible beyond this region by analytic continuation. For example, if $f(t)=1$, then $\tilde{f}(s)=1/s$, which is analytic everywhere except at $s = 0$, independent of the fact that the expression $\int_0^\infty e^{-st}dt\ (=1/s)$ is convergent for $\mathrm{Re}[s] > 0$. We can therefore perform analytic continuation beyond $s = 0$ to enable integration in negative real s (Figure 5.3).

Regarding the LT inversion, we bear in mind that $f(t) = 0$ for $t < 0$, and we ideally should write

$$p(t) = f(t)e^{-\gamma t} \text{ as } p(t) = f(t)e^{-\gamma t}H(t),$$

where H is the Heaviside step function. At any rate, γ is considered to be large enough to make $p(t)$ to be absolutely integrable. We can use a definition of the Fourier transform (FT)

$$\mathcal{F}((f(t)) = \frac{1}{\sqrt{2\pi}} \int_{-\infty}^{\infty} f(t)e^{i\lambda t}dt,$$

where λ is the FT variable to obtain the formula for LT inversion. With this formula, we write $p(t)$ as

$$p(t) = \mathcal{F}^{-1}(\mathcal{F}(p(t))) = \frac{1}{2\pi} \int_{-\infty}^{\infty} e^{-i\lambda t} \int_0^{\infty} f(\tau)e^{-\gamma\tau + i\lambda\tau}d\tau d\lambda,$$

where we have used the fact that $f = 0$ when $t < 0$ to change the lower limit on the inner integral from $-\infty$ to 0.

Substituting $f(t) = p(t)e^{\gamma t}$, we can write

$$f(t) = \frac{1}{2\pi} \int_{-\infty}^{\infty} e^{(\gamma t - i\lambda)t} \int_0^{\infty} f(\tau)e^{-\gamma\tau + i\lambda\tau}d\tau d\lambda.$$

Let us define $s = \gamma - i\lambda$, $ds = -id\lambda$ and note that $\lambda = \infty$ implies $s = \gamma - i\infty$ and $\lambda = -\infty$ implies $s = \gamma + i\infty$. If in addition we define $\tilde{f}(s) = \int_0^\infty f(t)e^{-st}dt$ and switch the limits of the integration over ds because of the negative sign in $ds = -id\lambda$, we obtain

$$f(t) = \frac{1}{2\pi i} \int_{\gamma - i\infty}^{\gamma + i\infty} \tilde{f}(s)e^{st}ds,$$

which is the required Laplace inversion formula.

A few elementary properties of the LT of relevance in this chapter are as follows:

a) $\int_0^\infty f^n(t)e^{-st}dt = s^n\tilde{f}(s) - s^{n-1}f(0^+) - \ldots - f^{(n-1)}(0^+)$, where $f^n \equiv d^nf/dt^n$. Note that $f(t)$ does not include 0.

b) $$\mathcal{L}\left\{\int_0^t f(\tau)d\tau\right\} = \frac{1}{s}\tilde{f}(s).$$

c) $\mathcal{L}\left\{\int_0^t f(t-\tau)g(\tau)d\tau\right\} \equiv f(t) * g(t) = g(t) * f(t) = \tilde{f}(s)\tilde{g}(s)$.

d) $\mathcal{L}\{e^{at}f(t)\} = \tilde{f}(s-a)$ (Shifting theorem).

e) $\mathcal{L}\{\frac{1}{c}f(\frac{t}{c})\} = \tilde{f}(cs), c > 0; c$ real, $(a+ib) > 0$, which makes no sense except when $b = 0$. If we make "a" positive, we can make $\tilde{f}$ less integrable, so we shift the plane of integrability by "a" to the right (Figure 5.4).

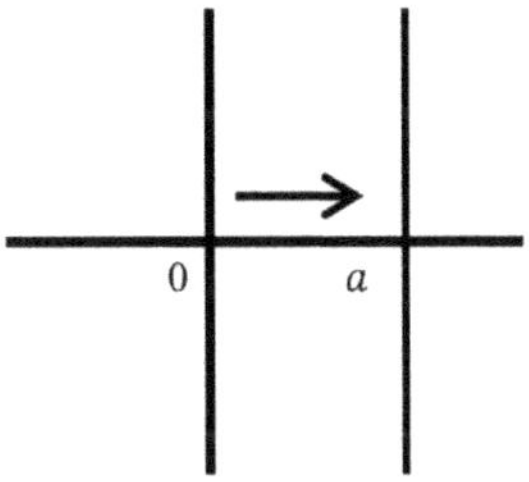

Figure 5.4: Shifting the plane of integrability to the right of the origin of s.

Consider the integral equation

$$\lambda\phi(t) + \int_0^t k(t-\tau)\phi(\tau)d\tau = f(t), \quad 0 < t < \infty,$$

where $f(t), k(t)$ are given; k is the kernel, and λ, ϕ are unknown. This is a particular type of kernel, as a general kernel is of the form $k = k(t,\tau)$. This special case is called translational kernel. To find ϕ, take the LT of both sides, noting that the LT is a linear operator:

$$\lambda\tilde{\phi} + \tilde{k}\tilde{\phi} = \tilde{f}, \quad \tilde{\phi} = \frac{\tilde{f}(s)}{\lambda + \tilde{k}(s)},$$

where λ could be manipulated to determine the region of validity. We need to invert $\tilde{\phi}$ to obtain ϕ, and the success of inversion may depend on λ. Sometimes $\tilde{\phi}$ may be difficult to invert, but if we can have certain information about ϕ (e.g., behavior at 0 and ∞), we may be able to invert.

Consider

$$\tilde{f}(s) = \int_0^\infty e^{-st}f(t)dt$$

on the half plane shown in Figure 5.5.

We may be able to analytically continue beyond the half plane if we have some poles outside of the region of analyticity, which is hatched in Figure 5.6. The dots in Figure 5.6 symbolize singularity.

The condition

$$\sup_{\delta > \gamma} \int_{-\infty}^{\infty} |F(\delta + i\tau)|^2 d\tau < \infty$$

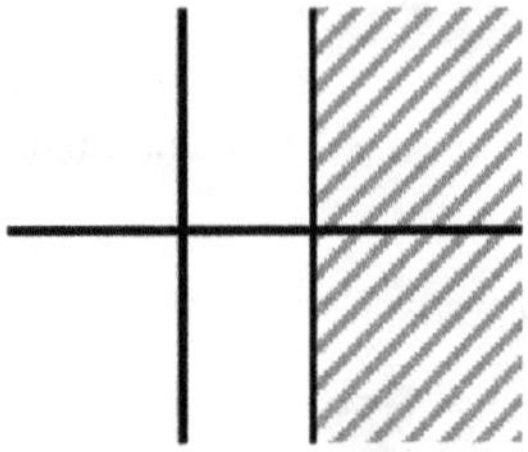

Figure 5.5: Domain of $\tilde{f}(s)$.

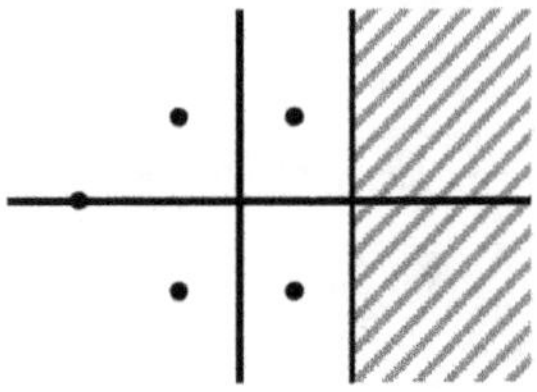

Figure 5.6: An illustration of analytic continuation beyond the region of formal analyticity.

is necessary and sufficient for a function $F(s)$, analytic in the half plane $\mathrm{Re}[s] > \gamma$, to be the LT of a function $f(t)$ satisfying the inequality

$$\int_{-\infty}^{\infty} |f(t)|^2 e^{-2\gamma t} dt < \infty. \tag{5.2}$$

Above, "sup" implies the maximum over all possibilities. The contour for LT integral is shown in Figure 5.7. It goes from $\gamma - i\infty$ to $\gamma + i\infty$.

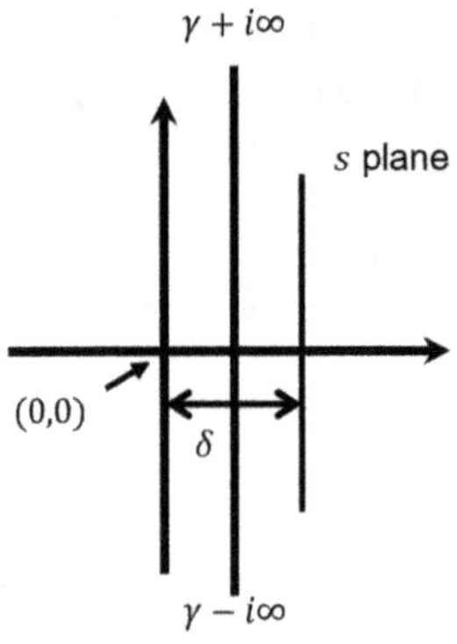

Figure 5.7: The $(\gamma - i\infty, \gamma + i\infty)$ contour for integration in the s plane.

Given an $F(s)$ that is analytic in the right half plane, we form an absolute value of $F(s)$ along any vertical line and integrate along the line, it is the inverse LT satisfying this criterion. As a counter example, consider $f(t) = t^{-1/2}$; $0 < t < \infty$. The condition in eq. (5.2) is violated because $1/t$ is not integrable at 0. The LT is

$$\tilde{f}(s) = \frac{\Gamma(1/2)}{s^{1/2}},$$

which is an analytical function on the right plane ($\mathrm{Re}[s] \geq \gamma$), but is multivalued so that a branch cut has to be made for the contour of integration for the inversion to not enclose the branch point, which is $s = 0$ (Figure 5.8). Note that any branch cut on the left is acceptable; we have analyticity on the right plane.

Theorem 5.1 (On inversion) Let $\int_0^\infty |f(t)|e^{-at}dt < \infty$ for all $a > \gamma$. Let $f(t)$ be a function of bounded variation (i.e., it can be integrated) in a neighborhood of a point t ($t \geq 0$), then

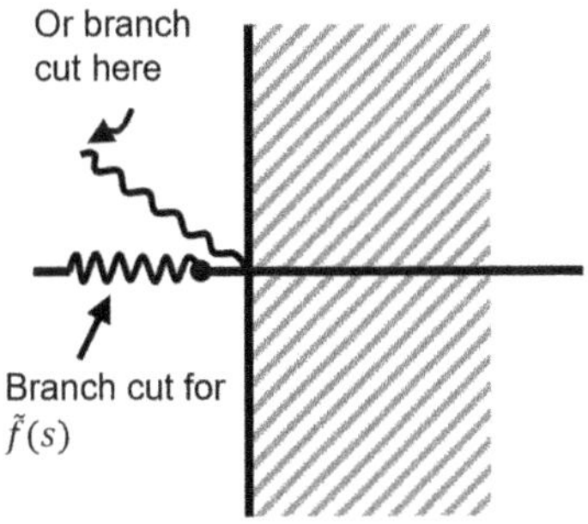

Figure 5.8: Possible branch cuts to prevent multivaluedness due to $s^{1/2}$.

$$\lim_{\omega\to\infty} \frac{1}{2\pi i} \int_{a-i\omega}^{a+i\omega} e^{st}\tilde{f}(s)ds = \begin{cases} \frac{f(t^+)+f(t-)}{2}, & t > 0 \\ \frac{f(0)}{2}, & t = 0 \\ 0, & t \leq 0. \end{cases}$$

Note in particular that the inverse transform is zero for $t < 0$. Given $\tilde{f}(s)$, we need to find where $\tilde{f}(s)$ $\left(\text{not } e^{st}\tilde{f}(s)\right)$ is analytic. For example, if $\tilde{f}(s) = 1/(s-1)$, the function has a pole at 1. This means that we have to select $a = 1$. A plausible contour ($a - i\omega < s < a + i\omega$) for inverting $\tilde{f}(s) = 1/(s-1)$ is shown in Figure 5.9. It is emphasized that there cannot be a singularity to the right of a.

Note that we can integrate along any vertical line $a + i\omega$, $a - i\omega$ as long as $a > \gamma$, $(\omega \to \infty)$.

If we define

$$g(t) = \frac{1}{2\pi i} \int_{a-i\infty}^{a+i\infty} e^{tz}\tilde{f}(z)dz,$$

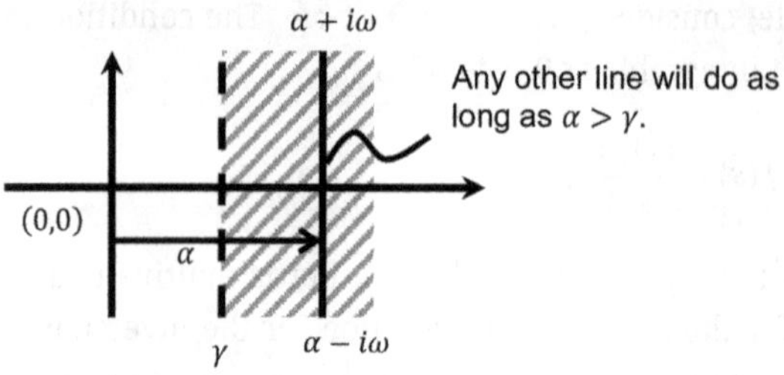

Figure 5.9: Plausible contour for inverting $\tilde{f}(s) = 1/(s-1)$. Here $\alpha = 1$.

we can show that $\tilde{g}(s) = \tilde{f}(s)$ as follows:

$$\tilde{g}(s) = \int_0^{\infty} g(t) e^{-st} dt = \frac{1}{2\pi i} \int_0^{\infty} e^{-st} dt \int_{\alpha - i\infty}^{\alpha + i\infty} e^{tz} \tilde{f}(z) dz.$$

Noting that

$$\frac{1}{2\pi i} \int_0^{\infty} e^{zt} e^{-st} dt \int_{\alpha - i\infty}^{\alpha + i\infty} \tilde{f}(z) dz = \frac{1}{2\pi i} \int_{\alpha - i\infty}^{\alpha + i\infty} \int_0^{\infty} e^{zt} e^{-st} \tilde{f}(s) dt dz$$

and assuming that we can interchange integrals:

$$\tilde{g}(s) = \frac{1}{2\pi i} \int_0^{\infty} e^{(z-s)t} dt \int_{\alpha - i\infty}^{\alpha + i\infty} \tilde{f}(z) dz.$$

We can write this as

$$\tilde{g}(s) = -\frac{1}{2\pi i} \int_{\alpha - i\infty}^{\alpha + i\infty} \frac{\tilde{f}(z)}{(z - s)} dz,$$

where we have assumed that $\mathrm{Re}[s] > \mathrm{Re}[z]$; otherwise $\int_0^{\infty} e^{(z-s)t} dt$ will blow up. Also, we are integrating vertically along α. At this juncture, we need to recall the Cauchy Integral Formula that

$$\phi(z) = \frac{1}{2\pi i} \oint_C \frac{\phi(z')}{z' - z} dz',$$

where C is a closed contour such as depicted in Figure 5.10, as a closed contour is required for the application of this theorem. On the other hand, the contour shown in Figure 5.8 goes from $\alpha - i\infty$ to $\alpha + i\infty$ and is therefore not closed (Figure 5.11). This contour can be closed as shown in Figure 5.12, and since there are no singularities to the right of the line $(\alpha - i\infty, \alpha + i\infty)$, the integral along the closed contour is zero in this case.

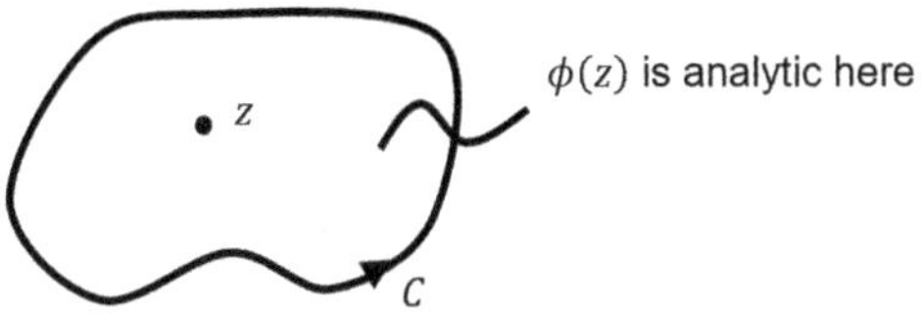

Figure 5.10: Closed contour for the application of Cauchy Integral Formula.

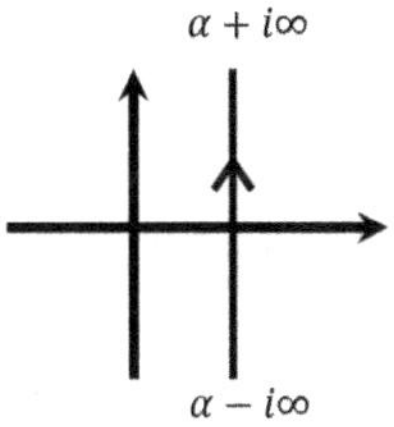

Figure 5.11: Open contour from $\alpha - i\infty$ to $\alpha + i\infty$.

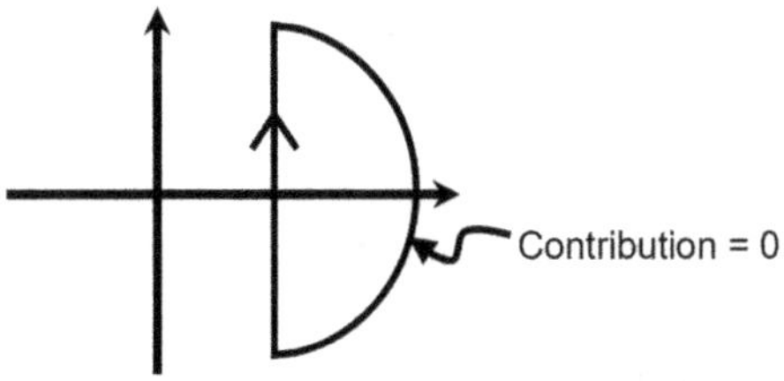

Figure 5.12: One closure of the contour $\alpha - i\infty$ to $\alpha + i\infty$..

However, if $\tilde{f}(s)$ has no singularity to the right of the line $\mathrm{Re}[z] = \alpha$ and $\tilde{f}(z) \to 0$ as $z \to \infty$ on the RHP of this line, then we can apply Cauchy theorem:

$$\frac{1}{2\pi i} \int_{\alpha - i\infty}^{\alpha + i\infty} \frac{\tilde{f}(z)}{(z - s)} dz = \tilde{f}(s)$$

for all s such that $\mathrm{Re}[s] > \alpha$. Note that if we have singularities inside the closed contour, which invariably will have to be located on the left of γ, we will have to include all the residues. We also need to be aware that any line is good to the right of $\mathrm{Re}[s] > \alpha$. We may also close on the left but may encounter singularities of $\tilde{f}(s)$ along the way so that the Cauchy Integral formula cannot be invoked and the LT $\tilde{g}(s)$ cannot be defined or set equal to $\tilde{f}(s)$.

Remark 5.1 The following itemized summary of the foregoing theoretical discussions on complex Laplace transform is in order:

a) If $f(t)$ is piecewise regular on $[0, \infty)$, with

$$\tilde{f}(s) = \int_0^\infty f(t)e^{-st}dt,$$

together with the integrability condition, we are ensured that $\tilde{f}(s) \to 0$ as $\text{Re}[s] \to \infty$.

b) For the inverse transform to be unique, we must limit $f(t)$ ($\tilde{f}(s)$) to functions that are piecewise continuous and discard those that are not defined on certain finite intervals.

c) A sufficient (but not necessary) condition for a function $f(t)$ to have a LT is that $f(t)$ must be piecewise regular on $[0, \infty)$ and of exponential order.

d) There must be a real value of s, $\text{Re}[s] \equiv \gamma$, such that $\tilde{f}(s)$ has no singularities in the complex plane on the right-hand side of γ. This is called the integrability condition and is mathematically expressed as $\left|\int_0^\infty f(t)e^{-\gamma t}dt\right| < M$.

e) To test if there is a γ and determine it, we will usually examine the singularities of $\tilde{f}(s)$. The singularity of $\tilde{f}(s)$ with the largest real part is the critical one in this context, as it allows us to determine $\gamma > s_0$. We cannot find $f(t)$ from $\tilde{f}(s)$ by Laplace inversion if no such γ exists.

f) In principle, we must also test for the condition $\tilde{f}(s) \to 0$ as $\text{Re}[s] \to \infty$ when we have a γ and that this condition is satisfied before we attempt to invert $\tilde{f}(s)$.

g) The contour $\gamma - i\infty < s < \gamma + i\infty$ is called the Bromwich contour.

h) Closing the inversion contour on the LHP is beneficial because $\text{Re}[s] < 0$, so that e^{st} goes to zero very quickly. We will not obtain convergence if we close on RHP.

i) If $\tilde{f}(s)$ has branch points, which may be different or the same as the poles of $\tilde{f}(s)$, we must introduce branch cuts since otherwise, we will not be able to close the inversion contour at infinity.

Example 5.1 Consider

$$\mathcal{L}^{-1}\left\{\frac{1}{s^2+1}\right\} = \sin t$$

and the contour in Figure 5.13.

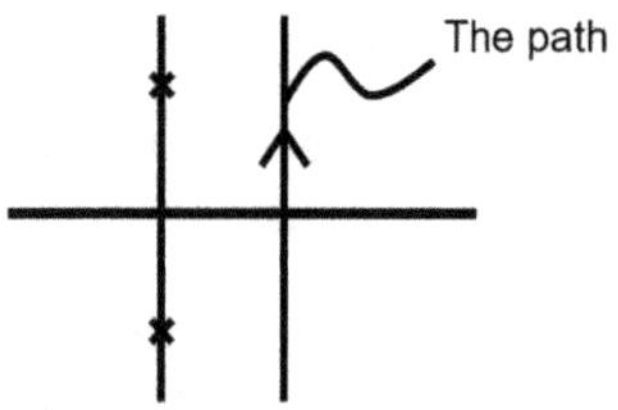

Figure 5.13: The required contour for inverting $\frac{1}{s^2+1}$.

Solution: The integral to be evaluated to obtain $f(t)$ is

$$I = \frac{1}{2\pi i} \int_{\gamma - i\infty}^{\gamma + i\infty} \frac{e^{st}}{s^2+1} ds. \tag{5.3}$$

The closed contour needed to apply the residue theorem is shown in Figure 5.14. Utilizing this theorem and noting that the contribution of the semicircular contour is zero or

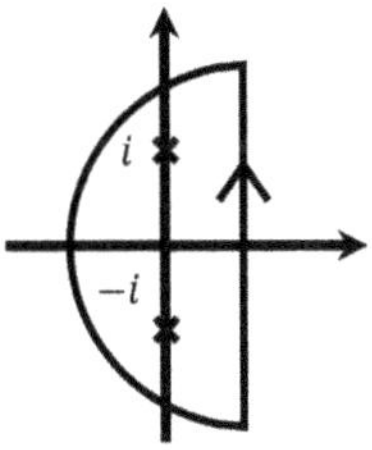

Figure 5.14: A closed contour for evaluating the integral in eq. (5.3).

$$\lim_{R\to\infty} \frac{e^{Rs}}{R^2} \pi R = \lim_{R\to 0} \frac{1}{R} e^{Rs} \to 0, \qquad s \text{ on left,}$$

we will have

$$I = \frac{2\pi i}{2\pi i} \cdot \operatorname{Res}\left(\frac{e^{st}}{s^2+1}; -i, i\right) = -\frac{e^{-it}}{2i} + \frac{e^{it}}{2i} = \sin t.$$

Note that all we need to do to obtain I is to integrate along $(a - i\infty, a + i\infty)$, no matter the contour. However, the left plane, which may harbor singularities, is best because the integral dies out exponentially fast, although residue calculus is required as shown above. The two contours implied are shown in Figure 5.15, but the one on the left of the figure is warranted and has been used.

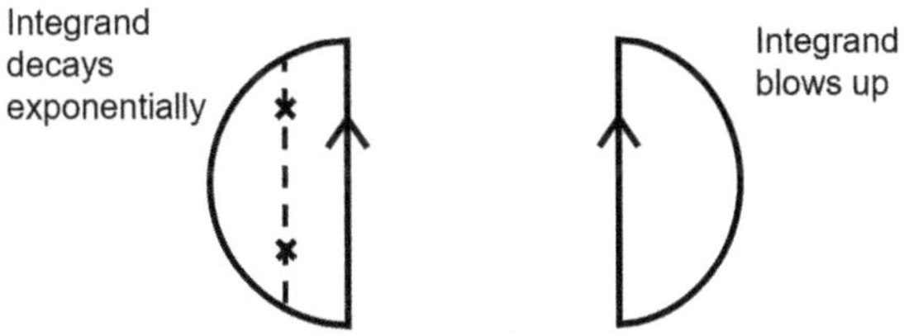

Figure 5.15: Two contours, one enclosing poles.

Example 5.2 Application to a finite integral problem. Use the LT method to obtain the solution to the eigenvalue problem:

$$y''+y=f(t), \qquad y(0)=y(\pi)=0.$$

Solution: We have a two-point boundary value (BV) problem, so we might not have a solution, as we must satisfy uniqueness and existence for the eigenvalue problem. Contrast this with an initial value (IV) problem, for which we always have a solution. Therefore, it is useful to be able to convert a BV problem to an IV problem. Another advantage of doing this is the availability of a wealth of powerful and accurate techniques for solving IV problems. We will use the shooting method for this purpose. Suppose the BCs were not given for the present problem, we would straight away determine the Laplace transform as if semi-infinite, i.e.,

$$s^2\tilde{y}-sy(0^+)-y'(0^+)+\tilde{y}=\tilde{f}(s).$$

The Dirichlet condition at $t=0$ is known $(y(0^+)=0)$; therefore, we will have:

$$s^2\tilde{y}-y'(0^+)+\tilde{y}=\tilde{f}(s).$$

To turn the BV problem to an IV problem, we could guess the slope of y at $t=0$, say, $\beta=y'(0^+)$, but we don't explicitly specify any hard conditions at $t=\pi$ because we are dealing with an IV problem. However, the goal is to obtain β that yields the desired BC at $t=\pi$. Obviously our initial guess for β will not do this, suggesting that some kind of iteration on β will be required. That is, by somehow adjusting β to shoot the solution to the second point (π) (Figure 5.16) where the BC is not known in an IV framework, we are able to generate the BC there. The governing equation in the LT space (for IV problem) becomes $s^2\tilde{y}-\beta+\tilde{y}=\tilde{f}(s)$, leading to

$$\tilde{y}=\frac{\tilde{f}+\beta}{s^2+1}=\frac{\tilde{f}}{s^2+1}+\frac{\beta}{s^2+1}.$$

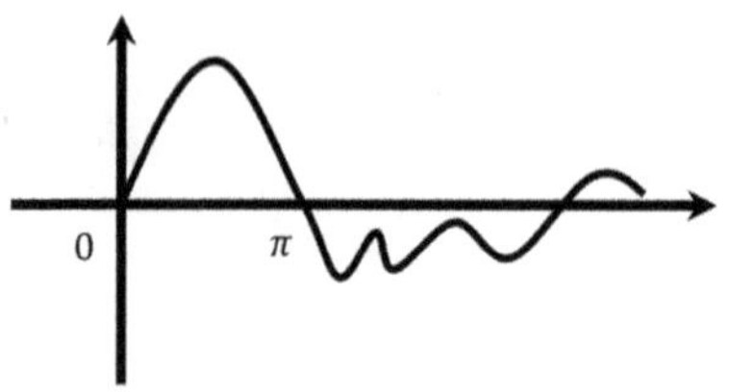

Figure 5.16: Shooting to a point in the s domain where a BC may not be known.

Using convolution we will have

$$y = \int_0^t f(t)\sin(t-\tau)d\tau + \beta \sin t.$$

We then use this equation to demand that $y(\pi) = 0$ for the IV problem, so that since $\beta \neq 0$ and $\sin\pi = 0$, we must satisfy the condition

$$\int_0^\pi f(t)\sin t\, dt = 0.$$

Thus, this problem has no solution unless f satisfies this condition, which gives an infinite number of solutions. This is an orthogonalization requirement: $f(t)$ has to be orthogonal to $\sin t$ (which is the solution of the homogeneous problem $y'' + y = 0$), where the inner product is here defined as

$$\langle f, g\rangle = \int_0^\pi f(\tau)g(\tau)d\tau.$$

Example 5.3 (Constant coefficient ODEs). Constant coefficient ODEs are relatively easy to solve. Take for example:

$$u'' + au' + bu = f(t),$$
$$u(0) = \alpha, u'(0) = \beta,$$

where a, b are constants. We can write the solution down immediately. Taking LT, we have

$$(s^2 + as + b)\tilde{u} = \tilde{f} + s\alpha + \beta + a\alpha,$$

$$\tilde{u} = \frac{\tilde{f}}{s^2 + as + b} + \frac{s\alpha + (\beta + a\alpha)}{s^2 + as + b}.$$

Suppose

$$\mathcal{L}^{-1}\left\{\frac{1}{s^2 + as + b}\right\} = k(t) = \frac{1}{2\pi i}\int_{\gamma - i\infty}^{\gamma + i\infty} \frac{e^{st}}{s^2 + as + b} ds.$$

Then, by convolution:

$$u = \underbrace{\int_0^t k(t-\tau)f(\tau)d\tau}_{A} + \underbrace{\mathcal{L}^{-1}\left\{\frac{sa+\beta+aa}{s^2+as+b}\right\}}_{B}. \tag{5.4}$$

Functions such as $k(t)$ above only have poles which are complex conjugates of each other. If the poles have negative real parts, the solutions will be stable. See Figure 5.17. A in eq. (5.4) is the particular solution or the nonhomogeneous part of the solution, while B is the homogeneous solution, which is exponential: $e^{s_1 t}, e^{s_2 t}$; $te^{s_1 t}$, $te^{s_2 t}$.

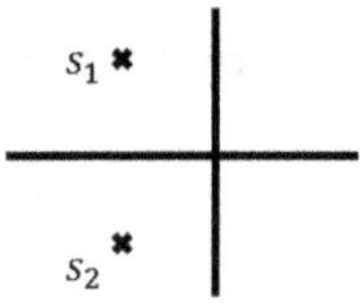

Figure 5.17: Location of poles s_1 and s_2 in the negative real s gives stable solutions.

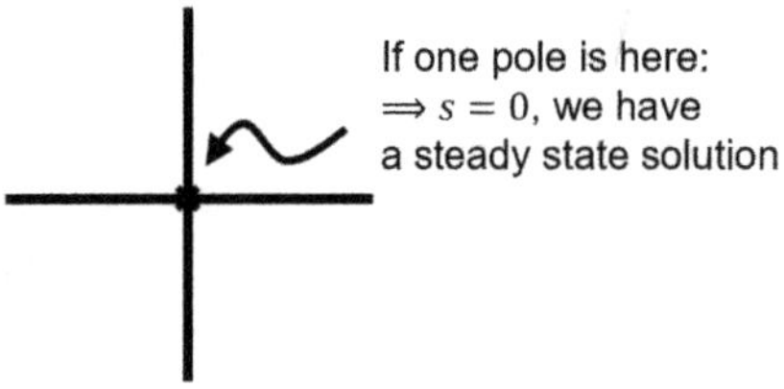

Figure 5.18: A pole at $s = 0$ implies a steady-state solution.

We have double roots (poles of order 2), hence the terms have $te^{s_1 t}$. If the roots are distinct, so are the poles. If a pole has $s = 0$ (Figure 5.18), then a steady solution is implied.

If the coefficients $a, b = 0$, we will have $1/s^2$, or double poles at zero, $e^{st} = 0$. In this case, we have a constant and a linear function of t when the integral is evaluated. The reader should verify these limits of the sample problem:

Example 5.4 Variable coefficient ODE problems can sometimes be solved using LT. Consider

$$(tu')' + tu = 0,$$

which is a Bessel equation of order zero. The solution proceeds as follows:

$$\widetilde{tu} = \mathcal{L}\{tu\} = -\frac{\partial}{\partial s}\tilde{u},$$

$$L\{tu'\} = -\frac{\partial}{\partial s}(\tilde{u}') = -\frac{\partial}{\partial s}[s\tilde{u} - u(0^+)].$$

Whereas constant coefficient ODEs give algebraic equations for s, variable coefficient ODEs give differential equations in s, as shown below. (A second-order ODE in s doesn't exactly sound like a simplification of the solution procedure.) We can continue the above procedure to obtain

$$(s^2+1)\frac{d\tilde{u}}{ds}+s\tilde{u}=0, \qquad \tilde{u}=\frac{1}{(s^2+1)^{1/2}}.$$

Although we have two solutions, we automatically get rid of the solution that is unbounded because we implicitly assume 0^+ exists. The solution $u(t)$ can be obtained from

$$u(t)=\frac{1}{2\pi i}\int_{\gamma-i\infty}^{\gamma+i\infty}\frac{e^{st}ds}{(s^2+1)^{1/2}}. \tag{5.5}$$

A plausible contour for inversion is shown in Figure 5.19.

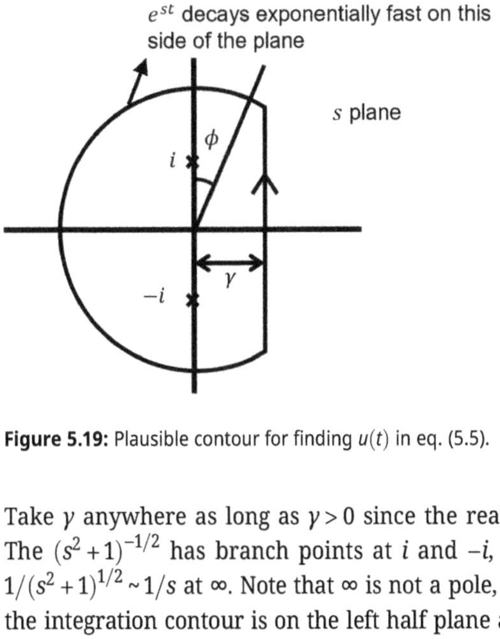

Figure 5.19: Plausible contour for finding $u(t)$ in eq. (5.5).

Take γ anywhere as long as $\gamma>0$ since the real part of the pole of $1/(s^2+1)$ is zero. The $(s^2+1)^{-1/2}$ has branch points at i and $-i$, while e^{st} is analytic everywhere and $1/(s^2+1)^{1/2}\sim 1/s$ at ∞. Note that ∞ is not a pole, and that e^{st} decays fast as $1/s$ because the integration contour is on the left half plane and γ only has to be greater than zero. We can use Jordan's lemma. (The angle ϕ becomes smaller and smaller.) However, note that $1/(s^2+1)^{1/2}$ is not analytic on the LHP because the function has branch points. We make it analytic by introducing a branch cut such as shown in Figure 5.20. The function $1/(s^2+1)^{1/2}$ is analytic everywhere outside the branch cut, and the nonanalyticity of this function is not due to the presence of singularities, but rather to the presence of branch points (at $\pm i$). The contour in Figure 5.21 is used to invert the integral in eq. (5.5). According to Cauchy's deformation theorem, we will have:

$$\frac{1}{2\pi i}\int_{\Gamma+C}\frac{e^{st}ds}{(s^2+1)^{1/2}}=\frac{1}{2\pi i}\oint\frac{e^{st}ds}{(s^2+1)^{1/2}}.$$

Figure 5.20: A plausible branch cut for $1/(s^2+1)^{1/2}$.

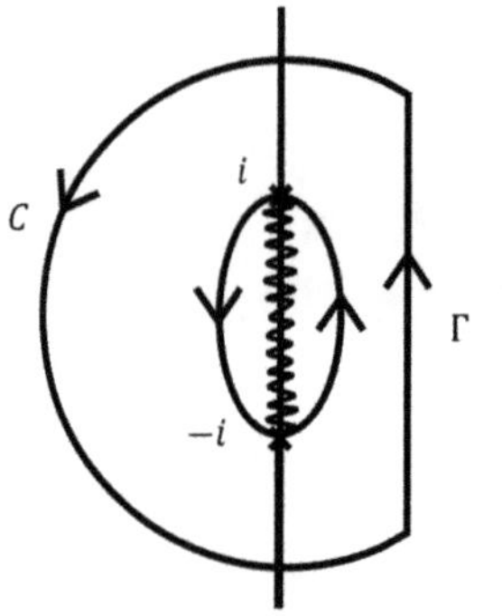

Figure 5.21: The closed contour with branch cut that we choose for determining $u(t)$ in eq. (5.5).

But

$$\frac{1}{2\pi i}\int_{\Gamma}\frac{e^{st}ds}{(s^2+1)^{1/2}}=\frac{1}{2\pi i}\int_{\gamma-i\infty}^{\gamma+i\infty}\frac{e^{st}ds}{(s^2+1)^{1/2}}=u(t)$$

and the integral over Curve C alone is

$$\frac{1}{2\pi i}\int_{C}\frac{e^{st}ds}{(s^2+1)^{\frac{1}{2}}}=0$$

by Jordan's Lemma. So,

$$u(t)=\oint\frac{e^{st}ds}{(s^2+1)^{\frac{1}{2}}}.$$

Let $s = iy$, based on the contour , so that

$$\vdots$$

$$u(t) = \frac{1}{\pi} \int_{-1}^{1} \frac{\cos(yt)dy}{\sqrt{1-y^2}} = J_0(t),$$

where $J_0(t)$ has no closed form.

The contour in Figure 5.21 above is not the only one possible. Another possibility is shown in Figure 5.22.

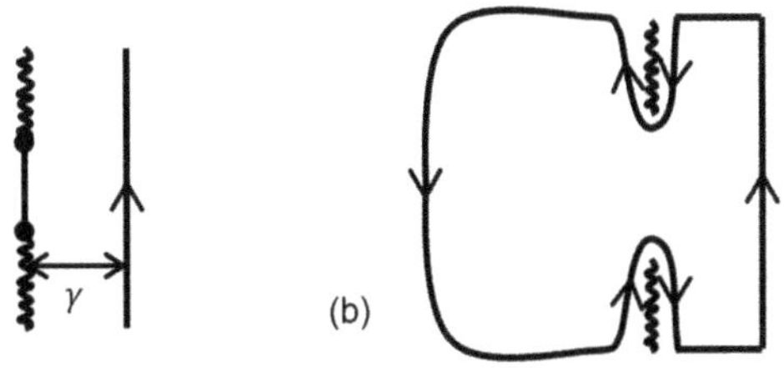

Figure 5.22: (a) The original contour, (b) an alternate closed contour with branch cuts that could be used to evaluate $u(t)$.

Note that the desired contour $(\gamma - i\infty, \gamma + i\infty)$ (Figure 5.22(a)) is now the union of the two indented contours in Figure 5.22(b) or

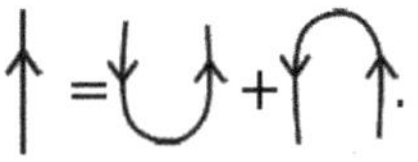

This is so because $\int$$f(s)ds = 0$ since there is no singularity inside the contour. Using this contour should give us the same answers as above, although we need to do one transformation of variables.

5.2 Applications to partial differential equations (PDES)

While there is no general technique for the complex LT solution of PDEs, this section illustrates the techniques for several PDEs of varying complexities.

5.2.1 Unsteady diffusion in an infinite domain

Example 5.5 Consider the unsteady heat diffusion equation in an infinite rod along the x-coordinate direction (Figure 5.23):

$$u_{xx} = u_t$$

with boundary and initial conditions

$$u(x = \pm\infty) = 0, u(x, t = 0) = f(x).$$

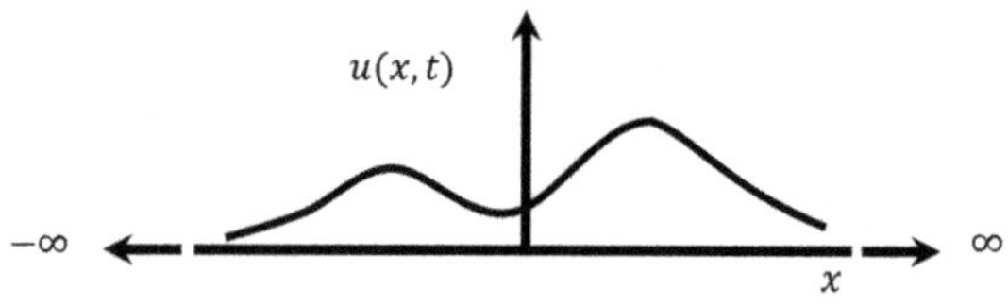

Figure 5.23: Sketch of a hypothetical distribution of $u(x, t)$ along a rod at an instant in time.

We want to find $u(x, t), t > 0$. We could also use the FT method to solve this problem. Which is better? We will solve the problem in a way that involves LT, FT, and the Green's functions to explore some relationship between the three approaches.

First, LT with respect to t gives

$$s\tilde{u} - u(x, t = 0) = \tilde{u}_{xx} \Rightarrow \tilde{u}_{xx} - s\tilde{u} = -f(x).$$

$$u(\pm\infty, t) = 0 \Rightarrow \tilde{u}(\pm\infty, s) = 0.$$

The general way of solving this kind of nonhomogeneous ODE is by Green's function:

$$\frac{d^2G}{dx^2} - sG = \delta(x - x') \qquad [-\infty, \infty].$$

Assume that we are placing a Green's function at x'. $\tilde{u}(x \to \pm\infty, s) \to 0$ due to the BC. The Green's function for this problem is

$$G(x, x', s) = \begin{cases} -\frac{1}{2\sqrt{s}} e^{\sqrt{s}(x-x')}, & -\infty < x < x' \\ -\frac{1}{2\sqrt{s}} e^{-\sqrt{s}(x-x')}, & x' < x < \infty. \end{cases} \qquad (5.6)$$

The general solution of the problem is

$$\tilde{u} = \int_{-\infty}^{\infty} G(x, x', s) f(x') dx',$$

where G can be computed from knowing two linearly independent (LI) solutions. Note that (a) G is continuous but has a kink at x'; (b) for $x > x'$, G decays exponentially.

We take the positive square root of s (from the two possible branches – positive or negative – of the square root of s) to gain exponential decay. The Green's function in eq. (5.6) for this problem can be combined as

$$G(x,x',s)=-\frac{1}{2\sqrt{s}}e^{-\sqrt{s}|x-x'|},$$

where the $|\cdot|$ symbol has been introduced to account for $x>x'$ and $x<x'$. We can solve for u as

$$u=\mathcal{L}^{-1}\{\tilde{u}\}=\mathcal{L}^{-1}\left\{\int_{-\infty}^{\infty}G(x,x',s)f(x')dx'\right\}=\int_{-\infty}^{\infty}\left[\mathcal{L}^{-1}G\right]f(x')dx'.$$

Here, $\mathcal{L}^{-1}$ operates only on G because only G has s:

$$\mathcal{L}^{-1}\{G\}=\mathcal{L}^{-1}\left\{\frac{e^{-\sqrt{s}|x-x'|}}{2\sqrt{s}}\right\}. \tag{5.7}$$

What is $\mathcal{L}^{-1}\left\{e^{-a\sqrt{s}}/\sqrt{s}\right\}$ since $|x-x'|$ is constant >0 or $a>0$? Toward evaluating

$$I\equiv\mathcal{L}^{-1}\left\{\frac{e^{-a\sqrt{s}}}{\sqrt{s}}\right\}=\frac{1}{2\pi i}\int_{\gamma-i\infty}^{\gamma+i\infty}\left(\frac{1}{\sqrt{s}}e^{-\sqrt{s}a}\right)e^{st}dt,$$

we take the positive branch of the square root function in order to have an exponential decay and note that the integral of $1/\sqrt{s}$ is 0 at ∞. The slit contour in Figure 5.24 is used.

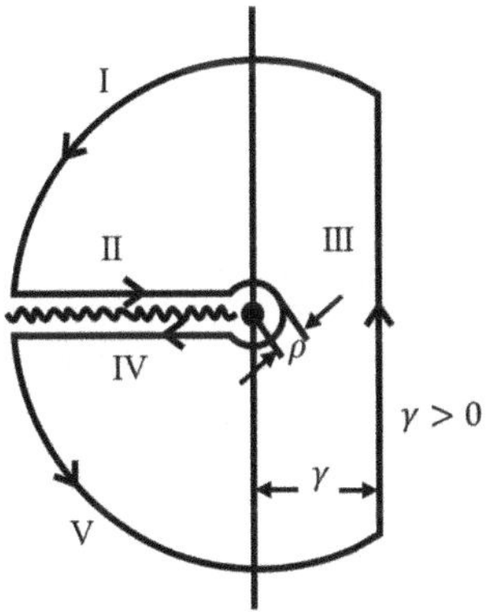

Figure 5.24: Slit contour for determining $\mathcal{L}^{-1}\{G\}$ in eq. (5.7).

$$\text{On I, V}: \quad \int = 0$$

$$\text{III:} \quad \int = 0 \text{ since } \frac{1}{\sqrt{s}} ds \sim \frac{1}{\sqrt{s}} \pi s = \sqrt{\rho}; \quad \rho \to 0 \Rightarrow \sqrt{\rho} = 0$$

Thus,

$$I = \frac{1}{2\pi i} \int_{\supset} \left(\frac{1}{\sqrt{s}} e^{-\sqrt{s}a} \right) e^{st} dt,$$

where the direction of the contour has been reversed to comply with Cauchy's theorem. On II and IV, s in purely negative ($s < 0$), so let's call it $s = -a^2$, where a is a positive real number. Then

$$\sqrt{s} = \sqrt{-a^2} = i\sqrt{a}$$

since we are taking the positive branch (on the top). Now

$$s = re^{i\theta}, \quad ds = e^{i\theta} dr + rie^{i\theta} d\theta.$$

At the top, $\theta = \pi$, $\sqrt{s} = \sqrt{r}e^{i\pi/2} = i\sqrt{r} = ia$. At the bottom, $\theta = -\pi$, $\sqrt{s} = \sqrt{r}e^{-i\pi/2} = -i\sqrt{r} = -ia$. Substituting and noting that: ↑ = −⤼ = ⤼, we have

$$\begin{aligned} I &= \frac{1}{2\pi i} \left[\int_0^{\infty} \frac{1}{ia} e^{-iaa} e^{-a^2 t} 2a \, da + \int_0^{\infty} \frac{1}{ia} e^{+iaa} e^{-a^2 t} 2a \, da \right] \\ &= +\frac{1}{\pi} \left[\int_0^{\infty} e^{-iaa} e^{-a^2 t} da + \int_0^{a} e^{iaa} e^{-a^2 t} da \right]. \end{aligned} \tag{5.8}$$

We then integrate or look up in integration tables to get

$$I \equiv \mathcal{L}^{-1} \left\{ \frac{e^{-a\sqrt{s}}}{\sqrt{s}} \right\} = \frac{e^{-a^2}/4t}{\sqrt{\pi t}}$$

so that

$$\tilde{u} = \int_{-\infty}^{\infty} \frac{e^{-|x-x'|^2/(4t)}}{2\sqrt{\pi t}} f(x') dx'$$

for any f. Note in passing that in physical space t for this problem:

$$u_{xx} = u_t, \quad u(x = \pm\infty, t) = 0, \quad u(x, t = 0) = f(x),$$

we could have placed an impulse, $\delta(\ )$, on the boundary where $f(x)$ is, and we would have found out that the Green's function is in the physical t-space is

$$G = \frac{e^{-|x-x'|^2/4t}}{2\sqrt{\pi t}}.$$

This Green's function could be used in other related problems for different functions $f(x)$. We can evaluate u in the present problem as

$$u(x,t) = \mathcal{L}^{-1} \int_{-\infty}^{\infty} \frac{1}{2\sqrt{s}} e^{-\sqrt{s}|x'-x|} f(x')dx'. \tag{5.9}$$

We substitute eq. (5.8) into eq. (5.9) to obtain

$$u(x,t) = \frac{1}{2\pi} \int_0^{\infty} e^{-a^2 t} da \left[\int_{-\infty}^{\infty} e^{-ia|x-x'|} f(x')dx' + \int_{-\infty}^{\infty} e^{ia|x-x'|} f(x')dx' \right].$$

By using the definition of absolute value, we can show that

$$\begin{aligned} \int_{-\infty}^{\infty} e^{-ia|x-x'|} f(x')dx' &= \int_{-\infty}^{x} e^{-ia(x-x')} f(x')dx + \int_{x}^{\infty} e^{-ia(x'-x)} f(x')dx' \\ &= e^{-iax} \int_{-\infty}^{x} e^{iax'} f(x')dx' + e^{iax} \int_{x}^{\infty} e^{-iax'} f(x')dx'. \end{aligned}$$

We do the same for the positive term and add them together to obtain

$$u(x,t) = \frac{1}{2\pi} \int_{-\infty}^{\infty} e^{-iax} e^{-a^2 t} \underbrace{\left[\int_{-\infty}^{\infty} e^{iax'} f(x')dx' \right]}_{\mathcal{F}\{f\}} da,$$

where $\mathcal{F}\{f\}$ is the FT of f. When we add, we can show that we have an even function:

$$u(x,t) = \frac{1}{2\pi i} \int_{-\infty}^{\infty} e^{-iax} e^{-a^2 t} \mathcal{F}\{f\} da.$$

To get the inversion formula back – just set $t = 0$ in the solution. That is, since $u(x, t = 0) = f(x)$, then

$$f(x) = \frac{1}{2\pi i} \int_{-\infty}^{\infty} e^{-iax} \mathcal{F}\{f\} da.$$

The reader is reminded that the notations used here for FT are:

$$\mathcal{F}\{\lambda\} = \frac{1}{\sqrt{2\pi}} \int_{-\infty}^{\infty} e^{i\lambda x} f(x)dx, \qquad f(x) = \frac{1}{\sqrt{2\pi}} \int_{-\infty}^{\infty} e^{-i\lambda x} \tilde{f}(\lambda)d\lambda.$$

Also note that the FT operator takes that the FT operator takes an L_2 function into an L_2 function. So also the inverse operator. These are unitary operations since the length of the function is preserved.

5.2.2 Unsteady diffusion with a source in a finite domain

The method of separation of variables (SOV) can be connected to LT (in time) and the Sturm-Liouville problems. The example below illustrates this.

Example 5.6 Unsteady heat conduction in a finite one-dimensional rod (Figure 5.25):

$$u_{xx} = u_t, \quad u(0,t) = u(a,t) = 0, \quad u(x,0) = f(x).$$

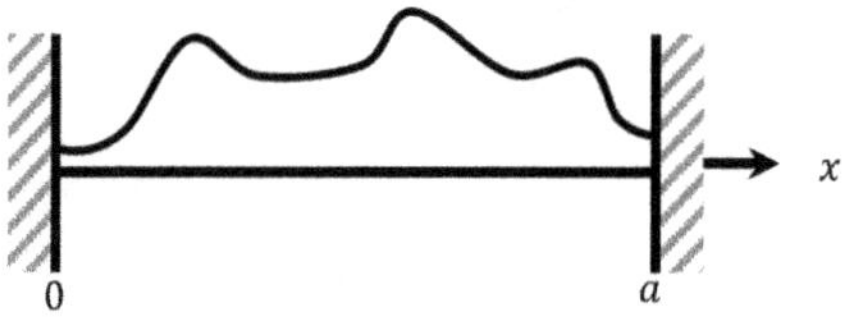

Figure 5.25: Sketch of the physical spatial domain for finite one-dimensional unsteady heat conduction in a rod.

For the SOV approach, we assume that the solution $u(x,t)$ can be separated into a product of functions, respectively, of x and t in this example:

$$u = h(x)g(t),$$

where $h(x)$ must have the form $h(x) = \sin\frac{n\pi x}{a}$ and $g(t)$ must have the form $g(t) = e^{-\frac{n^2\pi^2}{a^2}t}$. We have a linear problem so, we can add up the solutions:

$$u = \sum_{n=1}^{\infty} a_n e^{-\frac{n^2\pi^2 t}{a^2}} \sin\frac{n\pi x}{a}, \qquad f(x) = u(x,0) = \sum_{n=1}^{\infty} a_n \sin\frac{n\pi x}{a}.$$

Sine is complete so that

$$a_n = \frac{2}{a}\int_0^a f(x') \sin\frac{n\pi x'}{a} dx'.$$

For the LT approach, we take the LT with respect to time:

$$\tilde{u}_{xx} - s\tilde{u} = -f(x), \tilde{u}(0,s) = 0, \tilde{u}(a,s) = 0.$$

Let's examine a procedure for solving an ODE problem of the form:

$$y'' + p_1(x)y' + p_0(x)y = f(x), \tag{5.10}$$

$$\begin{aligned} \alpha_1 y(a) + \beta_1 y'(a) &= 0, \\ \alpha_2 y(b) + \beta_2 y'(b) &= 0. \end{aligned} \tag{5.11}$$

If α_1, β_1 are not zero, then we have mixed BCs at a. Similarly at b. This is a BV problem. Let y_1, y_2 be the homogeneous solution of eq. (5.10) and define the Green's functions as

$$G_1(x,x') = \frac{y_1(x')y_2(x)}{W(x')}, \quad a < x' < x$$
$$G_2(x,x') = \frac{y_1(x)y_2(x')}{W(x')}, \quad x' < x < b$$

where $W(x')$ is Wronskian:

$$W(x') = \begin{vmatrix} y_1 & y_2 \\ y'_1 & y'_2 \end{vmatrix} \neq 0,$$

that is, y_1, y_2 are linearly independent.

Furthermore, y_1 and y_2 must also satisfy the conditions in eq. (5.11). (Uniqueness and existence theorems guarantee that this is easily accomplished.) The solution of eqs. (5.10) and (5.11) is then

$$y(x) = \int_a^b G(x,x')f(x')dx'.$$

Clearly, the functions

$$y_1 = \sinh\sqrt{s}x, \quad y_2 = \sinh\sqrt{s}(a-x)$$

satisfy the BCs for this problem so that

$$W = -\sqrt{s}\sinh\sqrt{s}a,$$

which is independent of x. Returning to the unsteady heat diffusion problem,

$$\tilde{u}(x,s) = -\int_0^a G(x,x',s)f(x')dx'.$$

One of the Green's functions is

$$G(x,x',s)=\frac{\sinh\sqrt{s}x'\sinh\sqrt{s}(a-x)}{-\sqrt{s}\sinh\sqrt{s}a},\qquad 0<x'<x.$$

The second Green's function (for $x<x'<a$) can also be found. $u(x,t)$ can then be obtained as follows:

$$\mathcal{L}^{-1}\{\tilde{u}\}=\int_0^x f(x')dx'\left[\frac{1}{2\pi i}\int_{\gamma-i\infty}^{\gamma+i\infty}\frac{e^{st}\sinh\sqrt{s}x'\sinh\sqrt{s}(a-x)ds}{\sqrt{s}\sinh\sqrt{s}a}\right]$$
$$+\int_x^a f(x')dx'\left[\frac{1}{2\pi i}\int_{\gamma-i\infty}^{\gamma+i\infty}\frac{e^{st}\sinh\sqrt{s}x\sinh\sqrt{s}(a-x')ds}{\sqrt{s}\sinh\sqrt{s}a}\right].$$

We have split up the integral because of the partition $0<x'<x, x<x'<a$. Because of the appearance of $\sqrt{s}$ in this expression, one is tempted to treat $s=0$ as a branch point and introduce a branch cut. However, in the series expansion for $\sinh\sqrt{s}$, the first term is $\sqrt{s}$ (sinh is odd):

$$a_1\sqrt{s}+a_2\left(\sqrt{s}\right)^3+\ldots.$$

Therefore, we can factor out $\sqrt{s}\sqrt{s}=s$ so that there is no branch point or the need to introduce a branch cut for the purpose. In short: (a) $G(x,x',s)$ is not a multivalued function and (b) $s=0$ is not a singular point. At $s\to 0$, $\sqrt{s}$ behaves like s so that

$$\sqrt{s}\sinh\sqrt{s}a=sa[1+\ldots].$$

We have the same situation with the numerator so that s cancels out and we have no singularity. But we do need a singular point. The solution decays with time so that the singular point is on the left. Also, the singularity has to be on the real axis because we do not have an oscillatory solution (sin$h$ is not oscillatory). The singularity of $G(x,x',s)$ is where $\sqrt{s}\sinh\sqrt{s}a$ is $=0$, not at $s=0$. The points

$$s_n=-\frac{n^2\pi^2}{a^2},\qquad n\geq 1$$

are simple poles (Figure 5.26).

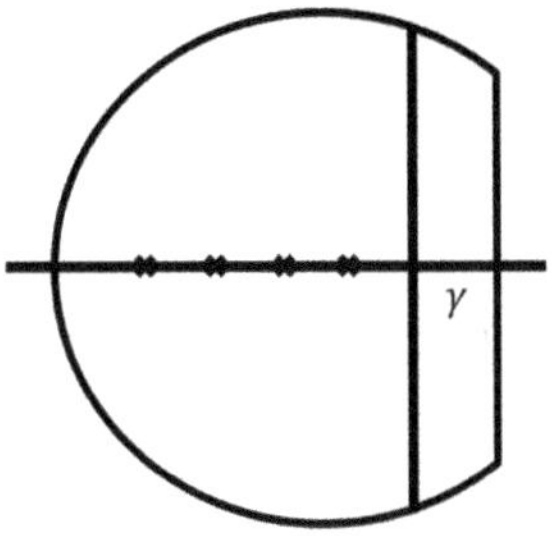

Figure 5.26: Contour for inverting $\tilde{u}(s)$ showing the poles on the real axis.

We can write

$$I_1 = \frac{1}{2\pi i} \int_{\gamma-i\infty}^{\gamma+i\infty} \frac{e^{st} \sinh \sqrt{s}x' \sinh \sqrt{s}(a-x)}{\sqrt{s} \sinh \sqrt{s}a} ds = \sum_{n=1}^{\infty} \operatorname{Res} \left[\frac{e^{st} \sinh \sqrt{s}x' \sinh \sqrt{s}(a-x)}{\sqrt{s} \sinh \sqrt{s}a}, s_j \right]$$

or after some algebra,

$$I_1 = \frac{2}{a} \exp\left[-\frac{n^2\pi^2 t}{a^2} \right] \sin\frac{n\pi x'}{a} \sin\frac{n\pi x}{a}, \quad 0 < x' < x,$$

where we have used $\sinh iz = i \sin z$.

Above the s_j's are the poles of $\sqrt{s} \sinh \sqrt{s}a$. The role of x and x' in the inversion is the same $\Rightarrow I_2 = I_1$ and

$$\begin{aligned} u(x,t) &= \frac{2}{a} \int_{f}^{a} f(x') \left[\sum_{n=1}^{\infty} \sin\frac{n\pi x'}{a} \sin\frac{n\pi x}{a} e^{-\frac{n^2\pi^2 t}{a^2}} \right] dx' \\ &= \sum_{n=1}^{\infty} \underbrace{\left[\frac{2}{a} \int_0^a f(x') \sin\frac{n\pi x'}{a} dx' \right]}_{\text{Fourier coefficient, } a_n} e^{-\frac{n^2\pi^2 t}{a^2}} \sin\frac{n\pi x}{a}. \end{aligned}$$

We have interchanged $\sum$ and $\int$ in the foregoing because the series is uniformly convergent. The solution above using LT agrees with that from SOV. We could as well solve the problem with Fourier series, which would be advantageous if we are interested in the large-time behavior, which could be given by the $n=1$ term, as the exponent decays fast for larger n. However, for the small-time behavior, we have to sum the sine series solution, as all the modes are important. With the LT method, we can obtain a good solution at small time if we so desire, expressing an advantage of the LT approach over the SOV.

5.2.3 Unsteady diffusion with periodic boundary conditions in a finite domain

Example 5.7 Solve the following unsteady heat equation for $(x,t) \in [-a,a] \times [0,\infty)$ using the LT method:

$$\phi_t = \phi_{xx}, \quad -a < x < a, \quad \phi(x,0) = 1, \quad \phi(-a,t) = \phi(a,t) = 0.$$

Solution: Take the LT in time:

$$-1 + s\tilde{\phi}(x,s) = \tilde{\phi},_{xx}(s,x), \quad \tilde{\phi}(-a,s) = 0, \quad \tilde{\phi}(a,s) = 0.$$

The particular (nonhomogeneous solution) is

$$\phi_p = \frac{1}{s}$$

so that the complete solution in the LT space is

$$\tilde{\phi} = \frac{1}{s} + A(s)\cosh\sqrt{s}x + B(s)\sinh\sqrt{s}x.$$

Using the given initial and BCs, we determine the constants $A(s), B(s)$ to obtain

$$\tilde{\phi} = \frac{1}{s} - \frac{\cosh\sqrt{s}x}{s\cosh\sqrt{s}a}, \quad \phi(x,t) = \mathcal{L}^{-1}\{\tilde{\phi}\}. \tag{5.12}$$

The singularities of $e^{st}\tilde{\phi}$

On a cursory look, the second term on the RHS of eq. (5.12) has singularities at $s = 0$ and $\cosh\sqrt{s}a = 0$, and a branch point at $s = 0$ on account of the square root operation in the denominator. However, this is not exactly the case. To see this, note that cosh is an even function ($\cosh(s) = \cosh(-s)$), and

$$\cosh\sqrt{s}x = 1 + \frac{(\sqrt{s}x)^2}{2} + \ldots = 1 + as + bs^2 + \ldots,$$

which is a Taylor series in s, without any singularities. Note that $s = 0$ is not a branch point of $\cosh\sqrt{s}$. Thus, $\tilde{\phi}$ is analytic at $s = 0$. Also, $1/s$ is a removable singularity: From eq. (5.12(a)) we have

$$\tilde{\phi} = \frac{1}{s} - \frac{\cosh\sqrt{s}x}{s\cosh\sqrt{s}a} \equiv \frac{1}{s} - \frac{1 + (\sqrt{s}x)^2 a_1 + \ldots}{s(1 + (\sqrt{s}a)^2 a_1 + \ldots}$$

$$\tilde{\phi} = \frac{1}{s}\left[1 - \left(1 + (\sqrt{s}x)^2 a_1 + \ldots\right)\left(1 - (\sqrt{s}a)^2 a_1 + \ldots\right)\right] = as^n; \quad n \geq 0.$$

Since $\cosh\sqrt{sa}$ has its zeros at $s_n = -(2n+1)^2\pi^2/4a^2$, by summing residues we can write

$$\phi(x,t) = \sum_{n=0}^{\infty} \frac{4(-1)^n}{(2n+1)\pi} \exp\left(\frac{-(2n+1)^2\pi^2}{4a^2}t\right) \cos\frac{(2n+1)\pi x}{2a}.$$

It is appropriate to sum this series when t is large but not for small time. Note that we can also write:

$$\mathcal{L}^{-1}\{\tilde{\phi}\} = \mathcal{L}^{-1}\left\{\frac{1}{s}\right\} - \mathcal{L}^{-1}\left\{\frac{\cosh\sqrt{s}x}{s\cosh\sqrt{sa}}\right\} = 1 - \mathcal{L}^{-1}\left\{\frac{\cosh\sqrt{s}x}{s\cosh\sqrt{sa}}\right\}$$

and can expand

$$I(s) \equiv \frac{\cosh\sqrt{s}x}{s\cosh\sqrt{sa}} = \frac{\exp\sqrt{s}x + \exp(-\sqrt{s}x)}{s\exp\sqrt{sa}\left[1+\exp(-2\sqrt{sa})\right]}$$

as

$$I(s) = \frac{e^{-\sqrt{sa}}}{s}\left[e^{\sqrt{s}x} + e^{-\sqrt{s}x}\right]\left[1 - \exp(-2\sqrt{s}a) + \exp(-4\sqrt{s}a) + \ldots\right].$$

We need to pick real $\sqrt{s}$ positive because of the way we pick the branch such that $\eta \equiv \left|e^{-\sqrt{sa}}\right| < 1$. (In fact, η converges absolutely.) With $\eta < 1$, we can write

$$\frac{1}{1+\eta} = 1 - \eta + \eta^2 - \eta^3 + \ldots.$$

Thus, we have

$$\begin{aligned} I(s) &= \frac{1}{s}\left[e^{-\sqrt{s}(a-x)} + e^{-\sqrt{s}(a+x)}\right][1 - \ldots] \\ &= \frac{1}{s}\left\{\exp\left[-\sqrt{s}(a-x)\right] - \exp\left[-\sqrt{s}(3a-x)\right] + \exp\left[-\sqrt{s}(5a-x)\right] + \ldots\right\} \\ &\quad + \frac{1}{s}\left\{\exp\left[-\sqrt{s}(a+x)\right] - \exp\left[-\sqrt{s}(3a+x)\right] + \ldots\right\}. \end{aligned}$$

We see that we need to invert functions of the form $e^{-\sqrt{sa}}/s$, where $a > 0$ in order to evaluate $I(s)$. The Laplace inverse of the function $e^{-\sqrt{sa}}/s$, $a > 0$ can be written as

$$\mathcal{L}^{-1}\left\{\frac{e^{-\sqrt{sa}}}{s}\right\} = \operatorname{erfc}\left(\frac{a}{2\sqrt{t}}\right),$$

where

$$\operatorname{erfc}(\eta) = \frac{2}{\sqrt{\pi}}\int_{\eta}^{\infty} \exp\left(-\xi^2\right) d\xi = \begin{cases} 1, & \eta = 0 \\ 0, & \eta \to \infty. \end{cases}$$

Thus,

$$\phi(x,t) = 1 - \left[\operatorname{erfc}\left(\frac{a-x}{2\sqrt{t}}\right) - \operatorname{erfc}\left(\frac{3a-x}{2\sqrt{t}}\right) + \dots\right] - \left[\operatorname{erfc}\left(\frac{a+x}{2\sqrt{t}}\right) - \operatorname{erfc}\left(\frac{3a+x}{2\sqrt{t}}\right) + \dots\right].$$

When t is small, $(a-x)/2\sqrt{t}$ becomes large and

$$\operatorname{erfc}(\eta) = \frac{1}{\sqrt{\pi}} e^{-\eta^2}\left[1 - \frac{1}{2\eta^3} + \dots\right], \qquad n \to \infty.$$

5.2.4 Unsteady diffusion with variable coefficients

Example 5.8 This is an example of an application that cannot be solved by separation of variables or directly by the series method. The forcing at $t = 0$ represents another source of complexity of this example. Consider the variable coefficient heat equation in a semi-infinite spatial domain

$$u_{xx} - xu_t = 0, \quad u(0,t) = 0 = u(\infty,t), \quad u(x,0) = f(x).$$

As a particular application, this problem could model heat conduction with spatially varying specific heat at constant pressure. The thermal diffusivity can be written as $\mathcal{D} \equiv k/\rho c$, where ρ is density $(\mathrm{kg/m^3})$, k is thermal conductivity $(\mathrm{W/m \cdot K})$, and c is the specific heat at constant pressure $(\mathrm{J/kg \cdot K})$. The equation could then also be written as

$$x\phi_t = \mathcal{D}\phi_{xx}$$

in dimensional form. However, we will set $\mathcal{D}$, the reference thermal diffusivity, to unity in this example.

We take LT with respect to time to obtain

$$\tilde{u}_{xx} - sx\tilde{u} = -xf(x), \quad \tilde{u}(0,s) = \tilde{u}(\infty,s) = 0$$

and find the Green's function so that

$$\tilde{u}(x,s) = \int_0^{\infty} G(x,x',s)[-x'f(x')]dx'.$$

Two LI solutions of the homogeneous problem are, say $v_1(x,s)$ and $v_2(x,s)$, so that

$$G(x,x',s) = \frac{v_1(x')v_2(x)}{W(x',s)}, \quad 0 < x' < x$$
$$= \frac{v_1(x)v_2(x')}{W(x',s)}, \quad x < x' < \infty \tag{5.13}$$

with $v_1(x=0,s)=0$, $v_2(x \to \infty, s) = 0$ and

$$W(x',s) = v_1(x',s)v_2{}'(x',s) - v_1{}'(x',s)v_2(x',s). \tag{5.14}$$

Recall Abel's theorem:

$$y'' + p_1(x)y' + p_0(x)y = 0 \Rightarrow W(x) = W(x_0)e^{-p(x)}, \text{ where } p(x) = \int_{x_0}^{x} p_1(t)dt.$$

If $p_1(t)=0$, $W(x) \equiv$ constant, which is our case: $p_0(x) = -sx$. We can write one of the homogeneous solutions as [15]

$$v_1 = x^{1/2} J_{1/3}\left[(-s)^{1/2}\frac{2}{3}x^{3/2}\right], \tag{5.15a}$$

where we take $-s = se^{i\pi}$, $-\pi < \arg s < \pi$, where [15]

$$J_p(\alpha) = \sum_{n=0}^{\infty} \frac{(-1)^n \left(\frac{x}{2}\right)^{2n+p}}{n!(p+n)!} = \sum_{n=0}^{\infty} \frac{(-1)^n \left(\frac{x}{2}\right)^{2n+p}}{n!\Gamma(n+p+1)}.$$

For p, n integers, $\Gamma(p+n+1) = (p+n)!$ and $\Gamma(\beta) = \int_0^\infty t^{\beta-1}e^{-t}dt$, $\mathrm{Re}[\beta] \geq 1$.

We can write the second homogeneous solution as

$$v_2 = x^{1/2} H^{(1)}_{1/3}\left[(-s)^{1/2}\frac{2}{3}x^{3/2}\right], \tag{5.15b}$$

where the Hankel function $H_\nu^{1,2} = J_\nu \pm iY_\nu$ and

$$Y_\nu = \frac{J_\nu(z)\cos\nu\pi - J_{-\nu}(z)}{\sin\nu\pi}.$$

Note that for small z,

$$H_\nu^{1,2}(z) \sim iY_\nu(z) = -i\frac{1}{\pi}\Gamma(\nu)\left(\frac{1}{2}z\right)^{-\nu}; \ \mathrm{Re}[\nu] > 0. \tag{5.16}$$

However,

$$H_\nu^{1,2}(z) \sim \sqrt{\frac{2}{\pi z}} e^{\pm i\left(z - \frac{1}{2}\nu\pi - \frac{1}{4}\pi\right)} \pm 1 \text{ as } z \to \infty$$

$$-\pi < \arg z < 2\pi \ (+), \quad -2\pi < \arg z < \pi \ (-).$$

This is the reason why we pick $(-s)^{1/2} = \left(se^{i\pi}\right)^{1/2} = e^{i\pi/2}s^{1/2} = i\sqrt{s}$. We take the positive branch for $s^{1/2}$ and $-\pi < \arg s < \pi$ so that $\mathrm{Re}\left[s^{1/2}\right] \geq 0$ and $e^{i\pi/2} = +i$ $\left(z = (-s)^{1/2}\frac{2}{3}x^{3/2}\right)$.

Thus,

$$H_\nu^1(z) \sim \frac{e^{iz}}{\sqrt{z}} = \frac{e^{i(i)s^{1/2}\frac{2}{3}x^{3/2}}}{\sqrt{z}} \to 0 \text{ as } x \to \infty.$$

The decay is exponentially fast. This is used to kill off $J_{1/3}(\ )$ as $x \to \infty$, see eq. (5.17), as

$$J_\nu(z) \sim z^{-1/2}\left[e^{iz} + e^{-iz}\right]$$

so that no matter which branch we take, the solution will blow up exponentially if z has nonzero imaginary part, unlike $e^{-s(\ )}$.

The Wronskian, as per Abel's theorem is

$$W(x', x) = \text{constant} = -\frac{3i}{\pi}.$$

This can easily be proved, as we now show. We can write

$$J_\nu(z) = \sqrt{\frac{2}{\pi z}}\left[\cos\left(z - \frac{1}{2}\nu\pi - \frac{1}{4}\pi\right)\right], z \to \infty \tag{5.17}$$

and

$$J_\nu(z) \cong \frac{(z/2)^\nu}{\Gamma(\nu+1)} \text{ as } z \to 0. \tag{5.18}$$

Combining eqs. (5.14)–(5.16), and (5.18), we get

$$\begin{aligned} v_1(x', s) &\sim x'^{1/2}\left[\tfrac{(-s)^{1/2}}{2}\left(\tfrac{2}{3}\right)x'^{3/2}\right]^{1/3}\Gamma\left(\tfrac{1}{3}+1\right) \\ &= x'\left[(-s)^{\frac{1}{2}}\tfrac{1}{3}\right]^{1/3}/\Gamma\left(\tfrac{1}{3}+1\right), \quad x' \sim 0. \end{aligned}$$

As for $v_2(x', s)$:

$$v_2(x', s) \sim -\frac{ix'^{1/2}}{\pi}\Gamma\left(\frac{1}{3}\right)\left[\frac{1}{2}(-s)^{1/2}\frac{2}{3}x^{3/2}\right]^{-1/3} + \frac{x\left[\frac{1}{3}(-s)^{1/2}\right]^{1/3}}{\Gamma(1/3+1)}$$
$$= \text{constant} + \frac{x\left[\frac{1}{3}(-s)^{1/2}\right]^{1/3}}{\Gamma(1/3+1)}, \quad x' \to 0.$$

For the Wronskian:

$$W(x', s) = v_1 v_2' - v_1' v_2|_0,$$

we can use the foregoing expressions for v_1, v_2 to obtain

$$v_1 v_2' \sim \frac{x'\left[(-s)^{1/2}\frac{1}{3}\right]^{1/3}}{\Gamma(1/3+1)}\frac{\left[\frac{1}{3}(-s)^{1/2}\right]^{1/3}}{\Gamma(1/3+1)} \to 0 \text{ as } x' \to 0$$

and

$$-v_1' v_2 \sim \frac{i(x')^{1/2}}{\pi}\Gamma\left(\frac{1}{3}\right)\left[\frac{1}{2}(-s)^{1/2}\frac{2}{3}x'^{3/2}\right]^{-1/3}\frac{\left[\left(\frac{1}{3}\right)(-s)^{1/2}\right]^{1/3}}{\Gamma(1/3+1)}$$
$$= \frac{i\Gamma(1/3)}{\pi\Gamma(1/3+1)}(x')^{1/2}\left[\frac{1}{3}(-s)^{1/2}\right]^{-1/3}(x')^{-1/2}\left[\frac{1}{3}(-s)^{1/2}\right]^{1/3}$$
$$= \frac{i\Gamma(1/3)}{\pi\Gamma(1/3+1)}.$$

Therefore

$$W(x', s) \sim -v_1' v_2 = \frac{i\Gamma(1/3)}{\pi\Gamma(1/3+1)} = \frac{i\Gamma(1/3)}{\pi\frac{1}{3}\Gamma(1/3)} = \frac{3i}{\pi}.$$

Thus, by eqs. (5.15a) and (5.15b),

$$G(x, x', s) = (x')^{1/2}x^{1/2}J_{1/3}\left[(-s)^{1/2}\frac{2}{3}(x')^{3/2}\right]H^{(1)}_{1/3}\left[(-s)^{1/2}\frac{2}{3}x^{3/2}\right]\frac{\pi}{3i}, x' < x$$
$$= (x)^{1/2}(x')^{1/2}J_{1/3}\left[(-s)^{1/2}\frac{2}{3}(x)^{3/2}\right]H^{(1)}_{1/3}\left[(-s)^{1/2}\frac{2}{3}(x')^{3/2}\right]\frac{\pi}{3i}, x' > x$$

That is, $G_1 = G_2$. The solution is the inverse of the transform

$$\tilde{u}(x,s) = \int_0^\infty G(x,x',s)[-x'f(x')]dx',$$

which we can write as

$$u(x,t) = -\int_0^\infty x'f(x')\left[\frac{1}{2\pi i}\int_{\gamma-i\infty}^{\gamma+i\infty} G(x,x',s)e^{st}ds\right]dx'.$$

From the expression for $G(x,x's)$ above, we see that the solution has one Bessel function and one Hankel function. Thus, the branch cut (Figure 5.27) that we need to make in order to evaluate the integral in this equation will be determined from our knowledge of the Bessel & Hankel functions. We note that $G(x,x',s)$ is in general a multivalued function and that $s = 0$ and $s = \infty$ are branch points.

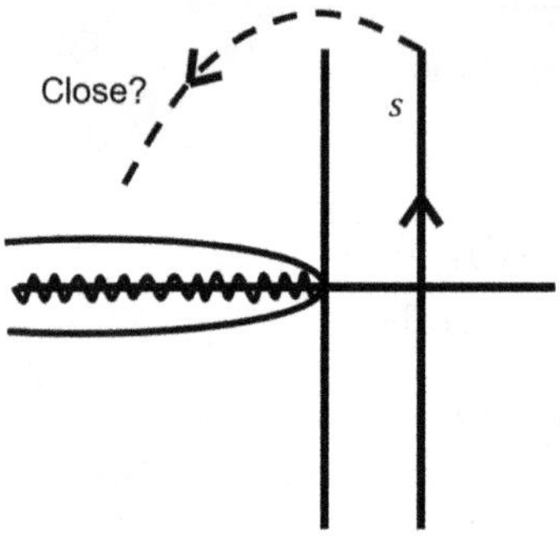

Figure 5.27: Branch cut on $(-\infty, 0]$.

The Bessel functions have no poles/singular points other than the branch points at 0 and ∞, and the Bessel function is analytic outside of the branch cut. We also have to check if it has an essential singularity at ∞, in which case, we may not be able to close the contour at ∞. So the branch cut is not the only consideration before we can say that $\uparrow = \supset$.

In order to close the LHP contour at ∞, we have to examine $G(x,x',s)$ as $s \to \infty$ in the LHP. However, we are not concerned with how it behaves on the RHP.

Behavior of $G(x,x',s)$ *as* $s \to \infty$

We are interested only in variable s, as x is not important in the integral

$$J_{1/3}\left[a(-s)^{1/2}\right]H^{(1)}_{1/3}\left[(-s)^{1/2}a\right]; \quad a > 0.$$

What happens as $s \to \infty$:

$$J_\nu(z) \sim \sqrt{\frac{2}{\pi z}} \cos\left(z - \frac{1}{2}\nu\pi - \frac{1}{4}\pi\right) \text{ as } z \to \infty.$$

For our interest,

$$J_\nu(z) = \frac{\text{constant}}{\sqrt{z}} \left[e^{iz} + e^{-iz}\right] \text{ as } z \to \infty.$$

We also know that

$$H_\nu^{(1)}(z) \sim \sqrt{\frac{2}{\pi z}} e^{i\left(z - \frac{1}{2}\nu\pi - \frac{1}{4}\pi\right)} = \frac{c}{\sqrt{z}} e^{iz}, \text{ as } z \to \infty.$$

We can multiply the Bessel and Hankel functions to see what happens:

$$J_\nu(z) H_\nu^{(1)}(z) \sim \frac{c}{z}\left[e^{2iz} + 1\right].$$

In our case, $z = a(-s)^{1/2} = ias^{1/2} = e^{-2as^{1/2}} \to 0$ as $s \to \infty$. Note that we have chosen $z = ias^{1/2}$; otherwise $\mathrm{Re}[s^{1/2}] > 0$ in the cut plane, leading to problems as $s \to \infty$. Thus,

$$J_{1/3}[a(-s)^{1/2}] H_{1/3}^{(1)}[a(-s)^{1/2}] \to c/(-s)^{1/2} = is^{\frac{1}{2}} \text{ (taking the positive branch) as } s \to \infty.$$

This goes to zero everywhere although all we really need is LHP. (Note that $x, x' > 0$.)

Further discussions on $(-s)^{1/2}$

From the foregoing, we have

$$v_2 = x^{1/2} H_{1/3}^{(1)}\left[(-s)^{1/2} \frac{2}{3} x^{3/2}\right], \quad H_\nu^{(1)} \equiv J_\nu + iY_\nu,$$

and

$$H_\nu^{(1)}(z) \sim iY_\nu(z) = -\frac{i}{\pi}\Gamma(\nu)\left(\frac{1}{2}z\right)^{-\nu}, \quad \mathrm{Re}[\nu] > 0;$$

that is,

$$H_{1/3}^{(1)} \sim z^{-1/3} \to \infty, z \to 0, \text{ and } z = x^{3/2}\left(\frac{2}{3}\right)(-s)^{1/2},$$

and

$$H_{1/3}^{(1)} \sim -\frac{1}{\pi}\Gamma(\nu)\left[\left(\frac{1}{2}\right)\left(\frac{2}{3}\right)(-s)^{1/2}\right]^{-1/3} x^{-1/2}.$$

The solution v_2 is bounded as $x \to 0$, and

$$H_\nu^{(1)}(z) \sim \sqrt{\frac{2}{\pi z}} e^{i\left(z - \frac{1}{2}\nu\pi + \frac{1}{4}\pi\right)}, \quad z \to \infty$$

so that

$$H_{1/3}^{(1)}\underbrace{\left(\frac{2}{3}(-s)^{\frac{1}{2}} x^{\frac{3}{2}}\right)}_{z} = \sqrt{\frac{2}{\pi z}}\, e^{i\left(\frac{2}{3}(-s)^{1/2} x^{3/2}\right)} e^{i\left(-\frac{1}{6}\pi + \frac{1}{4}\pi\right)}$$

$$= \sqrt{\frac{2}{\pi}\left(\frac{3}{2}\right)} (-s)^{-\frac{1}{2}} \left(x^{-\frac{3}{2}}\right) e^{i\left(\frac{2}{3}(-s)^{1/2} x^{3/2}\right)} e^{i\left(-\frac{1}{6}\pi + \frac{1}{4}\pi\right)},$$

where we have used $\nu = 1/3$ and $z = (2/3)(-s)^{1/2} x^{3/2}$.

Also,

$$J_\nu(z) = \sqrt{\frac{2}{\pi}\left(\frac{1}{z}\right)} \cos\left(z - \frac{1}{2}\nu\pi - \frac{1}{4}\pi\right), \quad z \to \infty$$

Recall that

$$G(x', x, s) = \frac{v_1(x', s) v_2(x, s)}{W(x', s)}, W(x', s) = (v_1 v_2{}' - v_2 v_1{}')|_0 = \frac{3i}{\pi},$$

which is an exact result. Also,

$$v_1(x', s) \sim \frac{x'\left((-s)^{1/2}\frac{1}{3}\right)^{1/3}}{\Gamma(1/3 + 1)}, \quad x' \to 0,$$

and

$$v_2 \sim -i\frac{(x')^{1/2}}{\pi}\Gamma\left(\frac{1}{3}\right)\left[\frac{1}{2}(-s)^{1/2}\frac{2}{3}x^{3/2}\right]^{-1/3} + \frac{x\left[\frac{1}{3}(-s)^{1/2}\right]^{1/3}}{\Gamma\left(\frac{1}{3} + 1\right)}, \quad x' \to 0,$$

where the first term on the RHS is a constant, and we have used a series expansion to obtain it. We see that

$$v_1 v_2' \to 0, x' \to 0, \quad -v_1' v_2 \to \frac{i\Gamma(1/3)}{\pi\Gamma(1/3 + 1)}, \quad x' \to 0.$$

Note that if we can get G to go to zero, and, having e^{st}, we can close the large contour and set it to zero by Jordan's Lemma. Therefore, the integral becomes

$$-\frac{1}{2\pi i}\int_{\gamma-i\infty}^{\gamma+i\infty} G(x',x,s)e^{st}ds \equiv I = -\frac{1}{2\pi i}\int G(x',x,s)e^{st}ds.$$

Two options for a contour to evaluate I are shown in Figure 5.28. That is, the slit contour and the combination of the slit contour with the contour $\gamma - i\infty < s < \gamma + i\infty$.

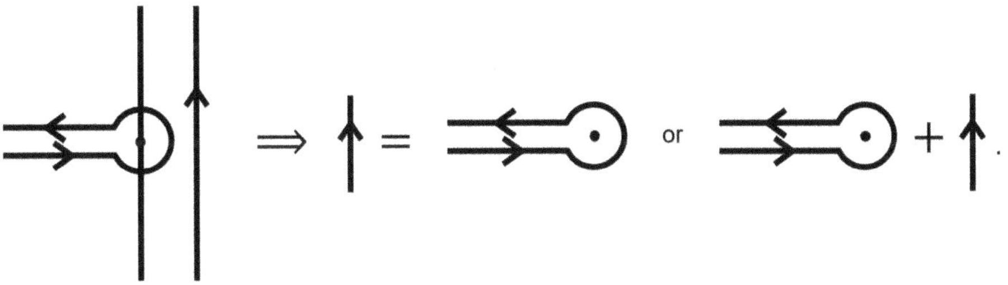

Figure 5.28: Two options for a contour.

Let's consider $x < x'$ and the former contour, which is also shown in Figure 5.29.

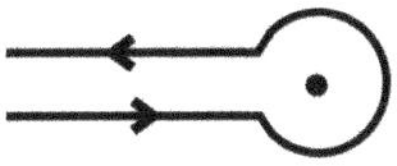

Figure 5.29: A plausible contour for evaluating I.

On the small circle in Figure 5.29, we have

$$J^{(1)}{}_{1/3}\left(a(-s)^{1/2}\right) \sim (-s)^{1/6} \to 0, \quad s \to 0,$$

$$H^{(1)}_{1/3}\left(a(-s)^{1/2}\right) \sim (-s)^{-1/6} \to \infty, \quad s \to 0.$$

When we multiply the two functions together, the product becomes bounded. Thus, to have s^{-a}, we need to have $a > 1$ to have a contribution because the integral here goes like s^1. Note that a pole does not have a meaning for a multivalued function such as $e^{a\sqrt{s}/s}$. However, you may miss some physics such as the steady-state solution if a pole is not accounted for in a multivaluedness-mandated slit contour that contains a pole (Figure 5.30).

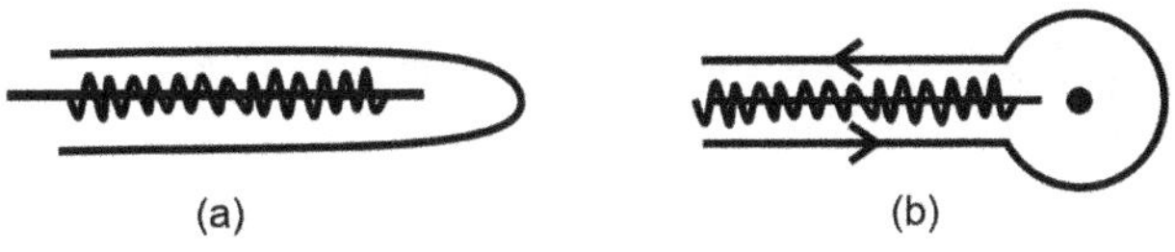

Figure 5.30: Slit contour without and with a pole included. In (a) the pole is excluded, while in (b), it is included.

Note that the "small" hole in Figure 5.29 eventually contributes $\int_c = 0$. Therefore, we don't need this hole as $\varepsilon \to 0$. The integral I can be written as

$$I = \frac{i\pi (x')^{1/2} (x)^{1/2}}{3(2\pi i)} \left[\int_0^{\infty} J_{1/3}\left(\frac{2\rho}{3}(x')^{\frac{3}{2}}\right) H_{1/3}^{(1)}\left(-\frac{2\rho}{2} x^{\frac{3}{2}}\right)(-2\rho)e^{-\rho^2 t} d\rho \right.$$

$$\left. - \int_0^{\infty} J_{1/3}\left(\frac{2\rho}{3}(x')^{\frac{3}{2}}\right) H_{1/3}^{(1)}\left(\frac{2\rho}{3} x^{\frac{3}{2}}\right)(-2\rho)e^{-\rho^2} d\rho \right].$$

On the argument $J_{1/3}\left(i(s)^{1/2}(2/3)x^{3/2}\right)$:

At the top part of the contour ($\theta = \pi$), we have $s^{1/2} = e^{i\pi/2} = i$. At the bottom ($\theta = -\pi$), we have $s^{1/2} = e^{-i\pi/2} = -i$. Let us introduce a change of variables: $s = \rho^2$.

Thus,

$$I = -\frac{(x')^{1/2} x^{1/2}}{3} \int_0^{\infty} \rho e^{-\rho^2 t} \left[J_{1/3}(-\beta') H_{1/3}^{(1)}(-\beta) - J_{1/3}(\beta') H_{1/3}^{(1)}(\beta) \right] d\rho,$$

where

$$\beta \equiv \frac{2\rho}{3}(x)^{\frac{3}{2}}; \; \beta' = \frac{2\rho}{3}(x')^{\frac{3}{2}}.$$

If we use the identities $J_\nu(ze^{i\pi}) = e^{i\pi\nu} J_\nu(z)$ and $H_\nu^{(1)}(z) = [J_{-\nu}(z) - e^{-i\pi\nu} J_\nu(z)]/(i \sin \nu\pi)$, we can write

$$\begin{aligned} J_{1/3}(-\beta') H_{1/3}^{(1)}(-\beta) &= e^{\frac{i\pi}{3}} J_{1/3}(\beta') \frac{[J_{-1/3}(-\beta) - e^{-i\pi/3} J_{1/3}(-\beta)]}{i \sin(\pi/3)} \\ &= -\frac{2i}{\sqrt{3}} J_{1/3}(\beta') \left[J_{-1/3}(\beta) - e^{i\pi/3} J_{1/3}(\beta) \right]. \end{aligned} \tag{5.19}$$

Similarly,

$$J_{1/3}(\beta') H_{1/3}^{(1)}(\beta) = -\frac{2i}{\sqrt{3}} J_{1/3}(\beta') \left[J_{-1/3}(\beta) - e^{-i\pi/3} J_{1/3}(\beta) \right]. \tag{5.20}$$

Adding eqs. (5.19) and (5.20), we have

$$J_{1/2}(-\beta') H_{1/3}^{(1)}(-\beta) = \frac{2i}{\sqrt{3}} \left(\underbrace{e^{i\pi/3} - e^{-i\pi/3}}_{2i \sin \pi/3} \right) [J_{1/3}(\beta') J_{1/3}(\beta)] = -2 J_{1/3}(\beta') J_{1/3}(\beta).$$

Therefore

$$I = \frac{2(x')^{1/2}(x)^{1/2}}{3}\int_0^\infty \rho e^{-\rho^2 t} J_{1/3}(\beta) J_{1/3}(\beta')d\rho, \quad x' < x.$$

Since x', x and β, β' appear in symmetric fashion, we do not need extra considerations for the case $x' > x$ and

$$I = \frac{2(x')^{1/2}x^{1/2}}{3}\int_0^\infty a e^{-a^2 t} J_{1/3}(\beta) J_{1/3}(\beta')da, \tag{5.21}$$

yielding

$$u(x,t) = \int_0^\infty x'(f(x')I(x,x')dx', \tag{5.22}$$

where $I(x,x')$ is a Green's function.

Equation (5.22) can be written as

$$u(x,t) = \int_0^\infty ac(a)x^{\frac{1}{2}} J_{1/3}\left(\frac{2}{3}ax^{3/2}\right)e^{-a^2 t}da,$$

where

$$c(a) \equiv \frac{2}{3}\int_0^\infty (x')^{3/2} J_{1/3}\left(\frac{2}{3}a(x')^{2/3}\right)f(x')dx'.$$

Let $t = 0$. Then

$$u(x,0) = f(x), \quad f(x) = \int_0^\infty ac(a)x^{\frac{1}{2}} J_{1/3}\left(\frac{2}{3}ax^{\frac{3}{2}}\right)da.$$

Let us adopt the following notations:

$$\beta \equiv \frac{2}{3}(x')^{\frac{3}{2}}, \quad g(\beta) \equiv \frac{f(x')}{\sqrt{x'}}, \quad \gamma \equiv \frac{2}{3}(x)^{\frac{3}{2}}.$$

Then we can write the Hankel transform and its inverse of order 1/3, respectively, as

$$g(\gamma) = \int_0^\infty ac(a)J_{\frac{1}{3}}(a\gamma)da, \quad c(a) = \int_0^\infty \beta g(\beta)J_{\frac{1}{3}}(\beta a)d\beta,$$

where γ is the Hankel transform variable, the same way we have used s and λ (sometimes ω) as the variables for the Laplace and FTs, respectively. In general, we don't have to use the Bessel function of order 1/3. The Hankel transform could be used instead. Recollect that the Hankel transform and its inverse are generally defined as

$$g_\nu(r) = \int_0^\infty \rho G_\nu(\rho) J_\nu(\rho r) d\rho, \quad G_\nu(\rho) = \frac{1}{2\pi}\int_0^\infty r g_\nu(r) J_\nu(r\rho) dr,$$

where $\mathrm{Re}[\nu] > -1$. This can be proved.

5.3 Generalized unsteady diffusion problems

The power of the LT method in physics and engineering analysis can be appreciated if we acknowledge its suitability for the analysis of the generalized linear PDE for transient diffusion problems. Let us interpret this with the heat conduction in a rod (bar), which can be written as [3]

$$\frac{\partial T}{\partial t} = \frac{k_0}{\rho c_\mathrm{p}} \frac{\partial}{\partial x}\left[(1+\beta T)\frac{\partial T}{\partial x}\right] - \left(U + \frac{sI}{\rho c_\mathrm{p} \omega}\right)\frac{\partial T}{\partial x} - T\left(\frac{hp}{\omega \rho c_\mathrm{p}} - \frac{a j I^2}{\rho c_\mathrm{p} \omega^2 \sigma_0}\right) + \frac{hpT_0}{\rho c_\mathrm{p} \omega} + \frac{jI^2}{\rho c_\mathrm{p} \omega^2 \sigma_0}, \tag{5.23}$$

where T is temperature, t is time, $k = k_0(1+\beta T)$ is the thermal conductivity, ρ is density, c_p is specific heat at constant pressure, x is the coordinate direction along the bar, U is the velocity of the bar as it moves in the direction of its length; s is the Thomson effect associated with current heating, which involves the direction of the current and is zero if alternating current is used to heat the rod; I is the current for electric heating, ω is the rod cross-sectional area, and h is the coefficient of convective heat transfer ($q_{\mathrm{radiation}} = h(T - T_\infty)p\,dx$), where T_∞ is the temperature of the medium with which the rod exchanges heat, p is the perimeter of the bar, dx is an element of the rod, a is the coefficient of dependence of electrical heating with temperature: $\frac{jI^2}{\rho c_\mathrm{p} \omega^2 \sigma_0} \Rightarrow \frac{jI^2}{\rho c_\mathrm{p} \omega^2 \sigma_0}(1 + aT)$, σ_0 is the electrical conductivity at zero temperature, T_0 is the initial temperature in the rod, j is a conversion factor, or the number of calories in 1 J. Equation (5.23) is equally applicable to molecular diffusion of the concentration of a species.

The above equation, which can be written in the more compact form

$$aT_{xx} + bT_x + cT + d = T_t,$$

where a, b, c, and d are constants and can be solved by the LT method in a manner that "treats all cases the same way," according to Carslaw and Jaeger [3]. The entire book by Carslaw and Jaeger [3] is devoted to the mathematics of heat conduction in solids, while the book by Crank [5] is dedicated to the analogous treatment of diffusion of concentration. The LT method is employed extensively to solve otherwise difficult problems in these texts.

Appendix A, which is a table of Laplace transform pairs that parallels a similar one in [3], may be convenient for solving complicated heat conduction and mass diffusion problems. Note that the inverse in the table can be obtained via the complex variable contour integration techniques presented in this chapter but have been provided in the appendix for the convenience. A few of the exercises dealing with LT at the end of this chapter may be conveniently solved by using the results in the table.

We will briefly show how the table in Appendix A could be used. Consider semi-infinite one-dimensional unsteady heat conduction in a rod, which is governed by the equation $\alpha T_{xx} - T_t = 0$, $x > 0$ with the initial condition $T_0(x)$. (α here is $k/\rho c_p$, the thermal diffusivity of the rod.) The LT of this problem with respect to time is

$$\alpha \tilde{T}_{xx} - s\tilde{T} + T_0(x), \quad x > 0,$$

where the BC

$$T = T_s(t), \quad x = 0, t > 0$$

transforms to

$$\tilde{T} = \tilde{T}_s(s).$$

Here, subscript s denotes surface which should not be confused with the LT variable. Consider the following three cases.

Case I:

$$T_s(t) \equiv T_\omega = \text{constant} \Rightarrow \tilde{T} = \frac{T_\omega}{s}$$

and

$$\tilde{T} = \frac{T_\omega}{s} e^{-\alpha' x},$$

where $\alpha' = \alpha/s$, $\alpha =$ thermal diffusivity of the solid.

Appendix A (5) shows that

$$T(x,t) = T_\omega \text{erfc} \frac{x}{2(\alpha t)^{\frac{1}{2}}}.$$

Case II: For arbitrary $T_s(t)$ at $x = 0$, Appendix A (3) gives the solution as

$$T(x,t) = \frac{x}{2\sqrt{(\pi\alpha)}} \int_0^t T_s(\lambda) \frac{e^{-\frac{x^2}{4\alpha(t-\lambda)}} d\lambda}{(t-\lambda)^{\frac{3}{2}}}.$$

Case III: For convective BC at $x = 0$:

$$\frac{\partial T}{\partial x} - h(T - T_\infty) = 0, \quad x = 0, t > 0,$$

where h is the film coefficient and T_∞ is the temperature of the fluid medium to which the surface $x = 0$ is exposed. The transform of the BC equation is

$$\frac{d\tilde{T}}{dx} - h\tilde{T} + \frac{hT_\infty}{s} = 0.$$

The (Laplace) transformed equation can be solved as

$$\tilde{T} = \frac{hT_\infty e^{-\alpha' x}}{s(\alpha' + h)}.$$

Appendix A (11) gives the solution as

$$T(x,t) = T_\infty \text{erfc} \frac{x}{2\sqrt{\alpha t}} - T_\infty e^{hx + h^2 \alpha t} \text{erfc} \left\{ \frac{x}{2\sqrt{\alpha t}} + h\sqrt{\alpha t} \right\}.$$

5.4 Miscellaneous examples

Example 5.9 Consider the linear flow of heat in the semi-infinite solid $x > 0$. Starting at $t = 0$, the boundary $x = 0$ is subjected to a fluctuating temperature $u = u_0 \cos \omega t$ (u_0 and ω const.). Determine u if initially it is zero throughout.

Solution: The equation for the temperature u at x at a time t is

$$\frac{\partial u}{\partial t} = \frac{\partial^2 u}{\partial x^2}, \quad x > 0, t > 0 \tag{5.24}$$

$$u = 0 \text{ when } t = 0, \ x > 0, \quad u = u_0 \cos \omega t \text{ when } x = 0, \ t > 0. \tag{5.25}$$

The LT of eq. (5.24) (using eq. (5.25)) is given as

$$\frac{d^2 \tilde{u}}{dx^2} - s\tilde{u} = 0, \quad \text{for} x > 0, \tag{5.26}$$

while

$$\tilde{u} = \frac{u_0 s}{s^2 + \omega^2} \tag{5.27}$$

when $x = 0$ by eq. (5.26). From eqs. (5.26) and (5.27) we find that

$$\tilde{u} = \frac{u_0 s}{s^2 + \omega^2} e^{-\sqrt{s}x}$$

so that

$$u = \frac{u_0}{2\pi i} \int\limits_{\gamma - i\infty}^{\gamma + i\infty} \frac{s}{s^2 + \omega^2} e^{st - \sqrt{s}x} ds. \tag{5.28}$$

There is a branch point at the origin, so we use the contour in Figure 5.31. The integrand is single-valued in the cut plane and has poles at $\pm i\omega$. The integrands over $B'F$ and $A'C$ tend to zero as $R \to \infty$, by Jordan's lemma, since

$$\left| \frac{s}{s^2 + \omega^2} e^{-\sqrt{s}x} \right| = O\left(\frac{1}{R} e^{-xRe\sqrt{s}}\right) = O\left(\frac{1}{R}\right).$$

On BB' and AA', the same estimate gives

$$\left| \int\limits_{\substack{BB' \\ \text{or } AA'}} \right| = e^{\gamma t} O\left(\frac{1}{R} \cdot R\beta\right) = e^{\gamma t} O\left(\sin^{-1} \frac{\gamma}{R}\right)$$

and these also tend to zero as $R \to \infty$.

Example 5.10 [16] When lateral displacements are taken into account in longitudinal oscillation of a slender rod and units are chosen suitably, the displacement $u(x, y)$ satisfies

$$\frac{\partial^2 u}{\partial t^2} - \frac{\partial^4 u}{\partial x^2 \partial t^2} - \frac{\partial^2 u}{\partial x^2} = 0,$$

and the stress is $\sigma(x, t) = \partial u / \partial x$. Starting at time $t = 0$ a constant stress σ_0 is applied to the end $x = 0$ of an infinite rod. Show that

$$\sigma(x,t)=\sigma_0\left[1-\frac{2}{\pi}\int_0^1 \cos at \sin\left(\frac{ax}{\sqrt{1-a^2}}\right)\frac{da}{a}\right].$$

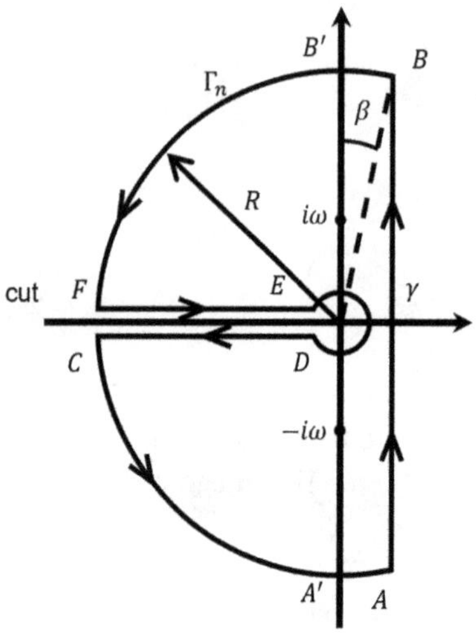

Figure 5.31: Contour for the inversion of u in eq. (5.28).

Solution: The differential equation for the stress, σ, is

$$\frac{\partial^2\sigma}{\partial t^2}-\frac{\partial^4\sigma}{\partial x^2\partial t^2}-\frac{\partial^2\sigma}{\partial x^2}=0,$$

and since σ and σ_t are initially zero, the Laplace transformation gives

$$\frac{\partial^2\tilde{\sigma}}{\partial x^2}-\frac{s^2}{1+s^2}\tilde{\sigma}=0, \tag{5.29}$$

with

$$\tilde{\sigma}(0,s)=\sigma_0/s. \tag{5.30}$$

The solution of eqs. (5.29) and (5.30) is

$$\tilde{\sigma}(x,s)=\frac{\sigma_0}{s}e^{-sx/\sqrt{1+s^2}}.$$

Hence on inversion we obtain

$$\sigma(x,t)=\frac{\sigma_0}{2\pi i}\int_{\gamma-i\infty}^{\gamma+i\infty} e^{st}e^{-sx/\sqrt{1+s^2}}\frac{ds}{s}.$$

The integrand has a simple pole at $s = 0$ and branch point at $s = \pm i$. We introduce a cut as shown (Figures 5.32 and 5.33) and define

$$\sqrt{s^2+1} = \rho_1^{1/2}\rho_2^{1/2}e^{\frac{i}{2}(\theta_1+\theta_2)}.$$

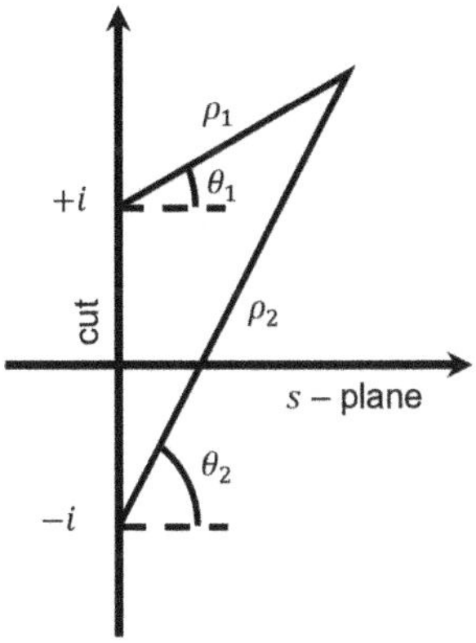

Figure 5.32: Branch cut on the region $-i \le s \le i$ in Sample Problem 5.10.

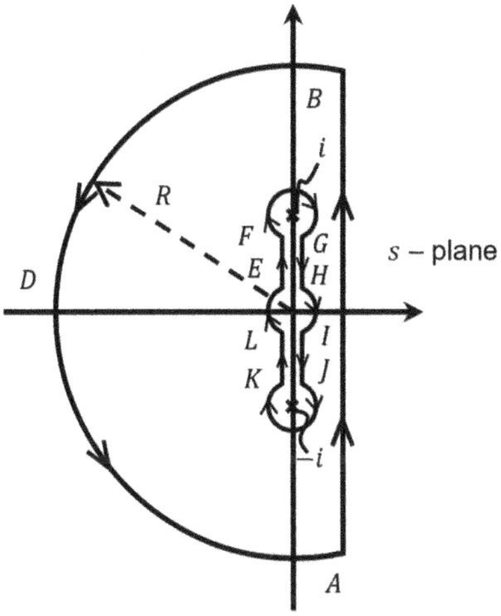

Figure 5.33: An integration contour for Sample Problem 5.10.

Now consider the combined contour, C, shown. The integrand is single-valued and has no singularities in a region containing C so that

$$\int_{ABDA} = \int_{LKJIHGFEL}.$$

By Jordan's lemma the integral over BDA tends to zero as $R \to \infty$. Various contours are shown in Figures 5.34 through 5.36.

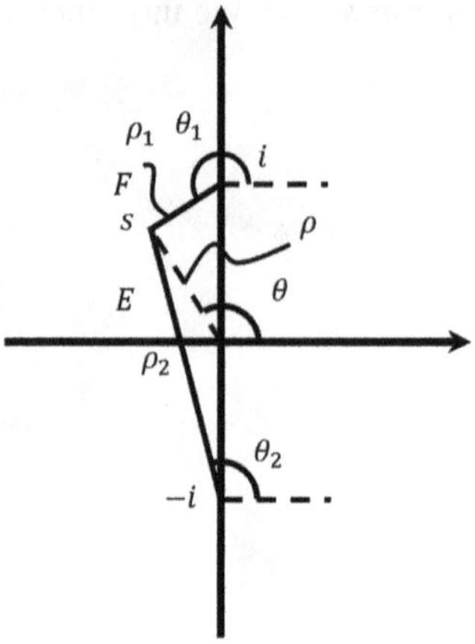

Figure 5.34: Related to the contour segments E-F, G-H, I-J, and K-L in Sample Problem 5.10.

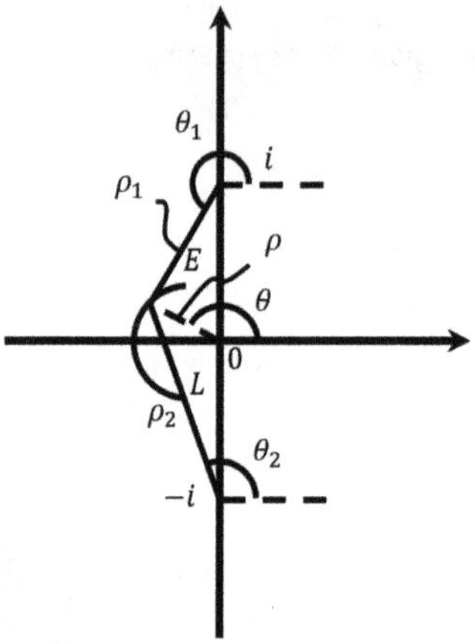

Figure 5.35: Related to the small circle at the origin in Sample Problem 5.10.

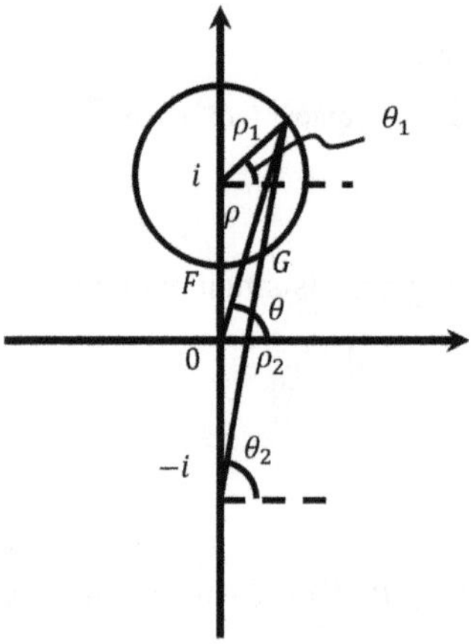

Figure 5.36: Related to the small circles at $s = \pm i$ in Sample Problem 5.10.

Along EF (Figure 5.34) $s = i\rho$, $\frac{ds}{s} = \frac{d\rho}{\rho}$, $\theta_1 + \theta_2 = 2\pi$, $\rho_1 = 1 - \rho$, $\rho_2 = 1 + \rho$;

$$\sqrt{s^2+1} = \sqrt{\rho_1\rho_2}e^{i\pi} = -\sqrt{1-\rho^2};$$

and

$$\int_{EF} = \int_0^1 e^{i\rho t} e^{\frac{i\rho x}{\sqrt{1-\rho^2}}} \frac{d\rho}{\rho}, \quad \int_{GH} = -\int_0^1 e^{i\rho t} e^{\frac{-i\rho x}{\sqrt{1-\rho^2}}} \frac{d\rho}{\rho},$$

$$\int_{IJ} = \int_0^1 e^{-i\rho t} e^{\frac{i\rho x}{\sqrt{1-\rho^2}}} \frac{d\rho}{\rho}, \quad \int_{KL} = -\int_0^1 e^{-i\rho t} e^{\frac{-i\rho x}{\sqrt{1-\rho^2}}} \frac{d\rho}{\rho}.$$

The small circle at the origin (Figure 5.35) gives $s = \rho e^{i\theta}$, $ds/s = i d\theta$, and $s/\sqrt{s^2+1} \to 0$ from either side of the cut. Hence

$$\int_{LEHI} = \int_{3\pi/2}^{-\pi/2} 1 \cdot 1 \cdot i d\theta = -2\pi i.$$

Now $s = i$ (see Figure 5.36) and

$$\frac{s}{\sqrt{1+s^2}} = \frac{\rho}{\rho_1^{1/2}\rho_2^{1/2}} e^{i\left(\theta - \frac{\theta_1}{2} - \frac{\theta_2}{2}\right)}$$

has argument between $-\pi/2$ and $+\pi/2$. Hence

$$\left| e^{-\frac{sx}{\sqrt{1+s^2}}} \right| = e^{-x\mathrm{Re}\left[\frac{s}{\sqrt{1+s^2}}\right]} \le 1,$$

and $\int_{FG} \to 0$. Now $s = -i$, $s/\sqrt{1+s^2}$ has argument between $-\pi/2$ and $\pi/2$ again so that $\int_{JK} \to 0$.

Thus, it follows that

$$\sigma(x,t) = -\frac{\sigma_0}{2\pi i}\left[\int_0^1 e^{i\rho t}\left[e^{\frac{ix\rho}{\sqrt{1-\rho^2}}} - e^{-\frac{ix\rho}{\sqrt{1-\rho^2}}}\right]\frac{d\rho}{\rho} + \int_0^1 e^{-i\rho t}\left[e^{\frac{ix\rho}{\sqrt{1-\rho^2}}} - e^{-\frac{ix\rho}{\sqrt{1-\rho^2}}}\right]\frac{d\rho}{\rho} - 2\pi i\right]$$

$$= \sigma_0\left[1 - \frac{2}{\pi}\int_0^1 \cos\rho t \sin\left(\frac{\rho x}{\sqrt{1-\rho^2}}\right)\frac{d\rho}{\rho}\right].$$

Example 5.11 A circular cylinder of radius a has its surface kept at constant temperature u_0, the initial temperature throughout the cylinder being zero. Find an approximate expression for the temperature at small time t. Is your result valid near the axis? If not, how would you obtain an expression?

Solution: The equations for the temperature, u, are

$$\frac{\partial u}{\partial t} = \frac{\partial^2 u}{\partial r^2} + \frac{1}{r}\frac{\partial u}{\partial r}, \quad 0 \le 0 \le r < a, \quad t < 0,$$

with $u = 0$ when $t = 0$, $0 \le r \le a$, $u = u_0$ when $t > 0$, $r = a$. Using LT we obtain

$$\frac{d^2\tilde{u}}{dr^2} + \frac{1}{r}\frac{d\tilde{u}}{dr} - s\tilde{u} = 0, \quad 0 \le r \le a$$

with

$$\tilde{u}(a, s) = \frac{u_0}{s}.$$

Hence the solution is

$$\tilde{u} = \frac{u_0}{s}\frac{I_0(qr)}{I_0(qa)}, \tag{5.31}$$

where $q = \sqrt{s}$.

In order to obtain u for small time we make asymptotic expansions of the Bessel functions in eq. (5.31) valid when r is not small:

$$\tilde{u} \sim \frac{u_0}{s}\sqrt{\frac{a}{r}}e^{-q(a-r)}\frac{\left[1 + \frac{1}{8qr} + \ldots\right] + O(e^{-2qr})}{\left[1 + \frac{1}{8qa} + \ldots\right] + O(e^{-2qa})}$$

$$\sim \frac{u_0}{s}\sqrt{\frac{a}{r}}e^{-q(a-r)}\left\{1 + \frac{a-r}{8qar}\right\} + \left(\text{Order } e^{-q(3a-r)} \text{and higher}\right).$$

Thus we obtain

$$u \sim u_0\sqrt{\frac{2a}{\pi r}}e^{-(a-r)^2/8t}\left\{D_{-1}\left(\frac{a-r}{\sqrt{2t}}\right) + \frac{a-r}{8ar}(2t)^{\frac{1}{2}}D_{-2}\left(\frac{a-r}{\sqrt{2t}}\right) + \ldots\right\}$$

When r is small we use the power series for $I_0(qr)$:

$$\tilde{u} \sim \frac{u_0}{s}\frac{1 + \frac{q^2r^2}{4} + \ldots}{\frac{e^{qa}}{\sqrt{2\pi qa}}\left[1 + \frac{1}{8qa} + \ldots\right]}$$

$$= u_0\sqrt{2\pi a}q^{-\frac{3}{2}}e^{-\frac{q}{a}}\left[1 + \frac{q^2r^2}{4} - \frac{1}{8}qa\ldots\right]$$

and

$$u \sim 2\sqrt{a}u_0 e^{-\frac{a^2}{8t}}\left[(2t)^{-\frac{1}{4}}D_{-\frac{1}{2}}\left(\frac{a}{\sqrt{2t}}\right) + \frac{r^2}{4}(2t)^{-\frac{5}{4}}D_{\frac{3}{2}}\left(\frac{a}{\sqrt{2t}}\right) - \frac{1}{8a}(2t)^{\frac{1}{4}}D_{-\frac{3}{2}}\left(\frac{a}{\sqrt{2t}}\right) + \dots\right].$$

Above, the $D_\nu(.)$'s are the parabolic cylinder functions.

Problems

5.1 At $t = 0$ the hinge $x = 0$ of a semi-infinite beam $(x > 0)$ is given a small displacement $u = u_0$. If the beam is initially straight and at rest, show that

$$u = u_0\left[1 - \frac{1}{\sqrt{\pi}}\int_0^\infty \{\cos(y^2/2) + \sin(y^2/2)\}dy\right].$$

Neglect gravity. Hint: $\frac{1}{\sqrt{\pi}}\int_0^\infty e^{-k^2 t}\sin kx\frac{dk}{k} = \int_0^{x/2\sqrt{t}} e^{-\lambda^2}d\lambda$, which follows from integrating

$$\int_0^\infty e^{-a^2k^2}\cos 2bk\,dk = \frac{\sqrt{\pi}}{2a}e^{-b^2/a^2} \quad (a > 0).$$

5.2 Consider linear flow of heat in the solid $0 < x < l$. The boundary $x = 0$ is insulated to prevent the escape of heat; the boundary $x = l$ is kept at a constant temperature $u = u_1$. Determine u if initially it has the same value u_0 throughout.

5.3 With suitably chosen units, the transverse vibrations of a uniform beam are governed by the equation $\partial^4 u/\partial x^4 + \partial^2 u/\partial t^2 = P$, where $u(x, t)$ is the displacement and $P(x, t)$ the externally applied force per unit length (including gravity when this is not neglected). At the ends $x = 0, l$ of a freely hinged beam the displacement and bending moment are zero: $u = \partial^2 u/\partial x^2 = 0$. Starting at $t = 0$ a concentrated load w moves from the end $x = 0$ across such a beam with uniform velocity v. Determine u if initially the beam is straight and at rest. Neglect gravity [16].

5.4 With suitably chosen units, the longitudinal vibrations of a uniform beam are governed by the equation $\partial^2 u/\partial x^2 - \partial^2 u/\partial t^2 = 0$, where $u(x, t)$ is the displacement. Two equal rods of length l are moving longitudinally with velocities $\pm U$ along the x-axis. At time $t = 0$ they collide at the origin. Determine the subsequent motion. [Hint: expand the transform in exponentials and note that the rods only stay in contact so long as the pressure between them is positive, i.e., $[\partial u/\partial x]_{x=0} < 0$.]

5.5 Consider the following radially symmetric vibration problem:

$$\left(\frac{\partial^2}{\partial r^2}+\frac{1}{r}\frac{\partial}{\partial r}\right)^2\varphi+\lambda^2\varphi_{tt}=0,$$

where λ is a real constant. If $\varphi(r,0)=f(r)$ and $\varphi_t(r,0)=0$:
(a) Find $\varphi(r,t)$ using the Hankel transform.
(b) Find $\varphi(r,t)$ using the LT.
(c) Find $\varphi(r,t)$ using both the Hankel and LTs at the same time.

5.6 Let $\varphi(r,\theta)$ satisfy [16]

$$r^2\varphi_{rr}+r\varphi_r+\varphi_{\theta\theta}-r^2\alpha^2\varphi=0;\quad \alpha>0$$

in the sector

$$0\le r<\infty,\quad -\theta_0<\theta<0,\quad \theta_0<\frac{\pi}{2}.$$

The BCs are

$$\varphi_\theta-r\varphi=0\quad\text{for}\quad\theta=0,$$

$$\varphi_\theta=0\quad\text{for}\quad\theta=-\theta_0.$$

Assume $r\varphi_r$ is bounded as $r\to 0$, although φ_r may not. Take LT to show that the new function $\psi=(\sinh\lambda)\tilde{\varphi}(s,\theta)$ must be harmonic in the variables λ, θ where $s=\alpha\cosh\lambda$, transforming the domain of interest to $\theta\in(-\theta_0,0)$, $\lambda\in(-\infty,\infty)$. A hint is given in [2]: "Rewrite the equation in terms of $(r\varphi_r)_r$ in order to avoid the appearance of the term $\varphi_r(0,\theta)$ in the transform formula. Note that a term-by-term transform of the equations would not yield the correct answer. Write $\psi=Re[f(\lambda+i\theta)]$ and express the boundary conditions for the string in terms of f." Analytical continuation could be explored to obtain f. This approach is credited to Peters [11].

5.7 Find the Laplace inverse of $e^{-a\sqrt{s}/s}$ and express the result in terms of the error function, where a is a positive real constant.

5.8 Evaluate

$$I(t)=\int_0^\infty \exp\left[-x^2-\frac{t^2}{x^2}\right]dx$$

by taking LT.

5.9 Consider

$$\varphi_{tt}+\omega_0^2\varphi=\sin\omega t,\quad t>0;$$

ω, ω_0 are real constants and $\varphi(0)=\varphi_t(0)=0$. How will resonance in this problem (i.e., $\omega=\omega_0$) show up in the LT method?

5.10 Consider the sinusoidally forced wave equation:

$$\omega_{tt} - 4\omega_{xx} + \omega = 16x + 20\sin x$$

subject to the initial condition

$$\omega(x,0) = 16x + 12\sin 2x - 8\sin 3x,$$
$$\omega_t(x,0) = 0,$$

and the BCs

$$\omega(0,t) = 0; \quad \omega(\pi,t) = 16\pi.$$

Solve for $\omega(x,t)$ using the LT method.

5.11 By taking LT, solve the integral equation

$$\varphi(x) = x^2 + \int_0^x K(\lambda)\sin(x-\lambda)d\lambda.$$

5.12 Solve for the unsteady temperature field $T(r,t)$ in an axisymmetric circular disk satisfying the following conditions:

$$T_t = k\left(T_{rr} + \frac{1}{r}T_r\right), \quad r \in (0,1),$$

$T(1,t) = 0$, $T(r,0) = T_0$, and $|T(r,t)|$ are bounded everywhere. Use the LT method.

5.13 Show that the temperature $T(x)$ satisfying the equation

$$aT_{xx} - UT_x - \nu(T - T_0) = 0$$

in a semi-infinite rod moving with velocity U in the direction of its length, with the points $x = 0$ and $x = 2l$ maintained at constant temperatures T_1 and T_2, respectively, and convective heat flux $h(T - T_0)$ along its length, has solution

$$T(x) = [\sinh 2\xi l]^{-1}\left[T_2 e^{-\frac{U(2l-x)}{2a}}\sinh \xi x + T_1 e^{\frac{Ux}{2a}}\sinh \xi(2l - x)\right],$$

where

$$a = \frac{k}{\rho c_p}, \quad \nu = \frac{hp}{\rho c_p \omega}, \quad \xi = \left[\left(\frac{U^2}{4a^2}\right) + \left(\frac{hp}{\omega k}\right)\right]^{\frac{1}{2}},$$

with the meanings of the variables as described in the text. Also explore solution of the unsteady problem with the LT method, assuming an initial temperature of T_0. Above, k is the thermal conductivity of the rod, ρ is its density, c_p is the specific heat at constant pressure, p is perimeter, and ω is the cross-sectional area.

5.14 Consider the problem of a semi-infinite rod, $x > 0$, with initial temperature $T_0 = 0$, with its $x = 0$ end placed in contact with a fluid medium, initial temperature T_{f_0}. Heat conduction in the rod is governed by

$$aT_{xx} = T_t,$$

where a is the thermal diffusivity of the solid. There is heat transfer between the fluid at $x = 0$ and the rod. If T_f is the temperature of the fluid, assuming the fluid is well-stirred such that there is no temperature gradient in the fluid itself, then the BCs at $x = 0$ can be written as

$$mc_{p_f}\frac{dT_f}{dt} + h(T_f - T) = 0, \quad x = 0, t > 0,$$

$$-k\frac{\partial T}{\partial x} = h(T_f - T), \qquad x = 0, t > 0,$$

where m is the mass of fluid, c_{p_f} is its specific heat at constant pressure, h is the heat transfer coefficient, and k is the thermal conductivity of the solid. Determine the temperature $T(x)$, $x \geq 0$. Use the LT method to solve this problem.

5.15 Consider a thin rod, $x \in (0, l)$, moving with velocity U_∞ along its length and satisfying the diffusion-convection equation:

$$aT_{xx} - U_\infty T_x - \nu T - T_t = 0, \quad x \in (0, l),$$

where a is the thermal diffusivity of the solid and ν is as defined in the text. The following initial and BCs are specified:

$$T(x, 0) = 0,$$

$$x = 0\colon\ T \equiv T_0 = 0, \quad t > 0,$$

$$x = l\colon\ T = T_l, \quad t > 0,$$

The longitudinal surface of the rod is exchanging heat with the surrounding which is maintained at temperature T_∞, with film coefficient h. Determine $T(x, t)$ for this problem. Use the LT method. Be aware that Appendix A does not contain the inversion formula required for this problem so that you will need to carry out the inversion yourself. Fortunately, the standard Bromwich contour can be used within the context of the residue theorem approach.

5.16 Suppose heat is produced at $t > 0$ in a semi-infinite rod, $x \in (0, \infty)$, at the rate of $\beta t^{\lambda+3}$ per unit time per unit volume, where β and λ are known positive constants. Formulate the unsteady heat conduction problem assuming the rod exchanges heat with its surrounding ($T_\infty = 0$) at $x = 0$ via $\partial T/\partial x = hT$, $t > 0$. Solve for $T(x, t)$ using the LT method.

5.17 Consider a slab, $x \in (0, l)$, in which an adiabatic condition is maintained at $x = 0$ and a zero temperature at $x = l$ and at $t = 0$. Heat is generated in the solid at the rate of $k(a + bT)$ per unit time per unit volume for $t > 0$, where k is the thermal conductivity of the solid and a and b are constants. Determine $T(x, t)$.

5.18 Consider the semi-infinite $(x > 0)$ problem of nonisothermal flow of a fluid parallel to a flat surface (Figure P5.18), with the design that the solid and the fluid exchange heat, such as in an industrial heat exchanger application.

Assuming that the thermal conductivity of the solid in the z-direction is infinite (i.e., no conductive resistance) but zero in the x-direction, determine $T_s(x, t)$ and $T_f(x, t)$, where $T_s(x, t)$ and $T_f(x, t)$ are, respectively, the temperatures in the solid and fluid. Note that the fluid is not well-mixed and that there is convective heat transfer between the solid and the fluid. Use the LT method. Hint: The governing equations are as follows:

Solid:

$$\frac{\partial T_s}{\partial t} + h_s'(T_s - T_f) = 0,$$

where $T_s = T_s(x, t)$ is the solid temperature, $T_f = T_f(x, t)$ is the fluid temperature, $h_s' \equiv h_s/m_s c_{p_s}$, is the convective heat transfer coefficient, m_s is the mass of solid per unit area of the wall, and c_{p_s} is the specific heat at constant pressure of the solid.

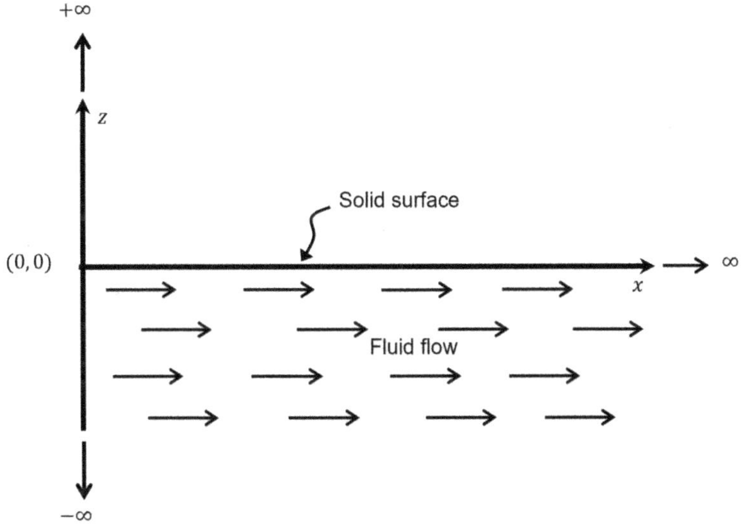

Figure P5.18: Convective fluid flow parallel to a solid surface.

Fluid:

$$\frac{\partial T_f}{\partial t} + U_\infty \frac{\partial T_f}{\partial x} + h_f'(T_f - T_s) = 0,$$

where $T_f = T_f(x,t)$ is the fluid temperature, $h_f' \equiv h_f/m_f c_{p_f}$, h_f is the convective heat transfer coefficient, m_f is the mass of the fluid per unit area of the wall, and c_{p_f} is the fluid specific heat at constant pressure. U_∞ is the velocity of the fluid, which is wholly in the x-direction. Note that molecular diffusion of heat is neglected in this problem, which means that there is no thermal boundary layer at the wall. Also note that $h_s = h_f$.

5.19 Revisit Problem 5.18 by assuming that there is a uniform heat generation rate of Q per unit time per unit volume.

5.20 Revisit Problem 5.18 by assuming a finite value for the thermal conductivity perpendicular to the direction flow and zero conductivity in the direction of flow. For this case, assume that thermal conduction is limiting relative to convective heat transfer.

Hint: The governing equations in this case are as follows:

$$\alpha(T_s)_{zz} - (T_s)_t = 0; \; z \in (0,\infty), \quad x \in (0,\infty), t \in (0,\infty).$$

$$k_f(T_s)_z = m_f c_{p_f}\left(\frac{\partial T_f}{\partial t} + U_\infty \frac{\partial T_f}{\partial x}\right), \quad x \in (0,\infty), t \in (0,\infty), z = 0.$$

$$T_s = T_f; \; x \in (0,\infty), \quad t \in (0,\infty), z = 0.$$

The same variable definition used in Problem 5.18 applies here as well. k_f is the fluid thermal conductivity.

5.21 A more realistic heat exchanger problem can be obtained if film coefficient h_f or h_s is brought back into Problem 5.20. That is, relaxing the assumption of infinite convective heat transfer coefficient so that both conductive and convective heat transfer resistances are comparably limiting. Obtain $T_f(x,t)$ for this case.

Suggested reading

[1] Abramowitz, M. and Stegnum, I.A. (Eds). Handbook of Mathematical Functions, Washington, WA, USA, United States Department Commerce, Bureau of Standards, 1964.

[2] Carrier, G.F., Krook, M. and Pearson, C.E. Functions of Complex Variables, Ithaca, NY, USA, Hod Book, 1983.

[3] Carslaw, H.S. and Jaeger, J.C. Conduction of Heat in Solids, Second Edition, Oxford, UK, Clarendon Press, 1959.

[4] Courant, R. and Hilbert, D. Methoden der Mathematischen Physic, Volume I, Berlin, Germany, Verlag von Julius, 1931.

[5] Crank, J. The Mathematics of Diffusion, Second Edition, Oxford, UK, Clarendon Press, 1975.

[6] Friedman, B. Principles and Techniques of Applied Mathematics, New York, NY, USA, Wiley and Sons, 1956.
[7] Miles, J. Integral Transforms in Applied Mathematics, New York, NY, USA, Cambridge University Press, 1971.
[8] Morse, P.M. and Feshbach, H. Methods of Theoretical Physics, Volume I, New York, NY, USA, McGraw-Hill, 1953.
[9] Nair, S. Advanced Topics in Applied Mathematics for Engineering and the Physical Sciences, New York, NY, USA, Cambridge University Press, 2011.
[10] Nettleton. Proc Phys Soc, 1916, 29, 50. From "Carslaw, H.S. and Jaeger, J.C., Conduction of Heat in Solids, Second Edition, Oxford, UK, Clarendon Press, 1959
[11] Peters, A.S. Water waves over sloping beaches and the solution of a mixed boundary value problem for $\Delta^2\phi - k^2\phi = 0$ in a sector, Comm Pure Appl Math, 1952, 5, 87–108.
[12] Sokonikoff, I.S. and Redheffer, R.M. Mathematics of physics and modern engineering, J Electrochem Soc, 1958, 105, 9.
[13] Sommerfeld, A. Partial Differential Equations in Physics, Volume VI of Lectures on Theoretical Physics, New York, NY, USA, Academic Press, 1949.
[14] Stakgold, I. Boundary Value Problems of Mathematical Physics, Volume I, New York, MacMillan Co, 1968.
[15] Watson, G.N. A Treatise on the Theory of Bessel Functions, Cambridge, UK, Cambridge University Press, 1964.
[16] Ludford, G. S. S., Personal Communication (1985).

6 Asymptotic behavior of complex integrals

Engineering problems governed by partial differential equations that are singular are commonplace but difficult to analyze. This is particularly so when the singularity is of the essential type, and no closed-form solutions are possible in general. Asymptotic methods offer an avenue for us to analyze such problems, and this is explored in this chapter. Since most differential equations of engineering relevance can be transformed to integral equations, and the converse is not generally true, we focus on integrals in this chapter and explore limiting behaviors at small time and longtime dynamics of engineering systems. The general treatment involves fairly complicated details of the complex variable theory. Asymptotic power series (APS) are discussed. The sectorial nature of asymptotic expansions is also presented. At the crux of the asymptotic analysis presented in this chapter are the complex variable inversion techniques, for example, within the framework of the Laplace transform (LT) method, and the characterization of the types of solutions (exponential decay, exponential growth, algebraic growth or decay, oscillatory solutions, etc.) as functions of the type of singularity involved in the transform space. The asymptotic methods of Laplace, steepest descent, saddle point, and stationary phase are presented, as are the complexities brought about by singularities and multivaluedness in closing inversion contours.

6.1 Introduction

Let us start with the integral

$$\tilde{\phi}(s) = \int_0^\infty e^{-st}\phi(t)dt, \tag{6.1}$$

which of course defines the LT of $\phi(t)$, where, in general, we are ultimately interested in obtaining $\phi(t)$. The LT gives good information on small time behavior. Large time behavior in the case of LT is usually given away by the poles.

In eq. (6.1), if $\mathrm{Re}[s] > 0$, as we approach from $+\infty$ on the right-hand plane (RHP), $\tilde{\phi}(s)$ goes to zero, unless we are looking at a time so small that "it kills" $s \to \infty$ in the product st in the exponent. The asymptotic behavior of $t \to 0$ is betrayed by the behavior of the LT as $s \to \infty$. That is to say that the only information left as $s \to \infty$ is when $t \to 0$; all other contributions are wiped out (since if t is large, $s \to \infty$ implies things go to zero even faster). Thus, we anticipate that the large s behavior of $\tilde{\phi}(s)$ implies the small time behavior of $\phi(t)$. Conversely, the large time behavior of LT is governed by the location of the poles and at times by small s. However, it is not in general true that the large time behavior of $\phi(t)$ implies the small s behavior of $\tilde{\phi}(s)$. This is especially so if the pole is positive.

https://doi.org/10.1515/9783111351179-008

As an example, for $\tilde{\phi}(s) = 1/(s+1), \phi(t) = e^t$, large time is governed by e^t and not $s \to 0$ (delta function). That is, the delta function going to 0 as $t \to \infty$ gives no information.

Example 6.1 Transient heat transfer with an impulsive thermal input (Figure 6.1): Consider the problem

$$u_{xx} = u_t, \qquad 0 < x < \infty, \qquad u(x,0) = 0, \qquad u(0, t > 0) = u_0.$$

The solution is $u(x,t) = u_0 \operatorname{erfc}\left(x/2\sqrt{t}\right)$, where erfc is the complimentary error function, which is defined as

$$\operatorname{erfc}(y) = \frac{2}{\sqrt{\pi}} \int_y^\infty e^{-\eta^2} d\eta = 1 - \operatorname{erf} y = 1 - \frac{2}{\sqrt{\pi}} \int_0^y e^{-\eta^2} d\eta.$$

Taking LT of the governing equation, we have:

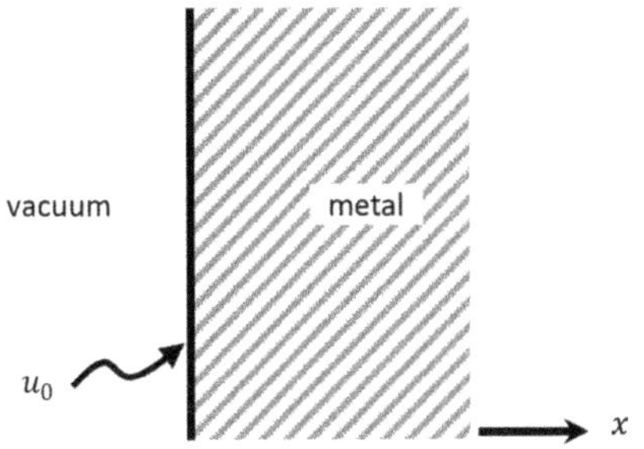

Figure 6.1: Physical domain for transient heat transfer with an impulsive heat input.

$$\tilde{u}(s,x) = \frac{u_0 e^{-\sqrt{s}x}}{s}.$$

Note that this is a rather difficult problem, whose behavior at $s = 0$ or $t = 0$ is singular (all of a sudden you impose a heat source), and no unique solution exists unless we bound the energy. A pole at $s = 0$ gives t^n at steady state and nothing like exponential.

6.1.1 Behavior for small time

The condition $t \to 0$ means that $y \equiv x/\sqrt{t}$ is large. There is no characteristic length scale for the problem, so we are interested in erfc $(y \to \infty)$:

$$\int_y^\infty e^{-\eta^2} d\eta = \int_y^\infty \frac{e^{-\eta^2}}{\eta} \eta\, d\eta = \frac{1}{2}\int_y^\infty \frac{e^{-\eta^2}}{\eta} d\eta^2 = -\frac{1}{2}\int_y^\infty \frac{de^{-\eta^2}}{\eta}$$

$$= +\frac{1}{2}\left[\frac{e^{-y^2}}{y} + \int_y^\infty \frac{1}{\eta^2} e^{-\eta^2} d\eta\right] = +\frac{1}{2}\left[\frac{e^{-y^2}}{y} - \frac{e^{-y^2}}{y^2} + \dots\right],$$

where we have used the substitution $x = -\eta^2$ and transformed back to $\eta = (-x)^{1/2}$. We see that the terms become successively smaller as $y \to \infty$. That is, the dominating behavior of this erfc y as $y \to \infty$ is $\frac{1}{2}\frac{e^{-y^2}}{y}$, leading to what we call an asymptotic series. However, we have this on the real line. If we analytically continue in the complex plane, we would have

$$\operatorname{erfc}(z) \sim \frac{1}{2}\left[\frac{e^{-z^2}}{z} - \dots\right].$$

This extension to the complex plane fails sometimes. That is, just because you get an asymptotic expansion doesn't mean it is valid everywhere in the complex plane, and validity may be restricted to a sector of the complex plane that may be defined by a range in $\arg z$. Consider the following example.

Example 6.2 What is the asymptotic behavior of the exponential integral $E_i(x)$:

$$E_i(x) = \int_x^\infty \frac{e^{-t}}{t} dt; \qquad x \text{ real},$$

in the limit of $x \to \infty$?

Solution: A trick to obtaining an asymptotic sequence is to integrate by parts:

$$E_i(x) = \frac{e^{-x}}{x} - \int_x^\infty \frac{e^{-t}}{t^2} dt$$

$$= \underbrace{e^{-x}\left[\frac{1}{x} - \frac{1}{x^2} + \frac{2!}{x^3} - \dots + \frac{(-1)^{n+1}(n-1)!}{x^n}\right]}_{S_n} + \underbrace{(-1)^n n! \int_x^\infty \frac{e^{-t}}{t^{n+1}} dt}_{R_n},$$

where S_n is the partial sum of the asymptotic sequence (not necessarily a convergent sequence) and R_n is the remainder, which satisfies the inequality that

$$|R_n| < \left|\frac{n!}{x^{n+1}} \int_x^\infty e^{-t} dt\right| = \frac{n!}{x^{n+1}} e^{-x}.$$

That is, for fixed n, $R_n(x)$ is of order $e^{-x}/x^{n+1} \longrightarrow 0$ as $x \longrightarrow \infty$, and $E_i(x) \sim S_n(x)$. For example, if $E_i(x) \approx e^{-x}/x$ (first term in the expansion), then

$$\left| E_i(x) - \frac{e^{-x}}{x} \right| < \frac{1}{x^2} e^{-x}.$$

The relative error is more important and is given by

$$\frac{\left| E_i(x) - \frac{e^{-x}}{x} \right|}{\frac{e^{-x}}{x}} < \frac{\frac{1}{x^2} e^{-x}}{\frac{1}{x} e^{-x}} \approx \frac{1}{x} \longrightarrow 0 \text{ as } x \longrightarrow \infty.$$

We see that we are not necessarily better off by keeping more terms. Moreover, we cannot sum the series. That is, for a fixed x, the partial sum of the asymptotic series is given by

$$\lim_{n-\infty} S_n = s = \infty,$$

which is divergent. A necessary condition for a series to converge is that $a_n \longrightarrow 0$ as $n \longrightarrow \infty$. Here for any fixed x,

$$a_n = \frac{(n-1)!(-1)^n}{x^n},$$

which blows up because of the magnitude of $n!$ relative to x^n. In general, we keep two terms. Asymptotic series don't usually converge but they are still very useful nevertheless. The idea here is that although the series diverges as $n \longrightarrow \infty$, we see that the integral is more and more closely approximated by the partial sum of the asymptotic series as $x \longrightarrow \infty$.

6.1.2 Asymptotic sequence

Definition 6.1 A sequence of functions $\langle \phi_n(z) \rangle$, all defined in an open region Ω, is said to be an asymptotic sequence as $z \longrightarrow z_0$, $z \in \Omega$, if

$$\phi_{n+1} = o(\phi_n), \quad z \longrightarrow z_0, \quad \text{or} \quad \lim_{z \to z_0} \frac{\phi_{n+1}(z)}{\phi_n(z)} = 0, \quad \text{for } n = 1, 2, \ldots.$$

There are an infinite number of ways of approaching 0 or ∞ in the complex plane (Figure 6.2). (There is only one direction in real space.) This is a main difference between asymptotic series in complex and real space. For example, for $z = 0$, Ω = the whole complex plane. A consequence of this is that asymptotic expansions are sometimes valid in a sector and not in the whole complex plane.

Example 6.3 Is $\langle\phi_n\rangle = 1/z^n$ an asymptotic sequence?

Solution:

$$\phi_n = \frac{1}{z^n}, \qquad \frac{\phi_{n+1}}{\phi_n} = \frac{1}{z} \to 0, \qquad z \to \infty.$$

Therefore, we have an asymptotic sequence as $z \to \infty$. On the other hand, the sequence $\langle\phi_n(z)\rangle = z^n$ will be asymptotic at $z \equiv z_0 \to 0$ if $\Omega = \{|z| < 1\}$. That is, we can see the domain Ω as defining the condition (region) of validity of a series in an asymptotic sequence.

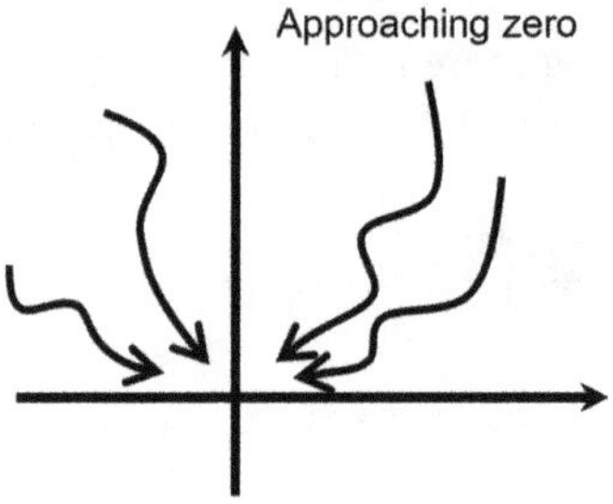

Figure 6.2: Sketch shows that there are an infinite number of ways to approach zero in the complex plane.

Example 6.4 Is $\langle z(\ln z)^{-n}\rangle$ an asymptotic sequence?

Solution:

$$\phi_n = z(\ln z)^{-n}, \qquad z_0 = 0, \qquad \Omega = s_{\pm\pi}, \qquad \frac{\phi_{n+1}}{\phi_n} = \frac{1}{\ln z},$$

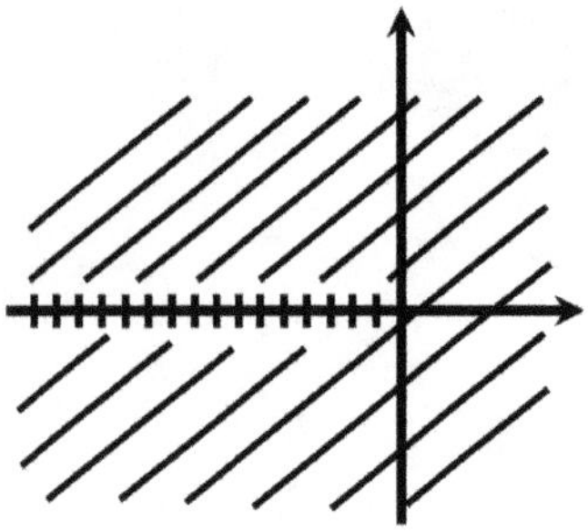

Figure 6.3: Definition of domain $\Omega \equiv s_{\pm\pi}$.

where $s_{\pm\pi}$ is the cut plane, and the notations are defined as follows:

$$s_R(\alpha,\beta) \equiv \{z|s_R(\alpha,\beta);\ 0 < |z| < R,\ \alpha < \arg z < \beta\}$$

$$\text{If } \alpha = \beta,\ s_R(\alpha,\beta) = s_{R\beta}, \text{if } R = \infty\ s_R(\alpha,\beta) = s_{\alpha\beta}.$$

Figure 6.3 shows the definition of the domain $\Omega = s_{\pm\pi}$.

Above, the subscript R denotes the radius of the complex variable z, while α and β denote the range of the angle of z. The definition of domain $s_R(\alpha,\beta)$ is depicted in Figure 6.4. The ϕ_ns are called gauge functions – they gauge how fast a function is decreasing or otherwise.

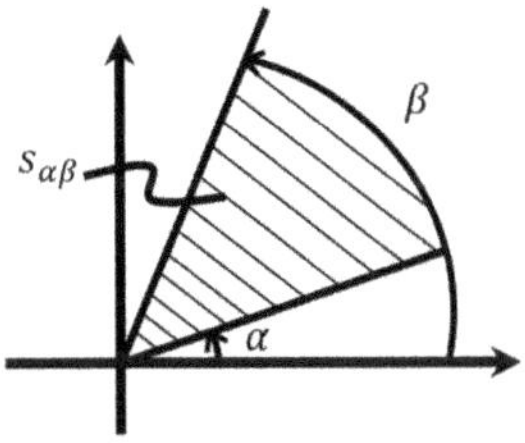

Figure 6.4: Definition of sectorial domains $s_{\alpha\beta}$.

6.2 Asymptotic expansion

Let $f(z)$ and $\langle\phi_n(z)\rangle$ be defined in the domain Ω (where the symbol $\langle\ \rangle$ implies a sequence of functions), $z_0 \in \bar{\Omega}$, and let $\langle\phi_n\rangle$ be an asymptotic sequence as $z \to z_0$. Then $f(z)$ is said to have an asymptotic development in $\langle\phi_n\rangle$ as $z \to z_0$ of order N if there exist constants C_k's such that

$$f(z) = \sum_{k=1}^{N} C_k\phi_k(z) + o(\phi_N), \tag{6.2}$$

which means that $o(\phi_N)/\phi_N \to 0$ as $z \to z_0$; or, equivalently:

$$f(z) \sim \sum_{k=1}^{N} C_k\phi_k(z), \quad \text{or} \quad \lim_{z\to z_0} \frac{\left|f(z) - \sum_{i=1}^{N} C_i\phi_i\right|}{\phi_N} \to 0,$$

where the denominator on the RHS is from the last term of eq. (6.2).

In a layperson's language, when we say that a function $F(x)$ is asymptotic to $G(x)$ (as $x \to 0$ or $x \to \infty$), it simply means that the remainder (the term neglected after $G(x)$ in a series) is insignificant (goes to zero), relative to the retained terms in $G(x)$, as $x \to 0$ or $x \to \infty$, as the case may be. So, basically, the idea is to get a series (anyhow) and show that wherever (in the series) you truncate, say at S_n, the remainder $R_n(x)$ goes to zero as $x \to 0$ or $x \to \infty$, independent of n. Above, Ω = Open region and $\bar{\Omega}$ = Boundary, where points on the boundary, including the origin, do not belong to $s_{\alpha\beta}$ (Figure 6.5). The approach is sectorial in the sense that the summation in eq. (6.2) is valid only for $\alpha < \arg(z - z_0) < \beta$.

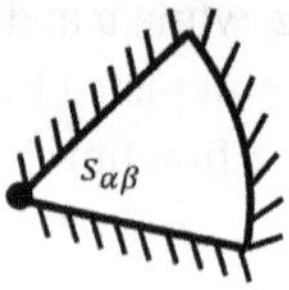

Figure 6.5: Depiction of the boundary of $s_{\alpha\beta}$. Note that $z = 0$ (or $z - z_0$) is part of the boundary.

Example 6.5 Investigate the asymptotic properties of $f(z) = \sin z + e^{-1/z^2}$ as $z_0 \to 0$ in the domain $s_{0,\pi/4}$, or

$$f(z) = \sin z + e^{-\frac{1}{z^2}}, \quad z_0 = 0, \quad \Omega = s_{0,\pi/4}. \tag{6.3}$$

Solution: The domain $\Omega = s_{0,\pi/4}$ is shown in Figure 6.6. Now,

$$\sin z = \sum_{k=0}^{\infty} \frac{z^{2k+1}}{(2k+1)!}(-1)^k \quad \forall z$$

has infinite radius of convergence (i.e., it is convergent for all z). Here, we are claiming that

$$f(z) \sim \sum_{k=0}^{N} \frac{z^{2k+1}}{(2k+1)!}(-1)^k + o(z^{2N+1}). \tag{6.4}$$

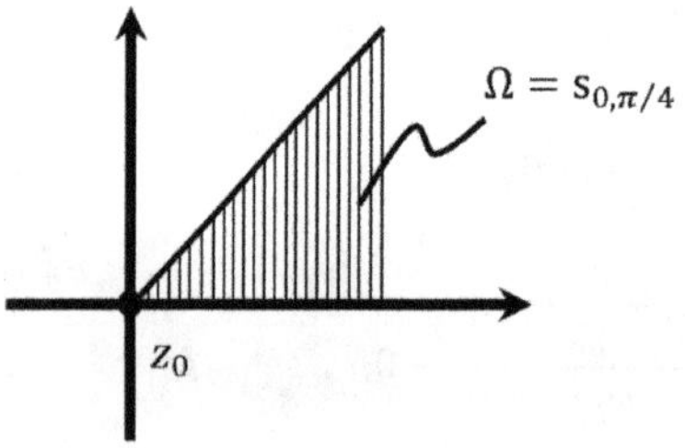

Figure 6.6: The domain $s_{0,\pi/4}$.

To see this, consider that, regarding the e^{-1/z^2} in eq. (6.3):

$$\frac{1}{z^2} = \frac{1}{r^2}e^{-2i\theta} = \frac{1}{r^2}[\cos(-2\theta) + i\sin(-2\theta)]$$

so that

$$\mathrm{Re}\left[\frac{1}{z^2}\right] = \frac{1}{r^2}\cos 2\theta, \qquad \left|e^{-\frac{1}{z^2}}\right| = e^{-\frac{1}{r^2}\cos 2\theta}$$

and realizing that $\cos\theta$ is positive for $-\pi/2 < \theta < \pi/2$ (Figure 6.7), we see that

$$\lim_{z\to 0} \frac{e^{-\frac{1}{r^2}\cos 2\theta}}{z^m} = 0 \text{ as } \theta < \frac{\pi}{4} \left(\text{or } 2\theta < \frac{\pi}{2} \text{ and } \cos 2\theta > 0\right).$$

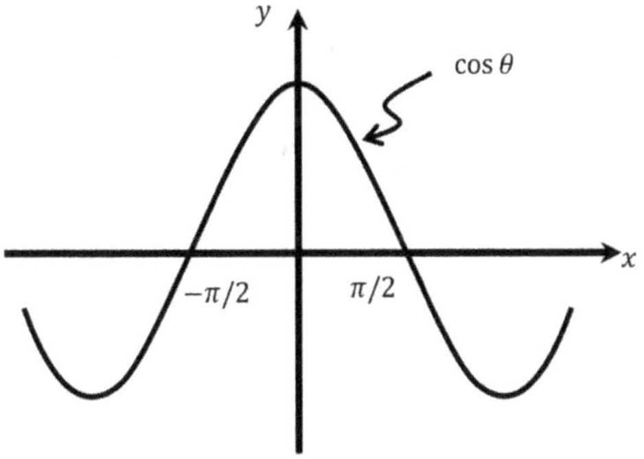

Figure 6.7: $\cos\theta$ is positive in $-\pi/2 < \theta < \pi/2$.

(Note that $\pi/4$ is not in the sector.) Thus, we will not see the exponential decay (the term e^{-1/z^2} in the given function), and we will have

$$\left|\frac{\sin z + e^{-1/z^2} - \sum_{k=1}^{N}\left[(-1)^k z^{k+1}/(2k+1)\right]}{z^{2N+1}}\right| \to 0,$$

where the denominator on the RHS is from the last term of eq. (6.4).

Suppose we consider $s_{\pi/4,\pi/2}$ instead so that $\pi/2 < 2\theta < \pi$ and hence $\cos 2\theta < 0$, then

$$\left|e^{-\frac{1}{z^2}}\right| = e^{-\frac{1}{r^2}\cos 2\theta} = e^{+\frac{1}{r^2}|\cos 2\theta|}, \qquad \frac{\pi}{4} < \theta < \frac{\pi}{2}. \tag{6.5}$$

So, we have a blow up when we cross the Stokes line, which is the $\pi/4$ line in this case. The sectors are shown in Figure 6.8. The result in eq. (6.5) clearly demonstrates how $\arg z$ or the domain (sector) could determine the existence or otherwise of an asymptotic expansion about a point.

In this case, the asymptotic behavior of the function changes as we cross the Stokes line, and eq. (6.4) is no longer valid because the exponential is dominated by e^{-1/z^2} and we have to use different gauge functions. The conclusion is that we have to be concerned about the Stokes line when we carry out an asymptotic expansion in the complex plane.

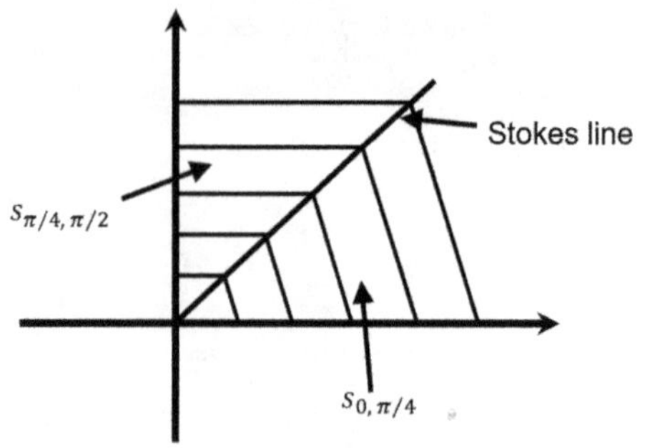

Figure 6.8: Sectors $s_{0,\pi/4}$ and $s_{\pi/4,\pi/2}$.

6.2.1 Definition of asymptotic power series

If gauge functions like z^n or z^{-n}, $n > 0$ are used, then the asymptotic series are called Asymptotic Power Series (APS). Normally, it is acceptable to integrate an asymptotic series term by term. However, we need to ensure that the resulting series is an APS, in the sense that

$$\int_z^\infty \left[f(z) - a_0 - \frac{a_1}{z}\right] dz \sim \frac{a_2}{z} + \frac{a_3}{2z^2} + \frac{a_4}{3z^3}$$

and

$$f(z) \sim a_0 + \frac{a_1}{z} + \frac{a_2}{z^2} + \dots.$$

The fact that a_1/z blows up at zero doesn't matter to our definition of APS since the other terms after a_1 are integrable and are asymptotic. Also, if $f'(z)$ possesses an asymptotic representation, it can be obtained by term differentiation of $f(z)$. However, for a counter example, consider

$$f(z) = e^{-\frac{1}{z}} \sin\left(e^{-\frac{1}{z}}\right); \quad z > 0. \tag{6.6}$$

We can see that

$\lim_{z\to 0}\left(e^{-\frac{1}{z}} \sin\left(e^{-\frac{1}{z}}\right)\right) z^{+n} = 0$ (from the definition of APS), so that $f(z) \sim 0$ to all orders $\langle z^{-n}\rangle$.

By differentiating eq. (6.6), we see that

$$f'(z) = \frac{e^{-1/z}}{z^2} \sin\left(e^{-\frac{1}{z}}\right) - z^{-2} \cos\left(e^{\frac{1}{z}}\right).$$

For the first term on the RHS, $f(z) \sim 0$ to all orders since $e^{-1/z} \ll z^2$ as $z \to 0$, while, for the second, there is no APS in terms of $\langle z^{-n}\rangle$ since we cannot expand in the neighborhood of $z = 0$.

Example 6.6 Investigate the asymptotic properties of the exponential integral $E_i(x) = \int_x^\infty \frac{e^{-t}}{t} dt$ as $x \to \infty$ (positive).

Solution:

$$E_i(x) = \int_x^\infty \frac{e^{-t}}{t} dt \quad \text{(Real } t \text{ space)}, \tag{6.7}$$

$$E_i(z) = \int_z^\infty \frac{e^{-t}}{t} dt \quad \text{(Complex } t \text{ space)}.$$

Given the limits of the integral (z, ∞), z doesn't pass through 0. Since e^{-z}/z is analytic as long as we skip $z = 0$, it is independent of path (Figure 6.9), although with paths passing through zero being unsuitable.

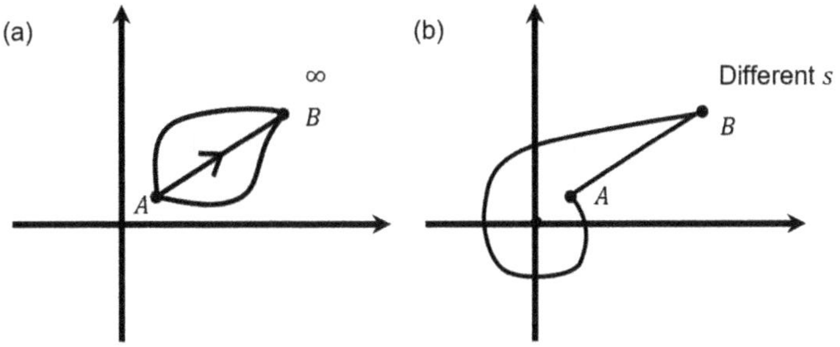

Figure 6.9: Path independence of $E_i(z)$ as long as we don't pass through $z = 0$. We have Condition (a) for the exponential integral problem in eq. (6.7), not (b), where $z = 0$ is enclosed by the contour.

We can start with $E_i(x)$ and put $z = x$ for the complex space exponential integral, assuming that we can do analytic continuation on the whole z plane. That is, we extend the real series

$$E_i(x) \sim e^{-x} \left[\frac{1}{x} - \frac{1}{x^2} + \frac{2!}{x^3} + \ldots + \frac{(-1)^{n+1}(n-1)!}{x^n} \right]$$

to the complex series

$$E_i(z) \sim e^{-z} \left[\frac{1}{z} - \frac{1}{z^2} + \frac{2!}{z^3} + \ldots + \frac{(-1)^{n+1}(n-1)!}{z^n} \right]. \tag{6.8}$$

By claiming analytic continuation, the "~" should be "=," but then the series is divergent. One possibility in order to extend to z is to realize that

$$E_i(x) = \int_x^\infty \frac{e^{-t}}{t}dt = -\int_0^x \left(\frac{e^{-t}}{t} - \frac{1}{t}\right)dt + \int_0^1 \left(\frac{e^{-t}}{t} - \frac{1}{t}\right)dt + \int_1^\infty \frac{e^{-t}}{t}dt - \int_1^x \frac{1}{t}dt,$$

which allows us to separate the singularity behavior. We can write this as

$$E_i(x) = -\gamma - \ln x - \int_0^x \left(\frac{e^{-t}-1}{t}\right)dt,$$

where

$$\gamma \equiv -\int_0^1 \frac{e^{-t} - 1 - e^{-1/t}}{t}dt,$$

which is just a number called the **Euler Constant**, which has a numerical value of 0.5772156649. The function $\int_0^x \left(\frac{e^{-t}-1}{t}\right)dt$ is analytic everywhere, independent of path, including at 0, but then we have $\ln z$, where we have replaced x with z. The problem with $z = 0$ is the multivaluedness of $\ln z$. When we make a branch cut to avoid the branch point, then the series for $E_i(z)$ in eq. (6.8) is analytic everywhere outside the branch cut. (Note that $\ln z$ is not defined if we go across cut.) A branch cut for this is shown in Figure 6.10. Thus, we can say that

$$E_i(z) = -\gamma - \ln z - \text{Entire function},$$

where an Entire function is one which is analytic for all finite values of the complex variable z. (Elementary examples of such functions include z^2, $\sin z$, and e^z.)

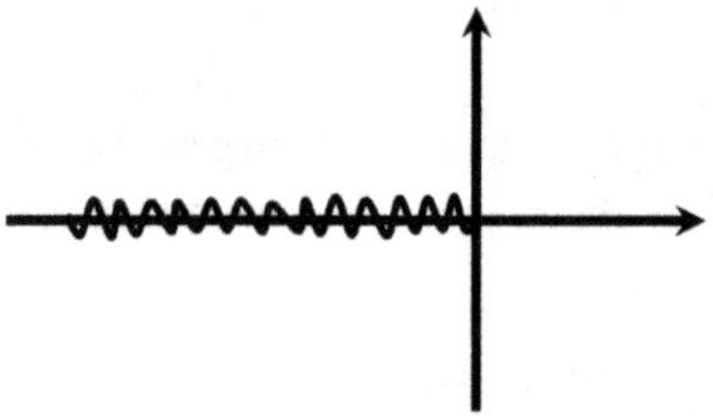

Figure 6.10: Branch cut for $E_i(z)$.

The APS for $E_i(z)$ is valid for $|\arg z| < 3\pi/2$ or $-3\pi/2 < \arg z < 3\pi/2$ (Figure 6.11(a)) since θ is positive for $-\pi/2 < \theta < \pi/2$, with relevance in $|e^{-1/z}| = e^{-(1/r)\cos\theta}$, where the exponent is positive for $\pi/2 < \theta < 3\pi/2$ and negative for $-\pi/2 < \theta < \pi/2$. Figure 6.11(b) shows the variation $\cos\theta$ with θ.

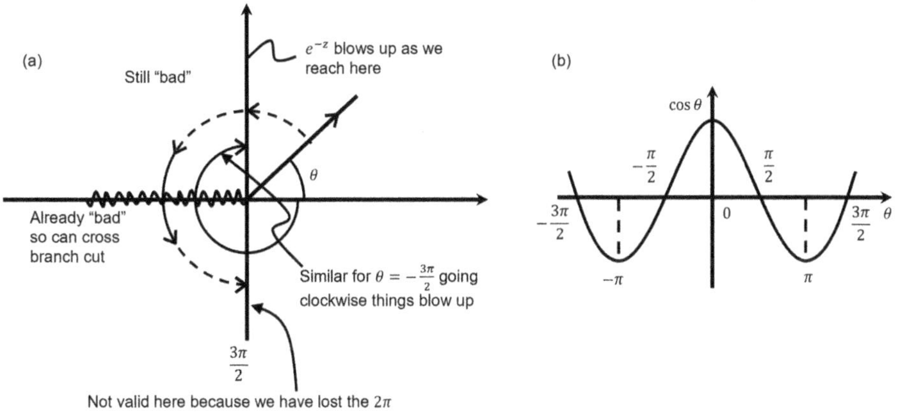

Figure 6.11: (a) Regions of validity of the APS for $E_i(z)$and (b) $\cos\theta$ versus θ.

6.2.2 The small time behavior of the inverse Laplace transform

6.2.2.1 Case 1: Independence of angle (Uniformity)

Theorem 6.1 If $\tilde{\phi}(s) \sim \sum_{n=1}^{N} a_n s^{-n}$ as $s \to \infty$ **uniformly** (independent of angle) and $\tilde{\phi}(s)$ is analytic outside a finite region in the s plane, then

$$\phi(t) \sim \sum_{n=1}^{N} \frac{a_n t^n}{(n-1)!}. \tag{6.9}$$

Equation (6.9) expresses the small-time asymptotic behavior of $\phi(t)$ because it has been obtained for $s \to \infty$. (Remember that $\tilde{\phi}(s)$ is the Laplace transform of $\phi(t)$.) The only possible singularity is essential singularity at infinity: because $\tilde{\phi}(s)$ is represented by the APS $\sum a_n s^{-n}$, it is bounded at ∞. The integration of the series in eq. (6.9) term by term toward Laplace transforming gives $\tilde{\phi}(s) \sim \sum a_n s^{-n}$.

6.2.2.2 Case 2: Nonuniformity

Nonuniformity refers to the dependence on angles in the complex plane. Let us consider an example in the sector $s_{(-\alpha,\beta)}$; $\alpha, \beta > 0$ (Figure 6.12) and the following requirements:

(1) $\phi(t)$ is analytic
(2) $\phi(t) = o(e^{\gamma t})$, $|t| \to \infty$, $\gamma = \text{constant}$
(3) $\phi(t) \sim \sum_{n=0}^{N} C_n t^{\gamma_n}$ as $|t| \to 0$, $t \in s_{(-\alpha,\beta)}$. (This is just the behavior of ϕ as $t \to 0$.)
(4) Define $\gamma_n \equiv \alpha_n + i\beta_n$. If $\alpha_n \le \alpha_{n+1}$, $\alpha_0 > -1$, or $\phi(t) \sim 1/t$, then the LT wouldn't exist (from $\sum_{n=0}^{N} C_n t^{\gamma_n}$ in Condition (3) above)

If the requirements implied by (1)–(4) are met and $|\arg(s - \beta)| < \pi/2$, then the large s behavior of $\tilde{\phi}(s)$ for the nonuniform sector-dependent case is given by:

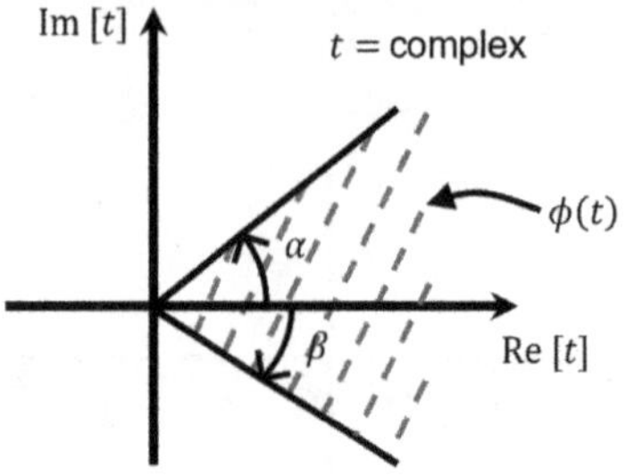

Figure 6.12: The sector $s_{(-\beta,\alpha)}$.

$$\tilde{\phi}(s) \sim \sum_{n=0}^{N} \Gamma(\gamma_n+1) C_n s^{-\gamma_n-1} \text{ as } |s| \to \infty \text{ in } s\left(-\frac{\pi}{2}-\beta+\delta, \frac{\pi}{2}+\alpha-\delta\right), \; \delta > 0.$$

See Figure 6.13 for the sector in this expression. The small time behavior of LT is given by this large s behavior in the shaded region. The unshaded region is/maybe the location where the APS is not valid. Requirements (2) and (4) above are those of the existence of LT. Requirement 3 is common, while Requirement 4 is asking for analyticity.

The series for $\tilde{\phi}(s)$ is called the generalized Watson Lemma on the complex plane. We can remove Requirement (1) if Requirements (2) and (3) are satisfied. If $\phi(t)$ is not analytic, we set $\alpha = \beta = 0$. That is, there is no extension (Figure 6.13), and we have $s_{-\pi/2,\pi/2}$ (Figure 6.14), which is the RHP.

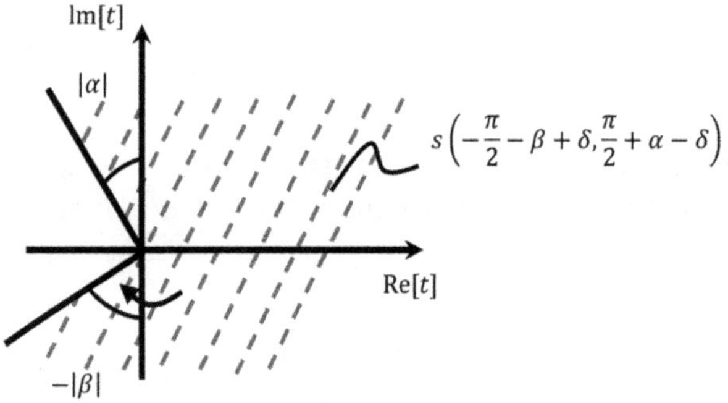

Figure 6.13: The region $s(-\frac{\pi}{2}-\beta+\delta, \frac{\pi}{2}+\alpha-\delta)$; $\delta > 0$.

6.2.2.3 Case 3: $\tilde{\phi}(s)$ has essential singularity at ∞

An example of this case is the function

$$\tilde{\phi}(s) = \frac{e^{-x\sqrt{s}}}{s}, \tag{6.10}$$

which is what we have for heat diffusion. In this case, we may need to do asymptotics on the LT inversion integral rather than directly on $\tilde{\phi}(s)$. A contour for this is shown in Figure 6.15.

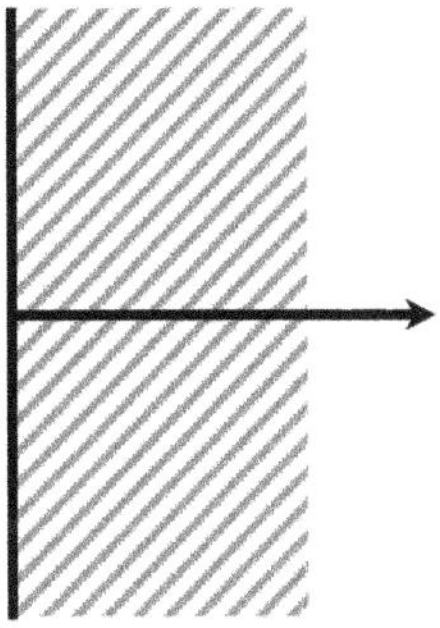

Figure 6.14: $S_{-\pi/2,\pi/2}$, $\alpha=\beta=0$ if $\phi(t)$ is not analytic.

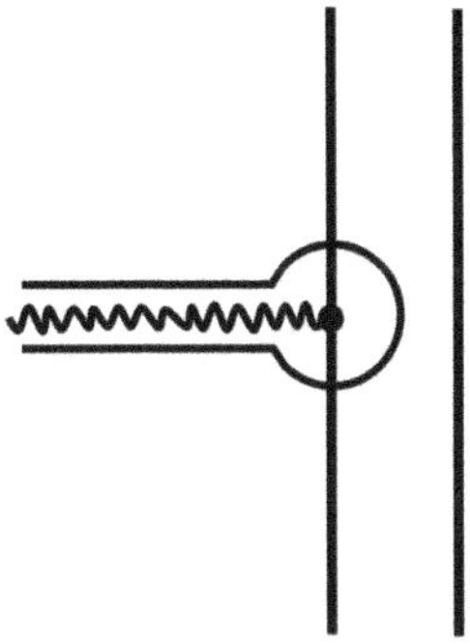

Figure 6.15: A slit contour for the inversion of the $\tilde{\phi}(s)$ in eq. (6.10).

We calculate the residues, etc., or whatever we can do approximately and apply asymptotics. We may not be able to evaluate the integral exactly. It may be possible to extend the limit beyond $\pi/2$ (i.e., the α, β above). The idea of this chapter is clearly espoused in this particular illustration. That is, even when we have an essential singularity, we can use the knowledge of asymptotics to obtain some useful results nonetheless.

6.2.2.4 Watson Lemma

To further motivate the small time behavior of the inverse LT, where small time implies large s, consider again the definition of LT:

$$\tilde{\phi}(s) = \int_0^\infty e^{-st} f(t)dt.$$

We can extend within the complex plane. In regions where $\mathrm{Re}[s] > 0$, e^{-st} decays so that $\tilde{\phi}(s)$ should be analytic (Figure 6.16).

Figure 6.16: Domain of $\text{Re}[s] > 0$.

Consider the simple function:

$$f(x) = \int_0^T e^{-xt} g(t)dt; \qquad 0 < T \leq \infty,$$

where x here is a surrogate for the LT variable s and x, t, g are real, and

$$g(t) = t^\lambda [a_0 + a_1 t + \ldots + a_n t^m] + t^\lambda R_{m+1}(t), \qquad \text{Re}[\lambda] > 0,\ \lambda > -1.$$

If there exists $A > 0$ such that $|R_{m+1}(t)| < Ct^{m+1}$ for all $t \in (0, A)$, then we can write

$$f(x) = \int_0^A e^{-xt} t^\lambda (a_0 + \ldots + a_m t^m)dt + \int_0^A e^{-xt} t^\lambda R_{m+1}(t)dt + \int_A^{T(T\text{ can be }\infty)} e^{-xt} g(t)dt.$$

(If $T = \infty$, then we require that $|g(t)| < Be^{bt}$ for some constant b and positive constant B.) Using

$$\int_0^A e^{-xt} t^\lambda a_k t^k dt = \frac{a_k \Gamma(\lambda + k + 1)}{x^{\lambda+k+1}} + O(e^{-Ax}),$$

which is easily obtained by the change of variables $-xt \equiv u$, we can write

$$f(x) = \underbrace{\frac{a_0 \Gamma(\lambda + 1)}{x^{\lambda+1}} + \ldots + \frac{a_m \Gamma(\lambda + m + 1)}{x^{\lambda+m+1}}}_{**} + O(e^{-Ax}), \tag{6.11}$$

where the remainder terms are assumed to follow the equation

$$\int_0^A e^{-xt} t^\lambda R_{m+1}(t)dt = O\left(\frac{1}{x^{\lambda+m+1}}\right) + O(e^{-Ax}).$$

Concerning the term $\int_A^T e^{-xt} g(t)dt$, consider $x > b$, so that the integral is $O(e^{-Ax})$, which goes to zero as $x \to \infty$, $A > 0$. We are looking for the proper x, with b fixed.

Thus, when we take x large enough, the leading balance will be given by the terms indicated by "$**$" in eq. (6.11). This equation is the Watson's lemma, which expresses, in this illustration, the inverse LT integral in the limit that the LT variable x goes to infinity, which is equivalent to the short-time behavior of the inverse LT. Therefore, it should be noted that the series in eq. (6.11) comes from the neighborhood of $z = 0$. It is therefore not surprising that the end point T does not appear in the series.

6.2.3 The large time behavior of the inverse Laplace transform

The behavior at large time is governed by the poles in the s – plane. There cannot be an infinite number of singularities unless we accept nonisolated singularity. In the previous heat diffusion problem, we had an infinite number of singularities but there was no finite plane that enclosed them.

Plausible poles and branch cuts in the complex s – plane for determining large time behavior are depicted in Figure 6.17.

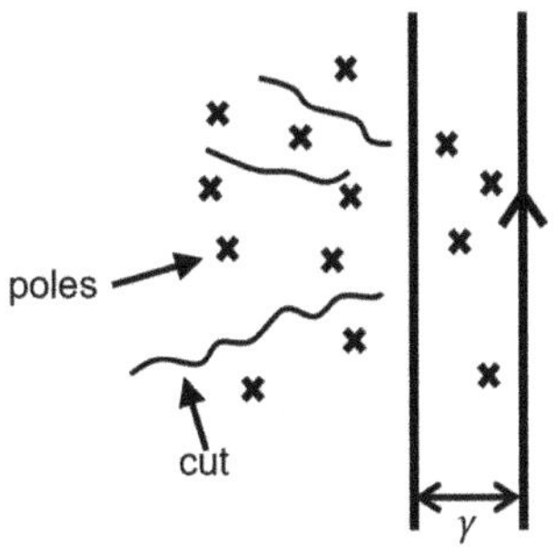

Figure 6.17: Conceptual contour for integration in the complex s plane with plausible poles and branch cuts.

6.2.3.1 A few remarks

a) If all the poles p_i are isolated singularities then the leading behavior of $f(t)$ is determined by the isolated singular point of $\int_{\gamma-i\infty}^{\gamma+i\infty}(\ )ds$ at p^* with $\text{Re}[p^*] \geq \text{Re}[p_i]\ \forall\ i$. Isolated singularities can be poles, and we could have essential singularity; but we still have to evaluate their residues. Since the dominant behavior is given by $\text{Re}[p^*]$, the approach to obtain this is to check where the poles are located and choose the biggest pole. We then evaluate $e^{st}\tilde{f}(s)$ at the largest pole $s = p^*$.
b) Suppose the p_is are branch points, what can we say about $f(t \to \infty)$? The assumption in (a) is that we can close the contour at infinity. Is the assumption necessary as far as examining the large time behavior of $f(t)$ is concerned? In general, it does not matter whether we can close the contour as far as $f(t)$ approaching infinity is concerned. Normally, the failure of our ability to close the contours at infinity is due to singularities at the origin $t = 0$, such as typical impulsive heat source at a boundary. We may also encounter difficulties in closing contours at

finite t. A delta source input or an error function could be disastrous at x' of $\delta(x-x')$, for example. It is emphasized that if we are interested in small s behavior, we will not in general be concerned with things that blow up at infinity, which again underlines the relevance of asymptotic analysis.

Example 6.7 (Heat diffusion problem) Let us examine the large-time behavior of $\tilde{u} = \frac{u_0}{s} e^{-x\sqrt{s}}$. Consider the solution

$$u(x,t) = u_0 \operatorname{erfc} \overbrace{\left(\frac{x}{2\sqrt{t}}\right)}^{y} = u_0 [1 - \frac{2}{\sqrt{\pi}} \int_0^y e^{-\eta^2} d\eta].$$

Large t implies small y, for which we can write

$$e^{-\eta^2} = 1 - \frac{\eta^2}{1!} + \frac{\eta^4}{2!} + \dots .$$

We can integrate term-by-term to obtain

$$u = u_0 \left[1 - \frac{x}{t^{1/2}} \frac{1}{\Gamma(1/2)} - \frac{t^{-3/2} x^3}{\Gamma(-1/2)} - \dots \right].$$

Let us try to expand the transformed u or $\tilde{u} = \frac{u_0}{s} e^{-x\sqrt{s}}$ about the only singularity ($s = 0$):

$$\tilde{u} = \frac{u_0}{s} \left[1 - x\sqrt{s} + \frac{x^2 s}{2!} + \dots \right].$$

This gives the behavior of $\tilde{u}$ at $s = 0$. Take $u_0 = 1$:

$$\mathcal{L}^{-1}\{\tilde{u}\} = 1 - \mathcal{L}^{-1}\left\{\frac{x}{\sqrt{s}}\right\} + \underbrace{\mathcal{L}^{-1}\left\{\frac{x^2}{2!}\right\}}_{\frac{x^2}{2!}\delta(t)} + \mathcal{L}^{-1}\left\{\frac{x^3\sqrt{s}}{3!}\right\} + \dots$$

Recall that

$$\mathcal{L}^{-1}\left\{s^{-\beta}\right\} = \frac{t^{+\beta-1}}{\Gamma(+\beta)}, \quad \beta > 0 \Rightarrow \mathcal{L}^{-1}\left\{\frac{x}{\sqrt{s}}\right\} = \frac{x}{t^{1/2}} \frac{1}{\Gamma(1/2)},$$

$$\mathcal{L}\{\delta(t)\} = \int_0^\infty \delta(t) e^{-st} dt = 1.$$

We are not concerned about $\frac{x^2}{2!}\delta(t)$ given our interest in $t \gg 0$, $\delta(t) = 0$.

Remarks on $\mathcal{L}^{-1}\{1\}$ ***by contour integration***

By definition, $\mathcal{L}^{-1}\{1\} = \int_0^\infty e^{st} ds$, and a plausible contour for this is shown in Figure 6.18.

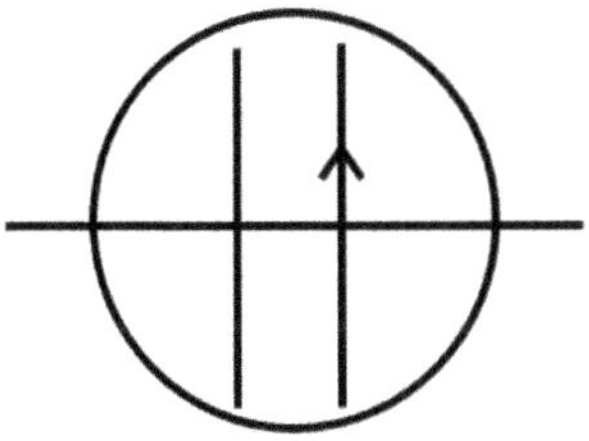

Figure 6.18: A plausible contour for determining $\mathcal{L}^{-1}\{1\}$.

However, we cannot close the contour at infinity (Figure 6.18) because of something within (a spike (δ) at 0). Therefore, we cannot evaluate $\mathcal{L}^{-1}\{1\}$ using contour integration. A heuristic argument will help evaluate $\mathcal{L}^{-1}\{1\}$.

6.2.3.2 Inverting $\tilde{u} = u_0\left(e^{-x\sqrt{s}}/s\right)$

Let us employ the key-hole contour in Figure 6.19 to invert $\tilde{u}(s)$, where the interest is to obtain the long-time behavior of the inverse; or $u(x,t)$, where

$$u = \frac{u_0}{2\pi i}\int_{-\infty}^{0^+} e^{st}\frac{e^{-x\sqrt{s}}}{s} ds,$$

and the notation used is $\int_{-\infty}^{0^+}$ = ⊂○, which is not just integrating from $-\infty$ to 0^+, as we need to include the small circle, at least initially, before we determine that it has zero contribution.

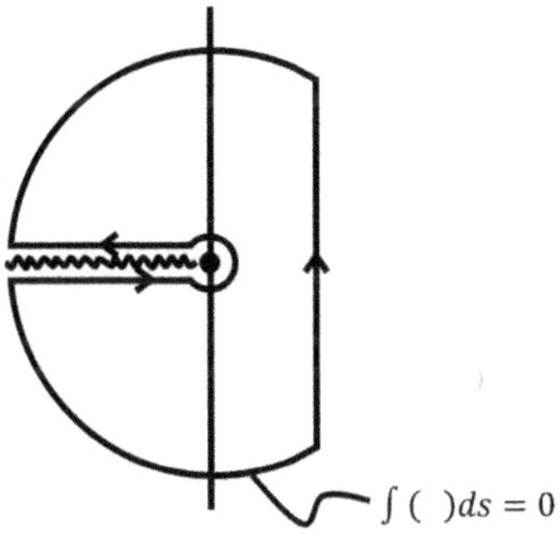

Figure 6.19: Contour for Laplace transform inversion of the heat diffusion problem.

The function $e^{-x\sqrt{s}}$ can be expanded for any values of x and s since $e^{-x\sqrt{s}}$ is analytic in the cut plane:

$$e^{-x\sqrt{s}} = 1 - x\sqrt{s} + \frac{x^2 s}{2!} + \ldots .$$

The ratio test says that the series converges everywhere (uniform convergence). However, $e^{-x\sqrt{s}}$ is not uniformly convergent, as it is convergent only in the cut plane. Because the series is uniformly convergent, we can integrate term-by-term to obtain

$$u = \frac{u_0}{2\pi i} \int_{-\infty}^{0^+} \frac{e^{st}}{s} \underbrace{\left(1 - x\sqrt{s} + \frac{x^2 s}{2!} + \dots\right)}_{\text{uniformly convergent series}} ds. \tag{6.12}$$

In order to integrate term-by-term, we will need to compute things like

$$J_a \equiv \frac{1}{2\pi i} \int_{-\infty}^{0^+} e^{st} s^a ds, \ \mathrm{Re}[a] \geq -1.$$

Since $a > 1$, the small circle does not contribute (Figure 6.20), but we have to integrate on both sides of the branch cut. We can write

$$J_a = -\frac{1}{\pi} \sin \pi a \int_0^\infty e^{-rt} r^a \, dr.$$

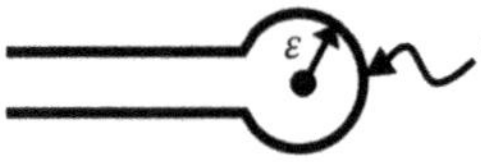

Figure 6.20: The small circle here doesn't contribute to integration because $a > 1$.

Introduce Γ:

$$J_a = -\frac{t^{-a-1}}{\pi} \sin \pi a \, \Gamma(1 + a).$$

But $\Gamma(z)\Gamma(1-z) = \pi \csc(\pi z)$, therefore

$$J_a = \frac{t^{-a-1}}{\Gamma(-a)} \quad (\text{where} - a = z). \tag{6.13}$$

If we substitute eq. (6.13) into eq. (6.12), we will obtain:

$$u(x,t) = u_0 \left[1 - \frac{xt^{-1/2}}{\Gamma(1/2)} - \frac{x^3 t^{-3/2}}{3!\Gamma(-3/2)} + \dots\right].$$

The terms with even x powers except 0 in eq. (6.12) vanish since a for this value of n are integers and $\Gamma(-a) = \infty$ for a even.

Define $y = xt^{-1/2}$ (i.e., $y = x/\sqrt{t}$). We can integrate eq. (6.12). The solution in eq. (6.13) is an exact solution as long as $xt^{-1/2} \nrightarrow \infty$ ($y < \infty$). **This is another way of writing the exact solution for large time.**

Suppose we say the behavior for small time is related to the behavior at large s and write

$$\mathcal{L}^{-1}\left\{\frac{u_0 e^{-\sqrt{s}x}}{s}\right\} = u_0 \mathcal{L}^{-1}\left\{\frac{1}{s} - \frac{x}{s^{\frac{1}{2}}} + \frac{x^2}{2!} - \frac{x^3\sqrt{s}}{3!} + \dots\right\},$$

we will immediately run into a difficulty because $\mathcal{L}^{-1}\{\sqrt{s}\}$ does not exist, as we cannot close the contour at ∞ (Figure 6.21).

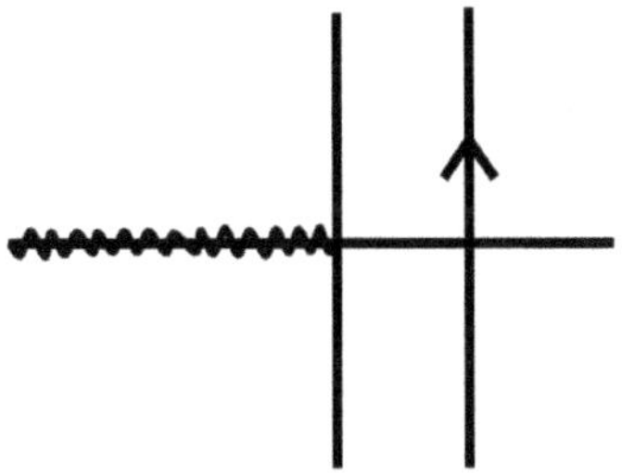

Figure 6.21: Branch cut to exclude branch point $s = 0$ extends to minus infinity.

This is unlike before where we used Jordan's Lemma to set $\int_{-\infty}^{0^+} = \text{⊂⊃}$, which is of course closable. That $\sqrt{s}$ cannot be inverted can be seen from the fact that it must be of the form $\frac{xt^{-1/2}}{\Gamma(1/2)}$ or $\frac{x^3 t^{-3/2}}{3!\Gamma(-3/2)}$, which has to be integrated, but the integration does not exist at 0.

Recall that for $n > 0$ and integer, $\int_0^\infty e^{-st}\delta(t)dt = 1$, $\int_0^\infty e^{-st}\delta'(t)dt = -s$, and $\int_0^\infty e^{-st}\delta^n(t)dt = (-1)^n s^n$ or

$$\mathcal{L}^{-1}\{(-1)^m s^m\} = \delta^m(t). \tag{6.14}$$

The reason why we cannot close the contour is that something happens in the finite region of the domain, not at ∞. However, it is still acceptable that small t implies big s.

Let us examine $\mathcal{L}^{-1}\{s^{n-\alpha}\}$ if $n > 0$, $1 > \alpha > 0$, $n =$ integer, $\beta > 0$, where $\beta \equiv n - \alpha$. Recall the expression in eq. (6.14) above to obtain, using the convolution theorem:

$$\mathcal{L}^{-1}\{s^{n-\alpha}\} = (-1)^n \int_0^t \delta^n(u) \frac{(t-u)^{\alpha-1}}{\Gamma(\alpha)} du$$

$$\mathcal{L}^{-1}\{s^n\} = (-1)^n \delta^n.$$

From a statement of convolution theorem, we will have $\mathcal{L}^{-1}\{s^{-\alpha}\} = t^{\alpha-1}/\Gamma(\alpha)$ and

$$\mathcal{L}^{-1}\{s^{n-\alpha}\} = \frac{(-1)^n}{\Gamma(\alpha)}\frac{d^n}{du^n}(t-u)^{\alpha-1}\bigg|_{u=0} = \frac{1}{\Gamma(\alpha)}t^{\alpha-n-1}(\alpha-1)\ldots(\alpha-n),$$

where

$$\Gamma(-\beta) = \Gamma(\alpha-n) = \frac{\Gamma(\alpha)}{(\alpha-1),(\alpha-2),\ \ldots\ (\alpha-n)}.$$

Thus,

$$\mathcal{L}^{-1}\{s^{n-\alpha}\} = \frac{t^{-\beta-1}}{\Gamma(-\beta)}, \qquad \beta > 0.$$

If β is an integer except 0, $\mathcal{L}^{-1}\{s^{n-\alpha}\}$ blows up, and the RHS of this equation is not integrable at 0. Therefore, there is no Laplace transformation. Note that we are considering s^{α} because we think that the large-time behavior of LT has to do with small s.

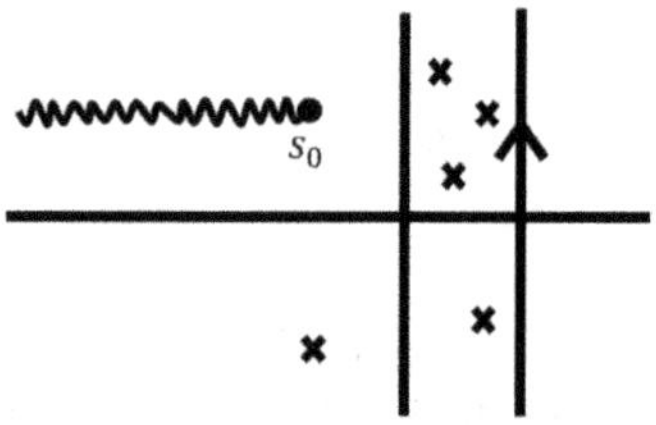

Figure 6.22: s – plane with multiple singularities and a branch cut, showing a particular branch point s_0.

The presence of poles (Figure 6.22) does not pose difficulties since **large time behavior is given by the largest pole**, though for essential singularity, it is a different matter. The behavior of the function $\tilde{\phi}(s)$ in the neighborhood of a particular singularity s_0 can be written as

$$\tilde{\phi}(s) \sim \sum_{n=1}^{N} a_n(s-s_0)^{\lambda_n},$$

where, focusing on the largest pole,

$$\mathrm{Re}[\lambda_n] < \mathrm{Re}[\lambda_{n+1}] < \ldots.$$

Therefore, we will ignore the fact that there are other singularities. That is, we will assume that the problem is well-posed and close the contour to the left. Also, we assume that we can close poles converging to ∞. We may still close if we have infinite singularity, or a branch cut, as in the cosh, sinh, example. Again, if we cannot immediately close the contour at ∞, it suggests that something is abrupt or impulsive at small time. For example, if the function has a delta behavior or we have the derivative of a delta function. For most problems we do not have a delta function behavior, so we can close. Actually, since we are not concerned with what happens as $s \to \infty$, which gives behaviors at small time, we could "force" a closure.

For inversion to obtain $\phi(t)$, we are interested in evaluating a loop integral of the form

$$I(t) = \frac{1}{2\pi i} \int_{\gamma} e^{st} \underbrace{\sum_{n=1}^{N} a_n (s - s_0)^{\lambda_n}}_{\tilde{\phi}(s)} \, ds \sim \phi(t),$$

where $\tilde{\phi}(s)$ has been expanded about s_0. We will have to do this for all poles and branch cuts and compare to obtain the one that gives the dominating (asymptotic) behavior.

Let $s' = s - s_0$ to shift the origin to s_0 (Figure 6.23), then

$$I(t) = \frac{1}{2\pi i} \int_{-\infty}^{0^+} e^{(s'+s_0)t} \tilde{\phi}(s' + s_0) ds',$$

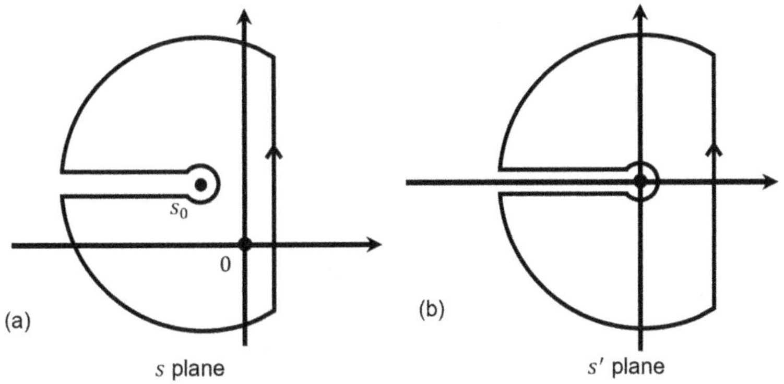

Figure 6.23: Contours without and with shifted origin.

which we can write as

$$I(t) = \left[\frac{1}{2\pi i} \int_{-\infty}^{0^+} e^{s't} \sum a_n (s')^{\lambda_n} ds \right] e^{s_0 t}.$$

Since we can integrate an asymptotic behavior, we will have

$$I(t) = \frac{\left[\sum a_n \int_{-\infty}^{0^+} e^{s't} (s')^{\lambda_n} ds \right]}{2\pi i} e^{s_0 t}.$$

The contour used is shown in Figure 6.24, where the contribution of the small hole at $s' = 0$ has been neglected. Also, note that the difference between A and B in the figure is given by the Γ function. If we assume $\mathrm{Re}[\lambda_n] > -1$ and evaluate the integral, we will get

$$I(t) = \left(\sum_{n=1}^{N} \frac{a_n}{\Gamma(-\lambda_n)t^{\lambda+1}}\right) e^{s_0 t}.$$

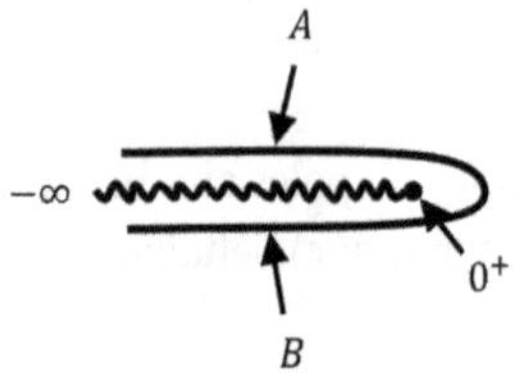

Figure 6.24: Contour used for integration. Note that the small circle at $s' = 0$ has been shrunk because its contribution is zero.

With a branch cut, we can still have an exponential behavior:

$$\mathrm{Re}[s_0] \begin{cases} < 0 & \text{exponential decay as } t \to \infty \\ > 0 & \text{exponential increase as } t \to \infty \\ = 0 & \text{sine and cosine in time as } t \to \infty. \end{cases}$$

For the case $\mathrm{Re}[s_0] = 0$, the term t^{λ_n+1} is very important because the factor $e^{s_0 t}$, which is controlling when $s_0 \neq 0$, becomes unity, transferring control to $z^{-(\lambda+1)}$. Suppose $\lambda = a + ib$ so that $z^{-(\lambda+1)} = z^{-(a+1)-ib}$, meaning that $\mathrm{Re}[\lambda_n] > -1 \Rightarrow \phi(t) \to 0$ because $a + 1 > 0$ for decaying solutions.

If s_0 is at the origin, we have algebraic decay. For branch cuts, there is a behavior that goes like $1/t$. Poles give exponential decay, or singularities of the function $t^m e^{s_0 t}$, where $m \geq 0$, integer. The case $m = 0$ implies simple poles of the function $\tilde{\phi}(x)$, and $m > 0$ implies multiple poles.

So for a purely imaginary pole, we will have a growth (Hopf bifurcation) as t^m since $e^{s_0 t}$ will be bounded by unity ($e^{s_0 t} = e^{i|\omega|t}$ so that $|e^{s_0 t}| \leq 1$). If $s = 0 + i\infty$, with multiple poles at the origin, we have an at^b behavior. The one that has the largest real part decides whether it is a branch point or not. A branch point behaves as

$$\frac{a_n}{\Gamma(-\lambda_n)t^{\lambda_n+1}},$$

giving a nonoscillatory behavior when ϕ decays to zero, and blowing up not as fast as an exponential behavior.

6.2.4 Logarithmic branch points

Since logarithmic functions are inherently multivalued, we need to consider logarithmic branch points (Figure 6.25), in conjunction with poles, and branch points due to other sources of multivaluedness. In this case, the asymptotic behavior of $\tilde{\phi}(s)$ close to s_0 can be written as

$$\tilde{\phi}(s) \sim (s - s_0)^n \ln(s - s_0).$$

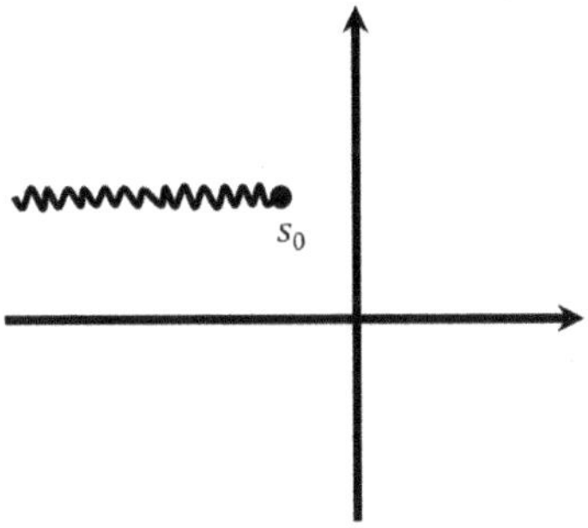

Figure 6.25: Branch cut to avoid multivaluedness due to a logarithm function.

We assume that we can close the contour (Figure 6.26) or use the other arguments that allow us to evaluate $\int_{-\infty}^{0^+}$. The $\phi(t)$ which we want to evaluate (by inversion) is asymptotic as follows:

$$\phi(t) \sim \frac{1}{2\pi i} \int e^{st} (s - s_0)^n \ln(s - s_0) ds.$$

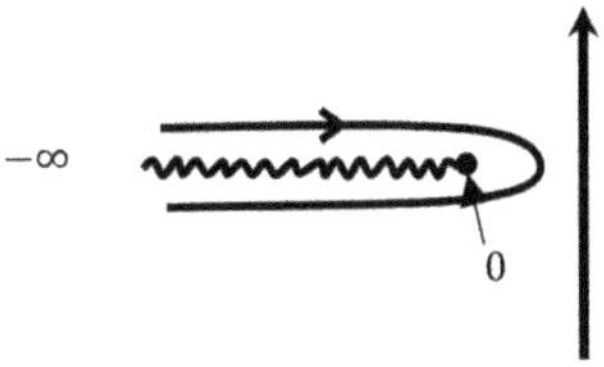

Figure 6.26: The contour for $\int_{-\infty}^{0^+} (\cdot) ds$.

If we shift the origin to $s' = 0$; $s' = s - s_0$, we will have

$$\begin{aligned} \phi(t) &\sim \frac{e^{s_0 t}}{2\pi i} \int_{-\infty}^{0^+} e^{s' t} (s')^n \ln s \, ds, \qquad n = \text{integer} \geq 0 \\ &= \frac{1}{2\pi i} \left[\int_0^\infty \underbrace{e^{-rt} (-r)^n (\ln r + i\pi)(-dr)}_{\text{on top: } s' = re^{i\pi} = -r} - \int_0^\infty e^{-rt} (-r)^n (\ln r - i\pi)(-dr) \right] e^{s_0 t}. \end{aligned}$$

Clearly the $\ln r$ cancels out and

$$\phi(t) \sim \left[- \int_0^\infty (-r)^n e^{-rt} \right] e^{s_0 t} = (-1)^{n+1} t^{-n-1} \Gamma(n+1).$$

So, the logarithm gives a result that is similar to that of a branch cut. Instead of having n as an integer, if $n = 1/2$, $(-r)^n$ will not cancel because it has its own singularity. We could use asymptotics to evaluate the behavior, using, for example, Watson's Lemma. We will have terms like $\int_0^\infty e^{-rt} r^a \ln r \, dr$.

To summarize, we have the rather interesting results:

$$\tilde{\phi}(s) \sim \sum_{n=0}^\infty a_n (s - s_0)^{n-1} + (s - s_0)^{\beta-1} \sum_{n=0}^\infty b_n (s - s_0)^n + \ln(s - s_0) \sum_{n=0}^\infty C_n (s - s_0)^n, \quad (6.15)$$

and

$$\phi(t) \sim e^{s_0 t} \left[a_0 + \sum_{n=0}^\infty b_n \frac{t^{-\beta-n}}{\Gamma(1 - \beta - n)} - e^{s_0 t} \sum_{n=0}^\infty (-1)^n C_n n! t^{-n-1} \right].$$

Many special functions have this type of behavior. Note that if we have the equal sign, "=," instead of the asymptotic sign, "~" in the equations above, we will encounter difficulties closing the contour.

Example 6.8 Consider the following to contrast with the function given earlier:

$$\mathcal{L}^{-1}\left\{ \frac{1}{\sqrt{s}(1-s)} \right\} = \phi(t).$$

In this case a pole is located at $s = 0$ and another one between 0 and γ. Thus, we are interested in the asymptotic behavior at $s = 0, 1$.

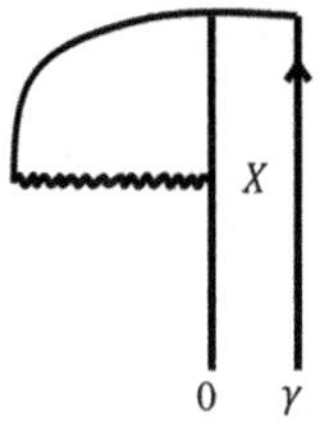

Figure 6.27: A pole situated between 0 and γ, or $s = 1$, and a branch point at $s = 0$.

We can close the contour at infinity because the integration on the large circle goes like $s^{1/2}$, $|e^s| \leq 1$; or, as per Jordan's Lemma (Figure 6.27):

$$\frac{1}{R^{1/2} R} \pi R \underbrace{e^{|Rt|}}_{\text{bounded}} \sim \frac{1}{R^{1/2}} \to 0 \text{ as } R \to \infty.$$

Note that

$$\mathcal{L}^{-1}\left\{\frac{1}{\sqrt{s}(1-s)}\right\} = \frac{1}{2\pi i}\int \frac{e^{st}ds}{(1-s)\sqrt{s}} = \phi(t).$$

The asymptotic behavior is given by

$$\phi(s) \sim \frac{1}{1-s}$$

or by e^{-at} near $s = 1$. However, near $s = 0$, we have

$$\tilde{\phi}(s) = \frac{1}{(1-s)\sqrt{s}} \sim \frac{1}{\sqrt{s}}\left[1+s+s^2+\ldots\right].$$

In the previous problem, near $s = 0$ we expanded the whole expression and the whole series is nonuniformly convergent. We cannot expand it beyond 1 because it has a radius of convergence of 1. This series is truly asymptotic. Other things remain the same, and this series is exactly of the form of eq. (6.15) above for s_0. In a real problem, we will not be bothered about the behavior at $s_0 = 0$, as the behavior at $s_0 = 1$ is far more important. Figure 6.28 depicts algebraic and asymptotic decay. In a realistic engineering or physics application, poles are found on the imaginary axis of the complex plane (Figure 6.29(a)), not at some random points such as in Figure 6.29(b). The latter will usually indicate instability in a realistic engineering problem.

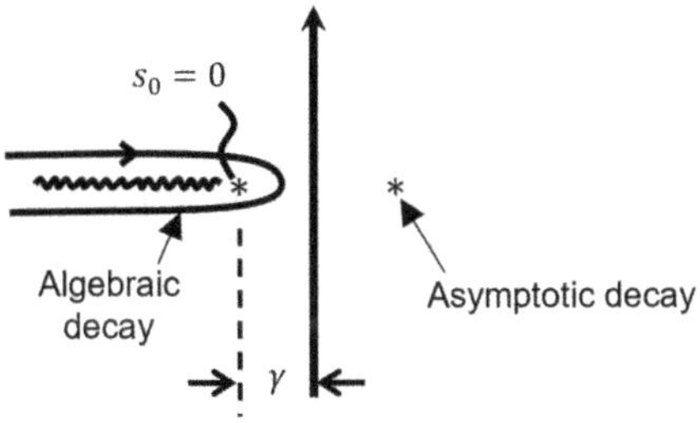

Figure 6.28: Algebraic decay versus asymptotic decay in the complex s-plane.

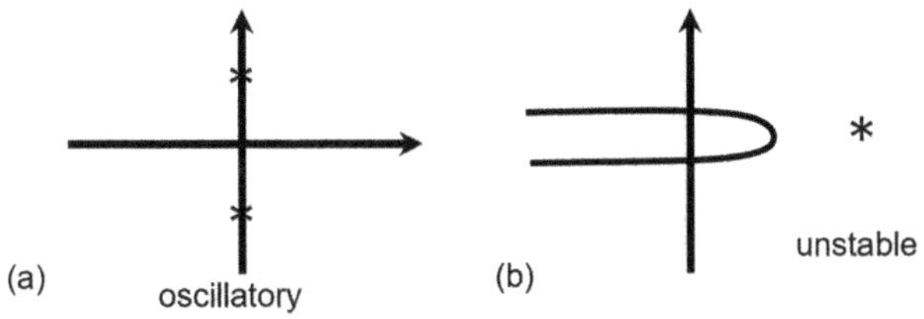

Figure 6.29: (a) Locations of poles in realistic engineering problems; (b) location of poles in unstable problems.

6.3 Asymptotics of integrals

The process of developing asymptotic series for integrals is taken up in the rest of this chapter. In Subsections 6.31. through 6.3.5, we give some relatively fundamental and direct analysis of the methods – integration by parts, Laplace method, method of stationary phase, and the method of steepest descent, while more advanced and heuristic analysis is presented in Subsections 6.36 through 6.3.10. Suggested reading for this topic includes References [1–10] at the end of this chapter.

6.3.1 Basic Analysis

We are generally interested in obtaining approximations or asymptotic expansions of integrals of the form

$$I(x) = \int_a^b \pm h(t) e^{\pm x p(t)} dt, \qquad x \to \infty,$$

where h, p are complex quantities and the limits "a", "b" are not necessarily "$\pm\infty$," in situations where the exponent in the integrand approaches a very large number. Asymptotic expansions are useful because we can drive the error $|f(z) - S_n(z)|$ to zero as $|z| \to \infty$ (independent of the number of terms n) even if the series of $f(z)$ that has the sum $S_n(z)$ is not convergent. Here, we contrast this with convergent series where $|f(z) - S_n(z)| \to 0$ as $n \to \infty$ but with z fixed.

This is important because using transforms, we can write the integral formulations of most of the differential equations of mathematical physics in the integral form, enabling us to obtain asymptotic solutions to the models represented by the differential equations. Take the zeroth-order Bessel's equation as an example:

$$x^2 y'' + xy' + x^2 y = 0, \qquad y(0) = 1.$$

(For a second-order ODE, we ordinarily need two initial conditions, but because we chose a singular point to investigate, only one arbitrary constant is needed for solution.) We can write the general solution of this Bessel's equation as

$$y = J_0(x) = \frac{1}{\pi} \int_0^\pi \cos(x \sin t) dt = \frac{1}{\pi} Re \int_0^\pi e^{ix \sin t} dt.$$

We are interested in determining how J_0 behaves with a large value of its argument. These equations will be written in the general form

$$I(x) = \int_a^b h(t) e^{x p(t)} dt,$$

where h, ρ are sometimes real, but $\rho(t) = \phi(t) + i\psi(t)$ in general. If $\psi(t) = 0$ so that $\rho(x)$ is real then $I(x)$ could have exponential growth. In this case, $I(x)$ is said to be a Laplace integral. The factor $h(t)$ could also be complex: $h(t) = f(t) + iq(t)$, but this is trivial to treat compared to ρ because it can always be broken up as the sum of two terms. The limits (a, b) can be anything, but will be restricted to the "zone of influence" later, and will often times be stretched to $(0, \infty)$.

We are going to explore four methods to obtain the asymptotic expansions: Integration by parts, Laplace method, Method of stationary phase, and the steepest descent method.

6.3.2 Integration by parts

In the case of Laplace integral, $I(x)$ evaluation involves real functions, and we can write the integral as follows:

$$I(x) = \int_a^b f(t) e^{x\phi(t)} dt.$$

The procedure is fairly elementary and is nothing more than integration-by-parts:

$$\begin{aligned} I(x) &= \int_a^b f(t) \underbrace{e^{x\phi(t)} dt}_{\frac{d\left(e^{x\phi(t)}\right)}{x\phi'(t)}} \\ &= \int_a^b \frac{f(t)}{x\phi'(t)} \cdot x\phi'(t) \cdot e^{x\phi(t)} dt, \quad u \equiv \frac{f}{x\phi'}, \quad v \equiv e^{x\phi} \end{aligned}$$

so that

$$dv = d\left(e^{x\phi(t)}\right) = x\phi'(t) e^{x\phi(t)} dt.$$

The basic integration-by-parts formula used is

$$\int u dv = uv - \int v du, \quad \text{or} \quad \int v\, du = uv - \int u\, dv.$$

Remark 6.1

a) x is just a parameter; t is the variable.

b) We have used the following:

$$d\left(e^{x\phi(t)}\right) = \frac{d}{dt} e^{x\phi(t)} dt = x\phi'(t) e^{x\phi(t)} dt \Rightarrow e^{x\phi(t)} dt = \frac{d\left(e^{x\phi(t)}\right)}{x\phi'(t)}.$$

c) Using $u = f$, $dv = e^{x\phi} dt$ would have led to the need to evaluate a complicated integral.

d) $\phi'(t)$ is assumed to not vanish in the interval of integration $[a, b]$, since it occurs in the denominator as shown here:

$$\int u\,dv = I(x) = \frac{f(t)}{x\phi'(t)} e^{x\phi(t)} \Bigg|_a^b - \frac{1}{x} \int_a^b e^{x\phi(t)} \frac{d}{dt}\left(\frac{f(t)}{\phi'(t)}\right) dt. \tag{6.16}$$

If the second integral exists and if it is asymptotically smaller than the first term (x is big $\Rightarrow 1/x \to 0$), then

$$I(x) \sim \frac{f(b)\overbrace{e^{x\phi(b)}}^{\circledast}}{x\phi'(b)} - \frac{f(a)e^{x\phi(a)}}{x\phi'(a)}.$$

We will assume that not both $f(a), f(b) = 0$; one will dominate. The larger of $\phi(a), \phi(b)$ in the exponentially growing term $\circledast$ will dominate because x is going to ∞. We will in practice need to find out whether integration-by-parts will work or whether a more rigorous method will be needed.

Example 6.9 Find the behavior for large x of the integral:

$$I(x) = \int_1^\infty e^{-xt^2} dt.$$

Solution:

$$I(x) = \int_1^\infty \underbrace{e^{-xt^2} dt}_{udv = \frac{d\left(e^{-xt^2}\right)}{-2xt}} = \underbrace{\frac{e^{-xt^2}}{-2xt}\Bigg|_{t=1}^{\infty}}_{uv} - \int_1^\infty \underbrace{e^{-xt^2}\left(-\frac{1}{2x}\right)\left(-\frac{dt}{t^2}\right)}_{v\,du} \tag{6.17}$$

\ vanishes at 0 (outside limits of integral)

Compared with the previous illustration, the following correspondence is evident:

$$\phi = -t^2;\ \phi' = -2t,\ \ f = 1,\ \ u = \frac{f}{x^2\phi'} = \frac{1}{-2xt} \Rightarrow du = +\frac{1}{2x} t^{-2} dt,$$

$$v = e^{x\phi} = e^{-xt^2},\quad dv = x\phi' e^{x\phi} dt = x(-2t)e^{-xt^2} dt.$$

Because of the appearance of t in the denominator, we would have gotten into trouble if the limits of the integral had involved $t=0$. At $x \to +\infty$, the first term in eq. (6.17) goes to zero. At $t=1$,

$$I(x) = \frac{e^{-x}}{2x} - \frac{1}{2x}\underbrace{\int_1^\infty \frac{1}{t^2} e^{-xt^2}\, dt}_{O(e^{-x}/x)}.$$

The claim is that $I(x) = e^{-x}/2x$. However, there is no reason to stop the evaluation at this level, and we wouldn't. In fact, we will build the asymptotic expansion by successively integrating $\int -v\, du$ by parts and picking up the "boundary" terms $(\cdot)|_a^b$ as the successive additions to the asymptotic series being developed for $I(x)$.

We now take the second term on the RHS of eq. (6.17) and integrate it by parts:

$$I_2(x) \equiv \frac{1}{2x}\int_1^\infty \frac{1}{t^2}\frac{d\left(e^{-xt^2}\right)}{-2xt}.$$

For this case, we have

$$f \equiv \tfrac{1}{t^2}, \quad \phi \equiv -t^2, \quad \phi' = -2t, \quad u \equiv \tfrac{f}{x\phi'} = \tfrac{1}{xt^2}\tfrac{1}{-2t},$$

$$uv| \equiv e^{-xt^2}\frac{1}{-2t^3 x}\Bigg|_1^\infty = -\frac{e^{-xt^2}}{e^{xt^3}},$$

so that

$$I_2(x) = -\frac{1}{2x}\frac{e^{-xt^2}}{2xt^3}\Bigg|_1^\infty + \frac{1}{2x}\underbrace{\int_1^\infty e^{-xt^2}\left(-3t^{-4}\right)dt}_{\text{hot}} = \frac{e^{-x}}{2x}\frac{1}{2x},$$

where "hot" stands for high order terms.

Therefore

$$\int_1^\infty e^{-xt^2}\, dt \approx \frac{e^{-x}}{2x}\left(1 - \frac{1}{2x} + O\left(\frac{1}{x^2}\right)\right).$$

We could continue the procedure whereby the new $f = -3t^4$, new $v =$ old v, new $du \equiv \frac{d}{dt}(u_{\text{old}})dt$, and so on. As a comment in passing, we note that standard integration

$$\int_1^\infty e^{-xt^2}\, dt = -\frac{1}{2xt}e^{-xt^2}\Bigg|_1^\infty = 0 + \frac{1}{2x}e^{-x} = \frac{1}{2x}e^{-x}$$

gives the leading order solution.

Failure of integration by parts

The approach based on integration by parts will fail under the conditions below:

(1) If $\phi'(t) = 0$ (i.e., a stationary point of ϕ in $[a, b]$), we get vanishing behavior.

(2) If $\frac{d}{dt}\left(\frac{f(t)}{\phi'(t)}\right) \to \infty$ in $[a, b]$ (see eq. (6.16)), we cannot have asymptotic behavior. For example, if $f(t) = t^{a}$, $0 < a < 1$ we will have problems as $t \to 0$. Thus, we go to an alternative, preferred, method: the Laplace method.

6.3.3 Laplace's method

In the integral

$$I(x) = \int_a^b f(t)e^{x\phi(t)}dt,$$

f and ϕ are real, which means that there are no issues with phases. The idea of the Laplace method is that as $x \to \infty$, $e^{x\phi(t)}$ has the largest contribution, where $\phi(t)$ is maximum and the integrand is restricted to that region. Obviously, the sharper the peak of the exponent in that region the more accurately it can represent the function. Both the f and ϕ are expanded about the maximum point.

6.3.3.1 Outline of the procedure

The Laplace method can be outlined as follows:

1. If $\phi(t)$ is max at $t = c$ (interior point, $c \in (a, b)$), then we replace $\int_a^b$ by $\int_{c-\varepsilon}^{c+\varepsilon}$. It is important that the critical point be interior for the Laplace method to work (see Figure 6.30). The exception is that if $c = a$, replace $\int_a^b$ by $\int_a^{a+\varepsilon}$, and if $c = b$, replace $\int_a^b$ by $\int_{b-\varepsilon}^b$.
2. Expand both $f(t)$ and $\phi(t)$ in series about $t = c$ and since this is an asymptotic process, truncate.
3. In order to evaluate the resulting integrals more easily, let $\varepsilon \to \infty$, that is, replace $\int_{c-\varepsilon}^{c+\varepsilon}$ by $\int_{-\infty}^{\infty}$, independent of the limits a and b in original integral; except when a or b coincides with c. If $c = a$, change $\int_a^{a+\varepsilon} \to \int_a^{\infty}$, without a change in the lower limit.

   The rational here is that $e^{x\phi}$ is the controlling factor because $x \to \infty$. Thus, the neighborhood where ϕ is maximum is where the main contribution is located. Another way of saying this is that (Figure 6.30)

$$\int_{-\varepsilon}^{\varepsilon} (\cdot)\, dt \gg \int_{-\infty}^{-\varepsilon} (\cdot)\, dt + \int_{\varepsilon}^{\infty} (\cdot)\, dt.$$

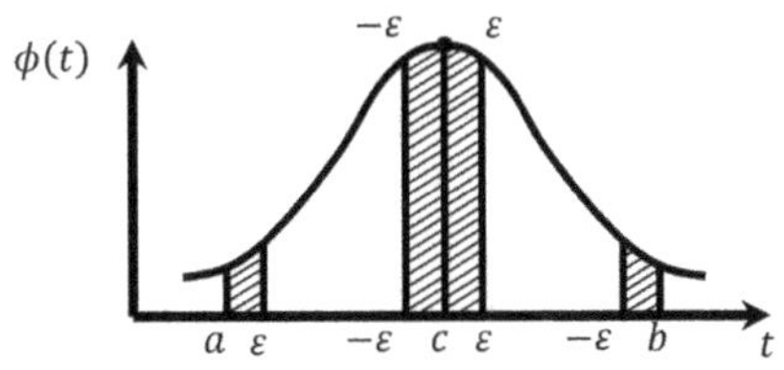

Figure 6.30: Integration limits in the Laplace method.

Example 6.10 Consider the problem of blood flow in the artery, which is governed by the Modified Bessel function, $K_n(x)$:

$$xy'' + xy' - (x^2 + n^2)y = 0, \quad y \to \infty, \; x \to 0.$$

The solution of this ODE can be written in terms of the modified Bessel function of order n of the second kind:

$$K_n(x) = \int_0^\infty \cosh nt \, e^{-x\cosh t} dt.$$

In terms of our notations, $\phi(t) = -\cosh t$ (which achieves its maximum at $t = 0$), $\phi'(t) = -\sinh t$, which will be zero in $[0.\infty)$, that is, at 0 so that Integration-by-Parts breaks down).

A sketch of the functions $\cosh t$ and $-\cosh t$ is shown in Figure 6.31.

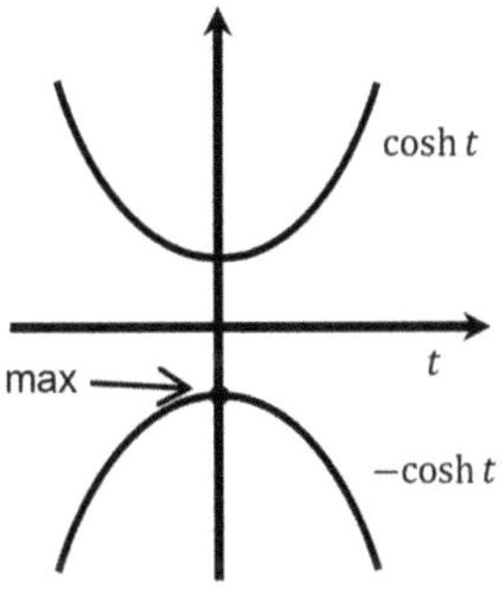

Figure 6.31: Sketch of $\cosh t$ and $-\cosh t$.

So we will have

$$K_n(x) \sim \int_{0 \,\leftarrow\, c \equiv a = 0}^{\varepsilon \,\leftarrow\, c+\varepsilon=\varepsilon} \underbrace{\cosh nt}_{f} \, e^{-x \overbrace{\cosh t}^{\phi}} dt.$$

The next step is to expand f and ϕ in series about c:

$$\cosh u = 1 + \frac{u^2}{2} + \ldots$$

so that

$$K_n(x) \sim \int_0^{\varepsilon} \left(1 + \frac{n^2t^2}{2} + \ldots\right) e^{-x\left(1+\frac{t^2}{2}+\ldots\right)} dt \sim e^{-x} \int_0^{\varepsilon} (1 + \ldots) e^{-\frac{xt^2}{2}} dt, \tag{6.18}$$

where we have used

$$e^{-x\left(1+\frac{t^2}{2}+\ldots\right)} = e^{-x} e^{-\frac{xt^2}{2}+\ldots}.$$

We will need to replace ε with ∞ because $\int_{\varepsilon}^{\infty}(\cdot)$ will be asymptotically smaller than $\int_0^{\varepsilon}$. It will prove to be useful to reduce eq. (6.18) to gamma function in order to carry out the resulting integration problem.

In the Laplace procedure, only the limit changes; based on c or where $\phi'(t) = 0$, and judicious extension of the limit of the integral to (say) ∞. In particular, we do not substitute $\phi(c)$ into the exponent $e^{\phi(t)}$. We are looking for a series solution. The quantity $n^2t^2/2!$ will be small compared to unity over the integration range $(0, \infty)$ because the evaluation is being done in the vicinity of $t = c = 0$ or $t \ll 1$.

Thus,

$$K_n(x) \sim e^{-x} \int_0^{\varepsilon} e^{-\frac{xt^2}{2}} dt.$$

Note that $e^{-xt^2/2}$ has its maximum at zero (Figure 6.32), so it is still relatively difficult to treat. The integrand suggests a gamma function:

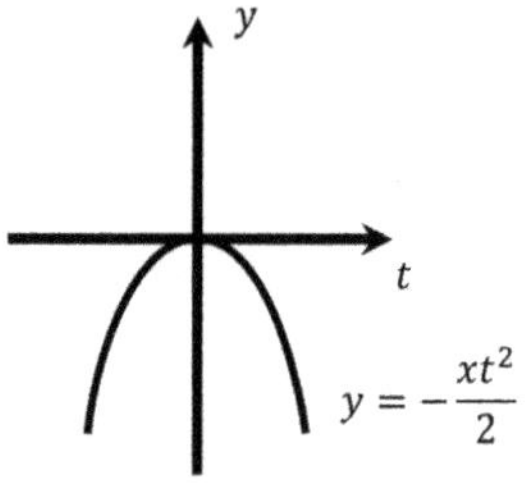

Figure 6.32: The function $y = -\frac{xt^2}{2}$.

$$K_n(x) \sim e^{-x} \int_0^{\infty} e^{-\frac{xt^2}{2}} dt,$$

where, again, we have added $\int_\varepsilon^\infty$, which is much less than $\int_0^\infty$.

The general problem will always have an integrand factor like $e^{-xt^2/2}$, maybe t^3 in place of t^2, so the approach is the same. As a result, we usually explore the introduction of the gamma function, whose integrand can usually be obtained from the one here by a simple change of variables.

For example, we could set $u = xt^2/2$ and replaced t by u so that

$$du = xt\ dt \Rightarrow\ t = \sqrt{\frac{2u}{x}} \Rightarrow du = x\left(\frac{2u}{x}\right)^{1/2} dt \Rightarrow dt = x^{-1}\left(\frac{2u}{x}\right)^{-1/2} du.$$

The limits of the integrand transform as follows: $t = 0 \Rightarrow u = 0$, $t = \infty \Rightarrow u = \infty$ so that

$$\int_0^\infty e^{-\frac{xt^2}{2}} dt = \int_0^\infty e^{-u} \frac{du}{x\sqrt{2u/x}} = \frac{1}{\sqrt{2x}} \int_0^\infty u^{-\frac{1}{2}} e^{-u}\, du.$$

6.3.3.2 A few elementary properties of the Gamma function

The Gamma function is defined as

$$\Gamma(z) \equiv \int_0^\infty u^{z-1} e^{-u} du.$$

(To help the reader remember the integrand of the Gamma function, note that the exponential factor is just sitting there, raised to the negative power of the dummy variable of integration, with absolutely nothing to do with the argument z of Γ.) The behavior of the integrands in the definition of $\Gamma(z)$ is shown in Figure 6.33.

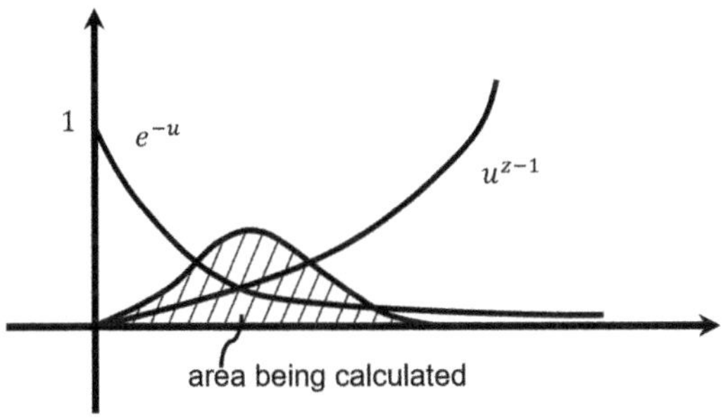

Figure 6.33: Graph of e^{-u} and u^{z-1} as a function of u.

The gamma function comes in most frequently in calculating factorials. From the definition

$$\Gamma(z+1) \equiv \int_0^\infty u^z e^{-u} du = -u^z e^{-u}|_0^\infty + \int_0^\infty e^{-u} z u^{z-1} du$$

we see that

$$\Gamma(z+1) = z\Gamma(z).$$

Note that z does not have to be an integer. For example,

$$\Gamma\left(\frac{3}{2}\right) = \Gamma\left(1+\frac{1}{2}\right) = \frac{1}{2}\Gamma\left(\frac{1}{2}\right), \qquad \Gamma(1) = \int_0^\infty e^{-u} du = -e^{-u}|_0^\infty = +1.$$

Also note that in $\int u^{-1/2} e^{-u} du$, $z-1=-1/2 \Rightarrow z=1/2 \Rightarrow \Gamma(1/2)$.

The following relations are obvious:

$$\Gamma(2)=1,\ \Gamma(3)=2\Gamma(2)=2,\ \Gamma(4)=3\Gamma(3)=3!,\ \Gamma(n+1)=n!\ \text{(Integers only)}.$$

For the last identity, note that

$$\Gamma(n+1) = n\Gamma(n) = n(n-1)\Gamma(n-1) = n(n-1)(n-2)\Gamma(n-2) = \ldots = n!$$

The variation of $\Gamma(n+1)$ as a function of n is shown in Figure 6.34.

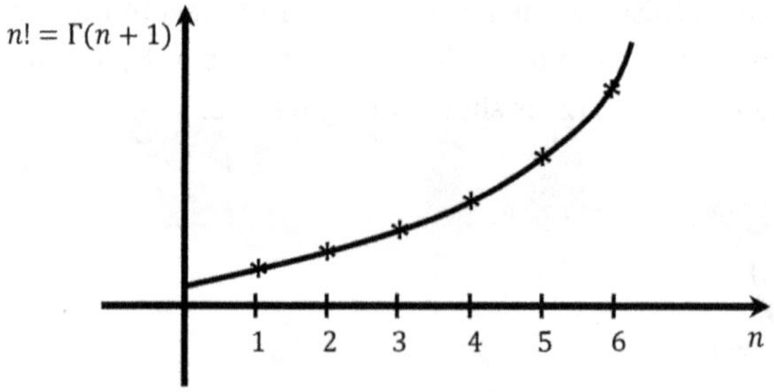

Figure 6.34: Graph of $\Gamma(n+1)$ as a function of n.

We can show that

$$\frac{K_n(x)}{e^{-x}} = \frac{1}{\sqrt{2x}}\Gamma\left(\frac{1}{2}\right)$$

and evaluate $\Gamma(1/2)$ as follows. From the definition of the gamma function, we have

$$\Gamma\left(\frac{1}{2}\right) = \int_0^\infty u^{-\frac{1}{2}} e^{(-u)} du.$$

With the change of variable, $u = s^2$, $s = \sqrt{u}$, $du = 2s\ ds$, we can write

$$\Gamma\left(\frac{1}{2}\right) = \int_0^\infty e^{-s^2} \frac{1}{s} 2s\ ds = 2\int_0^\infty e^{-s^2} ds \Rightarrow \Gamma\left(\frac{1}{2}\right) = \int_{-\infty}^{\infty} e^{-s^2}\ ds$$

since the integrand is an even function. Squaring both sides, we will have the following (Figure 6.35):

$$\left[\Gamma\left(\frac{1}{2}\right)\right]^2 = \left(\int_{-\infty}^{\infty} e^{-s^2} ds\right)\left(\int_{-\infty}^{\infty} e^{-s_1^2} ds_1\right)$$

$$= -\int_{-\infty}^{\infty}\int_{-\infty}^{\infty} e^{-(s^2+s_1^2)} \underbrace{ds ds_1}_{\substack{dA = dxdy \equiv r\,drd\theta \\ s^2+s_1^2 \equiv x^2+y^2 \equiv r^2}} = \int_0^{2\pi}\int_0^{\infty} e^{-r^2} r\,dr\,d\theta.$$

Use the transformation

$$v = r^2,\ dv = 2r\,dr.$$

Note that, not minding the transformation to polar coordinates, we still have the same $(-\infty < x < \infty,\ -\infty < y < \infty)$ domain since $|x| \to \infty$, $|y| \to \infty \Rightarrow r \to \infty$ and the angular measure θ has a range of 2π even at infinite radius.

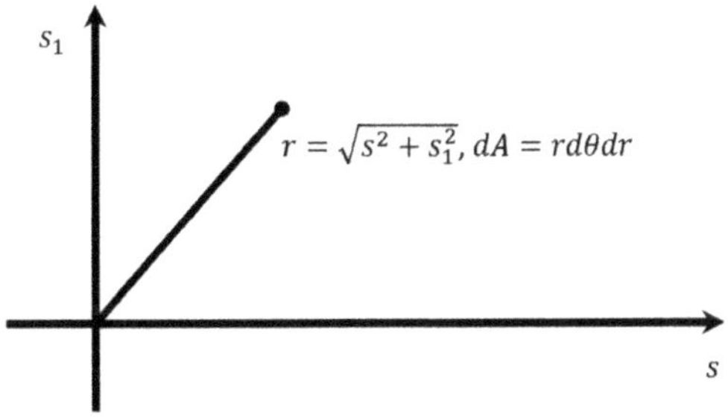

Figure 6.35: Elemental area is equal to the area of sector.

Thus,

$$\left[\Gamma\left(\frac{1}{2}\right)\right]^2 = 2\pi\int_0^\infty \frac{e^{-v}dv}{2} = \pi \Rightarrow \Gamma\left(\frac{1}{2}\right) = \sqrt{\pi}.$$

The graph of $\Gamma(z)$ is shown in Figure 6.36.

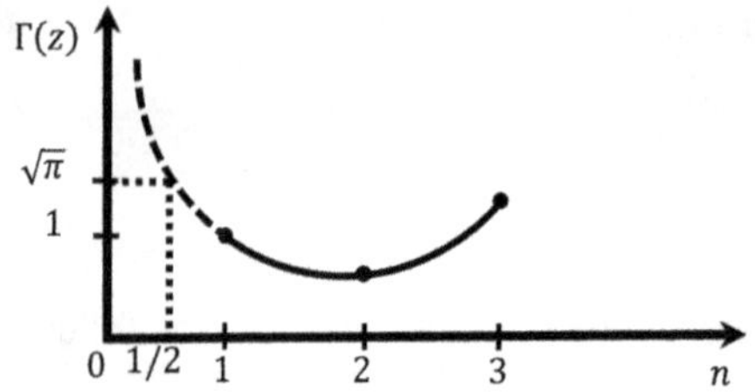

Figure 6.36: The graph of Γ(z).

It should be noted in passing that

$$\Gamma(0) = \int u^{0-1} d^{-u} du = \int_0^{\infty} \frac{1}{u} e^{-u} du = \infty.$$

\

blows up in the neighborhood of the lower limit

Coming back to the Modified Bessel function asymptotics, we have

$$K_n(x) \sim e^{-x} \sqrt{\frac{\pi}{2x}}.$$

Example 6.11 Express $I = \int_0^{\infty} e^{-x^4} dx$ in terms of the gamma function.

Solution:

$$I = \int_0^{\infty} e^{-x^4} dx = \int_0^{\infty} e^{-t} tx^{-\frac{3}{4}} \frac{dt}{4} = \frac{\Gamma(1/4)}{4},$$

where we have used $t = x^4$.

6.3.3.3 Watson's Lemma

Consider the integral

$$I(x) = \int_0^{b} f(t) e^{-xt}\, dt,$$

where

$$f(t) = t^{\alpha} \sum_{n=0}^{\infty} a_n\, t^{\beta_n}.$$

The reader should observe that $I(x)$ becomes the expression for the LT of $f(t)$ if the upper limit b is replaced by "∞." In that case, x is the LT variable which we often denote by the symbol s in this book. In Section 6.2.2.4 we obtained an asymptotic expansion, identified as Watson's Lemma, for $I(x)$ in the limit of $x \to \infty$, in order to determine the small time behavior of $f(t)$. In that section, $I(x)$ was identified as the LT of f, or $\mathcal{L}\{f(t)\}$. Here, we use the Laplace method to arrive at essentially the same result, or eq. (6.11). In the current procedure, the expansion of $f(t)$ is valid in $t \to 0$ because e^{-xt} has substantial value only in a neighborhood of 0 (and $f(t)$ dwarfs compared to e^{-xt} between 0 and b).

Where does $\phi(t) = -t$ achieve its maximum (between 0 and b)? Figure 6.37 shows that the maximum occurs at $t = 0$ for $t \in [0, b]$. That is, $f(t)$ does not show your usual type of maximum where $d/dt = 0$ since ϕ is a linear function t. Thus, $I \sim \int_0^\varepsilon f(t)e^{-xt}dt$.

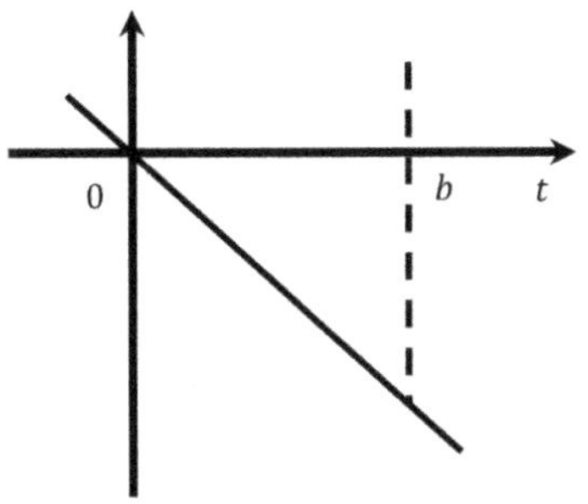

Figure 6.37: The function $f(t)$, with maximum at $t = 0$.

If we plug in the expression for $f(t)$, we will have

$$I(x) = \int_0^\varepsilon \left(t^\alpha \sum_{n=0}^{\infty} a_n t^{\beta_n}\right) e^{-xt}\, dt.$$

We need not be concerned about interchanging $\sum$ and $\int$ because we have convergence of the series $\sum a_n t^{\beta_n}$. Thus,

$$I(x) \sim \int_0^\varepsilon \left(t^\alpha \sum_{n=0}^{\infty} a_n t^{\beta_n}\right) e^{-xt} dt = \sum a_n \int_0^\infty t^{\alpha+\beta_n} e^{-xt}\, dt. \tag{6.19}$$

As noted previously, an important part of this analysis is the replacement of $\int_0^\varepsilon$ by $\int_0^\infty$, which we can do because $\int_\varepsilon^\infty \ll \int_0^\varepsilon$ asymptotically. We now move to introduce the Γ function. In eq. (6.19), let $u = xt \Rightarrow du = xdt$, $dt = du/x$, $t = u/x$, $t^{\alpha+\beta_n} = (u/x)^{\alpha+\beta_n} = u^{\alpha+\beta_n}/x^{\alpha+\beta_n}$ so that

$$\int_0^\infty t^{\alpha+\beta_n} e^{-xt} dt = \int_0^\infty \frac{u^{\alpha+\beta_n}}{x^{\alpha+\beta_n}} e^{-u} \frac{du}{x} = \frac{1}{x^{\alpha+\beta_n+1}} \Gamma(\alpha + \beta_n + 1).$$

Thus, $I(x) = \int_a^b f(t)e^{-xt}dt, f(t) = t^{\alpha}\sum_{n=0}^{\infty} a_n t^{\beta_n}$ has the asymptotic expansion

$$I(x) \sim \sum_{n=0}^{\infty} \frac{a_n}{x^{\alpha+\beta_n+1}}\Gamma(\alpha+\beta_n+1), \; x \to \infty.$$

This is Watson's Lemma. This is also a series for the LT of $f(t)$ when $f(t)$ itself is expressed in the series form above in the limit of the LT variable approaching infinity.

6.3.3.4 Using Laplace method to get high order terms

The Laplace method can be used to obtain high order terms in the asymptotic series. The following example illustrates this.

Example 6.12 Investigate the other solution (first kind, order n) to modified Bessel's equation:

$$x^2y'' + xy' - (x^2 + n^2)y = 0,$$

or

$$I_n(x) = \frac{1}{\pi}\int_0^{\pi} e^{x\cos t}\cos nt \, dt.$$

The function $\cosh t$ attains a maximum at $t = 0$ (Figure 6.38) within the range shown in the integral, and we cannot use integration by parts because $\phi'(t) = -\sin t$ blows up at $t = 0$. Now

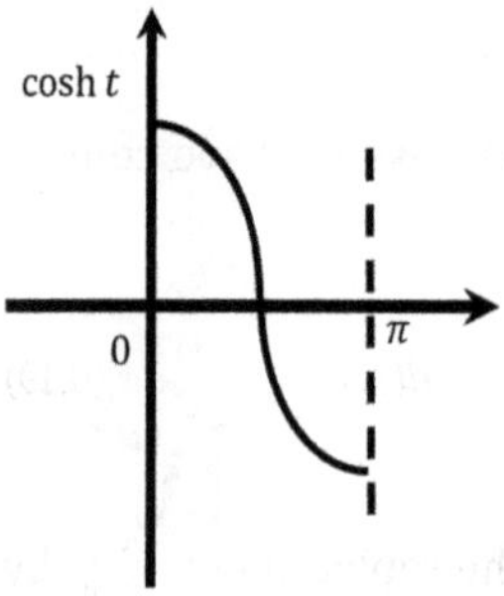

Figure 6.38: cosh t.

$$I_n(x) \sim \frac{1}{\pi}\int_0^{\varepsilon} e^{x\cos t}\cos nt \, dt, \qquad \cos u = 1 - \frac{u^2}{2!} + \frac{u^4}{4!} + \ldots$$

so that

$$I_n(x) \sim \frac{1}{\pi}\int_0^{\varepsilon} e^{x\left(1-\frac{t^2}{2}+\frac{t^4}{24}+\cdots\right)}\left(1-\frac{n^2t^2}{2}+\frac{n^4t^4}{24}+\ldots\right)dt$$

$$\sim \frac{e^x}{\pi}\int_0^{\varepsilon} e^{-\frac{xt^2}{2}}e^{\frac{xt^4}{24}}\ldots\left(1-\frac{n^2t^2}{2}+\frac{n^4t^4}{24}+\ldots\right)dt.$$

Since t has significant values around $t=0$, higher order t terms go to zero asymptotically. We will leave the leading term in exponential form and expand the other terms:

$$I(x) \sim \frac{e^x}{\pi}\int_0^{\varepsilon} e^{-x\frac{t^2}{2}}\left(1+\frac{xt^4}{24}+\ldots\right)\left(1-\frac{n^2t^2}{2}+\frac{n^4t^4}{24}\right)dt.$$

Let us change variables: $xt^2/2=u$, $t^2=2u/x$ so that asymptotically

$$I_n(x) \sim \frac{e^x}{\pi}\int_0^{\varepsilon} \frac{e^{-u}}{\sqrt{2ux}}\left(1+\frac{x}{24}\frac{4u^2}{x^2}+\ldots\right)\left(1-\frac{n^2 2u}{2x}+\frac{n^4 4u^2}{x^2}+\ldots\right)du,$$

where we have used $xt\,dt=du$, $dt=\frac{du}{x\sqrt{2u/x}}=\frac{du}{\sqrt{2ux}}$. The behavior of $\cosh nt$ in the vicinity of zero is shown in Figure 6.39.

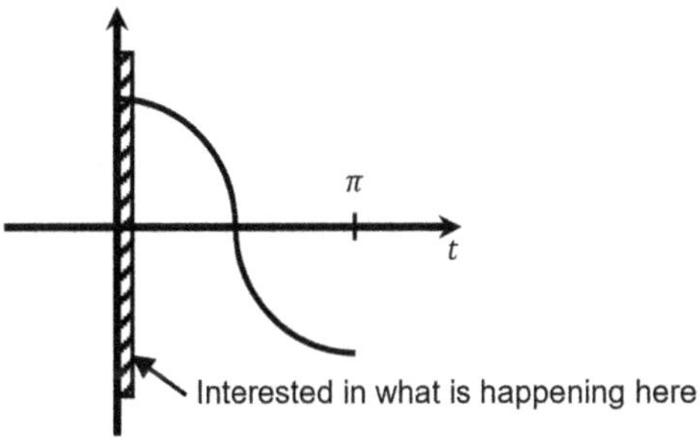

Figure 6.39: $\cosh nt$ in the vicinity of zero.

We want to get the same power of the two series, irrespective of how many terms we have to take.

$$I_n(x) \sim \frac{1}{\pi}\int_0^{\varepsilon} e^{x\left(1-\frac{t^2}{2!}+\frac{t^4}{4!}+\cdots\right)}\left(1-\frac{n^2t^2}{2!}+\frac{n^4t^4}{4!}\right)$$

$$\sim \frac{e^x}{\pi}\int e^{-\frac{xt^2}{2}}e^{\frac{xt^4}{24}-\frac{xt^6}{6!}}\left(1-\frac{n^2t^2}{2!}+\frac{n^4t^4}{4!}+\ldots\right)dt$$

$$\sim \frac{e^x}{\pi}\int_0^{\infty} e^{-\frac{xt^2}{2!}}\left(1+\frac{xt^4}{24}\right)\left(1-\frac{n^2t^2}{2}+\frac{n^4t^4}{24}+\ldots\right)\frac{du}{\sqrt{2ux}}.$$

With the change of variables: $e^{-\frac{xt^2}{2}} \equiv e^{-u} \Rightarrow u = xt^2/2,\, du = xtdt$, we obtain

$$I(x) \sim \frac{e^x}{\pi}\int_0^\infty e^{-u}\frac{du}{\sqrt{2ux}}\underbrace{\left(1+\frac{1}{24}\frac{4u^2}{x}+\dots\right)\left(1-\frac{n^2}{2}\frac{2u}{x}+\frac{n^4}{24}\frac{4u^2}{x^2}\right)}_{1+\frac{u^2}{6x}-\frac{n^2u}{x}+O\left(\frac{1}{x^2}\right)}dt.$$

Thus,

$$I_n(x) \sim \frac{e^x}{\sqrt{2x\pi}}\int_0^\infty e^{-4}\,du\left(u^{-\frac{1}{2}}+\frac{u^{3/2}}{6x}-\frac{n^2u^{1/2}}{x}+O\left(\frac{1}{x}\right)\right).$$

Noting that $\Gamma(\frac{5}{2}) = \Gamma(1+\frac{3}{2}) = \frac{3}{2}\Gamma(\frac{3}{2}) = \frac{3}{2}\Gamma(1+\frac{1}{2}) = \frac{3}{2}\cdot\frac{1}{2}\Gamma(\frac{1}{2})$ and $\Gamma(\frac{3}{2}) = \Gamma(1+\frac{1}{2}) = \frac{1}{2}\Gamma(\frac{1}{2})$, we can write

$$\begin{aligned} I_n(x) &\sim \frac{e^x}{\sqrt{2x\pi}}\left[\Gamma\left(\frac{1}{2}\right)+\frac{\Gamma(5/2)}{6x}-\frac{n^2\Gamma(3/2)}{x}+O\left(\frac{1}{x^2}\right)\right] \\ &\sim \frac{e^x}{\sqrt{2x\pi}}\left[(1+\frac{1}{x}\left(\frac{1}{8}-\frac{n^2}{2}\right)+\dots\right]. \end{aligned}$$

Just like integration by parts, we can push this method to the desired accuracy, but the leading order behavior is

$$I_n(x) \sim \frac{e^x}{\sqrt{2x\pi}}\left(1+O\left(\frac{1}{x}\right)\right).$$

We can always check our results against numerical integration.

6.3.3.5 Stirling's formula (the behavior of gamma function as $x \to \infty$)

The Stirling's formula

$$\Gamma(x) = x^x\int_0^\infty \frac{e^{x(\ln s - s)}}{s}\,ds$$

is used in testing convergence of series, for example, in conjunction with the ratio test. Its use here pertains to determining how the gamma function behaves as $x \to \infty$. To accomplish this, we write $\Gamma(x)$ in Laplace form so that we will be able to determine the behavior as $x \to \infty$. Note that because "$f(t)$" in the integrand has $e^{(\ )}$ in it, the behavior of $I(x)$ is no longer determined only by $e^{\phi(c)}$, where $\phi'(c) = 0$. One problem with this is that the maximum of $\phi(t)$ varies with x (i.e., we have a moving maximum). We resolve this by defining a new variable $s(x)$ with maximum $\phi(s)$ occurring at $s = 1$. We can write the gamma function as

$$\Gamma(x) = \int_0^\infty t^{x-1} e^{-t} dt = \int_0^\infty \frac{e^{-t}}{t} e^{x \ln t} \, dt.$$

We are interested in the value of t for which $x \ln t - t$ has its maximum:

$$\int_0^\infty \frac{e^{x \ln t - t}}{t} dt \equiv \int_0^\infty h(t) e^{x\phi(t)} dt \quad \left(\phi = \ln t - \frac{t}{x} = \phi(x,t)\right).$$

To determine where the maximum of the exponent occurs, we take derivative and set it to zero:

$$\frac{d}{dt}(x \ln t - t) = 0 \Rightarrow \frac{x}{t} - 1 = 0, \text{ or } t = x \equiv A(x).$$

We need to check that we have a maximum, not minimum. To do this, we examine the second derivative, noting that $x \neq x(t)$ and $x > 0$:

$$\frac{d^2}{dt^2}(x \ln t - t) = -\frac{x}{t^2} < 0,$$

which shows that we indeed have a maximum. Now we need to device a new variable s that will transform the integral into the Laplace form. The idea is to scale t so that the maximum of t (or $t = x$) occurs at $s = 1$ (so that it is now independent of x). Assume $s = k(x)t$ with the condition:

$$s_{\max} = k(x) t_{\max} \Rightarrow 1 = k(x)x \Rightarrow k(x) = \frac{1}{x} \Rightarrow s = \frac{t}{x} \Rightarrow ds = \frac{1}{x} dt.$$

This gives us the location of the maximum, so we don't need to calculate it. We will have

$$\begin{aligned} \int_0^\infty \frac{e^{x \ln t - t}}{t} dt &= \int_0^\infty \frac{e^{x \ln(sx) - sx}}{sx} x \, ds = \int_0^\infty \frac{e^{x(\ln x + \ln s) - sx}}{s} ds \\ &= e^{x \ln x} \int_0^\infty \frac{e^{x(\ln s - s)}}{s} ds = x^x \int_0^\infty \frac{e^{x(\ln s - s)}}{s} ds; \ \phi = \ln s - s. \end{aligned}$$

Thus, Stirling's formula is just the first term in an asymptotic expression. For more accuracy, we need to take more terms. The maximum ($\varphi'(s) = 1/s - 1 = 0$) occurs at $s = 1$; therefore the limits on the integral can be specified as follows:

$$\int_0^\infty \frac{e^{x \ln t - t}}{t} dt \sim x^x \int_{1-\varepsilon}^{1+\varepsilon} \frac{e^{x(\ln s - s)}}{s} ds.$$

We then carry out a Taylor series expansion about $s = 1$ (where s is maximum) and in powers of $(s-1)$:

$$\ln s = \ln 1 + (s-1) - \frac{(s-1)^2}{2} + \frac{2(s-1)^3}{2\cdot 3}.$$

We then plug it in for $\ln s$ above:

$$\int_0^\infty \frac{e^{x\ln t - t}}{t}\,dt \sim x^x \int_{1-\varepsilon}^{1+\varepsilon} \frac{e^{x\left(-1-\frac{(s-1)^2}{2}+\cdots\right)}}{1+(s-1)}\,ds \sim x^x e^{-x} \int_{1-\varepsilon}^{1+\varepsilon} \frac{e^{-\frac{x(s-1)^2}{2}}}{1+(s-1)}\,ds.$$

Using

$$s-1 = u,\ \ s = 1 \Rightarrow u = 0,\ \ s = 1-\varepsilon \Rightarrow\ u = -\varepsilon,\ \text{and}\ s = 1+\varepsilon \Rightarrow\ u = \varepsilon,$$

we will obtain

$$\int_0^\infty \frac{e^{x\ln t - t}}{t}\,dt \sim x^x e^{-x} \int_{-\varepsilon}^{\varepsilon} e^{-\frac{xu^2}{2}}\,du.$$

We simplify the computation of the integral by adding a term that is asymptotically smaller than the leading term $\left(\int_{-\infty}^{-\varepsilon} + \int_{\varepsilon}^{\infty} \ll \int_{-\varepsilon}^{\varepsilon}\right)$:

$$\int_0^\infty \frac{e^{x\ln t - t}}{t}\,dt \sim x^x e^{-x} \int_{-\infty}^{\infty} e^{-\frac{xu^2}{2}}\,du \sim 2x^x e^{-x} \int_0^\infty e^{-\frac{xu^2}{2}}\,du.$$

Note that we have used the fact that $f \equiv e^{-xu^2/2}$ is an even function $(f(u) = f(-u))$ in the above equation.

Let

$$\frac{xu^2}{2} = v \Rightarrow du = \frac{dv}{ux} = \frac{dv}{\sqrt{2vx}}.$$

Therefore,

$$\begin{aligned}\int_0^\infty \frac{e^{x\ln t - t}}{t}\,dt &\sim 2x^x e^{-x} \int_0^\infty e^{-v} \frac{dv}{\sqrt{2vx}} \\ &= \frac{2x^x e^{-x}}{\sqrt{2x}} \sqrt{\pi};\left(\Gamma\left(\frac{1}{2}\right) = \sqrt{\pi}\right) = \sqrt{2\pi x}x^x e^{-x} = \Gamma(x), \\ &x \to \infty \text{ (This is the Stirling formula).}\end{aligned}$$

That is,

$$\Gamma(x) = \int_0^\infty e^{-u}u^{x-1}du \ (\text{any}\, x) = x^x e^{-x}\sqrt{2\pi x} \ (\text{when}\, x \to \infty).$$

To recap, the expression

$$\Gamma(x) = \int_0^\infty \frac{1}{t} e^{x \ln t - t} dt$$

hits maximum when $t = x$, which is a moving maximum. We have used the transformation $s = k(x)t$, $s = 1$, $t = x \Rightarrow k = 1/x$ to put the equation in the form

$$\Gamma(x) = x^x \int_0^\infty \frac{1}{s} e^{x(\ln s - s)} \, ds = x^x \int_{1-\varepsilon}^{1+\varepsilon} \left(\frac{1}{s} e^{x(\ln s - s)}\right) ds.$$

This is in the Laplace form and $s = 1$ is the maximum point. A change of variables ($s = u + 1$) yields

$$\Gamma(x) \sim x^x e^{-x} \int_{-\varepsilon}^{\varepsilon} \frac{1}{u+1} e^{x(\ln(1+u)-u)} du.$$

Note that x has been factored out in the exponent: $e^{x[\cdot]}$, and $f(t)$ in the Laplace procedure is not of exponential form. Then we Taylor expand about $u = 0$ and change limits to $-\infty$ to ∞, keeping the leading term. We obtain

$$\Gamma(x) \sim \ x^x e^{-x} \int_{-\infty}^{\infty} \frac{1}{u+1} e^{x(\ln(1+u)-u)} du \sim \ x^x e^{-x} \int_{-\infty}^{\infty} \frac{1}{u+1} e^{x(\ln(1+u)-u)+\frac{u^2}{2}} du \ e^{-\frac{xu^2}{2}}$$

$$\equiv \int_{-\infty}^{\infty} e^{-\frac{xu^2}{2}} G(x,u) du; \ G \equiv \frac{1}{u+1} e^{x(\ln(1+u)-u)+\frac{u^2}{2}}$$

In the previous subsections, we analyzed the integral

$$\lim_{x\to\infty} \int_a^b h(t) e^{x\rho(t)} dt$$

for ρ real. However, we are ultimately interested in ρ complex, which is the case that we are going to treat next.

6.3.4 Method of stationary phase

Unlike the Laplace method, here we need the ψ term in $\rho(t) = \phi(t) + i\psi(t)$. Phase implies the existence of the sinusoidal behavior implied by raising e to a pure imaginary number, as per the Euler formula. The method of stationary phase is useful when the phase ψ is stationary or $\psi'(t) = 0$ in $[a, b]$; hence the description "stationary phase." As an illustration, consider the behavior of $J_0(x)$, the Bessel function of order zero of the first kind, as $x \to \infty$. The integral representation is

$$J_0(x) = \frac{1}{\pi} \int_0^{\pi} \cos(x \sin t) dt.$$

Using the Euler formula, $e^{ix\psi} = \cos x\psi + i \sin x\psi$, we have

$$J_0(x) = \frac{1}{\pi} Re \int_0^{\pi} e^{ix \sin t} dt,$$

which is of the form $\int_a^b f(t) e^{ix\,\psi(t)} dt$.

The Laplace method does not work for this because the problem is not of Laplace type. (Laplace is for real $h(\equiv f)$ and $\rho(\equiv \phi)$) As $x \to \infty$ the frequency will tend to infinity (Figure 6.40) and the waveform oscillates more and more rapidly. Note that we can talk of frequency because of the sinusoidal behavior implied by the phase (sine and cosine): $e^{ix\psi} = \cos x\psi + i \sin x\psi$ so that the frequency $f' = x\psi$ and the wavelength is $2\pi/x\psi$. We see that the frequency becomes very large when $x \to \infty$.

A high frequency of oscillation causes neighboring "ups and downs" to cancel out. We will not have cancelling areas where the frequency is very small or $\psi'(t) = 0$, as adjacent areas will not be close enough to cancel out. So the neighborhood of interest is where the frequency is small and things don't cancel out. The method of stationary phase makes use of the self-cancelling of oscillations (at high frequencies) rather than the decay of the exponential factor in the integrand.

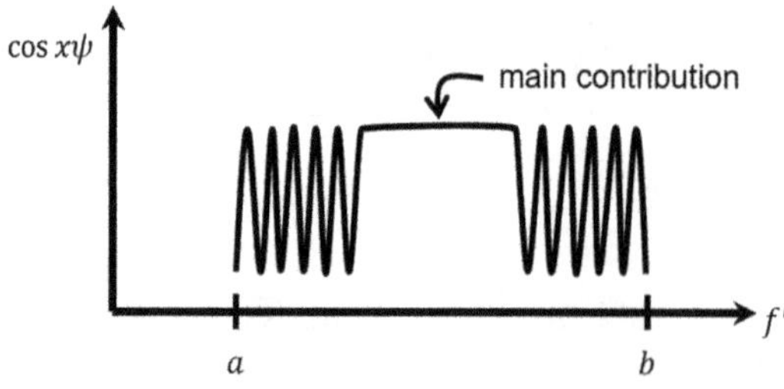

Figure 6.40: Exaggerated high and low frequencies of $\cos x\psi$, showing the main region of the contribution, which is at the low frequencies, where $\psi'(t) = 0$.

Why not use integration-by-parts? Well, integration-by-parts is applicable except when $\psi'(t) = 0$ in $[a, b]$. For example, with integration-by-parts:

$$\int_0^\pi e^{ix\sin t}dt = \int_0^\pi \frac{d(e^{ix\sin t})}{ix\cos t} = \left.\frac{e^{ix\sin t}}{ix\cos t}\right|_0^\pi - \int_0^\pi e^{ix\sin t}\frac{d}{dt}\left(\frac{1}{ix\cos t}\right)dt,$$

with $ix\cos t = 0$ when $t = \pi/2$, leading to problems. However, over some other intervals this method works. Above, we have used the following:

$$f(t) = 1,\ \psi = \sin t,\ u = \frac{f}{ix\psi'(t)} = \frac{1}{ix\psi'},\ v = e^{ix\psi},\ dv = ix\psi' e^{ix\psi},\ du = \frac{1}{ix}\frac{d}{dt}\left(\frac{1}{\psi'}\right)dt$$

in

$$\int udv = uv| - \int vdu.$$

In this method we can't throw away a term on the basis of large $\phi(a)$, $\phi(b)$ as $x \to \infty$. This represents a difference between the integration-by-parts here and before. Afterall, $|e^{ix\psi}| = 1$ even as $x \to \infty$.

Suppose that the oscillation is slowest, that is, it has smallest frequency at $t = t_0$, we will expand $\psi(t)$ in Taylor series about this point. Or

$$e^{ix\psi(t)} = e^{ix[\psi(t_0) + (t - t_0)\psi'(t_0) + \dots]}$$

$$\cos(wt + \phi)$$

Note that stationary phase implies that phase ψ' does not depend on t because $\psi \approx \psi(t_0) + (t - t_0)\psi' = \psi(t_0)$ since $\psi' = 0$. We emphasize that the frequency is lowest where $\psi' = 0$. So, the major contribution comes from the neighborhood of points where $\psi'(t) = 0$. The procedure is similar to the Laplace method otherwise.

Example 6.13 Investigate the large x behavior of

$$J_0(x) = \frac{1}{\pi} R_e \int_0^\pi e^{ix\sin t}\, dt$$

by using the method of stationary phase.

Three steps are involved:

1. In the Laplace method, we ask "where does the maximum (of ϕ) occur?" Here, we ask "where is the stationary phase, or where is $\psi'(t) = 0$?" ($\psi = 0$ for the Laplace method.) Here $d\psi/dt$ will be zero where $\frac{d}{dt}(\sin t) = 0$ or $\cos t = 0$. That is, at $t = \frac{\pi}{2}$,

or more generally $t = \pm(2n-1)\frac{\pi}{2}$. We focus on this point in the limits of the integral in order to obtain the asymptotic behavior:

$$J_0(x) \sim \frac{1}{\pi} R_e \int_{\pi/2-\varepsilon}^{\pi/2+\varepsilon} e^{ix\sin t}\, dt.$$

Note that the "$\pi/2$" in the limits of the integral was actually "c" in the Laplace method.

2. We Taylor expand $\psi = \sin t$ about $t = \pi/2$, the point of maximum contribution:

$$\sin t = \sin\frac{\pi}{2} - \sin\frac{\pi}{2}\frac{(t-\pi/2)^2}{2!} + \ldots = 1 - \frac{(t-\pi/2)^2}{2} + \ldots.$$

Therefore

$$J_0(x) \sim \frac{1}{\pi} Re \int_{\pi/2-\varepsilon}^{\pi/2+\varepsilon} e^{ix\left(1-\frac{1}{2}\left(t-\frac{\pi}{2}\right)^2\right)}\, dt.$$

We do change of variables: $s = t - \pi/2$ so that the limits $\pi/2 \pm \varepsilon$ become $\pm\varepsilon$:

$$J_0(x) \sim \frac{1}{\pi} R_e \int_{-\varepsilon}^{\varepsilon} e^{ix} e^{-\frac{ix}{2}s^2}\, ds \sim \frac{2}{\pi} Re \left[e^{ix} \int_{0}^{\varepsilon} e^{-\frac{ix}{2}s^2}\, ds \right].$$

3. We let $\varepsilon \to \infty$, just as in the Laplace method. We also want to get an e^{-u} toward getting the gamma function from the integral. So let $u = \frac{ix}{2}s^2$, $du = ixs\, ds$, $\int_0^{\varepsilon} \to \int_0^{\infty}$, $s = \sqrt{2u/ix}$.

   Therefore,

$$J_0\,(x) \sim \frac{2}{\pi} Re \left[e^{ix} \int_{0}^{\infty} e^{-u} \frac{du}{ix\sqrt{2u/ix}} \right].$$

Procedurally speaking, the method of stationary phase is the same as Laplace method. Only that we have the "i" in stationary phase, which, for all practical purposes, is pretty innocuous, other than telling us where to sharpen our focus on. We write the equation above as

$$J_0(x) \sim \frac{2}{\pi} Re \left[\frac{e^{ix}}{\sqrt{2ix}} \int_{0}^{\infty} e^{-u} u^{-\frac{1}{2}} du \right].$$

The integral here is just $\Gamma(1/2)=\sqrt{\pi}$ so that

$$J_0(x)\sim\frac{2}{\sqrt{\pi}}\frac{1}{\sqrt{2x}}\operatorname{Re}\left[e^{ix}i^{-\frac{1}{2}}\right].$$

Now, the principal value of $i^{1/2}$ is $i^{1/2}=\left(e^{i\pi/2}\right)^{1/2}=e^{+i\pi/4}$ so that $e^{ix}i^{-1/2}=e^{i(x-\pi/4)}$ and $\operatorname{Re}\left(e^{i(x-\pi/4)}\right)=\cos(x-\pi/4)$. Therefore,

$$J_0(x)\sim\sqrt{\frac{2}{\pi x}}\cos\left(x-\frac{\pi}{4}\right),\qquad x\to\infty.$$

Thus, for large x, $J_0(x)$ goes as shown in Figure 6.41. Also shown in the figure is $1/\sqrt{x}$, which models the boundary of $J_0(x)$ as x increases.

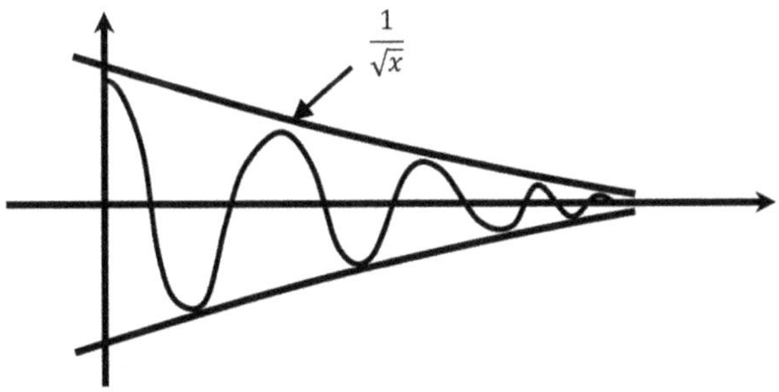

Figure 6.41: The behavior of $1/\sqrt{x}$ and $J_0(x)$ for large x.

Further comparison of Laplace's method with stationary phase

The method of stationary phase can only give leading order terms, and can't give useful extensions. On the other hand, Laplace integration can be extended to obtain arbitrary accuracy (high order terms). To see this, consider that, for $x\to\infty$, we found with the stationary phase method that

$$J_0(x)\sim\sqrt{\frac{2}{\pi x}}\cos\left(x-\frac{\pi}{4}\right).\tag{6.20}$$

If desired, we could extend (using Taylor expansion) as

$$\sqrt{\frac{2}{\pi x}}\cos\left(x-\frac{\pi}{4}\right)\left[1+\frac{a}{x}+\frac{b}{x^2}+\dots\right].$$

The next term to eq. (6.20) will be $O\left(\frac{1}{x^{3/2}}\right)$ from $\left(\sqrt{\frac{2}{\pi x}}\cdot\frac{a}{x}\right)$. On the other hand, when we integrate by parts, we see that

$$\int_0^\pi e^{ix\sin t}dt = \left.\frac{e^{ix\sin t}}{ix\cos t}\right|_0^\pi - \int_0^\pi \frac{d}{dt}\left(\frac{1}{ix\cos t}\right)e^{ix\sin t}\,dt. \tag{6.21}$$

Thus, there is no point in including the term $O(1/x^{3/2})$ if there have been other larger terms that have been neglected. That is, both the terms on the RHS of eq. (6.21) are $O(1/x)$, which is bigger than $O(1/x^{3/2})$. A caveat on the method of stationary phase, or where it wouldn't work, is in order. Consider

$$\int_a^b f(t)e^{ix\psi(t)}\,dt.$$

Suppose that $\psi'(t)=0$ at $t=t^*$. We need to be cautioned if $f(z^*)=0$. This does not mean the method of stationary phase will fail. However, the method may not give the leading behavior! The steepest descent method is another option.

6.3.5 Method of steepest descent

This is a more complicated procedure, where both $h(t)$ and $\rho(t)$ in

$$I(x) = \int_a^b h(t)e^{x\rho(t)}dt, \quad x\to\infty$$

are complex. The idea here is to think of t as a complex variable $t=u+iv$ (Figure 6.42), even though t is real, and integration is to be carried out between points a and b on the real line. Since there are no poles between a and b, we can compress the contour, pick a new contour so that the integral is easy to calculate using the Laplace method.

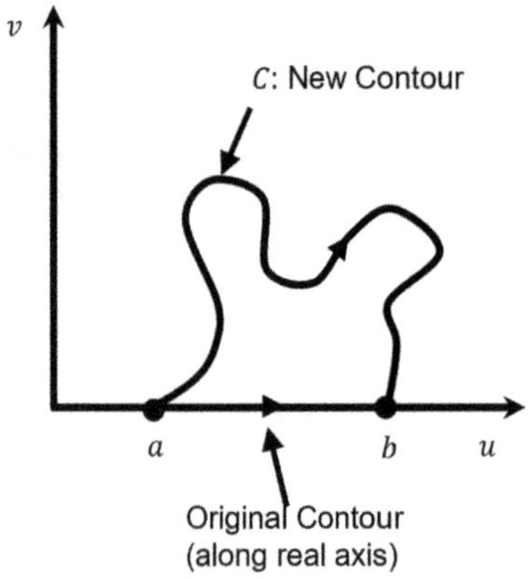

Figure 6.42: Complex-plane contour for integrating along a real line in time.

The original contour for integration goes from a to b. A main task in this approach is to pick a new contour (Figure 6.42) so that $\rho \equiv \phi + i\psi$ has a constant imaginary part. That is, the imaginary part $i\psi$ as a function of t is constant in the integral

$$\int_C h e^{x(\phi + i\psi)}\, dt.$$

Here, note that $h = f + ig$ and $\rho = \phi + i\psi$ and that $i\psi$ or ψ as a function of t is not necessarily real! An example will best illustrate the procedure for the steepest descent method.

Example 6.14 Consider the asymptotic expansion of the integral

$$I(x) = \int_0^1 e^{ixt^2}\, dt, \tag{6.22}$$

in the limit $x \to \infty$ so that

$$x\rho(t) \equiv x(\phi + i\psi) = ixt^2 \Rightarrow \psi = t^2$$

and the stationary point $(d\psi/dt = 0)$ is located at $t = 0$. We cannot analyze this problem with integration-by-parts, but we can with stationary phase, which is given as follows.

$$I(x) \sim \int_0^{\varepsilon} e^{ixt^2} dt \sim \int_0^{\infty} e^{ixt^2} dt.$$

Assuming $u = -ixt^2$, we have

$$I(x) \sim \frac{1}{2}\sqrt{\frac{i}{x}} \int_0^{\infty} e^{-u} u^{-\frac{1}{2}}\, du.$$

Thus,

$$I(x) \sim \frac{1}{2}\sqrt{\frac{\pi}{x}} e^{\frac{i\pi}{4}},$$

where $\Gamma(1/2) = \sqrt{\pi}$ has been used. With the steepest descent method, we carry out the following steps:

(1) We set $t = u + iv$ and set that in, with $t^2 = u^2 - v^2 + 2iv$ in eq. (6.22). The complex t-plane is shown in Figure 6.43.

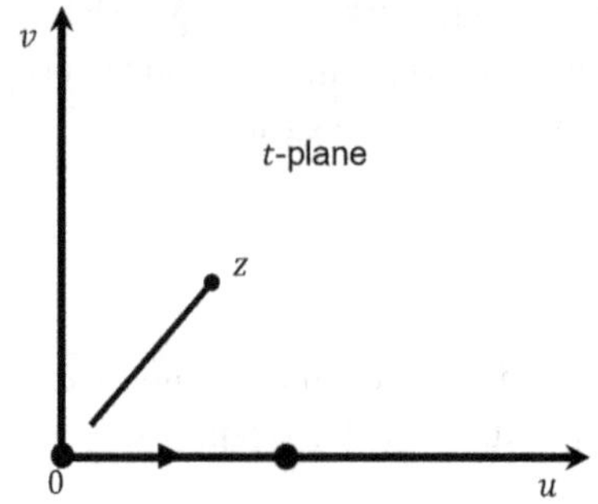

Figure 6.43: The complex t-plane, $t = u + iv$.

Note that in the steepest descent method, we don't allow the imaginary part (ψ), written in terms of u, v, to vary with t in the contour used to carry out the integration after setting $t = u + iv$.

In the integral

$$I(x) = \int_0^1 e^{ix\left(u^2 - v^2 + 2iuv\right)} dt,$$

isolate the imaginary part of ψ after setting $t = u + iv$ in ψ (the idea of the method) and require that it be constant relative to t:

$$I(x) = \int e^{-2xuv} e^{ix\left(u^2 - v^2\right)} dt.$$

(2) We pick a new contour such that $u^2 - v^2$, which is the imaginary part of ρ, is a constant, say c. In general, we will have

$$u^2 = c + v^2 \Rightarrow u = \pm\left(c + v^2\right)^{\frac{1}{2}}.$$

See Figures 6.44 through 6.47 and Tab 6.1 for the various behaviors. In the figures, we show the variation of v versus u for some values of c. For example, in Figure 6.44, v varies linearly with u, with slopes ± 1. This is the case when $c = 0$. The relationship between v and u is shown in Tab. 6.1 for three values of c and plotted in Figures 6.44 through 6.46, respectively. Figure 6.47 shows possible contours for carrying out the integration of $I(x)$ in the complex t-plane, instead of just on the real line $a \leq t \leq b$.

Let us recapitulate the procedure so far. The original integral problem is written with t in the real space as the dummy variable of the integral. But then we "extend" t to the complex plane and define it as $t = u + iv$. Regarding ψ, we start out with $\rho = \varphi + i\psi$, which suggests that ψ is real but then its dependence on t serves to extend it to the complex plane, just as t is. Thus, we have $\psi = \psi(t(u, v))$ as a complex number. Furthermore, we write $\rho = \rho(\varphi(t(u, v)))$. Asking that the imaginary part of ρ be constant implies that $\rho(\psi) = \rho(\psi(t(u, v))) = c$ so that $u = u(c, v)$, which exists entirely in the complex t-plane. Here, c is a constant. Thus, with $\rho = it^2$ in the example under consideration, $\psi = t^2 = u^2 - v^2 + 2uvi$ so that $\rho = -2uv + i(u^2 - v^2)$ and $\mathrm{Im}[\rho] = u^2 - v^2 = c$.

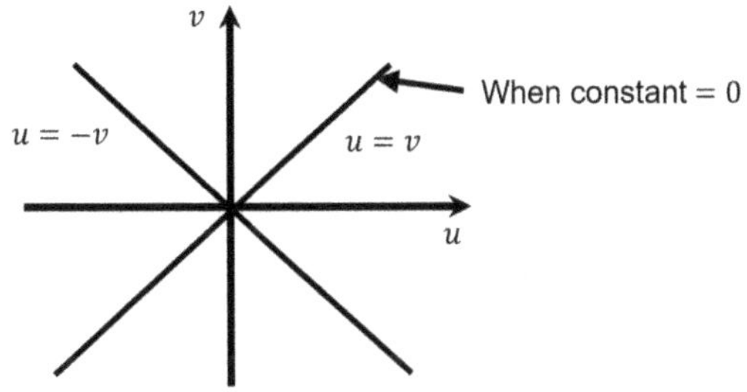

Figure 6.44: A new contour where $u^2 - v^2$ is a constant c equal to zero. The old contour is $a \le t \le b$ on the real line.

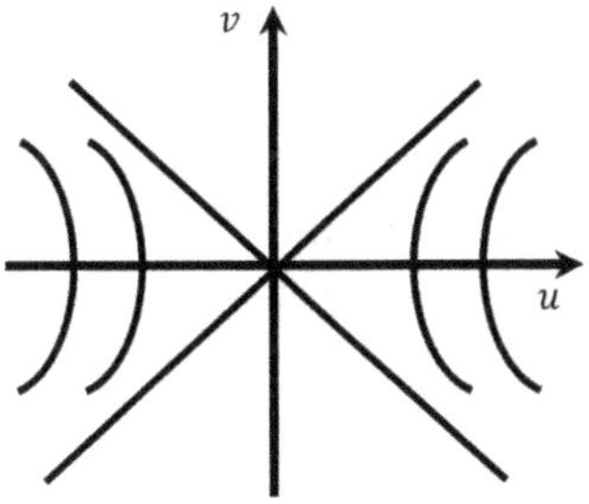

Figure 6.45: A sample of four nonlinear curves where $u^2 - v^2$ is a constant c and the constant c is greater than zero.

Table 6.1: Relationship between u and v for different values of c.

c	v
0	$-u, +u$
$\frac{1}{2}$	$\pm\sqrt{u^2 - \frac{1}{2}}$
c	$\pm\sqrt{u^2 - c}$

A sample of four contours is shown in Figure 6.45, where $u^2 - v^2$ is a constant, say, c, where $c > 0$. A subset of the curves that pass through $t = 0$ or $t = 1$, the limits of the integral $I(x)$, will be selected to obtain the closed contour that is needed for integration using the Cauchy theorem. The relevant contours will of course be obtained by judicious choice of the constant c. For example, $c = 0$ gives a curve that passes through a, where $a = 0$ for the illustration problem Figure (6.44).

Note that the limits must be $0 \to 1$ as given in the problem statement, no matter the contour that we choose. That is, we pick a constant such that the statement, imagi-

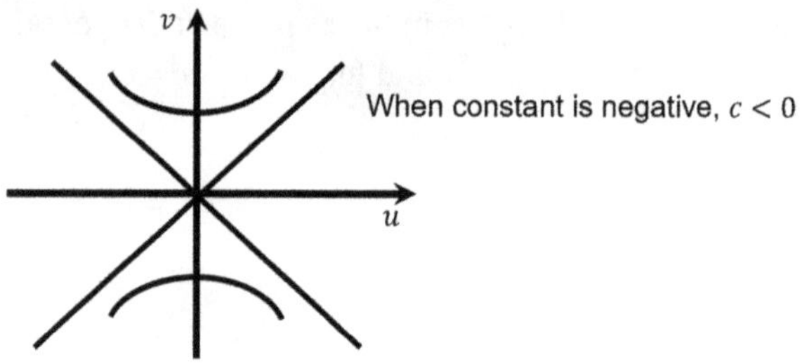

Figure 6.46: A new contour c where $u^2 - v^2$ is a constant c and the constant c is less than zero.

nary part of ρ or $u^2 - v^2 = \text{constant}$ goes through $t = 0$ and $t = 1$. That is, through $(u = 0, v = 0)$ and $(u = 1,\ v = 0)$, as per $t = u + iv$, and matching the limits of the integral. For example, $c = 1$ implies that

$$u^2 - v^2 = 1 \Rightarrow v = \pm\sqrt{-1 + u^2},$$

which also gives $u = \pm 1$ on the real axis of t (or $v = 0$).

Thus, based on the limits of the integral $I(x)$, we have two choices of contour: $u = v$ and $v = +\sqrt{-1 + u^2}$. Note that we will replace $\int_a^b (\cdot)dt$ with $\int_C (\cdot)d\mathcal{L}$, where the contour C is yet to be specified. Point a is located at $u = v = 0$, while point b is located at

$$v = \pm\sqrt{u^2 - 1} = 0,$$

which means that $u = 1$ from $u^2 - 1 = 0$.

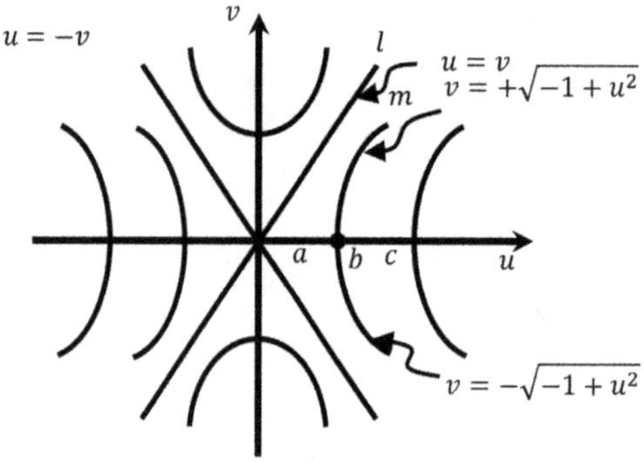

Figure 6.47: Possible contours for integrating $I(x)$ in the complex t-plane.

Whether we choose contours containing the segments $a - b$, $a - c$, or $b - c$, etc. depends on the integration limits for $I(x)$ in terms of t, vis-à-vis the transformation $t = u + iv$. Of course the other segments of the curves, such as $a - l$ and $b - m$, are given by the form of $\text{Im}[\rho]$. Finally whether you choose the contour below or above $v = 0$

depends on which of the two is descending. Let us examine the idea of ascending and descending curves, so we know which specific branch of a curve to pick as a *component* of the closed contour that will be used for integrating $I(x)$ in the complex plane.

Consider $u = \pm v$ first: Which shall we pick?

On $u = v$:

$$\psi = t^2 = u^2 - v^2 + 2iuv = 2iuv, \quad e^{ix(t^2)} = e^{ix(u^2 - v^2 + 2ix)} = e^{-2xv^2},$$
$$t = u + iv = v(1 + i)$$

so that

$$I(x) = \int_0^{v_1} e^{-2xv^2}(1 + i)dv,$$

where we have used $dt = du + idv = dv(1 + i)$ and $v_1 = (1 - i)/2$ or $v_1 = \infty$ if we consider that we close the contour at infinity, with the extension of t to the complex plane.

A plausible contour is shown in Figure 6.48. We can conceivably integrate up to where the line $u = v$ meets the curve that passes through $u = 1$ at infinity or integrate over the contour in dashed line and show convergence if we stop at D in Figure 6.48.

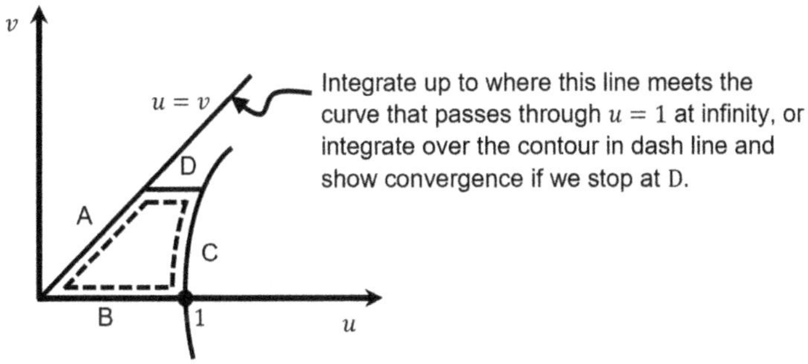

Figure 6.48: A plausible contour: ABCD, where D could either exist at infinity or is finite as shown but is able to give converged integral.

On $u = -v$:

$$\psi = t^2 = u^2 - v^2 + 2iuv = -2iv^2; \; dt = du + idv = (-1 + i)dv$$

and

$$I(x) = \int_0^{v_2} e^{+2xv^2}(-1 + i)dv,$$

where $v_2 = -(1+i)/2$ or $v_2 = -\infty$ if we consider that the contour is closed at $-\infty$, with the extension of t to the complex plane. Since it is based on integrating around the maximum point, we cannot use the Laplace method (even though that's what we're looking for) because the maximum occurs at 0, ∞. We usually want the descending branch. (The positive sign on the exponent $2xv^2$ suggests that we have an ascending branch.) We show descending and ascending branches in Figure 6.49.

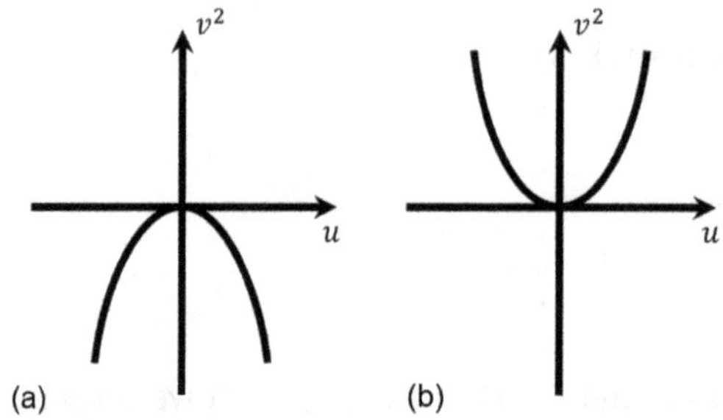

Figure 6.49: (a) Descending and (b) ascending branches.

Let us consider the two cases: (a) $v = +\sqrt{-1+u^2}$ and (b) $v = -\sqrt{-1+u^2}$, which correspond to the case with constant $c = 1$ in Tab. 6.1. The integral for Case (a) is

$$I(x) = \int_0^\infty e^{-2uvx} e^{ix(u^2-v^2)}\, dt = e^{ix} \int_{t=1}^\infty e^{-2u\sqrt{-1+u^2}}\, dt.$$

Again, the negative sign in front of the "2" in the exponent implies that we have the "descending" branch. The integral for Case (b) is

$$I(x) = e^{ix} \int_{t=1}^\infty e^{+2u\sqrt{-1+u^2}}\, dt.$$

This is an "ascending" branch, which is of no use here. Thus, we pick the branch $u = v$ and not the branch $u = -v$. The C_1 and C_3 contours are called steepest descent because $\left|e^{xp(t)}\right|$ decreases most rapidly along these paths as t moves up from the real t axis. C_1 and C_3 are joined at ∞ by C_2 (Figure 6.50).

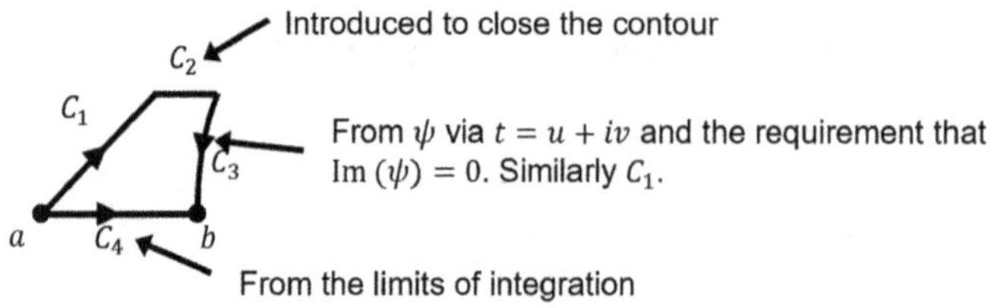

Figure 6.50: Plausible contour for integrating $I(x)$ in the complex t plane.

What happens on C_2? On C_2, v is a constant. In

$$I(x) = \int e^{-2xuv} e^{-ix(u^2-v^2)} (du + idv),$$

note that on C_2, $e^{-2xvu} \to 0$ as $v \to \infty$ since $u > 0, v > 0, x > 0$. This means that C_2 has no contribution to $I(x)$. Thus, in

$$\int_a^b = \int_{C_1} + \int_{C_2} - \int_{C_3},$$

and the contribution of $\int_{C_2}$ is zero. The C_i curves (excluding the curve from $t = 0$ to $t = 1$, which are the original limits of the integral) along which we integrate, satisfy $\mathrm{Re}[\rho] =$ constant in the t-plane:

$$I(x) = \int_0^\infty (1+i)e^{-2xv^2}\,dv - e^{ix}\int_0^\infty e^{-2xuv}(du + idv).$$

$$C_1\colon u = v, \quad C_3\colon iv = \sqrt{-1+u^2}, \quad it^2 = i(u^2 - v^2) - 2vu.$$

The contour chosen for integrating $I(x)$ is shown in Figure 6.51.

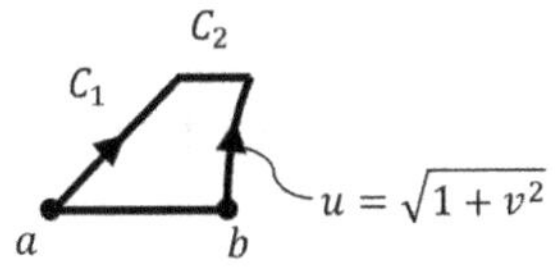

Figure 6.51: Contour chosen for integrating $I(x)$.

Let us evaluate the integrals. For $\int_{C_1}$, note that $u = v$ so that $dt = du + idv$ (general) or $= (1+i)dv$ on a particular line.

Thus, $\int_{C_1}$ can be evaluated as

$$\int_{C_1} \equiv \int_0^\infty e^{-2xv^2}(1+i)\,dv = \frac{e^{i\pi/4}}{2}\sqrt{\frac{\pi}{x}}.$$

The integral $I(x) = \int_{C_1} - \int_{C_3}$ can also be written as $I(x) = \int_0^\infty \left(\text{for } \int_{C_1}\right) - \int_1^\infty \left(\text{for } \int_{C_3}\right)$.

For $\int_{C_3}$, we can evaluate it using one of two approaches, direct and less-direct:

(1) *The direct approach to evaluate* $\int_{C_3}$

With $t = u + iv$, and on C_3, $u = \sqrt{1+v^2}$, $t = \sqrt{1+v^2} + iv$, and $t^2 = 1 + 2iv\sqrt{1+v^2}$, we have

$$2t\,dt = \frac{d}{dv}\left[1 + 2iv(1+v^2)^{\frac{1}{2}}\right]dv = \left[iv\left[1+v^2\right]^{-\frac{1}{2}} \cdot 2v + (1+v^2)^{\frac{1}{2}} \cdot 2i\right]dv$$

so that

$$dt = \left(\frac{v}{\sqrt{1+v^2}} + i\right) dv, \quad \int\limits_{C_3} = \int\limits_{t=1}^{1+\varepsilon} e^{ix\left[1+2iv\sqrt{1+v^2}\right]} \left(i + \frac{v}{\sqrt{1+v^2}}\right) dv.$$

This should be considered as a Laplace integral, with exponent having a maximum where $v = 0$. Therefore, we Taylor-expand the exponent about $v \equiv 0$ (or $t = 1$). We have

$$\int\limits_{C_3} \sim e^{ix} \int\limits_{t=1}^{t=1+\varepsilon} e^{\overbrace{-2xv\sqrt{1+v^2}}^{\text{TS about } v\,=\,0,\text{ with } \Delta v\,\equiv\,\varepsilon\,=\,v}} \left(\frac{v}{\sqrt{1+v^2}} + i\right) dv.$$

As in the Laplace method, we take Taylor series and change limits ($\varepsilon \to \infty$):

$$\Rightarrow \int\limits_{C_3} \sim e^{ix} \int\limits_{v=0}^{\infty} e^{-2xv\left(1+\frac{1}{2}v^2+\cdots\right)} (i + v + \ldots) dv.$$

On the Taylor series, it is in variable v about $v = 0$:

$$\left(\frac{1}{\sqrt{1+v^2}} = 1 - \frac{v^2}{2} + \ldots\right).$$

We leave the first term in exponential form and Taylor-series expand the next; that is, our usual procedure so that

$$\int\limits_{C_3} \sim e^{ix} \int\limits_{0}^{\infty} e^{-2xv} \underbrace{e^{-xv^3}}_{1-xv^3+\cdots} (i + v + \ldots) dv.$$

Let $s = 2xv$ so that

$$\int\limits_{C_3} \sim e^{ix} \int\limits_{0}^{\infty} e^{-2xv} (i + v) dv = e^{ix} \int\limits_{0}^{\infty} e^{-s} \left(i + \frac{s}{2x}\right) \frac{ds}{2x}.$$

With $\Gamma(z) = \int_0^\infty e^{-s} s^{z-1}\, ds$, we have

$$\int\limits_{C_3} \sim e^{ix} \int\limits_{0}^{\infty} e^{-s} \left(i + \tfrac{s}{2x}\right) \tfrac{ds}{2x} = e^{ix}\left[\tfrac{i}{2x}\Gamma(1) + \tfrac{1}{4x^2}\Gamma(2) + \ldots\right] = e^{ix}\left[\tfrac{i}{2x} + \tfrac{1}{4x^2} + \ldots\right]$$

$$= \tfrac{ie^{ix}}{2x}\left(1 - \tfrac{i}{2x} + \ldots\right).$$

We can already see an asymptotic series emerging. Contour C_3 is shown in Figure 6.52.

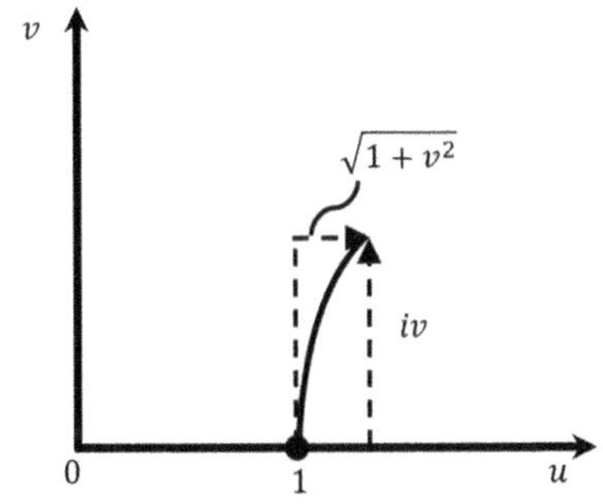

Figure 6.52: Contour C_3 in the vicinity of $t = u = 1$.

(2) *A less direct approach to evaluate* $\int_{C_3}$
In this approach, we approximate contour C_3 by its tangent at the maximum point (locally) which is at $t = u = 1$ (Figure 6.53). (Note that integrating along C_1 is easy by the direct approach so we had no need for this approach.)

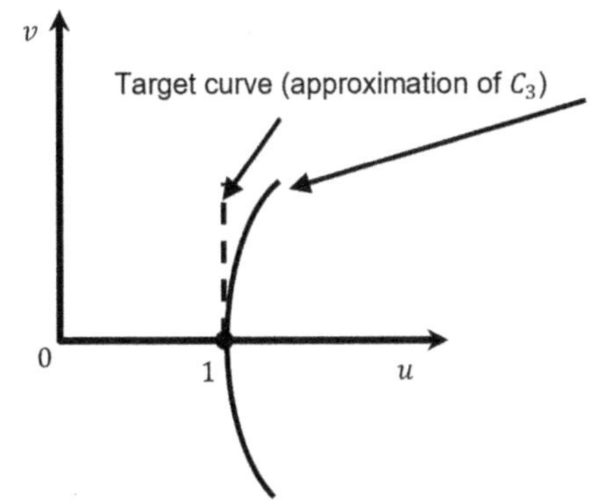

Figure 6.53: Using the tangent approach to evaluate $\int_{C_3}$.

To within the asymptotic approximation, the contour C_3 becomes $u = 1$ and $dt = du + idv = idv$ which is desirable, but then in $e^{ix(u^2 - v^2)}$, $u^2 - v^2 \neq 1$, because $u \equiv 1$, which is not so convenient. The approach is to truncate the integral $\int_{v=0}^{\varepsilon} e^{ixt^2}\, dt$ so that the contour becomes $t = 1 + iv$ (Figure 5.53):

$$\int_{C_3} \sim \int_0^{\varepsilon} e^{ix(1+iv)^2}\, idv \sim \int_0^{\infty} e^{ix(1 - v^2 + 2iv)}\, idv \sim ie^{ix} \int_0^{\infty} e^{-2xv} e^{-ixv^2}\, dv.$$

We then expand the third exponential in this relation, just as you would any term that is small compared to the leading term. Thus,

$$\int_{C_3} \sim ie^{ix} \int_0^{\infty} e^{-2xv} \underbrace{e^{-ixv^2}}_{1 - ixv^2 + \ldots} dv \sim ie^{ix} \int_0^{\infty} e^{-s} \left(1 - ix\left(\frac{s}{2x}\right)^2 + \ldots\right) \frac{ds}{2x}$$

$$\sim ie^{ix}\left[\frac{\Gamma(1)}{2x} - \frac{i}{8x^2}\Gamma(3) + \dots\right] \sim ie^{ix}\left(\frac{1}{2x} - \frac{i}{4x^2} + \dots\right).$$

Therefore,

$$I(x) \approx \frac{e^{i\pi/4}}{2}\sqrt{\frac{\pi}{x}} - \frac{ie^{ix}}{2x}\left(1 - \frac{i}{2x}\right).$$

Recall that the stationary phase approach gives only one term because it doesn't see anything apart from the point. Also, the method doesn't work (for $\int f(t)e^{ix\psi}dt$) when f vanishes at the point of stationary phase.

Why call the present procedure the method of steepest descent? Why not the method of constant imaginary part? To appreciate this, recall that the gradient vector ∇f of a scalar field $f(u,v)$ is expressed as $\nabla f = \frac{\partial f}{\partial u}e_u + \frac{\partial f}{\partial v}e_v$, where e_u and e_v are the unit vectors in the directions of the independent variables u, v. The ∇f points in the direction of the most rapid change in f. This direction, denoted here as n, is given by $\nabla f/|\nabla f|$. The derivative of f along n is given by the projection of ∇f in the direction of $\boldsymbol{n}$, or $\boldsymbol{n}\cdot\nabla f \equiv \frac{\partial f}{\partial n} = \frac{df}{ds}$, assuming that s is a direction along n. On a 2D plot (contour map) of f versus (u,v), the vector ∇f is perpendicular to the contours representing constant f, say $f_1, f_2, \dots, f_n$. These are called the level curves of f. The tangent $\hat{t}$ to a level surface, being perpendicular to the normal to the curve satisfies $\frac{\partial f}{\partial t} = 0$, where $\hat{t} = |\hat{t}|$.

Let $\rho(t) = \phi(t) + i\psi(t)$ be an analytic function such that $\phi = \phi(t(u,v))$, $\psi = \psi(t(u,v))$, and $t = u + iv$. Suppose $\frac{\partial \rho}{\partial u}, \frac{\partial \rho}{dv}$ are not zero. In the integrand $e^{x\rho(t)}$ in our procedure, $x > 0$, it can be shown that a steepest contour is defined as the contour whose tangent is parallel to $\nabla\left|e^{x\rho(t)}\right| = \nabla\left|e^{x\phi(t)}\right|$ since $\left|e^{ix\psi(t)}\right| = 1$. This is parallel to $\nabla\phi$ (Figure 6.54). The implication of this is that the steepest contour is that contour on which $\left|e^{x\rho(t)}\right|$ is changing the most rapidly with t.

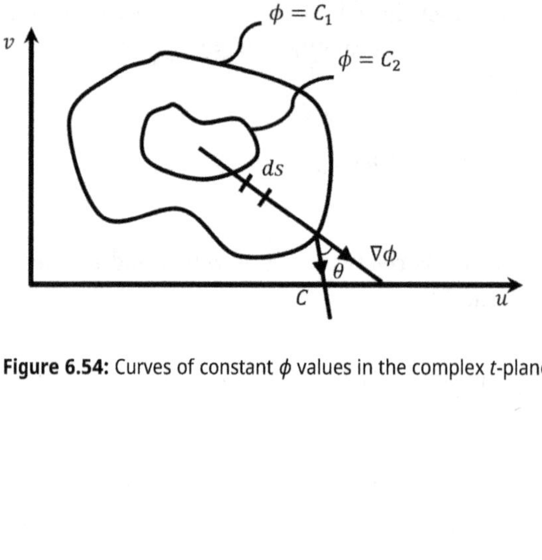

Figure 6.54: Curves of constant ϕ values in the complex t-plane.

We can show that the constant phase contours are also the steepest contours. To this end, on account of the analyticity of ρ, the C-R conditions are satisfied:

$$\frac{\partial \phi}{\partial u} = \frac{\partial \psi}{\partial v}, \frac{\partial \phi}{\partial v} = -\frac{\partial \psi}{\partial u} \text{ and } \phi_u \psi_u + \phi_v \psi_v = \begin{bmatrix} \frac{\partial \phi}{\partial u} & \frac{\partial \phi}{\partial v} \end{bmatrix} \begin{bmatrix} \frac{\partial \psi}{\partial u} \\ \frac{\partial \psi}{\partial v} \end{bmatrix} \equiv \nabla \phi \cdot \nabla \psi = 0.$$

Thus, $\nabla \phi \perp \nabla \psi$ and derivatives along $\nabla \phi$ satisfy $\frac{d\psi}{ds} = 0$, where s is along $\nabla \phi$ (Figure 6.54). Therefore, contours whose tangents are parallel to $\nabla \phi$ will have $\psi =$ constant. These are constant phase contours and are also the steepest contours (Figure 6.55).

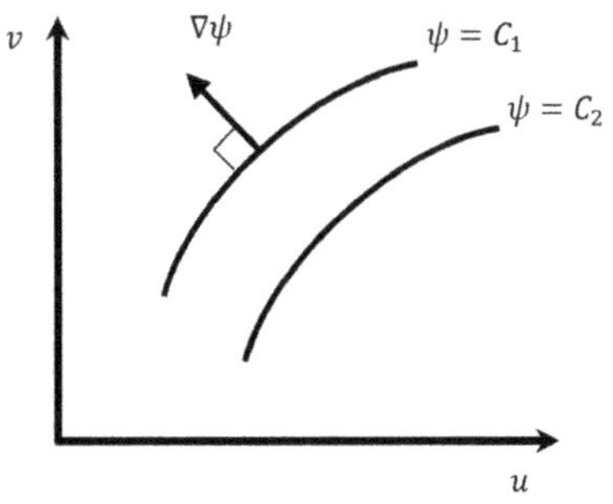

Figure 6.55: Curves of constant imaginary part and the gradient to them.

6.3.5.1 Example: Large-time behavior of the Bessel function

Recollect that the large x (i.e., the small-time) behavior of the Bessel function

$$J_0(x) = \frac{1}{\pi} Re \int_0^{\pi} e^{ix \sin t} \, dt$$

obtained by the method of stationary phase is given by

$$J_0(x) \sim \sqrt{\frac{2}{\pi x}} \cos\left(x - \frac{\pi}{4}\right).$$

Analyzing the large-time behavior of the Bessel function integral requires a different approach when compared with the approach used for the previous example. The present problem requires the saddle point approach. We will pick a new contour so that $\sin t \equiv \psi$ or $\mathrm{Im}[\rho]$ has a constant part on it. The imaginary part of $i \sin(u + iv)$ can be obtained as follows:

$$i \sin(u + iv) = i \sin u \cos iv + i \cos u \sin iv = i \sin u \cosh v - \cos u \sinh v, \tag{6.23}$$

where we have used $\cos iv = \cosh v$, $\sin iv = i \sinh v$. We will pick u and v such that the imaginary part, or $\sin u \cosh v$, is a constant. That is, the equation of the curve along which we integrate is $\sin u \cosh v =$ constant. This is completely analogous to $\mathrm{Im}(\rho(\psi(t(u, v)))) = c$ in the previous example. Also, as in the previous example, we

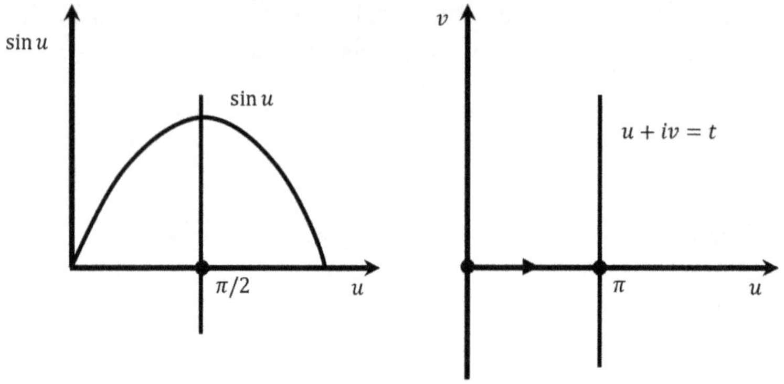

Figure 6.56: The sine wave and the contour $\pi + iv$.

will pick the curve that goes through the given end points (0 and π) of the integral (Figure 6.56), just like the $t = 0$ ($u = 0$) and $t = 1$ ($u = 1$) in the previous example. Note that if $u = v = 0$, then the constant $= 0$ or $\sin u \cosh v = 0$. The cosh function never vanishes $\Rightarrow \sin u = 0 \Rightarrow u \equiv 0$ is the point $d\psi/dt = 0$. Because we are interested in large values of the exponent, we need to know the (two) ways to go to infinity from $t = 0$ and π. We can go in the positive or negative v direction again, just as in the previous example.

If $(-\cos u \sinh v)$ attains its maximum at ∞, there is no use. We need to find the descending branch out of all these curves. (Note that the curves $u = 0$, $u = \pi$ are identical in concept to $u = \pm v$, $v = \sqrt{u^2 - 1}$ in the previous discussions, although the definition of ascending/descending is a little bit different.) We now need to determine the branch of the curves (passing through $u = 0$, π and defined by $i\psi =$ constant in the t-plane) where we have a descending behavior:

a) The branch of the curve through $u = 0$ has the integral

$$\int_0^\infty e^{-x \sinh v}(i\, dv); \; (t = iv).$$

Observe that $i \sin u \cosh v - \cos u \sinh v = \sinh v$ when $u = 0$. Whether we have a descending/ascending behavior depends on how $x \sinh v$ behaves. We need $v > 0$ for the maximum to occur (at $u = 0$ in this case). The function $\sinh v$ is shown in Figure 6.57.

b) The branch of the curve through $u \equiv \pi$ has the integral

$$\int_0^\infty e^{+x \sinh v}(i\, dv): \; (t = \pi + iv).$$

Here v has to be greater than 0 for the maximum to occur at $v = \pi$. So a new contour is shown in Figure 6.58. Two paths to go from $u = 0$ to $u = \pi$ are shown in this figure.

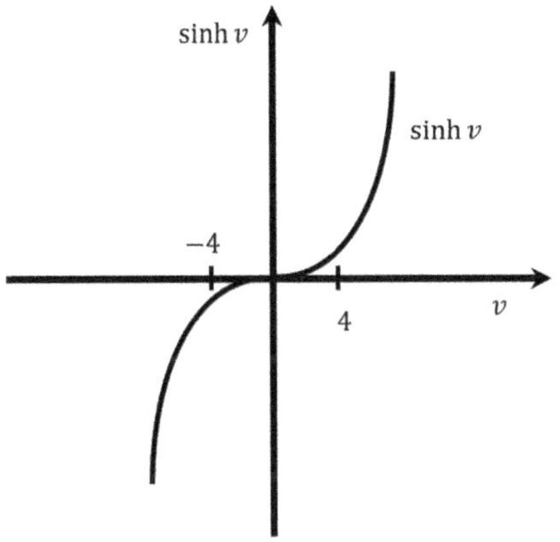

Figure 6.57: Graph of sinh *v*.

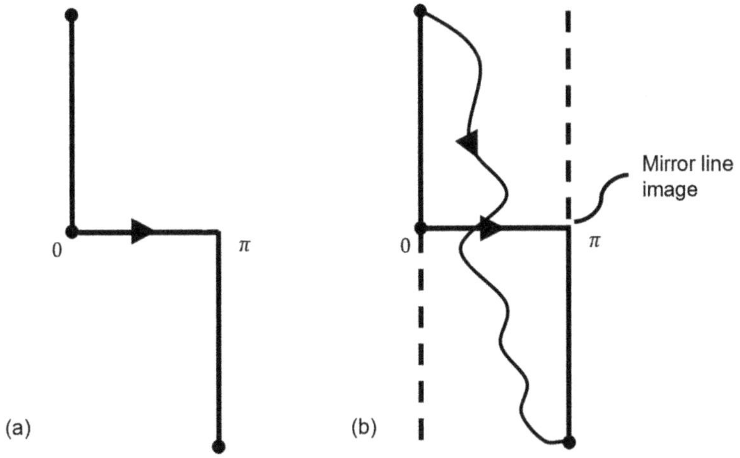

Figure 6.58: Two paths to go from $u = 0$ to $u = \pi$ for saddle point analysis.

The need to have the desired curve pass through the interior of the integration limits 0 and π is the main difference between this problem and the previous one. Thus, as shown below, we actually have a saddle point.

To proceed, we will look for singular points: We determine singular points only to obtain where the curve of steepest descent passes through.

$$\frac{\partial}{\partial u}(\sin u \cosh v) = 0 = \cos u \cosh v \Rightarrow u = \frac{\pi}{2},$$

$$\frac{\partial}{\partial v}(\sin u \cosh v) = 0 = \sin u \sinh v \Rightarrow v = 0.$$

So, the singular point is at $(\pi/2,\ 0)$. The curve that goes through the singular points has the equation (Figures 6.59 and 6.60) $\sin u \cosh v = 1$ or $\sin u = 1/\cosh v$. It is clear that $v \to \infty$, $\cosh v \to \infty \Rightarrow \sin u \to 0$.

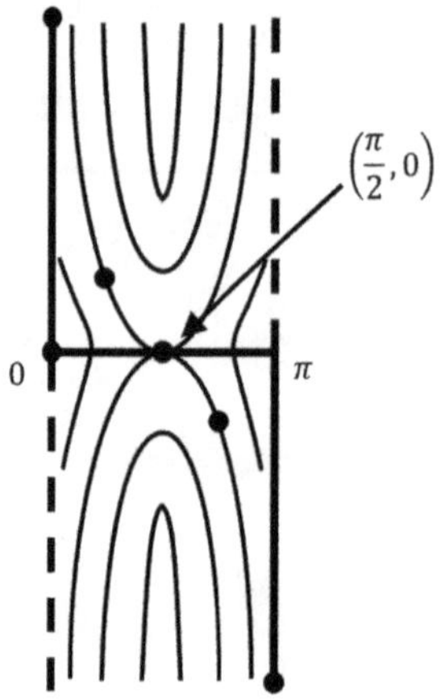

Figure 6.59: Paths from $u = 0$ to $u = \pi$. The curve that goes through the singularity $(\pi/2, 0)$ is shown with dots.

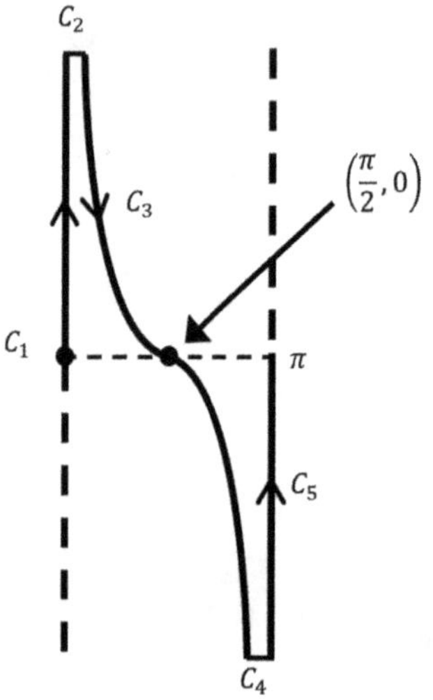

Figure 6.60: The final path from $u = 0$ to $u = \pi$ ($C_1 \to C_2 \to C_3 \to C_4 \to C_5$).

The only place that a curve of steepest descent can intersect is at an equilibrium point, with the realization that the normal at a singular point has to be nonunique in order to have an equilibrium point. The complete path from $u = 0$ to $u = \pi$ is shown in Figure 6.60. **We have a saddle point in this problem because the curve connecting the limits of the integral is not a straight line**, but as on Curve C_3 in Figure 6.60, which does look like a saddle!

We can evaluate the integral over C_2 as follows:

$$\int_{C_2} = \int e^{-x\cos u \sinh v} \underbrace{e^{ix\sin u \cosh v}}_{\text{Has modulus } \|\cdot\| = 1} du; \to 0 \text{ as } v \text{ (or } \sinh v) \to \infty.$$

A similar result can be obtained for integration along C_4. Thus, with $t = u + iv$, we can write

$$\int_0^{\pi} e^{ix\sin t}\,dt = \int_{C_1} + \int_{C_3} + \int_{C_5}$$

$$= i\int_{\substack{v=0\\u=0}}^{\infty} e^{-x\sinh v}dv + \int_{C_3} + i\int_{v=-\infty}^{0} e^{+x\sinh v}dv,$$

where the last term can be written as

$$-i\int_{v_1=\infty}^{0} e^{-x\sinh v_1}dv_1, \qquad (v_1=-v).$$

That is, C_1 and C_5 are really the same contour. So the contributions from C_1, C_5 are purely imaginary. Since we are going to take the real part at the end, we could neglect C_1 and C_5 now so that we only have $\int_{C_3}$ or

$$J_0(x) = \frac{Re}{\pi}\int_{C_3} e^{ix\sin t}\,dt.$$

$$\sin u\cosh v = 1.$$

6.3.5.2 Simplification

We could replace the curve with its tangent at the point where it has its tangent (Figure 6.61). We need to determine where the maximum of $ix\sin t$ occurs. Note that $\int OA = \int OB$. Using eq. (6.23), we can write

$$\int_{(0,\infty)}^{(\pi,-\infty)} e^{ix\cdot 1}e^{-x\cos u\sinh v}dt = 2\int_{(0,\infty)}^{(\frac{\pi}{2},0)} e^{ix\cdot 1}e^{-x\cos u\sinh v} = -2\int_{(\frac{\pi}{2},0)}^{(0,\infty)} e^{ix\cdot 1}e^{-x\cos u\sinh v}.$$

Note that the sinh function is odd. As you move toward O in the direction shown in Figure 6.62, $\sinh v$, $\cos u$ each grows, which means that the maximum of the integrand occurs at O or $(\pi/2, 0)$.

With the location of the maximum of the integrand known, we can introduce the ε parameter of the Laplace method. Or

$$J_0(x) \sim -\frac{2Re}{\pi}\int_{(\pi/2,0)}^{(\pi/2,0)+\varepsilon}. \tag{6.24}$$

Regarding the upper limit on the integrand in eq. (6.24), note that we are integrating along a curve. We will replace the curve with its tangent line (Figure 6.63), which

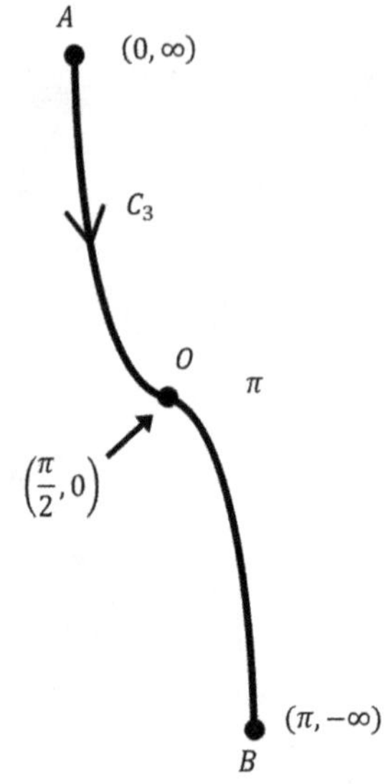

Figure 6.61: The C_3 contour. Note that the equation of the curve is $\sin u \cosh v = 1$.

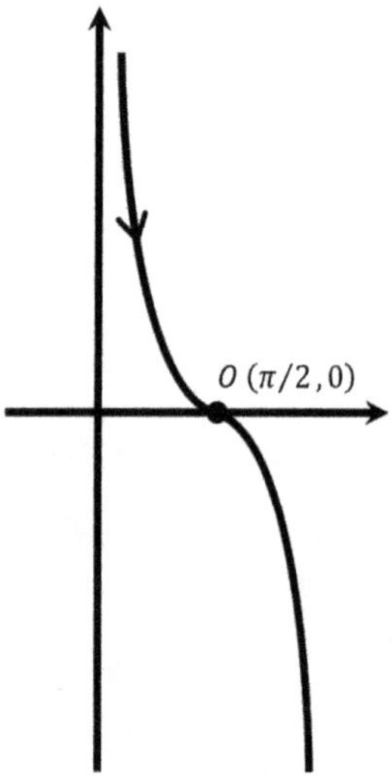

Figure 6.62: Traversing the curve C_3. Both $\sinh v$ and $\cos u$ grow along this curve so that the maximum of the integrand occurs at O (or$(\pi/2, 0)$).

has an angle of 45° to the horizontal. From $(\cosh v \cos u)du + (\sinh v \sin u)dv = 0$, we have $(dv/du)|_{(u=\pi/2, v=0)} = 0/0$.

Therefore, we cannot obtain the slope by differentiating! It's better to carry out a Taylor series expansion for the line (using the equation of the line) around the point. This procedure is shown below. We will introduce the real coordinate direction to be the same as u but with its origin at $u = \pi/2$. That is, we will introduce a new variable: $s = u - \pi/2$ so that $\sin(s + \pi/2)\cosh v = 1$, where $\sin s \cos(\pi/2) + \cos s \sin(\pi/2) \cos s$ and $(\cos s \cosh v) = 1 = (1 - s^2/2! + \ldots)(1 + v^2/2! + \ldots) = 1$, after a Taylor series expansion about $s = 0$, $v = 0$, respectively. Therefore, we will have $v^2 - s^2 + \ldots = 0$ as the equation of Curve C_3.

We have two branches: $(v + s)(v - s) + \ldots = 0$. The slope is $v = \pm s$. We select $v = -s$ since this is the path that better represents C_3 in the neighborhood of $s = 0$ along C_3 (Figure 6.64). Thus, with $v = -s$, $s = u - \pi/2$, we have $t = u + iv = (1 - i)s + \pi/2 \equiv (1 - i)\xi + \pi/2$. We replace C_3 by $t = \pi/2 + (i - 1)\xi$ so that

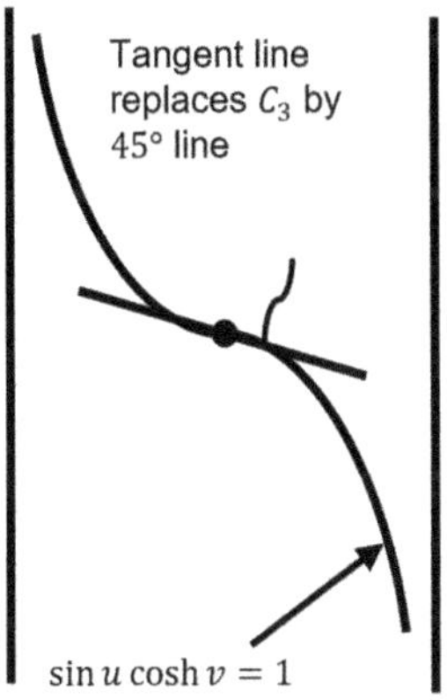

Figure 6.63: Curve C_3 is replaced by its tangent at $(\pi/2, 0)$. Equation of curve is $\sin u \cosh v = 1$.

$$J_0(x) \sim -\frac{2}{\pi} Re \int_{\xi=0}^{\varepsilon} e^{ix \overbrace{\sin\left[\frac{\pi}{2} + (i-1)\xi\right]}^{\cos(i-1)\xi}} (i-1)d\xi,$$

where we have used the formula for $\sin(A + B)$ with $\cos(\pi/2) = 0$. This is now a regular Laplace integral. So, we will carry out a Taylor series expansion with respect to ξ and pick $\xi \to \infty$ and take the real part when done. This is shown below.

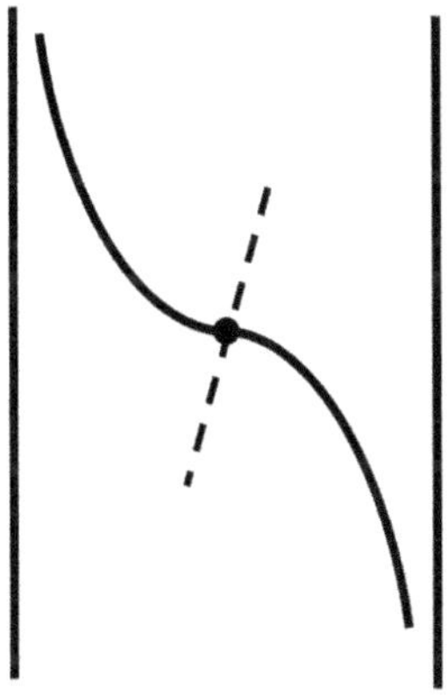

Figure 6.64: The slope $(v = +s)$ shown here in dash line is ignored and in favor of the slope $v = -s$.

We expand $\cos(i-1)\xi$:

$$\cos(i-1)\xi = 1 - \frac{(i-1)^2}{2}\xi^2 + \frac{(i-1)^4}{24}\xi^4 + \ldots = 1 + i\xi^2 - \frac{\xi^4}{6} + O\left(\xi^6\right).$$

Note that we are keeping extra terms in this expansion to get both the leading term and the correction.

Thus,

$$J_0(x) \sim -\frac{2}{\pi} Re\left[e^{ix}\int_0^\infty e^{-\xi x^2} e^{\left(-ix\xi^4/6\right)}(i-1)d\xi\right].$$

(The expression in the exponent $= 1 - ix\xi^4 + \ldots$)

Let $u = \xi^2 x$, $u^2/x^2 = \xi^4$, $du = 2\xi x d\xi$, $x\xi = \sqrt{ux}$.

Then

$$\begin{aligned} J_0(x) &\sim -\frac{2}{\pi}\text{Re}\left[e^{ix}(i-1)\int_0^\infty e^{-u}\left(1 - \frac{iu^2}{6x}\right)\frac{du}{2\sqrt{xu}}\right] \\ &\sim -\frac{2}{\pi}\text{Re}\left[e^{ix(i-1)}\frac{1}{2\sqrt{x}}\left[\sqrt{\pi} - \frac{i}{6x}\frac{3}{4}\sqrt{\pi}\right]\right]. \end{aligned}$$

We can now write $(i-1)$ in the last expression in Euler form, $re^{i\theta}$ (Figure 6.65): $i-1 = \sqrt{2}e^{-i\pi/4}$, to obtain

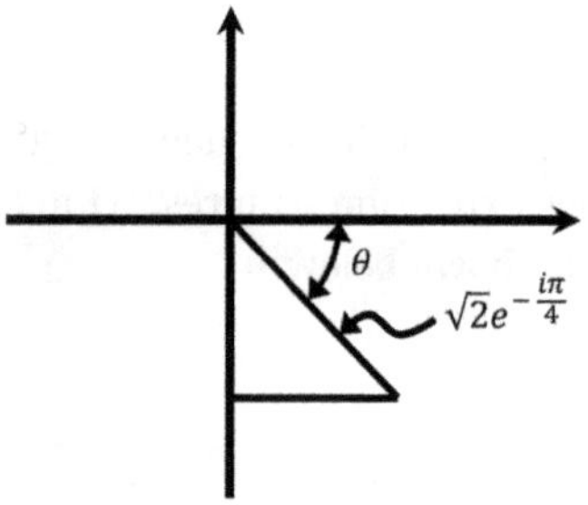

Figure 6.65: Euler form of $(i-1)$.

$$J_0(x) \sim \frac{2\sqrt{2}}{\pi}\text{Re}\left[e^{i\left(x-\frac{\pi}{4}\right)}\frac{\sqrt{\pi}}{2\sqrt{x}}\left(1 - \frac{i}{8x}\right)\right] \sim \sqrt{\frac{2}{\pi x}}\left[\cos\left(x - \frac{\pi}{4}\right) + \frac{1}{8x}\sin\left(x - \frac{\pi}{4}\right) + \ldots\right].$$

The first term agrees with the result from the stationary phase method, and the second term is the correction to this obtained by the current steepest descent method.

6.3.5.3 $\int h(t)e^{xp(t)}dt$: Summary of the methods of asymptotics of integrals

Table 6.2 provides a summary of the methods that we have used so far in this section to analyze the asymptotic of integrals. The rest of this chapter presents a more detailed and somewhat heuristic analysis.

Table 6.2: $\int h(t)e^{xp(t)}dt$: Summary of the methods for the asymptotics of integrals.

Method	$h = f + ig$	$p = \phi + i\psi$	Condition	Implementation
1. Integration by parts	We only consider the case $g = 0$.	We only consider the case $\psi = 0$.	Nothing special is done.	Regular integration by parts. No expansions.
2. Laplace method	f, g. We only considered the case $g = 0$.	We only consider the case $\psi = 0$.	Determine where ϕ is maximum (using $d\phi/dt = 0$), say at $t = t^*$.	Expand f and ϕ about t^*. Introduce ε based on t^* and stretch the integration limit(s) to $(-\infty, \infty)$ based on ε.
3. Stationary phase	f, g	ϕ, ψ	Find where (the phase) ψ is stationary, say $t = t^*$, by calculating $d\psi/dt = 0$.	Taylor expand ψ about t^* and follow the Laplace method.
4. Steepest descent	f, g	ϕ, ψ	Write $p = p(\psi(t(u, v)))$ and set $\text{Im}[p]$ equal to a constant, say, c. This leads to $u = u(c, v)$ in the complex t-plane. Find the value of the constant c that allows the curve $u(c, v)$ to pass through the original limits of the given integral. Determine the descending part of the chosen curve. Find where ψ is maximum by setting $d\psi/dt = 0$, say t^*. Use t^* to obtain the effective limit of the integral, via the ε parameter technique we use in the Laplace method. Expand exponent p about t^* and follow the Laplace method from here.	

6.3.6 More detailed analysis

6.3.6.1 Laplace method

The Laplace method in the complex space is essentially the same as for real space which is expected because the argument of the exponential factor is real in both cases. It is the case, however, that our interest includes Laplace inverse transformation in integrals such as

$$f(x,t) = \frac{1}{2\pi i}\int_{\gamma-i\infty}^{\gamma+i\infty} e^{st}\frac{e^{-x\sqrt{s}}}{s}\,ds,$$

where p is complex, or $p = \phi + i\psi$, s is large, and t is small or large. For the real limit p, or $p = \phi$, $\psi = 0$; we will present a slightly different, but almost equivalent, approach to the Laplace method discussed in the previous subsections. Suppose we are interested in the asymptotic behavior as $x \to \infty$ of the integral

$$I(x) = \int_a^b f(t) e^{x\phi(t)} dt, \quad I(x \to \infty),$$

where $f(t)$ can be imaginary, $\phi(t)$ is real, and x is real. The crux of the Laplace method is that the behavior as $x \to \infty$ is dominated where ϕ is maximum. Let's say the maximum of $\phi(t) = \phi^*$ and it occurs somewhere at $t = c$ in $[a, b]$, then

$$I(t) = \left[\int_a^b f(t) e^{x[\phi(t) - \phi^*]} dt \right] e^{x\phi^*}$$

since ϕ^* is maximum in the interval. We can write

$$I(t) = \left\{ \int_a^{c-\varepsilon} \left[f(t) e^{x[\phi(t)-\phi^*]} \right] dt + \int_{c-\varepsilon}^{c+\varepsilon} \left[f(t) e^{x[\phi(t)-\phi^*]} \right] dt + \int_{c+\varepsilon}^{b} \left[f(t) e^{x[\phi(t)-\phi^*]} \right] dt \right\} e^{x\phi^*}.$$

Now,

$$\int_a^{c-\varepsilon} f(t) e^{x[\phi(t)-\phi^*]} dt \sim O\left(e^{-x|\phi^* - \phi(c-\varepsilon)|} \right)$$

and

$$\int_{c+\varepsilon}^{b} f(t) e^{x[\phi(t)-\phi^*]} dt \sim O\left(e^{-x|\phi^* - \phi(c+\varepsilon)|} \right).$$

These last two expressions will exponentially decay since ϕ^* is greater than any ϕ. Therefore, these terms decay faster than anything else as $x \to \infty$. Thus, only the middle term $\int_{c-\varepsilon}^{c+\varepsilon} \left[f(t) e^{x[\phi(t)-\phi^*]} \right] dt$ matters. Let us consider two distinct cases depending on where c is located.

Case I: c is an interior point

A depiction of the location of the maximum point of ϕ, where $\phi'(c) = 0$, is shown in Figure 6.66 in the case where this is in the interior of the domain.

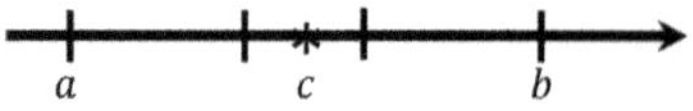

Figure 6.66: Depiction of the maximum of ϕ being located between the limits a and b.

It is only the neighborhood of c that matters, so we can set a, b to $-\infty$ and ∞, respectively. They exponentially decay. An expansion of $\phi(t)$ about $t = c$ gives, noting that $\phi'(c) = 0$:

$$\phi(t) = \phi^* + (1/2)\phi''(c)(t-c)^2 + \ldots.$$

Note that $\phi''(c) \le 0$ at the maximum point. The most general expansion is

$$\phi(t) = \phi^* + \frac{1}{p!}(t-c)^p \phi^p(c) + \ldots; \quad p \ge 2,\ \phi^p(c) < 0\ \{\max\}$$

$$\underset{x\to\infty}{I(x)} = e^{x\phi^*} \int_{c-\varepsilon}^{c+\varepsilon} f(c) e^{x\left[\frac{1}{2}\phi''(c)(t-c)^2\right]} dt = \left[f(c) \int_{c-\varepsilon}^{c+\varepsilon} e^{-\frac{1}{2}|\phi''(c)|(t-c)^2} dt\right] e^{x\phi''}.$$

(Again, ϕ'' is always negative.) Note that $f(c)$ is a first approximation to $f(t)$, while the quantity

$$(1/2)\phi''(c)(t-c)^2 \equiv (\phi(t) - \phi^*)$$

in the exponent is a truncated, second approximation. For large x, it is only when $t = c$ that matters, as all other t's go to zero very fast. That is,

$$\underset{x\to\infty}{I(x)} = f(c)\left[\int_{-\infty}^{\infty} e^{-\frac{x}{2}|\phi''(c)|(t-c)^2} dt\right] e^{x\phi^*}; \ \phi^* = \phi(c)$$

and

$$I(x) \sim \sqrt{2\pi} f(c) e^{x\phi^*} / \sqrt{-x\phi''(c)}.$$

Note that we will encounter a problem if $f(t)$ has a singularity in $[a, b]$, for example, $f(t)$ blowing up at $t = c$.

Case II: c is either a, b

The same argument as before holds, except now that

$$\phi(t) = \phi^* + \phi'(c)(t-c) + \ldots.$$

If we carry through as before, the result here, assuming $a = c$, is

$$I(x) \sim \frac{-f(a)e^{x\phi(a)}}{x\phi'(a)}.$$

Note again that since a is a maximum point, $\phi''(a) < 0$. The difference between Case I and Case II is the appearance of the square root in the denominator in Case I. This means that $I(x)$ goes to zero at a slower rate. This behavior is depicted in Figure 6.67, where we compare the behaviors when the maximum point is located at the boundary

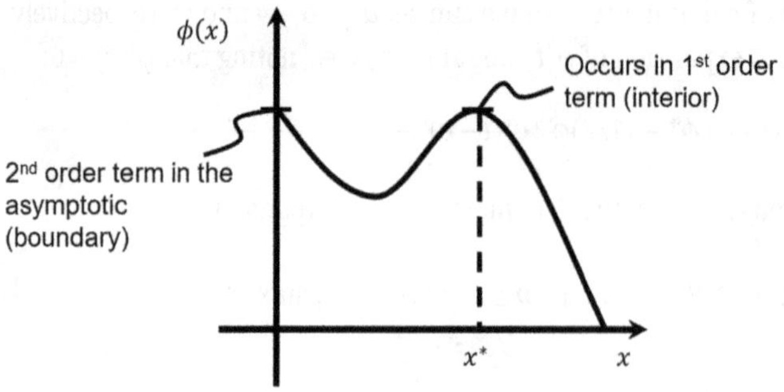

Figure 6.67: Comparison of the effects of having the maximum point at the boundary or in the interior in the Laplace method.

$(x = 0)$ or at the interior point $(x = x^*)$ of a function $\phi(x)$. We don't feel the boundary behavior as we do the interior because the interior (with the square root) dies down at a slower rate.

6.3.6.2 Method of steepest descent

We will give a rather heuristic discussion of the method of steepest descent starting from this subsection until Subsection 6.3.8. The reader should be able to skip these materials and proceed straight to Example 6.15 without missing much.

Consider the integral

$$I(x) = \int_a^b e^{xf(z)} g(z)\, dz,$$

where $f(z)$ is complex and analytic and a, b may be imaginary numbers, and $z \in \Omega$. Suppose a, b are real, we cannot use the Laplace method because $f(z)$ in the exponent is complex: $f = \phi + i\psi$. With this definition for f, we can write

$$I(x) = \int_a^b e^{x\phi} e^{ix\psi} g(z)\, dz.$$

Because of the complex $i\psi$, we cannot separate the maximum of ϕ from oscillations. So, the Laplace method does not work with complex f. A saddle point is a special case of steepest descent. The steepest descent method is focused on where the real part of f, or ϕ, is changing the fastest (steepest). That is, along $\nabla\phi$, which also has the direction of the path where $\frac{d\psi}{ds} = 0$, as discussed earlier in this chapter.

Definition 6.2 A path γ on which $\mathrm{Re}(f) \equiv \phi = \text{constant}$ is said to be a level curve of ϕ.

Definition 6.3 A path Γ on which $\mathrm{Im}(f) \equiv \psi = \text{constant}$ is called a path of stationary phase or a steepest path of the function $f; \Gamma =$ level curve of ψ. Note that the curves $\phi = \text{constant}$ and $\psi = \text{constant}$ must intersect perpendicularly as per the Cauchy's theorem (Figure 6.68). A path of stationary phase is also defined mathematically as that on which the gradient of the imaginary part of f, or $\mathrm{Im}[f]$, is zero.

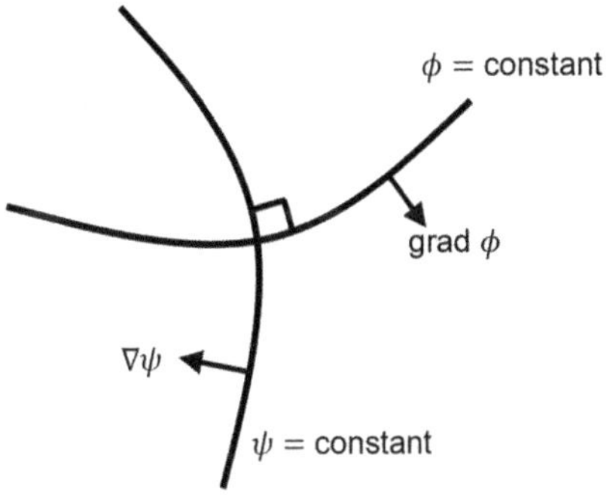

Figure 6.68: Curves of $\phi = \text{constant}$ and $\psi = \text{constant}$ must intersect at right angles as per Cauchy's theorem.

So, basically we want to find a path (Figure 6.69) where ψ is constant (or $d\psi/dz = 0$) and ϕ changes the fastest, which is also along $\nabla\phi$.

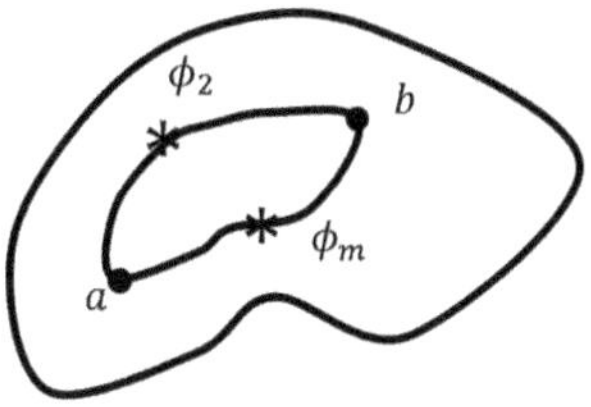

Figure 6.69: Hypothetical paths of $\phi_{\max}$.

Let the maximum of the real part of $f(z)$ occur at ϕ_m for any path Γ_{ab} in Ω. Different paths may have different ϕ_m's, but note that integration is independent of path as there are no poles. This implies that

$$|I(x)| < \int_{\Gamma_{ab}} |e^{xf(z)}||g(z)dz|$$
$$\leq \int_{\Gamma_{ab}} e^{x\phi_m}|g(z)dz|$$

on each path between a, b. Note that $|g(z)|$ is the same for all the paths between a and b. Furthermore, we will pick ϕ^*, which is the minimum of all the ϕ_m's in Ω. This will provide the best estimate of $\phi_{\max}$. The mapping of Ω to $f(\Omega)$ is shown in Figure 6.70. Note that $\mathrm{Re}[f] \equiv \phi$ runs horizontally in the sketches shown below.

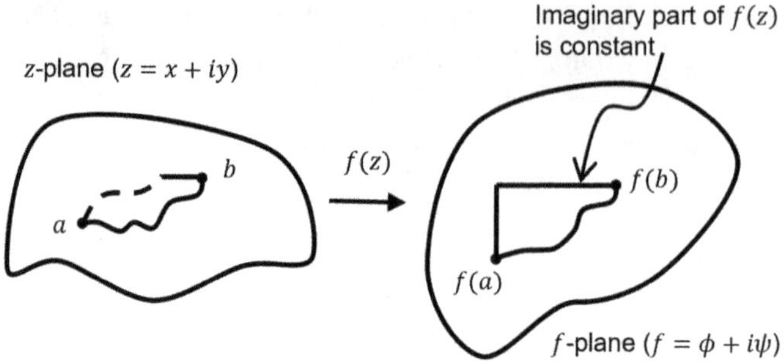

Figure 6.70: The mapping of Ω to $f(\Omega)$ by f. Note that $(a, b) \in \Omega$. Real part of f is maximum at b. We have a stationary phase procedure, where $\psi =$ constant and the gradient of ϕ is the highest.

The real part of f is maximum at b in the steepest descent method, but this is not so in the saddle point method (Figure 6.71).

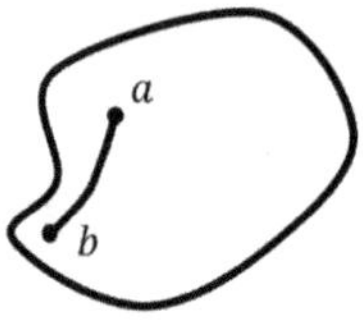

Figure 6.71: The real part of f is not maximum at b. So, we have a saddle point path.

Recollect that for stationary phase,

$$\int_a^b g(t)e^{ix\psi(t)}\,dt \equiv \int_a^b t^a e^{ixt^p}\,dt.$$

If $g(t) = t^a$ vanishes in $[a, b]$ when $\psi'(t) = 0$, then we need to be suspicious of the stationary phase approach and opt instead for the steepest descent method. Note that stationary phase ($\psi =$ constant) is valid for $a < p - 1$.

In the following, the method of steepest descent will be used to understand the asymptotic behavior of

$$I(x) = \int_{\Gamma_{ab}} g(z)e^{xf(z)}\,dz \text{ as } x \to \infty.$$

The points a, b can be anywhere in Ω, and both f, g are analytic. Let $f(z) = \phi(x, y) + i\psi(x, y)$. We are after paths from a to b. Two cases are considered (Figures 6.72 and 6.73):

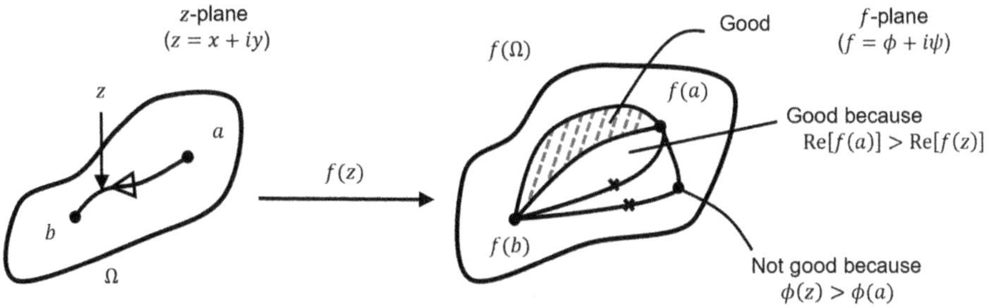

Figure 6.72: A few acceptable steepest descent paths.

Case I: There exists a path Γ_{ab} in Ω, such that $\phi(a) > \phi(z)\ \forall\ z \in \Gamma_{ab}$. That is, $\phi_{\max}$ occurs at a. (Without loss of generality, assume $\phi(a) > \phi(b)$. If smaller, reverse the roles of a and b.) The example of the steepest descent method given in the previous subsection (§6.3.5) is of this type. Note that Γ denotes a path in the domain Ω, while γ will be used to denote a path in the range $f(\Omega)$.

Case II: There is no such a path in the range of f, that is, $f(\Omega)$, in which $\phi(a) > \phi(b)$. This is depicted in Figure 6.73.

More details on Case I

Plausible paths in Ω and their images in $f(\Omega)$ are shown in Figure 6.74. Note that we can always deform a path while saving their end points. We have an open set, so consider a path – no matter how crooked – to consist of two paths (Figure 6.75). Again, by way of the notation for a path, Γ is in Ω while γ is in $f(\Omega)$.

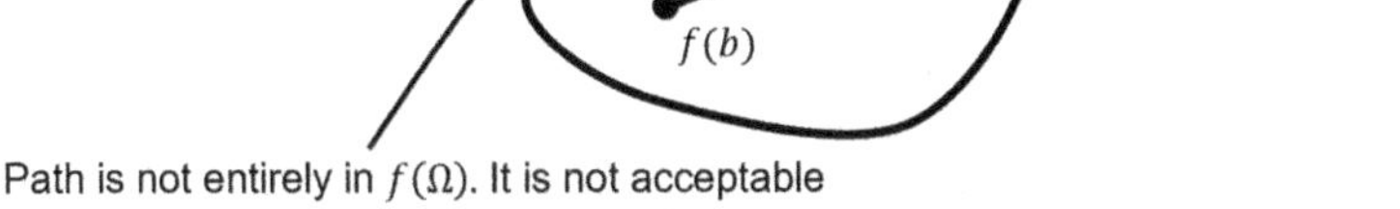

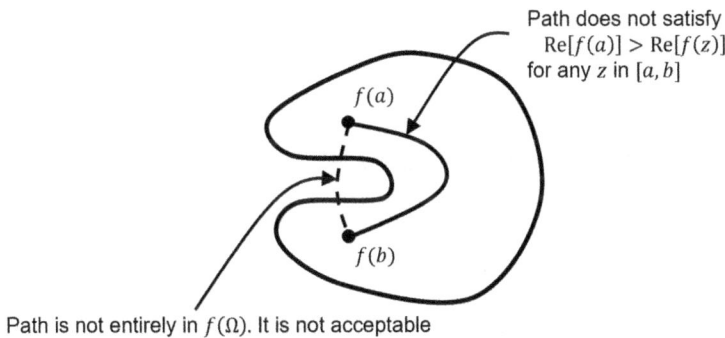

Figure 6.73: Example of an inappropriate steepest descent path.

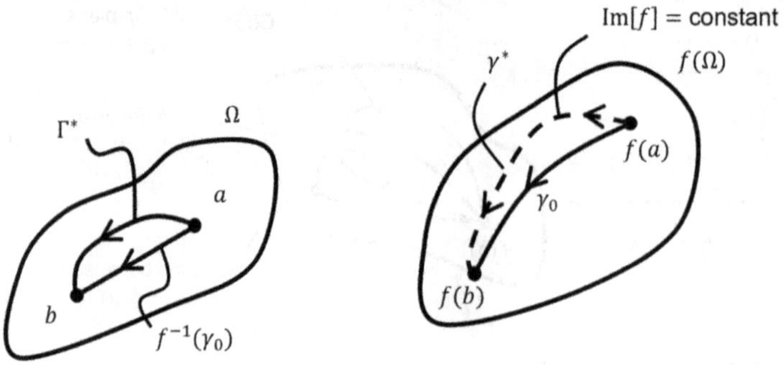

Figure 6.74: Plausible paths. Both the conditions $\text{Im}(f) = \text{constant}$ and $\text{Re}[f(a)] > \text{Re}[f(b)]$ are required for the steepest descent method.

$\Gamma^*_{aa'}$ is a path of stationary phase and hence it is a steepest path for $\phi(x,y)$, and ψ doesn't change along this path, while lines of constant ϕ and ψ are orthogonal. We can decompose Γ_{ab}, the steepest descent path, as we integrate over the path:

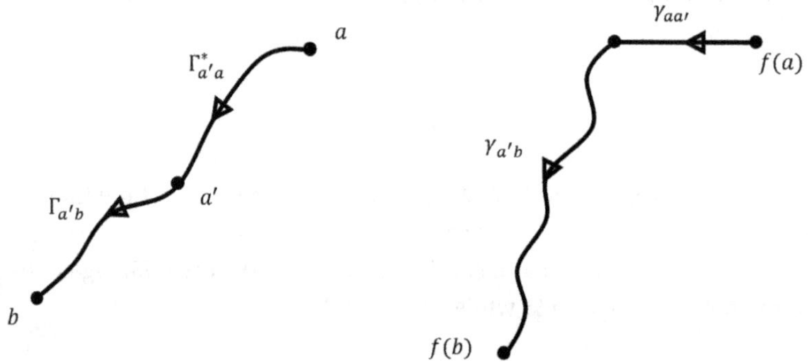

Figure 6.75: Partition of paths in Ω and $f(\Omega)$. Note that Γ is the notation for a path in Ω, while γ is the notation for a path in $f(\Omega)$.

$$I(x) = \int_{\Gamma_{ab}} e^{xf(z)} g(z)dz = \underbrace{\int_{\Gamma_{aa'}} e^{xf(z)} g(z)dz}_{\substack{I_1 \\ \text{Im}\,[f(z)]\text{ is constant} \\ \text{on}\,\Gamma_{aa'}}} + \underbrace{\int_{\Gamma_{a'b}} e^{xf(z)} g(z)dz}_{I_2}$$

Once we have found the maximum of ϕ along the steepest descent path, we can "localize" the exponential in the integrand, as in the Laplace method. To that end, consider the Taylor series, noting that ψ is constant on the path:

$$f(z)=f(a)+f'(a)(z-c)+\ldots=f(a)+\phi'(a)(z-c)+\ldots$$
$$\approx f(a)+\frac{\phi(z)-\phi(a)}{z-c}(z-c)+\ldots=f(a)+[\phi(z)-\phi(a)]$$
$$I_1(x)=e^{xf(a)}\int_{aa'} e^{-x[\phi(a)-\phi(z)]}g(z)dz.$$

Since Im$[f]$ is constant, we can use the Laplace method:

$$|I_2(x)|\le\int_{\Gamma_{a'b}}|e^{x\phi(z)}||g(z)dz|<e^{x\phi(a')}\int_{\Gamma_{a'b}}|g(z)dz|.$$

We have used the assumption that $\phi(a)$ is greater than any other ϕ (or $\phi(z)$) here. Along $\Gamma_{a'b}$, $\phi(z)<\phi(a)$. Thus,

$$|I_2(x)|<e^{x\phi*}\int_{\Gamma_\phi}(|g(z)dz|)$$

as long as $\phi^*<\phi(a)$. We are dealing with exponential comparison, so $I(x)=I_1(x)$.

Other possibilities in Case I include $\phi(b)>\phi(a)$, which can be clearly handled, but if $\phi(a)=\phi(b)$, we will pick Γ as shown in Figure 6.76. (Remember that a and b are the limits of the integral.)

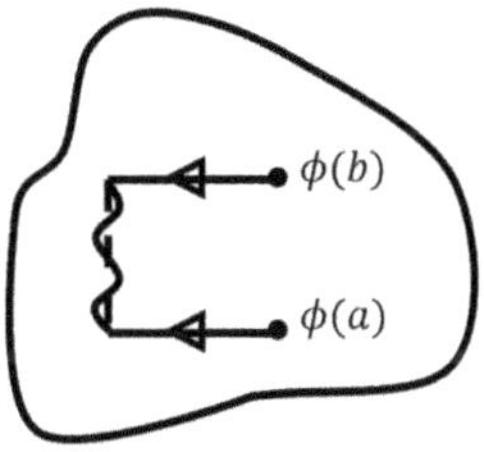

Figure 6.76: Case I when $\phi(a)=\phi(b)$.

More details on Case II

If Case 2 occurs, we will need a saddle point method. (Note that this case says that there is no path in the range of f in which $\phi(a)>\phi(b)$.) Take any path Γ_{ab} in Ω (Figures 6.77

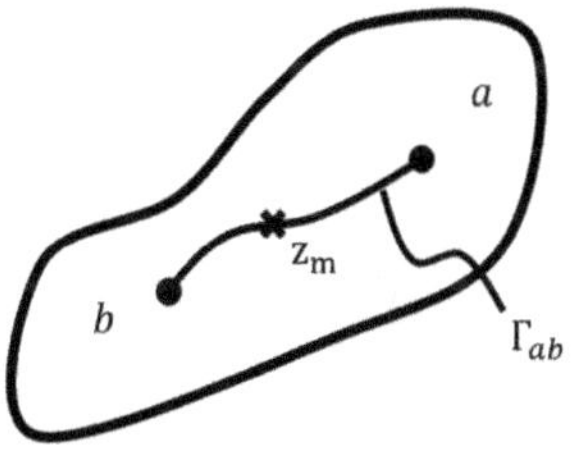

Figure 6.77: A sample path in Ω containing a maximum of ϕ.

through 6.79). Let the maximum of $\phi(z)$ along Γ_{ab} be ϕ_m. Let ϕ_m^* be the minimum of ϕ_m over all paths in Ω. (There could be several in different paths. Check all of them and take the local minimum.)

Let z^* be the location in Ω where $\phi(z^*) \equiv \phi_m^*$ on a certain path Γ_{ab}^*, that is, the particular path containing the minimum of the maximum ϕ_m's. Take a path of stationary phase $(\frac{d\psi}{ds} = 0)$ along z^*. This is possible, as we can determine it by solving a nonlinear equation. (Note that the stationary phase problem is in the space Ω, not the range space $f(\Omega)$.) The z^* is the maximum of $\mathrm{Re}[f]$ or ϕ on this path, Γ_{ab}^*, and minimum on neighboring paths of maximum ϕ_ms. In practical terms, implementation requires simultaneously satisfying the conditions for a stationary phase (steepest descent) and the condition for maximum $\mathrm{Re}[f]$ or ϕ. The first of this will be given by the paths $\alpha = \alpha(c, \beta)$, where $c = \mathrm{Im}[f(z(\alpha, \beta))]$, and is selected to determine the particular curves of the equation $\alpha = \alpha(c, \beta)$ that make the curves pass through the limits of the given integral. The second condition, $\phi \equiv \mathrm{Re}[f]$ is maximum and requires $\frac{d}{dz}\mathrm{Re}[f] = 0$. The $z \equiv z^*$ that satisfies the latter condition is the saddle point (Figure 6.79).

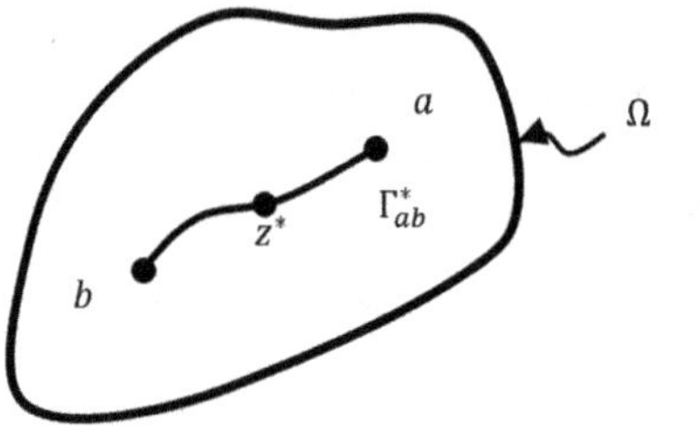

Figure 6.78: The path in Ω containing the minimum of the maximum ϕ_ms in Ω.

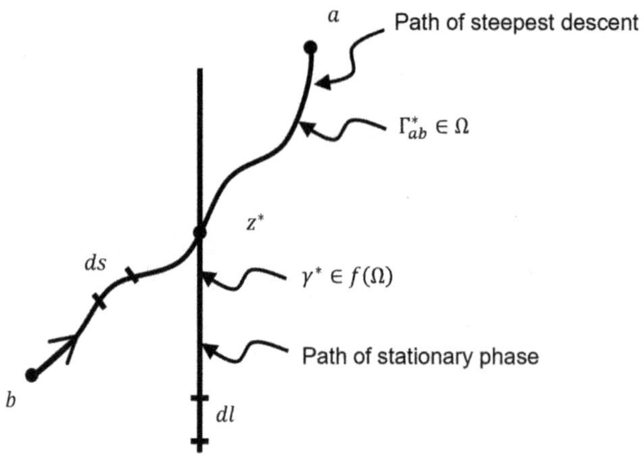

Figure 6.79: Intersection of the steepest descent path Γ_{ab}^* and the path of stationary phase.

6.3.7 Definition of ϕ^*

Referring to Figure 6.79, we have

$$\left.\frac{\partial\phi}{\partial s}\right|_{z*}=0,\quad \text{where } s=\text{arc length along } \Gamma^*_{ab}\ \left(s\in\Omega,\quad \phi^*\in\text{Range of }\Omega\right),\quad \text{and}$$

$$\left.\frac{\partial\phi}{\partial l}\right|_{z*}=0,\quad \text{where } l=\text{arc length on } \gamma^*\ \ (l\in\text{Range of }\Omega).$$

This means that the directional derivative along the two paths = 0 and that

$$\left.\frac{\partial\phi}{\partial x}\right|_{z*}=\left.\frac{\partial\phi}{\partial y}\right|_{z*}=0\Rightarrow z^* \text{ is a saddle point.}$$

If the two paths are coincident at a point, we still have a saddle point: Γ^*_{ab} tangent to γ^* at z^*. That $\partial\psi/\partial s=0$ implies stationary phase. We now also have $\partial\phi/\partial s=0$. That $\partial\phi/\partial s=\partial\phi/\partial l=0$ (or generically, $\partial\phi/\partial x=\partial\phi/\partial y=0$) implies the Cauchy-Riemann conditions, and we still have a saddle point. Note that $\partial\phi/\partial s=0$ implies that

$$\frac{\partial\phi}{\partial x}t_x+\frac{\partial\phi}{\partial y}t_y=0,\quad \frac{\partial\phi}{\partial y}t_x-\frac{\partial\phi}{\partial x}t_y=0.$$

The conclusion is that Case II is a saddle point problem. That is, $f'(z)=0$ at $z=z^*$. Let us assume that $f''(z^*)\neq 0=c$. With $f'(z^*)=0$, we can write

$$f(z)-f(z^*)=\frac{c}{2}(z-z^*)^2[1+c_1(z-z^*)+\ \dots\ ]\equiv\frac{c}{2}(z-z^*)^2h(z),$$

where $h(z=z^*)=1$. (Note that $h(z)$ cannot be zero in the neighborhood of z^*.) We can write

$$f(z)-f(z^*)=\left\{\pm\sqrt{h(z)}(z-z^*)\right\}^2\frac{c}{2}\equiv(\pm\beta)^2\frac{c}{2},$$

where $\pm\beta$ are two single-valued analytic functions. We will take the positive branch of these. We will also define an analytic function (near z^*) $\xi(z)$, such that

$$\xi(z)=+\lambda\equiv(z-z^*)\sqrt{h(z)}\ \text{so that}\ f(z)-f(z^*)=\frac{c}{2}\xi^2,$$

and change variables from the z plane to the ξ plane. Clearly $\frac{c}{2}\xi^2$ is much easier to plot than $h(z)$. Note that later we will define the ξ plane as $\xi=X_1+iX_2$. Recall that our task is to use the steepest descent method to look for the long time behavior of

$$I(t) = \int_{\Gamma_{ab}} e^{tf(z)} g(z) dz, \quad t \to \infty. \tag{6.25}$$

With the foregoing, eq. (6.25) becomes:

$$I(t) = e^{f^* t} \int_{\Gamma_{\alpha\beta}} e^{\overbrace{\left(\frac{c\xi^2}{2}\right)t}^{\text{We want to have bounded exponential and also be able to use Jordan's Lemma selectively on the positive branch}}} \chi(\xi) d\xi; \quad f^* = f(z^*),$$

where $\chi(\xi) \equiv g(z(\xi))(dz/d\xi)$ from the following change of variable formula:

$$\xi = (z - z^*)\sqrt{h(z)}, \quad g(z) \Rightarrow g(\xi),$$

$$d\xi = \frac{d\xi}{dz} dz \Rightarrow dz = \frac{dz}{d\xi} d\xi \Rightarrow g(z) dz = g(\xi) \frac{dz}{d\xi} d\xi.$$

Above, ξ locally maps the curve z, and a, b get mapped to α, β. The mapping is depicted in Figure 6.80. **Note that the z^* in the z-plane is mapped to the origin in the new ξ-plane. This is an important property of the transformation.**

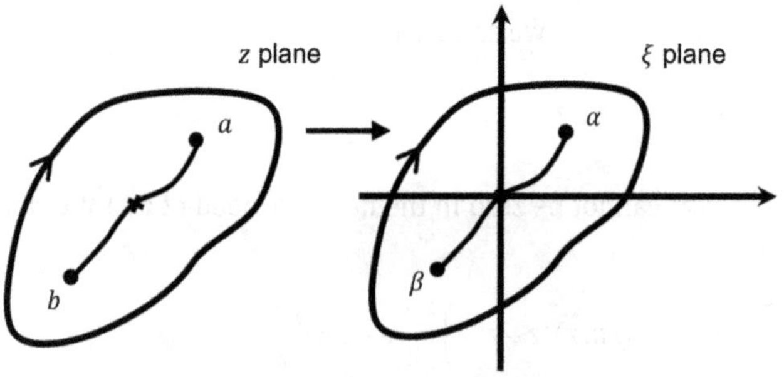

Figure 6.80: The mapping of z to ξ, both of which are complex planes.

Let us define $\phi + i\psi \equiv (c/2)\xi^2 = f(z) - f(z^*)$ in order to study what $e^{(c\xi^2/2)t}$ does locally. Consider

$$e^{tf(z)} = e^{tf^*} e^{t[f(z) - f^*]} = e^{tf^*} e^{t\left(\frac{c}{2}\xi^2\right)}. \tag{6.26}$$

We will look at the ξ plane and examine what the function looks like. Let c be real > 0 (easier to plot). In general, define $c \equiv f''(z^*)$ as

$$c = |c|e^{-2i\theta_0} = \frac{|c|}{2} r^2 e^{2i(\theta-\theta_0)},$$

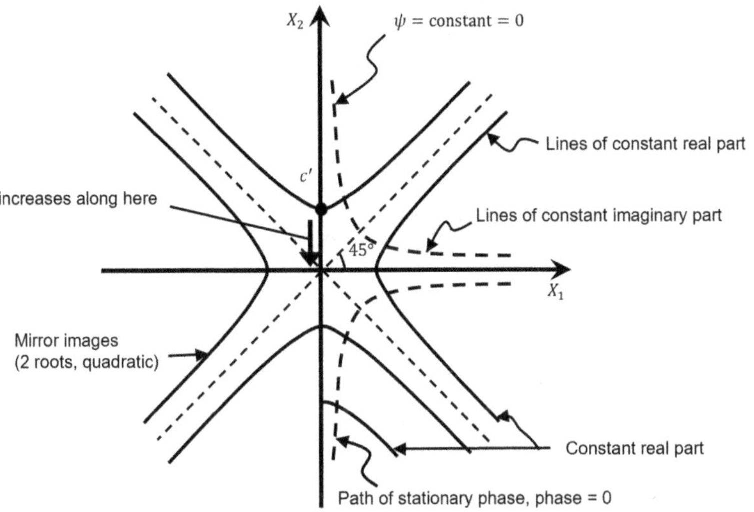

Figure 6.81: The ξ-plane showing the lines of constant real part, lines of rapidly increasing real part ($\nabla\phi$), and constant ψ lines.

where $\theta_0 = -(1/2)\arg c$, $\xi \equiv re^{i\theta}$. The plot of $(c/2)\xi^2$ is four-dimensional on a complex plane, so we cannot draw it. We will draw the contours (Figure 6.81):

$$\xi = X_1 + iX_2, \quad \xi^2 = (X_1^2 - X_2^2) + 2iX_1X_2.$$

The point 0 is the saddle point for the mapping. We have (the imaginary part) $\psi \equiv 0$ if $2(\theta-\theta_0) = \pi, 2\pi$, 3π and 4π, etc., from $\mathrm{Im}[e^{2i(\theta-\theta_0)}] = \sin 2(\theta-\theta_0) = 0$. The real part $\phi = 0$ when we have $\cos(2(\theta-\theta_0)) = 0$ or $2(\theta-\theta_0) = \pi/2$, $3\pi/2$, $5\pi/2$, $7\pi/2$, etc. In general, we do not know the location of α, β (Figure 6.82), where (α, β) are the images of the limits (a, b) on the integral when we map z to ξ. We can always deform our path and it will map to that shown in Figure 6.83.

We pick a path through the saddle point ($\xi = 0$) that goes through α as shown. The path will correspond to some path in a, b. We can pick any path in the plane and we will have a path such that the real part is constant. Figure 6.84 shows the scenario for Case II.

The path of steepest descent must pass through the saddle point in Case II. It doesn't have to in Case I. Therefore, in Case II, we have to determine the saddle point.

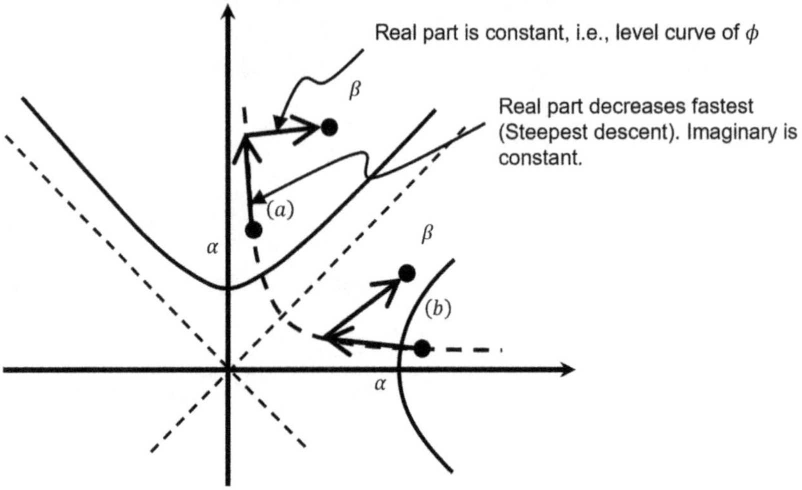

Figure 6.82: Possible paths. Paths (a) and (b) may exist.

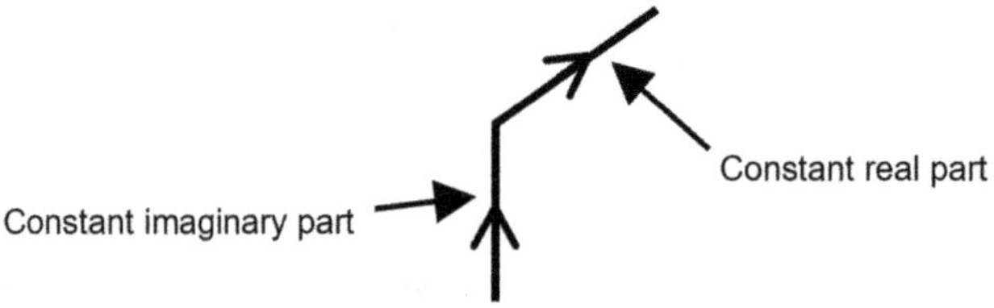

Figure 6.83: Deformation of path into a constant imaginary part and a constant real part.

Example 6.15 (Case II) Investigate $I(t)$ as $t \to \infty$:

$$I(t) = \frac{1}{2\pi i} \int_{\gamma - i\infty}^{\gamma + i\infty} \frac{\exp[t(z\xi - \sqrt{z}]\,dz}{z}, \quad t \to \infty \quad (x \equiv t),$$

where ξ is a real positive parameter (see [2]). This equation arises from the LT application to the heat diffusion equation. We have

$$f(z) = z\xi - \sqrt{z}.$$

One objective here is to replace the contour $(\gamma - i\infty, \gamma + i\infty)$ in Figure 6.85 with a convenient path such that there is no oscillation ($\psi = 0$). Let us look for a saddle point. That is, where $f'(z) = 0$ or where $\xi - 1/2\sqrt{z} = 0$ or $z^* = 1/4\xi^2$. If a path passes through the saddle point (Figure 6.86), it does so at a perpendicular angle. Let us find the path of steepest descent across the saddle point. If we know that the path passes through the saddle point, the problem is essentially solved because we know that the maximum of the function (f), or f^*, is at the saddle point. So, we will just integrate using the formula

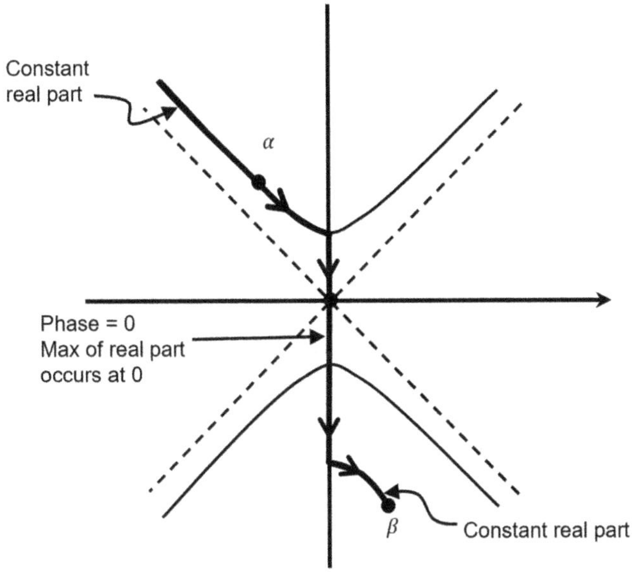

Figure 6.84: Case II graph.

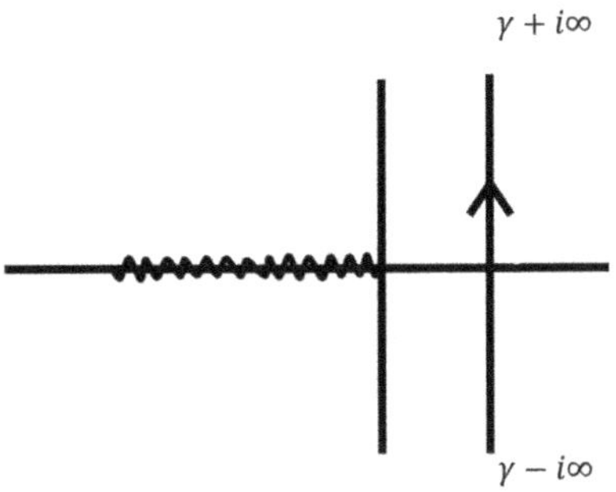

Figure 6.85: Standard path in the z plane for inverting Laplace transform for the heat diffusion equation.

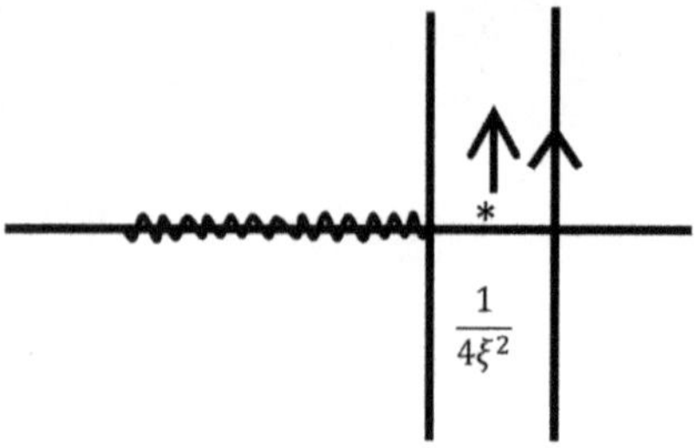

Figure 6.86: The saddle point for $f(z) = 2\xi - \sqrt{z}$ is at $z^* = \frac{1}{4\xi^2}$. The path through the saddle point will point northward from the saddle point, as shown by the arrow on the left in this figure.

$$I(t) = e^{f^* t} \int_{\Gamma_{\alpha\beta}} e^{\frac{tc\xi^2}{2}} \chi(\xi) d\xi,$$

where we have used $f(z) - f(z^*) = c\xi^2/2$. We just need to integrate along the imaginary axis. That is, $\chi(\xi) \equiv g$ needs to be evaluated at the saddle point, with the change of variables from the two-dimensional ξ to the one-dimensional y along the imaginary axis or

$$\frac{c\xi^2}{2} = -iy$$

so that

$$I(t) = e^{f^* t} \int_{-\infty}^{\infty} e^{-\frac{ty^2}{2}} g(z^*) dy.$$

If there are other saddle points, we will need to determine them since they could be more dominant. As pointed out earlier, when our function passes through the saddle point, we do not need to calculate the path of steepest descent. When it does not, as in Figure 6.87 (Case 2), we have to calculate the path of steepest descent.

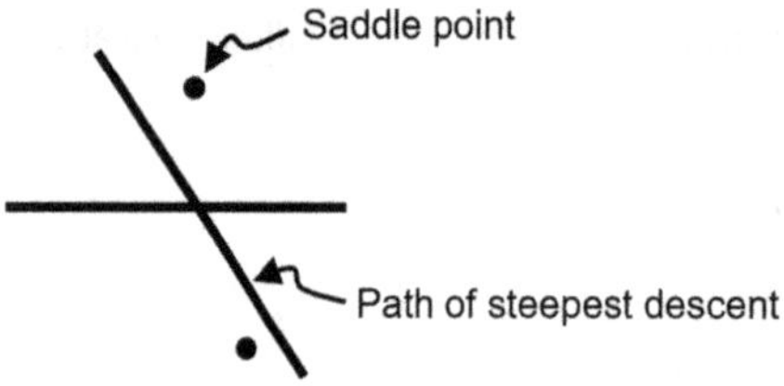

Figure 6.87: When the function does not pass through the saddle point.

We need to find $\text{Im}[f(z)]$ at the point (z^*). Note that $\text{Im}[f(z^*)] = 0 \Rightarrow f(z)$ is real at the saddle point. The equation for the steepest descent path is given by $\text{Im}\left[z\xi - \sqrt{z}\right] = 0$. Let $z = \alpha + i\beta$. Zero imaginary part implies $\beta = 0$ at $\alpha = 1/4\xi^2$ (Figure 6.88). Thus, $\alpha = 1/4\xi^2 - \xi^2\beta^2$ is the equation of the steepest descent path. The steepest descent path (occurs at

$\beta = \mathrm{Im}[f] = 0$); therefore $z^* = 1/4\xi^2 + i0$. The path is a parabola for each α and there are two values of β for each α.

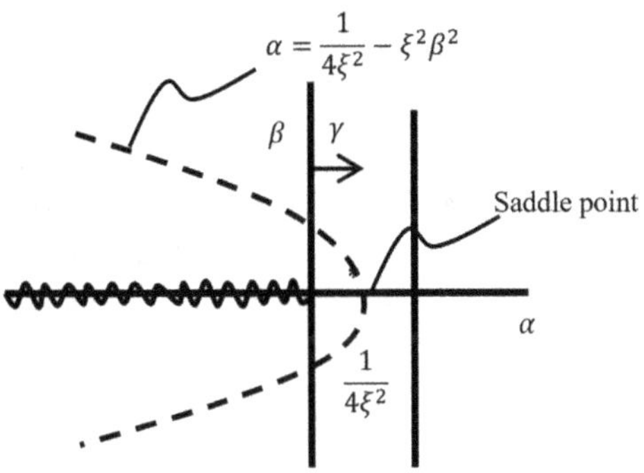

Figure 6.88: Steepest descent path $\alpha = 1/4\xi^2 - \xi^2\beta^2$ in the $z = \alpha + i\beta$ plane passes through the saddle point.

A Taylor series expansion of $f(z)$ about $z^* = 1/4\xi^2$ gives

$$f(z) = (z\xi - \sqrt{z}) = -\frac{1}{4\xi} + \xi^3 (z - z^*)^2 + \ldots.$$

Multiplying by x, we have

$$x(z\xi - \sqrt{z}) = -\frac{x}{4\xi} + x\xi^3 (z - z^*)^2. \tag{6.27}$$

It is easy to see that the steepest descent paths intersect at right angle. Everything is concentrated around the saddle point, the real part decreases just along 1, 2 in Figure 6.89. Note that Curve C cuts the real axis orthogonally. Let change variables by translating so that $t = 0$ becomes the saddle point:

$$z - z^* = it \tag{6.28}$$

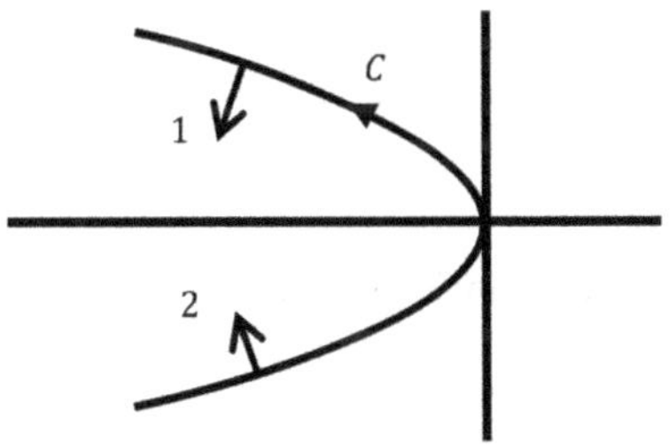

Figure 6.89: Intersection of $\phi =$ constant, $\psi =$ constant. Note that Curve C cuts the real axis orthogonally at the saddle point.

because we are only interested in the little segment after which things decay exponentially fast. Substituting eqs. (6.27) and (6.28) into $I(x)$:

$$I(x) \sim \underbrace{\frac{1}{2\pi i} e^{-\frac{x}{4\xi}} 4\xi^2}_{\text{constant}} \int_{-\infty}^{\infty} i \exp\left[-x\xi^3 t^2\right] dt,$$

where the integral is one-dimensional along the imaginary axis t.

Recall $1/z = g(z)$ since the first term replaces $g(z)$ by $g(z^*)$: $g(z^*) = 1/z^* = 4\xi^2$. Also, recall Watson's Lemma and look close to the point for everything that matters (Laplace method). That is, we have replaced $\int_{-\varepsilon}^{\varepsilon}$ by $\int_{-\infty}^{\infty}$ since everything beyond $\pm\varepsilon$ decays exponentially fast. We will have

$$I(x) \sim 2\sqrt{\frac{\xi}{\pi x}} e^{-\frac{x}{4\xi}}.$$

In most cases, we use numerical calculations to check the results from asymptotics. We need to prove convergence for proper work, but unfortunately, we can't in 99% of the cases. Asymptotics involve secular arguments.

To summarize the analysis of this example so far, we basically found the saddle point where steepest descent path, which is a parabola, also passes through. We use the point in $f(z)$ and merely Taylor-expand $f(z)$ inside the given integral. Then, we evaluate the integral by localizing it to the point with the maximum real part, ϕ, as per the Laplace method.

Remark 6.2 The following remarks can be made:
(1) The saddle point is given by $f'(z) = 0$.
(2) The steepest descent curve is given by $\mathrm{Im}[f(z)] =$ constant.
(3) The steepest descent method is also focused on where $\mathrm{Re}[f]$, or ϕ, is changing the fastest (steepest).
(4) $\mathrm{Im}[f(z)] =$ constant is also the path of stationary phase. Of course, this is also equivalent to setting gradient of $\mathrm{Im}[f(z)] = 0$.

6.3.8 Higher order terms

Toward generating high order terms in the asymptotic expansion, let us define τ by

$$-\tau^2 = \left(z\xi - \sqrt{z}\right) + \frac{1}{4\xi}.$$

This is in the form $f(z) - f(z^*) = (c\xi^2)/2(= -\tau^2)$, though the ξ here is different from the ξ in the previous subsection. Note that $-\tau^2$ is real on C, and $(f(z), f(z^*))$ must have the same imaginary parts on C:

$$-\tau^2 = \frac{1}{2} f''(z^*)(z - z^*)^2 h(z),$$

where $h = [1 + c_1(z - z^*) + c_2() + \ldots]$ is analytic and is not equal to zero at z^*, and the first term of the Taylor series is not necessarily zero because we are looking at the saddle point.

Thus,

$$\tau = +\sqrt{-\frac{1}{2} f''(z^*) h(z)}(z - z^*).$$

(This is the positive branch.) We substitute this expression into our equation (we haven't ignored any terms):

$$I(x) = \frac{1}{2\pi i} e^{-\frac{x}{4\xi}} \int_{-\infty}^{\infty} \frac{1}{z(\tau)} \exp(-x\tau^2) \frac{dz}{d\tau} d\tau.$$

We want to integrate this along the parabolic path, where τ is real along the path C from $-\infty$ to ∞. Note that the quantities $1/z(\tau)$, $dz/d\tau$ define the path. To evaluate these, use the equation for the path, which is given by

$$z = \left(\frac{1}{2\xi} + \frac{i\tau}{\sqrt{\xi}}\right)^2.$$

We have to evaluate $1/z(\tau), dz/d\tau$ even though we are interested in the local behavior at the saddle point. This is because of our interest in the higher order terms. Substituting for $1/z(\tau)$, $dz/d\tau$, expanding term by term, and integrating term by term, we obtain

$$I(x) \sim 2\sqrt{\frac{\xi}{\pi x}} e^{-\frac{x}{4\xi}} \left[1 - \frac{2\xi}{x} + \frac{12\xi^2}{x^2} + \ldots\right],$$

where the $1/x$, $1/x^2$, ... are the gauge functions. Note that quadratic $f(z)$ is good, for example, the parabolic profile here. If logarithmic, obtaining an inverse may be difficult.

6.3.9 (Case I) An example in which the steepest descent path does not pass through the saddle point

Consider the integral

$$I(x) \equiv \int_0^1 \frac{1}{\sqrt{z}} \exp\left[ix\left(z + z^2\right)\right] dz.$$

The path of integration is on the real axis by chance. Here, $f = i(z + z^2)$. The saddle point is given by

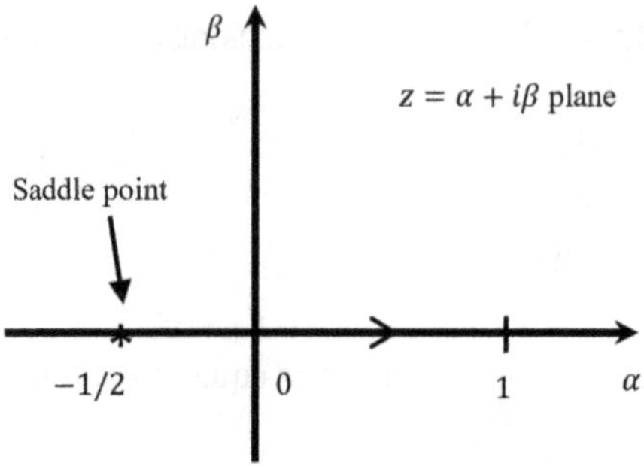

Figure 6.90: Steepest descent path doesn't pass through the saddle point $z^* = -1/2$.

$$f'(z^*) = 0 \Rightarrow i(1 + 2z^*) = 0 \Rightarrow z^* = -1/2.$$

Note that the limits of the integral are 0 and 1, and that $\mathrm{Im}[f(z = 0)]$=0 and $\mathrm{Im}[f(z = 1)]$= 2. They are not on opposite sides of the real axis. Therefore, there cannot be one steepest descent path that goes through both 0 and 1. So we have Case I. That no single steepest path joins the two points $z = 0, 1$ doesn't matter, as long as we follow the path of stationary phase in which the phase is constant along the path. We expect two steepest descent paths that emanate from paths $z = 0$ and $z = 1$. We could assume that the important behavior will be given by the neighborhood of the two points, but we could still find the equation of the steepest descent paths for the interest. For first-order behaviors, we will use the tangent at the saddle point.

The equation for the steepest path is given by $\mathrm{Im}[f(z(\alpha, \beta))] = c$ so that $\alpha = \alpha(c, \beta)$. We can also write this as

$$\mathrm{Im}[f(z)] = \text{constant} = c \text{ on steepest path } = \mathrm{Im}\left[i(z + z^2)\right] = \mathrm{Im}\left[i\left\{\alpha + i\beta + (\alpha + i\beta)^2\right\}\right].$$

Thus, $c = \alpha \pm (\alpha^2 - \beta^3)$, $c = 0 \Rightarrow \alpha + (\alpha^2 - \beta^2) = 0$, $c = 2 \Rightarrow 2 = \alpha + (\alpha^2 - \beta^2)$.

Steepest paths (not necessarily descent) that, respectively, pass through $(0, 0)$ and $(1, 0)$ have the equations

$$\beta^2 = \alpha^2 + \alpha \quad \text{(passes through } (0, 0)),$$
$$\beta^2 = \alpha^2 + \alpha - 2 \text{ (passes through } (1, 0)).$$

These are labeled as Γ_1 and Γ_2 in Figure 6.91. Note that we don't want to integrate along 0 and 1 because $\mathrm{Im}[f(z)]$ is changing, but the imaginary parts on Γ_1 and Γ_2 are constant.

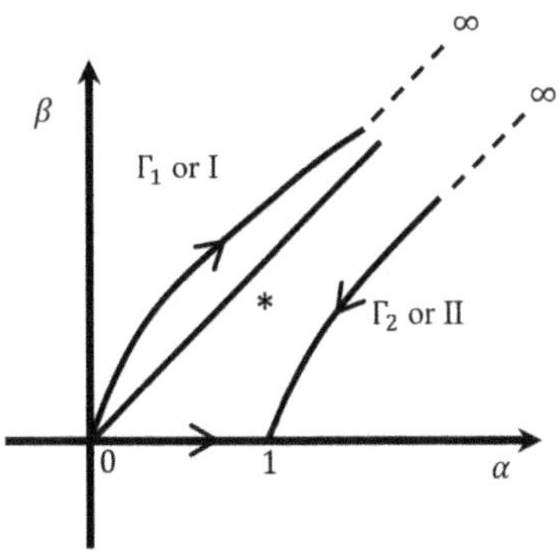

Figure 6.91: Two steepest paths, respectively, passing through (0, 0) and (1, 0).

If $f(z)$ has a pole at the saddle point z^*, we will evaluate the residue there. Then assume that there is no pole, do asymptotics (leading order only), and compare with the case with residue to see if we can throw the residue out.

On Curve I, $\alpha + (\alpha^2 - \beta^2) = 0$ and $\beta = \sqrt{\alpha^2 + \alpha}$, while on Curve II, $\alpha + (\alpha^2 - \beta^2) = 2$ everywhere and $\beta = \sqrt{\alpha^2 + \alpha - 2}$.

Define

$$\tau \equiv -i(z + z^2) \text{ (I)}, \quad \eta \equiv -i(z + z^2) + 2i \text{ (II)}.$$

The minus sign in these two equations has been used because we want τ and η to be positive at ∞. Note that with this change of variables, τ, η will both be real. We can write

$$d\tau = -i[1 + 2z]dz, \quad d\eta = -i[1 + 2z]dz.$$

Along Curve I, $\mathrm{Re}[f(z)] = -\beta - 2\alpha\beta < 0$; $\alpha, \beta > 0$. As we move further and further away from the two circles shown in Figure 6.92, with centers at 0 and 1, respectively, $\mathrm{Re}[f]$ decreases. So as in all Case I, all the action will be located at the centers of these two circles, where $\mathrm{Re}[f]$ is maximum.

Also, the maximum of $f(z) - f(0)$ or $f(z) - f(1)$ on I and II occurs at 0 and 1 since as we move further away, the function is decreasing. **Note that above, τ and η have been defined because we want to integrate along the two curves shown in Figure 6.91.**

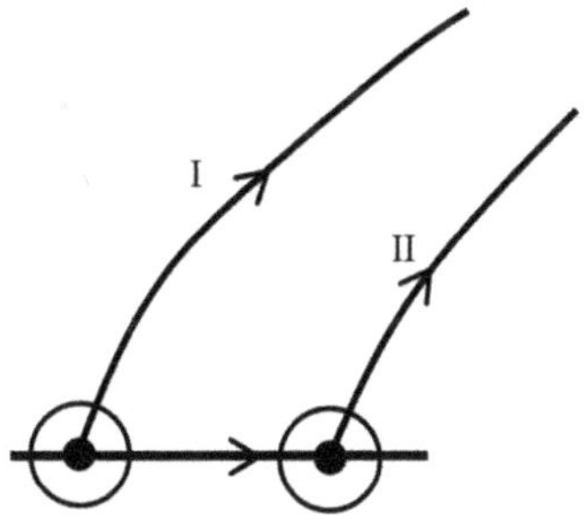

Figure 6.92: As we go further away from 0 and 1, contributions become negligible because $\mathrm{Re}[f]$ is maximum at 0, 1.

Now, on

$$\text{I: } z = \alpha + i\beta = \alpha + i\sqrt{\alpha^2 + \alpha}, \quad \text{II: } z = \alpha + i\sqrt{\alpha^2 + \alpha - 2}.$$

By Cauchy theorem, with no singularity on and between the two curves in Figure 6.92, and assuming that the two curves come closer and closer to each other at infinity, we will have

$$\int_0^1 = \int_{\text{I}} + \int_{\text{II}}.$$

Thus,

$$\int_0^1 \frac{e^{xf(z)}}{\sqrt{z}} dz = \int_0^\infty \frac{e^{-x\tau}}{\sqrt{z(\tau)}} \frac{dz}{d\tau} d\tau - \int_0^\infty \frac{e^{-x\eta}}{\sqrt{z(\eta)}} \frac{dz}{d\eta} d\eta. \tag{6.29}$$

The first integral on the RHS is along Curve I, while the minus sign after it implies that we are going downwards on Curve II. We have

$$\frac{dz}{d\tau} = \frac{i}{1+2z} = \frac{i}{\sqrt{1+4i\tau}}, \quad \frac{dz}{d\eta} = \frac{i}{\sqrt{9+4i\eta}}.$$

If we substitute these into eq. (6.29), we will have

$$\int_0^1 \frac{e^{xf(z)}}{\sqrt{z}} dz = \underbrace{\int_0^\infty \frac{e^{-x\tau} i d\tau}{\sqrt{-\frac{1}{2} + \frac{1}{2}\sqrt{1+4i\tau}}\sqrt{1+4i\tau}}}_{\text{along Curve 1}} - \underbrace{\int_0^\infty \frac{ie^{-x\eta} d\eta}{\sqrt{-\frac{1}{2} + \frac{1}{2}\sqrt{9+4i\eta}}\sqrt{9+4i\eta}}}_{\text{along Curve 2 in the reverse direction}}.$$

Note that $f(z)$ has been replaced by τ, η, which are real, because the imaginary part is constant. This is an important aspect of the method. Using Watson's Lemma, we have

$$I(x) \sim \sqrt{\frac{\pi}{x}} e^{\frac{i\pi}{4}} \left[1 - \frac{3i}{4x} + \dots\right] + \frac{i}{3} e^{2ix} \left(-\frac{1}{x} + \frac{7i}{18x^2} + \dots\right).$$

The dominant behavior is given by $\sqrt{\pi/x} e^{\frac{i\pi}{4}}$ from Curve I. **Note that in general when you have two or more saddle points, just take the one with the biggest real part.**

6.3.9.1 Method of stationary phase

In this case, we consider integrals of the form

$$I(x) = \int_a^b e^{ix\phi(t)} h(t) dt, \qquad x \to \infty$$

$$= \int_a^b e^{xf(z)} h(z)dz, \quad x \to \infty,$$

where a, b are real, $\phi(t)$ is real in $[a, b]$ but can be analytically continued. $h(t)$ can be a complex function. The $f(z)$ is general, with $i\phi(z)$ being a special case of f.

6.3.10 The Riemann-Lebesgue lemma

The Riemann-Lebesgue (RL) lemma states that

$$\lim_{x\to\infty} \int_a^b f(t)e^{ix\phi(t)}dt = 0 \text{ if } \int_a^b \left|f(t)\right|dt < \infty,$$

or $f(t)$ is integrable, $\phi(t)$ is continuously differentiable and is not constant in any subinterval of $[a, b]$. That is, the function $f(t)$ is integrable. We can recall area cancellations from the Riemann-Lebesgue lemma (Figure 6.93) (see Section 6.3.4 for a previous discussion in this book on cancelling high frequency oscillations). If we set $\phi(t) = t$, this integral is the Fourier transform of $f(t)$, which the lemma says goes to zero as $x \to \infty$, provided the integral of $f(t)$ exists. (Of course, x in this case is the Fourier transform variable.) Before analyzing the problem with the method of stationary phase, to which the RL lemma is fundamentally relevant, we will first examine the leading-order results from the method of integration by parts. With the latter, where x is a parameter, we will obtain

$$I(x) = \frac{1}{ix}\int_a^b \frac{h(t)}{\phi'(t)} \frac{de^{ix\phi(t)}}{dt} dt$$

$$= \underbrace{\frac{1}{ix}\frac{h(t)}{\phi'(t)}}_{\text{of } O(1/x)} \underbrace{e^{ix\phi(t)}}_{\text{of} O(1)}\Big|_a^b - \underbrace{\frac{1}{ix}\int_a^b \left[\frac{h(t)}{\phi'(t)}\right]' e^{ix\phi(t)}dt}_{\substack{\to 0 \text{according to RL lemma,} \\ \text{if the first factor in the} \\ \text{integrand is integrable}}}. \tag{6.30}$$

Note that this is acceptable provided $\phi'(t) \neq 0$ in $[a, b]$. With integration by parts, we only get$1/x$, $1/x^2$, $1/x^3$, . . ., so we do not expect the approach to work for Initial Value Problems. Thus, in

$$I(x) \equiv \int_a^b e^{ixf(t)} h(t)dt, \quad x \to \infty,$$

if $f'(t) \neq 0$ in $[a, b]$, then $I(x) \sim A/x$, $x \to \infty$, as per integration by parts. That is, $f(t)$ is either decreasing or increasing. Then locally it's like a linear function, with rapid variation (oscillation) (Figure 6.94). The analysis doesn't work if $f'(t) = 0$ in $[a, b]$; for example, $f'(t^*) = 0$. The c on the time (horizontal) axis in Figures 6.94 and 6.95 is the same as the t^* referred to here.

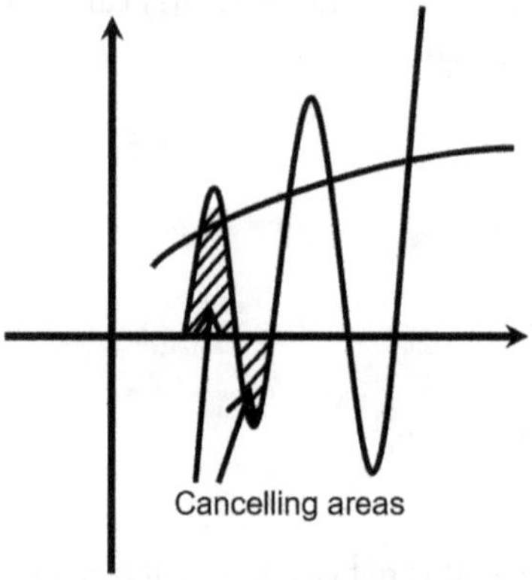

Figure 6.93: Cancellation of areas to give zero integral, as per the Riemann-Lebesgue lemma.

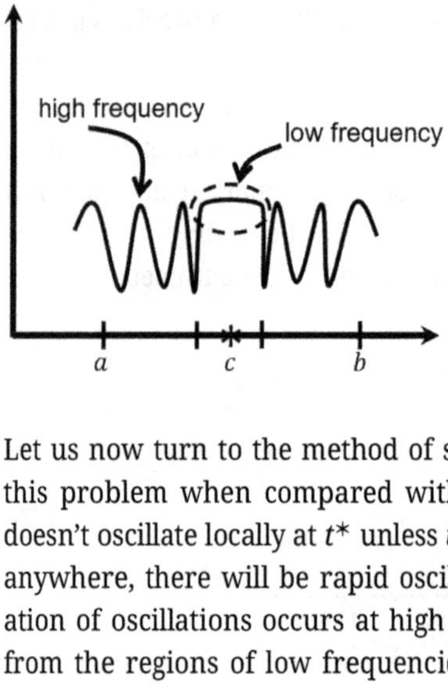

Figure 6.94: Behavior of $I(x)$ as predicted by integration by parts and use of Riemann-Lebesgue Lemma.

Let us now turn to the method of stationary phase, which gives a different result for this problem when compared with the method of integration by parts. The phase doesn't oscillate locally at t^* unless t changes, but t doesn't change locally. So if $f'(t^*) \neq 0$ anywhere, there will be rapid oscillations (Figure 6.95). As stated previously, cancelation of oscillations occurs at high frequencies so that the major contributions come from the regions of low frequencies, such as $t = c$ in Figure 6.95. Note that with the method of stationary phase, $I(x) \sim B/\sqrt{x}$, $x \to \infty$ at $t = c$.

Let us assume that $f^{(k)}(t = a) = 0$, $0 < k \leq p$, k integer. We assume that t^* occurs at a. If this is not located at the end, we will break the domain up so that t^* occurs at a 'new' end (Figure 6.96).

Note that we can derive p is an integer and is equal to unity in our case. From the Taylor series

$$f(t) \sim f(a) + \frac{f^{(p)}(a)}{p!}(t - a)^p + \dots \text{ locally,}$$

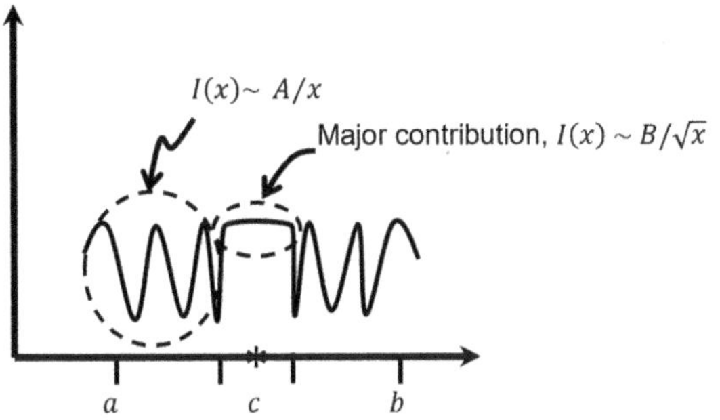

Figure 6.95: Rapid oscillations everywhere, due to $f'(t^*)\neq 0$ anywhere. Note the point "*" in the figure.

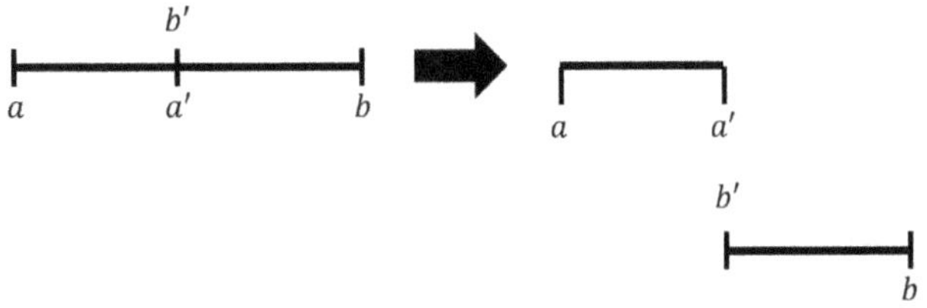

Figure 6.96: Break domain up when t^* (or c) occurs in the interior, say a'.

we can derive

$$I(x) = \int_a^{a+\varepsilon} e^{ixf(t)}h(t)dt + \underbrace{\int_{a+\varepsilon}^{b} e^{ixf(t)}h(t)dt}_{\substack{f'(t)\neq 0 \text{ in } [a+\varepsilon, b] \\ \sim \frac{1}{x}\to 0 \text{ as } x\to\infty}}$$

$$= \underbrace{h(a)}_{\substack{\text{outside because} \\ \varepsilon \text{ is small} \\ \text{and } h(a)\neq h(t)}} \int_a^{a+\varepsilon} e^{ix\left[f(a)+\frac{f^{(p)}(a)}{p!}(t-a)^p\right]} dt$$

$$= h(a)\int_a^{\infty} e^{ix\left[f(a)+\frac{f^{(p)}(a)}{p!}(t-a)^p\right]} dt$$

$$= h(a)e^{ixf(a)}\int_a^{\infty} \exp\left[\frac{ixf^{(p)}(a)}{p!}(t-a)^p\right] dt.$$

The result of the integral depends on whether $f^{(p)}(a) > 0$ or $f^{(p)}(a) < 0$. The only way to integrate is on a complex plane. Thus, we have to go from real to imaginary (Figure 6.97).

That is, this is one of the extensions of the real analysis. If we change variables, we will have

$$I(x) = h(a)e^{ixf(a)} \int_0^\infty \exp\left[\frac{ixf^{(p)}(a)}{p!} z^p\right] dz, \tag{6.31}$$

where

$$z = t - a, \quad z = re^{\frac{i\pi}{2p}}, \quad z^p = ir^p = e^{-x^p},$$

which is integrable. Thus, the integral in eq. (6.31) is reduced to $\int_0^\infty e^{-cx^{p'}} dx$ and

$$I(x) \sim h(a)e^{ixf(a) \pm \frac{i\pi}{2p}} \left[\frac{p'}{x|f^{(p)}(a)|}\right]^{1/p} \frac{\Gamma(1/p)}{p},$$

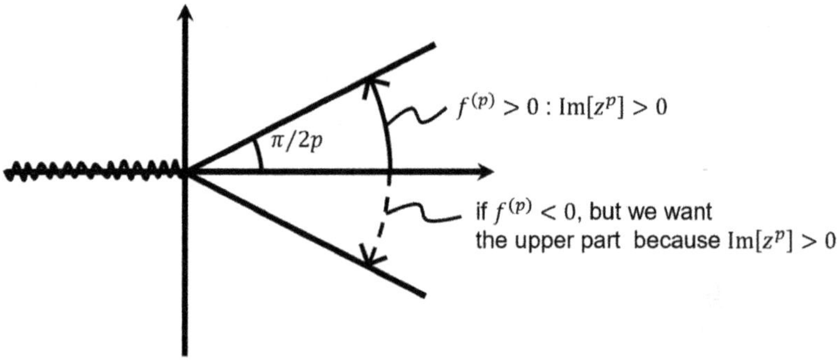

Figure 6.97: Integration in the complex plane is the only option for the problem at hand.

where the "+" sign implies $f^{(p)}(a) > 0$ and "–" sign implies $f^{(p)}(a) < 0$. If $p = 2$, that is, $f'(a) = 0$, $f''(a) \neq 0$ then $I(x) \sim 1/\sqrt{x}$. Note that we can only get one term from the method of stationary phase because of self-cancellations in the procedure. However, the method of steepest decent is more general; it has no assumptions.

Example 6.16 Investigate $J_0(x)$, $x \to \infty$ with the method of stationary phase.

Solution:

$$J_0(x) = -\frac{2}{\pi} \int_0^{\pi/2} \cos(x \cos\theta) d\theta = \frac{1}{\pi} \int_0^{\pi/2} e^{ix\cos\theta} d\theta + 1/\pi \int_0^{\pi/2} e^{-ix\cos\theta} d\theta$$

so that $f = \cos\theta$, $-\cos\theta$ in the two integrands. Stationary phase occurs at $d\cos\theta/d\theta = -\sin\theta = 0$, which is at the lower limit a of the integral. The result is

$$J_0(x) = \left(\frac{2}{\pi x}\right)^{1/2} \cos\left(x - \frac{\pi}{4}\right).$$

If the location of the stationary phase does not occur at the ends, we will need to break up the interval into two, as discussed previously. A cleaner way is to carry out $\int_{-\infty}^{\infty}$ instead of two integrations. If you forget to do this, you will miss the answer by a factor of two. It is appropriate to look at the relationship between the saddle point and steepest descent:

$$I(x) = \int_a^b e^{ixf(t)} h(t)\, dt = \int_a^b e^{xF(z)} h(z)\, dz,$$

where $F(z) = if(z)$. Along a, b, $F(z)$ is purely imaginary, that is, $\mathrm{Re}[F(z)] = 0$. Points where $f'(z) = 0$ in (a, b) or $F'(z) = 0$ is a saddle point. The conditions f'', f''', $\ldots = 0$ are more difficult to characterize as there could be more than 15 points, and we can't easily visualize this.

6.3.10.1 The saddle point method

In the previous problem, consider that the saddle point is located at $z = 0$ (Figure 6.98). (We can translate the saddle point to $z = 0$ if it is not at $z = 0$.) In the saddle point method, we usually consider

$$F(z) - F(z^*) = \frac{c}{2}\xi^2;\ z^* = 0,$$

and we work in the z-plane instead of the ξ-plane. Note that in the ξ-plane, the contour will look like a saddle point, whereas in the z-plane, the contour will look like a saddle point only in the neighborhood of a saddle point. The paths from a to b are depicted in Figure 6.99, in which the dashed curve and the solid line through the saddle point are the same paths except that the latter passes through the saddle point.

The results in the three regions in Figure 6.99 are as follows:

$$\text{I: } \frac{A}{x}, \quad \text{II: } \frac{B}{x}, \quad \text{III: } \frac{C}{\sqrt{x}} = h(a)e^{ixf(a)} \pm \frac{i\pi}{4}\left[\frac{2!}{x|f''(a)|}\right]^{1/2} \frac{\Gamma(1/2)}{2}.$$

First-term asymptotics must give the same answer for steepest descent and stationary phase since they are basically the same on a complex plane. Note that Case III result is for $p = 2$ or a sector angle of $\frac{\pi}{4}$ in Figure 6.97.

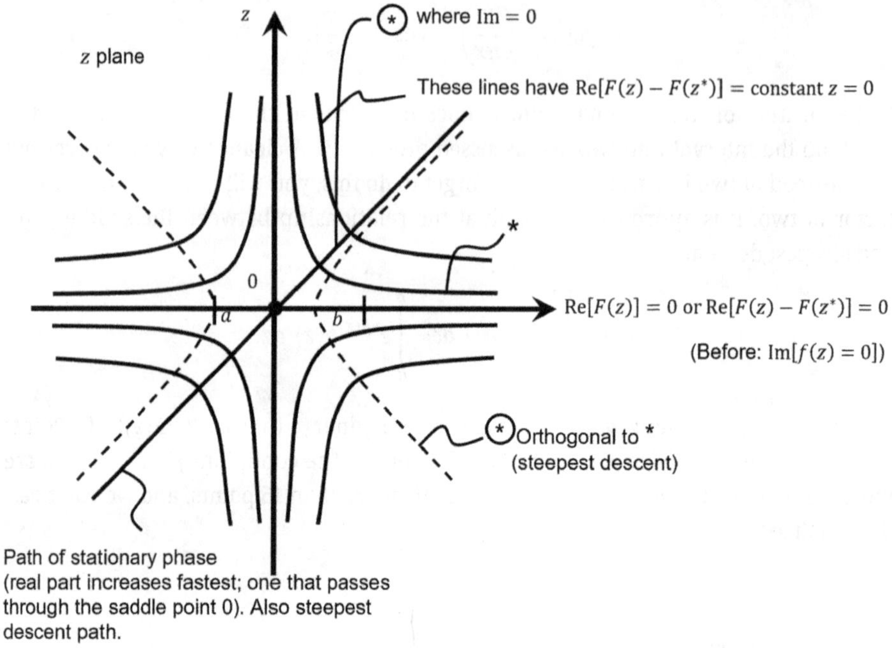

Figure 6.98: Contour plot showing lines of constant real and imaginary parts. Same as for steepest descent.

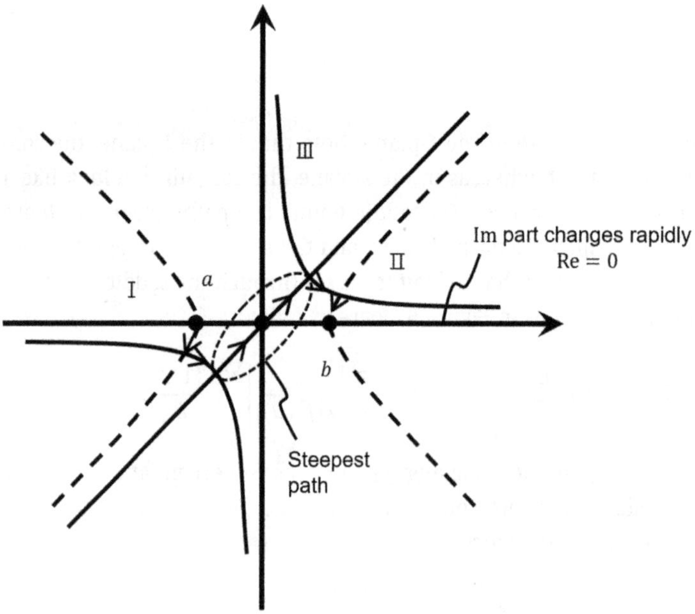

Figure 6.99: Simplified contour plot showing the curves I, II, and III along which asymptotic results are given.

6.4 Miscellaneous examples

Example 6.17 The Airy function is

$$Ai(z) = \frac{1}{\pi}\int_0^\infty \cos\left(\frac{1}{3}t^3 + zt\right)dt.$$

Determine its complete asymptotic expansion for z real, large, and positive, using the method of steepest descent [11].

Solution: Transform the Airy function

$$Ai(z) = \frac{1}{\pi}\int_0^\infty \cos\left(\frac{1}{3}t^3 + zt\right)dt = \frac{1}{2\pi}\int_{-\infty}^{\infty} e^{i\left(\frac{1}{3}t^3 + zt\right)}\,dt$$

to the imaginary axis by setting $t = it'$ (and dropping primes thereafter):

$$Ai(z) = \frac{1}{2\pi i}\int_{-i\infty}^{i\infty} e^{\left(zt - \frac{1}{3}t^3\right)}\,dt.$$

This path may be deformed into L_1 shown for real z (Figure 6.100). In point of fact, the definition of $Ai(z)$ may be taken as

$$Ai(z) = \frac{1}{2\pi i}\int_{L_1} e^{\left(zt + \frac{1}{3}t^3\right)}\,dt, \tag{6.32}$$

where L_1 begins at $t = \infty e^{-2\pi i/3}$ and ends at $t = \infty e^{2\pi i/3}$. Putting $t = 3^{1/2}z^{1/2}t'$ in eq. (6.32) and letting $r^2 = 3^{1/2}z^{3/2}$, we get (putting everything in terms of r^2)

$$Ai(z) = \frac{3^{1/2}r^{2/3}}{2\pi i}\int_{L_1} e^{r^2\left(t - t^3\right)}\,dt. \tag{6.33}$$

Note that for z positive, $r^2 > 0$.

See Irving and Mullineux [8] for a discussion of this point (p. 806). By an extension of Jordan's lemma the starting point of L_1 can be bent around to anywhere in the interval $\infty e^{i\pi/2}$, $\infty e^{-i5\pi/6}$, but no further. Similarly, for top end. We have

$$f(t) = t - t^3, \quad f'(t) = 1 - 3t^2.$$

There are two cols at $t = \pm 1/\sqrt{3}$. Writing $t = x + iy$, $\xi = u + iv$ (x, y, u, v real) we have

$$u(x,y) = x - x^3 + 3xy^2, \quad adv(x,y) = y - 3x^2y + y^3.$$

Through the col at $t=-1/\sqrt{3}$, which concerns us in this problem, we have the level curve

$$x-x^3+3xy^2=-2/3\sqrt{3}$$

and the curves of steepest descent or ascent are given by

$$y-3x^2y+y^3=0.$$

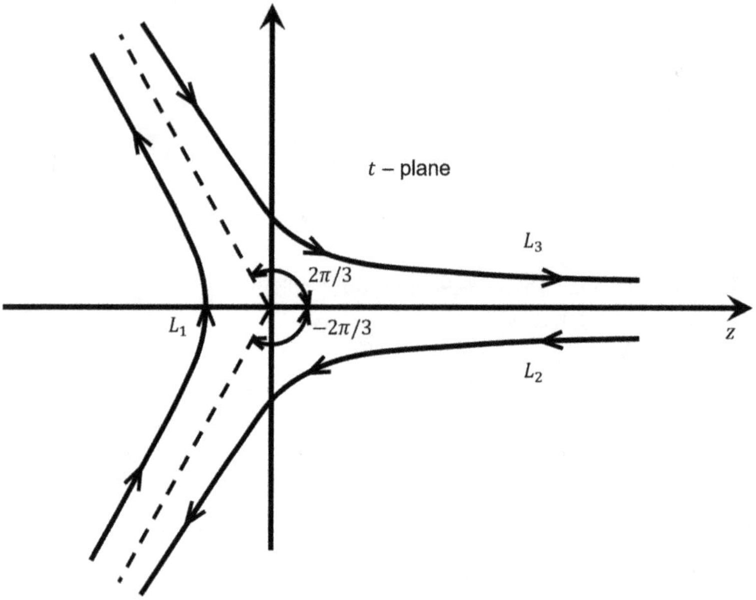

Figure 6.100: Path of integration.

The level curve is a cubic, but the steepest curves are the real axis (steepest ascent) and the hyperbola (steepest descent) $3x^2-y^2=1$. Figure 6.101 attempts to illustrate the form of the saddle area corresponding to this col.

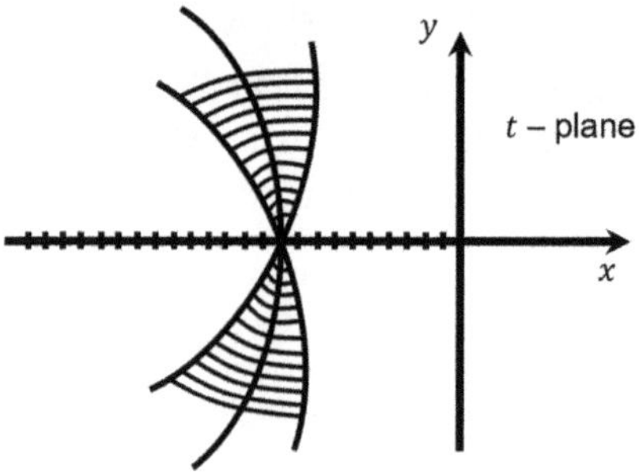

Figure 6.101: Two saddle points.

It is shown as a contour map: the full lines are level curves, and the hatched lines are steepest curves. Shaded areas are valleys and unshaded areas are hills. The path of integration can be deformed into the branch of the hyperbola $3x^2 - y^2 = 1$ which passes through the col $t = -1/\sqrt{3}$ since the latter goes off to infinitely in the same direction as L_1. Note that it cannot be deformed into the other branch, which is the steepest descent path through the col at $t = +1/\sqrt{3}$ since its ends cannot be swing around far enough.

The path of steepest descent from a col is obtained in parametric form by finding the solutions of

$$f(t) = \rho(t_0) - \tau$$

which satisfy the condition $t = t_0$ when $\tau = 0$, τ being so for regarded as an imaginary variable, and then making τ real and positive. The characteristic of the steepest descent path is that on it $\mathrm{Im}[\rho] = $ constant. Hence, on a steepest descent path, any deviation of $f(t)$ from $\rho(t_0)$ must be real. (If we took τ real and negative we would obtain the paths of steepest ascent.) In the present case, the equation to be solved is

$$t - t^3 = -\frac{2}{3\sqrt{3}} - \tau.$$

Writing

$$t = -\frac{1}{\sqrt{3}} + s, \quad \tau = \sigma^2$$

(since these are two paths of steepest descent from the col) we have

$$\begin{aligned} &\sqrt{3}s^2 - s^3 = -\sigma^2, \\ &\text{or } s\left(\sqrt{3} - s\right)^{1/2} = i\sigma. \end{aligned} \tag{6.34}$$

We want the solution of eq. (6.34) which vanishes when $\sigma = 0$, it being understood that $(\sqrt{3} - s)^{1/2}$ means that branch which has the value $3^{1/4}$ when $s = 0$.

By Lagrange's formula for the inversion of series, the required solution of eq. (6.34) is

$$s = \sum_{n=1}^{\infty} b_n (i\sigma)^n,$$

where nb_b is the residue of $\left\{s(\sqrt{3} - s)^{1/2}\right\}^{-n}$ at $s = 0$ which = coefficient of s^{n-1} in the expansion of

$$\left(\sqrt{3} - s\right)^{-\frac{n}{2}} = \frac{\Gamma\left(\frac{3}{2}n - 1\right)}{\Gamma\left(\frac{n}{2}\right)(n-1)!3^{(2n-2)/4}}.$$

Hence

$$t = -\frac{1}{\sqrt{3}} + \sqrt{3}\sum_{n=1}^{\infty} \frac{\Gamma\left(\frac{3}{2}n-1\right)}{\Gamma\left(\frac{n}{2}\right)n!}\left(\frac{\pm i\sigma}{3^{3/4}}\right)^n .$$

On the two paths of steepest descent τ is positive. But since

$$\left(\frac{dt}{d\sigma}\right)_{\sigma=0} = \frac{\pm i}{3^{1/4}},$$

the upper sign gives the path of steepest descent in the upper half plane, the lower sign the path in lower half plane.

From eqs. (6.33) and (6.34), we now have when $\sigma > 0$:

$$Ai(z) = \frac{3^{1/3}k^{2/3}}{2\pi i} e^{\frac{-2R^2}{3\sqrt{3}}} \left\{ \int_0^\infty e^{-k^2\tau} \frac{dt_1}{d\tau} d\tau - \int_0^\infty e^{-k^2\tau} \frac{dt_2}{d\tau} d\tau \right\}.$$

Watson's lemma now allows us to substitute the power series of $dt_1/d\tau$ and $dt_2/d\tau$ and proceed formally to integrate term by term to obtain the derived asymptotic expansion as $k \to +\infty$. Or more generally as $|k| \to \infty$ in $|\arg k^2| < \pi/2$:

$$Ai(z) \sim \frac{e^{-2r^2/3\sqrt{3}}}{2\pi z^{1/4}} \sum_{n=0}^{\infty} \frac{\Gamma\left(3n+\frac{1}{2}\right)(-1)^n}{(2n)!3^{\frac{3n}{2}}k^{2n}}.$$

Usually it is more convenient to take the slightly different variable $\xi = (2/3)z^{3/2}$. Then as $z \to +\infty$

$$\begin{aligned} Ai(z) &\sim \frac{e^{-\xi}}{2\pi z^{1/4}} \sum_{n=0}^{\infty} \frac{\Gamma(3n+1/2)}{(2n)!}\left(-\frac{2}{27\xi}\right)^n \\ &= \frac{e^{-\xi}}{2\sqrt{\pi}z^{1/4}}\left[1 - \frac{3 \cdot 5}{1!\,216\xi} + \frac{5 \cdot 7 \cdot 9 \cdot 11}{2!216^2\xi^2} - \dots\right]. \end{aligned}$$

Clearly these results hold also as $|z| \to \infty$ in $|\arg z| = \pi/3$.

Problems

6.1 Show that for kernels represented by

$$K(t,a) = \frac{1}{2\pi i}\int_{\gamma-i\infty}^{\gamma+i\infty} e^{[st-a\Psi(s)]}\Phi(s)ds,$$

where Φ and Ψ are independent of a, the integral equation

$$f(t) = \int_0^\infty K(t, a)u(a)da$$

can be solved explicitly. Using table, find two such K's.

6.2 Solve Abel's equation

$$f(t) = \int_0^t (t-a)^{-a}u(a)da, \quad 0 < a < 1,$$

making sure that the resulting integral is convergent.

6.3 Show that $u = Ai(z)$ satisfies the differential equation

$$u'' - zu = 0.$$

Hence, determine the asymptotic expansions in questions 6.2 by direct substitution (assuming only the general forms of the expansions). Also express $Ai(z)$ in terms of the Bessel function.

6.4 Derive the asymptotic behavior of $\int_0^1 e^{ixt^2}dt$ as $x \to \infty$ using
(a) Method of steepest descent
(b) Method of stationary phase.
Obtain at least one higher order term for part (a). Can you think of another method to solve this problem?

6.5 By exploring the possibility of expanding the integrands, find asymptotic expansions for the following integrals:
(a) $I(x) = \int_0^{\pi/2} (1 - x\cos^2\theta)^{-\frac{1}{2}}d\theta, \quad x \ll 1$
(b) $I(\varepsilon) = \int_0^{\pi/2} [1 - (1-\varepsilon)\cos^2\theta]^{-\frac{1}{2}}d\theta, \quad \varepsilon \to 0,$
(c) $I(x) = \int_0^\infty (1+t)^{-1}\exp[-xt]dt, \quad x \to +\infty.$

6.6 Assuming that

$$f(x) \sim a_0 + \frac{a_1}{x} + \frac{a_2}{x^2} + \frac{a_3}{x^3} + \ldots \text{ as } x \to \infty,$$

show that

$$\int_x^\infty \left[f(\lambda) - a_0 - \frac{a_1}{\lambda}\right]d\lambda \sim \frac{a_2}{x} + \frac{a_3}{2x^2} + \frac{a_4}{3x^3} + \ldots$$

as $x \to \infty$.

6.7 If

$$I(x) = \int_0^\infty \frac{e^{-\lambda}}{(1+x\lambda)^2}d\lambda,$$

obtain an asymptotic series for $I(x)$ by repeated application of integration by parts.

6.8 Investigate the steepest descent method to investigate the asymptotic behavior of

$$f(x) = \int_0^\infty \exp[ix\left(\frac{1}{3}s^3 + s\right)ds$$

in the limit $x \to \infty$ and compare the result with the previous expansion.

6.9 Consider the integral

$$I(x) = \int_a^\beta g(t)e^{ixh(t)}(t-a)^{\lambda-1}(\beta-t)^{\mu-1} \equiv A(x) + B(x),$$

where

$$\frac{dh}{dt} \equiv h_t(t) = (t-a)^{\rho-1}(\beta-t)^{\sigma-1}h_1(t).$$

Use the method of stationary phase to show that

$$A(x) \sim A_N(x) = -\sum_{n=0}^{N-1}\frac{k^{(n)}(0)}{n!\rho}\Gamma\left(\frac{n+\lambda}{\rho}\right)\exp\left[\frac{\pi i(n+\lambda)}{2\rho}\right]x^{-\frac{n+\lambda}{\rho}}e^{ixh(a)},$$

$$B(x) \sim B_N(x) = -\sum_{n=0}^{N-1}\frac{l^{(n)}(0)}{n!\sigma}\Gamma\left(\frac{n+\mu}{\sigma}\right)\exp\left[\frac{-\pi i(n+\mu)}{2\sigma}\right]x^{-\frac{n+\mu}{\sigma}}e^{ixh(\beta)}, \quad x \to \infty.$$

Note that for the integral

$$f(x) = \int_0^1 \exp\left[ixt^3\right]dt,$$

we have

$$a = 0,\ \beta = 1,\ h = t^3,\ \lambda = 1,\ \mu = 1.$$

We also have

$$\begin{aligned} h_t = \frac{d}{dt}(t^3) = 3t^2 &\equiv (t-a)^{\rho-1}(\beta-t)^{\sigma-1}h_1(t) \\ &= (t-0)^{\rho-1}(\beta-t)^{\sigma-1}h_1(t) \\ &\Rightarrow \rho = 3, h_1 = 3, \sigma = 1. \end{aligned}$$

These parameters can be substituted into $A_N(x)$ and $B_N(x)$ to obtain the asymptotic expansion of $f(x)$ as $x \to \infty$.

6.10 Use the method of steepest descent to show that the general integral

$$I(x) = \int_a^b f(z)\exp[xg(z)]dz, \qquad x \to \infty$$

can be expanded in an asymptotic series as

$$I(x) \sim -\frac{f(a)}{g'(a)}\frac{e^{xg(a)}}{x} + \frac{f(b)}{g'(b)}\frac{e^{xg(b)}}{x}.$$

6.11 Employ the method of steepest descent to determine the leading term for the integral

$$I(x) = \int_a^b z^2 \exp\left[ix\left(z^3 + 3z\right)\right]dz, \qquad x \to \infty,$$

where $a = -1 + i$, $b = 1 + i$. Take x as real.

6.12 Use the Laplace method to asymptotically expand

$$I(x) = \int_0^\infty \exp[-x(1+t^3)\left(1 + \frac{t^3}{3}\right)^{-\frac{1}{3}} dt.$$

6.13 Another Watson's Lemma is that the application of Laplace's method to asymptotically expand

$$f(x) = \int_0^A e^{-xt}t^\lambda h(t)dt$$

yields a true asymptotic expansion for $f(x)$ as $x \to \infty$, provided that $h(t)$ satisfies certain conditions. Note that $h(t)$ being of exponential order (or $|h(t)| < Ke^{\alpha t}$ in$(0, A)$) is important only if $A \to \infty$, as it guarantees the existence of the integral for sufficiently large values of x. Assume that in some interval $(0, T)$, $g(t)$ can be expressed as

$$g(t) = a_0 + a_1 t + a_2 t^2 + \ldots + a_m t^m + R_{m+1}^{(t)},$$

where m is a positive integer, and a constant c exists such that the remainder satisfies

$$|R_{m+1}(t)| < ct^{m+1}.$$

Under this condition, use the Laplace method to show that

$$f(x) \sim a_0 \frac{\Gamma(\lambda+1)}{x^{\lambda+1}} + a_1 \frac{\Gamma(\lambda+2)}{x^{\lambda+2}} + \ldots + a_m \frac{\Gamma(\lambda+m+1)}{x^{\lambda+m+1}} + O\left(\frac{1}{x^{l+m+2}}\right).$$

6.14 Employ the method of stationary phase to obtain an asymptotic expansion for the function

$$f(x) = \int_0^\infty \cos x\left(t^3 - t\right) dt$$

for $x \to +\infty$.

6.15 Suppose

$$I(s,\eta) = \frac{1}{2\pi i} \int_{\gamma-i\infty}^{\gamma+i\infty} \frac{\exp\left[s\left(\lambda\eta - \sqrt{\lambda}\right)\right]}{\lambda} d\lambda,$$

where η is real and $\eta > 0$, and the contour C_1 of integration is along a vertical line which is a distance γ to the right of the origin $(\lambda = \lambda_1 + i\lambda_2)$ (Figure P6.15). Use the method of steepest descent to obtain an asymptotic expansion for $I(s,\eta)$ in the limit $s \to \infty$.

6.16 Use the method of steepest descent to obtain an asymptotic expansion for

$$f(x) = \int_0^1 \frac{1}{\sqrt{\lambda}} \exp\left[ix\left(\lambda + \lambda^2\right)\right] d\lambda$$

for x real and large.

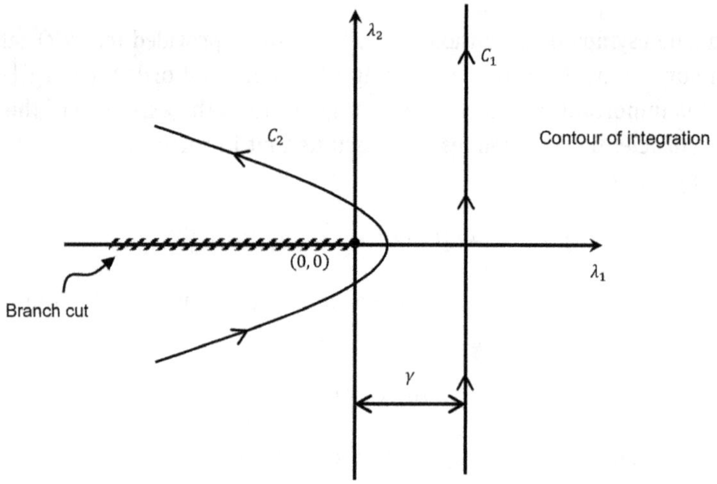

Figure P6.15: Contour of integration for $I(s,\eta)$..

6.17 Starting from an integral formula, obtain an asymptotic expansion for $\Gamma(z+1)$ in the limit $|z| \to \infty$.

6.18 By repeated application of integration by parts, develop an asymptotic expansion for the exponential integral (Ei)

$$-Ei(-x) = \int_{x}^{\infty} \left(\frac{e^{-t}}{t}\right) dt$$

in the limit $x \to \infty$. Show that the series thus obtained is asymptotic.

6.19 Consider the νth order first Hankel function:

$$H_{\nu}^{(1)}(z) = \left(\frac{e^{-\frac{i\nu\pi}{2}}}{\pi}\right) \int_{-\frac{1}{2}\pi+i\infty}^{\frac{1}{2}\pi-i\infty} e^{i(z\cos\lambda + \nu\lambda)}\, d\lambda.$$

Show that

$$H_{\nu}^{(1)}(\nu\sec\beta) \simeq \frac{e^{i\left[\nu(\tan\beta-\beta)+\left(\frac{\pi}{4}\right)\right]}}{\sqrt{\frac{1}{2}\nu\pi\tan\beta}} \left[1 - i\frac{1+\left(\frac{5}{3}\right)\cot^2\beta}{8\tan\beta} + \dots\right].$$

6.20 Show that

$$H_{\lambda}^{(1)}(\lambda) = \left(\frac{e^{-\frac{i\lambda\pi}{2}}}{\pi}\right) \int_{-\frac{1}{2}\pi+i\infty}^{\frac{1}{2}\pi-i\infty} e^{i\lambda(\cos t - t)}\, dt$$

can be asymptotically expanded as

$$H_{\lambda}^{(1)}(\lambda) \sim \left(\frac{1}{\pi}\right) e^{-\left(\frac{1}{2}\right)\pi i} \sqrt{3} \left(\frac{6}{\lambda}\right)^{\frac{1}{3}} \Gamma\left(\frac{4}{3}\right)$$

as $\lambda \to \infty$.

6.21 Obtain an asymptotic expansion for the integral

$$f(x) = \int_{0}^{\infty} \exp\left[i\left(\frac{1}{3}\xi^3 + \xi\right)\right] d\xi$$

as $x \to \infty$.

6.22 Obtain an asymptotic expansion for the integral

$$f(x) = \int_0^1 \exp\left[ixt^3\right] dt$$

as $x \to \infty$.

6.23 Given the integral

$$g(x) = \int_a^b f(t) e^{ix\varphi(t)} dt,$$

where x is a large positive variable and $\mathrm{Im}[\varphi(t)] = 0$. Assume that $f(t)$ is continuous and $\varphi(t)$ is twice differentiable. Let t_0, $a < t_0 < b$, be the only stationary point of φ, that is, where $\varphi'(t_0) = 0$ and $\varphi^N(t_0) > 0$. Show that

$$g(x) \sim \left[\frac{2\pi}{x\varphi''(t_0)}\right]^{\frac{1}{2}} f(t_0) \exp\left[ix\varphi(t_0) + \frac{i\pi}{4}\right].$$

6.24 Employ the technique of integration by parts to derive the following relations as $x \to \infty$:

(a) $\int_x^\infty e^{-t} t^{-\nu} dt \sim e^{-x} \sum_{m=0}^\infty (-1)^m (\nu)_m x^{-\nu-m}$

(b) $\int_b^x e^t t^{-\nu} dt \sim e^x \sum_{m=0}^\infty (\nu)_m x^{-\nu-m}$

6.25 The Stieltjes integral can be written as

$$I(x) = \int_0^\infty \frac{e^{-t}}{1+xt} dt.$$

(a) Find the behavior of $I(x)$ as $x \to 0$ using integration by parts.
(b) Show that difficulties are encountered when the method of integration by parts is applied to obtain the behavior of the function as $x \to \infty$.

6.26 Employ Watson's Lemma to expand the integral

$$I(x) = \int_0^8 \frac{e^{-xt}}{1+t^2} dt$$

as $x \to \infty$.

6.27 Derive the leading behavior of

$$I(x) = \int_0^1 \exp\left[ixe^{-\frac{1}{t}}\right] dt$$

as $x \to +\infty$. Use the method of steepest descent.

6.28 Show that

$$e^z z^{-a} \int_z^\infty e^{-x} x^{a-1} dx \sim \frac{1}{z} + \frac{a-1}{z^2} + \frac{(a-1)(a-2)}{z^3} + \ldots$$

as $z \to +\infty$.

6.29 Show that the function

$$f(x) = \int_0^\infty \left\{\log\lambda + \log\frac{1}{(1-e^{-\lambda})}\right\}\left(e^{-x\lambda}\right)\frac{d\lambda}{\lambda}$$

can be expanded asymptotically as

$$f(x) \sim \frac{1}{2x} - \frac{c_1}{2^2x^2} + \frac{c_2}{4^2x^4} - \frac{c_3}{6^2x^6} + \ldots$$

where the c_i's are constants and $x > 0$.

6.30 Derive the leading behavior of

$$I(x) = \int_0^1 \exp\left[ixe^{-1/t}\right] dt$$

as $x \to +\infty$. Use the method of stationary phase.

6.31 Consider

$$I(x) = \int_{-\infty}^\infty \frac{e^{ix(\lambda+\lambda^3/3)}}{\lambda^2+\lambda_0^2} d\lambda, \qquad x > 0,$$

where λ_0 is a real constant. Find the leading behavior for $x \to \infty$.

Suggested reading

[1] Abramowitz, M. and Stegnum, I. A. (Eds). Handbook of Mathematical Functions, Washington, WA, USA, United States Department Commerce, Bureau of Standards, 1964.

[2] Carrier, G. F., Krook, M., and Pearson, C. E. Functions of Complex Variables, Ithaca, NY, USA, Hod Book, 1983.

[3] Courant, R. and Hilbert, D. Methoden der Mathematischen Physic, Volume I, Berlin, Germany, Verlag von Julius, 1931.
[4] Erdelyi, A. Asymptotic Methods, Mineola, NY, USA, Dover Publications, 1956.
[5] Friedman, B. Principles and Techniques of Applied Mathematics, New York, NY, USA, Wiley and Sons, 1956.
[6] Miles, J. Integral Transforms in Applied Mathematics, New York, NY, USA, Cambridge University Press, 1971.
[7] Morse, P. M. and Feshbach, H. Methods of Theoretical Physics, Volume I, New York, NY, USA, McGraw-Hill, 1953.
[8] Irving, J. and Mullineaux, N. Mathematics in Physics and Engineering, Cambridge, MA, USA, Academic Press, 1959.
[9] Sirovich, L. Techniques of Asymptotic Analysis, New York, NY, USA, Springer-Verlag, 1971.
[10] Bender, C.M. and Orszag, S.A. Advanced Mathematical Methods for Scientists and Engineers, McGraw-Hill Book Company, New York (1978).
[11] Hui, H. Personal Communication, 1986.

7 The techniques of complex Fourier transform

This chapter examines the existence of the Fourier transform (FT) and its inverse from the context of the extension of the transform variable to the complex plane and analytic continuation. The extension of the domain of operation from semi-infinite to the fully infinite type that is typically required for FT operations is presented in substantial detail. Associated with this is the splitting of functions into parts that are analytic in the upper half and those that are analytic in the lower half of the complex plane of the transform variable. Several partial differential and integral equations of engineering interest requiring this decomposition for closed-form solution are used to illustrate the technique. Examples of function splitting are given, as are the systematic applications of the Wiener–Hopf technique to solve a mixed boundary value problem (BVP). The inversion of multidimensional FT is also contained in this chapter.

7.1 Generalization of Fourier transform

7.1.1 Introduction

In the real space, the FT and its inverse could be defined as follows:

$$\tilde{f}(\alpha) = \frac{1}{2\pi}\int_{-\infty}^{\infty} e^{-i\alpha x} f(x)dx, \qquad f(x) = \int_{-\infty}^{\infty} e^{i\alpha x}\tilde{f}(\alpha)d\alpha,$$

where α is the real space transform variable. We will extend the FT to the complex plane and represent the transform variable as $\lambda = \alpha + i\eta$. If a function is integrable (L_1), its FT exists. Also, if a function is square integrable (L_2), then there is a one-to-one mapping (unitary). That is, the mapping is preserved, as is the norm:

$$\|f\| = \int_{-\infty}^{\infty} |f|^2 dx, \qquad \|\tilde{f}\| = \|f\|.$$

Figure 7.1 shows a Fourier mapping from domain to range.

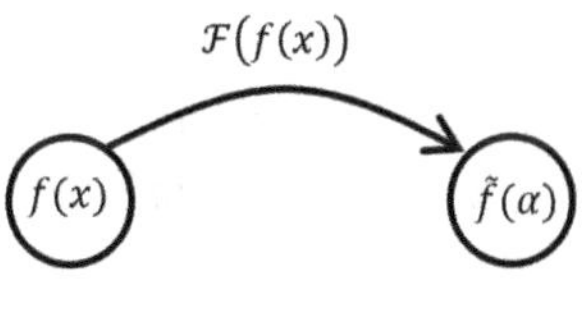

Figure 7.1: Fourier transformation of $f(x)$ to $\tilde{f}(\alpha)$.

The transform is the same as the Hilbert space itself. **We want to generalize the idea of FT because many functions do not have an FT in the traditional (real space)**

https://doi.org/10.1515/9783111351179-009

sense. For example, the function x^n, $n \geq 0$ for a reasonable n, is not L_2 (square integrable). So x^n does not have an FT in the traditional sense.

Also note that the number "1" is not square integrable but $f(x, t=0) = 1 \neq 0$ is of interest, even as $x \to \infty$. Thus, we need to generalize the idea of FT. In undergraduate, we did everything on the real plane, and limited the transform variable to α. This is very limiting and needs to be extended to the complex plane.

Example 7.1. Consider the function

$$f(x) = \begin{cases} 1, & x > 0 \\ 0, & x < 0. \end{cases}$$

Its transform is

$$\tilde{f}(\lambda) = \int_{-\infty}^{\infty} e^{-i\lambda x} dx = \int_{0}^{\infty} e^{-i\lambda x} dx = -\frac{1}{i\lambda} e^{-i\lambda x} \bigg|_0^{\infty} = +\frac{1}{i\lambda},$$

if we assume that $-(1/i\lambda)e^{-i\lambda x}|_{-\infty} = 0$. But if λ is real we cannot make this assumption, instead we will have oscillatory behavior and $|e^{-i\lambda x}| = 1$ is independent of the value of x. Hence, λ, the FT variable, must be complex in order for $\tilde{f}(\lambda) = 1/i\lambda$.

Does the backward (inverse) formula

$$f(x) = \frac{1}{2\pi} \int_{-\infty}^{\infty} e^{i\lambda x} \tilde{f}(\lambda) d\lambda,$$

where $\tilde{f}(\lambda) = 1/i\lambda$ work? The answer is "No," because the integral does not exist. Note that there are several possible results for the integral

$$I \equiv \frac{1}{2\pi i} \int_{-\infty}^{\infty} \frac{e^{i\lambda x}}{\lambda} d\lambda.$$

These are:

(a) The integral does not exist because $1/\lambda$ is not an integrable function

(b) The integral exists in the Cauchy principal value sense. In that case, the answer includes ½ of the residue of the function.

(c) The formula $\int_{-\infty}^{\infty}$, i.e., integration along the real axis, is not right. Let us suppose that we go along another line close to $(-\infty, \infty)$ on the real axis. The complex λ plane is shown in Figure 7.2. For x positive, we will close on the upper half-plane (UHP) ($b \to \infty$ in $\lambda = a + bi$), using Jordan's lemma. This gives a zero value for the integral because of the absence of poles (Figure 7.3). For x negative ($x < 0$), integration, again on the UHP, gives unity, given the residue at 0. This is not right.

(d) Let us try another line, below the real axis $(-\infty,\ \infty)$, or LHP. This gives a zero integral if $x < 0$, and unity if $x > 0$ (Figure 7.4).

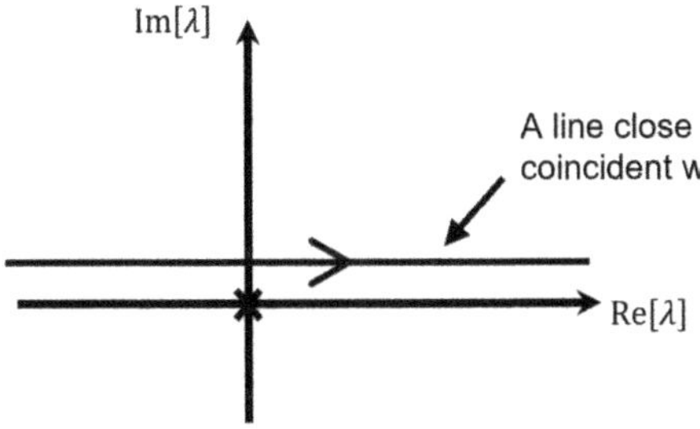

Figure 7.2: A line in UHP parallel and very close to the real line $(-\infty, \infty)$.

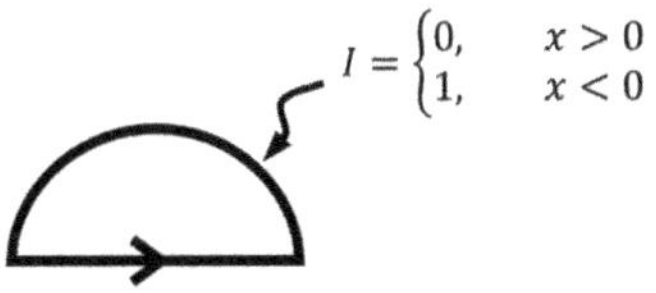

Figure 7.3: Integrating along a line parallel and very close to the real line gives a zero value for the integral because of the absence of poles on the line and in the UHP.

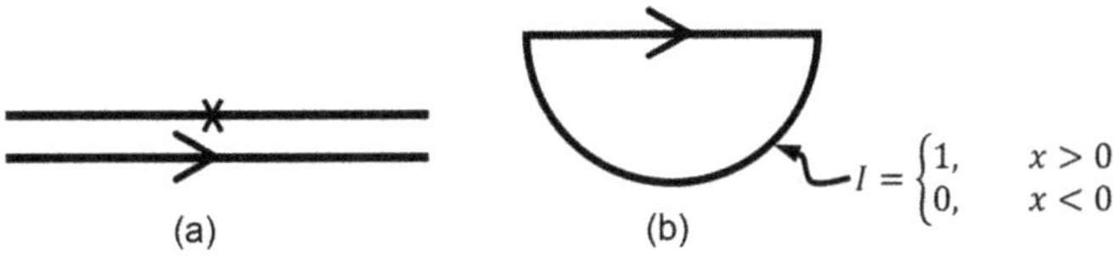

Figure 7.4: A line in the LHP parallel and very close to the real line $(-\infty, \infty)$.

The conclusion so far is that the integral along the real axis has no meaning. Let us look more closely. Let $f(x)$ be a function of $\beta_+,\ \beta_-$:

$$|f(x)| \le \left| \begin{array}{ll} Ae^{\beta_+ x}, & x \to +\infty \\ Be^{\beta_- x}, & x \to -\infty. \end{array} \right. \tag{7.1}$$

(The first part of eq. (7.1) is also required for Laplace transform.) The function $f(x)$ cannot grow faster than some exponential function as $x \to -\infty, \infty$. For example, the function $|f(x)| \sim e^{x^2}$ is unsuitable. (Note that β_+, β_- are real numbers.) On the other hand, $f(x) = x^n$ is suitable because it does not grow faster than exponential. For $f(x) = |x|^n,\ n > 0$,

$$\beta_+ = \varepsilon = \beta_-, \quad \varepsilon > 0.$$

Note that the bound on Laplace transform only deals with x positive, whereas the bound on FT deals with both x positive and x negative. For any given function satisfying eq. (7.1), we can define FT in certain ways:

$$\begin{aligned} f_1(x) &= e^{-\beta x} f(x), \quad \beta > \beta_+, \quad x > 0, \\ &= 0, \qquad\qquad\qquad\qquad\quad x < 0, \\ f_2(x) &= 0, \qquad\qquad\qquad\qquad\quad x > 0, \\ &= e^{\beta x} f(x), \quad \beta > \beta_-, \quad x < 0. \end{aligned}$$

We know that these functions are integrable at ∞, since

$$|f_1(x)| \to 0 \text{ as } |x| \to \infty \text{ (from } f_1(x) = e^{-\beta x} f(x) = Ae^{x(\beta_+ - \beta)}, \ \beta > \beta_+)$$

$$\text{and } |f_2(x)| \to 0 \text{ as } |x| \to \infty.$$

That is, the FT exists, with exponentially fast decay, depending of course on the choice of β.

Let us carry out an FT on $f_1(x)$, noting that there is nothing in $x < 0$; α is real:

$$\begin{aligned} \tilde{f}(\alpha) &= \int_0^\infty e^{-i\alpha x} e^{-\beta x} f(x) dx, \quad \alpha \text{ real} \\ &= \underbrace{\int_0^\infty e^{-i\lambda x} f_1(x)\, dx}_{\text{analytic; } \beta > \beta_+} = \tilde{f}_1(\lambda) = \tilde{f}(\lambda) \end{aligned}$$

if $\lambda = \alpha - i\beta$. Note: If $\eta = -\beta$, $\eta_+ = -\beta_+$, then $\beta > \beta_+ \Rightarrow \eta_+ > \eta$.

Therefore,

$$\tilde{f}(\lambda) = \int_0^\infty e^{-i\lambda x} f(x)\, dx; \ \lambda = \alpha + i\eta,$$

as long as λ stays below LHP, and we have analyticity for all $\eta < \beta_+$, or in the lower half-plane (LHP) $\eta < \beta_+$, i.e., $\text{Im}[\lambda] < -\beta_+$, where $-\beta_+ = \eta_+$ (Figure 7.5). Note that $\text{Im}[\lambda] = \eta$, $-i\lambda x = -i\alpha x + \eta x$, and that $\eta < -\beta^+$, which implies that $\eta < \eta_+$ and location in the LHP. The inverse of $\tilde{f}_1(\alpha)$ exists and is given by

$$f_1(x) = \frac{1}{2\pi} \int_{-\infty}^{\infty} e^{i\alpha x} \tilde{f}_1(\alpha)\, d\alpha.$$

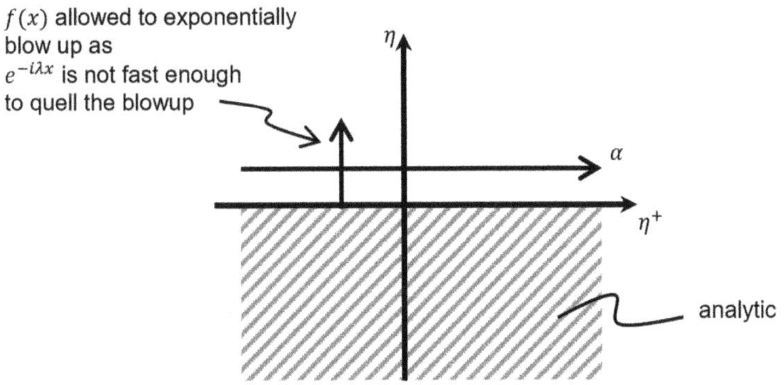

f(x) allowed to exponentially blow up as
e^{-iλx} is not fast enough
to quell the blowup
η
α
η+
analytic

Figure 7.5: Analyticity in the LHP of the complex λ plane ($\lambda \equiv \alpha + i\eta$).

For $x > 0$,

$$f_1 \equiv e^{\eta x} f(x) = \frac{1}{2\pi} \int_{-\infty}^{\infty} e^{i\alpha x} \tilde{f}_1(\alpha)\, d\alpha \Rightarrow f(x) = \frac{1}{2\pi} \int_{-\infty}^{\infty} e^{ix(\alpha+i\eta)} \underbrace{\tilde{f}_1(\alpha)}_{\tilde{f}(\lambda)}\, d\alpha.$$

Note that $d\alpha = d\lambda$. The inversion formula, if function exists in real axis, is

$$f(x) = \int_{-\infty+i\eta}^{\infty+i\eta} e^{i\lambda x} \tilde{f}(\lambda) d\lambda, \qquad \text{for } \mathrm{Re}[\lambda] < \eta^+ (\text{or } \alpha < \eta^+), \tag{7.2}$$

where any negative number $\eta < 0$ will work, or $\eta < \eta_+$. If $x < 0$ the theory of Fourier transformation says that the integral is 0. (Fourier inversion path works only for $x > 0$ (Figure 7.6).) $\tilde{f}$ is analytic in the LHP, which leads to zero inverse. Regarding the limits on the integral in eq. (7.2), note that a horizontal line in the λ plane that is offset by ω (from $\eta = 0$) along the imaginary axis ($i\eta$) runs from $-\infty + i\omega$ to $+\infty + i\omega$, analogous to the $\gamma - i\infty$ to $\gamma + i\infty$ for the vertical line offset by γ from $\alpha = 0$ in the LT inversion procedure. We will then use LHP if $x < 0$, since $i\lambda x = i\alpha x - \eta x$. (It should be noticed that the reference to lower (upper) half-plane does not imply $x < 0$ ($x > 0$) per se, but rather to $\eta < 0$ ($\eta > 0$). On the other hand, minus and plus functions often, respectively, refer to functions defined in $x < 0$ and $x > 0$.) We could also consider functions like

$$f_2(x) = \begin{cases} 0, & x > 0 \\ e^{\eta x} f(x), & x < 0, \end{cases}$$

$$\tilde{f}_2(\alpha) = \int_{-\infty}^{\infty} e^{-i\alpha x} f_2(x)\, dx = \int_{-\infty}^{\infty} e^{-ix(\alpha+i\eta)} f(x)\, dx = \int_{-\infty}^{\infty} e^{-i\lambda x} f(x)\, dx \equiv \tilde{f}(\lambda).$$

This function is analytic in the UHP, for $\eta > \eta_-$, or $\beta > \beta_-$ from $\lambda = \alpha + i\beta$, and

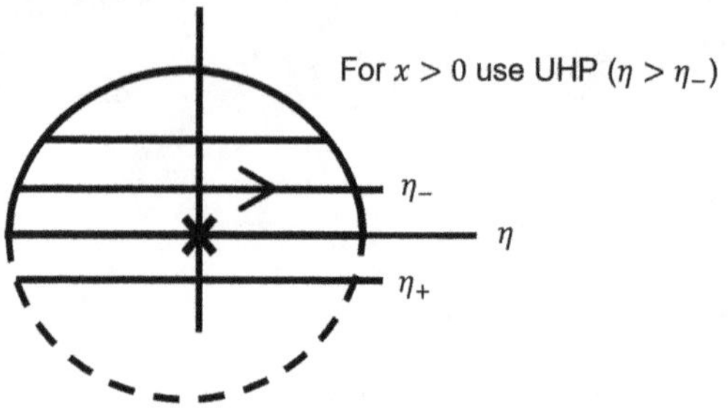

Figure 7.6: Inversion path for $x > 0$: use UHP.

$$f(x) = \frac{1}{2\pi} \int_{-\infty+i\eta}^{\infty+i\eta} e^{i\lambda x} \tilde{f}(\lambda) d\lambda, \qquad \mathrm{Im}[\lambda] \equiv \eta > \eta_-. \tag{7.3}$$

To generalize, any function can always be split. That is, we can always define:

$$\tilde{f}(\lambda) \equiv \int_{-\infty}^{\infty} e^{i\lambda x} f(x)\, dx = \int_{-\infty}^{0} e^{i\lambda x} f(x)\, dx + \int_{0}^{\infty} e^{i\lambda x} f(x)\, dx.$$

That is, the takeaway from the foregoing is that any function can be split as follows:

$$\tilde{f}(\lambda) \equiv \underbrace{\tilde{f}_-(\lambda)}_{f \text{ in eq.(7.2)}} + \underbrace{\tilde{f}_+(\lambda)}_{f \text{ in eq.(7.3)}} .$$

Thus, we need two inversion paths, as shown in Figure 7.7. One problem though is that we don't know the function we are inverting. That is, we don't know η_+, η_-, etc. The main idea pertains to how we should split the function into a part that is analytic in the UHP and another that is analytic in the LHP. This is also the idea behind the Wiener–Hopf technique that we shall see later in this chapter.

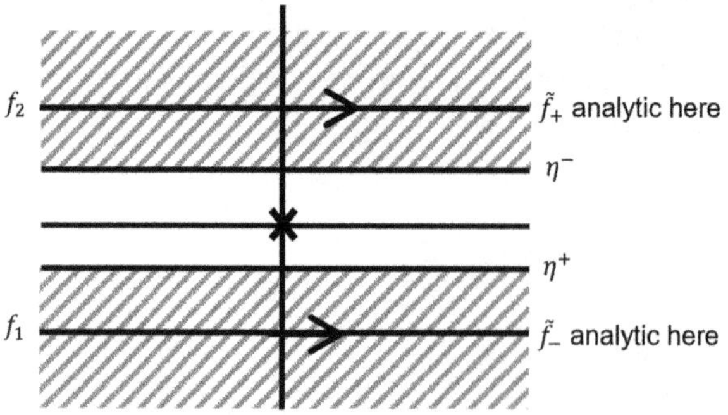

Figure 7.7: A function $\tilde{f}(\lambda)$ can be split into $\tilde{f}_+(\lambda)$ and $\tilde{f}_-(\lambda)$, which are respectively analytic in UHP and LHP.

The next section sheds some light on the precise meaning and magnitudes of η^+ and η^-.

7.2 Function splitting

In this section, we will discuss the behavior of the integrand of the inversion integral from the standpoint of analyticity and the selection of the contour for inversion.

7.2.1 The components of the Fourier transform of $e^{\pm ax}$ for real and complex transform variables

In this section, we examine the integral of the function $f(x) = e^{\pm ax}$ for real and complex versions of the FT variables. The real transform variable is denoted as α, while the complex one is denoted as $\lambda = \alpha + i\eta$, where α and η are real. We decompose the integral into $x \in (-\infty, 0]$ and $x \in [0, \infty)$ and examine the existence or otherwise of the integral and the required domain in the $i\eta$ direction. The eight cases considered are itemized as follows:

1) Case I: Transform of $e^{-ax}, x \in (-\infty, 0]$, real transform variable

$$I(x) = \int_{-\infty}^{0} e^{-ax} e^{-i\alpha x}\, dx = \int_{-\infty}^{0} e^{-(a+i\alpha)x}\, dx = -\frac{1}{(a+i\alpha)} e^{-(a+i\alpha)x}\Big|_{-\infty}^{0}.$$

Since x is negative between $(-\infty < x < 0)$, this integral does not exist at $x \to -\infty$.

2) Case II: Transform of $e^{-ax}, x \in (-\infty, 0]$, complex transform variable

$$I(x) = \int_{-\infty}^{0} e^{-ax} e^{-i\lambda x}\, dx = \int_{-\infty}^{0} e^{[(\eta - a) - i\alpha]x}\, dh.$$

This requires $\eta > a$ because x is negative in $(-\infty, 0)$. Therefore,

$$I(x) = \frac{1}{(\eta - a) - i\alpha}, \quad (\eta > a).$$

3) Case III: Transform of $e^{-ax}, x \in [0, \infty)$, real transform variable

$$I(x) = \int_{0}^{\infty} e^{-ax} e^{-i\alpha x}\, dx = \int_{0}^{\infty} e^{(-a-i\alpha)x}\, dx = \int_{0}^{\infty} e^{-(a+i\alpha)x}\, dx = -\frac{1}{(a+i\alpha)} e^{-(a+i\alpha)x}\Big|_{0}^{\infty} = +\left[\frac{1}{(a+i\alpha)}\right]$$

since x is positive.

4) Case IV: Transform of $e^{-ax}, x \in [0, \infty)$, complex transform variable

$$\int_0^\infty e^{-ax} e^{-i\lambda x}\, dx = \int_0^\infty e^{[(\eta-a)-i\alpha]x}\, dx = \frac{1}{[(\eta-a)-i\alpha]} e^{[(\eta-a)-i\alpha]x}\Big|_0^\infty = -\frac{1}{[(\eta-a)-i\alpha]}.$$

This requires that $\eta - a < 0$, i.e., $\eta < a$, since x is positive.

5) Case V: Transform of $e^{ax}, x \in (-\infty, 0]$, real transform variable

$$I(x) = \int_{-\infty}^0 e^{ax} e^{-i\alpha x}\, dx = \int_{-\infty}^0 e^{(a-i\alpha)x}\, dx = \frac{1}{(a-i\alpha)} e^{(a-i\alpha)x}\Big|_{-\infty}^0 = \frac{1}{(a-i\alpha)}$$

since x is negative.

6) Case VI: Transform of $e^{ax}, x \in (-\infty, 0]$, complex transform variable

$$I(x) = \int_{-\infty}^0 e^{ax} e^{-i\lambda x}\, dx = \int_{-\infty}^0 e^{[(a+\eta)-i\alpha]x}\, dx = \frac{1}{(a+\eta)-i\alpha} e^{[(a+\eta)-i\alpha]x}\Big|_{-\infty}^0 = \frac{1}{[(a+\eta)-i\alpha]}.$$

This requires that $a + \eta > 0$, or $\eta > -a$, since x is negative.

7) Case VII: Transform of $e^{ax}, x \in [0, \infty)$, real transform variable

$$I(x) = \int_0^\infty e^{ax} e^{-i\alpha x}\, dx = \int_0^\infty e^{(a-i\alpha)x}\, dx = \frac{1}{(a-i\alpha)} e^{(a-i\alpha)x}\Big|_0^\infty.$$

This goes to infinity since x is positive.

8) Case VIII: Transform of $e^{ax}, x \in [0, \infty)$, complex transform variable

$$I(x) = \int_0^\infty e^{ax} e^{-i\lambda x}\, dx = \int_0^\infty e^{[(a+\eta)-i\alpha]x}\, dx = \frac{1}{[(a+\eta)-i\alpha]} e^{[(a+\eta)-i\alpha]x}\Big|_0^\infty = -\frac{1}{[(a+\eta)-i\alpha]}.$$

This requires $(a + \eta) < 0$ or $\eta < -a$, since x is positive.

Table 7.1 summarizes the result. The last two columns in the table, "Condition on η" and "Sketch of the domain of existence of integral," clearly define the η^+ and η^- referred to in the previous sections. The condition on η for the integrals above has been written in terms of the constant a. Certainly this constant will be different for functions other than $e^{\pm ax}$. Obviously, an analogous analysis would apply to the inverse transformation as well. That is, with $e^{i\alpha x}$ and $e^{i\lambda x}$ replacing $e^{-i\alpha x}$ and $e^{-i\lambda x}$ in the integrand.

Table 7.1: The components of the Fourier transform of $e^{\pm ax}$ for real and complex transform variables.

	Integral	Transform variable	Value of integral	Condition on η	Sketch of the domain of existence of integral
1)	$\int_{-\infty}^{0} e^{-ax} e^{-i\alpha x}\,dx$	α	Infinity	–	None
2)	$\int_{-\infty}^{0} e^{-ax} e^{-i\lambda x}\,dx$	$\lambda = \alpha + i\eta$	$\frac{1}{[(\eta - a) - i\alpha]}$	$\eta > a$	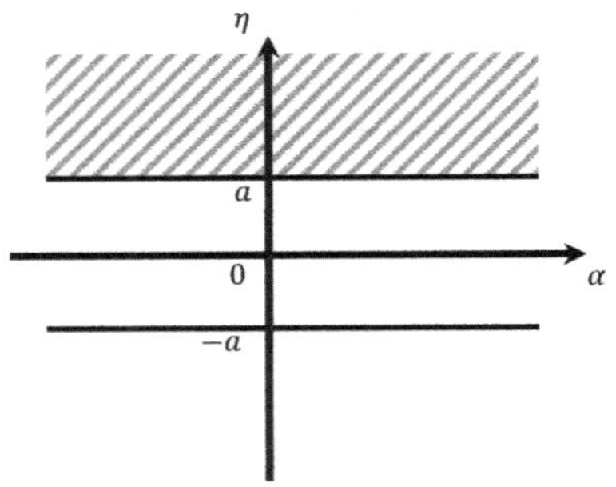
3)	$\int_{0}^{\infty} e^{-ax} e^{-i\alpha x}\,dx$	α	$\frac{1}{(a + i\alpha)}$	–	No restriction
4)	$\int_{0}^{\infty} e^{-ax} e^{-i\lambda x}\,dx$	$\lambda = \alpha + i\eta$	$\frac{-1}{[(\eta - a) - i\alpha]}$	$\eta < a$	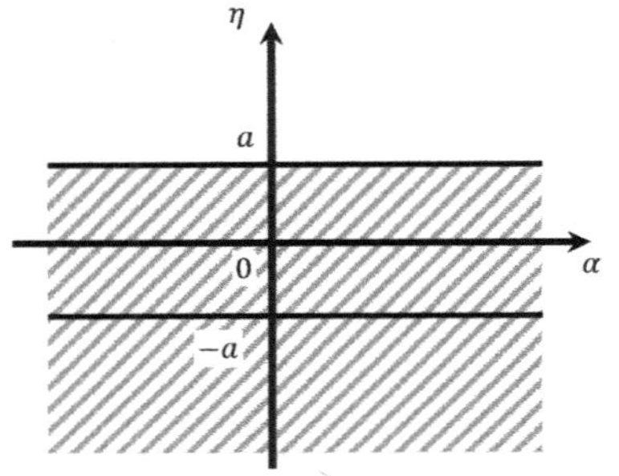

(continued)

Table 7.1 (continued)

	Integral	Transform variable	Value of integral	Condition on η	Sketch of the domain of existence of integral
5)	$\int_{-\infty}^{0} e^{ax}e^{-i\alpha x}\,dx$	α	$\frac{1}{(a-i\alpha)}$	–	No restriction
6)	$\int_{-\infty}^{0} e^{ax}e^{-i\lambda x}\,dx$	$\lambda=\alpha+i\eta$	$\frac{1}{[(\eta+a)-i\alpha]}$	$\eta>-a$	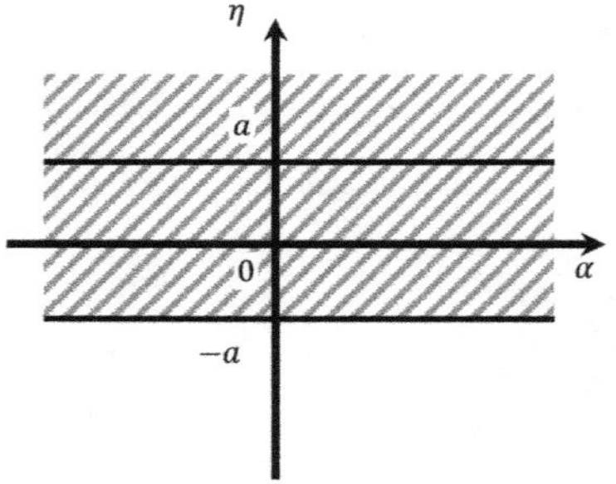
7)	$\int_{0}^{\infty} e^{ax}e^{-i\alpha x}\,dx$	α	Infinity	–	None
8)	$\int_{0}^{\infty} e^{ax}e^{-i\lambda x}\,dx$	$\lambda=\alpha+i\eta$	$\frac{-1}{[(\eta+a)-i\alpha]}$	$\eta<-a$	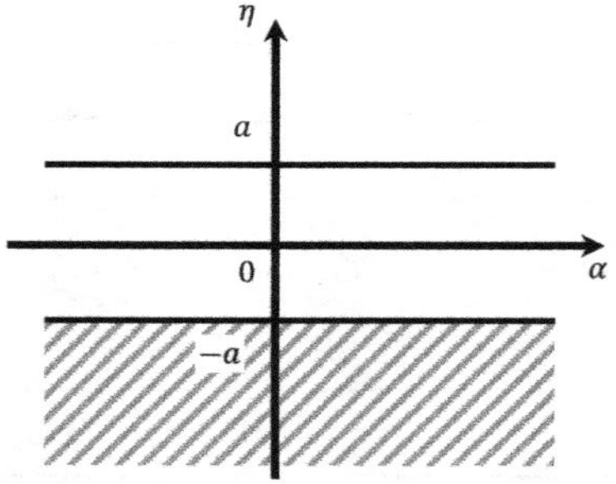

Note that in all cases $a>0$, α real, η real.

7.2.2 Location of inversion contour

This is related to the sign of x. To see this, consider $I = \int_{-\infty}^{\infty} e^{i\lambda x} d\lambda$, $\lambda = a + i\eta$. Only the case $\pm\eta \to \infty$ is of interest. Thus, since e^{iax} is bounded independent of the sign and magnitude of x, we can write

$$\left|e^{i\lambda x}\right| = \left|e^{i(a+i\eta)x}\right| = \left|e^{-\eta x}\right|.$$

So, when $x > 0$, we choose the horizontal inversion contour in UHP, where $\eta > 0$ since $\left|e^{i\lambda x}\right| \to 0$ as $\eta \to \infty$. Conversely, when $x < 0$, we close contour in LHP, where $\eta < 0$ since $\left|e^{i\lambda x}\right| \to 0$ as $\eta \to -\infty$. The foregoing results are irrespective of the presence of poles that may be due to $\tilde{f}(\lambda)$.

7.2.3 Where to close inversion contours versus plane of analyticity

In addition to the concern about the boundedness of $f(x)e^{i\lambda x}$, in relation to the condition on η, we also have to deal with the boundedness of $\tilde{f}(\lambda)$ and the type of singularity that it has. While isolated singularities of $\tilde{f}(\lambda)$ may be needed for nontrivial results of integration, we have to be mindful of multivaluedness and any needs to introduce branch cuts that may result from the nature of $\tilde{f}(\lambda)$. Note that once η^+ and η^- have been determined as the constant "horizontal" η line to integrate along in $a \in (-\infty, \infty)$, we may choose to close the inversion contour on the UHP or LHP of the η^+ or η^- line, depending on the nature of the integrand in the inversion formula. Invariably we will close over the plane where the integrand is not analytic. Thus, the exponential growth of the integrand is used to determine η^+, η^-, while other singularities – poles and multivaluedness – are used to determine where to close the contour.

7.2.4 More details on inversion contours

Let us define a split as follows:

$$\tilde{f}(\lambda) = F_+(\lambda) + F_-(\lambda), \qquad F_+(\lambda) = \int_{-\infty}^{0} e^{-i\lambda x} f(x)\, dx, \qquad F_-(\lambda) = \int_{0}^{\infty} e^{-i\lambda x} f(x)\, dx,$$

where $\lambda = a + i\eta$.

Suppose we are given η_-, η_+ then $F_+(\lambda)$ is analytic in some UHP, $\eta > \eta_-$, and $F_-(\lambda)$ is analytic in some LHP, $\eta < \eta_+$, where η_+ and η_- are defined by

$$|f(x)| < \begin{cases} e^{\eta_+ x}, & x \to \infty,\ x > 0 \\ e^{\eta_- x}, & x \to -\infty,\ x < 0 \end{cases}.$$

The paths of inversion are shown in Figure 7.8. The inversion formula is

$$f(x) = \frac{1}{2\pi}\int_{\Gamma_1} F_-(\lambda)e^{+i\lambda x}\,d\lambda + \frac{1}{2\pi}\int_{\Gamma_2} F_+(\lambda)e^{i\lambda x}\,d\lambda.$$

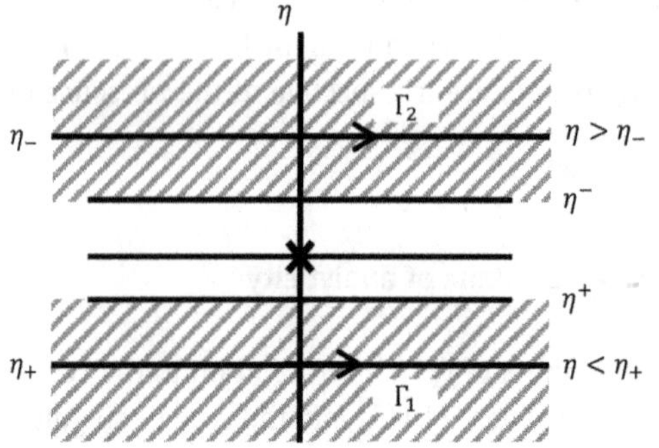

Figure 7.8: Distinct paths of inversion.

Two typical situations exist:

a) $\eta_- > 0,\ \eta_+ < 0$, such as in Figure 7.8. In this case, the two planes of analyticity of F_+ and F_- do not intersect (except at ∞).

b) $\eta_- < \eta_+$. In this case, we have an overlapping strip Ω of analyticity, as shown in Figures 7.9 and 7.10.

Suppose $F_+ = F_- = F(\lambda)$, where the last equal sign is as a result of analytic continuation, as a special case. So in this case, we can pick any contour in Ω for inversion, e.g., Γ, as long as Γ is in Ω. We shall have

$$f(x) = \frac{1}{2\pi}\int_{\Gamma} F(\lambda)e^{i\lambda x}\,d\lambda, \quad \Gamma \in \Omega.$$

Note that even if $F_+ \neq F_-$, Γ_1, Γ_2 are replaced by Γ. Also note that $F(\lambda) = 1$, $F(\lambda) = \lambda$ are analytic except at ∞.

An example of Case (b) is $f(x) = e^{-|a|x}$, or $\tilde{f}(\lambda) \propto \frac{1}{\lambda^2 + a^2}$ (from $\tilde{f}(\lambda) \propto \frac{1}{a+i\lambda} + \frac{1}{a-i\lambda} = \frac{2}{(a^2+\lambda^2)}$). Note in Table 7.1 that $(\eta + a) - ia = a - i\lambda$ and $(\eta - a) - ia = -(a + i\lambda)$.

We have an overlapping strip in this case (Figure 7.10).

F_+ is analytical in the UHP from plane $-ia$ upward while F_- is analytic in the LHP from ia downward. Note that, in general, $\tilde{f}$ is what we know.

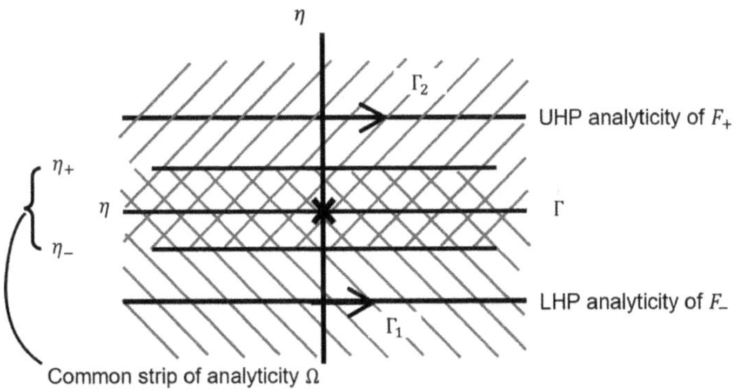

Figure 7.9: Overlapping strip of analyticity.

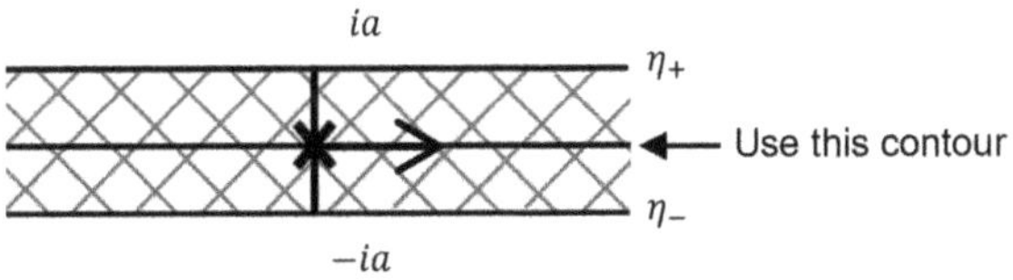

Figure 7.10: Overlapping strip of analyticity for $\tilde{f}(\lambda) \propto 1/\left(\lambda^2 + a^2\right)$.

Example 7.2. Invert the function $\tilde{f}(\lambda) = 1/i\lambda$.
We consider the contour in Figure 7.11. From the definition of FT inverse:

$$f(x) = \int_{-\infty}^{\infty} \frac{e^{+i\lambda x}}{i\lambda} d\lambda.$$

Heuristically, we see that

$$\left.\begin{array}{ll} f(x) = 1, & x > 0 \\ f(x) = 0, & x < 0 \end{array}\right\} \Rightarrow \tilde{f}(\lambda) = \frac{1}{i\lambda}.$$

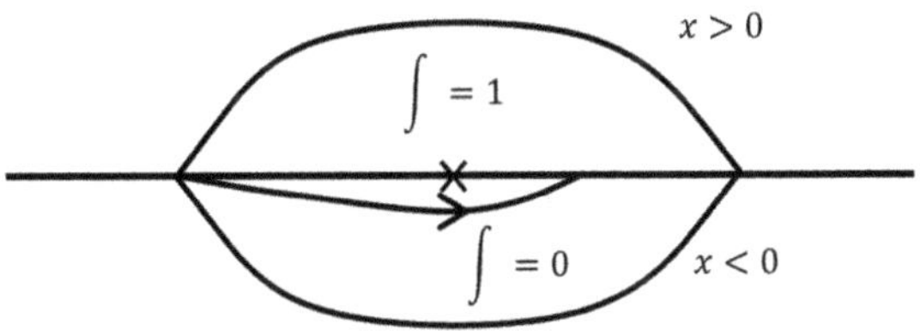

Figure 7.11: Integration domain influence.

Heuristically, because $\frac{d}{d\lambda}\left(\frac{e^{i\lambda x}}{i\lambda}\right)=1\cdot e^{i\lambda x}$ so that $f(x)=1$ when $x>0$. However, $f(x)=0$ when $x<0$ because the formula for inversion involves $e^{i\lambda x}$, not $e^{-i\lambda x}$. We do not have the issue of things going to infinity here because of the purely imaginary exponent.

Thus, we have a step function, so that a nonunique solution is imminent. We can consider splitting $\tilde{f}(\lambda)$ into $F_-(\lambda)$ and $F_+(\lambda)$, or

$$\tilde{f}(\lambda)=\frac{1}{i\lambda}=\int_{-\infty}^{0} e^{-i\lambda x}f(x)\,dx+\int_{0}^{\infty} e^{-i\lambda x}f(x)\,dx\equiv F_-(\lambda)+F_+(\lambda).$$

Note that this equation does suggest an approach to decompose a function. The domain dependence of step function $\tilde{f}(\lambda)$ is shown in Figure 7.12. The decomposition is $F_+(\lambda)=\frac{1}{i\lambda}$, $F_-(\lambda)=0$, which is analytic everywhere.

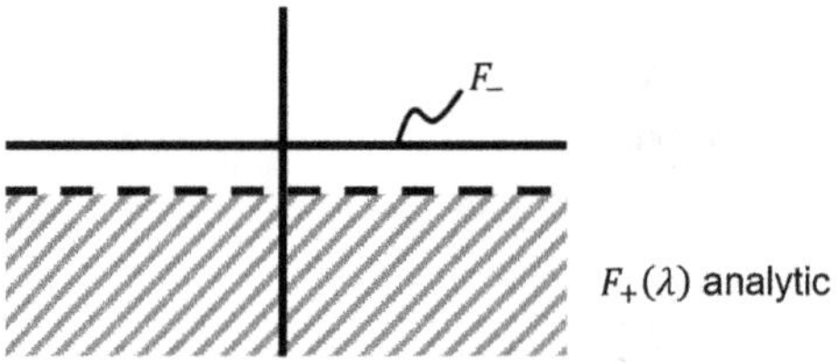

Figure 7.12: Depiction of the domain dependence of step-function $\tilde{f}(\lambda)$. Note that F_- is analytic everywhere, but F_+ has a pole at $\lambda=0$.

We could obtain an overall $f(x)$ as $f(x)=\frac{1}{2}[f(0^-)+f(0^+)]=\frac{1}{2}$ and recover the $\tilde{f}(\lambda)$ from

$$\tilde{f}(\lambda)=\frac{1}{2}\left[\underbrace{\int_{-\infty}^{0} e^{-i\lambda x}\,dx}_{1/i\lambda}+\underbrace{\int_{0}^{\infty} e^{-i\lambda x}\,dx}_{1/i\lambda}\right]=\frac{1}{i\lambda}.$$

Two possible paths for the FT inversion of $\tilde{f}(\lambda)$ are shown in Figure 7.13.

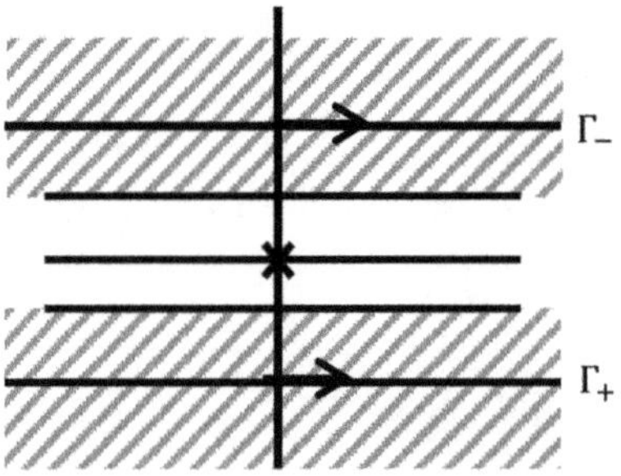

Figure 7.13: Two possible paths for the Fourier transform inversion integration of $\tilde{f}(\lambda)$.

We need to be careful here as the result is not unique, but dependent on the path taken. This happens if something goes wrong at ∞ that leads to $\tilde{f}(\lambda) \neq 0$ or some singularity as $\lambda \to \infty$. We do not have this problem for $\tilde{f} = 1/\left(\lambda^2 + a^2\right)$, as an example.

Example 7.3. On uniqueness (due to an impulse at $x = 0$) (Figure 7.14). Consider the transient diffusion equation

$$\varphi_t = \varphi_{xx}, \qquad t > 0, \tag{7.4}$$

with the boundary conditions (BCs):

$$\varphi(x,0) = \begin{Bmatrix} 2, & x > 0 \\ 0, & x < 0 \end{Bmatrix} = 2H(x).$$

FT in time (instead of x) is not natural, since t is defined for positive time only. The half-plane situation in the time domain creates some complications. Let $\varphi(x,t) = 0$, for all $t < 0$. Then,

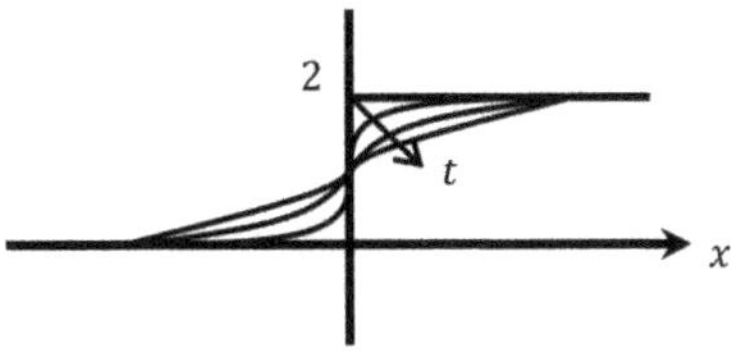

Figure 7.14: Sketch of the transient results for an impulsive application of heat.

$$\tilde{\varphi} = \int_{-\infty}^{\infty} e^{-i\lambda t}\varphi(x,t)\,dt = \int_{0}^{\infty} e^{-i\lambda t}\varphi(x,t)\,dt.$$

Taking the FT in time of eq. (7.4) above:

$$\int_0^{\infty} e^{-i\lambda t}\varphi_t\,dt = \int_0^{\infty} \varphi_{xx}e^{-i\lambda t}\,dt = \int_{-\infty}^{\infty} \varphi_{xx}e^{-i\lambda t}\,dt \equiv \tilde{\varphi}_{xx}. \tag{7.5}$$

Integrate LHS by parts:

$$\int_0^{\infty} e^{-i\lambda t}\varphi_t\,dt = \varphi(x,t)e^{-i\lambda t}\Big|_0^{\infty} + i\lambda\int_0^{\infty} \varphi e^{-i\lambda t}\,dt$$
$$= -\varphi(x,0) + i\lambda\tilde{\varphi},$$

where we have assumed that $\varphi(x, t \to \infty)e^{-i(t\to\infty)\lambda} \to 0$.

Note that λ has to be imaginary and negatively large. Substituting eq. (7.6) into (7.5) and noting that $-\varphi(x,0) = -2H(x)$, we will have

$$-2H(x) + i\lambda\tilde{\varphi} = \tilde{\varphi}_{xx} \quad \forall x.$$

For $x < 0$: Homogeneous solution:

$$\tilde{\varphi}(x,\lambda) = A(\lambda)\exp\left(x\sqrt{i\lambda}\right) + B(\lambda)\exp\left(-x\sqrt{i\lambda}\right).$$

Regarding $\sqrt{i\lambda}$, there are two choices, but it does not matter which branch we take since the role of the two terms interchange. We only need to be consistent with $x > 0$. We will take $\mathrm{Re}\left[\sqrt{i\lambda}\right] > 0$ as the branch if λ is real and positive. We can take $\mathrm{Re}\left[\sqrt{i\lambda}\right] = e^{i\pi/4}$ when dealing with $x < 0$.

As $x \to -\infty$ – we are dealing with $x < 0$ – the second term in $\tilde{\varphi}(x,\lambda)$ blows up for any λ that has a positive real part. So we will take $B(\lambda) = 0$. We want $\mathrm{Re}\left[\sqrt{i\lambda}\right] > 0$ when λ is real and < 0. (We want convergence because we will be integrating over all λ.) A problem to address is how we can make $x\sqrt{i\lambda}$ positive in the whole plane. Note that if $\lambda < 0$, this appears as positive.

Because of the BCs, we have to pick the branch cut in UHP (Figure 7.15). We could use one of the branch cuts (1) and (2). Let us take (2).

For $x > 0$: Nonhomogeneous solution:

$$\tilde{\varphi} = C(\lambda)\exp\left(-x\sqrt{i\lambda} + \frac{2}{i\lambda}\right). \tag{7.7}$$

In writing eq. (7.7), we have thrown away a D term. The branch cuts in Figure 7.15 are also consistent because $x\sqrt{i\lambda}$ is positive, and we obtain exponential decay as $x \to \infty$. We use continuity of $\tilde{\varphi}$ and $\tilde{\varphi}_{,x}$ at $x = 0$, so that

$$A(\lambda) = -C(\lambda) = \frac{1}{i\lambda}$$

and

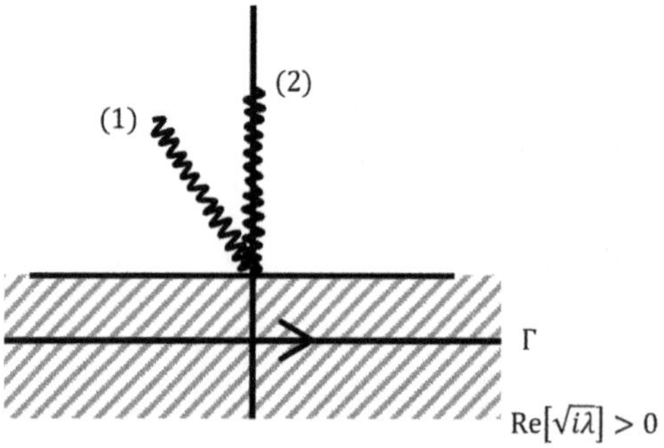

Figure 7.15: Analyticity in LHP and two branch cuts in the UHP.

$$\tilde{\varphi}(x,\lambda) = \frac{-|x|}{i\lambda x}\exp\left(-|x|\sqrt{i\lambda} + \frac{2}{i\lambda}H(x)\right).$$

The term $-|x|/x$ has been introduced just for the sign. The inverse of this equation can be written as

$$\varphi(x,t) = \frac{-|x|}{2\pi x}\int_{\Gamma} e^{i\lambda t}\left\{\left[\exp\left(-|x|\sqrt{i\lambda}\right)\right] + \frac{2}{i\lambda}H(x)\right\}d\lambda.$$

For $t<0$; we close at the bottom (from Jordan's lemma), since we have to take top or bottom so only y coordinate is relevant, sort of, and the fact that $e^{i\lambda t} = e^{it[x+iy]} \leq |e^{-ty}| = |e^{+ty}|$ for $t<0$. We also have to have y negative to get exponential decay. Note that $\varphi(x,t) = 0 \;\; \forall t<0$, or analyticity below.

For $t>0$; close on top (Figure 7.16). The solution is

$$\varphi(x,t) = 1 + \operatorname{erf}\frac{x}{2\sqrt{t}},$$

where

$$\operatorname{erf}(z) = \frac{2}{\sqrt{\pi}}\int_0^z e^{-t^2}dt = -\operatorname{erf}(-z).$$

A few comments on solution uniqueness: Is the solution obtained by Fourier series for the present diffusion problem unique? There is no simple answer to this question. Consider derivatives in space and time of the solution:

$$\varphi_{,t} = \varphi_{,xx}, \quad \varphi_{,tx} = \varphi_{,xxx}, \quad \varphi_{,x} = \frac{2}{\sqrt{\pi}}e^{-x^2/(2\sqrt{t})}.$$

Let

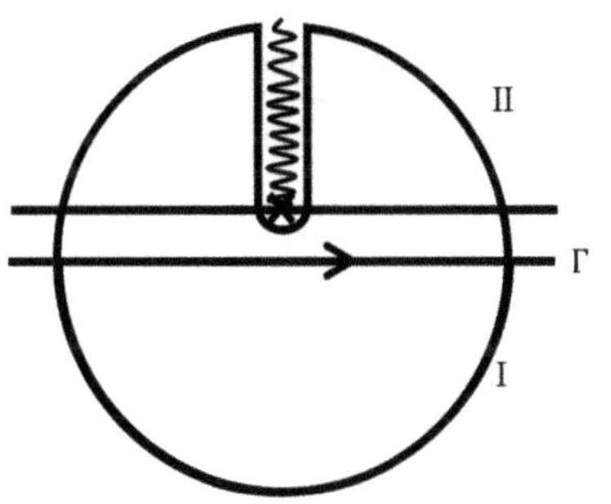

Figure 7.16: Fourier inversion contour for $t>0$.

$$\varphi(x,t) = \Phi(x,t) + A\Phi_{,x},$$

where the first term on the RHS satisfies the BCs and ICs (initial conditions). The function

$$\varphi(x,t) = \Phi(x,t) + A\Phi_{,x} + B\Phi_{,t} + C\Phi_{,xt} + D\Phi_{,xxt} \tag{7.8}$$

is a solution of the problem that satisfies the ICs and BCs and are bounded at origin. Note that you do not need each component of the solutions to satisfy the IC and BC, only the overall solution needs to. The first term on the RHS of eq. (7.8) gives bounded energy.

The FT solution has ruled out all the higher order singularities, although we could get them if we use other methods. We need to use extra conditions to exclude the other terms.

Why do we rule out all these terms? We have to seek arguments at points of singularity. Some rate-dependent processes occurring at the singularity point may tell us that there are extra terms that are important. The method of separation of variables (SOVs) works only for eigenvalue problems. That is, it works only for noninfinite domains (bounded domains).

Eigenfunctions have bounds: If we have an infinite dimension domain, we can get or specify any type of wave, and fit any waveform, and there is virtually no restriction because of the infinite dimension. On the other hand, for finite dimension, we have a definite spectrum, discrete spectrum; we can only fit certain modes (definite modes and eigenmodes). The SOV is used when we want to obtain expansion theorems in terms of eigenfunctions. Table 7.2 shows plausible solution methods for partial differential equations (PDEs) in different domains.

Table 7.2: Plausible solution methods for PDEs in different domains.

Domain	**PDE solution method**
Time: semi-infinite	Laplace transform
(x, y, z) bounded	Series, e.g., Fourier series, definite modes; SOV
(x, y, z) infinite	Fourier transform, Fourier integral
(x, y, z) semi-infinite	Laplace transform, Fourier transform with function splitting

7.3 Convolution in Fourier transform

Given the definition

$$f(x) * g(x) \equiv f * g = \int_{-\infty}^{\infty} f(\xi)g(x-\xi)d\xi,$$

it is simple to show that

$$\frac{1}{2\pi}\int_{-\infty}^{\infty} e^{i\lambda x}\left[\tilde{f}(\lambda)\tilde{g}(\lambda)\right]d\lambda = f(x) * g(x)$$

or

$$\mathcal{F}^{-1}\left\{\tilde{f}(\lambda)\tilde{g}(\lambda)\right\} = f(x) * g(x).$$

This is a statement of convolution in FT. As you can see, the idea of convolution is quite simple: start with a product of two transformed functions, invert the two transforms separately, and put them together under an integral sign with one of them having a translated kernel. Note that $f * g \neq f \times g$, and that $f * g = g * f$.

Example 7.4. A simple transient heat conduction equation can be used to illustrate the application of convolution. Consider the problem

$$T_{xx} - T_t = 0, \quad T(x,0) = f(x).$$

The $f(x)$ is on the time boundary and not in the spatial boundary or volume, i.e., in the equation. So a unit impulse is possible only on the boundary, at least from the standpoint of application in engineering or physics. The BC for this problem is

$$T(x = \pm\infty, t) = 0.$$

Let us take the FT in x to obtain

$$(i\lambda)^2\tilde{T} - \tilde{T}_t = 0.$$

The IC transforms as $\tilde{T}(\lambda,0) = \tilde{f}(\lambda)$ so that $\tilde{T}$ can be solved as

$$\tilde{T}(\lambda) = \tilde{f}(\lambda)e^{-\lambda^2 t} \equiv \tilde{f}(\lambda)\tilde{g}(\lambda), \text{ where } \tilde{g}(\lambda) \equiv e^{-\lambda^2 t}, \text{ and}$$

$$g(x,t) = \mathcal{F}^{-1}\left\{e^{-\lambda^2 t}\right\} = \frac{1}{2\pi}\int_{-\infty}^{\infty} e^{i\lambda x}e^{-\lambda^2 t}d\lambda = \frac{1}{\sqrt{4\pi t}}e^{-\frac{x^2}{4t}}$$

and

$$T(x,t) = \mathcal{F}^{-1}\{\tilde{g}(\lambda)\tilde{f}(\lambda)\} = \int_{-\infty}^{\infty} f(\xi)\frac{1}{\sqrt{4\pi t}}e^{-\frac{(x-\xi)^2}{4t}}d\xi.$$

If $f(x)$ is $\delta(x)$, which is a delta function at the origin (in time), we will have

$$T(x,t) = \frac{e^{-x^2/4t}}{\sqrt{4\pi t}}.$$

This is the Green's function of the problem. From the FT we can easily see the Green's function. We will illustrate the procedure above with a more complicated example, with solutions that are not as precise as the one we have just presented.

Example 7.5. Consider the damped wave equation

$$\varphi_{xx} = \frac{1}{c^2}\varphi_{tt} - \varepsilon\varphi_t \equiv p(x,t). \tag{7.9}$$

If $\varepsilon \to 0$, this gives the governing equation for strain with lateral loading, where φ is the lateral displacement of a string. If $\varepsilon > 0$, $\varepsilon\phi_t$ could represent the presence of wind resistance. The domain is $(-\infty,\infty)$, so we will use FT. (If $x \in [0,l]$, use SOV.) Let us look for a steady-state solution by assuming that

$$p(x,t) = f(x)e^{i\omega t}$$

in order to extract the time dependence. Thus, we anticipate that the solution for long time is in the form of

$$\phi(x,t) = \psi(x)e^{i\omega t} \quad \text{as } t \to \infty. \tag{7.10}$$

We will substitute eq. (7.10) into eq. (7.9) to obtain

$$\psi_{xx}(x) + \left(\frac{\omega^2}{c^2} - i\varepsilon\omega\right)\psi(x) = f(x). \tag{7.11}$$

The FT of eq. (7.11) with respect to x gives

$$-\lambda^2\tilde{\psi} + i\lambda\left(\frac{\omega^2}{c^2} - i\varepsilon\omega\right)\tilde{\psi} = \tilde{f},$$

from which we have

$$\tilde{\psi} = \frac{-\tilde{f}}{\lambda^2 - (\omega^2/c^2 - i\varepsilon\omega)}, \text{ and } \psi = -\frac{1}{2\pi}\int_{-\infty}^{\infty} \frac{\tilde{f}(\lambda)e^{i\lambda x}\,d\lambda}{\lambda^2 - (\omega^2/c^2 - i\varepsilon\omega)}.$$

Let λ be such that we can invert on the real axis, with convolution

$$\psi = \mathcal{F}^{-1}\left\{\tilde{f}(\lambda)\tilde{g}(\lambda)\right\} = -\int_{-\infty}^{\infty} f(\xi)g(x-\xi,t)\,d\xi,$$

where

$$g(x) \equiv \mathcal{F}^{-1}\left\{\frac{1}{\lambda^2 - (\omega^2/c^2 - i\varepsilon\omega)}\right\} = \frac{1}{2\pi}\int_{-\infty}^{\infty} e^{i\lambda x}\frac{d\lambda}{\lambda^2 - (\omega^2/c^2 - i\varepsilon\omega)}.$$

Singularity of $g(\lambda)$:

$$\lambda^2 = \frac{\omega^2}{c^2}\left(1 - \frac{i\varepsilon c^2}{\omega}\right) \Rightarrow \lambda = \pm\frac{\omega}{c}\left(1 - \frac{i\varepsilon c^2}{\omega}\right)^{1/2}.$$

Let us carry out the asymptotic expansion about $\varepsilon = 0$:

$$\lambda \sim \pm\frac{\omega}{c}\left[1 - \frac{1}{2}\frac{\varepsilon c^2}{\omega}\right] \quad \text{for } \varepsilon \to 0.$$

Independent of the singularities of $g(\lambda)$, as discussed previously, the presence of $e^{i\lambda x}$ means that we need to close in the UHP if we want to use Jordan's lemma for x positive. For $x < 0$, we will use LHP (Figure 7.17).

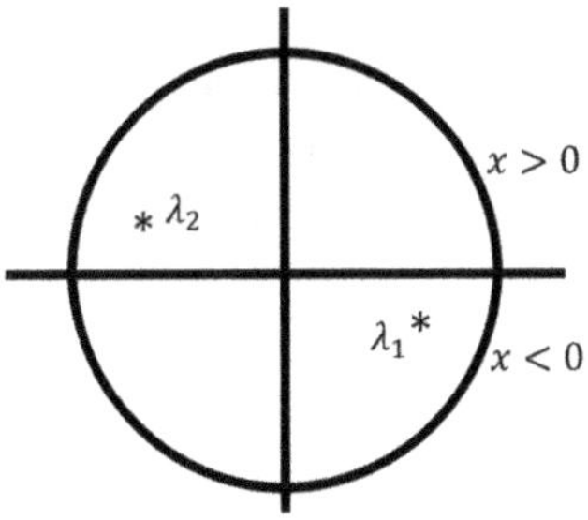

Figure 7.17: Integration contours depending on the sign of x, or where we are in the domain $x \in (-\infty, \infty)$.

$\psi(x)$ can be written as

$$\psi(x) = \int_{-\infty}^{\infty} \frac{ic}{2\omega}\left\{\frac{f(\xi)\exp\left[-(i\omega/c)|\xi - x|\sqrt{1 - (i\varepsilon c^2/\omega)}\right]}{\sqrt{1 - (i\varepsilon c^2/\omega)}}\right\}d\xi.$$

For $x > 0$, we will use the UHP (Figure 7.17) and evaluate the residue at $\lambda = \lambda_2$:

$$\text{Res}\left[\frac{1}{2\pi}\frac{e^{i\lambda x}}{\lambda^2 - (\omega^2/c^2)(1 - i\varepsilon c^2/\omega)}\right] \quad \text{at } \lambda = \lambda_2 \equiv -\frac{\omega}{c}\left(1 - \frac{i\varepsilon c^2}{\omega}\right)^{1/2}.$$

The result is

$$\frac{1}{2\pi}\frac{e^{i\lambda_2 x}}{(\lambda_2 - \lambda_1)} = \frac{1}{2\pi}\frac{\exp\left[-\frac{i\omega}{c}\left(1 - \frac{i\varepsilon c^2}{\omega}\right)x\right]}{\lambda_2 - \lambda_1}.$$

For $x < 0$, we will use the LHP (Figure 7.17) and evaluate the residue at $\lambda = \lambda_1$, with the result:

$$\text{Res}\left[\frac{1}{2\pi}\frac{e^{i\lambda x}}{\lambda^2 - (\omega^2/c^2)(1 - i\varepsilon c^2/\omega)}\right] \quad \text{at } \lambda_1 = \frac{1}{2\pi}\frac{e^{i\lambda_1 x}}{\lambda_1 - \lambda_2} = \frac{1}{2\pi}\frac{e^{i\lambda_2 |x|}}{\lambda_1 - \lambda_2}.$$

Take $\lim \varepsilon \to 0$:

$$\psi(x) = \frac{ic}{2\omega}\int_{-\infty}^{\infty} f(\xi)\exp\left[-\frac{i\omega}{c}\,|\xi - x|\right]d\xi.$$

Note that we could run into some problems without invoking the small parameter, ε. This is the reason why some people prefer to perturb the system. That is, if we set $i\varepsilon\omega = 0$ in $g(x)$, we will have

$$g(x) = \frac{1}{2\pi}\int_{-\infty}^{\infty} e^{i\lambda x}\frac{d\lambda}{\lambda^2 - \omega^2/c^2}.$$

Poles are located on the real axis. There are several possible contours, as follows:

① Contour passes through the two singularity points, as in Figure 7.18.

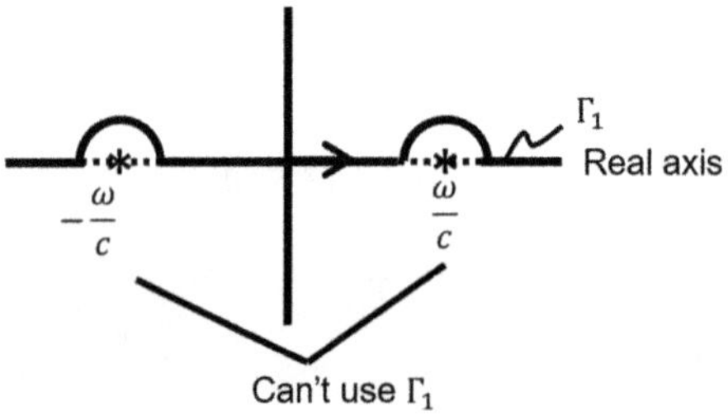

Figure 7.18: Location of poles on the real axis. We may need to use the Cauchy principal value approach.

We can carry out the Cauchy principal integral solution, which we have seen gives 1/2 the residues at each of the two singularity points.

② For $x > 0$, we could use a contour displaced upward in the UHP (Figure 7.18(a)) which does not pass through the poles.

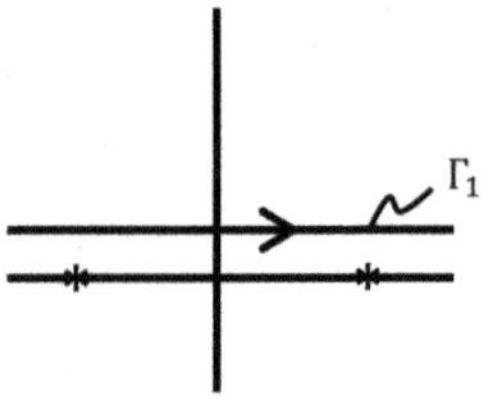

Figure 7.18(a): Contour for $x > 0$.

③ For $x < 0$, we could use a contour displaced downward in the LHP (Figure 7.18(b)), which does not pass through the poles.

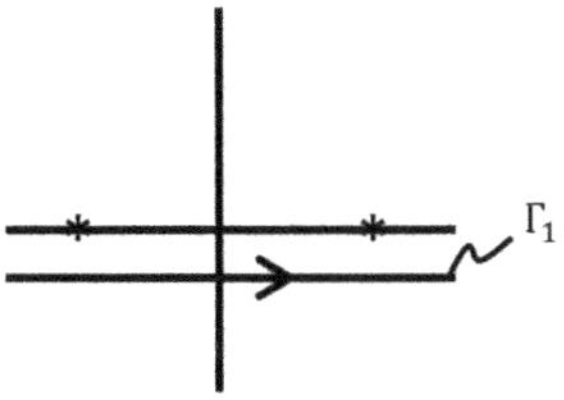

Figure 7.18(b): Contour for $x < 0$.

④ For $x > 0$, we could also use a contour approaching from the LHP that does not pass through the poles (Figure 7.18(c)).

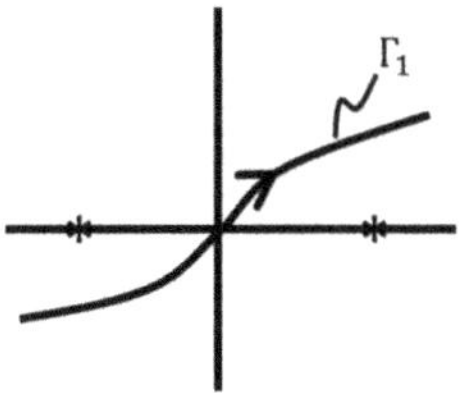

Figure 7.18(c): Alternate contour for $x > 0$.

⑤ Similarly, for $x < 0$, we could use a contour approaching from the UHP that does not pass through the poles (Figure 7.18(d)).

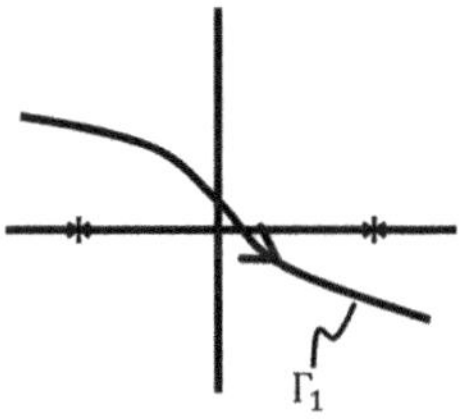

Figure 7.18(d): Alternate contour for $x < 0$.

The solution from each of the contours in Figures 7.18 satisfies the differential equations but only some of the solutions decay at infinity. Some of the solutions will be useful under different conditions.

Contour number 4 (Figure 7.18(d)) corresponds to what we picked earlier to get (noting that the poles λ_1 and λ_2 are now located on the real axis due to setting $i\varepsilon\omega \to 0$ in $g(\lambda)$):

$$\psi(x) = \frac{ic}{2\omega} \int_{-\infty}^{\infty} f(\xi) \exp\left[-\frac{i\omega}{c} |\xi - x| \right] d\xi; \ \left(\phi(x,t) = \psi e^{i\omega t}\right),$$

$$\phi(x,t) = \frac{ic}{2\omega} \int_{-\infty}^{\infty} f(\xi) d\xi \exp\left[-\frac{i\omega}{c} \overbrace{(|\xi - x| - ct)}^{\text{outgoing wave}} \right].$$

In this case, there is no incoming wave for each element, as the solution allows only outgoing wave. This means that energy is not pumped in from infinity. This is called **Somerfield's radiation condition**. So, if you want only outgoing waves, choose this solution. Each of these contours gives several combinations of incoming and outgoing waves. So, when you have poles on real axis, things are not as clear!

7.4 Multidimensional Fourier transform

The FT procedure for multidimensional problems is discussed in this section. Consider the transform of a function f of two independent variables $f(x,y)$, in x first:

$$f(x,y) \Rightarrow \tilde{f}(\lambda_1, y) = \int_{-\infty}^{\infty} e^{-i\lambda x} f(x,y) dx,$$

and then the transform in y:

$$\approx{f}(\lambda_1, \lambda_2) = \int_{-\infty}^{\infty} e^{-i\lambda_2 y} \tilde{f}(\lambda_1, y) dy.$$

For simplicity of notations, we will use $\tilde{f}(\lambda_1, \lambda_2) \equiv \overset{\approx}{f}(\lambda_1, \lambda_2)$. We can write this procedure as

$$\tilde{f}(\lambda_1, \lambda_2) = \int_{-\infty}^{\infty} \int_{-\infty}^{\infty} \underbrace{e^{-i\lambda_1 x_1 - i\lambda_2 x_2}}_{e^{-i\lambda \cdot \mathbf{r}}} \underbrace{f(x,y) dx dy}_{f(r,\theta) r d\theta dr}$$

or

$$\tilde{f}(\lambda_1, \lambda_2) = \int_0^{2\pi} \int_0^{\infty} e^{-i\lambda \cdot \mathbf{r}} f(r,\theta) r d\theta dr,$$

where $r \in [0, \infty)$ and $\theta \in [0, 2\pi)$.

The formula for inversion is

$$f(x,y) = \frac{1}{4\pi^2} \int_{-\infty}^{\infty} \int_{-\infty}^{\infty} e^{i\lambda_1 x + i\lambda_2 y} \tilde{f}(\lambda_1, \lambda_2) d\lambda_1 \, d\lambda_2.$$

Convolution in two dimensions (2D) is simply

$$\begin{aligned} f(x,y) * g(x,y) \equiv f * g &= \int_{-\infty}^{\infty}\int_{-\infty}^{\infty} g(x-\xi, y-\eta) f(\xi,\eta) d\xi\, d\eta \\ &= \mathcal{F}^{-1}\left\{\tilde{f}(\lambda_1,\lambda_2)\tilde{g}(\lambda_1,\lambda_2)\right\}. \end{aligned} \tag{7.12}$$

It should be noted in eq. (7.12) that the kernel translation is not shared between g and f, as only g is translated in both coordinate directions. However, we could also have f carry the two translations and obtain the same results. We need to invert in the complex plane in 2D.

Example 7.6. Solve the Poisson equation

$$\nabla^2\varphi = f(x,y,z), \quad -\infty < x,y,z < \infty$$

for the electric potential ϕ.

Solution: We will combine the methods of Green's function and FT to solve this problem. Recall that the Green's function of an inhomogeneous PDE $\mathcal{L}u = f$, where $u = u(x,y,z)$ and $f = f(x,y,z)$, is that function G that satisfies the PDE $\mathcal{L}G = \delta(x-x^o)(y-y^o)(z-z^o)$. Knowing G, by solving this equation, we can calculate

$$u = \iiint f(x^o,y^o,z^o)\, G(x-x^o, y-y^o, z-z^o) dx^o dy^o dz^o.$$

The Green's function from $\nabla^2 G = \delta(x-x^0)\delta(y-y^0)\delta(z-z^0)$ is proportional to $\int 1/r$. The solution can therefore be written as

$$\varphi = \frac{1}{4\pi}\int \frac{f(x',y',z')\, dx'dy'dz'}{(x-x')^2 + (y-y')^2 + (z-z')^2}.$$

Applying the FT operation

$$\tilde{\varphi} = \iiint_{-\infty}^{\infty} e^{-i(\lambda_1 x + \lambda_2 y + \lambda_3 z)}\varphi(x,y,z)\, dxdydz$$

to the given equation, we will obtain

$$-\lambda^2\tilde{\varphi} = \tilde{f}, \quad \lambda^2 \equiv \lambda_1^2 + \lambda_2^2 + \lambda_3^2.$$

To use the convolution theorem, factorize $\tilde{\varphi}$ into $\tilde{f}$ and $h \equiv -1/\lambda^2$:

$$\tilde{\varphi} = -\frac{\tilde{f}}{\lambda^2} \Rightarrow \text{Find } \mathcal{F}^{-1}\left\{\frac{-1}{\lambda^2}\right\} \text{ and } \mathcal{F}^{-1}\left\{\tilde{f}\right\},$$

so that

$$\varphi = \mathcal{F}^{-1}(\tilde{\varphi}) = \mathcal{F}^{-1}\left(\tilde{f}\tilde{h}\right) = \left(\mathcal{F}^{-1}\left(\tilde{f}\right)\right) * \left(\mathcal{F}^{-1}\left(\tilde{h}\right)\right) = f * h.$$

We can also write

$$\varphi = \iiint\limits_{-\infty}^{\infty} f(x', y', z') G(x - x', y - y', z - z')\, dx' dy' dz'.$$

Note that

$$\mathcal{F}^{-1}\left\{\frac{-1}{\lambda^2}\right\} = G(x, y, z) = \frac{1}{8\pi^3} \iiint\limits_{-\infty}^{\infty} \frac{e^{i\boldsymbol{\lambda}\cdot\mathbf{r}}}{\lambda^2} dV_\lambda,$$

where $\boldsymbol{\lambda} = \lambda_1 \mathbf{i} + \lambda_2 \mathbf{j} + \lambda_3 \mathbf{k}$, $\mathbf{r} = x\mathbf{i} + y\mathbf{j} + z\mathbf{k}$. The expression in the middle of the equation above results from $\nabla^2 G = \delta \Rightarrow (i\lambda)^2 \tilde{G} = 0$.

The polar coordinate representation is shown in Figure 7.19.

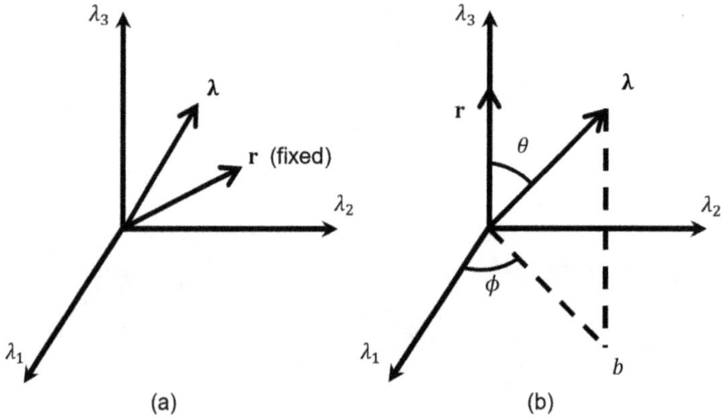

Figure 7.19: Coordinate system in the Fourier space: (a) basic system and (b) placement of **r** to coincide with **k**.

The δ function is independent of the orientation in space and the solution is independent of where the **r** vector is placed. Let us place it along the **k** unit vector, as in Figure 7.19(b), where **r** is fixed in the space in which we are integrating, and r is a constant. Also, let us do this in spherical coordinates, and integrate over λ_1, λ_2, λ_3: $\boldsymbol{\lambda} \cdot \mathbf{r} = \lambda r \cos\theta$, where $\lambda = \|\boldsymbol{\lambda}\|$, $r = \|\mathbf{r}\|$:

$$G(x, y, z) = \frac{1}{8\pi^3} \int\limits_0^{2\pi} d\phi \int\limits_0^{2\pi} \sin\theta\, d\theta \int \frac{e^{i\lambda r\cos\theta}}{\lambda^2} \lambda^2\, d\lambda = \frac{1}{8\pi^3} \frac{2\pi^2}{r} = \frac{1}{4\pi r}.$$

Example 7.7. Solve the PDE

$$\nabla^2\phi = \frac{1}{c^2}\phi_{,tt} + f(\mathbf{r},t) \quad (\text{in 3D}). \tag{7.13}$$

Solution: We need to find the Green's function for this problem. Let us set

$$f(\mathbf{r},t) = \delta(x-x')\delta(y-y')\delta(z-z')e^{i\omega t}$$

in order to obtain a steady-state solution which is sinusoidal (because of $e^{i\omega t}$). We will for a solution of the form $\phi = \psi e^{i\omega t}$, where ψ is the Green's function. Taking the FT, we obtain

$$-\lambda^2\tilde{\psi} + \frac{\omega^2}{c^2}\tilde{\psi} = \mathcal{F}\{\delta(x-x')\delta(y-y')\delta(z-z')\}$$

$$= e^{-i\lambda\cdot\mathbf{r}'}\left(\text{since} \int g(x)\delta(x-x') = g(x')\right),$$

where $\mathbf{r}' = x'\mathbf{i} + y'\mathbf{j} + z'\mathbf{k}$, and

$$\tilde{\psi}(\lambda) = \frac{-e^{-i\lambda\cdot\mathbf{r}}}{\lambda^2 - k^2} = \tilde{G},$$

where $k^2 = \omega^2/c^2$.

$$\psi(x,y,z) = -\frac{1}{8\pi^3}\iiint_{-\infty}^{\infty}\frac{e^{i\lambda\cdot(\mathbf{r}-\mathbf{r}')}}{(\lambda^2-k^2)}dV_\lambda.$$

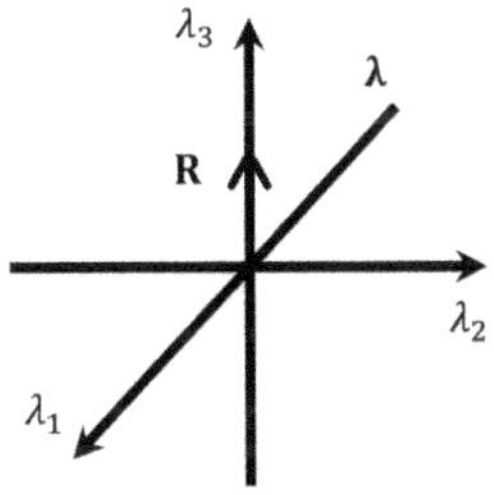

Figure 7.20: Alignment of **R** with λ_3.

Let $(\mathbf{r}-\mathbf{r}') \equiv \mathbf{R}$, where $|\mathbf{R}| \equiv R$ and $\mathbf{R}$ is aligned with λ_3 (Figure 7.20). We can then write

$$\psi(x,y,z) = \frac{1}{2\pi^2 R}\int_0^\infty \frac{\lambda\sin(\lambda R)d\lambda}{\lambda^2 - k^2}.$$

The $(0,\infty)$ limit on the integral involves the real axis, and the integral describes the outgoing and incoming waves. Note that we have integrated over θ, ϕ. Recognize that the integrand is an even function so that

$$\psi(x,y,z) = \frac{1}{4\pi^2 R} \int_{-\infty}^{\infty} \frac{\lambda \sin(\lambda R)\, d\lambda}{\left(\lambda^2 - k^2\right)}.$$

As seen previously, we have a choice of many solutions when the poles are all located on the real axis: incoming, outgoing waves, etc., depending on how we close the contour (Figures 7.21 and 7.22). The two cases of incoming and outgoing waves are presented as follows:

1) Outgoing waves: The contour for this case is shown in Figure 7.21.

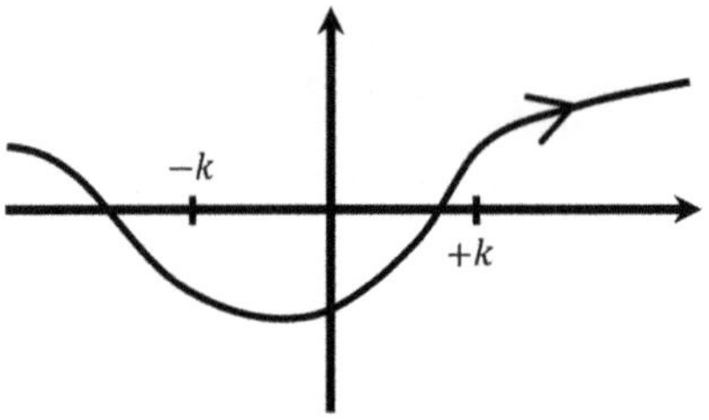

Figure 7.21: Contour for outgoing waves.

The solution in this case is

$$\varphi = \frac{1}{4\pi R} e^{-ik(R-ct)}.$$

Note that we have multiplied by $e^{i\omega t}$, etc.

2) Incoming waves: The contour for the case of incoming waves is shown in Figure 7.22.

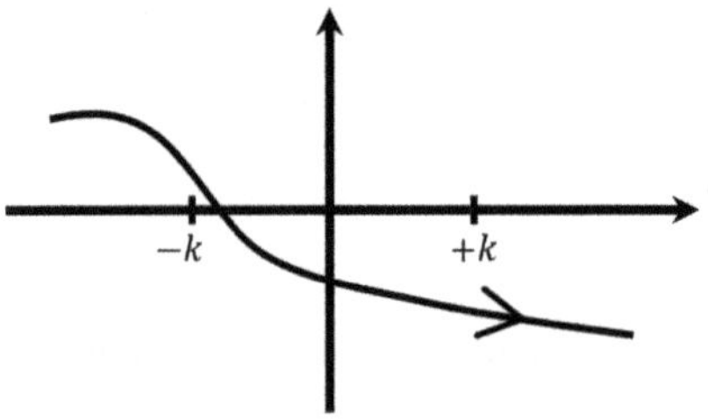

Figure 7.22: Contour for incoming waves.

The solution is

$$\varphi = \frac{1}{4\pi R} e^{-ik(R+ct)}.$$

If we add damping into the problem, we will remove the ambiguity in the foregoing because we will have poles below and above the real axis. The three-dimensional (3D) problem is, in a sense, easier to solve.

7.4.1 The acoustic speaker problem

This interesting problem consists of finding the sound amplitude perceived by an observer located at an arbitrary point (x, y, z) in a 3D space when an acoustic speaker is placed at the boundary $z = 0$. The governing equation is $\nabla^2\phi = \frac{1}{c^2}\phi_{,tt}$, as in eq. (7.13) without the forcing. We also assume that

$$\phi = \psi e^{i\omega t} \Rightarrow \nabla^2\psi + k^2\psi = 0, \quad \text{where } k^2 = \frac{\omega^2}{c^2}.$$

We could place a δ function on the RHS of eq. (7.13) if $f \neq 0$, and on a boundary if f is located on it. Convolution takes care of this case automatically. In the present problem, leading to the BC, the speaker is placed on the z – axis (Figure 7.23):

$$\psi_{,z} = f(x, y), \text{ at the boundary } z = 0.$$

Thus, this is an example of a situation where the driving force is located on a boundary, not inside (the equation). We effectively have a 2D analysis. We take FT in the x, y coordinate directions:

$$\tilde{\psi}_{zz} - \left(\lambda^2 - k^2\right)\tilde{\psi} = 0, \quad \lambda^2 \equiv \lambda_1^2 + \lambda_2^2,$$

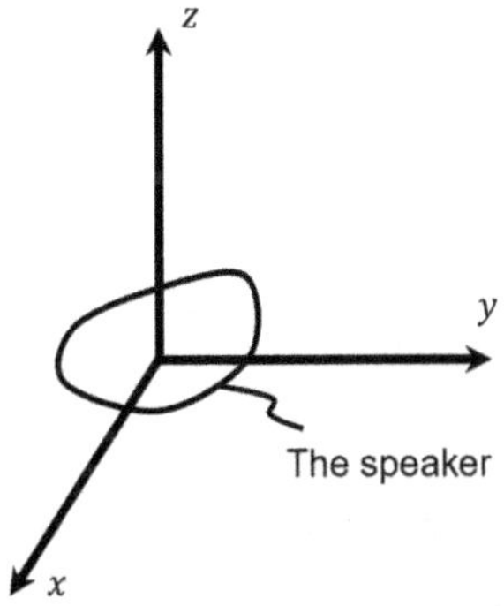

Figure 7.23: The placement of the speaker (at $z = 0$) with respect to the coordinate system.

and solve for $\tilde{\psi}$:

$$\tilde{\psi} = A(\lambda_1, \lambda_2)\exp\underbrace{\left[-z\sqrt{\lambda^2 - k^2}\right]}_{a} + B(\lambda_1, \lambda_2)\exp\underbrace{\left[z\sqrt{\lambda^2 - k^2}\right]}_{b}.$$

We have a choice to make on the sign of z in order to have bounded solutions from the terms "a" and "b" in this equation. We do not know whether "a" or "b" grows or dies unless we determine our branch cut, which is something we are not going to do here. That is, we will not be concerned with what the branch cut looks like. We set $B = 0$ for boundedness:

$$\psi_z|_{z=0} \Rightarrow \tilde{\psi}_z|_{z=0} = -\sqrt{\lambda^2 - k^2}e^{-z\sqrt{\lambda^2 - k^2}}|_{z=0} = \tilde{f}(\lambda_1, \lambda_2) \Rightarrow A(\lambda_1, \lambda_2) = \frac{-\tilde{f}(\lambda_1, \lambda_2)}{\sqrt{\lambda^2 - k^2}},$$

giving

$$\tilde{\psi} = \frac{-\tilde{f}}{\sqrt{\lambda^2 - k^2}} e^{-z\sqrt{\lambda^2 - k^2}}.$$

Employ the convolution theorem:

$$\psi = -\int f(x', y', z') G(x - x', y - y', z)\, dx' dy',$$

where

$$G(x, y, z) = \mathcal{F}^{-1}\left\{\frac{e^{-z\sqrt{\lambda^2 - k^2}}}{\sqrt{\lambda^2 - k^2}}\right\} = \left(\mathcal{F}^{-1}\left\{e^{-z\sqrt{(\lambda^2 - k^2)}}\right\}\right) * \left(\mathcal{F}^{-1}\left\{\frac{1}{\sqrt{\lambda^2 - k^2}}\right\}\right).$$

The Green's function depends on z, which is the third physical space coordinate direction in Figure 7.23. Convolution then tells us where the Green's function is located and where the driving force is. Meanwhile, if we have to find the Green's function, we will just apply convolution and pick up the function. In general, the Green's function, written as $G(\mathbf{x}|\mathbf{x}_s)$, tells us where the "observer" (or listener) is located, or $\mathbf{x}$, and where the sound source is placed, or $\mathbf{x}_s$.

Now,

$$G = \mathcal{F}^{-1}\left\{\frac{e^{-z\sqrt{\lambda^2 - k^2}}}{\sqrt{\lambda^2 - k^2}}\right\} = -\frac{1}{4\pi^2}\iint_{-\infty}^{\infty} \frac{e^{-z\sqrt{\lambda^2 - k^2}} e^{i\lambda \cdot \mathbf{r}} d\lambda_1 d\lambda_2}{\sqrt{\lambda^2 - k^2}}, \tag{7.14}$$

where the RHS term has been obtained from the definition of Fourier inverse. We cannot take FT of this because we do not know the definition of z, whether positive or negative. This involves some trick. Define

$$a(\lambda, z) \equiv \frac{e^{-|z|\sqrt{\lambda^2 - k^2}}}{\sqrt{\lambda^2 - k^2}} \text{ for all } z, \tag{7.15}$$

and note that a $G(\lambda, z)$ for $z > 0$ is the same as the first factor in the integrand of eq. (7.14), as $z = |z|$ when $z > 0$. So, we can take the FT of eq. (7.15), and invert it to obtain the result for $z > 0$. This is called the **method of images**. Now, the FT of $a(\lambda_1, \lambda_2, \lambda_3)$ is as follows:

$$\tilde{a}(\lambda_1, \lambda_2, \lambda_3) \equiv \int_{-\infty}^{\infty} a(\lambda, z) e^{-i\lambda_3 z} dz$$
$$= \underbrace{\int_{-\infty}^{0} \frac{e^{+z\sqrt{\lambda^2 - k^2}}}{\sqrt{\lambda^2 - k^2}} e^{-i\lambda_3 z} dz}_{z<0} + \underbrace{\int_{0}^{\infty} \frac{e^{-z\sqrt{\lambda^2 - k^2}}}{\sqrt{\lambda^2 - k^2}} e^{-i\lambda_3 z} dz}_{z>0}.$$

These integrals are easy to evaluate. The result is

$$\tilde{a}(\lambda_1, \lambda_2, \lambda_3) = \frac{1}{\sqrt{\lambda^2 - k^2}} \left[\frac{1}{-i\lambda_3 + \sqrt{\lambda^2 - k^2}} - \frac{1}{-i\lambda_3 - \sqrt{\lambda^2 - k^2}} \right]$$
$$= \frac{2}{\lambda^2 + \lambda_3^2 - k^2}.$$

Note that $\sqrt{\lambda^2 - k^2} = \text{constant} \equiv a$, and $z\left(\sqrt{\lambda^2 - k^2} - i\lambda_3\right) \equiv zb$. Inverting $\tilde{a}(\lambda_1, \lambda_2, \lambda_3)$ in $\lambda_1, \lambda_2, \lambda_3$ (3D space) and considering the resulting function for $z > 0$, we have

$$G(x, y, z) = \frac{1}{8\pi^3} \iiint_{-\infty}^{\infty} \frac{e^{i\lambda \cdot \mathbf{r}} dV_\lambda}{\lambda_1^2 + \lambda_2^2 + \lambda_3^2 - k^2} = \frac{1}{2\pi r} e^{-ikr}.$$

We can invoke convolution as usual to obtain:

$$\psi(x, y, z) = \frac{1}{2\pi} \iint_{-\infty}^{\infty} \frac{f(x', y') e^{-ik\sqrt{(x-x')^2 + (y-y')^2 + z^2}}}{\sqrt{(x-x')^2 + (y-y')^2 + z^2}} dx' dy'.$$

7.4.2 Two-dimensional wave equation

In this section, we will consider the procedure to analyze a spatially infinite 2D wave problem with a temporal sinusoidal forcing. The governing equation can be written as

$$\nabla^2 \phi - \frac{1}{c^2} \phi_{tt} = H(x, y) e^{i\omega t}.$$

As before, let us look for steady-state solutions of the type $\phi = \psi e^{i\omega t}$ so that

$$\nabla^2\psi + k^2\psi = H(x,y), \quad k \equiv \frac{\omega}{c}.$$

This is the Helmholtz equation. No BCs are specified, as we are dealing with a hyperbolic mathematical problem in an infinite space. That is, we are solving for $0 < |x|,\ |y| < \infty$. Whenever you see this kind of problem (infinite domain), try to find the Green's function. There are at least two ways of solving this equation.

(a) First approach – Green's function
Here we determine G from

$$\nabla^2 G + k^2 G = \delta(x)\delta(y). \tag{7.16}$$

We need to know that the solution will only depend on r. We look for this, and write the governing equation in polar coordinates:

$$\frac{1}{r}\frac{d}{dr}\left(r\frac{dG}{dr}\right) + k^2 G = \delta(x)\delta(y) = \infty \text{ at } (0,0).$$

Neglecting the origin $(0,0)$ for the time being, this equation reduces to

$$\frac{1}{r}\frac{d}{dr}\left(r\frac{dG}{dr}\right) + k^2 G = 0,$$

which has the solution

$$G = a_1 H_0^1(kr) + a_2 H_0^2(kr),$$

which are the Hankel functions of order zero, first, and second kind, respectively. Toward determining a_1, a_2, we note that

$$H_0^2(kr) \sim -\frac{2i}{\pi}\ln r$$

as $r \to 0$. Differentiation gives $1/r$ variation with r. H_0^1 has the same properties at the origin, but also has

$$H_0^2(kr) \sim \sqrt{\frac{2}{\pi kr}}\, e^{-i\left(kr-\frac{\pi}{4}\right)} \quad \text{as } r \to \infty. \tag{7.17}$$

If we multiply eq. (7.17) by $H(x,y)e^{i\omega t}$, we get outgoing waves. If we use H_0^1, we get incoming waves, which is not wanted, so we set $a_1 = 0$, so that

$$G = a_2 H_0^2(kr).$$

To obtain a_2, we integrate eq. (7.16) over the whole space to obtain unity on the RHS, i.e., $\int_{A_\varepsilon} \nabla^2 G\, dA = 1$, since $\int_{A_\varepsilon} G\, dA = 0$ as $\varepsilon \to 0$, where ε is the radius of the circle shown in Figure 7.24. We are carrying out this analysis in the neighborhood of the origin, noting that we are solving the differential equation outside of the origin in that case.

Thus,

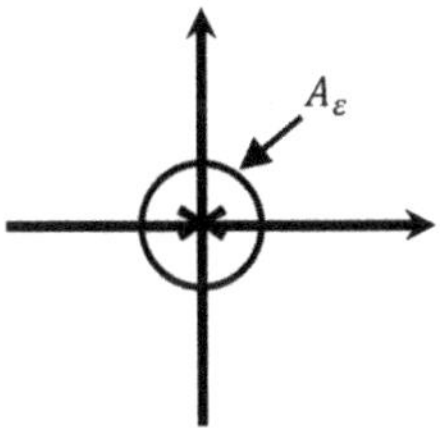

Figure 7.24: Elemental area A_ε at the origin of a circular domain.

$$1 = \int_{A_\varepsilon} \nabla^2 G\, dA \equiv \int_{|\varepsilon|=r} \frac{\partial G}{\partial r} ds \left(\text{or} \int \mathbf{n} \cdot \nabla G\, ds \right),$$

where $s = \varepsilon\, d\theta$ is an elemental arc length on A_ε. Thus, by using $G = -(2/\pi) i \ln r$ in the neighborhood of the origin, we have $A = i/4$. We can use this method to solve for many types of related second-order linear differential equations as long as we have infinite domain. We have solved the current differential equation outside the origin.

(b) Second approach – Fourier transform:
We take the FT of the governing equation to obtain

$$\left(\lambda_1^2 + \lambda_2^2 - k^2\right) \tilde{\psi} = -\tilde{H}(\lambda_1, \lambda_2),$$

and employ convolution:

$$\psi = -\frac{1}{4\pi^2} \iint_{-\infty}^{\infty} H(x', y') dx' dy' \underbrace{\iint_{-\infty}^{\infty} \frac{e^{i(x-x')\lambda_1 + i(y-y')\lambda_2}}{\lambda_1^2 + \lambda_2^2 - k^2} d\lambda_1 d\lambda_2}_{\text{Have to find this } \equiv I}.$$

Let $x' - x = a$, $y' - y = b$, so that

$$I = \iint_{-\infty}^{\infty} \frac{e^{ia\lambda_1} e^{ib\lambda_2}}{\lambda^2 - k^2} d\lambda_1 d\lambda_2,$$

where $\lambda^2 = \lambda_1^2 + \lambda_2^2$ (Figure 7.25). Note that a, b are now the physical coordinates, whereas λ_1, λ_2 are the transform variables. We transform the λ domain to the polar coordinates, where $\lambda_1 \equiv x$, $\lambda_2 \equiv y$, and $\lambda =$ the radius r, $dxdy \equiv d\lambda_1 d\lambda_2 \equiv rdrd\theta \equiv \lambda d\lambda d\theta$, and $\lambda_1 = \lambda\cos\theta$. I in the new coordinate system can be written as

$$I = \int_0^{2\pi} d\theta \int_0^{\infty} \frac{\lambda e^{i\rho\lambda\cos\theta}}{\lambda^2 - k^2} d\lambda,$$

where $\rho = a\mathbf{i} + b\mathbf{j}$.

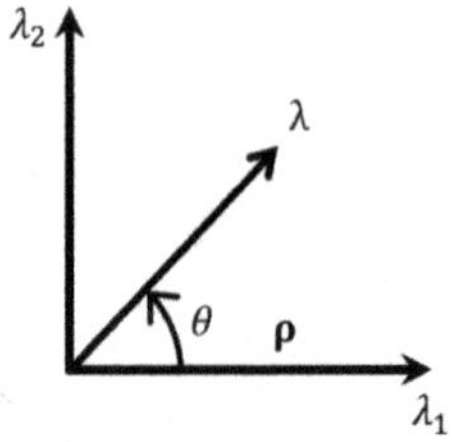

Figure 7.25: The λ coordinate directions.

Recollect that in the 3D problem, $\boldsymbol{\rho}$ (as $\mathbf{r}$) was lined up with the vertical axis (λ_3), but here it can be the horizontal axis, although we can put it on any axis. The integral can be written as

$$I = \int_0^{\infty} 2\pi \underbrace{\left[\frac{1}{2\pi}\int_0^{2\pi} e^{i\rho\lambda\cos\theta} d\theta\right]}_{J_0(\lambda\rho)} \lambda\left(\lambda^2 - k^2\right)^{-1} d\lambda = 2\pi \int_0^{\infty} \frac{\lambda J_0(\lambda\rho)d\lambda}{\lambda^2 - k^2}.$$

This is expected to give a Hankel function. So some manipulation is necessary. We know that

$$\begin{aligned} H_0^2(z) &= J_0(z) - iY_0(z) \\ &= \text{Even function in } z - \underbrace{\frac{2i}{\pi}\ln\left(\frac{z}{2}\right)\underbrace{J_0(z)}_{\text{even}}}_{\text{not even, not odd}}. \end{aligned} \tag{7.18}$$

For odd function $f(z)$, we have

$$\int_{-\infty}^{\infty} f(z)H_0^2(za)dz = -2\int_0^{\infty} f(z)J_0(az)dz. \tag{7.19}$$

We substitute the RHS of eq. (7.18) into the LHS of eq. (7.19), where the LHS vanishes for even function i.e., $\int_{-\infty}^{\infty} f(z) \times [\text{Even function in } z]dz = \int_{-\infty}^{\infty}[\text{Odd function}]dz = 0$. Thus, the term

$$-\frac{2i}{\pi}\int_{-\infty}^{\infty} f(z)\ln\left(\frac{za}{2}\right)J_0(az)dz \tag{7.20}$$

remains on the LHS of eq. (7.19), where $J_0(az)$ is analytic everywhere. We will have a branch cut as shown in Figure 7.26. Note that the log term cancels at 1, 2 in the figure, so that we are left with the J_0 term.

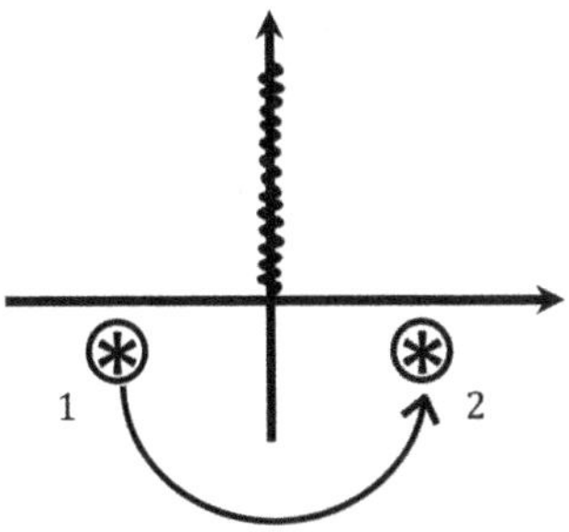

Figure 7.26: Branch cut to integrate eq. (7.20).

Thus,

$$\psi = \frac{1}{4\pi}\int_{-\infty}^{\infty}\frac{\lambda H_0^2(\lambda\rho)d\rho}{\lambda^2 - k^2},$$

where we have used

$$z \leftrightarrow \lambda, \quad f(z) \leftrightarrow \frac{\lambda}{\lambda^2 - k^2}. \tag{7.21}$$

Note that we have a singularity in eq. (7.21), but our f above is nonsingular, so we have a different meaning. We therefore shift the origin as shown in Figure 7.27.

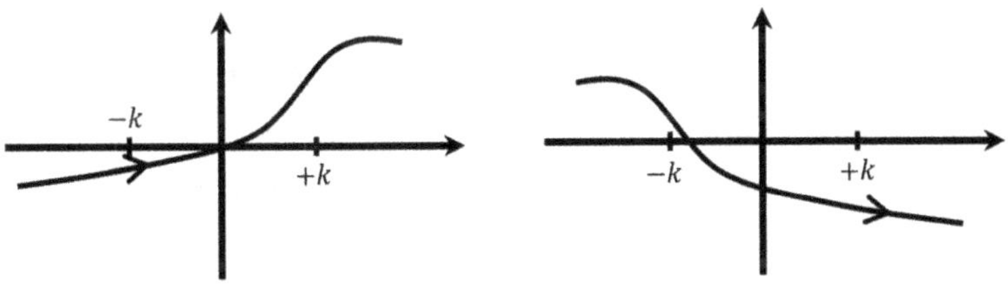

Figure 7.27: Shifting of origin.

If we do not use this trick, we would have a large integral to evaluate. Needless to say that we need to be familiar with working with the Bessel functions, to appreciate some of the solution techniques employed in this section. We will now attack the problem in a direct approach so we can appreciate the problem with inverting multidimensional FT problems. Starting from eq. (7.22):

$$I = \iint_{-\infty}^{\infty} \frac{e^{ia\lambda_1} e^{ib\lambda_2}}{\lambda_1^2 + \lambda_2^2 - k^2} d\lambda_1 d\lambda_2, \tag{7.22}$$

we will integrate in λ_1 first using the contour in Figure 7.28. That is, we hold λ_2 fixed, so that, for example, $\lambda_2^2 - k^2$ is treated as a constant. Thus, the poles associated with integrating with respect to λ_1 are given by

$$\lambda_1^2 = \pm i\sqrt{\lambda_2^2 - k^2}.$$

We do not know where the λ_i's are located, but we will put them at the two points shown in Figure 7.28. (We do not want them on $\text{Re}[\lambda_1]$.) Some guess work is involved here, which is acceptable because we do know that one pole is above and the other is below the real axis.

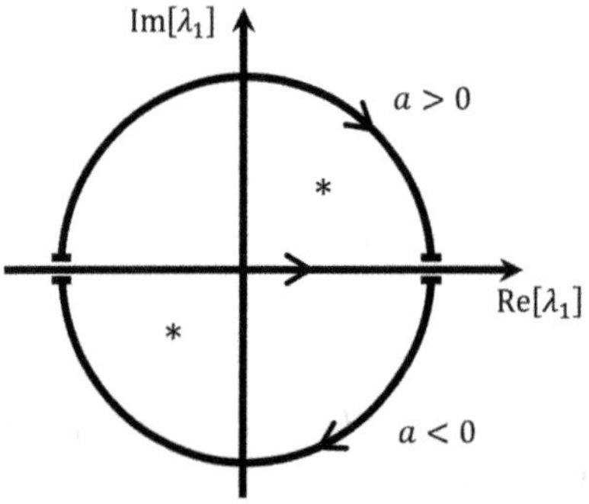

Figure 7.28: Contour in the complex λ_1 plane for evaluating λ_1 integral.

For $a > 0$ in $ia\lambda_1 = ia\text{Re}[\lambda_1] - a(\text{Im}[\lambda_1])$, or $ia\lambda_1 = iaa - a\eta$ if $\lambda_1 = a + i\eta$, we close in the UHP (or $\text{Im}[\lambda_1] > 0$), whereas for $a < 0$, we close in the LHP. The λ_1 integral is

$$I(\lambda_1) \equiv \int_{-\infty}^{\infty} e^{ia\lambda_1} \left[\frac{e^{ib\lambda_2}}{\lambda_1^2 + \lambda_2^2 - k^2}\right] d\lambda_1 = 2\pi i \left[e^{-|a|\sqrt{\lambda_2^2 - k^2}} \Big/ \left(2i\sqrt{\lambda_2^2 - k^2}\right)\right].$$

We will then substitute this into I and **integrate again**, but this time with respect to λ_2. We face choices again. We have two branch cuts and the problem becomes that of branch cuts.

We want to evaluate

$$I = \int_{-\infty}^{\infty} e^{ib\lambda_2} 2\pi i \left[\frac{e^{-|a|\sqrt{\lambda_2^2 - k^2}}}{2i\sqrt{\lambda_2^2 - k^2}} \right] d\lambda_2. \tag{7.23}$$

Because of the factor $\xi \equiv \sqrt{\lambda_2^2 - k^2}$ in the exponent in eq. (7.23), the branch cut must be such that the real part of ξ is positive along the integration path, otherwise you get a blow up. So, we will ask that along the integration path, $\mathrm{Re}\left[\sqrt{\lambda_2^2 - k^2}\right] > 0$. We have four options for the integration path, as shown in Figure 7.29. We can write

$$\left(\lambda_2^2 - k^2\right) \approx |\lambda| e^{\frac{i(\theta_1+\theta_2)}{2}} = |\lambda| e^{-i\pi},$$

of which we have to select the option that gives $\mathrm{Re}\left[\sqrt{\lambda_2^2 - k^2}\right] > 0$. Options (c) and (d) in Figure 7.29 violate this and we will get a blow up. Options (a) and (b) are acceptable: (a) gives outgoing waves, while (b) gives incoming waves.

If we use (a) we will obtain one solution, but it will be very complicated. It is the path integral for Hankel function, H_0^2. However, it is easy to do asymptotics with this contour. Path (b) will give H_0^1, or the path of integration for H_0^1. Previously (see Examples 7.5 and 7.7), instead of having branch cuts we have poles. It is easy to see outgoing and incoming waves in the latter.

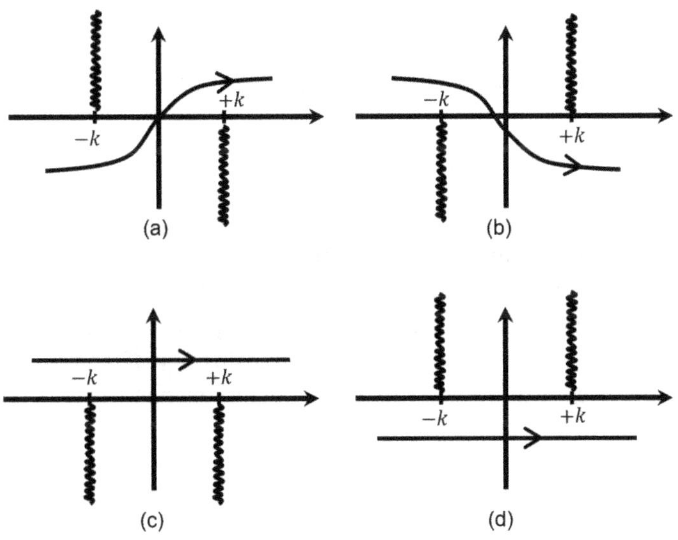

Figure 7.29: Four options for integration path in the complex λ_2 plane.

The asymptotics will go like Hankel asymptotics. The first approach (Green's function) may not work for more complicated BCs, and so we might need to opt for FT in that case.

7.5 The Wiener–Hopf techniques

The Wiener–Hopf technique is a method of solving certain linear PDEs subject to mixed BCs on semi-infinite physical domains. It is also equally applicable to integral equations of convolution type. To paraphrase Noble [11], the authority on this method, "When we say that a method of solution of a PDE is based on the Wiener–Hopf technique we mean that at some stage of the solution a function $K(\lambda)$, $\lambda = \alpha + i\eta$, is given which is regular and non-zero in a strip $\eta_- < \eta < \eta_+$, $-\infty < \alpha < \infty$, of the complex λ-plane." The method of solution requires that $K(\lambda)$ be decomposed in the form $K(\lambda) = K_+(\lambda)K_-(\lambda)$, where K_+ is regular and nonzero in $\eta > \eta_-$ and K_- is regular and nonzero in $\eta < \eta_+$. So, we see that the notion of plus or minus functions primitively concerns analyticity and existence in the upper or lower half of the transform variable plane λ. However, we shall see that the region of the physical domain we are considering, $x < 0$, $x > 0$, may affect whether a function is plus or minus.

In some applications, a mixed BV problem can be cast in integral equations such as

$$\int_{-\infty}^{\infty} K(x-t)f(t)dt = g(x),$$

which by convolution leads to

$$\tilde{K}\tilde{f} = \tilde{g}$$

which can easily be solved. However, we cannot directly to do so in other cases, such as

$$\int_{0}^{\infty} K(x-t)f(t)dt = g(x).$$

The reason is that we cannot use FT directly because the domain is in $[0, \infty)$ instead of the standard $(-\infty, \infty)$. We have to devise some means to convert the problem to one in $(-\infty, \infty)$. The application of complex FT can be significantly expanded if techniques can be devised to decompose functions into the two subdomains $(-\infty, 0]$ and $[0, \infty)$. Several examples are now given to illustrate plausible decomposition methods.

Example 7.8. Consider the equation

$$\int_{0}^{\infty} e^{-a|x-t|} f(t)dt = g(x). \tag{7.24}$$

(Note that if $f(x)=g(x)$, we have an eigenvalue problem.) The function $K\equiv e^{-a|x|}$ is a translation kernel because of the appearance of $x-t$. Toward extending the domain to $(-\infty,\infty)$ from $(0,\infty)$, we could try to set

$$f(x)=0 \quad \forall x<0.$$

The problem that arises is we now do not know what $f(x)$ is in the subdomain $x>0$. Let us consider

$$g(x)=0, \quad \forall x<0, \quad \text{and} \quad g(x)=g(x), \quad \forall x>0.$$

Let us rename $g(x)$ for $x<0$:

$$m(x)=0, \quad \forall x<0 \tag{7.25}$$

$$m(x)=g(x), \quad \forall x>0.$$

Can we now write our original equation as

$$\int_{-\infty}^{\infty} e^{-a|x-t|}f(t)dt=m(x)? \tag{7.26}$$

The answer is No, although we can write

$$\int_{-\infty}^{0} e^{-a|x-t|}f(t)dt=0.$$

We will not be able to define negative x since if $x<0$, eq. (7.26) will be equal to zero instead of being the given function in eq. (7.24). Therefore, we need to add something to $m(x)$, or

$$\int_{-\infty}^{\infty} e^{-a|x-t|}f(t)dt=m(x)+h(x),$$

where

$$h(x)=\begin{cases} 0 & \forall x>0 \\ \int_0^{\infty} e^{-a|x-t|}f(t)dt & x<0. \end{cases}$$

If $g=e^{-a|x|}$ then

$$\int_{-\infty}^{\infty} e^{-a|x-t|}f(t)dt=g(x)*f(x)$$

and

$$\tilde{g} = \int_{-\infty}^{\infty} e^{-a|x|} e^{-i\lambda x} = \int_{-\infty}^{0} e^{+ax-i\lambda x} dx + \int_{0}^{\infty} e^{-ax-i\lambda x} dx$$

$$= \int_{-\infty}^{0} e^{x(a-i\lambda)} dx + \int_{0}^{\infty} e^{-x(a+i\lambda)} dx \equiv \int_{-\infty}^{0} e^{xb} dx + \int_{0}^{\infty} e^{-xc} dx,$$

where $b = a - i\lambda$, $c = a + i\lambda$, and

$$\tilde{g} = \frac{1}{b} e^{xb}\Big|_{-\infty}^{0} - \frac{1}{c} e^{-xc}\Big|_{0}^{\infty} = \frac{2a}{\lambda^2 + a^2}.$$

We can now apply convolution:

$$\left(\frac{2a}{a^2 + \lambda^2}\right)\tilde{f}_- = \tilde{m}_- + \tilde{h}_+,$$

where f_-, m_- are analytic in some LHP, and h_+ is analytic in some UHP. This is the case because we can regulate λ so that it becomes negative enough to have analyticity.

Let $g(x) = f(x)$:

$$\frac{2a}{a^2 + \lambda^2}\tilde{f}_- = \tilde{f}_- + \tilde{h}_+ \Rightarrow \left(\frac{2a}{a^2 + \lambda^2} - 1\right)\tilde{f}_- = \tilde{h}_+$$

$$\underbrace{\frac{2a - a^2 - \lambda^2}{\lambda - ia}}_{(A)} \underbrace{\tilde{f}_-}_{(B)} = \underbrace{(\lambda + ia)}_{(C)} \underbrace{\tilde{h}_+}_{(D)} = \text{constant}, c. \tag{7.27}$$

The regions of analyticity of the different factors in eq. (7.27) are shown in Figure 7.30. Note that $2a - a^2 - \lambda^2$ is analytic everywhere, $\tilde{f}_-$ is analytic in the LHP, $(\lambda + ia)$ on the RHS is analytic everywhere, and $\tilde{h}_+$ is analytic on the UHP. Also the whole LHS of this equation is not analytic at ia.

So both the terms *AB*, *CD* agree in the strip. These two quantities must be equal to the same function which is analytic everywhere in the whole plane by virtue of analytic continuation. Let this function be $E(\lambda)$. Therefore, $AB = CD \equiv E(\lambda)$. This procedure is reminiscent of the kind of thing we do in SOVs where we look for the separation constant, which is invariably an eigenvalue.

To determine $E(\lambda)$, we have to assume some behaviors of $\tilde{f}$ or $\tilde{h}$. We could then solve and find out if the assumed behaviors are obeyed. If we assume that $f(x)$ is bounded, then

$$\tilde{f}(\lambda) = \int_{0}^{\infty} e^{-i\lambda x} f(x)\, dx.$$

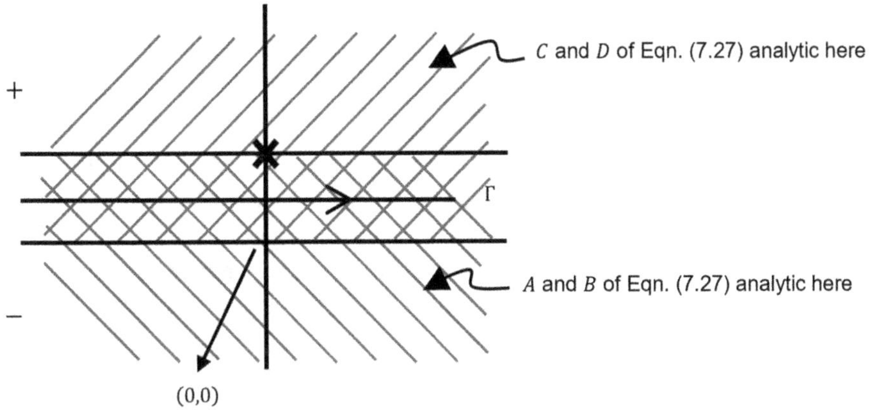

Figure 7.30: Regions of analyticity of the different factors in eq. (7.27).

If $\tilde{h}(\lambda) \to 0$ as $\lambda \to \infty$, then if $E(\lambda) =$ constant anywhere it must be constant everywhere, even at ∞. Then from eq. (7.27) we can solve for

$$\tilde{h}_+ = \frac{c}{\lambda + ia},$$

where c is a constant (see eq. (7.27)). Also from eq. (7.27):

$$\tilde{f}_- = \frac{(\lambda - ia)c}{2a - a^2 - \lambda^2}. \tag{7.28}$$

With this kind of analysis, we have to be adventurous and must not hesitate to guess. We must always have a strip. Equation (7.24) is an eigenvalue problem $(g(x) = f(x))$. We could pick $E(\lambda) = \lambda^2, \ldots$, or anything, although the solution may be unbounded. So, after we obtain the solution for h and f, we must check to see that the strip exists.

Inversion: We choose the inversion contour Γ shown in Figure 7.31. With $\tilde{h}_+ = \frac{c}{\lambda + ia}$, the pole is at $-ia$. We can invert h to get $h = 0$ (homolytic) for $x > 0$, or $h = ce^{+ax}$ for $x < 0$.

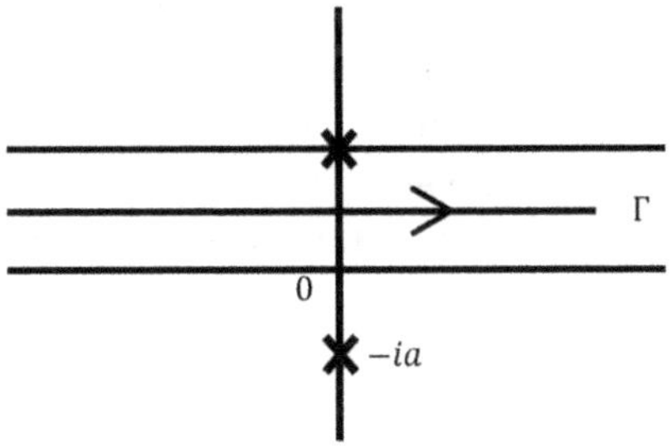

Figure 7.31: Contour Γ for inversion, with the pole at $-ia$.

Similarly,

$$f = \begin{cases} c\left[-\frac{ai}{b}\sin bx - i\cos bx\right], & x > 0, \\ 0, & x < 0. \end{cases}$$

The poles of $\tilde{f}_-$ are located at $\lambda^2 = \pm\sqrt{2a - a^2}$, which must be located in the UHP since f is analytic in the LHP. We will have some failure for some values of a. If $2a < a^2$, we will have imaginary roots. Thus, we need to check this rigorously for certain roots. This is because if the poles are not located at the right places, we may not obtain zero when we close the contour, and we may violate the step function assumption for f, h. Let us now examine a few examples.

Example 7.9. Determine the FT of

$$f(x) = e^{-a|x|}, \quad a > 0, \quad -\infty < x < \infty$$

and decompose this into a plus and a minus function.

Solution: Defining the FT of $f(x)$ as

$$\tilde{f}(\lambda) \equiv \mathcal{F}(f(x)) = \int_{-\infty}^{\infty} f(x)e^{-i\lambda x}\,dx,$$

we will have

$$\begin{aligned} \tilde{f}(\lambda) &= \int_{-\infty}^{\infty} e^{-a|x|-i\lambda x}\,dx = \int_{0}^{\infty} e^{-(a+i\lambda)x}\,dx + \int_{-\infty}^{0} e^{(a-i\lambda)x}\,dx \\ &= (a+i\lambda)^{-1} + (a-i\lambda)^{-1} = 2a\left(a^2+\lambda^2\right)^{-1}. \end{aligned}$$

Note that the procedure above automatically decomposes $\tilde{f}(\lambda)$ into $\tilde{f}_+(\lambda)$ and $\tilde{f}_-(\lambda)$:

$$\tilde{f}_+(\lambda) = (a+i\lambda)^{-1}, \quad \tilde{f}_-(\lambda) = (a-i\lambda)^{-1}.$$

Also note that the decomposition above is informed by the requirement for boundedness that the real part of the exponent in the integrand be negative independent of the value of x. That is, $g \equiv -a|x| > 0$. Thus, $g = -ax$ for $x > 0$, and $g = +ax$ for $x < 0$. **The function $\tilde{f}_+(\lambda)$ is analytic in some UHP, meaning that its singularities are located in the LHP (or at $i\lambda = -a$). Similarly, $f_-(\lambda)$ is analytic in some LHP, so that its singularities are located in the UHP (or at $i\lambda = a$).**

Example 7.10. Using the following definition for the FT of $f(x)$:

$$\tilde{f}(\xi) = \int_{-\infty}^{\infty} f(x)e^{i\xi x}dx,$$

show that the real function

$$f(x) = \begin{cases} e^{ax}, & x \geq 0,\ a > 0 \\ 0, & x < 0 \end{cases}$$

can only be Fourier-transformed if the transform variable is imaginary, as opposed to the transform variable ξ, which is real.

Solution: First, notice that $f(x)$ as given is not absolutely integrable because

$$\int_0^{\infty} e^{ax}dx = \frac{1}{a}e^{ax}\Big|_0^{\infty},$$

which is unbounded as $x \to \infty$. On the other hand, if we define a q such that

$$q = f(x)e^{-\eta x}, \quad \text{where } \eta > a,$$

we can try to obtain the FT of q, $\tilde{q}(\lambda)$, as follows:

$$\tilde{q}(\lambda) \equiv \mathcal{F}(q(x)) = \int_{-\infty}^{\infty} q(x)e^{i\xi x}\,dx = \int_{-\infty}^{\infty} e^{ax}e^{i(\xi+i\eta)x}\,dx \equiv \int_{-\infty}^{\infty} f(x)e^{i\lambda x}\,dx; \quad \lambda = \xi + i\eta.$$

Thus, whereas the given function $f(x)$ is not integrable, the function $q(x) = f(x)e^{-\eta x}$ is, provided the real part of the exponent in $f(x)e^{i\lambda x}$, or $(a - \eta)x$, is negative, meaning that $\eta > a$ if x is positive and $a > \eta$ if x is negative. Another result from this rather trivial analysis is that we need to operate in the complex plane, $\lambda \equiv \xi + i\eta$, as the real transform variable ξ will not enable the evaluation of the integral associated with the FT. This is consistent with previous discussions in this chapter. This example and a few others below have been included partly to remind the reader that the FT can be defined differently from the approach that have been used so far in this book.

Example 7.11. Decompose the function $f(x) = x^2$ into $f_+(x)$ and $f_-(x)$ using the FT definition given in the previous example problem.

Solution: We can obtain $f_+(x)$ and $f_-(x)$ by first determining $F_+(\lambda)$ and $F_-(\lambda)$, which are the respective FTs. These are obtained as follows:

$$F_+(\lambda) = \int_0^\infty x^2 e^{i\lambda x}\, dx = -\frac{2i}{\lambda^3}, \quad F_-(\lambda) = \int_{-\infty}^0 x^2 e^{i\lambda x}\, dx = \frac{2i}{\lambda^3}.$$

$f_+(x)$ and $f_-(x)$ can easily be obtained, respectively, from $\int_{-\infty}^{\infty} -\frac{2i}{\lambda^3} e^{-i\lambda x} d\lambda$ and $\int_{-\infty}^{\infty} \frac{2i}{\lambda^3} e^{-i\lambda x} d\lambda$. This approach directly gives $F_-(\lambda)$ and $F_+(\lambda)$ from $f(x)$ by applying the FT operation in $x \in (-\infty, 0]$ and $x \in [0, \infty)$, respectively. It is reminded that UHP and LHP do not refer to $x > 0$, $x < 0$ but rather to regions in the η coordinate direction in $\lambda = \alpha + i\eta$.

Example 7.12. Obtain $F_+(\lambda)$ and $F_-(\lambda)$ for the function

$$f(x) = e^{|x|}.$$

Use the FT definition in Example 7.10.

Solution: We can decompose as follows in the physical space: $f(x) = f_+(x) + f_-(x) = e^{|x|}$, where

$$f_+(x) = \int_{-\infty}^{\infty} F_+(\lambda) e^{-i\lambda x} d\lambda, \quad f_-(x) = \int_{-\infty}^{\infty} F_-(\lambda) e^{-i\lambda x} d\lambda,$$

and

$$F_+(\lambda) = \int_0^\infty e^{(1+i\lambda)x} dx = \frac{-1}{(1+i\lambda)}, \quad F_-(\lambda) = \int_{-\infty}^0 e^{(-1+i\lambda)x} dx = \frac{1}{(-1+i\lambda)}.$$

We note in the foregoing that a positive exponent is required everywhere in $x \in (-\infty, \infty)$ because of the absolute value sign. Therefore, $g \equiv e^{|x|}$ must be e^x when $x > 0$ and e^{-x}, when $x < 0$).

Example 7.13. Use the results in the previous example to solve the integral equation

$$\varphi(x) - \int_{-\infty}^{\infty} e^{-|x-t|} \varphi(t) dt = x^2 \equiv F(x). \tag{7.29}$$

Solution: Suppose $k \equiv e^{-|x|}$ is the kernel of the given equation and $\tilde{k}(\lambda)$ is its FT. On inspection we see that convolution gives

$$k(t) * \varphi(t) = \mathcal{F}^{-1}\left(\tilde{k}(\lambda)\tilde{\varphi}(\lambda)\right) = \int_{-\infty}^{\infty} \tilde{k}(\lambda)\tilde{\varphi}(\lambda) e^{-i\lambda x}\, d\lambda = \int_{-\infty}^{\infty} k(x-t)\varphi t(t)\, dt.$$

The penultimate example, with $F(x) \equiv x^2$, shows that $F_+(\lambda) = -2i/\lambda^3$, $F_-(\lambda) = 2i/\lambda^3$. Moreover, the FT of the kernel is $\tilde{k}(\lambda) = 2/\left(\lambda^2+1\right)$. We see that $\tilde{k}(\lambda)$ is analytic within the strip $-1 \le \eta \le 1$ in the complex plane $\lambda = \xi + i\eta$. Let us suppose that $\varphi(x)$ is decomposed into $\varphi_+(x)$ and $\varphi_-(x)$, with respective FTs of $\tilde{\varphi}_+(\lambda)$ and $\tilde{\varphi}_-(\lambda)$. Consistent with the current notations, we can determine $\varphi_+(x)$ and $\varphi_-(x)$ as follows:

$$\varphi_+(x) = \int_{C_+} \tilde{\varphi}_+(\lambda) e^{-i\lambda x}\, d\lambda, \quad \varphi_-(x) = \int_{C_-} \tilde{\varphi}_-(\lambda) e^{-i\lambda x}\, d\lambda.$$

The horizontal contour C_+ (along ξ) that we use in the λ plane for inversion must be chosen so that it runs through the region in which $\tilde{k}(\lambda)$ is analytic ($-1 \le \eta \le 1$). The transform of $\varphi_+(x)$, or $\tilde{\varphi}_+(\lambda)$, must be analytic in the analytic region of $\tilde{k}(\lambda)$ and also be analytic above it. In a similar fashion, $\tilde{\varphi}_-(\lambda)$ must be analytic in the analytic region of $\tilde{k}(\lambda)$ and also in the region below it.

We can now write the inversion formula for eq. (7.29), in which we separate the "+" functions from the "–" functions:

$$\int_{C_+} \left\{ \tilde{\varphi}_+(\lambda) - \frac{2}{\lambda^2+1}\tilde{\varphi}_+(\lambda) - F_+(\lambda) \right\} e^{-i\lambda x}\, d\lambda + \int_{C_-} \left\{ \tilde{\varphi}_-(\lambda) - \frac{2}{\lambda^2+1}\tilde{\varphi}_-(\lambda) - F_-(\lambda) \right\} e^{-i\lambda x}\, d\lambda$$

which can be simplified to

$$\int_{C_+} \left\{ \tilde{\varphi}_+(\lambda) \left[\frac{\lambda^2-1}{\lambda^2+1}\right] - F_+(\lambda) \right\} e^{-i\lambda x}\, d\lambda + \int_{C_-} \left\{ \tilde{\varphi}_-(\lambda) \left[\frac{\lambda^2-1}{\lambda^2+1}\right] - F_-(\lambda) \right\} e^{-i\lambda x}\, d\lambda = 0. \quad (7.30)$$

The contours C_+ and C_-, together with the region of analyticity of $\tilde{k}(\lambda)$ in the λ plane, the strip $-1 \le \eta \le 1$, are shown in Figure 7.32. Suppose we define a closed contour $C \equiv (C_-) \cup (\Gamma_1) \cup (-C_+) \cup (\Gamma_2)$ as shown in Figure 7.33, with the length of C_{-1} and C_+ being infinite and the contributions of Γ_1 and Γ_2 are negligible. If we integrate a function $R(\lambda)$ that is analytic inside the contour C, then by the Cauchy–Goursat theorem

$$\int_C R(\lambda) e^{-i\lambda x}\, d\lambda = 0.$$

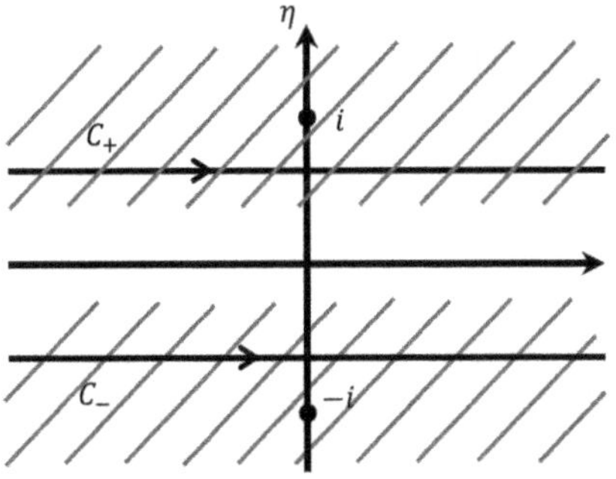

Figure 7.32: Region of analyticity of $\tilde{k}(\lambda)$, or $-1 \le \eta \le 1$, and the contours C_+ and C_- for $\tilde{\varphi}_+$ and $\tilde{\varphi}_-$, respectively.

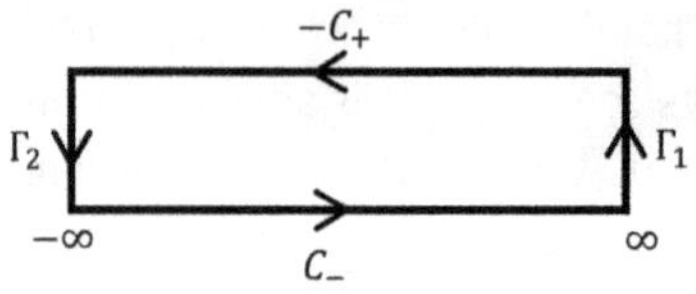

Figure 7.33: Contour for LT inversion.

If we decompose $R(\lambda)$ into $R_+(\lambda)$ and $R_-(\lambda)$ and consider that

$$\int_{C_+} R_+(\lambda)e^{-i\lambda x}d\lambda + \int_{C_-} R_-(\lambda)e^{-i\lambda x}d\lambda = 0,$$

where $R_+ = 0$ in C_- and $R_- = 0$ in C_+, we see upon comparison with eq. (7.30) that

$$-\left[\frac{\lambda^2-1}{\lambda^2+1}\varphi_+ - F_+\right] = \left[\frac{\lambda^2-1}{\lambda^2+1}\varphi_- - F_-\right] = R(\lambda).$$

Thus,

$$\tilde{\varphi}_+ = \left[\frac{\lambda^2+1}{\lambda^2-1}\right](F_+ - R),\ \ \tilde{\varphi}_- = \left[\frac{\lambda^2+1}{\lambda^2-1}\right](F_- + R).$$

The functions $\tilde{\varphi}_+(\lambda)$ and $\tilde{\varphi}_-(\lambda)$ are singular at $\lambda = \pm 1$, or $\xi = \pm 1$. There is also a singularity at $\lambda = 0$ from $F_\pm(\lambda)$, and the kernel $\tilde{k}(\lambda)$ is analytic in the strip $-1 \le \eta \le 1$. Therefore, C_+ must be chosen to be below $\eta = 1$ and C_- must be located above $\eta = -1$ but below $\eta = 0$. Introducing R into eq. (7.30), we can obtain $\varphi(x)$ as follows:

$$\begin{aligned}\varphi(x) &= \int_{-C_+} \tilde{\varphi}_+(\lambda)e^{-i\lambda x}d\lambda + \int_{C_-} \tilde{\varphi}_-(\lambda)e^{-i\lambda x}d\lambda \\ &= \int_{-C_+}\left[\frac{\lambda^2+1}{\lambda^2-1}\right]\left[-\frac{2i}{\lambda^3}\right]e^{-i\lambda x}d\lambda - \int_{-C_+}\left[\frac{\lambda^2+1}{\lambda^2-1}\right]Re^{-i\lambda x}d\lambda + \int_{C_-}\left[\frac{\lambda^2+1}{\lambda^2-1}\right]\left[\frac{2i}{\lambda^3}\right]e^{-i\lambda x}d\lambda \\ &\quad + \int_{C_-}\left[\frac{\lambda^2+1}{\lambda^2-1}\right]Re^{-i\lambda x}d\lambda \\ &= \int_{-C_+} A\left[-\frac{2i}{\lambda^3}\right]e^{-i\lambda x}d\lambda + \int_{C_-} A\left[\frac{2i}{\lambda^3}\right]e^{-i\lambda x}d\lambda + \int_C A\cdot Re^{-i\lambda x}d\lambda,\end{aligned}$$

where $A = \left[\frac{\lambda^2+1}{\lambda^2-1}\right]$, $C = (C_-) \cup (-C_+)$, and R is as defined above. We can write this equation as

$$\varphi(x) = I + I_+ + I_-.$$

The integrals above can easily be evaluated using the residue theorem. In specifics, with the contour C presented above, the evaluation of I is straightforward, with residues evaluated at $\lambda = \pm 1$. The singularities for I_+ and I_- are located at $\lambda = 0$ (third-order pole) and $\lambda = \pm 1$ (two first-order poles). Because these latter poles have been used in the evaluation of I, their contributions to I_+ and I_- can be omitted. In closing the C_+ and C_- contours, we will use the LHP if $x > 0$, since $e^{-i\lambda x} = e^{-i\xi x} e^{\eta x} \to 0$ as $\eta \to -\infty$. The contours for C_+ and C_- in this case, are shown in Figure 7.34.

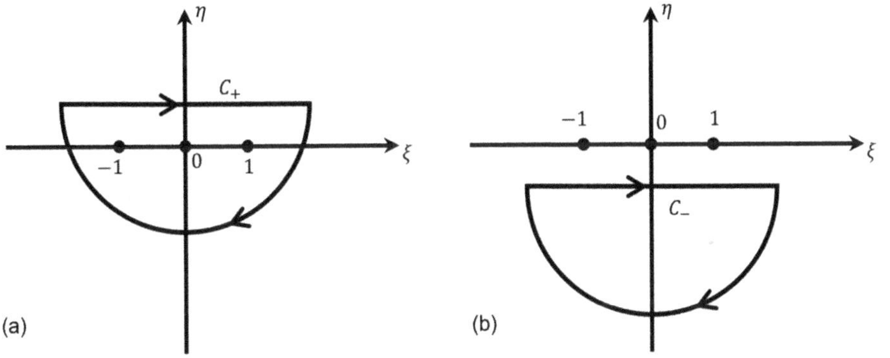

Figure 7.34: The contours for evaluating (a) I_+ and (b) I_-, when $x > 0$.

Similarly, when $x < 0$, $e^{-i\lambda x} = e^{-i\xi x} e^{\eta x} \to 0$ as $\eta \to \infty$, so that we will need to close the two contours on the UHP, as shown in Figure 7.35.

The results are

$$I = C_1 \cos x + C_2 \sin x,$$

$$I_+ = \left(4 - x^2\right) H(x), \qquad I_- = \left(4 - x^2\right) H(-x),$$

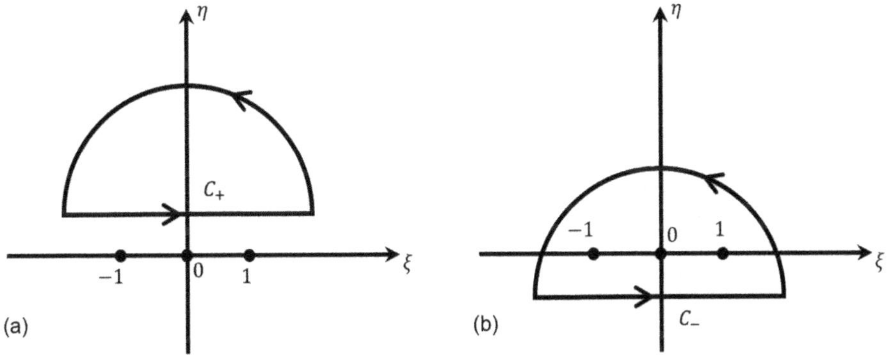

Figure 7.35: The contours for evaluating (a) I_+ and (b) I_-, when $x < 0$.

where $H(x)$ is the Heaviside step function. The combined solution is

$$\varphi(x) = C_1 \cos x + C_2 \sin x + 4 - x^2.$$

Example 7.14. Consider a convective-diffusive problem in transport phenomenon such as occurs in temperature or species transport. The governing equation associated with the flow from a semi-infinite plate to a fluid moving past the plate with a normalized constant velocity of magnitude unity can be written as [17]

$$\nabla^2 \phi - \phi_{,x} = 0 \text{ in } \mathcal{D},$$

where $\mathcal{D}$ is the infinite plane outside the half-line $x > 0$, i.e., the whole x,y plane $(-\infty,\ \infty)$ except line $x = 0 \to \infty$. That is, we have a discontinuity. See Figures 7.36 and 7.37. The condition

$$\phi(x, y = 0) = e^{-ax}, \quad a > 0$$

is imposed in $x \in (0, \infty)$, with $y = 0$. We do not know anything else about the problem. The equation is valid outside the cut plane. This condition is additionally imposed: $\phi(x,y) \to 0$ as $x, y \to \pm\infty$. This problem is also discussed in Carrier, Krook, and Pearson [3], but the approach presented here differs significantly. The Wiener-Hopf technique, though popularized by the authors in Ref [17], was reportedly invented by Carleman [19].

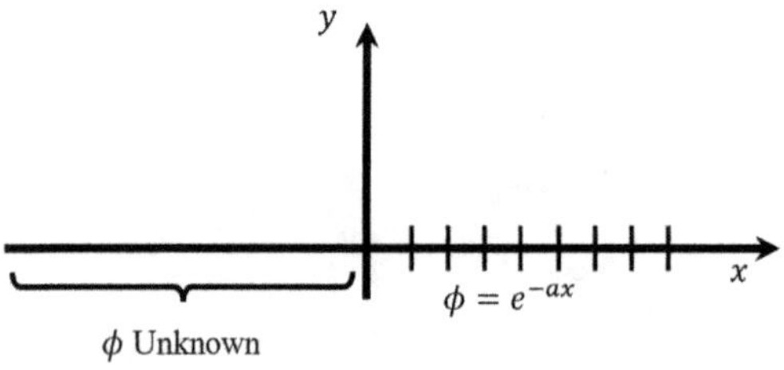

Figure 7.36: Domain $x \in (0, \infty)$ for transformation to $x \in (-\infty, \infty)$, with an imposed ϕ in $(x,y) \in [0,\ \infty) \times [0,\ 0]$.

We can change the problem to a mixed BVP, as opposed to the purely Dirichlet one described so far. That is, we could additionally specify the gradient of ϕ on the same boundary region $(x, y) \in [0,\ \infty) \times [0,\ 0]$.

With brute-force we will take the FT in x, even though we do not have enough information on the BCs. We only know the BC on the positive x – axis. The FT of the governing equation can be written as

$$-\lambda^2 \tilde{\phi} + \tilde{\phi}_{yy} - i\lambda \tilde{\phi} = 0, \tag{7.31}$$

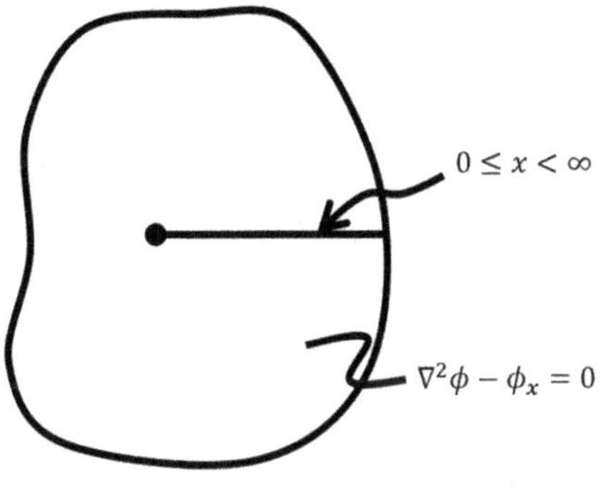

Figure 7.37: The physical domain, with BC e^{-ax}, $a > 0$ along the line $0 \le x < \infty$.

while the FT of the BC on $y = 0$ can be written as follows, using our original definition of FT:

$$\tilde{\phi} = \underbrace{\int_{-\infty}^{0} e^{-i\lambda x} \underbrace{\phi(x, y=0)}_{u(x),\,\text{unknown}}}_{\tilde{u}(\lambda)} + \underbrace{\int_{0}^{\infty} e^{-i\lambda x} e^{-ax} dx.}_{\substack{B \equiv \left(\frac{1}{a+i\lambda}\right) \\ \text{known}}} \tag{7.32}$$

In eq. (7.31), FT is taken only in the x-coordinate direction. The introduction of the first term on the RHS of eq. (7.32) is our attempt to extend the problem (BC domain) in x from $x \in [0, \infty)$ to $x \in (-\infty, \infty)$. Note that $-i\lambda x - ax = -iax + (\eta - a)x$ for the term B. With $x > 0$ in this subdomain, the "horizontal" contour for inversion must satisfy $\eta < a$. Furthermore, with $B = \frac{1}{a + i\lambda}$, the pole is located at $a + i\lambda = 0$ or $\lambda = ai$. Therefore, B is analytic in the LHP and so we will close the inversion contour in the UHP.

Let us define

$$u(x) = \begin{cases} 0, & x > 0 \\ \phi(x, y=0), & x < 0. \end{cases}$$

The $\tilde{u}(\lambda)$ is a "plus function" (analytic in some UHP) as $\eta > a$, but we do not know $\tilde{u}(\lambda)$ yet. From eq. (7.31) we have

$$\tilde{\phi}_{yy} - \left(\lambda^2 + i\lambda\right)\tilde{\phi} = 0,$$

or

$$\tilde{\phi} = A(\lambda)e^{-\sqrt{\lambda^2 + i\lambda}y} + B(\lambda)e^{\sqrt{\lambda^2 + i\lambda}y}. \tag{7.33}$$

We have to throw away some things because we need solutions that are convergent for all y. For positive y, the first term goes to zero exponentially, whereas the second grows exponentially. For negative y, we have the opposite behavior for the two terms.

We have branch points where $\lambda^2 + i\lambda = 0$, or $\lambda = 0, -i$. There is a way to pick $\tilde{\phi}$ such that $[\lambda^2 + i\lambda]$ in eq. (7.33) is always positive with a particular suitable branch cut. As we shall see in this procedure, this problem shows the significance of the proper choice of branch cut and the fact that coming up with the right cuts could be a difficult task. If we can find this contour such that $\mathrm{Re}\left[\sqrt{\lambda^2 + i\lambda}\right]$ is always positive independent of whether y is positive or negative, then the solution of $\tilde{\phi}(\lambda, y)$ of the problem can be written as

$$\tilde{\phi} = c(\lambda) e^{-|y|\sqrt{\lambda^2 + i\lambda}}, \tag{7.34}$$

where $c(\lambda)$, which applies when $y = 0$, is defined below. We have introduced $|y|$ in order to have validity for both $y \to \infty$, $y \to -\infty$, $x > 0$, $x < 0$ are primarily concerned with the behavior at $\eta \to \infty, -\infty$, where $\lambda = a + i\eta$, but then we have to ensure that the contribution from the real part does not lead to exponential growth. So we will need to pick the BC so that the assumption is right. The requirement is that, going along the real axis (imaginary =0),

$$\sqrt{\lambda^2 + i\lambda} \to +\infty \text{ as } \lambda \to \pm\infty,$$

otherwise the integration will not converge. Thus, we will choose branch cuts such that $\mathrm{Re}\left[\sqrt{\lambda^2 + i\lambda}\right]$ is positive. Figure 7.38 shows contours that could be considered for inversion. We could determine c via the BC statement in eq. (7.32) (see some details in Figure 7.39):

$$\tilde{\phi}(\lambda, 0) = \tilde{u}(\lambda) + \frac{1}{a + i\lambda} = c(\lambda) \text{ on } y = 0.$$

This expression for $c(\lambda)$ is used in eq. (7.34) to obtain the solution $\tilde{\phi}(\lambda, y)$. The $\tilde{u}(\lambda)$ in this solution must be determined prior to inversion to obtain $\phi(x, y)$. We note that

$$\phi_y(x, y = 0^+) \neq \phi_y(x, y = 0^-), \text{ on } x > 0,$$

so that

$$\phi_y(x, y = 0^+) - \phi_y(x, y = 0^-) \equiv f(x).$$

$$f(x) = ?, \quad x > 0.$$

Also, $f(x) = 0$ for $x < 0 \Rightarrow \phi_y$ is continuous at $y = 0$, $x < 0$. Let us take the FT of the BC statement above in the x-direction:

$$\tilde{\phi}_y(\lambda, 0^+) - \tilde{\phi}_y(\lambda, 0^-) = \tilde{f}(\lambda),$$

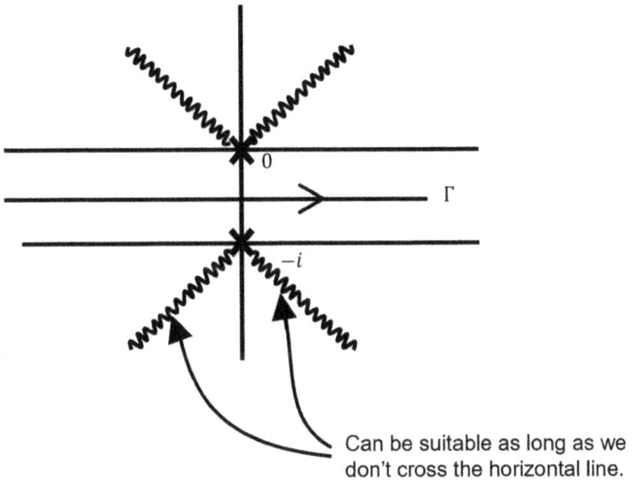
0
Γ
$-i$
Can be suitable as long as we don't cross the horizontal line.

Figure 7.38: Options for integration contours, showing branch cuts, such that $\mathrm{Re}\left[\lambda^2 + i\lambda\right] > 0$ on the real line.

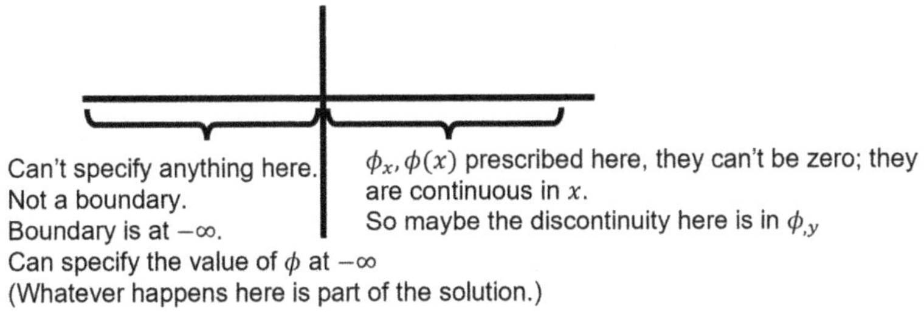
Can't specify anything here.
Not a boundary.
Boundary is at $-\infty$.
Can specify the value of ϕ at $-\infty$
(Whatever happens here is part of the solution.)
$\phi_x, \phi(x)$ prescribed here, they can't be zero; they are continuous in x.
So maybe the discontinuity here is in $\phi_{,y}$

Figure 7.39: More details on the conditions along the physical x-coordinate direction.

where $\tilde{f}(\lambda)$ is a minus function and the y-jump in the y-gradient of $\tilde{\varphi}$ in $x \in [0, \infty)$. Because of the $|y|$ absolute value symbol in eq. (7.34), ϕ_y is discontinuous. We will differentiate eq. (7.34) with respect to y for $y > 0$ to obtain $\tilde{f}(\lambda) = -2c(\lambda)\sqrt{\lambda^2 + i\lambda}$, where $c(\lambda) = \tilde{u}(\lambda) + 1/(a + i\lambda)$ so that

$$\tilde{f}(\lambda) = -2\left[\tilde{u}(\lambda) + \frac{1}{a + i\lambda}\right]\sqrt{\lambda^2 + i\lambda}, \qquad \tilde{f}_- = -2\left[\tilde{u}_+ + \frac{1}{(a + i\lambda)_-}\right]\sqrt{\lambda^2 + i\lambda},$$

where u_+ is not known on the negative real axis. The inversion path Γ must pass through as shown in Figure 7.38 because we cannot go through the branch cut. We will need to do something along the line of:

$(\text{Something})_{+\text{ve}} = (\text{Something})_{-\text{ve}}$ via analytic continuation = another function, say $E(\lambda)$, which decays. Factorizing $\sqrt{\lambda^2 + i\lambda}$, we can write the expression for $\tilde{f}(\lambda)$ as follows:

$$\sqrt{\lambda^2+i\lambda}=\sqrt{\lambda}\left(\sqrt{\lambda+i}\right) \Rightarrow \frac{\tilde{f}}{\sqrt{\lambda}}=-2\left(\tilde{u}+\frac{1}{a+i\lambda}\right)\sqrt{\lambda+i}.$$

A branch cut for $\tilde{f}/\sqrt{\lambda}$ is shown in Figure 7.40. This function is negative, as is $\tilde{f}$.

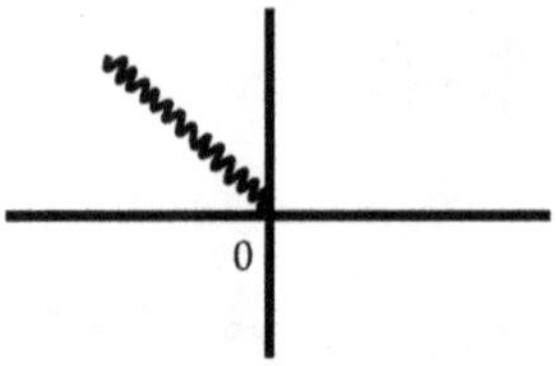

Figure 7.40: The branch cut for $\tilde{f}/\sqrt{\lambda}$.

We can write $\tilde{f}/\sqrt{\lambda}$ as follows:

$$\underbrace{\frac{\tilde{f}}{\sqrt{\lambda}}}_{-\text{ve}}=-2\left[\tilde{u}\underbrace{\sqrt{\lambda+i}}_{+\text{ve}}+\underbrace{\frac{\sqrt{\lambda+i}}{a+i\lambda}}_{\substack{\text{one}+\text{ve}\\ \text{one}-\text{ve}}}\right], \tag{7.35}$$

in which the function $\sqrt{\lambda}$ is analytic everywhere except at the branch cut, but $\tilde{f}$ is analytic in some LHP. Let this plane be arbitrary as shown in Figure 7.41. We want to write the expression on the LHS below as the **sum** of two functions L_+ and L_-:

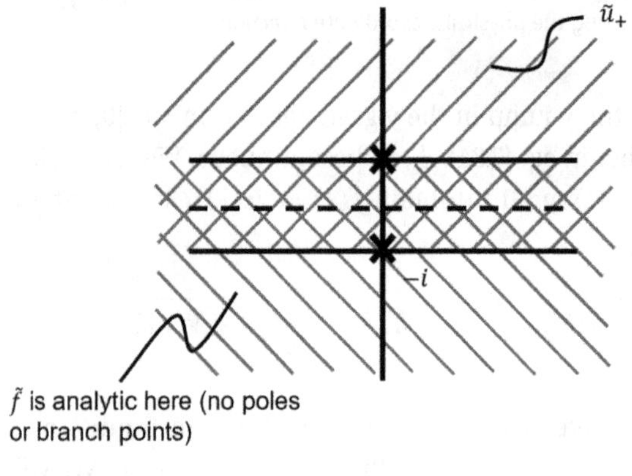

Figure 7.41: Region of analyticity of $\tilde{u}_+$ and $\tilde{f}$.

$$\frac{\sqrt{\lambda+i}}{a+i\lambda}=L_{+}+L_{-}.$$

We know that $\sqrt{\lambda+i}$ is a positive function that is analytic in UHP; but not in LHP, because it has a branch point at $\lambda=-i$. The function $\frac{1}{a+i\lambda}$ is a negative function that is not analytic in the UHP because it has a pole at $\lambda=ia$. We can easily get rid of the pole:

$$\frac{\sqrt{\lambda+i}}{a+i\lambda}=\overbrace{\frac{\sqrt{\lambda+i}-\sqrt{ia+i}}{a+i\lambda}}^{\text{Analytic because we subtracted the residue}}+\overbrace{\frac{\sqrt{ia+i}}{\underbrace{a+i\lambda}_{\text{-ve, has a pole}}}}^{\text{constant, residue}}.$$

No pole but has branch cut $\lambda=-i \Rightarrow$ +ve function, analytic in some UHP

The reader should observe that in the foregoing we have used the fact that the residue of $\sqrt{\lambda+i}/[a+i\lambda]$ is given by $\lim_{z\to ia}(a+i\lambda)\frac{\sqrt{\lambda+i}}{a+i\lambda}=\sqrt{ia+i}$.

Therefore, substituting all these into eq. (7.35) we will have:

$$\left[\left(\frac{\tilde{f}}{2\sqrt{\lambda}}\right)+\frac{\sqrt{ia+i}}{a+i\lambda}\right]_{-}=\left[\frac{\sqrt{\lambda+i}-\sqrt{ia-i}}{a+i\lambda}+\tilde{u}\sqrt{\lambda+i}\right]_{+}=E(\lambda),$$

where the function on the LHS is an analytic continuation of the one on the RHS. They are equal in the strip. Analogous to the idea of separation constant in the method of SOVs, the two terms define a function $E(\lambda)$, which is entire. It can be argued that each of the two terms also tends to zero as $|\lambda|\to\infty$, implying by Lioville's theorem, that $E(\lambda)$ is zero.

We can obtain $\tilde{f}$ by integrating in the LHP:

$$\tilde{f}=\int_{0}^{\infty}e^{-i\lambda x}f(x)dx.$$

Recall that although $\tilde{f}$ is equal to a jump of $\tilde{\phi}_y$, we assume that it is integrable. Also, we assume that $\tilde{f}\to 0$ as $\lambda\to\infty$; i.e., $f=\sqrt{\lambda}\varepsilon(\lambda)\to 0$ as $\lambda\to\infty$ so that $\varepsilon(\lambda)=0$ (LHP). Also note that $1/2\sqrt{\lambda}\to 0$ on LHP, $E(\lambda)\to 0$ on LHP, and that on LHP, we will have

$$\left[\frac{\sqrt{\lambda+i}-\sqrt{ia-i}}{a+i\lambda}+\tilde{u}\sqrt{\lambda+i}\right]\to 0.$$

On the UHP, $-i\lambda x=-iax+\eta x$, $x<0$ so that $\eta>0$. If u is continuous we will have $\tilde{u}=\frac{1}{i\lambda}\int_{-\infty}^{0}u(x)de^{-i\lambda x}=-\frac{e^{-i\lambda x}u(x)}{i\lambda}\Big|_{-\infty}^{0}+\int\dots$ We will assume that the integrand of the first term on the RHS of this equation exponentially goes to zero as $\lambda\to-\infty$.

Also on the UHP,

$\tilde{f} = \frac{-2\sqrt{\lambda}\sqrt{ia+i}}{a+i\lambda}$, which is negative (with a jump in the positive x-axis),

and

$$\tilde{u} = \frac{-\sqrt{\lambda+i}+\sqrt{ia-i}}{(a+i\lambda)\sqrt{\lambda+i}}. \tag{7.36}$$

This $\tilde{u}(\lambda)$ is required in $c(\lambda)$ in order to invert eq. (7.34) to obtain $\phi(x,y)$. In eq. (7.36), $(a+i\lambda)$ does not give a pole, but $\sqrt{\lambda+i}=0$ does at $\lambda=-i$. We need to check that $\tilde{f}$ is indeed a negative function. For x negative, it must be 0 (analytic in some LHP). To evaluate $\tilde{f}$, for $x<0$, we close in the LHP where the integral decays with η (in $\lambda=\xi+i\eta$), since $|e^{i\lambda x}|=|e^{\eta x}|$. Refer to Figure 7.42.

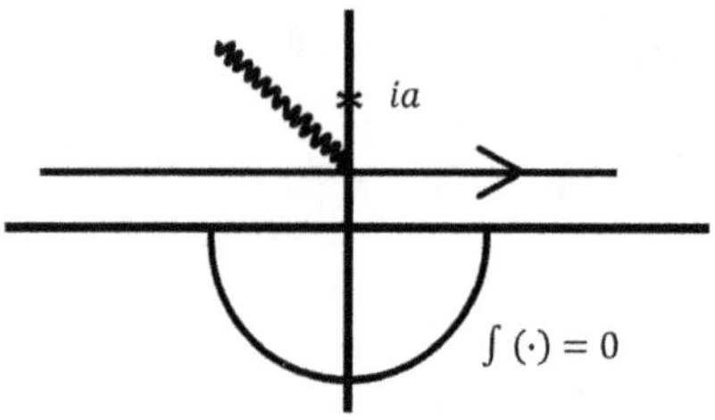

Figure 7.42: Region of analyticity of $\tilde{f}$, and the integration path.

Regarding $\tilde{u}$, it has a singularity at $-i$ (Figure 7.43), as illustrated in eq. (7.36).

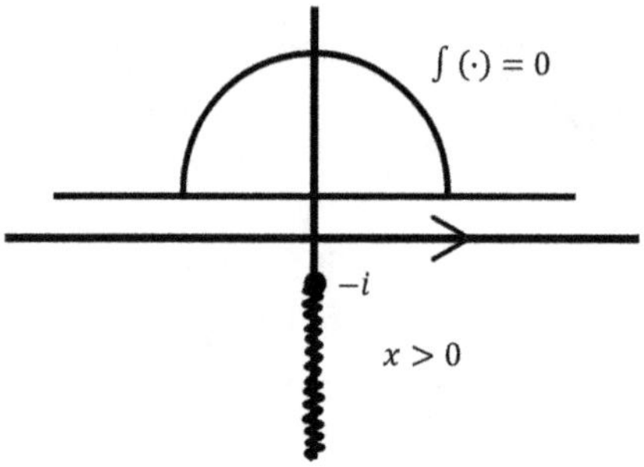

Figure 7.43: Region of analyticity of $\tilde{u}(x)$, and the integration path.

7.5.1 Summary of the Wiener–Hopf technique

Suppose we want to solve the integral equation

$$\int_0^\infty \underbrace{K(x-t)}_{\text{known}}\ \underbrace{f(t)}_{\text{unknown}}\ dt = \underbrace{g(x)}_{\text{known}}$$

for $f(x)$. We let

$$f(x) = \begin{cases} 0, & x < 0 \\ ?, & x > 0, \end{cases}$$

$$m(x) = \begin{cases} g(x), & x > 0 \\ ?, & x < 0, \end{cases}$$

and

$$h(x) = \begin{cases} \int_0^\infty K(x,t)f(t)dt, & x < 0 \\ 0, & x > 0, \end{cases}$$

and rewrite the integral equation as

$$\int_{-\infty}^{\infty} K(x-t)f(t)dt = m(x) + h(x),$$

where K, f, m, h are all defined over the whole range of x. The idea of the Wiener–Hopf technique is to extend the domain $(0, \infty)$ to the domain $(-\infty, \infty)$ so we can use FT. It is basically a domain extension technique.

The FT of the above equation is

$$\tilde{K}\tilde{f} = \tilde{m} + \tilde{h}.$$

Since f is unknown in $x > 0$ and 0 in $x < 0$, then $\tilde{f}$ is a minus function ($|e^{i\lambda x}| = |e^{\eta x}|$), and so it is analytic in some LHP. $\tilde{m}$ is a minus function and $\tilde{h}$ is a plus function. We will assume that there exists a λ such that we can make the imaginary part of λ large (negative enough) to dominate so that $\int_0^\infty e^{-i\lambda x} f(x)dx \to 0$:

$$\tilde{K}\tilde{f}_- = \tilde{m}_- + \tilde{h}_+ (\text{Wiener} - \text{Hopf equation}). \tag{7.37}$$

We will decompose the kernel $\tilde{K}$ into two parts to use the Wiener–Hopf technique. Let us take

$$\tilde{K} = \frac{\tilde{K}_-}{\tilde{K}_+}, \tag{7.38}$$

so that

$$\tilde{K}_-\tilde{f}_- = \tilde{K}_+\tilde{m}_- + \tilde{K}_+\tilde{h}_+.$$

The splitting of $\sqrt{\lambda(\lambda + i)}$, as done in eq. (7.38), is easy. $\tilde{f}$ is the function we want, which is why we separate it out. We will try to split $K_+ m_-$ additively into a function that is positive and another that is minus: ($K_+\tilde{m}_- \equiv L_+ + X_-$) so that

$$\tilde{K}_-\tilde{f}_- = L_+ + X_- + \tilde{K}_+ h_+.$$

We then group all minuses on one side, etc.:

$$\tilde{K}_-\tilde{f}_- - X_- = \tilde{K}_+\tilde{h}_+ + L_+ = E(\lambda).$$

Here, $E(\lambda)$ is being determined based on an assumed behavior of f, e.g., at 0, or from some asymptotics.

If we can write an eq. (7.37), which is the Wiener–Hopf equation, we can always get an integral equation which corresponds to it for the diffusion equation. The solution of the PDE is the solution of the integral equation. In fact, any second-order linear differential equation can always be written as an integral equation, although the converse is not necessarily true. This is the essence of the boundary integral method numerics. The classical reference for the Wiener–Hopf procedure is Noble [11]. A limited coverage is available in Carrier et al. [3], Nair [10], Morse and Feshbach [9], and Crighton et al. [5].

7.5.2 A standard way of kernel decomposition

One difficult aspect of the Wiener–Hopf method is the decomposition of the kernel $\tilde{K}(\lambda)$ into $\tilde{K}_+(\lambda)$ and $\tilde{K}_-(\lambda)$. A relatively easy approach is based on additive decomposition:

$$\tilde{K}(\lambda) = \tilde{K}_+ + \tilde{K}_-.$$

An assumption in this approach is that $\tilde{K}(\lambda)$ is analytic in some strip $\Omega\,(\xi, i\eta) \in (-\infty, \infty) \times [\beta, \alpha]$, as shown in Figure 7.44. Using the Cauchy integral theorem, with any suitable closed contour such as $\Gamma = \cup\, \Gamma_i$, $i = 1, 2, \ldots, 4$, we can write

$$\tilde{K}(\lambda) = \frac{1}{2\pi i}\int_{\Gamma} \frac{\tilde{K}(t)dt}{t-\lambda},$$

or

$$2\pi i\tilde{K}(\lambda) = \int_{\Gamma_1} \frac{\tilde{K}(t)dt}{t-\lambda} + \int_{\Gamma_3} \frac{\tilde{K}(t)dt}{t-\lambda} + \int_{\Gamma_2+\Gamma_4} \frac{\tilde{K}(t)dt}{t-\lambda}. \tag{7.39}$$

Above, λ, which is enclosed inside Γ is not a point of singularity of $\tilde{K}(\lambda)$. Let $AB \to \infty$ and $\tilde{K}(t)$ is such that the contributions from $\Gamma_2, \Gamma_4 = 0$. That is, they cancel each other, or they go to zero individually as $AB \to \infty$. With $\int_{\Gamma_2+\Gamma_4} \to 0$ and $K(\lambda) \to 0$ in the strip as $\mathrm{Re}[\lambda] \to \infty$, we are left with $\int_{\Gamma_1}$, $\int_{\Gamma_3}$ (Figure 7.45). The integral over Γ_1 defines a function in the λ plane which is analytic everywhere above Γ_1, while the integral over Γ_3 is for a function that is analytic everywhere below Γ_3. Therefore,

$$\tilde{K}(\lambda) = \tilde{K}_+(\lambda) - \tilde{K}_-(\lambda),$$

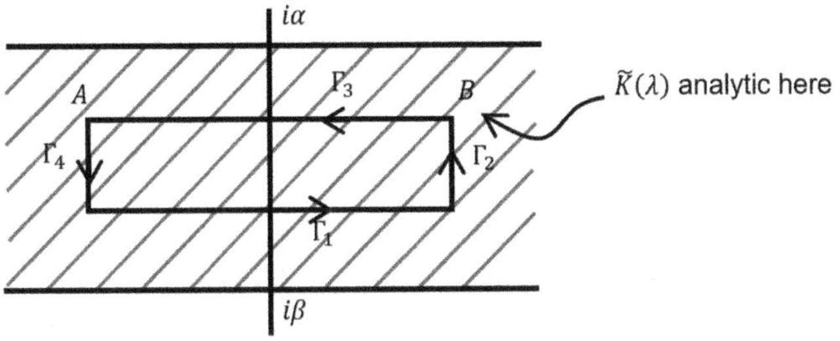

Figure 7.44: Strip of analyticity and contour for eq. (7.39).

where

$$\int_{\Gamma_1} \frac{\tilde{K}(t)dt}{t-\lambda} \equiv \tilde{K}_+, \qquad \int_{\Gamma_3} \frac{\tilde{K}(t)dt}{t-\lambda} \equiv \tilde{K}_-.$$

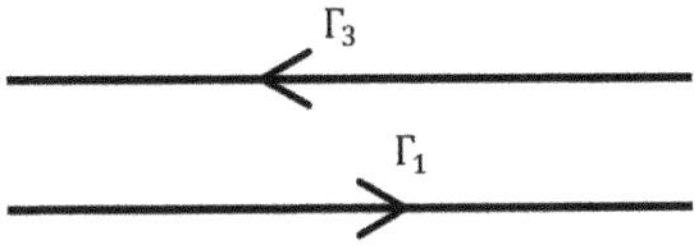

Figure 7.45: The integration contour boils down to $\Gamma_1 \cup \Gamma_3$ as the contributions of $\Gamma_2 \cup \Gamma_4$ are assumed to be zero as $\mathrm{Re}(\lambda) \to \infty$.

See Figure 7.46 for the depiction of the overlap strip.

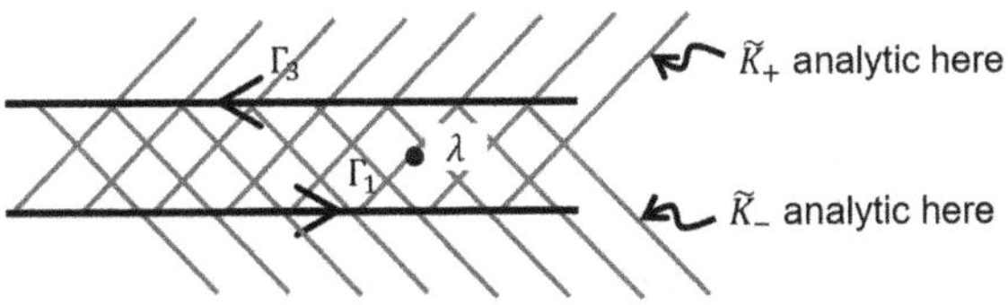

Figure 7.46: The overlap strip.

We will then choose suitable regions to evaluate the integrals and obtain $\tilde{K}_+, \tilde{K}_-$. Note that if

$$g(\lambda) = \int_{\Gamma_1} \frac{\tilde{K}(t)dt}{t-\lambda},$$

then $g(\lambda)$ is analytic everywhere as long as λ is not on Γ_1. If we can split $\tilde{K}(\lambda)$, we can also split $\left(\tilde{K}(\lambda) - 1\right)$ automatically.

7.5.3 Other ways of kernel decomposition

Since the quotient decomposition $\tilde{K}(\lambda) = \tilde{K}_-(\lambda)/\tilde{K}_+(\lambda)$ is often difficult to accomplish, we could reduce this to a sum decomposition by taking the logarithm, to obtain

$$\ln \tilde{K} = \ln \tilde{K}_- - \ln \tilde{K}_+.$$

Thus, the quotient decomposition reduces to sum decomposition for $\ln \tilde{K}$, albeit with a negative sign as shown. Obviously, in like manner, product decomposition directly reduces to sum decomposition for $\ln \tilde{K}$, or $\ln \tilde{K} = \ln \tilde{K}_- + \ln \tilde{K}_+$ when $\tilde{K} = \tilde{K}_- \tilde{K}_+$.

Algebraic functions are easy to split, whereas transcendental functions are not. The function $\tan^{-1}(\ )$ cannot easily be integrated. In this case, one would approximate the kernel into first, second, and third, etc., kernels. Suppose $\tilde{K}$ has no zero in the interval (strip), we may not have a problem (see Figure 7.47). But $\tilde{K}$ may have a singularity outside. This means that because of the branch cuts associated with log functions, the value of $\tilde{K}$ as we go to ∞ in A, B may be different. In general, we can do the following trick, theoretically. Suppose there are many zeroes inside the strip (Figure 7.48), and suppose the zeroes are simple zeroes, and we can define

$$k_1(\lambda) \equiv \frac{\tilde{K}(\lambda)}{\prod_{j=1}^{n}(\lambda - \lambda_j)}$$

such that this function has no zeroes inside the strip, allowing us to split $k_1(\lambda)$ (Figure 7.48).

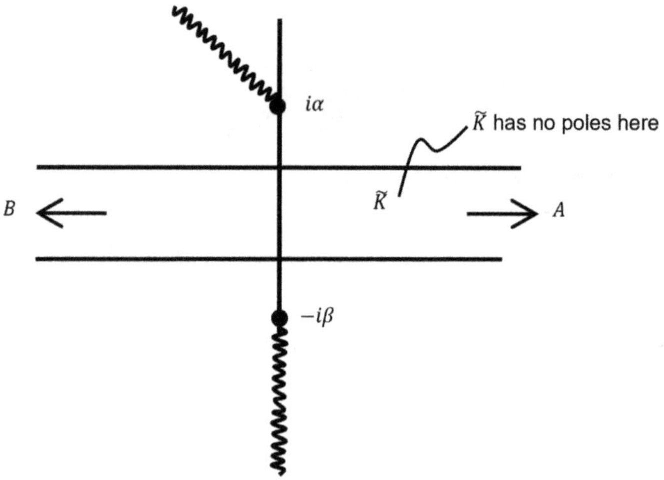

Figure 7.47: $\tilde{K}$ has no singularities inside the strip but has branch points outside.

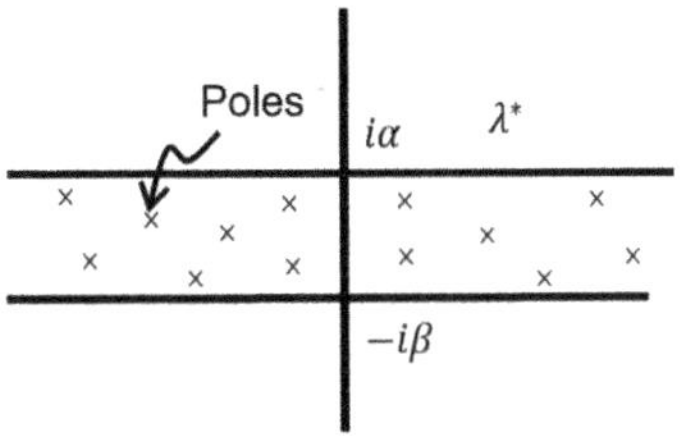

Figure 7.48: $\tilde{K}$ has many (countable number of) singularities inside strip.

We introduce a point λ^* outside the strip, where λ^* can be anything, e.g., $i\beta$, to obtain unity for $k_1(\lambda)$ as $\lambda \to \infty$:

$$k_1(\lambda) = \frac{\tilde{K}(\lambda)(\lambda - \lambda^*)^n}{\prod_{j=1}^{n}(\lambda - \lambda_j)}.$$

We could pick $\tilde{K} = (\lambda - i\beta)^m / (\lambda - i\alpha)^m$, so that

$$k_2(\lambda) \equiv \frac{(\lambda - i\beta)^{m+n}\tilde{K}(\lambda)}{(\lambda - i\alpha)^m \prod_{j=1}^{n}(\lambda - \lambda_j)}, \quad \lambda^* = i\beta, \quad k_2(\lambda) \to 1 \text{ as } \lambda \to \infty.$$

Let us assume that we can split $k_2 = k_2^- / k_2^+$ using the procedure described above. Then

$$\tilde{K} = \frac{k_2(\lambda)(\lambda - i\alpha)^m \prod_{j=1}^{n}(\lambda - \lambda_j)}{(\lambda - i\beta)^{m+n}}, \text{ where } k_2 = \frac{k_2^-}{k_2^+}.$$

7.6 Miscellaneous examples

Example 7.15. Use the Wiener–Hopf method to determine $\varphi(x)$ from the equation

$$\varphi(x) = \beta \int_0^\infty k(x - t)\varphi(t)\, dt,$$

where $x \in (-\infty, \infty)$, $\varphi(x)$ is bounded as $|x| \to \infty$, and $k(x) = e^{-|x|}$ (Ref [20]).

Solution: Let us decompose $\varphi(x) = \varphi_+(x) + \varphi_-(x)$, such that

$$\varphi_+(x) = \begin{cases} \varphi(x), & x > 0 \\ 0, & x < 0 \end{cases}; \qquad \varphi_-(x) = \begin{cases} 0, & x > 0 \\ \varphi(x), & x < 0. \end{cases}$$

Realizing that $\varphi(x) = \varphi_+(x)$ in $[0, \infty)$ and $\varphi_-(x) = 0$ in $(-\infty, 0]$, we can write

$$\varphi_+(x) + \varphi_-(x) = \beta \int_{-\infty}^{\infty} k(x-t)\varphi_+(t)\, dt.$$

Suppose the FT of $k(x)$ is $\tilde{k}(\lambda)$, and that this is analytic in the strip $-a < \eta < b$, where $\lambda = \xi + i\eta$. This means that $k \sim e^{-ax}$ for $x \to +\infty$ and $k \sim e^{bx}$ for $x \to -\infty$. Suppose $\varphi_+(x) \sim e^{-cx}$ for $x \to +\infty$, then the integrand $k(x-t)\varphi_+(t) \sim e^{-ax} e^{(a-c)t}$, so that $c > a$ for convergence if $x \to \infty$. Similarly, from $k(x-t)\varphi_-(t) \sim e^{bx} e^{-(b+c)t}$ as $x \to -\infty$, we require that $b + c > 0$.

Under the foregoing conditions, $\tilde{\varphi}_+(\lambda)$ will be analytic above $\eta = -c$, while $\tilde{\varphi}_-(\lambda)$ will be analytic below $\eta = b$. Following the procedure in the text, the FT of the integral equation can be written as

$$\left[1 - \sqrt{2\pi}\beta\tilde{k}(\lambda)\right] \tilde{\varphi}_+(\lambda) = -\tilde{\varphi}_-(\lambda).$$

Then, the brute-force assumption is made that we can quotient-decompose as

$$\left[1 - \sqrt{2\pi}\beta\tilde{k}(\lambda)\right] = \frac{\tilde{k}_+}{\tilde{k}_-}, \quad \text{or} \quad \tilde{k}_+(\lambda)\tilde{\varphi}_+(\lambda) = -\tilde{k}_-(\lambda)\tilde{\varphi}_-(\lambda).$$

As shown in the text, we can solve as follows:

$$\tilde{\varphi}_+(\lambda) = \frac{H(\lambda)}{\tilde{k}_+(\lambda)}, \qquad \tilde{\varphi}_-(\lambda) = -\frac{H(\lambda)}{\tilde{k}_-(\lambda)},$$

with $H(\lambda)$ being analytic in $-c \le \eta \le b$. As stated above, the functions $\tilde{\varphi}_+(\lambda)$ and $\tilde{\varphi}_-(\lambda)$ will be analytic, respectively, above $\eta = -c$ and below $\eta = b$. The domains are depicted in Figure 7.49.

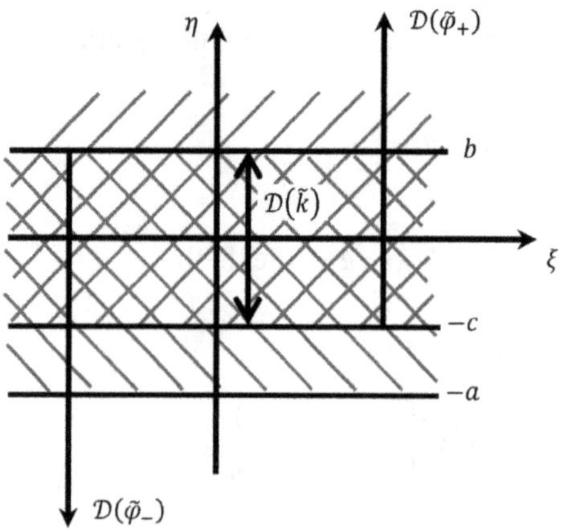

Figure 7.49: The domain of analyticity of $\tilde{\varphi}_+(\lambda)$, $\tilde{\varphi}_-(\lambda)$, and $\tilde{k}(\lambda)$. The symbol $D(\cdot)$ stands for "domain" of analyticity of its argument.

If we use $k(x) = e^{-|x|}$, with $\tilde{k}(\lambda) = \sqrt{2/\pi}\left[1/\left(\lambda^2+1\right)\right]$, the poles of $\tilde{k}(\lambda)$ are located at $\lambda = \pm i$, so that $a = b = 1$ in the expression $-a < \eta < b$. The decomposition introduced earlier implies that

$$\left[1-\sqrt{2\pi}\lambda\tilde{k}(\lambda)\right] = \frac{\lambda^2+1-2\beta}{\lambda^2+1} \equiv \frac{\tilde{k}_+(\lambda)}{\tilde{k}_-(\lambda)}.$$

With $\tilde{k}_+$ in the numerator, $\tilde{k}_-$ in the denominator, $\tilde{k}_+$ analytic above $\eta = -i$ and $\tilde{k}_-$ analytic below $\eta = i$, and the factorization of the denominator as

$$\frac{\lambda^2+1-2\lambda}{\lambda^2+1} = \frac{\lambda^2+1-2\beta}{(\lambda-i)(\lambda+i)},$$

we see that one way to decompose $\tilde{k}(\lambda)$ is

$$\tilde{k}_+ = \frac{\lambda^2+1-2\beta}{\lambda+i}, \qquad \tilde{k}_- = \lambda - i.$$

Obviously, this is not a rigorous approach, and there could be other ways to decompose $\tilde{k}(\lambda)$. Allowing $\tilde{\varphi}_+(\lambda) \to 0$ as $\mathrm{Re}[\lambda] \to \infty$ as it should, we will have $H(\lambda) =$ a constant, say C_O, so that

$$\tilde{\varphi}_+(\lambda) = \frac{C_O(\lambda+i)}{\lambda^2+1-2\beta}, \qquad \tilde{\varphi}_-(\lambda) = -\frac{C_O}{(\lambda-i)}.$$

If we use the inversion formula

$$\varphi_+(x) = \frac{1}{(2\pi)^{1/2}}\int_C e^{-i\lambda x}\frac{H(\lambda)}{\tilde{k}_+(\lambda)}d\lambda, \qquad \varphi_-(x) = -\frac{1}{(2\pi)^{\frac{1}{2}}}\int_C e^{-i\lambda x}\frac{H(\lambda)}{\tilde{k}_-(\lambda)}d\lambda,$$

we can obtain the solutions

$$\varphi_+(x) = D_O\left[\cos\beta' x + \frac{\sin\beta' x}{\beta'}\right], \qquad \varphi_-(x) = D_o e^x, \text{ where } \beta' = \sqrt{2\beta-1}.$$

Given the homogeneous problem in this example, additional conditions are needed in order to evaluate the constant D_O.

Example 7.16. Determine the solution of

$$\frac{\partial^2\psi}{\partial x^2} + \frac{\partial^2\psi}{\partial y^2} \pm k^2\psi = H(x,y),$$

which vanish as $x^2+y^2 \to \infty$, and identify the two Green's functions by using polar coordinates in the transform plane. Assume that the time factor $e^{i\omega t}$ has been extracted.

Solution: We have

$$\frac{\partial^2\psi}{\partial x^2}+\frac{\partial^2\psi}{\partial y^2}\pm k^2\psi=H(x,y),$$

and the 2D FT given in the usual notation gives

$$\left(\xi^2+\eta^2\mp k^2\right)\bar{\psi}=-\bar{H}.$$

Thus

$$\bar{\psi}=\frac{-\bar{H}}{\xi^2+\eta^2\mp k^2},$$

and so

$$\psi=-\frac{1}{4\pi^2}\iint\limits_{-\infty}^{\infty}H(x',y')dx'dy'\iint\limits_{-\infty}^{\infty}\frac{e^{i[\xi(x-x')+\eta(y-y')]}}{\xi^2+\eta^2\mp k^2}d\xi d\eta.$$

In polar coordinates (Figure 7.50), the inner integral becomes

$$\int_0^{2\pi}d\theta\int_0^{\infty}\frac{R\,dR\,e^{irR\cos\theta}}{R^2\mp k^2}=4\int_0^{\pi/2}d\theta\int_0^{\infty}\cos(rR\cos\theta)\frac{R\,dR}{R^2\mp k^2}$$

$$=2\pi\int_0^{\infty}\frac{RJ_0(rR)dR}{R^2\mp k^2}\quad\text{(see Watson [18, p. 21]).}$$

Now from a table of Hankel transforms (or Watson [18, p. 425]) we find

$$\int_0^{\infty}\frac{RJ_0(rR)}{R^2+z^2}dR=K_0(zr),\ \text{provided } \mathrm{Re}[z]>0.$$

This formula can be derived from contour integration of $H_0^{(1)}(rR)$.

With this result, the Green's function for $\partial^2/\partial x^2+\partial^2/\partial y^2-k^2$ is

$$\bar{G}(x,y;x',y')=-\frac{1}{2\pi}K_0(kr);\quad r^2=(x-x')^2+(y-y')^2.$$

The Green's function for $\partial^2/\partial x^2+\partial^2/\partial y^2+k^2$ is obtained in the limit $z=ik$ with $\mathrm{Im}\,k=0^-$ (so that $\mathrm{Re}[z]=0^+$). Then

$$G^+(x,y;x',y')=-\frac{1}{2\pi}K_0(ikr)=\frac{i}{4}H_0^{(2)}(kr)$$

(see Watson [18, pp. 75 and 78]).

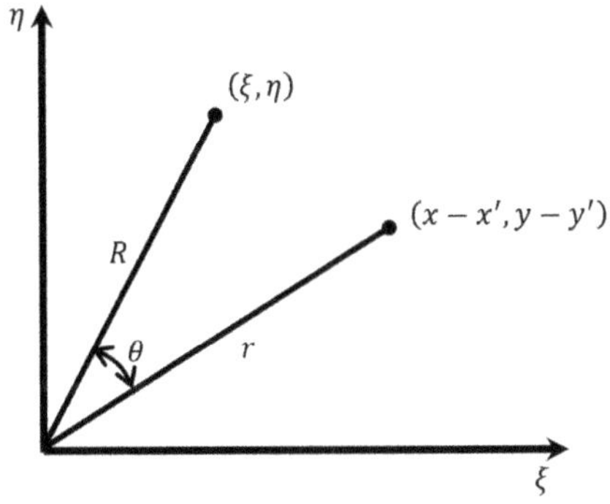

Figure 7.50: Polar coordinate system.

$\bar{G}$ is well behaved, and that it is exponentially small for large r. The corresponding equation comes from Laplace's equation in 3D on separating out $\exp \pm ikz$ – as is required in certain water wave problems, and also from Oseen's weak flow equations in 2D on taking out a factor e^{kx}.

Problems

7.1 Evaluate the integral

$$I = (x, y, a, k) = \int_{-\infty}^{\infty} \frac{|y|}{y} \frac{e^{i\xi x - |y|[(\xi+ia)(\xi-ik)]^{1/2}}}{(\xi - ik)^{1/2}} d\xi,$$

where a, k, x, y are real numbers.

7.2 Obtain the Green's functions for the following operators (L) in $L\varphi = Q$:

(a) $L = \frac{\partial^2}{\partial t^2} - c^2\nabla^2$ (2D and 3D)

(b) $L = \frac{\partial^2}{\partial t^2} - c^2\nabla_x^2$

(c) $L = \frac{\partial^2}{\partial t^2} - c^2\nabla_h^2$

where ∇_h^2 is the Laplacian in the horizontal coordinates, and c is a constant.

(d) $L = \frac{\partial^2}{\partial x^2} + k^2$

(e) $L = \frac{\partial^2}{\partial t^2} + \omega_0^2$

(f) $L = \frac{\partial^2}{\partial t^2} - \lambda\nabla^2 \frac{\partial}{\partial t} - c^2\nabla^2$ (2D and 3D)

(g) $L = \frac{\partial^2}{\partial t^2} + \lambda^2\nabla^4$

(h) $L = \frac{\partial^2}{\partial t^2} - \lambda^2 \frac{\partial^2}{\partial x^2} + c^2$.

Give the expressions for the solution $\varphi(\mathbf{x};\, t)$, $\varphi(x;\, t)$, etc., as the case may be for each of (a) through (h) above, assuming the homogeneous solution is φ_h in each case.

7.3 Consider, in free space,

$$\varphi_{xx} + \varphi_{yy} - \frac{1}{c^2}\varphi_{tt} - \frac{\varepsilon}{c^2}\varphi_t = -4\pi s(x, y, t),$$

where the $4\pi s$ on the RHS stands for a source term. By applying the FT in x, y, t in the domain $(x,y,t) \in (-\infty,\infty) \times (-\infty,\infty) \times (-\infty,\infty)$, determine $\varphi(x,y,t)$.

7.4 Evaluate

$$I(\theta) = 2i \sin\frac{1}{2}\theta \int_{-\infty}^{\infty} \frac{\exp(ikr\cosh t)\sin\frac{1}{2}(\theta + it)}{\cos(\theta + it) + \cos\theta}\, dt.$$

7.5 Consider

$$\varphi_{xx} + \varphi_{yy} + k^2\varphi = 0$$

$$\varphi = f(x) \text{ on } y = 0, \qquad 0 < x < \infty$$

$$\varphi_y = g(x) \text{ on } y = 0, \qquad -\infty < x < \infty.$$

Use the Wiener–Hopf technique to obtain $\varphi(x,y)$.

7.6 Consider the problem

$$(T_t)_{xx} + (T_t)_{yy} + a^2 T_t = \dot{Q}(x,y),$$

where $\dot{Q}(x,y)$ is a time-independent source of T. In a half-plane in $y = 0$, $-\infty < x < \infty$, we have $(T_t)_y = 0$. Otherwise, the domain is $(x,y) \in (-\infty,\infty) \times (-\infty,\infty)$.

Determine $T(x,t)$.

7.7 Consider the problem of two semi-infinite plates located at $y = \pm b$, $-\infty < x \le 0$, $\varphi_{xx} + \varphi_{yy} + k^2\varphi = 0$ (Figure P7.7), that is, an incident potential field φ_i and $\varphi_t \equiv \varphi + \varphi_i = 0$ on the two plates. Assume that φ can be produced by a distribution of line sources $f_1(x)$ and $f_2(x)$ on $y = +b, -b$, respectively. Given the linearity of the problem, we can use superposition to obtain the total potential at any point (x,y) as

$$\varphi_t(x,y) = \varphi_i(x,y) + \pi i \int_{-\infty}^{0} \left\{ H_0^{(1)}(kR_1) f_1(\xi) + H_0^{(1)}(kR_2) f_2(\xi) \right\} d\xi,$$

where

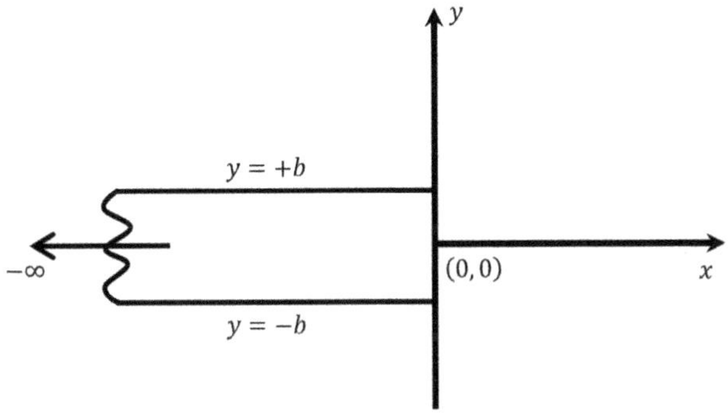

Figure Problem 7.7: Two semi-infinite plates.

$$R_1^2 = (x-\xi)^2 + (y-b)^2$$
$$R_2^2 = (x-\xi)^2 + (y+b)^2.$$

As (x,y) becomes $(x, \pm b)$, we have two integral equations:

(a) $\pi i \int_{-\infty}^{0} \left\{ H_0^{(1)} \left(k|x-\xi| f_1(\xi) + H_0^{(1)}(kR) f_2 \xi \right) \right\} d\xi + \varphi_i(x,b) = 0, \ (x \in (-\infty, 0]),$

(b) $\pi i \int_{-\infty}^{0} \left\{ H_0^{(1)}(kR) f_1(\xi) + H_0^{(1)}(k|x-\xi|) f_2(\xi) \right\} d\xi + \varphi_i(x,-b) = 0, \ (x \in (-\infty, 0]),$

where $R^2 = (x-\xi)^2 + 4b^2$.

We can add and subtract (a) and (b) and define $s(\xi) = f_1(\xi) + f_2(\xi)$, $d(\xi) = f_1(\xi) - f_2(\xi)$, to obtain two independent integral equations, each valid in $x \in (-\infty, 0]$:

(c) $\pi i \int_{-\infty}^{0} \left\{ H_0^{(1)}(k|x-\xi|) + H_0^{(1)}(kR) \right\} s(\xi) d\xi + \varphi_i(x,b) + \varphi_i(x,-b) = 0,$

(d) $\pi i \int_{-\infty}^{0} \left\{ H_0^{(1)}(k|x-\xi|) - H_0^{(1)}(kR) \right\} d(\xi) d\xi + \varphi_i(x,b) - \varphi_i(x,-b) = 0.$

Equations (c) and (d) are of Wiener–Hopf type. Solve them and hence obtain $\varphi_t(x,y)$.

7.8 Solve for $f(t)$ in this equation:

$$f(x) - \int_{-\infty}^{\infty} e^{-|x-t|} f(t) dx = x^2.$$

7.9 Find $\varphi(x,y)$ for the BVP

$$\nabla^2 \varphi + k^2 \varphi = 0$$

in $\mathcal{D}$, which is the domain exterior to the half-line $y = 0$, $x \geq 0$. These conditions apply:

$$\varphi_y = e^{i\beta x} \text{ on } y = 0, \ x \geq 0,$$

$$r\left(\varphi_y + ik\varphi\right) \to 0 \text{ as } r^2 = x^2 + y^2 \to \infty.$$

Note that $0 < \beta < k$ is assumed and k is real.

7.10 (a) Find the inverse FT of

$$F(\lambda) = \beta\left[(\alpha + i\lambda)^2 + \beta^2\right]^{-1},$$

where α and β are constants and λ is the transform variable.

(b) Find the inverse FT of the function

$$H(\lambda) = \frac{1}{2}\left[F(\lambda) + (i\lambda c)^{-1}G(\lambda)\right]e^{i\lambda ct} + \frac{1}{2}\left[F(\lambda) - (i\lambda c)^{-1}G(\lambda)\right]e^{-i\lambda ct},$$

where

$$\mathcal{F}[f(x)] = F(\lambda), \ \ \mathcal{F}[g(x)] = G(\lambda);$$

c is the speed of sound and t is time.

7.11 Consider the Laplacian operator in cylindrical coordinates

$$\Delta_n \equiv \left(\frac{\partial}{\partial r}\right)^2 + \frac{1}{r}\frac{\partial}{\partial r} - \left(\frac{n}{r}\right)^2.$$

Show that the Hankel transform of $\Delta_n \varphi$ can be written as

$$H_n\{\Delta_n\varphi\} = -\lambda^2 H_n\varphi$$

if we assume that the partially integrated terms vanish at both 0 and ∞, and the Bessel's equation

$$\left(\Delta_n + \lambda^2\right)J_n(\lambda r) = 0$$

is invoked.

7.12 Consider

$$\Delta\Delta\varphi - \Delta\varphi_x = 0 \text{ in domain } \mathcal{D},$$

where $\mathcal{D}$ is the xy plane exterior to the half-line $y = 0$, $x > 0$, φ is odd in y,

$$\varphi(x, 0) = 0,$$

$$\varphi_y(x, 0) = \begin{cases} ?, & x < 0 \\ 1, & x > 0, \end{cases}$$

and $\varphi \to 0$, as $x^2 + y^2 \to \infty$ in any sector excluding the line $y = 0$, $x > 0$. φ and $\nabla\varphi$ are assumed bounded for all x, y. Determine $\varphi(x, y)$.

7.13 Solve the equation

$$\nabla^2\varphi = 0$$

in $y < 0$, with $\varphi = 1$ on $y = 0$, $x < 0$; $\varphi_y + \varphi = 0$ on $y = 0$, $x > 0$.

7.14 A strip of metal occupies the region $-b \le y \le b$, $-\infty < x < \infty$. An imposed temperature profile $\theta(x)$ (above some reference level) is a function of x alone. As a consequence of this temperature distribution, certain stresses are set up in the strip; the stresses are given by the various second derivatives of a function $\varphi(x,y)$ satisfying the equation

$$\varphi_{xxxx} + 2\varphi_{xxyy} + \varphi_{yyyy} = -E\alpha\theta_{xx},$$

where the constants α and E are the linear coefficients of thermal expansion and the elastic modulus, respectively. The BCs on φ are $\varphi(x, \pm b) = \varphi_y(x, \pm b) = 0$. Let $\theta = \theta_0 e^{\epsilon x}$ for $x < 0$ and $\theta = 0$ for $x > 0$, where ϵ is a small positive constant (eventually allowed to approach zero). Take an FT in x to show that

$$\varphi = -\frac{E\alpha\theta_0}{\pi}\int_{-\infty}^{\infty} e^{i\lambda x}\frac{1}{i\lambda^3}\left[A\cosh\lambda y + B\lambda y\sinh\lambda y - \frac{1}{2}\right]d\lambda,$$

where

$$A = \frac{\sinh b\lambda + \lambda b(\cosh b\lambda)}{2\lambda b + \sinh 2\lambda b}$$

and

$$B = -\frac{\sinh b\lambda}{2\lambda b + \sinh 2\lambda b}.$$

Obtain the explicit form of the longitudinal stress given by $\tau = \varphi_{yy}$.

To find numerical values of τ, either we can use direct numerical integration – often a tedious process with transform formulas, because of the oscillating integrands – or we can try to sum the residues. Determine at least approximately where the poles of the denominator are. What curve is asymptotic to their locations? If we approximate the denominator by $\sinh 2\lambda b$ alone, the residues may be summed explicitly; show that this leads to

$$\frac{\tau}{E\alpha\theta_0} = \frac{e^{\frac{\pi x}{b}} + \cos\left(\frac{\pi y}{b}\right)}{2\cos\left(\frac{\pi y}{b}\right) + 2\cosh\left(\frac{\pi x}{b}\right)} - 1 + \frac{1}{\pi}\arctan\frac{2e^{\frac{\pi x}{2b}}\cos\left(\frac{\pi y}{2b}\right)}{e^{\frac{\pi x}{b}} - 1} - \frac{y}{2b}\left[\frac{\sinh\left(\frac{\pi x}{2b}\right)\sin\left(\frac{\pi y}{2b}\right)}{\cos^2\left(\frac{\pi y}{2b}\right) + \sinh^2\left(\frac{\pi x}{2b}\right)}\right]$$

for the case $x < 0$. (The function arctan is an angle in the second quadrant.) By examining the difference between a typical exact and approximate residue, show how correction terms could be obtained, and comment on their magnitudes [3].

7.15 Small deflection $\varphi(x,y)$ of an elastic membrane subjected to a loading distribution $f(x,y)$ is governed by the equation

$$\varphi_{xx}+\varphi_{yy}-a^2\varphi=f(x,y),$$

where a is a constant, and $a^2\varphi$ represents an elastic restraint on the membrane. The domain is $(x,y)\in(-\infty,\infty)\times(-\infty,\infty)$. The BCs are:

$$\varphi,\ \varphi_x,\ \varphi_y\to 0 \text{ as } r^2=x^2+y^2\to\infty.$$

Solve for $\varphi(x,y)$ under the foregoing conditions.

7.16 The Green's function $G(x,y;\xi,\eta)$ for an infinite strip problem satisfying $\varphi_{xx}+\varphi_{yy}=0$ in the region $(x,y)\in(-\infty,\infty)\times[0,l]$ and satisfying $G=0$ for $y=0$, $y=l$, can be written as

$$G=\frac{1}{4\pi}\ln\frac{\cosh\left(\frac{\pi}{l}\right)(\xi-x)-\cos\left(\frac{\pi}{l}\right)(\eta-y)}{\cosh\left(\frac{\pi}{l}\right)(\xi-x)-\cos\left(\frac{\pi}{l}\right)(\eta+y)}\equiv\frac{1}{4\pi}\ln A.$$

Expand A in the form of an infinite product, so as to show how this result can be interpreted in terms of the method of images. In this case, the strip is replaced by the whole plane with positive unit sources at $\ldots(\xi,4l+\eta)$, $(\xi,2l+\eta)$, $(\xi,\eta),(\xi,-2l+\eta)$, $\ldots$ and negative unit sources at $\ldots(\xi,4l-\eta),(\xi,2l-\eta),(\xi,-\eta),(\xi,-2l-\eta),\ldots$ Note that the lines $y=l$, $y=0$ are lines of symmetry where the effects of all sources cancel. Obtain the final result by the use of a Fourier sine series [3].

7.17 Use the Wiener–Hopf procedure to find the eigensolution of the integral equation

$$f(x)=\frac{1}{2}\int_0^\infty f(t)\exp\{-|x-t|\}dt,\ x\in(0,\infty).$$

7.18 Consider the vibrating string problem with a discontinuity in properties (densities). The governing equations can be written as

$$\frac{d^2u_-}{dx^2}+k_1^2u_-=0,\quad x<0,$$

$$\frac{d^2u_+}{dx^2}+k_2^2u_+=0,\quad x>0,$$

where $k_i^2\equiv\omega^2\rho_i/T$, ρ_i is the density, which is different at $x<0$ (ρ_1) from $x>0$ (ρ_2). T is the tension in the string. The equilibrium position of the string is $y=0$. ω is the (temporal) frequency from the time factor $e^{i\omega t}$.

The BCs are:

$$u_- \sim Te^{-ik_1x} \text{ as } x \to -\infty$$

$$u_+ \sim Te^{+ik_2x} \text{ as } x \to +\infty$$

$$(u_+ - u_-) = 1, \ (u'_+ - u'_-) = ik, \text{ at } x = 0.$$

Determine u_+ and u_- by using the Wiener–Hopf equation.

7.19 Solve the (one-sided) integral equation

$$f(x) = \lambda \int_0^\infty K(x-t)f(t)dt,$$

where $x \in (-\infty, \infty)$. We express $f(x)$ as follows:

$$f_+ = \begin{cases} f(x), & x > 0 \\ 0, & x < 0 \end{cases}; \qquad f_- = \begin{cases} 0, & x > 0 \\ f(x), & x < 0 \end{cases}.$$

7.20 Calculate $\varphi(x,y)$ for the system:

$$\nabla^2\varphi - \varphi_x = 0 \text{ in } \mathcal{D}$$

with

$$\varphi(x,y) \to 0 \text{ as } x^2 + y^2 \to \infty$$

$$\varphi = e^{-ax} \text{ on } y = 0, \ x \geq 0.$$

$\mathcal{D}$ is the domain exterior to the half-line $y = 0$, $x \geq 0$. It is required that φ be bounded everywhere and φ_y be integrable at all points.

Suggested reading

[1] Abramowitz, M. and Stegnum, I.A. (Eds). Handbook of Mathematical Functions, Washington, WA, USA, United States Department Commerce, Bureau of Standards, 1964.

[2] Brown, J.W. and Churchill, R.V. Complex Variables and Applications, 6th Edition, New York, NY, USA, McGraw-Hill, 1996.

[3] Carrier, G.F., Krook, M., and Pearson, C.E. (1983) Functions of Complex Variables. Ithaca, NY, USA, Hod Book.

[4] Courant R. and Hilbert D. Methoden der Mathematischen Physic, Volume I. Berlin, Germany, Verlag von Julius, 1931.

[5] Crighton, D.G., Dowling, A.P., Ffowcs Williams, J.E., Heckl, M., and Leppington, F.G. Modern Methods in Analytic Acoustics, London, UK, Springer-Verlag, 1992.

[6] Erdelyi, A. Asymptotic Methods. Mineola, NY, USA, Dover Publications, 1956.

[7] Friedman, B. Principles and Techniques of Applied Mathematics, New York, NY, USA, Wiley and Sons, 1956.

[8] Miles, J. Integral Transforms in Applied Mathematics, New York, NY, USA, Cambridge University Press, 1971.

[9] Morse, P.M. and Feshbach, H. Methods of Theoretical Physics, Volume I, New York, NY, USA, McGraw-Hill, 1953.

[10] Nair, S. Advanced Topics in Applied Mathematics for Engineering and the Physical Sciences, New York, NY, USA, Cambridge University Press, 2011.

[11] Noble, B. Methods based on the Wiener–Hopf Technique, London, UK, Pergamon Press, 1958.

[12] Sirovich, L. Techniques of Asymptotic Analysis, New York, NY, USA, Springer-Verlag, 1971.

[13] Sokonikoff, I.S. and Redheffer, R.M. Mathematics of physics and modern engineering, J Electrochem Soc, 1958, 105, 9.

[14] Sommerfeld, A. Partial Differential Equations in Physics, Volume VI of Lectures on Theoretical Physics, New York, NY, USA, Academic Press, 1949.

[15] Stakgold, I. Boundary Value Problems of Mathematical Physics, Volume I, New York, NY, USA, MacMillan Co., 1968.

[16] Wiener, N. The Fourier Integral and Certain of Its Applications, Mineola, NY, USA, Dover Publications, 1958.

[17] Wiener, N. and Hopf, E.S.B. Preuss Akad Wiss, 1931, 696.

[18] Watson, G.N. A Treatise on the Theory of Bessel Functions, Cambridge, UK, Cambridge University Press, 1966.

[19] Carleman, T. Cber die Abelische Integralgleichung mit konstanten Integrationsgrenzen, Math Z, 1922–23, 15, 111–120.

[20] Ludford, G.S.S. Personal Communication, 1985.

8 Modern applications of complex variables

In previous chapters of this book, we have seen that quite a few classical topics in complex variable theory attract the attention of the engineer and physicist. Certainly, one of these is conformal transformations, which when carried out successively in multiples enable closed-form solution of fairly realistic engineering problems, particularly those governed by Laplace equation in two-dimensional coordinate directions. Perfect fluid flows and purely diffusive transport benefit from the approach. We have also seen that the Schwarz-Christoffel (SC) transformation, when combined with the hodograph, log-hodograph, and other transformations, enables closed form solutions of complicated potential flows with unknown flow boundaries. An expanded use of the Fourier transform technique in engineering analysis has also been demonstrated in earlier chapters when the transform variable is allowed to be complex. The application to a class of mixed boundary value partial differential equations, or their integral equation equivalents, has been demonstrated in the use of the Wiener–Hopf technique [57]. Mathematical physics and engineering models with essential singularities cannot benefit from classical mathematical analysis such as separation of variables, series solutions, transform methods, etc., and one is left with procedures such as the method of dominant balance, and asymptotic and perturbation analysis. However, we have demonstrated in this book that the asymptotic behavior of integrals resulting from physics and engineering modeling provides a powerful means of analyzing engineering problems with irregular singular points.

The topics alluded to above are the classical ones in applied complex variables. However, there are modern developments and applications, some components of which are relevant to engineering and physics. While some of these new developments are strictly for the academic excitement of the mathematician, others have implications for mathematical physics. It is the latter that this chapter is mostly concerned with. The developments in applied complex variables in the past few decades include the following:

1) Efficient computational algorithms for obtaining and applying conformal transformations. Computational complex variable analysis is perhaps one of the most extensive recent developments. Whereas classical methods are mostly based on closed-form solutions, there is now an extensive amount of work on the development of the computer algorithms to solve conformal mapping problems [1–3].
2) Advances in the development and application of Schottky-Klein prime functions. A significant amount of the new developments in this area comes from the numerous publications by Crowdy and his colleagues. They treat geometries of varying complexities from unit disk to multiply connected domains. The latter application utilizes Schottky doubling process wherein a domain is reflected from its outer boundary curve, following by "gluing" with a holomorphic identification of circular boundaries by means of the linear fractional transformation (LFT). (The LFT procedure is presented in detail in Chapter 3 of this book.)

https://doi.org/10.1515/9783111351179-010

3) Extension of the Schwarz-Christoffel transformations from simply connected to arbitrarily connected multiple domains. New methods have extended the SC procedure to handle arbitrarily connected domains. We also now have the "unified method" of Fokas [4, 58, 59], and complex-variable based approaches to solve the six Painlevé equations have received a lot of attention lately [5–7].
4) Advanced image processing and computer visualization. General conformed mappings have recently found extensive applications in image processing and computer visualization, including applications in medicine. Wegert [3], has written a book on the visualization of complex functions using a variety of colored phase plots to describe the behaviors of functions, especially at the poles of the functions. His MATLAB-based computer code "Complex Function Explorer," which is described in his book, is freely available.
5) More extensive applications of conformal mappings in mechanics. This is particularly the case for fluid flows, including bubble and general free surface flows. There is now quite a sizable amount of work on the application of complex variables to two-phase (gas/liquid) systems [8–10], an application that is of immense importance in mathematical physics. A significant amount of work has also been carried out recently on hydrodynamic stability that is based on complex variable theory [11].
6) Ordinary differential equations in the complex domain. These are nonlinear, scalar, second-order ordinary differential equations in which the independent variable is complex. They have extensive applications, including in rotating systems and dynamics. New solution methods are continuously being developed.
7) Advances in the solution and application of the Riemann-Hilbert problem. The new developments and applications in this area are the subject of the book by Trogdon and Olver [2], to which the interested reader could consult for the state-of-the-art.

The remainder of this chapter will focus on prime functions, the Painlevé equations, recent developments in the Schwarz-Christoffel mapping, and recent computer applications and algorithm developments in applied complex variable.

8.1 Recent advances in prime function-based mappings

As engineers and scientists, our interest in mapping domains lies more on the ability to solve some partial differential equations that describe certain physics in the domains, than on the geometric aspect of the transformations, of course with the acknowledgement that the two are inextricably linked. This is obvious from the various illustrations given in Chapters 3 and 4 of this book. However, the most advanced examples in those chapters pertain to doubly connected domains.

Research efforts within the framework of complex variables have recently been focused on the mapping of multiply connected domains. Although several independent groups worldwide are involved, the brief survey in this section will focus exclusively on the extensive amount of work that have been undertaken in this area which is led by Darren Crowdy. A monograph by the same author [12] summarizes the recent work in this area, with numerous references that the interested reader could consult.

The primary analytical vehicle for Crowdy's work is the Schottky-Klein prime functions, of which Bottazzini and Gray [60] have provided a historical perspective. The simplest instance of a prime function can be written as

$$w(z,a) = (z-a),$$

where w is the mapping, $z = x + iy$, and a is a complex-valued point that is allowed to move around in the complex plane. The function w has a simple zero when $z = a$, and a simple pole at $z = \infty$. This particular expression of the mapping is for a unit disc, such as shown in Figure 8.1, with the interior region represented as $\mathcal{D}$ and the boundary $|z| = 1$ as C_0. The ratio of prime functions

$$R(z;a,b) = \frac{w(z,a)}{w(z,b)},$$

where a and b are arbitrary nonzero and noncoincident points, is important in the theory. For example, the complex conjugate of R for $z \in C_0$, where $z = 1/\bar{z}$, can be written as

$$\overline{R(z;a,b)} = \frac{\bar{a}}{\bar{b}} R\left(z; \frac{1}{\bar{a}}, \frac{1}{\bar{b}}\right).$$

This ratio, for the unit disc, when $w(z,a) = (z-a)$, is the same as the linear fractional transformation (LFT) presented in Chapter 3 of this book and written there as $w(z) = (az+b)/(ac+d)$.

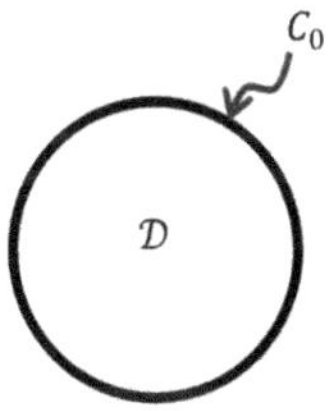

Figure 8.1: The unit disc $\mathcal{D}$ with boundary C_0.

Associated with $w(z,a)$ for the unit disc are the Green's functions

$$G'_0(z,a) = \frac{1}{2\pi i}\log\left[\frac{1}{|a|}R\left(z;a,\frac{1}{\bar{a}}\right)\right] = \frac{1}{2\pi i}\log\left[\frac{w(z,a)}{|a|w(z,1/\bar{a})}\right]$$

and

$$G_0(z,a) = \mathrm{Im}[G'_0(z,a)] = -\frac{1}{2\pi}\log\left[\frac{w(z,a)}{|a|w(z,1/\bar{a})}\right],$$

with the property that $G_0(z_0,a)=0$, where $z_0 \in C_0$. More precisely, $G_0(z,a)$ is the Green's function of the unit disc, while $G'_0(z,a)$ is the analytic extension of $G_0(z,a)$ as a function of z in $\mathcal{D}$.

Multiply connected domains are obtained in the procedure by employing the Schottky doubling process wherein the domain $\mathcal{D}$ is reflected from its outer boundary C_0. According to Crowdy [12], the "gluing" is done by using a holomorphic identification of circular boundaries by means of the LFT.

A more general version of $w(a,b)$ is required for more complex mapping:

$$w(a,b) = X(a,b)^{\frac{1}{2}},$$

where the sign of the square root is chosen to satisfy $w(a,b) \sim (a-b)$ as $a \to b$. Here, X is defined as

$$X(a,b) = \lim_{\substack{z\to a\\ w\to b}}\left[-(z-a)(w-b)e^{-\Pi_{a,b}^{z,w}}\right],$$

$$\Pi_{a,b}^{z,w} = \int_b^a d\Pi^{z,w}(a'),$$

and

$$d\Pi^{z,w}(a') = 2\pi i\left[\frac{\partial G'_0(z,a')}{\partial a'} - \frac{\partial G'_0(w,a')}{\partial a'}\right]da'.$$

The ratio of prime functions that is associated with this can be written as

$$\overline{R(z;a,b)} = \begin{cases} \frac{\bar{a}}{\bar{b}}R\left(z;\frac{1}{\bar{a}},\frac{1}{\bar{b}}\right) & |z \in C_0 \\ e^{-2\pi i\left(\overline{v_j(a)}-\overline{v_j(b)}\right)}\frac{\bar{a}}{\bar{b}}R\left(z;\frac{1}{\bar{a}},\frac{1}{\bar{b}}\right) & |z \in C_j, \quad j=1,2,\ldots,M \end{cases}$$

for M LFT maps.

Crowdy [12], gives $v_j(z)$ as follows:

$$v_j(z) = \frac{1}{2\pi i}\log\left[\frac{w(z,\theta_j(1/\bar{a}))}{w(z,1/\bar{a})}\right] - \frac{1}{2\pi i}\log\left[-\frac{(1-\bar{\delta}_j/\bar{a})}{q_j}\right] + v_j\left(\frac{1}{\bar{a}}\right) + \frac{\tau_{jj}}{2},$$

where $\tau_{jk} = v_j(\theta_k(a)) - v_j(a)$ is the period matrix, δ_j is the radius of circle C_j, $q_j^2 \equiv |z - \delta_j|^2 = (z - \delta_j)(\bar{z} - \bar{\delta}_j)$, and $\theta_j(z) = \delta_j + q_j^2 z/(1 - \bar{\delta}_j z)$, $j = 1, 2, \ldots, M$, is the number of LFT transformations. The foregoing description of the models for mapping multiply connected domains is for the most part geometric. However, as engineers and scientists we are more interested in using the mappings to solve PDEs in realistic geometries. This connection is provided by the discussion of Schwarz problems in $\mathcal{D}$, which pertain to the solution of boundary value problems with specified (Dirichlet) values of the dependent variable, its gradient, or mixed Dirichlet and gradient conditions on the boundary of the domain. Note that we are still limited to harmonic functions, without yet a facility to handle strongly hyperbolic problems.

Crowdy [10], presents several applications of this approach to multiply connected domains including electric transport theory and Hall effect, laminar flow in ducts, torsion of hollow prismatic rods, laminar convective heat transfer in flow through a doubly connected duct, mixed-type boundary value problems, slow viscous flow, plane elasticity, and vortex patch equilibria of the Euler equations of fluid mechanics. It is Crowdy's application of his approach to free surface Euler flow that we will summarize in this book.

The details of perfect fluid flow as presented in Chapter 4 of this book are relevant to this example, with $F \equiv \phi + i\psi$ and $W \equiv u - iv$ being the complex potential and complex velocity, respectively, in the notations used in Chapter 4. The u and v are the components of velocity in the x- and y-coordinate directions, respectively.

The application involves a staggered hollow vortex street comprising an L-periodic row of hollow vortices of circulation $+\Gamma$ located next to another L-periodic row of hollow vortices of circulation $-\Gamma$, as shown in Figure 8.2. The hollow vortices travel without deformation in their shape at a constant speed of U units in the x-coordinate direction. We move with the vortices at this speed, where the flow and vortices will be steady. In the far-field, $|y| \to \infty$, the fluid velocity will be $-U$ units.

The transformation considered is

$$z = f(\zeta),$$

where z is in the vortex plane and ζ is in the plane of the concentric annulus, with $\rho < |\zeta| < 1$. There are two logarithmic singularities located at $\zeta = \alpha, \beta$ in the interior of the annulus. These will serve as branch points and the preimages of the points $\pm\infty$ in the vortex plane, with α being the preimage of $+\infty$ and β the preimage of $-\infty$. The mapping $f(\zeta)$ will have the form

$$f(\zeta) = -\frac{iL}{2\pi} \log(\zeta - \alpha) + \text{a locally analytic function}$$

near $\zeta = \alpha$, while near $\zeta = \beta$, the form of the mapping is

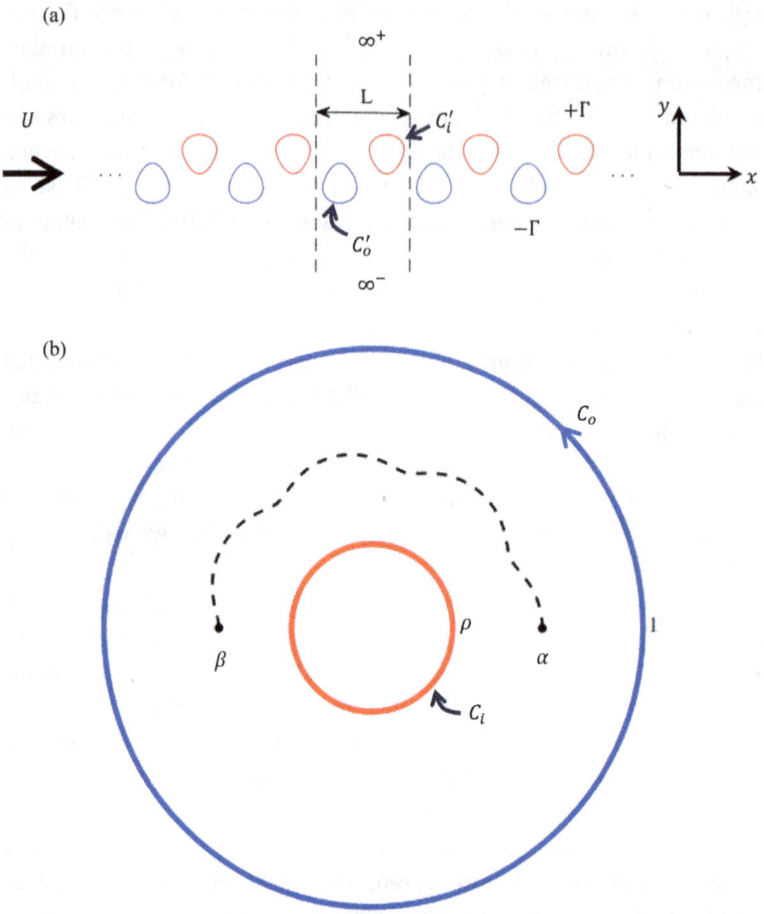

Figure 8.2: Conformal mapping from a concentric annulus with two logarithmic branch points at α and β mapped to a single period of a staggered periodic hollow vortex street in which all vortices have the same shape. The circles C_o and C_i map to the boundaries C'_o and C'_i of the two vortices in each period window [12].

$$f(\zeta) = +\frac{iL}{2\pi}\log(\zeta - \beta) + \text{a locally analytic function.}$$

In these expressions, L is the separation between the vortices as shown in Figure 8.2. At $y \to \pm\infty$ in the z-plane, the complex velocity $\frac{dW}{dz} \to -U$. The complex potential $F(\zeta)$ with the imposed circulation Γ is derived as

$$F(\zeta) = \frac{iLU}{2\pi}\log\left[\frac{|\beta| w(\zeta,\alpha) w(\zeta, 1/\bar{\beta})}{|\alpha| w(\zeta,\beta) w(\zeta, 1/\bar{\alpha})}\right] - \frac{i\Gamma}{2\pi}\log\zeta,$$

which explicitly shows the appearance of the prime function w. The conformal transformation is obtained as

$$z = f(\zeta) = -\frac{iL}{2\pi}\left\{\log\left[\frac{\beta w(\zeta,\alpha)}{\alpha w(\zeta,\beta)}\right] - \chi \log\left[\frac{\alpha w(\zeta,1/\alpha)}{\beta w(\zeta,1/\beta)}\right]\right\} + \frac{iL}{4\pi}\log\left(\frac{\beta}{\alpha}\right),$$

where χ is defined by

$$\sqrt{\chi} = \frac{\rho}{\beta|\gamma_1\gamma_2|^2}\frac{w(\beta,\gamma_1)w(\beta,\gamma_2)}{w(\beta,1/\bar{\gamma}_1)w(\beta,1/\bar{\gamma}_2)},$$

and $\zeta = \gamma_1$ and $\zeta = \gamma_2$ are the preimages of the stagnation points of the flow in the annulus $\rho < |\zeta| < 1$. The streamlines representing lines of constant $\psi(x,y)$, from the complex potential $F = \phi + i\psi$, is shown in Figure 8.3.

Figure 8.3: Distributions of the stream function for a typical hollow vortex street obtained for doubly connected domain via the use of prime functions [12].

8.2 Ordinary differential equations in the complex plane

The subject of ordinary differential equations (ODEs) in the complex plane, in which the independent variable is the complex variable $z = x + iy$, has recently received significant amount of attention, although more from the mathematics community than from the applied mathematics or mathematical physics communities. There are a few classes of these equations, including the Painlevé, Korteweg-de Vries, vortex, and bilinear equations. We will limit the description in this chapter to the Painlevé equations.

The Painlevé equations, first introduced by Painlevé [13], are a set of second-order nonlinear ODEs in which the independent variable is complex, $z = x + iy$. They are also referred to as Painlevé transcendents. One attractive property of the Painlevé equations is that they have only movable singularities, which means that their solutions do not have any singularities that are fixed in space. This property makes the Painlevé equations ideal for studying nonlinear systems, enabling closed-form solutions in certain limits. The equations are relevant in a number of fields, such as statistical mechanics, random processes, fluid dynamics, and nonlinear wave theory, to name a few.

The ODEs in the Painlevé class, with $w(z)$ as the dependent variable, are of the form $d^2w/dz^2 = F(z, w, dw/dz)$, where F is a rational function of its arguments. At least fifty of these equations have been investigated, out of which 44 were found to be expressible in terms of standard special functions. The remaining six continue to receive attention. Denoted as $P_I, P_{II}, \ldots, P_{VI}$, they can be written as follows:

$$P_I: \quad \frac{d^2w}{dz^2} = 6w^2 + z$$

$$P_{II}: \quad \frac{d^2w}{dz^2} = 2w^3 + zw + \alpha$$

$$P_{III}: \quad \frac{d^2w}{dz^2} = \frac{1}{w}\left(\frac{dw}{dz}\right)^2 - \frac{1}{z}\frac{dw}{dz} + \frac{\alpha w^2 + \beta}{z} + \gamma w^3 + \frac{\delta}{w}$$

$$P_{IV}: \quad \frac{d^2w}{dz^2} = \frac{1}{2w}\left(\frac{dw}{dz}\right)^2 + \frac{3}{2}w^3 + 4zw^2 + 2(z^2 - \alpha)w + \frac{\beta}{w}$$

$$P_V: \quad \frac{d^2w}{dz^2} = \left(\frac{1}{2w} + \frac{1}{w-1}\right)\left(\frac{dw}{dz}\right)^2 - \frac{1}{z}\frac{dw}{dz} + \frac{(w-1)^2}{z^2}\left(\alpha w + \frac{\beta}{w}\right) + \frac{\gamma}{z} + \frac{\delta w(w+1)}{w-1}$$

$$P_{VI}: \quad \frac{d^2w}{dz^2} = \frac{1}{2}\left(\frac{1}{w} + \frac{1}{w-1} + \frac{1}{w-z}\right)\left(\frac{dw}{dz}\right)^2 - \left(\frac{1}{z} + \frac{1}{z-1} + \frac{1}{w-z}\right)\frac{dw}{dz}$$
$$+ \frac{w(w-1)(w-z)}{z^2(z-1)^2}\left[\alpha + \frac{\beta z}{w^2} + \frac{\gamma(z-1)}{(w-1)^2} + \frac{\delta z(z-1)}{(w-z)^2}\right],$$

where α, β, γ, and δ are arbitrary constants.

Fornberg and Weideman [6, 7], Raeger and Fornberg [15], Clarkson [15–17], Clarkson and Jordan [5], Clarkson and Mansfield [18], Filipuk and Clarkson [19], and Clarkson [61] have investigated the PEs extensively in recent years. The equations have interesting properties. For example, research has shown that the six equations can be linear integral equations. The transformation procedure for this is embodied in the Riemann-Hilbert boundary value problems [2]. The overview in this book will be limited to P_{II} and P_{IV}.

Regarding P_{II}, or $d^2w/dz^2 = 2w^3 + zw + \alpha$, where α is an arbitrary constant and $z = x + iy$, there are a couple of relevant theorems:

Theorem 8.1 [62, 63] P_{II} has rational solutions if and only if $\alpha = n$ with $n \in \mathbb{C}$.

Theorem 8.2 [62, 63] Suppose $Q_n(z)$ satisfies the recursion relation

$$Q_{n+1}Q_{n-1} = zQ_n^2 - 4\left[Q_n\frac{d^2Q_n}{dz^2} - \left(\frac{dQ_n}{dz}\right)^2\right],$$

where $Q_0(z) = 1$ and $Q_1(z) = z$. Then the rational function

$$w(z;n) = \frac{d}{dz}\ln\left[\frac{Q_{n-1}(z)}{Q_n(z)}\right] = \frac{Q'_{n-1}(z)}{Q_{n-1}(z)} - \frac{Q'_n(z)}{Q_n(z)}$$

satisfy P_{II}, with $\alpha = n \in \mathbb{C}^+$. In addition, $w(z;0) = 0$ and $w(z;-n) = -w(z;n)$. The $Q_n(z)$ polynomials introduced by [62, 63], are monic polynomials with degree $[n(n+1)]/2$. These polynomials are as follows:

$$Q_0(z) = 1,\ Q_1(z) = z,\ Q_2(z) = z^3 + 4,$$

$$Q_3(z) = z^6 + 20z^3 - 80,\ Q_4(z) = z^{10} + 60z^7 + 11200z,$$

$$Q_5(z) = z^{15} + 140z^{12} + 2800z^9 + 78400z^6 - 313600z^3 - 6272000,$$

$$Q_6(z) = z^{21} + 280z^{18} + 18480z^{15} + 627200z^{12} - 17248000z^9 + 1448832000z^6 + 19317760000z^3 - 38635520000.$$

Kajiwara and Ohta [20] have reported on the solutions of P_{II} in terms of a determinant. If $u_j(z)$ is the polynomial defined by

$$\sum_{i=0}^{\infty} u_i(z)\lambda^i = \exp\left(z\lambda - \frac{4}{3}\lambda^3\right),$$

and $\beta_n(z)$ is an $n \times n$ determinant given by the Wronskian $\beta_n(z) = W(u_1, u_3, \ldots, u_{n-1})$, then

$$w_n(z) = \frac{d}{dz}\ln\left[\frac{\beta_{n-1}(z)}{\beta_n(z)}\right]$$

is a solution P_{II} if α in P_{II} is set to n.

For P_{IV}:

$$\frac{d^2w}{dz^2}=\frac{1}{2w}\left(\frac{dw}{dz}\right)^2+\frac{3}{2}w^3+4zw^2+2(z^2-\alpha)w+\frac{\beta}{w},$$

the following theorem, attributable to Kajiwara and Ohta [21] and Noumi and Yamada [22], establishes the solution of the equation in terms of the generalized Hermite polonomials.

Theorem 8.3 (Kajiwara and Ohta [21], Noumi and Yamada [22]) Define the generalized Hermite polynomial $H_{m,n}(z)$, which has degree mn, by

$$H_{m,n}(z)=a_{m,n}W(H_m(z),H_{m+1}(z),\ldots,\ H_{m+n-1}(z)),\quad m,n\geq 1,$$

where $W(\varphi_1,\varphi_2,\ldots,\varphi_n)$ is the Wronskian, $H_n(z)$ is the nth Hermite polynomial and $a_{m,n}$ is a constant. Then

$$w_{m,n}^{(i)}(z)=w\left(z;\alpha_{m,n}^{(i)},\beta_{m,n}^{(i)}\right)=\frac{d}{dz}\ln\frac{H_{m+1,n}(z)}{H_{m,n}(z)}$$

$$w_{m,n}^{(ii)}(z)=w\left(z;\alpha_{m,n}^{(ii)},\beta_{m,n}^{(ii)}\right)=\frac{d}{dz}\ln\frac{H_{m,n}(z)}{H_{m,n+1}(z)}$$

$$w_{m,n}^{(iii)}(z)=w\left(z;\alpha_{m,n}^{(iii)},\beta_{m,n}^{(iii)}\right)=-2z+\frac{d}{dz}\ln\frac{H_{m,n+1}(z)}{H_{m+1,n}(z)}$$

are solutions of P_{IV}, respectively, for

$$\left(\alpha_{m,n}^{(i)},\beta_{m,n}^{(i)}\right)=(2m+n+1,\ -2n^2),$$
$$\left(\alpha_{m,n}^{(ii)},\beta_{m,n}^{(ii)}\right)=(-m-2n-1,-2m^2)$$
$$\left(\alpha_{m,n}^{(iii)},\beta_{m,n}^{(iii)}\right)=\left(n-m,-2(m+n+1)^2\right),$$

where the generalized Hermite polynomials can be expressed as follows [5]:

$$H_{m,n}(z)=\frac{\pi^{m/2}\prod_{k=1}^{m}k!}{2^{m(m+2n-1)/2}}\underbrace{\int_{-\infty}^{\infty}\cdots\int_{-\infty}^{\infty}}_{n}\prod_{i=1}^{n}\prod_{j=i+1}^{n}(x_i-x_j)^2\prod_{k=1}^{n}(z-x_k)^m$$
$$\times\exp\left(-x_1^2-x_2^2-\ldots-x_n^2\right)dx_1dx_2\ldots dx_n.$$

The monic orthogonal polynomials on the real axis with respect to the Hermite weight

$$w(x;t)=x^{2m}\exp\left(-x^2+tx\right),\quad x,t\in\mathbb{R},\quad m\in\mathbb{N},$$

satisfy the recurrence relation

$$xP_n(x;t) = P_{n+1}(x;t) + \alpha_n(t)P_n(x;t) + \beta_n(t)P_{n-1}(x;t),$$

where the coefficients are given by

$$\alpha_n(t) = \frac{1}{2}t + \frac{d}{dt}\ln\frac{H_{n+1,2m}\left(\frac{1}{2}t\right)}{H_{n,2m}\left(\frac{1}{2}t\right)}, \quad \beta_n(t) = \frac{1}{2}n + \frac{d^2}{dt^2}\ln H_{n,2m}\left(\frac{1}{2}t\right),$$

and $H_{m,n}(z)$ is the generalized Hermite polynomial [5]. For the Hermite weight, $w(x;t) = x^{2m}\exp(-x^2 + tx)$, $t \in \mathbb{R}, N \in \mathbb{Z}^+$, the moment $\mu_k(t)$ can be written as

$$\mu_k(t) = \int_{-\infty}^{\infty} x^{k+2m}\exp(-x^2 + tx)\,dx = \sqrt{\pi}\left(-\frac{1}{2}i\right)^{2m+k} H_{2m+k}\left(\frac{1}{2}it\right)\exp\left(\frac{1}{4}t^2\right),$$

while the Hankel determinant $\Delta_n(t)$ takes the form

$$\Delta_n(t) = \det\left[\mu_{j+k}(t)\right]_{j,k=0}^{n-1} = \frac{\prod_{k=1}^{n-1} k!}{\prod_{k=1}^{2m-1} k!}\frac{\pi^{n/2}}{2^{(n+2m-1)(n+2m)}} H_{n,2m}\left(\frac{1}{2}t\right)\exp\left(\frac{1}{4}nt^2\right).$$

The coefficients in the recurrence relation are given by

$$\alpha_n(t) = \frac{d}{dt}\ln\frac{\Delta_{n+1}(t)}{\Delta_n(t)} = \frac{1}{2}t + \frac{d}{dt}\ln\frac{H_{n+1,2m}\left(\frac{1}{2}t\right)}{H_{n,2m}\left(\frac{1}{2}t\right)},$$

$$\beta_n(t) = \frac{\Delta_{n+1}(t)\Delta_{n-1}(t)}{\Delta_n^2(t)} = \frac{1}{2}m + \frac{d^2}{dt^2}\ln H_{n,2m}\left(\frac{1}{2}t\right).$$

The generalized Okamoto polynomial solutions of P_{IV} are described in the following theorem:

Theorem 8.4 (Kajiwara and Ohta [21], Noumi and Yamada [22], and Clarkson [15]) Let $\varphi_k(z) = 3^{k/2}e^{-k\pi i/2}H_k\left(\frac{1}{3}\sqrt{3}iz\right)$, with $H_k(\zeta)$ the kth Hermite polynomial, and define the generalized Okamoto polynomial $\Omega_{m,n}(z)$ by

$$\Omega_{m,n}(z) = W(\varphi_1, \varphi_4, \ldots, \varphi_{3m+3n-5}; \varphi_2, \varphi_5, \ldots, \varphi_{3n-4})$$

with $m, n \geq 1$, where $W(\varphi_1, \varphi_2, \ldots, \varphi_n)$ is the Wronskian. Then

$$\tilde{w}_{m,n}^{(i)}(z) = w\left(z; \tilde{\alpha}_{m,n}^{(i)}, \tilde{\beta}_{m,n}^{(i)}\right) = -\frac{2}{3}z + \frac{d}{dz}\ln\frac{\Omega_{m+1,n}(z)}{\Omega_{m,n}(z)}$$

$$\tilde{w}_{m,n}^{(ii)}(z) = w\left(z; \tilde{\alpha}_{m,n}^{(ii)}, \tilde{\beta}_{m,n}^{(ii)}\right) = -\frac{2}{3}z + \frac{d}{dz}\ln\frac{\Omega_{m,n}(z)}{\Omega_{m,n+1}(z)}$$

$$\tilde{w}^{(iii)}_{m,n}(z) = w\left(z; \tilde{\alpha}^{(iii)}_{m,n}, \tilde{\beta}^{(iii)}_{m,n}\right) = -\frac{2}{3}z + \frac{d}{dz}\ln\frac{\Omega_{m,n+1}(z)}{\Omega_{m+1,n}(z)}$$

are the solutions of P_{IV}, respectively, for

$$\left(\tilde{\alpha}^{(i)}_{m,n}, \tilde{\beta}^{(i)}_{m,n}\right) = \left(2m+n, -2\left(n-\frac{1}{3}\right)^2\right)$$

$$\left(\tilde{\alpha}^{(ii)}_{m,n}, \tilde{\beta}^{(ii)}_{m,n}\right) = \left(-m-2n, -2\left(m-\frac{1}{3}\right)^2\right)$$

$$\left(\tilde{\alpha}^{(iii)}_{m,n}, \tilde{\beta}^{(iii)}_{m,n}\right) = \left(n-m, -2\left(m+n+\frac{1}{3}\right)^2\right).$$

The roots of the Okamoto polynomials are shown in Figure 8.4 for the case $\Omega_{10,10}(z)$.

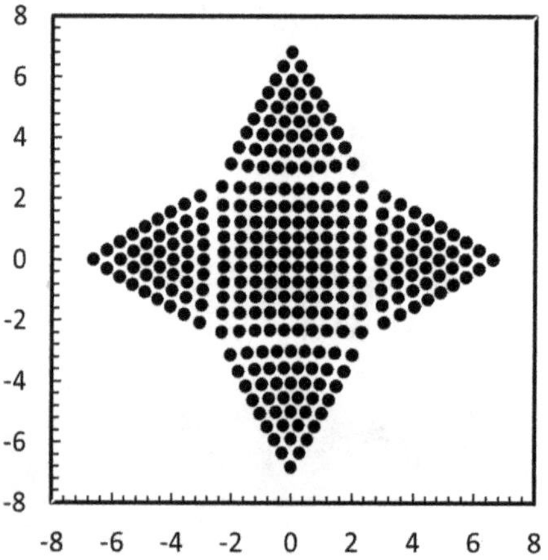

Figure 8.4: The roots of the Okamoto polynomials for the case $\Omega_{10,10}(z)$.

Regarding the movable singularity of the PEs, we point out that there are two kinds of singularity: fixed and movable. Fixed singularity occurs in some functions when the independent variable (z) attains a certain value. For example, the function $u(z) = ce^{-1/z}$ becomes singular when $z = 0$ independent of the initial condition. This function could be the solution of $u'/u = 1/z^2$. The singularity of this ODE does not depend on the initial condition $u_0 \equiv u(z=0)$. On the hand the ODE $u'/u^2 = 1$, which has the solution $-u_0/(zu_0 - 1)$, where $u_0 \equiv u(z=0)$, has a singularity when $z = 1/u_0$, so that the initial condition for the ODE determines the location z where a singularity is to be found. The PEs have this kind of (movable) singularities.

8.3 Advances in relevant Schwarz-Christoffel (SC) mappings

In this section we will briefly present the state-of-the-art of multiply connected Schwarz-Christoffel (MCSC) transformations for unbounded and bounded domains with a focus on the work of DeLillo and his colleagues [14, 23–25, 64]; DeLillo [26]; [27–29]; and those by Crowdy and his coworkers [30–34]. Impressive mappings have been reported from these groups, including methods based on the Fast Fourier Transform (FFT), extensions of Fornberg's [35] method for the disk, MCSC maps for unbounded and bounded domains, and the comparison of the transformations with those reported by Crowdy who used the Schottky-Klein's prime functions. The use of Laurent series to generate MCSC mappings, and the application to electric resistance calculations have also been undertaken.

DeLillo and Elcrat [36–38]; DeLillo, Elcrat, and Pfaltzgraff [39–41]; and DeLillo, Horn, and Pfaltzgraff [42] have developed Fourier series-based conformal mappings with applications to exterior regions, exterior regions with corners, and numerical conformal mapping of multiply connected regions using [35]-like methods, and the use of Faber (polynomial) series and Chebyshev polynomials for mapping, plus a host of other interesting transformations using different approaches. The contents of this section are driven more by the bias toward mathematical physics, and not necessarily the excitement generated by the results, although the mappings reported here are nonetheless interesting in their own right.

DeLillo et al. (2001) re-derived the SC map for an annulus with motivation coming from previous derivations for a disk by others. The finite annulus radius case, $\mu < z \leq 1$, was extended to the case of an annulus with infinite outer radius, leading to the transformation:

$$f(z) = A \int^{z} \prod_{k=1}^{m} \left[\Theta\left(\frac{\zeta}{\mu z_{0,k}}\right) \right]^{-\beta_{0,k}} \prod_{k=1}^{n} \left[\Theta\left(\frac{\mu \zeta}{z_{1,k}}\right) \right]^{\beta_{1,k}} d\zeta + B,$$

where $z_{v,k}$ are the prevertices and $\beta_{v,k}$ are the turning parameters, and

$$\Theta(w) = \prod_{v=0}^{\infty} \left(1 - \mu^{2v+1} + w\right) \left(1 - \frac{\mu^{2v+1}}{w}\right).$$

Table 8.1 contains the definitions of some of the symbols used in this section.

The $\mu^k z_{0,k}$'s and $\mu^k z_{1,k}$'s are reflections of the prevertices on the outer and inner circles, respectively, through concentric reflected circles. The researchers require that SC maps satisfy a geometric boundary condition, which they express as follows:

Table 8.1: The symbols used in Section 8.3.

Symbol	Meaning
K_j	Number of vertices on the jth polygon, Γ_j
m	Number of polygons
$w_{k,j}$	The kth vertex on Γ_j
$f(z_{k,j})$	The mapping of $z_{k,j}$. Satisfies $f(z_{k,j}) = w_{k,j}$
$z_{k,j}$	$c_j + r_j e^{i\theta_{k,j}} = k^{th}$ prevertex on circle C_j
c_j	Center of circle j
Γ_j	$f(C_j)$, the images of the circle boundaries for the jth polygon
$\pi\alpha_{k,j}$	Interior angles of the polygons at the vertices, $0 < \alpha_{k,j} \leq 2$
$\pi\beta_{k,j}$	Turning angle of the tangent vector on the polygonal boundaries at the vertices. Note that $\pi(\alpha_{k,j} + \beta_{k,j}) = \pi$.
$S(z)$	Singularity function; $f(z) = \int \exp[\int S(z)]$
μ	Normalized inner radius of annulus. Outer radius is 1.
$z_{k,j}$	The value of z at the kth vertex on the jth polygon
ψ	Tangent angle
$z_{k,vi}$	Reflection of prevertices
s_{vi}	Reflection of centers
$w(z, z_{k,j})$	The Schottky prime function, where $z_{k,j}$ is the value of z at the kth vertex of polygon Γ_j.

The tangent angle $\psi(t) = \arg\{ie^{it}f'(e^{it})\}$ is constant in the arc between the prevertices and so for a disk we can write

$$\psi'(t) = 1 + \text{Re}\left[e^{it}\frac{f''(e^{it})}{f'(e^{it})}\right] = 0.$$

The pre-Schwarzian of f, or $f''(z)/f'(z) \equiv S(z)$, which was also referred to as the singularity function, was introduced into the formulation, and DeLillo et al. [64] extended the transformation $f(z)$ to global multivalued function using Schwarz reflection. The $S(z)$ is invariant under affine maps $w \mapsto aw + b$ and, hence, single-valued. Thus,

$$\frac{(af(z) + b)''}{(af(z) + b')} = \frac{f''(z)}{f'(z)}.$$

The behavior near corner $f(z_{k,i})$ is represented by an $\alpha_{k,i}$ root, or

$$f(z) - f(z_{k,i}) = (z - z_{k,i})^{\alpha_{k,i}} h_{k,i}(z),$$

where $h_{k,i}(z)$ is analytic and nonvanishing near z_k. Local expansion:

$$\frac{f''(z)}{f'(z)} = \frac{\beta_{k,i}}{z - z_{k,i}} + H_{k,i}(z), \quad \beta_{k,i} = \alpha_{k,i} - 1,$$

where $H_{k,i}(z)$ is analytic in a neighborhood of $z_{k,i}$. The singularity function $S(z)$ was obtained by Delillo et al. as

$$S(z) = \sum_{k=1}^{m} -\beta_{0,k}\left\{\frac{1}{z-z_{0,k}} + \frac{1}{z-\mu^2 z_{0,k}} + \frac{1}{z-\mu^{-2}z_{0,k}} + \ldots\right\}$$
$$+ \sum_{k=1}^{n} \beta_{1,k}\left\{\frac{1}{z-z_{1,k}} + \frac{1}{z-\mu^{-2}z_{1,k}} + \frac{1}{z-\mu^{2}z_{1,k}} + \ldots\right\}$$
$$= -\sum_{k=1}^{m} \beta_{0,k}\sum_{v=0}^{\infty}\left[\frac{-\mu^{2v}/z_{0,k}}{1-\mu^{2v}(z/z_{0,k})} + \frac{\mu^{2(v+1)}(z_{0,k}/z^2)}{1-\mu^{2(v+1)}(z_{0,k}/z)}\right]$$
$$+ \sum_{k=1}^{n} \beta_{1,k}\sum_{v=0}^{\infty}\left[\frac{\mu^{2v}(z_{1,k}/z^2)}{1-\mu^{2v}(z_{1,k}/z)} - \frac{\mu^{2(v+1)}/z_{1,k}}{1-\mu^{2(v+1)}(z/z_{1,k})}\right].$$

Since the singularity function satisfies the boundary conditions, and $S(z) = \frac{f''(z)}{f'(z)} = \frac{1}{f'}\frac{df'}{dz} = \frac{d}{dz}\log f'$, we can obtain:

$$\frac{d}{dz}\log f'(z) = \sum_{k=1}^{m} -\beta_{0,k}\frac{d}{dz}\log\prod_{v=0}^{\infty}\left(1-\frac{\mu^{2v}z}{z_{0,k}}\right)\left(1-\frac{\mu^{2(v+1)}z_{0,k}}{z}\right)$$
$$+ \sum_{k=1}^{n} \beta_{1,k}\frac{d}{dz}\log\prod_{v=0}^{\infty}\left(1-\frac{\mu^{2(v+1)}z}{z_{1,k}}\right)\left(1-\frac{\mu^{2v}z_{1,k}}{z}\right)$$
$$= \frac{d}{dz}\sum_{k=1}^{m} -\beta_{0,k}\log\Theta\left(\frac{z}{\mu z_{0,k}}\right) + \frac{d}{dz}\sum_{k=1}^{n}\beta_{1,k}\log\Theta\left(\frac{\mu z}{z_{1,k}}\right).$$

Integrating and taking the exponent gives

$$f'(z) = A\prod_{k=1}^{m}\left[\Theta\left(\frac{z}{\mu z_{0,k}}\right)\right]^{-\beta_{0,k}}\prod_{k=1}^{n}\left[\Theta\left(\frac{\mu z}{z_{1,k}}\right)\right]^{\beta_{1,k}}.$$

It is important to note that the derivation of DeLillo and his coworkers was based on the invariance of the pre-Schwarzian under reflections, yielding solutions that ensure straight sides by the method of images. The interested reader should consult the references for details.

The transformation for the MCSC mapping for the unbounded case is reported by DeLillo et al. [4] as

$$f'(z) = A\prod_{i=1}^{m}\prod_{k=1}^{K_i}\left[\prod_{\substack{j=0\\ v\in\sigma_j(i)}}^{\infty}\left(\frac{z-z_{k,vi}}{z-s_{vi}}\right)\right]^{\beta_{k,i}}, \quad \sum_{k=1}^{K_i}\beta_{k,i} = 2,\ i = 1, \ldots, m,$$

where $z_{k,\nu i}$ and $s_{\nu i}$ are the reflections of the prevertices and centers, respectively. In addition, $\sigma_n(i) = \{\nu \in \sigma_n : \nu_n \neq i\}$ denotes sequences in σ_n whose last factor never equals i. Equivalently, the set of multi-indices ν of length $|\nu| = n > 0$ is denoted by $\sigma_n = \{\nu_1 \nu_2 \cdots \nu_n : 1 \leq \nu_i < m$ for $i = 1, 2, \ldots, n$; and $\nu_i \neq \nu_{i+1}$ for $i = 1, 2, \ldots, n-1\}$, with $\sigma_0 := \phi$. If $\nu \in \sigma_0$, the authors write $\nu_j = j$. Integration of the equation above gives

$$f(z) = A \int^{z} \prod_{i=1}^{m} \prod_{k=1}^{K_i} \left[\prod_{\substack{j=0 \\ \nu \in \sigma_j(i)}}^{\infty} \left(\frac{\zeta - z_{k,\nu i}}{\zeta - s_{\nu i}} \right) \right]^{\beta_{k,i}} d\zeta + B.$$

The singularity function for the unbounded multiply connected case is single valued, as is obtained by reflecting the poles $\beta_{k,i}/(z - z_{k,i})$ and $-2/(z - c_i)$ using the method of images. The noncovergent form of the function is [4]:

$$S(z) = \sum_{j=0}^{\infty} \sum_{i=1}^{m} \sum_{\nu \in \sigma_j(i)} \left(\sum_{k=1}^{K_i} \frac{\beta_{k,i}}{z - z_{k,\nu i}} - \frac{2}{z - s_{\nu i}} \right)$$

The authors compared their derivation with the mapping reported by Crowdy using the Schottky-Klein prime functions, which are implemented as the generalization of the elliptic Θ functions to finitely connected domains bounded by circles with invariance under the group of reflections in circles. This approach is more "algebraic" based on the fact that certain ratios of prime functions are constant on the circles.

Crowdy's formula for the interior of the unit disk C_1 can be written as

$$f'(z) = AS_\infty(z) \prod_{k=1}^{K_1} [w(z, z_{k,1})]^{\beta_{k,1}} \prod_{i=2}^{m} \prod_{k=1}^{K_i} [w(z, z_{k,i})]^{\beta_{k,i}},$$

where w is the Schottky-Klein prime function. DeLillo's group also extended their analysis to the bounded case, which pertains to mapping the interior of unit circle C_0 minus m interior disks with boundary circles C_{0i}, $i = 1, \ldots, m$ to a domain interior to a finite polygonal domain minus the m nonoverlapping, interior polygonal domains. The transformation is similar to the unbounded case with the reflections of centers $s_{\nu i}$ replaced by reflections $z_{k,\nu i0}$ of the prevertices $z_{k,0}$ on C_0. The formula reported for the bounded, multiply connected case is:

$$f(z) = A \int^{z} \prod_{k=1}^{K_0} (\zeta - z_{k,0})^{\beta_{k,0}} \prod_{i=1}^{m} \prod_{\substack{j=0 \\ \nu \in \sigma_j(i)}}^{\infty} \left[\prod_{k=1}^{K_0} (\zeta - z_{k,\nu i0})^{\beta_{k,0}} \prod_{k=1}^{K_i} (\zeta - z_{k,\nu i})^{\beta_{k,i}} \right] d\zeta + B.$$

For comparison, Crowdy [30] used the prime function approach to obtain the following mapping:

$$f(z) = A \int^{z} S_c(\zeta) \prod_{k=1}^{K_0} [w(\zeta, z_{k,0})]^{\beta_{k,0}} \prod_{i=1}^{m} \prod_{k=1}^{K_i} [w(\zeta, z_{k,i})]^{\beta_{k,i}} d\zeta + B,$$

where

$$S_c(\zeta) := \frac{w_\zeta(\zeta, a) w(\zeta, \bar{a}^{-1}) - w_\zeta(\zeta, \bar{a}^{-1}) w(\zeta, a)}{\prod_{j=1}^{m} w\left(\zeta, \gamma_1^j\right) w\left(\zeta, \gamma_2^j\right)},$$

where γ_1 and γ_2 are preimage points, and $w(\zeta, \gamma)$ are the Schottky-Klein prime functions,

$$w(\zeta, \gamma) := (\zeta - \gamma) \prod_{\theta_i \in \Theta''} \frac{(\theta_i(\zeta) - \gamma)(\theta_i(\gamma) - \zeta)}{(\theta_i(\zeta) - \zeta)(\theta_i(\gamma) - \gamma)},$$

where Θ'' is a certain subset of the Schottky group. The transformations obtained using the prime functions were shown to be reducible to those reported by DeLillo's group. The prime functions are the natural generalizations to the multiply connected case of the (Jacobi) theta functions above for the doubly connected case.

In concluding this section, the interested reader must consult the original references for the details of the various procedures as well as for relevant visualization and/or graphical depictions of the transformations.

8.4 Numerical complex variable simulation

In this section, we present an overview of selected numerical methods that have been used in connection with complex variable analysis. In the first subsection, we briefly summarize the use of conformal mapping for grid generation in computational fluid dynamics and solid mechanics. However, the emphasis in this section is on the numerical calculation of conformal mapping, which is the subject of subsection two.

8.4.1 Use of Schwarz-Christoffel transformation in mesh generation

A major task in the use deployment of the computational fluid dynamics or solid mechanics approach to solving highly nonlinear and multidimensional mechanical problems is the generation of the computational grids on which the solution will be obtained. The Schwarz-Christoffel mapping has been investigated as a candidate for this task [1]. For the elliptic grid generation method [43], we could consider the general conformal mapping $z = f(\zeta)$, for which the Laplacian Δz transforms as $\Delta\Phi = |f(\zeta)|^{-2}\Delta\phi$ as presented in Section 3.2 of this book. We have also shown that Dirichlet and homogeneous Neumann boundary conditions are invariant under a conformal transformation.

This suggests that fast Poisson solvers may be used to take advantage of this invariance after transformation. However, there are potential issues with the use of conformal transformation in grid generation, including the high possibility that $|f(\zeta)|^{-2}$ may be undefined at the corners of the meshed region, given the complex geometries that we deal with in engineering. Another issue with the use of conformal mapping for grid generation is that the procedure is the inherent two-dimensionality of the procedure, whereas realistic engineering models are mostly three-dimensional. Although Ives [44] reported possible roles for conformal mapping in grid generation, it is not clear that a revolutionary contribution in this direction is imminent.

8.4.2 Numerical conformal mapping

The state-of-the-art in numerical conformal mapping (NCM) in the mid-1960s was reviewed by Gaier [45]. Significant advances have been made since then, and improvements in the efficiency of some of the new methods were reported with the introduction of the FFT algorithm by Cooley and Tukey [46] shortly afterwards [47]. The most important method at the time appeared to center on different approximations of the Theodorsen's integral equation [47] and the approach of Menikoff and Zemack [48]. The method of the former authors concerns numerical solution of the Cauchy-Riemann equations in conjunction with optimization techniques.

In one of the earlier studies of NCM that precedes Fornberg [35], Menikoff and Zemach [48] developed and analyzed nonlinear integral equations for the boundary functions that determine conformal transformations in two dimensions. One of their equations has a nonsingular logarithmic kernel that is reported to be well suited for numerical computation of conformal maps, including those that deal with regions having highly distorted boundaries. The authors describe numerical procedures based on interspersed Gaussian quadrature for approximating the integrals, combined with the Newton-Raphson approach to solve the resulting nonlinear algebraic equations that result from their discretization. Applications to regions bounded by spike curves that are common in the Rayleigh-Taylor instability phenomena are reported. The authors also report a differential equation that relates changes in the mapping function to changes of the boundary, with the target application being potential problems for regions with time-dependent boundaries.

The classical paper by Fornberg [35] calculates the Taylor coefficients of the analytic function which maps the unit circle onto a region bounded by any smooth simple connected curve. The approach uses an outer iteration that is quadratically convergent and an inner one that is super-linearly convergent. An operation count of $O(N \log N)$) is reported for obtaining the images, on the edge of a mapped region, and the approximations for the $N/2$ first Taylor coefficients, when N complex points are distributed equidistantly around the periphery of a unit circle.

Bisshopp [49] derived and implemented, using fast Fourier transforms (FFTs), a numerical method for determining least-square approximations of an arbitrary complex mapping function. His method factors a discrete Hilbert transform into a pair of Fourier transforms to reduce the operation count of the longest computation to $O(N \log N)$, apparently to be competitive with the approach by Fornberg 35. By comparison, a similar factoring of the discrete Poisson integral formula allows an explicit inversion of it in $O(N \log N)$ operations instead of $O(N^3)$. Bisshopp concluded that the resulting scheme for analytic continuation appears to be considerably more reliable than the evaluation of polynomials.

In Murid and Nasser [50] and Wegmann et al. [51], the Riemann-Hilbert problem on simply connected regions is reduced to Fredholm integral equations of the second kind with a kernel which can be interpreted as a generalization of the Neumann kernel. However, the latter researchers have observed that the issue of solvability and uniqueness must be properly addressed to make the approach useful. This prompted Wegmann and Nasser [52] to extend the integral equation approach to multiply connected regions. It is shown that there are several characteristic differences between the simply connected and the multiply connected cases. The authors determined the eigenspaces to the eigenvalues of the integral operator with generalized Neumann kernel and its adjoint. The dimensions of these spaces yield the number of solutions of the homogeneous Fredholm equations, as well as the number of constraints. The authors' conclusion is that the information about the existence and uniqueness of their solutions makes these Fredholm integral equations a useful tool for the numerical solution of Riemann-Hilbert problems on multiply connected regions. A comprehensive presentation of the numerical solution of the Riemann-Hilbert problem is covered in Chapters 4 through 7 of Trogdon and Olver [2].

Murid and Hu [53] present a boundary integral equation for conformal mapping of multiply connected regions onto an annulus. Their (boundary integral) equation, which involves unknown circular radii, were discretized to obtain a system of nonlinear equations which were solved iteratively with a modified Gauss-Newton method named Lavenberg-Marquardt with the Fletcher's algorithm for solving the nonlinear least squares problems. The Cauchy's integral formula is used to determine the mapping function in the interior of the region once the boundary values of the mapping function have been calculated.

Trefethen [54] presented new algorithms for numerical conformal mapping based on rational approximations and the solution of Dirichlet problems by least-squares fitting on the boundary. The methods are targeted at regions with corners, where the Dirichlet problem is solved by the "lightning Laplace solver," wherein poles are exponentially clustered near each singularity. The possibility of further simplifications is suggested for circular polygons and noncircular polygons.

Numerical procedures for analyzing the Schwarz-Christoffel transformation for unbounded multiply connected domains were developed by DeLillo et al. [29]. In their approach, the infinite product representation for the derivative of the mapping function is replaced by a finite factorization in which the inner factors are made to satisfy certain boundary conditions. The Laurent series is used to develop least squares approximations that also satisfy some other boundary conditions, leading to what is reported to be a more efficient method than the original method based on reflections. The authors report that their procedure makes accurate mapping of domains of higher connectivity feasible.

Motivated by the desire to have a very fast, but noniterative method, in contrast to Wegman [55] and Wegman and Nasser [52] whose methods involve iterations, Ivanshin and Shirokova [56], applied the method of boundary curve re-parametrization to construct an approximate analytical conformal mapping of the unit disk onto an arbitrary finite domain with smooth boundary but with a finite number of points with acute angles. Their method combines the solution of a Fredholm equation and spline-interpolation and consists of an approximate solution of a linear system with unknown Fourier coefficients and construction of correction splines. The approximate mapping function has the form of a Cauchy integral. The method was demonstrated for multiply connected domains with boundary angle points. Finally, the numerical evaluation of the Schottky-Klein prime functions, including those for mapping multiply connected domains, is covered in Chapter 14 of Crowdy [12].

Suggested reading

[1] Driscoll, T.A. and Trefethen, L.N. Schwarz-Christoffel Mapping, Cambridge, UK, Cambridge University Press, 2002.

[2] Trogdon, T. and Olver, S. Riemann-Hilbert Problems, Their Numerical Solution and Computation of Nonlinear Special Functions, Philadelphia, USA SIAM, 2016.

[3] Wegert, E. Visual Complex Functions, Freiberg, Germany, Birkhauser, 2012.

[4] DeLillo, T., Elcrat, A.R. and Pfaltzgraff, J.A. Schwarz-Christoffel mapping of multiply connected domains, J d'Analyse Math 2004, 94, 17–47.

[5] Clarkson, P.A. and Jordaan, K. The relationship between semiclassical Laguerre polynomials and the fourth Painlevé equation, Constr Approx 2014, 39, 223–254.

[6] Fornberg, B. and Weideman, J.A.C. A numerical methodology for the Painlevé equations, J Comp Phys 2011, 230, 5957–5973.

[7] Fornberg, B. and Weideman, J.A.C. A computational exploration of the second Painlevé equation, Comp Math 2014, 14, 985–1016.

[8] Alimov, M.M. and Kornev, K.G. Meniscus on a shaped fibre: Singularities and hodograph formulation, Proc Roy Soc A 2014, 470, 113.

[9] Gardiner, B.P.J., McCue, S.W., Dallastonand, M.C. and Moroney, T.J. Saffman–Taylor fingers with kinetic undercooling, Phys Rev E 2015, 91, 023016.

[10] McDonald, N.R. Generalised Hele-Shaw flow: A Schwarz function approach, Eur J Appl Math 2011, 22, 517–532.

[11] Elcrat, A. and Protas, B. A framework for linear stability analysis of finite-area vortices, Proc Roy Soc A 2013, 469, 20120709.

[12] Crowdy, D. Solving Problems in Multiply Connected Domains, Society for Industrial and Applied Mathematics (SIAM), Philadelphia, USA, 2020.

[13] Painlevé, P. Sur les équationsdifférentielles du second ordre et d'ordresupérieurdontl'intégralegénéraleestuniforme, Acta Math 1902, 1–85.
[14] DeLillo, T., Elcrat, A.R. and Kropf, E.H. Calculation of resistances for multiply connected domains using the Schwarz-Christoffel transformations, CMFT J 2011, 11, 725–745.
[15] Clarkson, P.A. Special polynomials associated with rational solutions of the Painlevé equations and applications to soliton equations, Comp Methods Func Theory 2006, 6, 329–401.
[16] Clarkson, P.A. The fourth Painlevé equation and associated special polynomials, J Math Phys 2003, 44, 5350–5374.
[17] Clarkson, P.A. Vortices and polynomials, Stud Appl Math 2009, 123, 37–62.
[18] Clarkson, P.A. and Mansfield, E.L. The second Painlevé equation, its hierarchy and associated special polynomials, Nonlinearity 2003, 16, R1–R26.
[19] Filipuk, G.V. and Clarkson, P.A. The symmetric fourth Painlevé hierarchy and associated special polynomials, Stud Appl Math 2008, 121, 157–188.
[20] Kajiwara, K. and Ohta, Y. Determinant structure of the rational solutions for the Painlevé II equation, J of Math Phys 1996, 37, 9.
[21] Kajiwara, K. and Ohta, Y. Determinant structure of the rational solutions for the Painlevé IV equation, J Phys A: Math Gen 1998, 31, 2431. DOI: 10.1088/0305-4470/31/10/017.
[22] Naoumi, M. and Yamada, Y. Higher Order Painleve Equations of Type $A_l^{(1)}$, Funkcialaj Ekvacioj 1998, 41, 483–503.
[23] DeLillo, T., Driscoll, T.A., Elcrat, A.R. and Pfaltzgraff, J.A. Computation of multiply connected Schwarz-Christoffel maps for exterior domains, CMFT J 2006, 6, 301–315.
[24] DeLillo, T., Driscoll, T.A., Elcrat, A.R. and Pfaltzgraff, J.A. Radial and circular slit maps of unbounded multiply connected circle domains, Proc R Soc A 2008, 464, 1719–1737.
[25] DeLillo, T., Driscoll, T.A., Elcrat, A.R. and Pfaltzgraff, J.A. Efficient calculation of Schwarz-Christoffel transformations for multiply connected domains using Laurent series, CMFT J 2013, 13, 307–336.
[26] DeLillo, T. Schwarz-Christoffel mapping of bounded, multiply connected domains, CMFT J 2006, 6, 275–300.
[27] DeLillo, T. and Kropf, E.H. Slit maps and Schwarz-Christoffel maps for multiply connected domains, ETNA 2010, 36, 195–223.
[28] DeLillo, T. and Kropf, E.H. Numerical computation of the Schwarz-Christoffel transformation for multiply connected domains, SISC 2011, 33, 1369–1394.
[29] DeLillo, T.K., Elcrat, A.R., Kropf, E.H. and Pfaltzgraff, J.A. Efficient Calculation of Schwarz–Christoffel Transformations for Multiply Connected Domains Using Laurent Series, Comput Methods Funct Theory 2013, 13, 307–336. DOI: 10.1007/s40315-013-0023-1.
[30] Crowdy, D. The Schwarz-Christoffel mapping to bounded multiply-connected polygonal domains, Proc R Soc A 2005, 461, 2653–2678.
[31] Crowdy, D. Schwarz-Christoffel mapping to unbounded multiply connected polygonal regions, Math Proc Camb Phil Soc 2007, 142, 319–339.
[32] Crowdy, D. and Marshall, J. Conformal mapping between canonical multiply connected domains, Comput Methods Funct Theory 2006, 6(1), 59–76.
[33] Crowdy, D. and Marshall, J. Computing the Schottky-Klein prime function on the Schottky double of planar domains, Comput Methods Funct Theory 2007, 7(1), 293–308.
[34] Crowdy, D. The Schottky-Klein prime function on the Schottky double of planar domains, Comput Methods Funct Theory 2010, 10(2), 501–517.
[35] Fornberg, B. A numerical method for conformal mapping, SIAM J Sc Comput 1980, 1(3), 386–400.
[36] DeLillo, T. and Elcrat, A.R. A comparison of some numerical conformal mapping methods for exterior regions, SISC 1991, 12, 399–422.
[37] DeLillo, T. and Elcrat, A.R. Numerical conformal mapping methods for exterior regions with corners, Jcp 1993, 108, 199–208.

[38] DeLillo, T. and Elcrat, A.R. A Fornberg-like conformal mapping method for slender regions, JCAM 1993, 46, 49–64.
[39] DeLillo, T. and Pfaltzgraff, J.A. Extremal distance, harmonic measure and numerical conformal mapping, JCAM 1993, 46, 103–113.
[40] DeLillo, T., Elcrat, A.R. and Pfaltzgraff, J.A. Numerical conformal mapping methods based on Faber series, JCAM 1997, 83, 205–236.
[41] DeLillo, T. and Pfaltzgraff, J.A. Numerical conformal mapping methods for simply and doubly connected regions, SISC 1998, 19, 155–171.
[42] DeLillo, T., Horn, M.A. and Pfaltzgraff, J.A. Numerical conformal mapping of multiply connected regions by Fornberg-like methods, Numer Math 1999, 83, 205–230.
[43] Thompson, J.F., Soni, B.K. and Weatherhill, N.P. Handbook of Grid Generation, Boca Raton, Florida, USA, CRC Press, 1999.
[44] Ives, D.C. Conformal Grid Generation, Appl Math Comput 1982, 10(11), 107–135. In "Numerical Grid Generation," Thompson JF Editor, 107–35. North-Holland, New York, NY, USA, 1982.
[45] Gaier, D. Konstruktive Methoden der konformen Abbildung, Berlin, Germany, Springer, 1964.
[46] Cooley, J.W. and Tukey, J.W. An algorithm for the machine calculation of complex Fourier series, Math Comp 1965, 19, 297–301.
[47] Henrici, P. Fast Fourier methods in computational complex analysis, SIAM Rev 1979, 21, 481–527.
[48] Menikof, R. and Zemach, C. Methods for Numerical Conformal Mapping, Los Alamos Report Number LA-7836-MS/UC-32 1979.
[49] Bisshopp, F. Numerical conformal mapping and analytic continuation, Q Appl Math 1983, 125–142.
[50] Murid, A.H.M. and Nasser, M.M.S. Eigenproblem of the generalized Neumann kernel, Bull Malaysian Math Sci Soc 2003, 26(2), 13–33.
[51] Wegmann, R., Murid, A.H.M. and Nasser, M.M.S. The Riemann–Hilbert problem and the generalized Neumann kernel, J Comput Appl Math 2005, 182, 388–415.
[52] Wegman, R. and Nasser, M.M.S. The Riemann-Hilbert problem and the generalized Neumann kernel on multiply connected regions, J Comput Appl Math 2008, 214, 36–57.
[53] Murid, H.M. and Hu, L.N. Numerical conformal mapping of bounded multiply connected regions by an integral equation method, Int J Con- Temp Math Sci 2009, 4, 1121–1147.
[54] Trefethen, L.N. arXiv:1911.03696 [math.CV], https://doi.org/10.48550/arXiv.1911.03696.
[55] Wegmann, R. Fast conformal mapping of multiply connected regions, J Comput Appl Math 2001, 130, 119–138.
[56] Ivanshin, P.N. and Shirokova, E.A. The approximate conformal mapping of a disk onto domain with an acute angle, arXiv:2307.03516v1 [math.CV], 2023.
[57] Noble, B. Methods based on the Wiener-Hopf Technique, UK, Pergamon Press, 1958.
[58] Deconinck, B. Trogdon, T., Vasan, V. The method of Fokas for solving linear partial differential equations, SIAM Rev, 2014, 56, 159–186.
[59] Fokas, A.S. A unified approach to boundary value problems. Soc Ind Appl Math, 2008, Publication 78.
[60] Bottazzini, U. and Gray, J. Hidden Harmony – Geometric Fantasies. The Rise of Complex Function Theory, New York, Springer Science + Business Media, LLC, 2013.
[61] Clarkson, P.A. Vortices and Polynomials, Presented at "Modern Applications of Complex variables: Modeling, Theory and Computation," Banff, Canada, January 2013.
[62] Vorob'ev, A.P. On the rational solutions of the second Painlevé equation. Differ Uravn, 1965, 1, 79–81.
[63] Yablonskii, A. I. On rational solutions of the second Painlevé equation. Vest AN BSSR, Ser Fiz-Tech Nauk, 1959, 3, 30–35.
[64] Delillo, T.K., Elcrat, A.R. and Pfaltzgraff, J.A. Schwarz-Christoffel mapping of annulus, SIAM Rev, 2001, 43, 469–477.

Appendix A
Table of Laplace transforms pairs in diffusion analysis

We write $\alpha' = \sqrt{s/\alpha}$, α and x are always real and positive, β and h are unrestricted, and s is the Laplace transform variable. This appendix is intended for Section 5.3 of the text, where α is the thermal diffusivity.

	$\tilde{T}(s)$	$T(t)$
1.	$\frac{1}{s^{\nu+1}}, \nu > -1$	$\frac{t^{\nu}}{\Gamma(\nu+1)}$
2.	$\frac{1}{s+\beta}$	$e^{-\beta t}$
3.	$e^{-\alpha' x}$	$\frac{x}{2\sqrt{\pi \alpha t^3}} e^{-\frac{x^2}{4\alpha t}}$
4.	$\frac{e^{-\alpha' x}}{\alpha'}$	$\left(\frac{\alpha}{\pi t}\right)^{\frac{1}{2}} e^{-\frac{x^2}{4\alpha t}}$
5.	$\frac{e^{-\alpha' x}}{s}$	$\operatorname{erfc}\frac{x}{2\sqrt{(\alpha t)}}$
6.	$\frac{e^{-\alpha' x}}{\alpha' s}$	$2\left(\frac{\alpha t}{\pi}\right)^{\frac{1}{2}} e^{-\frac{x^2}{4\alpha t}} - x\operatorname{erfc}\frac{x}{2\sqrt{(\alpha t)}}$
7.	$\frac{e^{-\alpha' x}}{s^2}$	$\left(t+\frac{x^2}{2\alpha}\right)\operatorname{erfc}\frac{x}{2\sqrt{(\alpha t)}} - x\left(\frac{t}{\pi \alpha}\right)^{\frac{1}{3}} e^{-\frac{x^2}{4\alpha t}}$
8.	$\frac{e^{-\alpha' x}}{s^{1+\frac{1}{2}n}}, n = 0, 1, 2, \ldots$	$(4t)^{\frac{1}{2}n} i^n \operatorname{erfc}\frac{x}{2\sqrt{(\alpha t)}}$
9.	$\frac{e^{-\alpha' x}}{\alpha' + h}$	$\left(\frac{\alpha}{\pi t}\right)^{\frac{1}{2}} e^{-\frac{x^2}{4\alpha t}} - h\alpha e^{hx+\alpha t h^2} \times \operatorname{erfc}\left\{\frac{x}{2\sqrt{(\alpha t)}} + h\sqrt{(\alpha t)}\right\}$
10.	$\frac{e^{-\alpha' x}}{\alpha'(\alpha' + h)}$	$\alpha e^{hx+\alpha t h^2} \operatorname{erfc}\left\{\frac{x}{2\sqrt{(\alpha t)}} + h\sqrt{(\alpha t)}\right\}$
11.	$\frac{e^{-\alpha' x}}{s(\alpha' + h)}$	$\frac{1}{h}\operatorname{erfc}\frac{x}{2\sqrt{(\alpha t)}} - \frac{1}{h} e^{hx+\alpha t h^2} \times \operatorname{erfc}\left\{\frac{x}{2\sqrt{(\alpha t)}} + h\sqrt{(\alpha t)}\right\}$

https://doi.org/10.1515/9783111351179-011

(continued)

	$\tilde{T}(s)$	$T(t)$
12.	$\dfrac{e^{-\alpha' x}}{s\alpha'(\alpha+h)}$	$\dfrac{2}{h}\left(\dfrac{\alpha t}{\pi}\right)^{\frac{1}{2}} e^{-\frac{x^2}{4\alpha t}} - \dfrac{(1+hx)}{h^2} \times \operatorname{erfc}\dfrac{x}{2\sqrt{\alpha t}} + \dfrac{1}{h^2} e^{hx+\alpha th^2} \times \operatorname{erfc}\left\{\dfrac{x}{2\sqrt{\alpha t}} + h\sqrt{\alpha t}\right\}$
13.	$\dfrac{e^{-\alpha' x}}{\alpha'^{n+1}(\alpha'+h)}$	$\dfrac{\alpha}{(-h)^n} e^{hx+\alpha th^2} \operatorname{erfc}\left\{\dfrac{x}{2\sqrt{\alpha t}} + h\sqrt{\alpha t}\right\} - \dfrac{\alpha}{(-h)^n} \sum_{r=0}^{n-1} \left[-2h\sqrt{\alpha t}\right]^r i^r \operatorname{erfc}\dfrac{x}{2\sqrt{\alpha t}}$
14.	$\dfrac{e^{-\alpha' x}}{(\alpha'+h)^2}$	$-2h\left(\dfrac{\alpha^3 t}{\pi}\right)^{\frac{1}{2}} e^{-\frac{x^2}{4\alpha t}} + \alpha(1+hx+2h^2\alpha t) e^{hx+\alpha th^2} \times \operatorname{erfc}\left\{\dfrac{x}{2\sqrt{\alpha t}} + h\sqrt{\alpha t}\right\}$
15.	$\dfrac{e^{-\alpha' x}}{s(\alpha'+h)^2}$	$\dfrac{1}{h^2}\operatorname{erfc}\dfrac{x}{2\sqrt{\alpha t}} - \dfrac{2}{h}\left(\dfrac{\alpha t}{\pi}\right)^{\frac{1}{2}} e^{-\frac{x^2}{4\alpha t}} - \dfrac{1}{h^2}(1-hx-2h^2\alpha t) e^{hx+\alpha th^2} \times \operatorname{erfc}\left\{\dfrac{x}{2\sqrt{(\alpha t)}} + h\sqrt{(\alpha t)}\right\}$
16.	$\dfrac{e^{-\alpha' x}}{s-\beta}$	$\dfrac{1}{2} e^{\beta t} \left\{ e^{-x\sqrt{\left(\frac{\beta}{\alpha}\right)}} \operatorname{erfc}\left[\dfrac{x}{2\sqrt{(\alpha t)}} - \sqrt{(\beta t)}\right] + e^{x\sqrt{\left(\frac{\beta}{\alpha}\right)}} \operatorname{erfc}\left[\dfrac{x}{2\sqrt{(\alpha t)}} + \sqrt{(\beta t)}\right] \right\}$
17.	$\dfrac{1}{s^{3/4}} e^{-\alpha' x}$	$\dfrac{1}{\pi}\left(\dfrac{x}{2t\alpha^{1/2}}\right)^{1/2} \left(e^{-\frac{x^2}{8\alpha t}}\right) K_{1/2}\left(\dfrac{x^2}{8\alpha t}\right)$
18.	$\dfrac{1}{s^{3/4}} K_{2\nu}(\alpha' x)$	$\dfrac{1}{2\sqrt{(\pi t)}} e^{-\frac{x^2}{8\alpha t}} K_\nu\left(\dfrac{x^2}{8\alpha t}\right)$
19.	$\left.\begin{matrix} I_\nu(\alpha' x') K_\nu(\alpha' x), x > x' \\ I_\nu(\alpha' x) K_\nu(\alpha' x'), x < x' \end{matrix}\right\}$	$\dfrac{1}{2t} e^{-\frac{x^2+x'^2}{4\alpha t}} I_\nu\left(\dfrac{xx'}{2\alpha t}\right), \nu \geq 0$
20.	$K_0(\alpha' x)$	$\dfrac{1}{2t} e^{-\frac{x^2}{4\alpha t}}$
21.	$\dfrac{1}{s} e^{\frac{x}{s}}$	$I_0\left[2\sqrt{(xt)}\right]$
22.	$\dfrac{\exp\left\{xs - x[(s+a)(s+b)]^{\frac{1}{2}}\right\}}{[(s+a)(s+b)]^{\frac{1}{2}}}$	$e^{-\frac{1}{2}(a+b)(t+x)} I_0\left\{\dfrac{1}{2}(a-b)[t(t+2x)]^{\frac{1}{2}}\right\}$
23.	$s^{\frac{1}{2}\nu-1} K_\nu(x\sqrt{s})$	$x^{-\nu} 2^{\nu-1} \int_{x^2/4t}^{\infty} e^{-u} u^{\nu-1} du$
24.	$\left[s - \sqrt{s^2 - x^2}\right]^\nu, \nu > 0$	$\dfrac{\nu x^\nu I_\nu(xt)}{t}$

(continued)

	$\tilde{T}(s)$	$T(t)$
25.	$\dfrac{\exp\left\{x\left[(s+a)^{\frac{1}{2}}-(s+b)^{\frac{1}{2}}\right]^2\right\}}{(s+a)^{\frac{1}{2}}(s+b)^{\frac{1}{2}}\left[(s+a)^{\frac{1}{2}}+(s+b)^{\frac{1}{2}}\right]^{2\nu}}$, $\nu>0$	$\dfrac{t^{\frac{1}{2}\nu}e^{-\frac{1}{2}(a+b)t}I_\nu\left[\frac{1}{2}(a-b)t^{\frac{1}{2}}(t+4x)^{\frac{1}{2}}\right]}{(a-b)^\nu(t+4x)^{\frac{1}{2}\nu}}$
26.	$\dfrac{e^{-\alpha' x}}{(s-\beta)^2}$	$\frac{1}{2}e^{\beta t}\left\{\left(t-\dfrac{x}{2\sqrt{\alpha\beta}}\right)e^{-x\sqrt{\frac{\beta}{\alpha}}}\operatorname{erfc}\left[\dfrac{x}{2\sqrt{\alpha t}}-\sqrt{\beta t}\right]+\left(t+\dfrac{x}{2\sqrt{\alpha\beta}}\right)e^{x\sqrt{\frac{\beta}{\alpha}}}\operatorname{erfc}\left[\dfrac{x}{2\sqrt{\alpha t}}+\sqrt{\beta t}\right]\right\}$
27.	$\dfrac{e^{-\alpha' x}}{\alpha'(s-\beta)}$	$\frac{1}{2}e^{\beta t}\left(\dfrac{\alpha}{\beta}\right)^{\frac{1}{2}}\left\{e^{-x\sqrt{\frac{\beta}{\alpha}}}\operatorname{erfc}\left[\dfrac{x}{2\sqrt{\alpha t}}-\sqrt{\beta t}\right]-e^{x\sqrt{\frac{\beta}{\alpha}}}\operatorname{erfc}\left[\dfrac{x}{2\sqrt{\alpha t}}+\sqrt{\beta t}\right]\right\}$
28.	$\dfrac{e^{-\alpha' x}}{(s-\beta)(\alpha'+h)}$, $\beta\neq\alpha h^2$	$\frac{1}{2}e^{\beta t}$ $\left\{\dfrac{\alpha^{\frac{1}{2}}}{h\alpha^{\frac{1}{2}}+\beta^{\frac{1}{2}}}e^{-x\sqrt{\frac{\beta}{\alpha}}}\operatorname{erfc}\left[\dfrac{x}{2\sqrt{\alpha t}}-\sqrt{\beta t}\right]+\dfrac{\alpha^{\frac{1}{2}}}{h\alpha^{\frac{1}{2}}-\beta^{\frac{1}{2}}}e^{x\sqrt{\frac{\beta}{\alpha}}}\operatorname{erfc}\left[\dfrac{x}{2\sqrt{\alpha t}}+\sqrt{\beta t}\right]\right\}$ $-\dfrac{h\alpha}{h^2\alpha-\beta}e^{hx+h^2\alpha t}\operatorname{erfc}\left[\dfrac{x}{2\sqrt{\alpha t}}+h\sqrt{\alpha t}\right]$
29.	$\dfrac{1}{s}\ln s$	$-\ln(Ct)$, $\ln C=\gamma=0.5772\ldots$ (Euler constant)
30.	$s^{\frac{1}{2}\nu}K_\nu(x\sqrt{s})$	$\dfrac{x^\nu}{(2t)^{\nu+1}}e^{-\frac{x^2}{4t}}$

(continued)

Appendix B
Table of general properties of Laplace transforms

$$f(s) = \int_0^\infty e^{-st} F(t)\,dt$$

	$\tilde{f}(s)$	$f(t)$
1.	$a\tilde{f}_1(s) + b\tilde{f}_2(s)$	$af_1(t) + bf_2(t)$
2.	$\tilde{f}\left(\frac{s}{a}\right)$	$af(at)$
3.	$\tilde{f}(s-a)$	$e^{at}f(t)$
4.	$e^{-as}\tilde{f}(s)$	$u(t-a) = \begin{cases} f(t-a) & t > a \\ 0 & t < a \end{cases}$
5.	$s\tilde{f}(s) - f(0)$	$f'(t)$
6.	$s^2\tilde{f}(s) - sf(0) - f'(0)$	$f''(t)$
7.	$s^n\tilde{f}(s) - s^{n-1}f(0) - s^{n-2}f'(0)$ $- \ldots - f^{(n-1)}(0)$	$f^{(n)}(t)$
8.	$\tilde{f}'(s)$	$-tf(t)$
9.	$\tilde{f}''(s)$	$t^2f(t)$
10.	$\tilde{f}^{(n)}(s)$	$(-1)^n t^n f(t)$
11.	$\frac{\tilde{f}(s)}{s}$	$\int_0^t f(u)du$
12.	$\frac{\tilde{f}(s)}{s^n}$	$\int_0^t \ldots \int_0^t f(u)du^n = \int_0^t \frac{(t-u)^{n-1}}{(n-1)!} f(u)du$
13.	$\tilde{f}(s)\tilde{g}(s)$	$\int_0^t f(u)g(t-u)du$
14.	$\int_s^\infty \tilde{f}(u)\,du$, where $\mathcal{L}\{f(t)\} = \tilde{f}(s)$	$\frac{f(t)}{t}$
15.	$\frac{1}{1-e^{-sT}} \int_0^T e^{-su} f(u)du$	$f(t) = f(t+T)$
16.	$\frac{\tilde{f}(\sqrt{s})}{s}$	$\frac{1}{\sqrt{\pi t}} \int_0^\infty e^{-u^2/4t} f(u)du$

https://doi.org/10.1515/9783111351179-012

(continued)

	$\tilde{f}(s)$	$f(t)$
17.	$\frac{1}{s}\tilde{f}\left(\frac{1}{s}\right)$	$\int_0^\infty J_0\left(2\sqrt{ut}\right)f(u)du$
18.	$\frac{1}{s^{n+1}}\tilde{f}\left(\frac{1}{s}\right)$	$t^{\frac{n}{2}}\int_0^\infty u^{-n/2}J_n\left(2\sqrt{ut}\right)f(u)du$
19.	$\frac{\tilde{f}(s+1/s)}{s^2+1}$	$\int_0^t J_0\left(2\sqrt{u(t-u)}\right)f(u)du$
20.	$\frac{1}{2\sqrt{\pi}}\int_0^\infty u^{-3/2}e^{-s^2/4u}\tilde{f}(u)du$, where $\mathcal{L}\{f(t)\}=\tilde{f}(s)$	$f(t^2)$
21.	$\frac{\tilde{f}(\ln s)}{s\ln s}$	$\int_0^\infty \frac{t^u f(u)}{\Gamma(u+1)}du$
22.	$P(s)$ = polymonial of degree less than n, $Q(s)=(s-\alpha_1)(s-\alpha_2)\ldots(s-\alpha_n)$, where $\alpha_1, \alpha_2, \ldots, \alpha_n$ are all distinct.	$\sum_{k=1}^{n}\frac{P(\alpha_k)}{Q'(\alpha_k)}e^{\alpha_k t}$

Appendix C
Table of common Laplace transform pairs

	$\tilde{f}(s)$		$f(t)$	
1.	$\frac{1}{s}$		1	
2.	$\frac{1}{s^2}$		t	
3.	$\frac{1}{s^n}$	$n = 1, 2, 3, \ldots$	$\frac{t^{n-1}}{(n-1)!}$,	$0! = 1$
4.	$\frac{1}{s^n}$	$n > 0$	$\frac{t^{n-1}}{\Gamma(n)}$	
5.	$\frac{1}{s-a}$		e^{at}	
6.	$\frac{1}{(s-a)^n}$	$n = 1, 2, 3, \ldots$	$\frac{t^{n-1}e^{at}}{(n-1)!}$,	$0! = 1$
7.	$\frac{1}{(s-a)^n}$	$n > 0$	$\frac{t^{n-1}e^{at}}{\Gamma(n)}$	
8.	$\frac{1}{s^2+a^2}$		$\frac{\sin at}{a}$	
9.	$\frac{s}{s^2+a^2}$		$\cos at$	
10.	$\frac{1}{(s-b)^2+a^2}$		$\frac{e^{bt}\sin at}{a}$	
11.	$\frac{s-b}{(s-b)^2+a^2}$		$e^{bt}\cos at$	
12.	$\frac{1}{s^2-a^2}$		$\frac{\sinh at}{a}$	
13.	$\frac{s}{s^2-a^2}$		$\cosh at$	
14.	$\frac{1}{(s-b)^2-a^2}$		$\frac{e^{bt}\sinh at}{a}$	
15.	$\frac{s-b}{(s-b)^2-a^2}$		$e^{bt}\cosh at$	
16.	$\frac{1}{(s-a)(s-b)}$	$a \neq b$	$\frac{e^{bt}-e^{at}}{b-a}$	
17.	$\frac{s}{(s-a)(s-b)}$	$a \neq b$	$\frac{be^{bt}-ae^{at}}{b-a}$	

https://doi.org/10.1515/9783111351179-013

(continued)

	$\tilde{f}(s)$	$f(t)$
18.	$\frac{1}{(s^2+a^2)^2}$	$\frac{\sin at - at\cos at}{2a^3}$
19.	$\frac{s}{(s^2+a^2)^2}$	$\frac{t\sin at}{2a}$
20.	$\frac{s^2}{(s^2+a^2)^2}$	$\frac{\sin at + at\cos at}{2a}$
21.	$\frac{s^3}{(s^2+a^2)^2}$	$\cos at - \frac{1}{2}at\sin at$
22.	$\frac{s^2-a^2}{(s^2+a^2)^2}$	$t\cos at$
23.	$\frac{1}{(s^2-a^2)^2}$	$\frac{at\cosh at - \sinh at}{2a^3}$
24.	$\frac{s}{(s^2-a^2)^2}$	$\frac{t\sinh at}{2a}$
25.	$\frac{s^2}{(s^2-a^2)^2}$	$\frac{\sinh at + at\cosh at}{2a}$
26.	$\frac{s^3}{(s^2-a^2)^2}$	$\cosh at + \frac{1}{2}at\sinh at$
27.	$\frac{s^2+a^2}{(s^2-a^2)^2}$	$t\cosh at$
28.	$\frac{1}{(s^2+a^2)^3}$	$\frac{(3-a^2t^2)\sin at - 3at\cos at}{8a^5}$
29.	$\frac{s}{(s^2+a^2)^3}$	$\frac{t\sin at - at^2\cos at}{8a^3}$
30.	$\frac{s^2}{(s^2+a^2)^3}$	$\frac{(1+a^2t^2)\sin at - at\cos at}{8a^3}$
31.	$\frac{s^3}{(s^2+a^2)^3}$	$\frac{3t\sin at + at^2\cos at}{8a}$
32.	$\frac{s^4}{(s^2+a^2)^3}$	$\frac{(3-a^2t^2)\sin at + 5at\cos at}{8a}$
33.	$\frac{s^5}{(s^2+a^2)^3}$	$\frac{(8-a^2t^2)\cos at - 7at\sin at}{8}$

(continued)

	$\tilde{f}(s)$	$f(t)$
34.	$\dfrac{3s^2-a^2}{(s^2+a^2)^3}$	$\dfrac{t^2 \sin at}{2a}$
35.	$\dfrac{s^3-3a^2s}{(s^2+a^2)^3}$	$\dfrac{1}{2}t^2 \cos at$
36.	$\dfrac{s^4-6a^2s^2+a^4}{(s^2+a^2)^4}$	$\dfrac{1}{6}t^3 \cos at$
37.	$\dfrac{s^3-a^2s}{(s^2+a^2)^4}$	$\dfrac{t^3 \sin at}{24a}$
38.	$\dfrac{1}{(s^2-a^2)^3}$	$\dfrac{(3+a^2t^2)\sinh at-3at\cosh at}{8a^5}$
39.	$\dfrac{s}{(s^2-a^2)^3}$	$\dfrac{at^2\cosh at-t\sinh at}{8a^3}$
40.	$\dfrac{s^2}{(s^2-a^2)^3}$	$\dfrac{at\cosh at+(a^2t^2-1)\sinh at}{8a^3}$
41.	$\dfrac{s^3}{(s^2-a^2)^3}$	$\dfrac{3t\sinh at+at^2\cosh at}{8a}$
42.	$\dfrac{s^4}{(s^2-a^2)^3}$	$\dfrac{(3+a^2t^2)\sinh at+5at\cosh at}{8a}$
43.	$\dfrac{s^5}{(s^2-a^2)^3}$	$\dfrac{(8+a^2t^2)\cosh at+7at\sinh at}{8}$
44.	$\dfrac{3s^2+a^2}{(s^2-a^2)^3}$	$\dfrac{t^2\sinh at}{2a}$
45.	$\dfrac{s^3+3a^2s}{(s^2-a^2)^3}$	$\dfrac{1}{2}t^2\cosh at$
46.	$\dfrac{s^4+6a^2s^2+a^4}{(s^2-a^2)^4}$	$\dfrac{1}{6}t^3\cosh at$
47.	$\dfrac{s^3+a^2s}{(s^2-a^2)^4}$	$\dfrac{t^3\sinh at}{24a}$
48.	$\dfrac{1}{s^3+a^3}$	$\dfrac{e^{at/2}}{3a^2}\left\{\sqrt{3}\sin\dfrac{\sqrt{3}at}{2}-\cos\dfrac{\sqrt{3}at}{2}+e^{-3at/2}\right\}$
49.	$\dfrac{s}{s^3+a^3}$	$\dfrac{e^{at/3}}{3a}\left\{\cos\dfrac{\sqrt{3}at}{2}+\sqrt{3}\sin\dfrac{\sqrt{3}at}{2}-e^{-3at/2}\right\}$

(continued)

	$\tilde{f}(s)$	$f(t)$
50.	$\dfrac{s^2}{s^3+a^3}$	$\dfrac{1}{3}\left(e^{-at}+2e^{at/2}\cos\dfrac{\sqrt{3}at}{2}\right)$
51.	$\dfrac{1}{s^3-a^3}$	$\dfrac{e^{-at/2}}{3a^2}\left\{e^{3at/2}-\cos\dfrac{\sqrt{3}at}{2}-\sqrt{3}\sin\dfrac{\sqrt{3}at}{2}\right\}$
52.	$\dfrac{s}{s^3-a^3}$	$\dfrac{e^{-at/2}}{3a}\left\{\sqrt{3}\sin\dfrac{\sqrt{3}at}{2}-\cos\dfrac{\sqrt{3}at}{2}+e^{3at/2}\right\}$
53.	$\dfrac{s^2}{s^3-a^3}$	$\dfrac{1}{3}\left(e^{at}+2e^{-at/2}\cos\dfrac{\sqrt{3}at}{2}\right)$
54.	$\dfrac{1}{s^4+4a^4}$	$\dfrac{1}{4a^3}(\sin at\cosh at-\cos at\sinh at)$
55.	$\dfrac{s}{s^4+4a^4}$	$\dfrac{\sin at\sinh at}{2a^2}$
56.	$\dfrac{s^2}{s^4+4a^4}$	$\dfrac{1}{2a}(\sin at\cosh at+\cos at\sinh at)$
57.	$\dfrac{s^3}{s^4+4a^4}$	$\cos at\cosh at$
58.	$\dfrac{1}{s^4-a^4}$	$\dfrac{1}{2a^3}(\sinh at-\sin at)$
59.	$\dfrac{s}{s^4-a^4}$	$\dfrac{1}{2a^2}(\cosh at-\cos at)$
60.	$\dfrac{s^2}{s^4-a^4}$	$\dfrac{1}{2a}(\sinh at+\sin at)$
61.	$\dfrac{s^3}{s^4-a^4}$	$\dfrac{1}{2}(\cosh at+\cos at)$
62.	$\dfrac{1}{\sqrt{s+a}+\sqrt{s+b}}$	$\dfrac{e^{-bt}-e^{-at}}{2(b-a)\sqrt{\pi t^3}}$
63.	$\dfrac{1}{s\sqrt{s+a}}$	$\dfrac{\operatorname{erf}\sqrt{at}}{\sqrt{a}}$
64.	$\dfrac{1}{\sqrt{s}(s-a)}$	$\dfrac{e^{at}\operatorname{erf}\sqrt{at}}{\sqrt{a}}$
65.	$\dfrac{1}{\sqrt{s-a}+b}$	$e^{at}\left\{\dfrac{1}{\sqrt{\pi t}}-be^{b^2t}\operatorname{erfc}\left(b\sqrt{t}\right)\right\}$
66.	$\dfrac{1}{\sqrt{s^2+a^2}}$	$J_0(at)$

(continued)

	$\tilde{f}(s)$	$f(t)$
67.	$\dfrac{1}{\sqrt{s^2-a^2}}$	$I_0(at)$
68.	$\dfrac{\left(\sqrt{s^2+a^2}-s\right)^n}{\sqrt{s^2+a^2}}$ $n>-1$	$a^n J_n(at)$
69.	$\dfrac{(s-\sqrt{(s^2-a^2)})^n}{\sqrt{s^2-a^2}}$ $n>-1$	$a^n I_n(at)$
70.	$\dfrac{e^{b\left(s-\sqrt{s^2+a^2}\right)}}{\sqrt{s^2+a^2}}$	$J_0\left(a\sqrt{t(t+2b)}\right)$
71.	$\dfrac{e^{-b\sqrt{s^2+a^2}}}{\sqrt{s^2+a^2}}$	$\begin{cases} J_0\left(a\sqrt{t^2-b^2}\right) & t>b \\ 0 & t>b \end{cases}$
72.	$\dfrac{1}{(s^2+a^2)^{3/2}}$	$\dfrac{tJ_1(at)}{a}$
73.	$\dfrac{s}{(s^2+a^2)^{3/2}}$	$tJ_0(at)$
74.	$\dfrac{s^2}{(s^2+a^2)^{3/2}}$	$J_0(at)-atJ_1(at)$
75.	$\dfrac{1}{(s^2-a^2)^{3/2}}$	$\dfrac{tI_1(at)}{a}$
76.	$\dfrac{s}{(s^2-a^2)^{3/2}}$	$tI_0(at)$
77.	$\dfrac{s^2}{(s^2-a^2)^{3/2}}$	$I_0(at)+atI_1(at)$
78.	$\dfrac{1}{s(e^s-1)}=\dfrac{e^{-s}}{s(1-e^{-s})}$	$F(t)=n,\ n\leqq t<n+1,\ n=0,\ 1,\ 2,\ \ldots$
79.	$\dfrac{1}{s(e^s-r)}=\dfrac{e^{-s}}{s(1-re^{-s})}$	$F(t)=\sum_{k=1}^{[t]} r^k$ where $[t]$ = greatest integer $\leqq t$
80.	$\dfrac{e^s-1}{s(e^s-r)}=\dfrac{1-e^{-s}}{s(1-re^{-s})}$	$F(t)=r^n,\ n\leqq t<n+1,\ n=0,\ 1,\ 2,\ \ldots$
81.	$\dfrac{e^{-a/s}}{\sqrt{s}}$	$\dfrac{\cos 2\sqrt{at}}{\sqrt{\pi t}}$
82.	$\dfrac{e^{-a/s}}{s^{3/2}}$	$\dfrac{\sin 2\sqrt{at}}{\sqrt{\pi a}}$

(continued)

	$\tilde{f}(s)$	$f(t)$
83.	$\frac{e^{-a/s}}{s^{n+1}}$ $n > -1$	$\left(\frac{t}{a}\right)^{n/2} J_n\left(2\sqrt{at}\right)$
84.	$\frac{e^{-a\sqrt{s}}}{\sqrt{s}}$	$\frac{e^{-a^2/4t}}{\sqrt{\pi t}}$
85.	$e^{-a\sqrt{s}}$	$\frac{a}{2\sqrt{\pi t^3}} e^{-a^2/4t}$
86.	$\frac{1-e^{-a\sqrt{s}}}{s}$	$\operatorname{erf}\left(\frac{a}{2\sqrt{t}}\right)$
87.	$\frac{e^{-a\sqrt{s}}}{s}$	$\operatorname{erfc}\left(\frac{a}{2\sqrt{t}}\right)$
88.	$\frac{e^{-a\sqrt{s}}}{\sqrt{s}\left(\sqrt{s}+b\right)}$	$e^{b(bt+a)} \operatorname{erfc}\left(b\sqrt{t}+\frac{a}{2\sqrt{t}}\right)$
89.	$\frac{e^{-a/\sqrt{s}}}{s^{n+1}}$ $n > -1$	$\frac{1}{\sqrt{\pi t a^{2n+1}}} \int_0^\infty u^n e^{-u^2/4a^2t} J_{2n}\left(2\sqrt{u}\right) du$
90.	$\ln\left(\frac{s+a}{s+b}\right)$	$\frac{e^{-bt}-e^{-at}}{t}$
91.	$\frac{\ln\left[\left(s^2+a^2\right)/a^2\right]}{2s}$	$\operatorname{Ci}(at)$
92.	$\frac{\ln[(s+a)/a]}{s}$	$\operatorname{Ei}(at)$
93.	$-\frac{(\gamma+\ln s)}{s}$ $\gamma =$ Euler's constant $= .5772\ldots$	$\ln t$
94.	$\ln\left(\frac{s^2+a^2}{s^2+b^2}\right)$	$\frac{2(\cos at - \cos bt)}{t}$
95.	$\frac{\pi^2}{6s} + \frac{(\gamma+\ln s)^2}{s}$ $\gamma =$ Euler's constant $= .5772\ldots$	$\ln^2 t$
96.	$\frac{\ln s}{s}$	$-(\ln t+\gamma)$ $\gamma =$ Euler's constant $= .5772\ldots$
97.	$\frac{\ln^2 s}{s}$	$(\ln t+\gamma)^2 - \frac{1}{6}\pi^2$ $\gamma =$ Euler's constant $= .5772\ldots$
98.	$\frac{\Gamma'(n+1)-\Gamma(n+1)\ln s}{s^{n+1}}$ $n > -1$	$t^n \ln t$

(continued)

	$\tilde{f}(s)$	$f(t)$
99.	$\tan^{-1}(a/s)$	$\frac{\sin at}{t}$
100.	$\frac{\tan^{-1}(a/s)}{s}$	$\mathrm{Si}(at)$
101.	$\frac{e^{a/s}}{\sqrt{s}}\,\mathrm{erfc}\left(\sqrt{\frac{a}{s}}\right)$	$\frac{e^{-2\sqrt{at}}}{\sqrt{\pi t}}$
102.	$e^{s^2/4a^2}\,\mathrm{erfc}\left(\frac{s}{2a}\right)$	$\frac{2a}{\sqrt{\pi}}e^{-a^2t^2}$
103.	$\frac{e^{s^2/4a^2}\,\mathrm{erfc}(s/2a)}{s}$	$\mathrm{erf}(at)$
104.	$\frac{e^{as}\,\mathrm{erfc}\sqrt{as}}{\sqrt{s}}$	$\frac{1}{\sqrt{\pi(t+a)}}$
105.	$e^{as}\,\mathrm{Ei}(as)$	$\frac{1}{t+a}$
106.	$\frac{1}{a}\left[\cos as\left\{\frac{\pi}{2}-\mathrm{Si}(as)\right\}-\sin as\,\mathrm{Ci}(as)\right]$	$\frac{1}{t^2+a^2}$
107.	$\sin as\left\{\frac{\pi}{2}-\mathrm{Si}(as)\right\}+\cos as\,\mathrm{Ci}(as)$	$\frac{t}{t^2+a^2}$
108.	$\frac{\cos as\left\{\frac{\pi}{2}-\mathrm{Si}(as)\right\}-\sin as\,\mathrm{Ci}(as)}{s}$	$\tan^{-1}\left(\frac{t}{a}\right)$
109.	$\frac{\sin as\left\{\frac{\pi}{2}-\mathrm{Si}(as)\right\}+\cos as\,\mathrm{Ci}(as)}{s}$	$\frac{1}{2}\ln\left(\frac{t^2+a^2}{a^2}\right)$
110.	$\left[\frac{\pi}{2}-\mathrm{Si}(as)\right]^2+\mathrm{Ci}^2(as)$	$\frac{1}{t}\ln\left(\frac{t^2+a^2}{a^2}\right)$
111.	0	$\mathcal{N}(t)$
112.	1	$\delta(t)$
113.	e^{-as}	$\delta(t-a)$
114.	$\frac{e^{-as}}{s}$	$u(t-a)$
115.	$\frac{\sinh sx}{s\sinh sa}$	$\frac{x}{a}+\frac{2}{\pi}\sum_{n=1}^{\infty}\frac{(-1)^n}{n}\sin\frac{n\pi x}{a}\cos\frac{n\pi t}{a}$
116.	$\frac{\sinh sx}{s\cosh sa}$	$\frac{4}{\pi}\sum_{n=1}^{\infty}\frac{(-1)^n}{2n-1}\sin\frac{(2n-1)\pi x}{2a}\sin\frac{(2n-1)\pi t}{2a}$
117.	$\frac{\cosh sx}{s\sinh sa}$	$\frac{t}{a}+\frac{2}{\pi}\sum_{n=1}^{\infty}\frac{(-1)^n}{n}\cos\frac{n\pi x}{a}\sin\frac{n\pi t}{a}$

(continued)

	$\tilde{f}(s)$	$f(t)$
118.	$\dfrac{\cosh sx}{s\cosh sa}$	$1+\dfrac{4}{\pi}\sum_{n=1}^{\infty}\dfrac{(-1)^n}{2n-1}\cos\dfrac{(2n-1)\pi x}{2a}\cos\dfrac{(2n-1)\pi t}{2a}$
119.	$\dfrac{\sinh sx}{s^2\sinh sa}$	$\dfrac{xt}{a}+\dfrac{2a}{\pi^2}\sum_{n=1}^{\infty}\dfrac{(-1)^n}{n^2}\sin\dfrac{n\pi x}{a}\sin\dfrac{n\pi t}{a}$
120.	$\dfrac{\sinh sx}{s^2\cosh sa}$	$x+\dfrac{8a}{\pi^2}\sum_{n=1}^{\infty}\dfrac{(-1)^n}{(2n-1)^2}\sin\dfrac{(2n-1)\pi x}{2a}\cos\dfrac{(2n-1)\pi t}{2a}$
121.	$\dfrac{\cosh sx}{s^2\sinh sa}$	$\dfrac{t^2}{2a}+\dfrac{2a}{\pi^2}\sum_{n=1}^{\infty}\dfrac{(-1)^n}{n^2}\cos\dfrac{n\pi x}{a}\left(1-\cos\dfrac{n\pi t}{a}\right)$
122.	$\dfrac{\cosh sx}{s^2\cosh sa}$	$t+\dfrac{8a}{\pi^2}\sum_{n=1}^{\infty}\dfrac{(-1)^n}{(2n-1)^2}\cos\dfrac{(2n-1)\pi x}{2a}\sin\dfrac{(2n-1)\pi t}{2a}$
123.	$\dfrac{\cosh sx}{s^3\cosh sa}$	$\dfrac{1}{2}(t^2+x^2-a^2)-\dfrac{16a^2}{\pi^3}\sum_{n=1}^{\infty}\dfrac{(-1)^n}{(2n-1)^3}\cos\dfrac{(2n-1)\pi x}{2a}\cos\dfrac{(2n-1)\pi t}{2a}$
124.	$\dfrac{\sinh x\sqrt{s}}{\sinh a\sqrt{s}}$	$\dfrac{2\pi}{a^2}\sum_{n=1}^{\infty}(-1)^n n e^{-n^2\pi^2 t/a^2}\sin\dfrac{n\pi x}{a}$
125.	$\dfrac{\cosh x\sqrt{s}}{\cosh a\sqrt{s}}$	$\dfrac{\pi}{a^2}\sum_{n=1}^{\infty}(-1)^{n-1}(2n-1)e^{-(2n-1)^2\pi^2 t/4a^2}\cos\dfrac{(2n-1)\pi x}{2a}$
126.	$\dfrac{\sinh x\sqrt{s}}{\sqrt{s}\cosh a\sqrt{s}}$	$\dfrac{2}{a}\sum_{n=1}^{\infty}(-1)^{n-1}e^{-(2n-1)^2\pi^2 t/4a^2}\sin\dfrac{(2n-1)\pi x}{2a}$
127.	$\dfrac{\cosh x\sqrt{s}}{\sqrt{s}\sinh a\sqrt{s}}$	$\dfrac{1}{a}+\dfrac{2}{a}\sum_{n=1}^{\infty}(-1)^n e^{-n^2\pi^2 t/a^2}\cos\dfrac{n\pi x}{a}$
128.	$\dfrac{\sinh x\sqrt{s}}{s\sinh a\sqrt{s}}$	$\dfrac{x}{a}+\dfrac{2}{\pi}\sum_{n=1}^{\infty}\dfrac{(-1)^n}{n}e^{-n^2\pi^2 t/a^2}\sin\dfrac{n\pi x}{a}$
129.	$\dfrac{\cosh x\sqrt{s}}{s\cosh a\sqrt{s}}$	$1+\dfrac{4}{\pi}\sum_{n=1}^{\infty}\dfrac{(-1)^n}{2n-1}e^{-(2n-1)^2\pi^2 t/4a^2}\cos\dfrac{(2n-1)\pi x}{2a}$
130.	$\dfrac{\sinh x\sqrt{s}}{s^2\sinh a\sqrt{s}}$	$\dfrac{xt}{a}+\dfrac{2a^2}{\pi^3}\sum_{n=1}^{\infty}\dfrac{(-1)^n}{n^3}(1-e^{-n^2\pi^2 t/a^2}\sin\dfrac{n\pi x}{a}$
131.	$\dfrac{\cosh x\sqrt{s}}{s^2\cosh a\sqrt{s}}$	$\dfrac{1}{2}(x^2-a^2)+t-\dfrac{16a^2}{\pi^3}\sum_{n=1}^{\infty}\dfrac{(-1)^n}{(2n-1)^3}e^{-(2n-1)^2\pi^2 t/4a^2}\cos\dfrac{(2n-1)\pi x}{2a}$
132.	$\dfrac{J_0(ix\sqrt{s})}{sJ_0(ia\sqrt{s})}$	$1-2\sum_{n=1}^{\infty}\dfrac{e^{-\lambda_n^2 t/a^2}J_0(\lambda_n x/a)}{\lambda_n J_1(\lambda_n)}$, where $\lambda_1, \lambda_2, \ldots$ are the positive roots of $J_0(\lambda)=0$
133.	$\dfrac{J_0(ix\sqrt{s})}{s^2J_0(ia\sqrt{s})}$	$\dfrac{1}{4}(x^2-a^2)+t+2a^2\sum_{n=1}^{\infty}\dfrac{e^{-\lambda_n^2 t/a^2}J_0(\lambda_n x/a)}{\lambda_n^3 J_1(\lambda_n)}$, where $\lambda_1, \lambda_2, \ldots$ are the positive roots of $J_0(\lambda)=0$

(continued)

	$\tilde{f}(s)$	$f(t)$
134.	$\frac{1}{as^2}\tanh\left(\frac{as}{2}\right)$	Triangular wave function 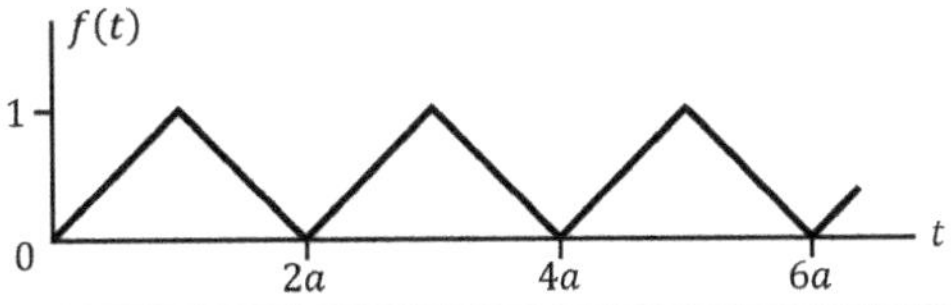
135.	$\frac{1}{s}\tanh\left(\frac{as}{2}\right)$	Square wave function 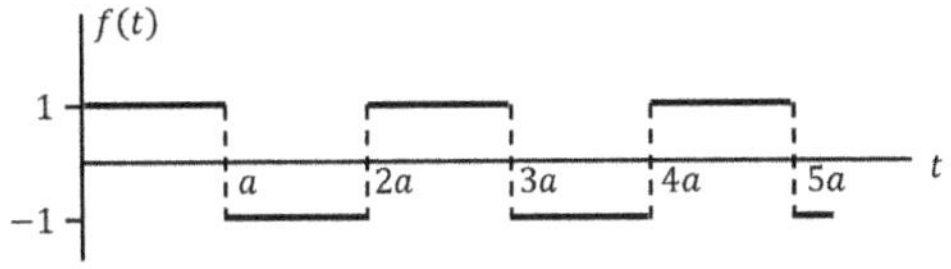
136.	$\frac{\pi a}{a^2s^2+\pi^2}\coth\left(\frac{as}{2}\right)$	Rectified sine wave function 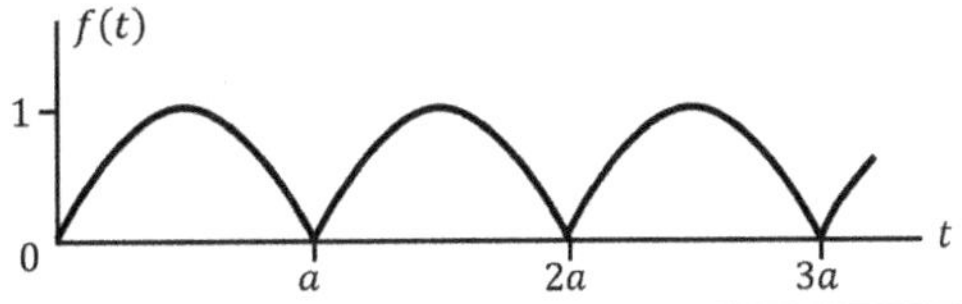
137.	$\frac{\pi a}{(a^2s^2+\pi^2)(1-e^{-as})}$	Half rectified sine wave function 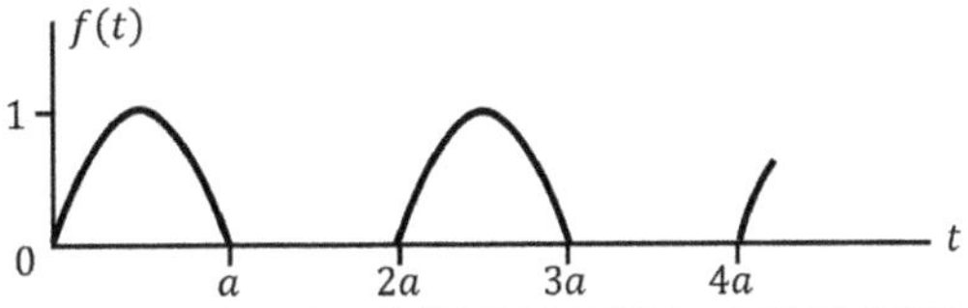
138.	$\frac{1}{as^2}-\frac{e^{-as}}{s(1-e^{-as})}$	Saw tooth wave function 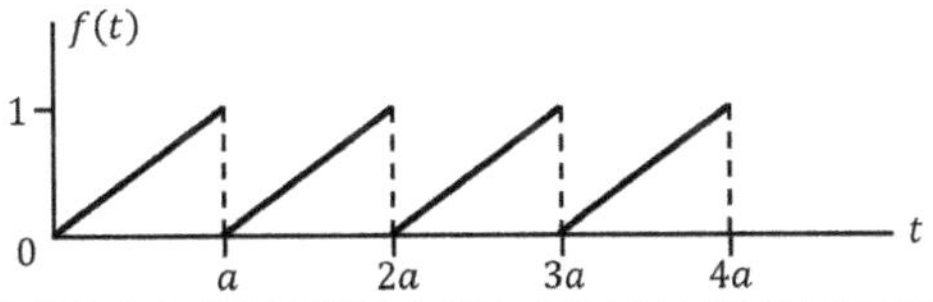
139.	$\frac{e^{-as}}{s}$	Heaviside's unit function $u(t-a)$ 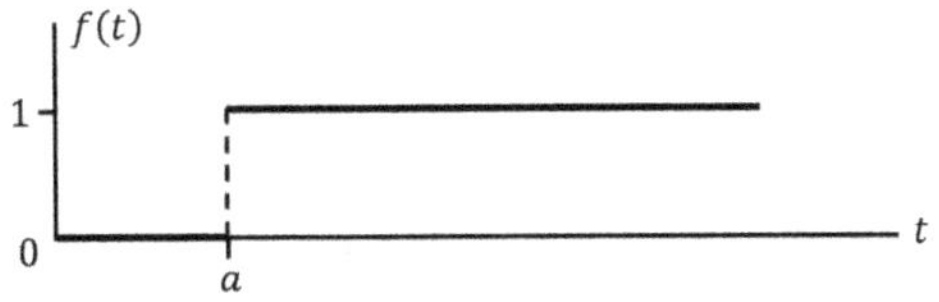

(continued)

	$\tilde{f}(s)$	$f(t)$
140.	$\dfrac{e^{-as}(1-e^{-\varepsilon s})}{s}$	Pulse function 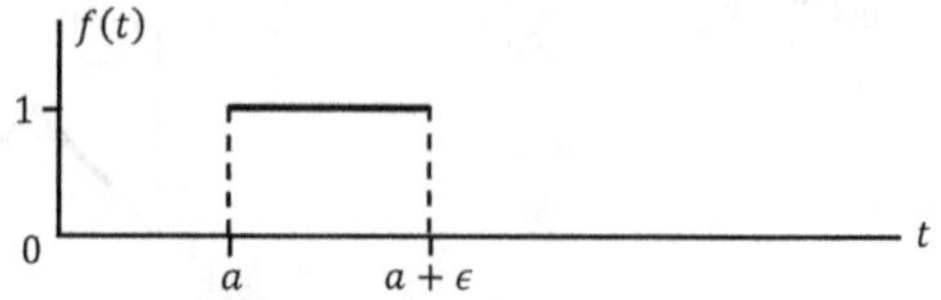
141.	$\dfrac{1}{s(1-e^{-as})}$	Step function 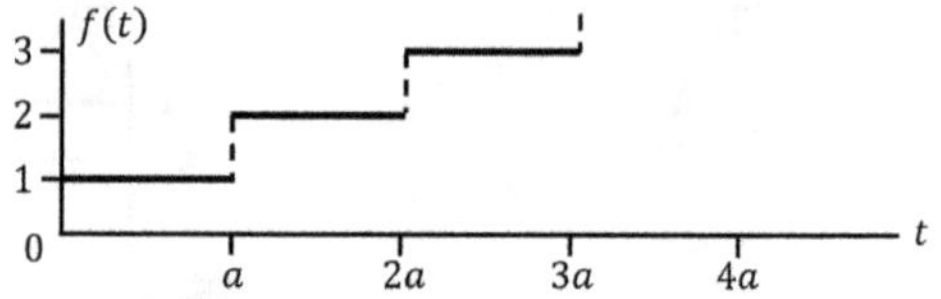
142.	$\dfrac{e^{-s}+e^{-2s}}{s(1-e^{-s})^2}$	$f(t)=n^2,\ n\leqq t<n+1,\ n=0,\ 1,\ 2,\ \ldots$ 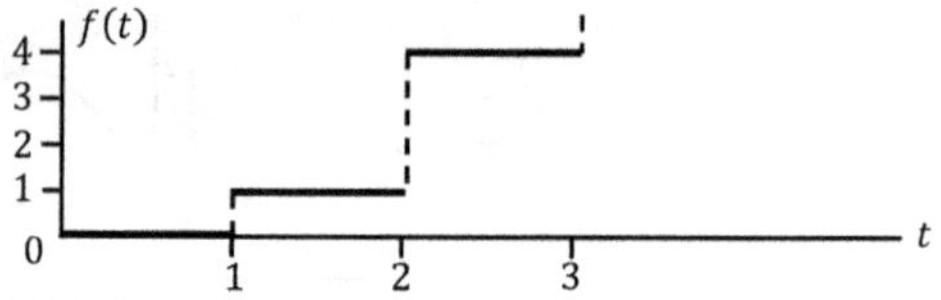
143.	$\dfrac{1-e^{-s}}{s(1-re^{-s})}$	$f(t)=r^n,\ n\leqq t<n+1,\ n=0,\ 1,\ 2,\ \ldots$ 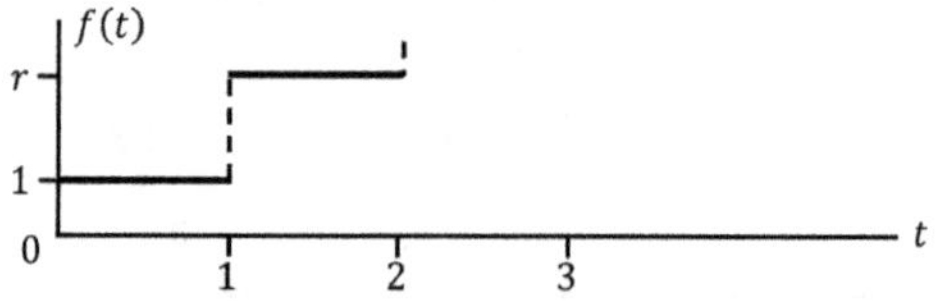
144.	$\dfrac{\pi a(1+e^{-as})}{a^2s^2+\pi^2}$	$f(t)=\begin{cases}\sin\left(\frac{\pi t}{a}\right) & 0\leqq t\leqq a\\ 0 & t>a\end{cases}$ 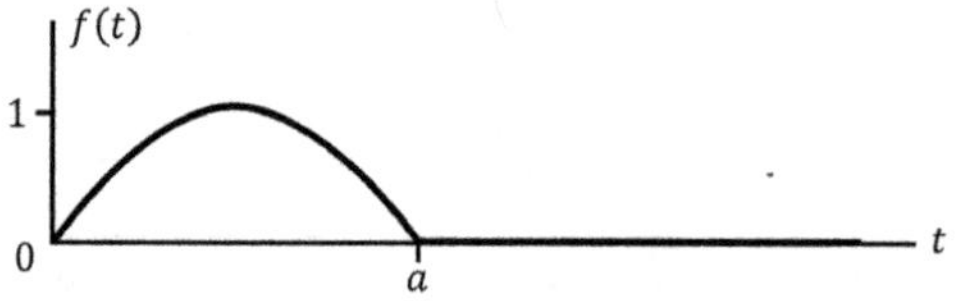

Appendix D
Table of special functions

1. Gamma function

$$\Gamma(n) = \int_0^\infty u^{n-1} e^{-u} du, \quad n > 0$$

2. Beta function

$$B(m,n) = \int_0^1 u^{m-1}(1-u)^{n-1} du = \frac{\Gamma(m)\Gamma(n)}{\Gamma(m+n)}, \quad m,n > 0$$

3. Bessel function

$$J_n(s) = \frac{x^n}{2^n \Gamma(n+1)} \left\{ 1 - \frac{x^2}{2(2n+2)} + \frac{x^4}{2 \cdot 4(2n+2)(2n+4)} - \dots \right\}$$

4. Modified Bessel function

$$I_n(x) = i^{-n} J_n(ix) = \frac{x^n}{2^n \Gamma(n+1)} \left\{ 1 + \frac{x^2}{2(2n+2)} + \frac{x^4}{2 \cdot 4(2n+2)(2n+4)} + \dots \right\}$$

5. Error function

$$\operatorname{erf}(t) = \frac{2}{\sqrt{\pi}} \int_0^t e^{-u^2} du$$

6. Complementary error function

$$\operatorname{erfc}(t) = 1 - \operatorname{erf}(t) = \frac{2}{\sqrt{\pi}} \int_t^\infty e^{-u^2} du$$

7. Exponential integral

$$\operatorname{Ei}(t) = \int_t^\infty \frac{e^{-u}}{u} du$$

8. Sine integral

$$\operatorname{Si}(t) = \int_0^t \frac{\sin u}{u} du$$

9. Cosine integral

$$\operatorname{Ci}(t) = \int_t^\infty \frac{\cos u}{u} du$$

10. Fresnel sine integral

$$S(t) = \int_0^t \sin u^2 du$$

https://doi.org/10.1515/9783111351179-014

(continued)

11. Fresnel cosine integral $C(t) = \int_0^t \cos u^2 du$
12. Laguerre polynomials $L_n(t) = \frac{e^t}{n!} \frac{d^n}{dt^n}\left(t^n e^{-t}\right), \; n = 0, \; 1, \; 2, \; \ldots$

Appendix E
Bessel Functions

E1 Remarks on Bessel Functions

The Bessel equation of order v:

$$\frac{d^2y}{dz^2} + \frac{1}{z}\frac{dy}{dz} + \left(1 - \frac{v^2}{z^2}\right)y = 0 \tag{E1}$$

is satisfied by the Bessel function $J_v(z)$ which can be defined as

$$J_v(z) = \sum_{r=0}^{\infty} \frac{(-1)^r \left(\frac{1}{2}z\right)^{v+2r}}{r!\Gamma(v+r+1)},$$

where v is real and z may be complex, with the principal value of the argument assumed. If v is not an integer, $J_v(z)$ and $J_{-v}(z)$ are independent solutions of eq. (E1), but if v is an integer, n, then:

$$J_n(z) = (-1)^n J_{-n}(z).$$

To obtain a second linearly independent solution of eq. (E1) that is available for all values of v, we define:

$$Y_v(z) = \frac{J_v(z)\cos v\pi - J_{-v}(z)}{\sin v\pi},$$

where the definition for $Y_n(z)$, n integer, are given by $\lim_{n\to v} Y_v(z)$. Also,

$$\frac{1}{2}\pi Y_0(z) = \left\{\ln\left(\frac{1}{2}z\right) + \gamma\right\}J_0(z) + \left(\frac{1}{2}z\right)^2 - \left(1+\frac{1}{2}\right)\frac{\left(\frac{1}{2}z\right)^4}{(2!)^2} + \left(1+\frac{1}{2}+\frac{1}{3}\right)\frac{\left(\frac{1}{2}z\right)^6}{(3!)^2} - \ldots,$$

where $\gamma = 0.5772\ldots$ is the Euler's constant. When n is any positive integer, we have:

$$\pi Y_n(z) = 2\left\{\ln\left(\frac{1}{2}z\right) + \gamma\right\}J_n(z) - \sum_{r=0}^{\infty}(-1)^r \frac{\left(\frac{1}{2}z\right)^{n+2r}}{r!(n+r)!}\left[\sum_{m=1}^{n+r} m^{-1} + \sum_{m=1}^{r} m^{-1}\right]$$
$$- \sum_{r=0}^{n-1}\left(\frac{1}{2}z\right)^{-n+2r}\frac{(n-r-1)!}{r!}.$$

If $r = 0$ we must replace $\left(\sum_{m=1}^{n+r} m^{-1} + \sum_{m=1}^{r} m^{-1}\right)$ with $\sum_{m=1}^{n} m^{-1}$.

https://doi.org/10.1515/9783111351179-015

Asymptotic expansions: For large values of z the functions $J_\nu(z)$ and $Y_\nu(z)$ can be approximated as follows:

$$J_\nu(z) = w_1(z)\left(\frac{2}{\pi z}\right)^{1/2}\cos\left(z-\frac{1}{2}\nu\pi-\frac{1}{4}\pi\right) - w_2(z)\left(\frac{2}{\pi z}\right)^{1/2}\sin\left(z-\frac{1}{2}\nu\pi-\frac{1}{4}\pi\right),$$

$$Y_\nu(z) = w_2(z)\left(\frac{2}{\pi z}\right)^{1/2}\cos\left(z-\frac{1}{2}\nu\pi-\frac{1}{4}\pi\right) + w_1(z)\left(\frac{2}{\pi z}\right)^{1/2}\sin\left(z-\frac{1}{2}\nu\pi-\frac{1}{4}\pi\right),$$

$$w_1(z) \sim \sum_{n=0}^{\infty}(-1)^n c_{2n} z^{-2n}, \quad z\to\infty;\ |\arg z|<\pi,$$

$$w_2(z) \sim \sum_{n=0}^{\infty}(-1)^n c_{2n+1} z^{-2n-1}, \quad z\to\infty;\ |\arg z|<\pi,$$

$$c_n = \frac{(4\nu^2-1^2)(4\nu^2-3^2)\cdots\left(4\nu^2-(2n-1)^2\right)}{8^n n!}, \quad c_0 = 1.$$

The modified Bessel equation

$$\frac{d^2y}{dz^2}+\frac{1}{z}\frac{dy}{dz}-\left(1+\frac{\nu^2}{z^2}\right)y=0 \tag{E2}$$

is satisfied by:

$$I_\nu(z) = \sum_{r=0}^{\infty}\frac{\left(\frac{1}{2}z\right)^{\nu+2r}}{r!\,\Gamma(\nu+r+1)}.$$

If ν is not an integer, $I_{-\nu}(z)$ is an independent solution of eq. (E2). However, for a second solution valid for all values of ν we can use:

$$K_\nu(z) = \frac{1}{2}\pi\frac{I_{-\nu}(z)-I_\nu(z)}{\sin\nu\pi},$$

where the relations for $K_n(z)$, n integer, are given by $\lim_{\nu\to n}K_\nu(z)$. The zeroth-order function can be written as

$$K_0(z) = -\left\{\ln\left(\frac{1}{2}z\right)+\gamma\right\}I_0(z)+\left(\frac{1}{2}z\right)^2+\left(1+\frac{1}{2}\right)\frac{\left(\frac{1}{2}z\right)^4}{(2!)^2}+\left(1+\frac{1}{2}+\frac{1}{3}\right)\frac{\left(\frac{1}{2}z\right)^6}{(3!)^2}+\ldots.$$

When n is any positive integer, we can write:

$$K_n(z) = (-1)^{n+1}\left\{\ln\left(\frac{1}{2}z\right) + \gamma\right\} I_n(z) + \frac{1}{2}(-1)^n \sum_{r=0}^{\infty} \frac{\left(\frac{1}{2}z\right)^{n+2r}}{r!(n+r)!}\left[\sum_{m=1}^{n+r} m^{-1} + \sum_{m=1}^{r} m^{-1}\right]$$

$$+ \frac{1}{2}\sum_{r=0}^{n-1}(-1)^r\left(\frac{1}{2}z\right)^{-n+2r}\frac{(n-r-1)!}{r!}.$$

If, for $r=0$, $\left(\sum_{m=1}^{n+r} m^{-1} + \sum_{m=1}^{r} m^{-1}\right)$ is replaced with $\sum_{m=1}^{n} m^{-1}$.

Asymptotic expansions: For large values of z the functions $K_\nu(z)$ and $I_\nu(z)$ can be approximated as follows:

$$K_\nu(z) = \left(\frac{\pi}{2z}\right)^{\frac{1}{2}} e^{-z}\left\{1 + \frac{4\nu^2 - 1^2}{1!8z} + \frac{(4\nu^2 - 1^2)(4\nu^2 - 3^2)}{2!(8z)^2} + O\left(\frac{1}{z^3}\right)\right\},$$

$$I_\nu(z) = \frac{e^z}{\sqrt{2\pi z}}\left\{1 - \frac{4\nu^2 - 1^2}{1!8z} + \frac{(4\nu^2 - 1^2)(4\nu^2 - 3^2)}{2!(8z)^2} + O\left(\frac{1}{z^3}\right)\right\}$$

$$+ \frac{e^{-z \pm \left(\nu + \frac{1}{2}\right)\pi i}}{\sqrt{2\pi z}}\left\{1 + O\left(\frac{1}{z}\right)\right\}.$$

Above, the positive sign is used if $-\frac{1}{2}\pi < \arg z < \frac{3}{2}\pi$, and the negative sign if $-\frac{3}{2}\pi < \arg z < \frac{1}{2}\pi$.

E2 Additional Bessel function related results

Bessel functions

1. Functional relations:

$$J_\nu(z) = e^{i\nu\pi/2} I_\nu\left(ze^{-i\pi/2}\right), \quad -\frac{\pi}{2} < \arg z \le \pi,$$

$$Y_\nu(z) = ie^{i\nu\pi/2} I_\nu\left(ze^{-i\pi/2}\right) - \frac{2}{\pi} e^{-i\nu\pi/2} K_\nu\left(ze^{-i\pi/2}\right), \quad -\frac{\pi}{2} < \arg z \le \pi.$$

2. Integral representations:

$$J_\nu(z) = \frac{1}{\pi}\int_0^\pi \cos(z\sin t - \nu t)dt - \frac{\sin(\nu\pi)}{\pi}\int_0^\infty e^{-z\sinh t - \nu t}dt, \quad |\arg z| < \frac{1}{2}\pi,$$

$$Y_\nu(z) = \frac{1}{\pi}\int_0^\pi \sin(z\sin t - \nu t)dt - \frac{1}{\pi}\int_0^\infty \left[e^{\nu t} + e^{-\nu t}\cos(\nu\pi)\right]e^{-z\sinh t}dt, \quad |\arg z| < \frac{1}{2}\pi.$$

3. Difference equations [$y_\nu(x)$ is either $J_\nu(x)$ or $Y_\nu(x)$]:

$$y_{\nu-1}(x) + y_{\nu+1}(x) = \frac{2\nu}{x} y_\nu(x),$$

$$2y'_\nu(x) = y_{\nu-1}(x) - y_{\nu+1}(x),$$

$$J'_0(x) = -J_1(x),$$

$$Y'_0(x) = -Y_1(x).$$

4. Generating function:

$$e^{z(t-1/t)/2} \sum_{k=-\infty}^{\infty} t^k J_k(z).$$

Modified Bessel functions

1. Functional relations:

$$I_\nu(z) = \pm \frac{i}{\pi} K_\nu\left(ze^{\pm i\pi}\right) \mp \frac{ie^{\mp i\nu\pi}}{\pi} K_\nu(z),$$

$$K_\nu(z) = 2\cos(\pi\nu) K_\nu\left(ze^{\pm i\pi}\right) - K_\nu\left(ze^{\pm 2i\pi}\right).$$

2. Integral representations:

$$I_\nu(z) = \frac{1}{\pi} \int_0^\pi e^{z\cos t} \cos(\nu t)\, dt - \frac{\sin(\nu\pi)}{\pi} \int_0^\infty e^{-z\cosh t - \nu t}\, dt, \quad |\arg z| < \frac{1}{2}\pi,$$

$$K_\nu(z) = \int_0^\infty e^{-z\cosh t} \cosh(\nu t)\, dt, \quad |\arg z| < \frac{1}{2}\pi.$$

3. Difference equations [$y_\nu(x)$ is either $I_\nu(x)$ or $K_\nu(x)$]:

$$y_{\nu-1}(x) - y_{\nu+1}(x) = \frac{2\nu}{x} y_\nu(x),$$

$$2y'_\nu(x) = y_{\nu-1}(x) + y_{\nu+1}(x),$$

$$I'_0(x) = I_1(x),$$

$$K'_0(x) = -K_1(x).$$

4. Generating function:

$$e^{z\cos t} = I_0(z) + 2\sum_{k=1}^{\infty} I_k(z)\cos(kt).$$

Equations Related to the Bessel Equation

The following differential equations have the given respective complete solutions:

a) $$y'' + a^2x^{k-2}y = 0,$$

$$y = \sqrt{x}\left[AJ_{1/k}\left(2ax^{k/2}/k\right) + BY_{1/k}\left(2ax^{k/2}/k\right)\right].$$

b) $$\frac{d^{2n}y}{dx^{2n}} = \left(-a^2\right)^n x^{-n}y,$$

$$y = x^{n/2}\left[AJ_n\left(2a\omega x^{1/2}\right) + BY_n\left(2a\omega x^{1/2}\right)\right],$$

where $\omega^n = 1$.

c) ***Theorem E.1*** If $\left(1-g^2\right) \geq 4c$ and $d \neq 0$, $p \neq 0$, $q \neq 0$, then, except in the special case where it reduces to Euler's equation, the equation:

$$x^2y'' + x(g + 2bx^p)y' + \left[c + dx^{2q} + b(g + p - 1)x^p + b^2x^{2p}\right]y = 0$$

has a complete solution that can be written as

$$y = x^{\eta}e^{-\beta x^p}\left[AJ_\nu(\lambda x^q) + BY_\nu(\lambda x^q)\right],$$

where A and B are constants, and

$$\eta = \frac{1-g}{2}, \quad \beta = \frac{b}{p}, \quad \lambda = \frac{\sqrt{|d|}}{q}, \quad \nu = \frac{\sqrt{(1-g)^2 - 4c}}{2q}.$$

If $d < 0$, J_ν and Y_ν should be replaced by I_ν and K_ν, respectively. If ν is not an integer, then Y_ν and K_ν could optionally be replaced by $J_{-\nu}$ and $K_{-\nu}$, respectively. To prove this theorem, transform the independent and dependent variables x and y in the given equation to a new independent variable x^* and a dependent variable y^* as follows:

$$x = \left(\frac{qx^*}{\sqrt{|d|}}\right)^{1/q}, \quad y = x^{(1-g)/2}e^{-\left(\frac{b}{p}\right)x^p}y^*.$$

The resulting Bessel's equation will appear in the variables x^* and y^*. A special case stated in the corollary below may be found useful.

Corollary E.1 If $(1-\xi)^2 \geq 4b$, then except in the special case when $g=0$, $\xi=2$, $s=0$, $b=0$; when it reduces to Euler's equation, the ODE

$$\left(x^{\xi}y'\right)' + \left(gx^{s} + bx^{\xi-2}\right)y = 0$$

has the complete solution:

$$y = x^{\eta}\left[AJ_{v}(\lambda x^{\gamma}) + BY_{v}(\lambda x^{\gamma})\right],$$

where

$$\eta = \frac{1-\xi}{2}, \quad \gamma = \frac{2-\xi+s}{2}, \quad \lambda = \frac{2\sqrt{|g|}}{2-\xi+s},$$

$$v = \frac{\sqrt{(1-\xi)^2 - 4b}}{(2-\xi+s)}.$$

If $g<0$, J_v and Y_v should be replaced by I_v and K_v, respectively. If v is not an integer, Y_v and K_v could optionally be replaced by J_{-v} and I_{-v}.

Appendix F
Special functions

Airy functions

1. The differential equation:

$$y'' = xy.$$

has solutions that are linear combinations of $\mathrm{Ai}\,(x)$ and $\mathrm{Bi}\,(x)$.

2. Taylor series:

$$\mathrm{Ai}\,(x) = 3^{-2/3} \sum_{n=0}^{\infty} \frac{x^{3n}}{9^n n! \Gamma\left(n + \frac{2}{3}\right)} - 3^{-4/3} \sum_{n=0}^{\infty} \frac{x^{3n+1}}{9^n n! \Gamma\left(n + \frac{4}{3}\right)},$$

$$\mathrm{Bi}\,(x) = 3^{-1/6} \sum_{n=0}^{\infty} \frac{x^{3n}}{9^n n! \Gamma\left(n + \frac{2}{3}\right)} + 3^{-5/6} \sum_{n=0}^{\infty} \frac{x^{3n+1}}{9^n n! \Gamma\left(n + \frac{4}{3}\right)},$$

$$\mathrm{Ai}\,(0) = \mathrm{Bi}\,(0)/\sqrt{3} = 3^{-2/3}/\Gamma\,(2/3) \doteqdot 0.335\ 028,$$

$$\mathrm{Ai}'(0) = -\mathrm{Bi}'(0)/\sqrt{3} = -3^{-1/3}/\Gamma(1/3) \doteqdot -0.258819.$$

3. Functional relations:

$$\mathrm{Ai}\,(z) + \omega\,\mathrm{Ai}\,(\omega z) + \omega^2\,\mathrm{Ai}\,(\omega^2 z) = 0,$$

$$\mathrm{Bi}\,(z) = i\omega\,\mathrm{Ai}\,(\omega z) - i\omega^2\,\mathrm{Ai}\,(\omega^2 z),$$

where $\omega = e^{-2i\pi/3}$.

4. Relation to Bessel functions:

$$\mathrm{Ai}\,(z) = \pi^{-1}\sqrt{z/3}K_{1/3}\left(2z^{3/2}/3\right),$$

$$\mathrm{Bi}\,(z) = \sqrt{z/3}\left[I_{-1/3}\left(2z^{3/2}/3\right) + I_{1/3}\left(2z^{3/2}/3\right)\right].$$

5. Asymptotic expansions:

$$\mathrm{Ai}\,(z) \sim \frac{1}{2}\pi^{-1/2} z^{-1/4} e^{-2z^{3/2}/3} \sum_{n=0}^{\infty} (-1)^n c_n z^{-3n/2},\ z \to \infty;\ |\arg z| < \pi,$$

$$\mathrm{Bi}\,(z) \sim \pi^{-1/3} z^{-1/4} e^{2z^{3/2}/3} \sum_{n=0}^{\infty} c_n z^{-3n/2},\ z \to \infty;\ |\arg z| < \frac{1}{3}\pi,$$

https://doi.org/10.1515/9783111351179-016

$$\mathrm{Ai}\,(z) = w_1(z)\sin\left[\frac{2}{3}(-z)^{3/2} + \frac{\pi}{4}\right] - w_2(z)\cos\left[\frac{2}{3}(-z)^{3/2} + \frac{\pi}{4}\right],$$

$$\mathrm{Bi}\,(z) = w_2(z)\sin\left[\frac{2}{3}(-z)^{3/2} + \frac{\pi}{4}\right] + w_1(z)\cos\left[\frac{2}{3}(-z)^{3/2} + \frac{\pi}{4}\right],$$

$$w_1(z) \sim \pi^{-1/2}(-z)^{-1/4}\sum_{n=0}^{\infty} c_{2n}z^{-3n}, \quad z \to \infty; \quad \frac{\pi}{3} < \arg z < \frac{5\pi}{3},$$

$$w_2(z) \sim \pi^{-1/2}(-z)^{-7/4}\sum_{n=0}^{\infty} c_{2n+1}z^{-3n}, \quad z \to \infty; \quad \frac{\pi}{3} < \arg z < \frac{5\pi}{3},$$

$$c_n = \frac{(2n+1)(2n+3)\cdots(6n-1)}{144^n n!}$$

$$= \frac{1}{2\pi}\left(\frac{3}{4}\right)^n \frac{\Gamma\left(n+\frac{5}{6}\right)\Gamma\left(n+\frac{1}{6}\right)}{n!}, \qquad c_0 = 1.$$

6. Integral representations:

$$\mathrm{Ai}(x) = \frac{1}{\pi}\int_0^{\infty} \cos\left(\frac{1}{3}t^3 + xt\right)dt,$$

$$\mathrm{Bi}(x) = \frac{1}{\pi}\int_0^{\infty} \left[e^{-t^3/3+xt} + \sin\left(\frac{1}{3}t^3 + xt\right)\right]dt.$$

Parabolic cylinder functions

1. The solutions of the differential equation:

$$y'' + \left(v + \frac{1}{2} - \frac{1}{4}x^2\right)y = 0.$$

are written in terms of the parabolic cylinder funcitions $D_v(\pm x)$ and $D_{-v-1}(\pm ix)$. Only two of these functions are linearly independent.
2. Taylor series:

$$D_v(x) = \frac{\pi^{1/2}2^{v/2}}{\Gamma\left(\frac{1}{2}-\frac{1}{2}v\right)}\sum_{n=0}^{\infty}\frac{a_{2n}x^{2n}}{(2n)!} - \frac{\pi^{1/2}2^{(v+1)/2}}{\Gamma\left(-\frac{1}{2}v\right)}\sum_{n=0}^{\infty}\frac{a_{2n+1}x^{2n+1}}{(2n+1)!},$$

where $a_0 = a_1 = 1$ and $a_{n+2} = -\left(v + \frac{1}{2}\right)a_n + \frac{1}{4}n(n-1)a_{n-2}$.

$$D_\nu(0)=\pi^{1/2}2^{\nu/2}/\Gamma\left(\frac{1}{2}-\frac{1}{2}\nu\right).$$

$$D_\nu{}'(0)=-\pi^{1/2}2^{(\nu+1)/2}/\Gamma\left(-\frac{1}{2}\nu\right).$$

3. Functional relation:

$$D_\nu(z)=e^{i\nu\pi}D_\nu(-z)+\frac{(2\pi)^{1/2}}{\Gamma(-\nu)}e^{i(\nu+1)\pi/2}D_{-\nu-1}(-iz).$$

4. Asymptotic expansions:

$$D_\nu(z)\sim z^\nu e^{-z^2/4}\sum_{n=0}^{\infty}(-1)^n c_n z^{-2n},\quad z\to\infty;\ |\arg z|<\frac{3}{4}\pi,$$

$$D_\nu(z)\sim z^\nu e^{-z^2/4}\sum_{n=0}^{\infty}(-1)^n c_n z^{-2n}-\frac{(2\pi)^{1/2}}{\Gamma(-\nu)}e^{i\pi\nu}z^{-\nu-1}e^{z^2/4}\sum_{n=0}^{\infty}d_n z^{-2n},$$

$$z\to\infty;\ \frac{1}{4}\pi<\arg z<\frac{5}{4}\pi,$$

$$c_n=\frac{\nu(\nu-1)\cdots(\nu-2n+1)}{2^n n!},\quad c_0=1,$$

$$d_n=\frac{(\nu+1)(\nu+2)\cdots(\nu+2n)}{2^n n!},\quad d_0=1.$$

5. Integral representation:

$$D_\nu(x)=\sqrt{\frac{2}{\pi}}e^{x^2/4}\int_0^\infty e^{-t^2/2}t^\nu\cos\left(xt-\frac{\nu\pi}{2}\right)dt,\ \operatorname{Re}\nu>-1.$$

6. Difference equations:

$$xD_\nu(x)=D_{\nu+1}(x)+\left(\nu+\frac{1}{2}\right)D_{\nu-1}(x),$$

$$D_\nu'(x)=-\frac{1}{2}xD_\nu(x)+\left(\nu+\frac{1}{2}\right)D_{\nu-1}(x).$$

7. Relation to Hermite polynomials:

$$D_n(x)=\mathrm{He}_n(x)e^{-x^2/4}.$$

Gamma and Digamma (PSI) functions

1. Integral representation:

$$\Gamma(z) = \int_0^\infty t^{z-1} e^{-t} dt, \quad \text{Re } z > 0.$$

2. Difference equation:

$$\Gamma(x+1) = x\Gamma(x).$$

3. Special values:

$$\Gamma(0) = 1, \quad \Gamma\left(\frac{1}{2}\right) = \sqrt{\pi}, \quad \Gamma(n+1) = n!.$$

4. Stirling's asymptotic formula:

$$\Gamma(z) \sim (z/e)^z \sqrt{2\pi/z}\left[1 + \frac{1}{12z} + \frac{1}{288z^2} - \frac{139}{51840z^3} - \dots\right], \quad z \to \infty; \ |\arg z| < \pi.$$

5. Other formulas:

$$\Gamma(z)\Gamma(1-z) = \pi/\sin(\pi z),$$

$$\Gamma(2z) = \frac{1}{2}\pi^{-1/2}4^z\Gamma(z)\Gamma\left(z + \frac{1}{2}\right),$$

$$\int_0^1 t^{x-1}(1-t)^{y-1} dt = \Gamma(x)\Gamma(y)/\Gamma(x+y), \quad \text{Re } x > 0, \text{Re } y > 0.$$

6. Psi function:

$$\psi(z) = \Gamma'(z)/\Gamma(z).$$

7. Difference equation:

$$\psi(z+1) = \psi(z) + \frac{1}{z}.$$

8. Special values:

$$\psi(1) = -\gamma, \quad \psi(n+1) = -\gamma + \sum_{k=1}^{n} 1/k,$$

where $\gamma \doteq 0.5772$ is Euler's constant.

9. Taylor series:

$$\psi(1+z) = -\gamma - \sum_{n=2}^{\infty} \zeta(n)(-z)^{n-1},$$

where $\zeta(n) = \sum_{k=1}^{\infty} k^{-n}$ is the Riemann zeta function.

10. Asymptotic expansion:

$$\psi(z) \sim \ln z - \frac{1}{2z} - \frac{1}{12z^2} + \frac{1}{120z^4} - \frac{1}{252z^6} + \ldots, \quad z \to \infty; \ |\arg z| < \pi.$$

Exponential integrals

1. Integral representation:

$$E_n(z) = \int_1^{\infty} \frac{e^{-zt}}{t^n} dt, \quad \operatorname{Re} z > 0.$$

2. Series expansion:

$$E_n(z) = \frac{(-z)^{n-1}}{(n-1)!}[-\ln z + \psi(n)] - \sum_{\substack{m=0 \\ m \neq n-1}}^{\infty} \frac{(-z)^m}{(m-n+1)m!}, \quad |\arg z| < \pi,$$

where $\psi(n)$ is the psi function.

3. Recurrence relations:

$$E_{n+1}(z) = \frac{1}{n}[e^{-z} - zE_n(z)], \quad \frac{dE_n(z)}{dz} = -E_{n-1}(z).$$

4. Asymptotic expansions:

$$E_n(z) \sim \frac{e^{-z}}{z}\left[1 - \frac{n}{2} + \frac{n(n+1)}{z^2} - \ldots\right], \quad z \to \infty; \ |\arg z| < \frac{3}{2}\pi.$$

Appendix G
Miscellaneous functions

G1 Trigonometric and hyperbolic functions (Angles in radians)

x	$\sin x$	$\cos x$	$\tan x$	$\sinh x$	$\cosh x$	$\tanh x$	e^x	e^{-x}
0.0	0.0000	1.0000	0.0000	0.0000	1.0000	0.0000	1.0000	1.0000
0.2	0.1987	0.9801	0.2027	0.2013	1.0201	0.1974	1.2214	0.8187
0.4	0.3894	0.9211	0.4228	0.4108	1.0811	0.3799	1.4918	0.6703
0.6	0.5646	0.8253	0.6841	0.6367	1.1855	0.5370	1.8221	0.5488
0.8	0.7174	0.6967	1.0296	0.8881	1.3374	0.6640	2.2255	0.4493
1.0	0.8415	0.5403	1.5574	1.1752	1.5431	0.7616	2.7183	0.3679
1.2	0.9320	0.3624	2.5722	1.5095	1.8106	0.8337	3.3201	0.3012
1.4	0.9854	+0.1700	+5.7979	1.9043	2.1509	0.8854	4.0552	0.2466
1.6	0.9996	−0.0292	−34.233	2.3756	2.5775	0.9217	4.9530	0.2019
1.8	0.9738	−0.2272	−4.2863	2.9422	3.1075	0.9468	6.0496	0.1653
2.0	0.9093	−0.4161	−2.1850	3.6269	3.7622	0.9640	7.3891	0.1353
2.2	0.8085	−0.5885	−1.3738	4.4571	4.5679	0.9757	9.0250	0.1108
2.4	0.6755	−0.7374	−0.9160	5.4662	5.5569	0.9837	11.023	0.0907
2.6	0.5155	−0.8569	−0.6016	6.6947	6.7690	0.9890	13.464	0.0742
2.8	0.3350	−0.9422	−0.3555	8.1919	8.2527	0.9926	16.445	0.0608
3.0	+0.1411	−0.9900	−0.1425	10.018	10.068	0.9951	20.086	0.0498
3.2	−0.0584	−0.9983	+0.0585	12.246	12.287	0.9967	24.533	0.0407
3.4	−0.2555	−0.9668	0.2643	14.965	14.999	0.9978	29.964	0.0331
3.6	−0.4425	−0.8968	0.4935	18.285	18.313	0.9985	36.598	0.0273
3.8	−0.6119	−0.7910	0.7736	22.339	22.362	0.9990	44.701	0.0223
4.0	−0.7568	−0.6536	1.1578	27.290	27.308	0.9993	54.598	0.0183
4.2	−0.8716	−0.4903	1.7778	33.335	33.351	0.9996	66.686	0.0150
4.4	−0.9516	−0.3073	3.0963	40.719	40.732	0.9997	81.451	0.0123
4.6	−0.9937	−0.1122	+8.8602	49.737	49.747	0.9998	99.484	0.0100
4.8	−0.9962	+0.0875	−11.385	60.751	60.759	0.9999	121.51	0.0082
5.0	−0.9589	0.2837	−3.3805	74.203	74.210	0.9999	148.41	0.0067
5.2	−0.8835	0.4685	−1.8856	90.633	90.639	0.9999	181.27	0.0055
5.4	−0.7728	0.6347	−1.2175	110.70	110.71	1.0000	221.41	0.0045
5.6	−0.6313	0.7756	−0.8139	135.21	135.22	1.0000	270.43	0.0037
5.8	−0.4646	0.8855	−0.5247	165.15	165.15	1.0000	330.30	0.0030
6.0	−0.2794	0.9602	−0.2910	201.71	201.71	1.0000	403.43	0.0025
6.2	−0.0831	0.9965	−0.0834	246.37	246.37	1.0000	492.75	0.0020
6.4	+0.1165	0.9932	+0.1173	300.92	300.92	1.0000	601.85	0.0016
6.6	0.3115	0.9502	0.3279	367.55	367.55	1.0000	735.10	0.0013
6.8	0.4941	0.8694	0.5683	448.92	448.92	1.0000	897.85	0.0011

https://doi.org/10.1515/9783111351179-017

(continued)

x	$\sin x$	$\cos x$	$\tan x$	$\sinh x$	$\cosh x$	$\tanh x$	e^x	e^{-x}
7.0	0.6570	0.7539	0.8714	548.32	548.32	1.0000	1096.6	0.0009
7.2	0.7937	0.6084	1.3046	669.72	669.72	1.0000	1339.4	0.0007
7.4	0.8987	0.4385	2.0493	817.99	817.99	1.0000	1636.0	0.0006
7.6	0.9679	0.2513	3.8523	999.10	999.10	1.0000	1998.2	0.0005
7.8	0.9985	+ 0.0540	+ 18.507	1220.3	1220.3	1.0000	2440.6	0.0004
8.0	0.9894	– 0.1455	– 6.7997	1490.5	1490.5	1.0000	2981.0	0.0003

G2 Trigonometric and hyperbolic functions

x	$\sin \pi x$	$\cos \pi x$	$\tan \pi x$	$\sinh \pi x$	$\cosh \pi x$	$\tanh \pi x$	$e^{\pi x}$	$e^{-\pi x}$
0.00	0.0000	1.0000	0.0000	0.0000	1.0000	0.0000	1.0000	1.0000
0.05	0.1564	0.9877	0.1584	0.1577	1.0124	0.1558	1.1701	0.8546
0.10	0.3090	0.9511	0.3249	0.3194	1.0498	0.3042	1.3691	0.7304
0.15	0.4540	0.8910	0.5095	0.4889	1.1131	0.4392	1.6020	0.6242
0.20	0.5878	0.8090	0.7265	0.6705	1.2040	0.5569	1.8745	0.5335
0.25	0.7071	0.7071	1.0000	0.8687	1.3246	0.6558	2.1933	0.4559
0.30	0.8090	0.5878	1.3764	1.0883	1.4780	0.7364	2.5663	0.3897
0.35	0.8910	0.4540	1.9626	1.3349	1.6679	0.8003	3.0028	0.3330
0.40	0.9511	0.3090	3.0777	1.6145	1.8991	0.8501	3.5136	0.2846
0.45	0.9877	0.1564	6.3138	1.9340	2.1772	0.8883	4.1112	0.2432
0.50	1.0000	0.0000	∞	2.3013	2.5092	0.9172	4.8105	0.2079
0.55	0.9877	– 0.1564	– 6.3138	2.7255	2.9032	0.9388	5.6287	0.1777
0.60	0.9511	– 0.3090	– 3.0777	3.2171	3.3689	0.9549	6.5861	0.1518
0.65	0.8910	– 0.4540	– 1.9626	3.7883	3.9180	0.9669	7.7063	0.1298
0.70	0.8090	– 0.5878	– 1.3764	4.4531	4.5640	0.9757	9.0170	0.1109
0.75	0.7071	– 0.7071	– 1.0000	5.2280	5.3228	0.9822	10.5507	0.0948
0.80	0.5878	– 0.8090	– 0.7265	6.1321	6.2131	0.9870	12.3453	0.0810
0.85	0.4540	– 0.8910	– 0.5095	7.1879	7.2572	0.9905	14.4451	0.0692
0.90	0.3090	– 0.9511	– 0.3249	8.4214	8.4806	0.9930	16.9020	0.0592
0.95	0.1564	– 0.9877	– 0.1584	9.8632	9.9137	0.9949	19.7769	0.0506
1.00	0.0000	– 1.0000	0.0000	11.5487	11.5920	0.9963	23.1407	0.0432
1.05	– 0.1564	– 0.9877	0.1584	13.5199	13.5568	0.9973	27.0767	0.0369
1.10	– 0.3090	– 0.9511	0.3249	15.8253	15.8568	0.9980	31.6821	0.0316
1.15	– 0.4540	– 0.8910	0.5095	18.5219	18.5489	0.9985	37.0709	0.0270
1.20	– 0.5878	– 0.8090	0.7265	21.6766	21.6996	0.9989	43.3762	0.0231
1.25	– 0.7071	– 0.7071	1.0000	25.3672	25.3869	0.9992	50.7540	0.0197
1.30	– 0.8090	– 0.5878	1.3764	29.6849	29.7018	0.9994	59.3867	0.0168

(continued)

x	$\sin \pi x$	$\cos \pi x$	$\tan \pi x$	$\sinh \pi x$	$\cosh \pi x$	$\tanh \pi x$	$e^{\pi x}$	$e^{-\pi x}$
1.35	−0.8910	−0.4540	1.9626	34.7367	34.7511	0.9996	69.4877	0.0144
1.40	−0.9511	−0.3090	3.0777	40.6473	40.6596	0.9997	81.3068	0.0123
1.45	−0.9877	−0.1564	6.3138	47.5628	47.5733	0.9998	95.1362	0.0105
1.50	−1.0000	0.0000	∞	55.6544	55.6634	0.9998	111.3178	0.0090
1.55	−0.9877	0.1564	−6.3138	65.1220	65.1297	0.9999	130.2517	0.0077
1.60	−0.9511	0.3090	−3.0777	76.1997	76.2063	0.9999	152.4060	0.0066
1.65	−0.8910	0.4540	−1.9626	89.1615	89.1671	0.9999	178.3286	0.0056
1.70	−0.8090	0.5878	−1.3764	104.3277	104.3325	1.0000	208.6603	0.0048
1.75	−0.7071	0.7071	−1.0000	122.0735	122.0776	1.0000	244.1511	0.0041
1.80	−0.5878	0.8090	−0.7265	142.8375	142.8410	1.0000	285.6784	0.0035
1.85	−0.4540	0.8910	−0.5095	167.1331	167.1361	1.0000	334.2691	0.0030
1.90	−0.3090	0.9511	−0.3249	195.5610	195.5636	1.0000	391.1245	0.0026
1.95	−0.1564	0.9877	−0.1584	228.8241	228.8263	1.0000	457.6504	0.0022
2.00	0.0000	1.0000	0.0000	267.7449	267.7468	1.0000	535.4917	0.0019

G3 Hyperbolic tangent of complex quantity

$$\tanh[\pi(\alpha - i\beta)] = \theta - i\chi = |\zeta| e^{-i\varphi}$$

α	$\tanh \pi\alpha$	θ	χ	$\|\zeta\|$	φ	θ	χ	$\|\zeta\|$	φ
		$\beta = 0.00$				$\beta = 0.05$			
0.0000	0.00	0.0000	0.0000	0.0000	0–90°	0.0000	0.1584	0.1584	90.00°
0.0159	0.05	0.0500	0.0000	0.0500	0.00	0.0512	0.1580	0.1660	72.03
0.0319	0.10	0.1000	0.0000	0.1000	0.00	0.1025	0.1567	0.1872	56.82
0.0481	0.15	0.1500	0.0000	0.1500	0.00	0.1537	0.1547	0.2180	45.20
0.0645	0.20	0.2000	0.0000	0.2000	0.00	0.2048	0.1519	0.2549	36.56
0.0813	0.25	0.2500	0.0000	0.2500	0.00	0.2558	0.1482	0.2956	30.09
0.0985	0.30	0.3000	0.0000	0.3000	0.00	0.3068	0.1438	0.3388	25.12
0.1163	0.35	0.3500	0.0000	0.3500	0.00	0.3577	0.1386	0.3836	21.18
0.1349	0.40	0.4000	0.0000	0.4000	0.00	0.4084	0.1325	0.4293	17.98
0.1543	0.45	0.4500	0.0000	0.4500	0.00	0.4589	0.1256	0.4758	15.32
0.1748	0.50	0.5000	0.0000	0.5000	0.00	0.5093	0.1181	0.5228	13.05
0.1968	0.55	0.5500	0.0000	0.5500	0.00	0.5596	0.1096	0.5702	11.08
0.2207	0.60	0.6000	0.0000	0.6000	0.00	0.6095	0.1005	0.6177	9.36
0.2468	0.65	0.6500	0.0000	0.6500	0.00	0.6593	0.0905	0.6654	7.82
0.2761	0.70	0.7000	0.0000	0.7000	0.00	0.7088	0.0798	0.7133	6.43

(continued)

α	$\tanh \pi\alpha$	θ	χ	$\lvert\zeta\rvert$	φ	θ	χ	$\lvert\zeta\rvert$	φ
		$\beta = 0.00$				$\beta = 0.05$			
0.3097	0.75	0.7500	0.0000	0.7500	0.00	0.7581	0.0683	0.7612	5.15
0.3497	0.80	0.8000	0.0000	0.8000	0.00	0.8070	0.0561	0.8090	3.97
0.3999	0.85	0.8500	0.0000	0.8500	0.00	0.8558	0.0432	0.8569	2.88
0.4686	0.90	0.9000	0.0000	0.9000	0.00	0.9041	0.0295	0.9047	1.87
0.5831	0.95	0.9500	0.0000	0.9500	0.00	0.9523	0.0151	0.9524	0.91
		$\beta = 0.10$				$\beta = 0.15$			
0.0000	0.00	0.0000	0.3249	0.3249	90.00°	0.0000	0.5095	0.5095	90.00°
0.0159	0.05	0.0553	0.3240	0.3286	80.32	0.0629	0.5079	0.5118	82.93
0.0319	0.10	0.1104	0.3213	0.3398	71.03	0.1256	0.5031	0.5186	75.98
0.0481	0.15	0.1655	0.3169	0.3575	62.43	0.1878	0.4951	0.5296	69.22
0.0645	0.20	0.2202	0.3106	0.3808	54.67	0.2493	0.4841	0.5445	62.75
0.0813	0.25	0.2746	0.3027	0.4087	47.78	0.3099	0.4700	0.5629	56.61
0.0985	0.30	0.3286	0.2929	0.4402	41.72	0.3692	0.4531	0.5845	50.82
0.1163	0.35	0.3820	0.2815	0.4745	36.39	0.4273	0.4333	0.6085	45.40
0.1349	0.40	0.4349	0.2684	0.5110	31.68	0.4838	0.4110	0.6347	40.35
0.1543	0.45	0.4871	0.2537	0.5492	27.51	0.5385	0.3860	0.6626	35.63
0.1748	0.50	0.5386	0.2374	0.5886	23.79	0.5914	0.3589	0.6917	31.25
0.1968	0.55	0.5893	0.2196	0.6289	20.44	0.6423	0.3295	0.7219	27.16
0.2207	0.60	0.6390	0.2003	0.6697	17.41	0.6911	0.2982	0.7527	23.34
0.2468	0.65	0.6880	0.1796	0.7110	14.63	0.7378	0.2652	0.7840	19.76
0.2761	0.70	0.7358	0.1576	0.7525	12.08	0.7822	0.2305	0.8155	16.43
0.3097	0.75	0.7827	0.1342	0.7941	9.73	0.8243	0.1945	0.8469	13.28
0.3497	0.80	0.8235	0.1096	0.8357	7.54	0.8642	0.1573	0.8783	10.32
0.3999	0.85	0.8731	0.0838	0.8771	5.48	0.9015	0.1191	0.9094	7.52
0.4686	0.90	0.9166	0.0569	0.9184	3.56	0.9367	0.0800	0.9401	4.88
0.5831	0.95	0.9589	0.0289	0.9594	1.73	0.9695	0.0403	0.9704	2.38
		$\beta = 0.20$				$\beta = 0.25$			
0.0000	0.00	0.0000	0.7265	0.7265	90.00°	0.0000	1.0000	1.000	90.00°
0.0159	0.05	0.0763	0.7238	0.7278	83.98	0.0998	0.9950	1.000	84.28
0.0319	0.10	0.1520	0.7155	0.7315	78.01	0.1980	0.9802	1.000	78.58
0.0481	0.15	0.2265	0.7019	0.7375	72.12	0.2934	0.9560	1.000	72.93
0.0645	0.20	0.2993	0.6831	0.7458	66.34	0.3846	0.9230	1.000	67.38
0.0813	0.25	0.3698	0.6593	0.7560	60.72	0.4706	0.8824	1.000	61.93
0.0985	0.30	0.4376	0.6312	0.7680	55.27	0.5504	0.8348	1.000	56.60
0.1163	0.35	0.5023	0.5989	0.7816	50.01	0.6236	0.7818	1.000	51.42
0.1349	0.40	0.5635	0.5627	0.7964	44.96	0.6896	0.7241	1.000	46.40
0.1543	0.45	0.6212	0.5235	0.8123	40.12	0.7484	0.6632	1.000	41.55

(continued)

			$\beta = 0.20$					$\beta = 0.25$	
0.1748	0.50	0.6749	0.4814	0.8290	35.50	0.8000	0.6000	1.000	36.87
0.1968	0.55	0.7247	0.4370	0.8462	31.09	0.8446	0.5355	1.000	32.38
0.2207	0.60	0.7703	0.3907	0.8637	26.89	0.8824	0.4706	1.000	28.07
0.2468	0.65	0.8120	0.3430	0.8815	22.91	0.9139	0.4060	1.000	23.95
0.2761	0.70	0.8497	0.2943	0.8992	19.11	0.9395	0.3423	1.000	20.02
0.3097	0.75	0.8835	0.2451	0.9169	15.51	0.9600	0.2800	1.000	16.27
0.3497	0.80	0.9136	0.1955	0.9343	12.08	0.9757	0.2195	1.000	12.68
0.3999	0.85	0.9401	0.1459	0.9514	8.82	0.9869	0.1611	1.000	9.27
0.4686	0.90	0.9632	0.0967	0.9681	5.73	0.9945	0.1050	1.000	6.03
0.5831	0.95	0.9831	0.0480	0.9843	2.79	0.9986	0.0512	1.000	2.93
			$\beta = 0.30$					$\beta = 0.35$	
0.0000	0.00	0.0000	1.3764	1.3764	90.00°	0.0000	1.9626	1.9626	90.00°
0.0159	0.05	0.1440	1.3664	1.3740	83.98	0.2403	1.9391	1.9539	82.93
0.0319	0.10	0.2841	1.3373	1.3671	78.01	0.4672	1.8708	1.9283	75.98
0.0481	0.15	0.4164	1.2904	1.3559	72.12	0.6697	1.7653	1.8882	69.22
0.0645	0.20	0.5382	1.2282	1.3410	66.34	0.8408	1.6326	1.8365	62.75
0.0813	0.25	0.6470	1.1537	1.3228	60.72	0.9776	1.4829	1.7762	56.61
0.0985	0.30	0.7419	1.0701	1.3021	55.27	1.0809	1.3261	1.7109	50.82
0.1163	0.35	0.8223	0.9803	1.2794	50.01	1.1538	1.1701	1.6432	45.40
0.1349	0.40	0.8885	0.8873	1.2556	44.96	1.2007	1.0200	1.5755	40.35
0.1543	0.45	0.9413	0.7933	1.2311	40.12	0.2267	0.8793	1.5092	35.63
0.1748	0.50	0.9820	0.7005	1.2063	35.50	1.2359	0.7500	1.4457	31.25
0.1968	0.55	1.0120	0.6102	1.1819	31.09	1.2324	0.6323	1.3852	27.16
0.2207	0.60	1.0326	0.5237	1.1578	26.89	1.2198	0.5263	1.3284	23.34
0.2468	0.65	1.0449	0.4415	1.1344	22.91	1.2002	0.4313	1.2755	19.76
0.2761	0.70	1.0507	0.3640	1.1121	19.11	1.1762	0.3467	1.2262	16.43
0.3097	0.75	1.0509	0.2916	1.0906	15.51	1.1493	0.2713	1.1808	13.28
0.3497	0.80	1.0465	0.2240	1.0703	12.08	1.1202	0.2039	1.1386	10.32
0.3999	0.85	1.0387	0.1612	1.0511	8.82	1.0902	0.1440	1.0996	7.52
0.4686	0.90	1.0278	0.1032	1.0330	5.73	1.0599	0.0905	1.0637	4.88
0.5831	0.95	1.0148	0.0494	1.0160	2.79	1.0296	0.0428	1.0305	2.38
			$\beta = 0.40$					$\beta = 0.45$	
0.0000	0.00	0.0000	3.0777	3.0777	90.00°	0.0000	6.3138	6.3138	90.00°
0.0159	0.05	0.5115	2.9990	3.0423	80.32	1.3580	5.7272	6.0211	72.03
0.0319	0.10	0.9565	2.7833	2.9431	71.03	2.9217	4.4691	5.3394	56.82
0.0481	0.15	1.2048	2.4799	2.7976	62.43	3.2313	3.2535	4.5855	45.20
0.0645	0.20	1.5189	2.1427	2.6265	54.67	3.1500	2.3362	3.9217	36.56
0.0813	0.25	1.6444	1.8124	2.4473	47.78	2.9200	1.6953	3.3816	30.09
0.0985	0.30	1.6960	1.5119	2.2720	41.72	2.6722	1.2524	2.9511	25.12

(continued)

				$\beta = 0.40$				$\beta = 0.45$	
0.1163	0.35	1.6966	1.2501	2.1074	36.39	2.4310	0.9417	2.6070	21.18
0.1349	0.40	1.6652	1.0277	1.9569	31.68	2.2154	0.7188	2.3291	17.98
0.1543	0.45	1.6149	0.8411	1.8208	27.51	2.0269	0.5550	2.1015	15.32
0.1748	0.50	1.5547	0.6853	1.6989	23.79	1.8633	0.4319	1.9126	13.05
0.1968	0.55	1.4901	0.5554	1.5901	20.44	1.7210	0.3371	1.7538	11.08
0.2207	0.60	1.4247	0.4466	1.4932	17.41	1.5972	0.2632	1.6187	9.36
0.2468	0.65	1.3609	0.3553	1.4065	14.63	1.4887	0.2044	1.5026	7.82
0.2761	0.70	1.2994	0.2781	1.3289	12.08	1.3931	0.1568	1.4019	6.43
0.3097	0.75	1.2412	0.2129	1.2593	9.73	1.3083	0.1180	1.3137	5.15
0.3497	0.80	1.1862	0.1569	1.1966	7.54	1.2331	0.0857	1.2361	3.97
0.3999	0.85	1.1349	0.1090	1.1401	5.48	1.1655	0.0587	1.1670	2.88
0.4686	0.90	1.0867	0.0674	1.0888	3.56	1.1047	0.0360	1.1053	1.87
0.5831	0.95	1.0418	0.0314	1.0423	1.73	1.0499	0.0167	1.0500	0.97
				$\beta = 0.475$				$\beta = 0.50$	
0.0000	0.00	0.0000	12.706	12.706	90.00°	∞	0.0000	∞	0–90.00°
0.0159	0.05	5.787	9.030	10.725	57.35	20.000	0.0000	20.000	0.00
0.0319	0.10	6.213	4.811	7.859	37.75	10.000	0.0000	10.000	0.00
0.0481	0.15	5.260	2.682	5.905	27.01	6.6667	0.0000	6.6667	0.00
0.0645	0.20	4.356	1.636	4.653	20.58	5.0000	0.0000	5.0000	0.00
0.0813	0.25	3.662	1.075	3.817	16.35	4.0000	0.0000	4.0000	0.00
0.0985	0.30	3.138	0.7445	3.225	13.35	3.3333	0.0000	3.3333	0.00
0.1163	0.35	2.736	0.5367	2.789	11.10	2.8571	0.0000	2.8571	0.00
0.1349	0.40	2.422	0.3979	2.454	9.33	2.5000	0.0000	2.5000	0.00
0.1543	0.45	2.170	0.3008	2.190	7.90	2.2222	0.0000	2.2222	0.00
0.1748	0.50	1.9638	0.2304	1.9773	6.70	2.0000	0.0000	2.0000	0.00
0.1968	0.55	1.7927	0.1778	1.8014	5.67	1.8182	0.0000	1.8182	0.00
0.2207	0.60	1.6486	0.1375	1.6543	4.76	1.6667	0.0000	1.6667	0.00
0.2468	0.65	1.5257	0.1060	1.5293	3.97	1.5385	0.0000	1.5385	0.00
0.2761	0.70	1.4194	0.0809	1.4217	3.27	1.4286	0.0000	1.4286	0.00
0.3097	0.75	1.3269	0.0605	1.3284	2.61	1.3333	0.0000	1.3333	0.00
0.3497	0.80	1.2458	0.0438	1.2465	2.02	1.2500	0.0000	1.2500	0.00
0.3999	0.85	1.1737	0.0300	1.1741	1.46	1.1765	0.0000	1.1765	0.00
0.4686	0.90	1.1095	0.0183	1.1097	0.94	1.1111	0.0000	1.1111	0.00
0.5831	0.95	1.0518	0.0085	1.0519	0.46	1.0526	0.0000	1.0526	0.00

G4 Inverse hyperbolic tangent of complex quantity

$$\pi(\alpha - i\beta) = \tanh^{-1}(\theta - i\chi)$$

θ	α	β	α	β	α	β	α	β	α	β
	$\chi = 0$		$\chi = 0.2$		$\chi = 0.4$		$\chi = 0.6$		$\chi = 0.8$	
0.0	0.0000	0.0000	0.0000	0.0628	0.0000	0.1211	0.0000	0.1720	0.0000	0.2148
0.2	0.0645	0.0000	0.0619	0.0653	0.0552	0.1250	0.0468	0.1762	0.0386	0.2186
0.4	0.1349	0.0000	0.1281	0.0738	0.1118	0.1379	0.0931	0.1894	0.0760	0.2302
0.6	0.2206	0.0000	0.2041	0.0936	0.1703	0.1640	0.1373	0.2135	0.1103	0.2500
0.8	0.3497	0.0000	0.2955	0.1426	0.2255	0.2110	0.1749	0.2500	0.1386	0.2776
1.0	∞	0–0.5	0.3672	0.2659	0.2593	0.2814	0.1985	0.2964	0.1576	0.3106
1.2	0.3816	0.5000	0.3271	0.3894	0.2562	0.3524	0.2041	0.3436	0.1661	0.3445
1.4	0.2852	0.5000	0.2681	0.4394	0.2322	0.4013	0.1962	0.3826	0.1655	0.3750
1.6	0.2334	0.5000	0.2255	0.4610	0.2060	0.4307	0.1823	0.4111	0.1593	0.3999
1.8	0.1994	0.5000	0.1950	0.4723	0.1832	0.4488	0.1674	0.4312	0.1505	0.4198
2.0	0.1748	0.5000	0.1721	0.4792	0.1644	0.4605	0.1536	0.4454	0.1409	0.4341
2.2	0.1561	0.5000	0.1542	0.4837	0.1490	0.4686	0.1411	0.4557	0.1317	0.4454
2.4	0.1412	0.5000	0.1399	0.4868	0.1361	0.4743	0.1302	0.4634	0.1230	0.4541
2.6	0.1291	0.5000	0.1281	0.4890	0.1252	0.4786	0.1208	0.4692	0.1151	0.4610
2.8	0.1188	0.5000	0.1181	0.4908	0.1159	0.4819	0.1124	0.4737	0.1079	0.4665
3.0	0.1103	0.5000	0.1097	0.4921	0.1080	0.4845	0.1052	0.4773	0.1016	0.4709
3.2	0.1029	0.5000	0.1025	0.4931	0.1010	0.4865	0.0988	0.4802	0.0959	0.4744
3.4	0.0965	0.5000	0.0961	0.4940	0.0950	0.4881	0.0931	0.4826	0.0907	0.4774
3.6	0.0908	0.5000	0.0905	0.4947	0.0895	0.4895	0.0880	0.4845	0.0860	0.4799
3.8	0.0858	0.5000	0.0855	0.4953	0.0847	0.4906	0.0834	0.4862	0.0817	0.4820
4.0	0.0813	0.5000	0.0812	0.4958	0.0804	0.4916	0.0793	0.4876	0.0778	0.4838
	$\chi = 0$		$\chi = 0.2$		$\chi = 0.4$		$\chi = 0.6$		$\chi = 0.8$	
0.0	0.0000	0.2500	0.0000	0.2789	0.0000	0.3026	0.0000	0.322	0.0000	0.3524
0.2	0.0316	0.2532	0.0259	0.2814	0.0213	0.3046	0.0178	0.3238	0.0127	0.3534
0.4	0.0619	0.2627	0.0506	0.2890	0.0417	0.3106	0.0348	0.3285	0.0249	0.3564
0.6	0.0892	0.2783	0.0729	0.3012	0.0602	0.3201	0.0503	0.3360	0.0362	0.3612
0.8	0.1118	0.2993	0.0916	0.3173	0.0760	0.3326	0.0639	0.3459	0.0464	0.3675
1.0	0.1281	0.3238	0.1058	0.3360	0.0885	0.3472	0.0749	0.3574	0.0552	0.3750
1.2	0.1373	0.3493	0.1150	0.3558	0.0974	0.3628	0.0832	0.3699	0.0623	0.3833
1.4	0.1403	0.3734	0.1197	0.3750	0.1028	0.3783	0.0890	0.3826	0.0679	0.3920
1.6	0.1386	0.3944	0.1207	0.3926	0.1054	0.3930	0.0924	0.3949	0.0719	0.4008
1.8	0.1341	0.4120	0.1190	0.4080	0.1056	0.4064	0.0938	0.4064	0.0745	0.4093
	$\chi = 0$		$\chi = 0.2$		$\chi = 0.4$		$\chi = 0.6$		$\chi = 0.8$	
2.0	0.1281	0.4262	0.1157	0.4211	0.1042	0.4182	0.0937	0.4169	0.0760	0.4174
2.2	0.1216	0.4376	0.1114	0.4321	0.1017	0.4284	0.0926	0.4262	0.0766	0.4249

(continued)

θ	α	β	α	β	α	β	α	β	α	β
	χ = 0		χ = 0.2		χ = 0.4		χ = 0.6		χ = 0.8	
2.4	0.1150	0.4468	0.1067	0.4412	0.0985	0.4372	0.0906	0.4344	0.0764	0.4318
2.6	0.1087	0.4542	0.1019	0.4488	0.0950	0.4446	0.0883	0.4416	0.0756	0.4381
2.8	0.1027	0.4602	0.0973	0.4551	0.0913	0.4510	0.0855	0.4478	0.0743	0.4437
3.0	0.0974	0.4652	0.0927	0.4604	0.0878	0.4564	0.0828	0.4531	0.0729	0.4488
3.2	0.0924	0.4693	0.0884	0.4648	0.0843	0.4610	0.0799	0.4579	0.0712	0.4533
3.4	0.0877	0.4727	0.0844	0.4686	0.0808	0.4650	0.0771	0.4619	0.0695	0.4573
3.6	0.0835	0.4756	0.0807	0.4718	0.0776	0.4650	0.0743	0.4655	0.0676	0.4609
3.8	0.0796	0.4781	0.0773	0.4746	0.0745	0.4714	0.0717	0.4686	0.0657	0.4641
4.0	0.0760	0.4802	0.0739	0.4769	0.0716	0.4740	0.0691	0.4713	0.0639	0.4670

G5 Logarithmic and inverse hyperbolic functions

x	ln x	$\sinh^{-1}x$	$\cosh^{-1}x$	x	ln x	$\sinh^{-1}x$	$\cosh^{-1}x$
0.0	−∞	0.0000	......	4.0	1.3863	2.0947	2.0634
0.1	−2.3026	0.0998	......	4.2	1.4351	2.1421	2.1137
0.2	−1.6094	0.1987	......	4.4	1.4816	2.1874	2.1616
0.3	−1.2040	0.2957	......	4.6	1.5261	2.2308	2.2072
0.4	−0.9163	0.3900	......	4.8	1.5686	2.2724	2.2507
0.5	−0.6931	0.4812	......	5.0	1.6094	2.3124	2.2924
0.6	−0.5108	0.5688	......	5.2	1.6487	2.3509	2.3324
0.7	−0.3567	0.6527	......	5.4	1.6864	2.3880	2.3709
0.8	−0.2231	0.7327	......	5.6	1.7228	2.4238	2.4078
0.9	−0.1054	0.8089	......	5.8	1.7579	2.4584	2.4435
1.0	0.0000	0.8814	0.0000	6.0	1.7918	2.4918	2.4779
1.1	0.0953	0.9503	0.4436	6.2	1.8245	2.5241	2.5111
1.2	0.1823	1.0160	0.6224	6.4	1.8563	2.5555	2.5433
1.3	0.2624	1.0785	0.7564	6.6	1.8871	2.5859	2.5744
1.4	0.3365	1.1380	0.8670	6.8	1.9169	2.6154	2.6046
1.5	0.4055	1.1948	0.9624	7.0	1.9459	2.6441	2.6339
1.6	0.4700	1.2490	1.0470	7.2	1.9741	2.6720	2.6624
1.7	0.5306	1.3008	1.1232	7.4	2.0015	2.6992	2.6900
1.8	0.5878	1.3504	1.1929	7.6	2.0281	2.7256	2.7169
1.9	0.6419	1.3980	1.2572	7.8	2.0541	2.7514	2.7431
2.0	0.6931	1.4436	1.3170	8.0	2.0794	2.7765	2.7687
2.1	0.7419	1.4875	1.3729	8.2	2.1041	2.8010	2.7935
2.2	0.7885	1.5297	1.4254	8.4	2.1282	2.8249	2.8178
2.3	0.8329	1.5703	1.4750	8.6	2.1518	2.8483	2.8415

(continued)

x	$\ln x$	$\sinh^{-1}x$	$\cosh^{-1}x$	x	$\ln x$	$\sinh^{-1}x$	$\cosh^{-1}x$
2.4	0.8755	1.6094	1.5221	8.8	2.1748	2.8711	2.8647
2.5	0.9163	1.6472	1.5668	9.0	2.1972	2.8934	2.8873
2.6	0.9555	1.6837	1.6094	9.2	2.2192	2.9153	2.9094
2.7	0.9933	1.7191	1.6502	9.4	2.2407	2.9367	2.9310
2.8	1.0296	1.7532	1.6892	9.6	2.2618	2.9576	2.9522
2.9	1.0647	1.7863	1.7267	9.8	2.2824	2.9781	2.9729
3.0	1.0986	1.8184	1.7627	10.0	2.3026	2.9982	2.9932
3.1	1.1314	1.8496	1.7975	10.2	2.3224	3.0179	3.0131
3.2	1.1632	1.8799	1.8309	10.4	2.3418	3.0373	3.0326
3.3	1.1939	1.9093	1.8633	10.6	2.3609	3.0562	3.0518
3.4	1.2238	1.9379	1.8946	10.8	2.3795	3.0748	3.0705
3.5	1.2528	1.9657	1.9248	11.0	2.3979	3.0931	3.0890
3.6	1.2809	1.9928	1.9542	11.2	2.4159	3.1110	3.1071
3.7	1.3083	2.0193	1.9827	11.4	2.4336	3.1287	3.1248
3.8	1.3350	2.0450	2.0104	11.6	2.4510	3.1460	3.1423
3.9	1.3610	2.0702	2.0373	11.8	2.4681	3.1630	3.1594
4.0	1.3863	2.0947	2.0634	12.0	2.4849	3.1798	3.1763

G6 Spherical harmonic functions

$$P_0^0(x) = 1.0000; \ P_1^0(x) = x$$

x	$P_1^1(x)$	$P_2^0(x)$	$P_2^1(x)$	$P_2^2(x)$	$P_3^0(x)$	$P_3^1(x)$	$P_3^2(x)$	$P_3^3(x)$
1.00	0.0000	1.0000	0.0000	0.0000	1.0000	0.0000	0.0000	0.0000
0.95	0.3122	0.8538	0.8899	0.2925	0.7184	1.6452	1.3894	0.4567
0.90	0.4359	0.7150	1.1769	0.5700	0.4725	1.9942	2.5650	1.2423
0.85	0.5268	0.5838	1.3433	0.8325	0.2603	2.0643	3.5381	2.1927
0.80	0.6000	0.4600	1.4400	1.0800	0.0800	1.9800	4.3200	3.2400
0.75	0.6614	0.3438	1.4882	1.3125	−0.0703	1.7983	4.9219	4.3407
0.70	0.7141	0.2350	1.4997	1.5300	−0.1925	1.5533	5.3550	5.4632
0.65	0.7599	0.1338	1.4819	1.7325	−0.2884	1.2681	5.6306	6.5829
0.60	0.8000	+0.0400	1.4400	1.9200	−0.3600	0.9600	5.7600	7.6800
0.55	0.8352	−0.0462	1.3780	2.0925	−0.4091	0.6420	5.7544	8.7379
0.50	0.8660	−0.1250	1.2990	2.2500	−0.4375	0.3248	5.6250	9.7428
0.45	0.8930	−0.1962	1.2056	2.3925	−0.4472	+0.0167	5.3831	10.683
0.40	0.9165	−0.2600	1.0998	2.5200	−0.4400	−0.2750	5.0400	11.548
0.35	0.9367	−0.3162	0.9836	2.6325	−0.4178	−0.5445	4.6069	12.330
0.30	0.9539	−0.3650	0.8585	2.7300	−0.3825	−0.7870	4.0950	13.021

(continued)

x	$P_1^1(x)$	$P_2^0(x)$	$P_2^1(x)$	$P_2^2(x)$	$P_3^0(x)$	$P_3^1(x)$	$P_3^2(x)$	$P_3^3(x)$
0.25	0.9682	−0.4062	0.7262	2.8125	−0.3359	−0.9985	3.5156	13.616
0.20	0.9798	−0.4400	0.5879	2.8800	−0.2800	−1.1758	2.8800	14.109
0.15	0.9887	−0.4662	0.4449	2.9325	−0.2166	−1.3162	2.1994	14.497
0.10	0.9950	−0.4850	0.2985	2.9700	−0.1475	−1.4179	1.4850	14.776
0.05	0.9987	−0.4962	0.1498	2.9925	−0.0747	−1.4794	0.7481	14.944
0.00	1.0000	−0.5000	0.0000	3.0000	0.0000	−1.5000	0.0000	15.000

G7 Legendre functions for large arguments

x	$P_0^0(x)$	$P_1^0(x)$	$iP_1^1(x)$	$P_2^0(x)$	$iP_2^1(x)$	$-P_2^2(x)$
1.0	1.0000	1.0000	0.0000	1.0000	0.0000	0.0000
1.2	1.0000	1.2000	0.6633	1.6600	2.3880	1.3200
1.4	1.0000	1.4000	0.9798	2.4400	4.1151	2.8800
1.6	1.0000	1.6000	1.2490	3.3400	5.9952	4.6800
1.8	1.0000	1.8000	1.4967	4.3600	8.0820	6.7200
2.0	1.0000	2.0000	1.7321	5.5000	10.392	9.0000
2.2	1.0000	2.2000	1.9596	6.7600	12.933	11.520
2.4	1.0000	2.4000	2.1817	8.1400	15.708	14.280
2.6	1.0000	2.6000	2.4000	9.6400	18.720	17.280
2.8	1.0000	2.8000	2.6153	11.260	21.969	20.520
3.0	1.0000	3.0000	2.8284	13.000	25.456	24.000
3.5	1.0000	3.5000	3.3541	17.875	35.218	33.750
4.0	1.0000	4.0000	3.8730	23.500	46.476	45.000
4.5	1.0000	4.5000	4.3875	29.875	59.231	57.750
5.0	1.0000	5.0000	4.8990	37.000	73.485	72.000
5.5	1.0000	5.5000	5.4083	44.875	89.237	87.750
6.0	1.0000	6.0000	5.9161	53.500	106.49	105.00
6.5	1.0000	6.5000	6.4226	62.875	125.24	123.75
7.0	1.0000	7.0000	6.9282	73.000	145.49	144.00
7.5	1.0000	7.5000	7.4330	83.875	167.24	165.75
8.0	1.0000	8.0000	7.9372	95.500	190.49	189.00
x	$Q_0^0(x)$	$Q_1^0(x)$	$Q_1^1(x)$	$Q_2^0(x)$	$Q_2^1(x)$	$Q_2^2(x)$
1.0	1.5223	0.6745	1.7028	0.3518	1.2549	8.1352
1.2	1.1990	0.4387	1.0138	0.19025	0.6345	3.4372
1.4	0.8959	0.2542	0.5511	0.08595	0.2733	1.2968
1.6	0.7332	0.17307	0.3653	0.04878	0.15215	0.6825
1.8	0.6264	0.12749	0.2652	0.03102	0.09574	0.4164

(continued)

x	$Q_0^0(x)$	$Q_1^0(x)$	$Q_1^1(x)$	$Q_2^0(x)$	$Q_2^1(x)$	$Q_2^2(x)$
2.0	0.5493	0.09861	0.2033	0.02118	0.06495	0.2771
2.2	0.4904	0.07891	0.16167	0.01520	0.04640	0.19541
2.4	0.4437	0.06476	0.13210	0.01132	0.03446	0.14375
2.6	0.4055	0.05421	0.11022	0.00868	0.02636	0.10922
2.8	0.3736	0.04610	0.09350	0.00682	0.02066	0.08513
3.0	0.3466	0.03972	0.08040	0.00546	0.01651	0.06777
3.5	0.2939	0.02863	0.05775	0.00334	0.01009	0.04112
4.0	0.2554	0.02165	0.04359	0.00220	0.00663	0.02691
4.5	0.2260	0.01697	0.03411	0.00153	0.00460	0.01860
5.0	0.2027	0.01366	0.02744	0.00110	0.00332	0.01341
5.5	0.1839	0.01124	0.02256	0.00082	0.0028	0.00999
6.0	0.1682	0.00942	0.01889	0.00063	0.00190	0.00765
6.5	0.1551	0.00800	0.01605	0.00050	0.00149	0.00599
7.0	0.1438	0.00689	0.01380	0.00040	0.00119	0.00478
7.5	0.1341	0.00599	0.01200	0.00032	0.00096	0.00387
8.0	0.1257	0.00526	0.01053	0.00026	0.00079	0.00318

G8 Legendre functions for imaginary arguments

x	$P_0^0(ix)$	$-iP_1^0(ix)$	$iP_1^1(x)$	$-P_2^0(ix)$	$-iP_2^1(ix)$	$P_2^2(ix)$
0.0	1.0000	0.0000	1.0000	0.5000	0.0000	3.0000
0.2	1.0000	0.2000	1.0198	0.5600	0.6119	3.1200
0.4	1.0000	0.4000	1.0770	0.7400	1.2924	3.4800
0.6	1.0000	0.6000	1.1662	1.0400	2.0991	4.0800
0.8	1.0000	0.8000	1.2806	1.4600	3.0735	4.9200
1.0	1.0000	1.0000	1.4142	2.0000	4.2426	6.0000
1.2	1.0000	1.2000	1.5621	2.6600	5.6234	7.3200
1.4	1.0000	1.4000	1.7205	3.4400	7.2260	8.8800
1.6	1.0000	1.6000	1.8868	4.3400	9.0566	10.680
1.8	1.0000	1.8000	2.0591	5.3600	11.119	12.720
2.0	1.0000	2.0000	2.2361	6.5000	13.416	15.000
2.5	1.0000	2.5000	2.6926	9.8750	20.194	21.750
3.0	1.0000	3.0000	3.1623	14.000	28.461	30.000
3.5	1.0000	3.5000	3.6401	18.875	38.221	39.750
4.0	1.0000	4.0000	4.1231	24.500	49.477	51.000
4.5	1.0000	4.5000	4.6098	30.875	62.232	63.750
5.0	1.0000	5.0000	5.0990	38.000	76.485	78.000
5.5	1.0000	5.5000	5.5902	45.875	92.238	93.750

(continued)

x	$P_0^0(ix)$	$-iP_1^0(ix)$	$iP_1^1(x)$	$-P_2^0(ix)$	$-iP_2^1(ix)$	$P_2^2(ix)$
6.0	1.0000	6.0000	6.0828	54.500	109.49	111.00
6.5	1.0000	6.5000	6.5765	63.875	128.24	129.75
7.0	1.0000	7.0000	7.0711	74.000	148.49	150.00
7.5	1.0000	7.5000	7.5664	84.875	170.24	171.75
x	$iQ_0^0(ix)$	$-Q_1^0(ix)$	$-Q_1^1(ix)$	$-iQ_2^0(ix)$	$-iQ_2^1(ix)$	$-iQ_2^2(ix)$
0.0	1.5708	1.0000	1.5708	0.7854	2.0000	4.7124
0.2	1.3734	0.7253	1.2045	0.4691	1.2385	3.3004
0.4	1.1903	0.5239	0.9106	0.2808	0.7642	2.2526
0.6	1.0304	0.3818	0.6871	0.17159	0.4782	1.5216
0.8	0.8961	0.2832	0.5228	0.10824	0.3070	1.0330
1.0	0.7854	0.2146	0.4036	0.07079	0.20337	0.7124
1.2	0.6947	0.1663	0.3170	0.04800	0.13919	0.5019
1.4	0.6202	0.13165	0.2534	0.03366	0.09826	0.3619
1.6	0.5586	0.10624	0.2060	0.02432	0.07137	0.2670
1.8	0.5071	0.08722	0.1700	0.01805	0.05316	0.2012
2.0	0.4636	0.07270	0.14232	0.01371	0.04050	0.15471
2.5	0.3805	0.04873	0.09607	0.00750	0.02227	0.08636
3.0	0.3218	0.03475	0.06878	0.00451	0.01342	0.05251
3.5	0.2783	0.02595	0.05150	0.00291	0.00867	0.03411
4.0	0.2450	0.02009	0.03993	0.00198	0.00591	0.02332
4.5	0.2187	0.01599	0.03183	0.00140	0.00420	0.01662
5.0	0.19740	0.01302	0.02594	0.00103	0.00309	0.01224
5.5	0.17985	0.01081	0.02154	0.00078	0.00233	0.00927
6.0	0.16515	0.00911	0.01816	0.00060	0.00180	0.00718
6.5	0.15265	0.00778	0.01552	0.00047	0.00143	0.00567
7.0	0.14190	0.00672	0.01341	0.00038	0.00114	0.00456
7.5	0.13255	0.00586	0.01171	0.00031	0.00093	0.00372

G9 Legendre functions of half-integral degree

x	$P^0_{-1/2}(x)$	$-iP^1_{-1/2}(x)$	$-P^2_{-1/2}(x)$	$P^0_{1/2}(x)$	$iP^1_{1/2}(x)$	$P^0_{3/2}(x)$
1.0	1.0000	0.0000	0.0000	1.0000	0.0000	1.0000
1.2	0.9763	0.07447	0.02536	1.0728	0.2344	1.3910
1.4	0.9549	0.09968	0.04613	1.1416	0.3283	1.8126
1.6	0.9355	0.11603	0.06340	1.2070	0.3986	2.2630
1.8	0.9177	0.12778	0.07794	1.2694	0.4567	2.7406

(continued)

x	$P^0_{-1/2}(x)$	$-iP^1_{-1/2}(x)$	$-P^2_{-1/2}(x)$	$P^0_{1/2}(x)$	$iP^1_{1/2}(x)$	$P^0_{3/2}(x)$
2.0	0.9013	0.1367	0.0903	1.3291	0.5072	3.2439
2.2	0.8861	0.1436	0.1009	1.3866	0.5523	3.7719
2.4	0.8719	0.1491	0.1101	1.4419	0.5933	4.3236
2.6	0.8587	0.1536	0.1181	1.4954	0.6311	4.8979
2.8	0.8463	0.1572	0.1251	1.5472	0.6664	5.4941
3.0	0.8346	0.1602	0.1313	1.5974	0.6996	6.1113
3.5	0.8082	0.1657	0.1438	1.7169	0.7753	7.7427
4.0	0.7850	0.1692	0.1533	1.8290	0.8432	9.4930
4.5	0.7643	0.1715	0.1606	1.9349	0.9052	11.355
5.0	0.7457	0.1728	0.1663	2.0356	0.9627	13.322
5.5	0.7289	0.1736	0.1708	2.1316	1.0165	15.389
6.0	0.7136	0.1739	0.1744	2.2237	1.0673	17.552
6.5	0.6995	0.1739	0.772	2.3122	1.1156	19.806
7.0	0.6864	0.1737	0.1795	2.3975	1.1616	22.148
7.5	0.6743	0.1734	0.1813	2.4799	1.2058	24.575
8.0	0.6631	0.1729	0.1828	2.5598	1.2482	27.083

x	$Q^0_{-1/2}(x)$	$Q^1_{-1/2}(x)$	$Q^2_{-1/2}(x)$	$Q^0_{1/2}(x)$	$Q^1_{1/2}(x)$	$Q^0_{3/2}(x)$
1.0	2.8612	2.3661	10.644	0.9788	1.9471	0.4818
1.2	2.5010	1.7349	5.6518	0.6996	1.2524	0.2856
1.4	2.1366	1.2918	3.1575	0.4598	0.7618	0.14609
1.6	1.9229	1.0943	2.3230	0.3430	0.5501	0.09080
1.8	1.7723	0.9748	1.9018	0.2720	0.4285	0.06214
2.0	1.6566	0.8918	1.6454	0.2240	0.3489	0.04516
2.2	1.5634	0.8293	1.4712	0.18932	0.29263	0.03422
2.4	1.4856	0.7798	1.3441	0.16312	0.25076	0.02676
2.6	1.4193	0.7391	1.2465	0.14266	0.21842	0.02143
2.8	1.3617	0.7048	1.1687	0.12628	0.17274	0.01751
3.0	1.3110	0.6753	1.1048	0.11289	0.17189	0.01454
3.5	1.2064	0.6163	0.9846	0.08824	0.13380	0.00966
4.0	1.1242	0.5713	0.8990	0.07154	0.10819	0.00682
4.5	1.0572	0.5353	0.8339	0.05957	0.08993	0.00503
5.0	1.0011	0.5057	0.7820	0.05063	0.07634	0.00384
5.5	0.9532	0.4806	0.7393	0.04374	0.06588	0.00301
6.0	0.9117	0.4591	0.7033	0.03829	0.05764	0.00241
6.5	0.87524	0.44025	0.67231	0.03389	0.05099	0.00197
7.0	0.84288	0.42362	0.64530	0.03028	0.04553	0.00163
7.5	0.81389	0.40877	0.62144	0.02727	0.04099	0.00137
8.0	0.78772	0.39542	0.60015	0.02473	0.03716	0.00116

G10 Bessel functions for cylindrical coordinates

x	$J_0(x)$	$N_0(x)$	$J_1(x)$	$N_1(x)$	$J_2(x)$	$N_2(x)$
0.0	1.0000	$-\infty$	0.0000	$-\infty$	0.0000	$-\infty$
0.1	0.9975	−1.5342	0.0499	−6.4590	0.0012	−127.64
0.2	0.9900	−1.0811	0.0995	−3.3238	0.0050	−32.157
0.4	0.9604	−0.6060	0.1960	−1.7809	0.0197	−8.2983
0.6	0.9120	−0.3085	0.2867	−1.2604	0.0437	−3.8928
0.8	0.8463	−0.0868	0.3688	−0.9781	0.0758	−2.3586
1.0	0.7652	+0.0883	0.4401	−0.7812	0.1149	−1.6507
1.2	0.6711	0.2281	0.4983	−0.6211	0.1593	−1.2633
1.4	0.5669	0.3379	0.5419	−0.4791	0.2074	−1.0224
1.6	0.4554	0.4204	0.5699	−0.3476	0.2570	−0.8549
1.8	0.3400	0.4774	0.5815	−0.2237	0.3061	−0.7259
2.0	0.2239	0.5104	0.5767	−0.1070	0.3528	−0.6174
2.2	0.1104	0.5208	0.5560	+0.0015	0.3951	−0.5194
2.4	+0.0025	0.5104	0.5202	0.1005	0.4310	−0.4267
2.6	−0.0968	0.4815	0.4708	0.1884	0.4590	−0.3364
2.8	−0.1850	0.4359	0.4097	0.2635	0.4777	−0.2477
3.0	−0.2601	0.3768	0.3391	0.3247	0.4861	−0.1604
3.2	−0.3202	0.3071	0.2613	0.3707	0.4835	−0.0754
3.4	−0.3643	0.2296	0.1792	0.4010	0.4697	+0.0063
3.6	−0.3918	0.1477	0.0955	0.4154	0.4448	0.0831
3.8	−0.4026	+0.0645	+0.0128	0.4141	0.4093	0.1535
4.0	−0.3971	−0.0169	−0.0660	0.3979	0.3641	0.2159
4.2	−0.3766	−0.0938	−0.1386	0.3680	0.3105	0.2690
4.4	−0.3423	−0.1633	−0.2028	0.3260	0.2501	0.3115
4.6	−0.2961	−0.2235	−0.2566	0.2737	0.1846	0.3425
4.8	−0.2404	−0.2723	−0.2985	0.2136	0.1161	0.3613
5.0	−0.1776	−0.3085	−0.3276	0.1479	+0.0466	0.3677
5.2	−0.1103	−0.3312	−0.3432	0.0792	−0.0217	0.3617
5.4	−0.0412	−0.3402	−0.3453	+0.0101	−0.0867	0.3439
5.6	+0.0270	−0.3354	−0.3343	−0.0568	−0.1464	0.3152
5.8	0.0917	−0.3177	−0.3110	−0.1192	−0.1989	0.2766
6.0	0.1507	−0.2882	−0.2767	−0.1750	−0.2429	0.2299
6.2	0.2017	−0.2483	−0.2329	−0.2223	−0.2769	0.1766
6.4	0.2433	−0.2000	−0.1816	−0.2596	−0.3001	0.1188
6.6	0.2740	−0.1452	−0.1250	−0.2858	−0.3119	+0.0586
6.8	0.2931	−0.0864	−0.0652	−0.3002	−0.3123	−0.0019
7.0	0.3001	−0.0259	−0.0047	−0.3027	−0.3014	−0.0605
7.2	0.2951	+0.0339	+0.0543	−0.2934	−0.2800	−0.1154

(continued)

x	$J_0(x)$	$N_0(x)$	$J_1(x)$	$N_1(x)$	$J_2(x)$	$N_2(x)$
7.4	0.2786	0.0907	0.1096	– 0.2731	– 0.2490	– 0.1645
7.6	0.2516	0.1424	0.1592	– 0.2428	– 0.2097	– 0.2063
7.8	0.2154	0.1872	0.2014	– 0.2039	– 0.1638	– 0.2395
8.0	0.1716	0.2235	0.2346	– 0.1581	– 0.1130	– 0.2630

G11 Hyperbolic Bessel functions

$$I_m(z) = i^{-m} J_m(iz)$$

x	$I_0(z)$	$I_1(z)$	$I_2(z)$
0.0	1.0000	0.0000	0.0000
0.1	1.0025	0.0501	0.0012
0.2	1.0100	0.1005	0.0050
0.4	1.0404	0.2040	0.0203
0.6	1.0921	0.3137	0.0464
0.8	1.1665	0.4329	0.0843
1.0	1.2661	0.5652	0.1358
1.2	1.3939	0.7147	0.2026
1.4	1.5534	0.8861	0.2876
1.6	1.7500	1.0848	0.3940
1.8	1.9895	1.3172	0.5260
2.0	2.2796	1.5906	0.6890
2.2	2.6292	1.9141	0.8891
2.4	3.0492	2.2981	1.1342
2.6	3.5532	2.7554	1.4338
2.8	4.1574	3.3011	1.7994
3.0	4.8808	3.9534	2.2452
3.2	5.7472	4.7343	2.7884
3.4	6.7848	5.6701	3.4495
3.6	8.0278	6.7926	4.2540
3.8	9.5169	8.1405	5.2325
4.0	11.302	9.7594	6.4222
4.2	13.443	11.705	7.8683
4.4	16.010	14.046	9.6259
4.6	19.093	16.863	11.761
4.8	22.794	20.253	14.355

(continued)

x	$I_0(z)$	$I_1(z)$	$I_2(z)$
5.0	27.240	24.335	17.505
5.2	32.584	29.254	21.332
5.4	39.010	35.181	25.978
5.6	46.738	42.327	31.621
5.8	56.039	50.945	38.470
6.0	67.235	61.341	46.788
6.2	80.717	73.888	56.884
6.4	96.963	89.025	69.141
6.6	116.54	107.31	84.021
6.8	140.14	129.38	102.08
7.0	168.59	156.04	124.01
7.2	202.92	188.25	150.63
7.4	244.34	227.17	182.94
7.6	294.33	274.22	222.17
7.8	354.68	331.10	269.79
8.0	427.57	399.87	327.60

G12 Bessel functions for spherical coordinates

$$j_n(x) = \sqrt{\pi/2x}J_{n+1/2}(x);\ n_n(x) = \sqrt{\pi/2x}N_{n+1/2}(x)$$

x	$j_0(x)$	$n_0(x)$	$j_1(x)$	$n_1(x)$	$j_2(x)$	$n_2(x)$
0.0	1.0000	$-\infty$	0.0000	$-\infty$	0.0000	$-\infty$
0.1	0.9983	9.9500	0.0333	−100.50	0.0007	−3005.0
0.2	0.9933	4.9003	0.0664	−25.495	0.0027	−377.52
0.4	0.9735	2.3027	0.1312	−6.7302	0.0105	−48.174
0.6	0.9411	1.3756	0.1929	−3.2337	0.0234	−14.793
0.8	0.8967	0.8709	0.2500	−1.9853	0.0408	−6.5740
1.0	0.8415	0.5403	0.3012	−1.3818	0.0620	−3.6050
1.2	0.7767	0.3020	0.3453	−1.0283	0.0865	−2.2689
1.4	0.7039	0.1214	0.3814	−0.7906	0.1133	−1.5728
1.6	0.6247	0.0183	0.4087	−0.6133	0.1416	−1.1682
1.8	0.5410	0.1262	0.4268	−0.4709	0.1703	−0.9111
2.0	0.4546	0.2081	0.4354	−0.3506	0.1985	−0.7340
2.2	0.3675	0.2675	0.4346	−0.2459	0.2251	−0.6028

(continued)

x	$j_0(x)$	$n_0(x)$	$j_1(x)$	$n_1(x)$	$j_2(x)$	$n_2(x)$
2.4	0.2814	0.3072	0.4245	−0.1534	0.2492	−0.4990
2.6	0.1983	0.3296	0.4058	−0.0715	0.2700	−0.4121
2.8	0.1196	0.3365	0.3792	+0.0005	0.2867	−0.3359
3.0	+0.0470	0.3300	0.3457	0.0630	0.2986	−0.2670
3.2	−0.0182	0.3120	0.3063	0.1157	0.3054	−0.2035
3.4	−0.0752	0.2844	0.2623	0.1588	0.3066	−0.1442
3.6	−0.1229	0.2491	0.2150	0.1921	0.3021	−0.0890
3.8	−0.1610	0.2082	0.1658	0.2158	0.2919	−0.0378
4.0	−0.1892	0.1634	0.1161	0.2300	0.2763	+0.0091
4.2	−0.2075	0.1167	0.0673	0.2353	0.2556	0.0514
4.4	−0.2163	0.0699	+0.0207	0.2321	0.2304	0.0884
4.6	−0.2160	+0.0244	−0.0226	0.2213	0.2013	0.1200
4.8	−0.2075	−0.0182	−0.0615	0.2037	0.1691	0.1456
5.0	−0.1918	−0.0567	−0.0951	0.1804	0.1347	0.1650
5.2	−0.1699	−0.0901	−0.1228	0.1526	0.0991	0.1781
5.4	−0.1431	−0.1175	−0.1440	0.1213	0.0631	0.1850
5.6	−0.1127	−0.1385	−0.1586	0.0880	+0.0278	0.1856
5.8	−0.0801	−0.1527	−0.1665	0.0538	−0.0060	0.1805
6.0	−0.0466	−0.1600	−0.1678	+0.0199	−0.0373	0.1700
6.2	−0.0134	−0.1607	−0.1629	−0.0124	−0.0654	0.1547
6.4	+0.0182	−0.1552	−0.1523	−0.0425	−0.0896	0.1353
6.6	0.0472	−0.1440	−0.1368	−0.0690	−0.1094	0.1126
6.8	0.0727	−0.1278	−0.1172	−0.0915	−0.1243	0.0875
7.0	0.0939	−0.1077	−0.0943	−0.1092	−0.1343	0.0609
7.2	0.1102	−0.0845	−0.0692	−0.1220	−0.1391	0.0337
7.4	0.1215	−0.0593	−0.0429	−0.1294	−0.1388	+0.0068
7.6	0.1274	−0.0331	−0.0163	−0.1317	−0.1338	−0.0189
7.8	0.1280	−0.0069	+0.0095	−0.1289	−0.1244	−0.0427
8.0	0.1237	+0.0182	0.0336	−0.1214	−0.1111	−0.0637

G13 Legendre functions for spherical coordinates

ϑ	$P_{-1}=P_0$	$P_1(\cos\vartheta)$	$P_2(\cos\vartheta)$	$P_3(\cos\vartheta)$	$P_4(\cos\vartheta)$
0°	1.0000	1.0000	1.0000	1.0000	1.0000
5	1.0000	0.9962	0.9886	0.9773	0.9623
10	1.0000	0.9848	0.9548	0.9106	0.8532
15	1.0000	0.9659	0.8995	0.8042	0.6847
20	1.0000	0.9397	0.8245	0.6649	0.4750
25	1.0000	0.9063	0.7321	0.5016	0.2465
30	1.0000	0.8660	0.6250	0.3248	0.0234
35	1.0000	0.8192	0.5065	0.1454	−0.1714
40	1.0000	0.7660	0.3802	−0.0252	−0.3190
45	1.0000	0.7071	0.2500	−0.1768	−0.4063
50	1.0000	0.6428	0.1198	−0.3002	−0.4275
55	1.0000	0.5736	−0.0065	−0.3886	−0.3852
60	1.0000	0.5000	−0.1250	−0.4375	−0.2891
65	1.0000	0.4226	−0.2321	−0.4452	−0.1552
70	1.0000	0.3420	−0.3245	−0.4130	−0.0038
75	1.0000	0.2588	−0.3995	−0.3449	+0.1434
80	1.0000	0.1736	−0.4548	−0.2474	0.2659
85	1.0000	0.0872	−0.4886	−0.1291	0.3468
90	1.0000	0.0000	−0.5000	0.0000	0.3750

ϑ	$P_5(\cos\vartheta)$	$P_6(\cos\vartheta)$	$P_7(\cos\vartheta)$	$P_8(\cos\vartheta)$	$P_9(\cos\vartheta)$
0°	1.0000	1.0000	1.0000	1.0000	1.0000
5	0.9437	0.9216	0.8962	0.8675	0.8358
10	0.7840	0.7045	0.6164	0.5218	0.4228
15	0.5471	0.3983	0.2455	0.0962	−0.0428
20	0.2715	0.0719	−0.1072	−0.2518	−0.3517
25	0.0009	−0.2040	−0.3441	−0.4062	−0.3896
30	−0.2233	−0.3740	−0.4102	−0.3388	−0.1896
35	−0.3691	−0.4114	−0.3096	−0.1154	+0.0965
40	−0.4197	−0.3236	−0.1006	+0.1386	0.2900
45	−0.3757	−0.1484	+0.1271	0.2983	0.2855
50	−0.2545	+0.0564	0.2854	0.2947	0.1041
55	−0.0868	0.2297	0.3191	0.1422	−0.1296
60	+0.0898	0.3232	0.2231	−0.0736	−0.2679
65	0.2381	0.3138	0.0422	−0.2411	−0.2300
70	0.3281	0.2089	−0.1485	−0.2780	−0.0476
75	0.3427	0.0431	−0.2731	−0.1702	+0.1595
80	0.2810	−0.1321	−0.2835	+0.0233	0.2596
85	0.1577	−0.2638	−0.1778	0.2017	0.1913
90	0.0000	−0.3125	0.0000	0.2734	0.0000

G14 Periodic Mathieu functions

	$x = 0°$	10°	20°	30°	40°	50°	60°	70°	80°	90°
					$Se_0(h, \cos x)$					
$h^2 = 0$	1.0000	1.0000	1.0000	1.0000	1.0000	1.0000	1.0000	1.0000	1.0000	1.0000
1	1.0000	1.0080	1.0313	1.0674	1.1126	1.1617	1.2089	1.2480	1.2739	1.2829
2	1.0000	1.0170	1.0666	1.1448	1.2445	1.3550	1.4633	1.5549	1.6162	1.6379
3	1.0000	1.0268	1.1057	1.2319	1.3956	1.5812	1.7667	1.9262	2.0344	2.0728
4	1.0000	1.0373	1.1481	1.3279	1.5657	1.8408	2.1212	2.3663	2.5345	2.5946
5	1.0000	1.0485	1.1935	1.4323	1.7542	2.1340	2.5286	2.8789	3.1221	3.2094
6	1.0000	1.0601	1.2415	1.5445	1.9604	2.4610	2.9906	3.4679	3.8029	3.9238
7	1.0000	1.0721	1.2917	1.6638	2.1840	2.8222	3.5092	4.1376	4.5831	4.7447
8	1.0000	1.0845	1.3439	1.7899	2.4248	3.2183	4.0870	4.8927	5.4696	5.6799
9	1.0000	1.0972	1.3979	1.9226	2.6828	3.6501	4.7268	5.7387	6.4701	6.7379
					$Se_1(h, \cos x)$					
$h^2 = 0$	1.0000	0.9848	0.9397	0.8660	0.7660	0.6428	0.5000	0.3420	0.1736	0.0000
1	1.0000	0.9886	0.9539	0.8943	0.8076	0.6927	0.5499	0.3825	0.1963	0.0000
2	1.0000	0.9927	0.9693	0.9250	0.8535	0.7486	0.6066	0.4289	0.2225	0.0000
3	1.0000	0.9971	0.9858	0.9585	0.9042	0.8112	0.6711	0.4822	0.2527	0.0000
4	1.0000	1.0018	1.0037	0.9951	0.9603	0.8815	0.7443	0.5434	0.2877	0.0000
5	1.0000	1.0069	1.0230	1.0351	1.0224	0.9604	0.8275	0.6138	0.3282	0.0000
6	1.0000	1.0123	1.0438	1.0786	1.0910	1.0489	0.9220	0.6944	0.3748	0.0000
7	1.0000	1.0180	1.0662	1.1261	1.1668	1.1479	1.0292	0.7867	0.4286	0.0000
8	1.0000	1.0242	1.0902	1.1777	1.2503	1.2584	1.1502	0.8921	0.4904	0.0000
9	1.0000	1.0307	1.1160	1.2335	1.3419	1.3815	1.2866	1.0119	0.5610	0.0000
					$Se_2(h, \cos x)$					
$h^2 = 0$	1.0000	0.9397	0.7660	0.5000	0.1736	−0.1736	−0.5000	−0.7660	−0.9397	−1.0000
1	1.0000	0.9467	0.7917	0.5496	0.2451	−0.0882	−0.4103	−0.6794	−0.8582	−0.9208
2	1.0000	0.9530	0.8147	0.5944	0.3098	−0.0110	−0.3298	−0.6026	−0.7869	−0.8522
3	1.0000	0.9586	0.8355	0.6348	0.3681	+0.0581	−0.2589	−0.5366	−0.7275	−0.7956
4	1.0000	0.9638	0.8544	0.6716	0.4209	0.1200	−0.1968	−0.4810	−0.6797	−0.7512
5	1.0000	0.9686	0.8720	0.7057	0.4696	0.1764	−0.1420	−0.4345	−0.6424	−0.7178
6	1.0000	0.9731	0.8885	0.7379	0.5154	0.2285	−0.0929	−0.3955	−0.6142	−0.6942
7	1.0000	0.9774	0.9044	0.7689	0.5593	0.2779	−0.0481	−0.3627	−0.5937	−0.6790
8	1.0000	0.9816	0.9200	0.7992	0.6023	0.3256	−0.0062	−0.3346	−0.5798	−0.6710
9	1.0000	0.9857	0.9355	0.8294	0.6450	0.3727	+0.0338	−0.3103	−0.5715	−0.6695
					$Se_3(h, \cos x)$					
$h^2 = 0$	1.0000	0.8660	0.5000	0.0000	−0.5000	−0.8660	−1.0000	−0.8660	−0.5000	0.0000
1	1.0000	0.8732	0.5242	+0.0407	−0.4530	−0.8261	−0.9753	−0.8562	−0.4981	0.0000
2	1.0000	0.8802	0.5481	0.0815	−0.4052	−0.7842	−0.9484	−0.8443	−0.4952	0.0000
3	1.0000	0.8871	0.5717	0.1221	−0.3566	−0.7410	−0.9195	−0.8307	−0.4912	0.0000
4	1.0000	0.8938	0.5949	0.1625	−0.3078	−0.6965	−0.8891	−0.8155	−0.4863	0.0000
5	1.0000	0.9004	0.6176	0.2023	−0.2590	−0.6513	−0.8572	−0.7991	−0.4807	0.0000
6	1.0000	0.9067	0.6398	0.2415	−0.2104	−0.6057	−0.8245	−0.7820	−0.4747	0.0000

(continued)

					$Se_3(h, \cos x)$					
7	1.0000	0.9128	0.6612	0.2798	-0.1625	-0.5602	-0.7916	-0.7643	-0.4683	0.0000
8	1.0000	0.9187	0.6820	0.3171	-0.1154	-0.5151	-0.7587	-0.7467	-0.4620	0.0000
9	1.0000	0.9244	0.7020	0.3533	-0.0694	-0.4709	-0.7263	-0.7295	-0.4558	0.0000
					$Se_4(h, \cos x)$					
$h^2 = 0$	1.0000	0.7660	0.1736	0.5000	0.9397	0.9397	0.5000	0.1736	0.7660	1.0000
1	1.0000	0.7730	0.1944	0.4727	0.9214	0.9410	0.5181	0.1504	0.7465	0.9835
2	1.0000	0.7798	0.2153	0.4448	0.9019	0.9407	0.5349	0.1279	0.7274	0.9671
3	1.0000	0.7867	0.2361	0.4165	0.8813	0.9390	0.5505	0.1061	0.7085	0.9510
4	1.0000	0.7935	0.2570	0.3876	0.8595	0.9358	0.5648	0.0851	0.6900	0.9350
5	1.0000	0.8002	0.2779	0.3584	0.8367	0.9311	0.5778	0.0648	0.6716	0.9190
6	1.0000	0.8069	0.2988	0.3287	0.8127	0.9249	0.5895	0.0451	0.6535	0.9031
7	1.0000	0.8136	0.3197	0.2986	0.7877	0.9173	0.6000	0.0262	0.6356	0.8872
8	1.0000	0.8201	0.4305	0.2681	0.7617	0.9083	0.6093	0.0080	0.6179	0.8714
9	1.0000	0.8267	0.8267	0.2374	0.7348	0.8979	0.6173	0.0096	0.6004	0.8555
					$So_1(h, \cos x)$					
$h^2 = 0$	0.0000	0.1736	0.3420	0.5000	0.6428	0.7660	0.8660	0.9397	0.9848	1.0000
1	0.0000	0.1743	0.3471	0.5159	0.6769	0.8242	0.9507	1.0484	1.1104	1.1317
2	0.0000	0.1750	0.3523	0.5325	0.7129	0.8864	1.0424	1.1675	1.2490	1.2773
3	0.0000	0.1757	0.3577	0.5498	0.7507	0.9528	1.1415	1.2976	1.4013	1.4378
4	0.0000	0.1764	0.3632	0.5676	0.7905	1.0235	1.2484	1.4393	1.5684	1.6141
5	0.0000	0.1771	0.3688	0.5861	0.8321	1.0985	1.3634	1.5933	1.7511	1.8075
6	0.0000	0.1778	0.3745	0.6051	0.8757	1.1782	1.4869	1.7602	1.9505	2.0188
7	0.0000	0.1785	0.3804	0.6248	0.9213	1.2625	1.6192	1.9409	2.1675	2.2494
8	0.0000	0.1793	0.3864	0.6451	0.9688	1.3517	1.7609	2.1360	2.4033	2.5005
9	0.0000	0.1800	0.3925	0.6660	1.0184	1.4460	1.9122	2.3464	2.6590	2.7733
					$So_2(h, \cos x)$					
$h^2 = 0$	0.0000	0.1710	0.3214	0.4330	0.4924	0.4924	0.4330	0.3214	0.1710	0.0000
1	0.0000	0.1714	0.3246	0.4422	0.5098	0.5172	0.4610	0.3460	0.1854	0.0000
2	0.0000	0.1719	0.3278	0.4517	0.5279	0.5434	0.4910	0.3725	0.2010	0.0000
3	0.0000	0.1723	0.3312	0.4616	0.5470	0.5712	0.5230	0.4010	0.2179	0.0000
4	0.0000	0.1728	0.3346	0.4718	0.5669	0.6006	0.5572	0.4318	0.2362	0.0000
5	0.0000	0.1733	0.3382	0.4824	0.5877	0.6316	0.5937	0.4648	0.2560	0.0000
6	0.0000	0.1738	0.3418	0.4934	0.6095	0.6644	0.6326	0.5004	0.2774	0.0000
7	0.0000	0.1743	0.3455	0.5047	0.6322	0.6990	0.6741	0.5386	0.3005	0.0000
8	0.0000	0.1748	0.3494	0.5164	0.6560	0.7356	0.7183	0.5796	0.3254	0.0000
9	0.0000	0.1753	0.3533	0.5285	0.6807	0.7741	0.7654	0.6236	0.3523	0.0000
					$So_3(h, \cos x)$					
$h^2 = 0$	0.0000	0.1667	0.2887	0.3333	0.2887	0.1667	0.0000	-0.1667	-0.2887	-0.3333
1	0.0000	0.1671	0.2916	0.3411	0.3017	0.1822	+0.0135	-0.1587	-0.2866	-0.3337
2	0.0000	0.1675	0.2945	0.3489	0.3147	0.1977	0.0270	-0.1511	-0.2852	-0.3349

(continued)

					$So_3(h, \cos x)$					
3	0.0000	0.1679	0.2974	0.3567	0.3277	0.2134	0.0404	– 0.1438	– 0.2844	– 0.3369
4	0.0000	0.1683	0.3003	0.3644	0.3409	0.2292	0.0540	– 0.1368	– 0.2843	– 0.3398
5	0.0000	0.1687	0.3032	0.3722	0.3542	0.2452	0.0676	– 0.1300	– 0.2848	– 0.3434
6	0.0000	0.1691	0.3060	0.3801	0.3676	0.2615	0.0814	– 0.1233	– 0.2859	– 0.3479
7	0.0000	0.1695	0.3089	0.3881	0.3813	0.2781	0.0955	– 0.1168	– 0.2876	– 0.3532
8	0.0000	0.1699	0.3118	0.3961	0.3953	0.2951	0.1098	– 0.1104	– 0.2900	– 0.3593
9	0.0000	0.1704	0.3148	0.4042	0.4095	0.3126	0.1246	– 0.1040	– 0.2929	– 0.3662
					$So_4(h, \cos x)$					
$h^2 = 0$	0.0000	0.1607	0.2462	0.2165	0.0855	– 0.0855	– 0.2165	– 0.2462	– 0.1607	0.0000
1	0.0000	0.1611	0.2489	0.2228	0.0941	– 0.0783	– 0.2137	– 0.2476	– 0.1630	0.0000
2	0.0000	0.1615	0.2516	0.2292	0.1028	– 0.0709	– 0.2108	– 0.2490	– 0.1653	0.0000
3	0.0000	0.1619	0.2542	0.2355	0.1116	– 0.0634	– 0.2079	– 0.2504	– 0.1677	0.0000
4	0.0000	0.1623	0.2569	0.2419	0.1205	– 0.0558	– 0.2048	– 0.2517	– 0.1701	0.0000
5	0.0000	0.1627	0.2596	0.2483	0.1294	– 0.0480	– 0.2016	– 0.2532	– 0.1726	0.0000
6	0.0000	0.1631	0.2622	0.2547	0.1385	– 0.0401	– 0.1983	– 0.2546	– 0.1751	0.0000
7	0.0000	0.1635	0.2649	0.2612	0.1476	– 0.0320	– 0.1950	– 0.2561	– 0.1778	0.0000
8	0.0000	0.1639	0.2675	0.2676	0.1569	– 0.0238	– 0.1915	– 0.2576	– 0.1805	0.0000
9	0.0000	0.1643	0.2701	0.2741	0.1662	– 0.0155	– 0.1881	– 0.2593	– 0.1834	0.0000

Appendix H
Table of transformations of regions

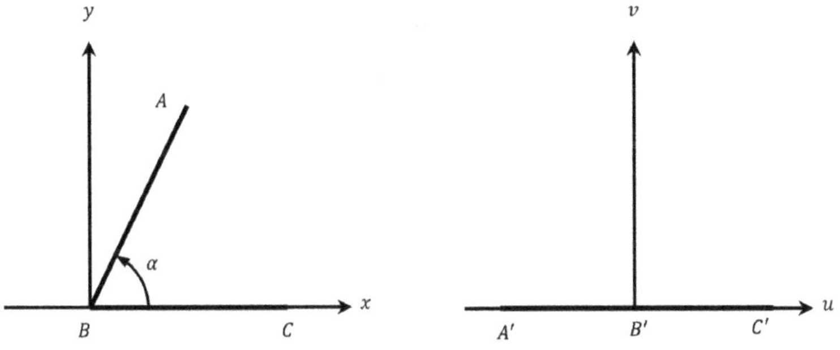

1. $w = z^{\pi/\alpha}$

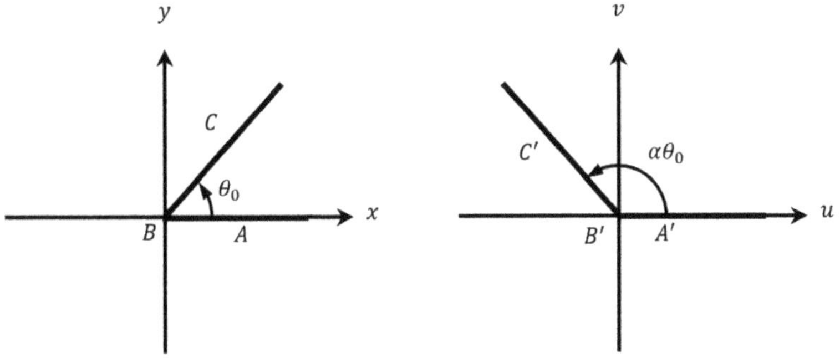

2. $w = z^{\alpha},\ \alpha > 0$

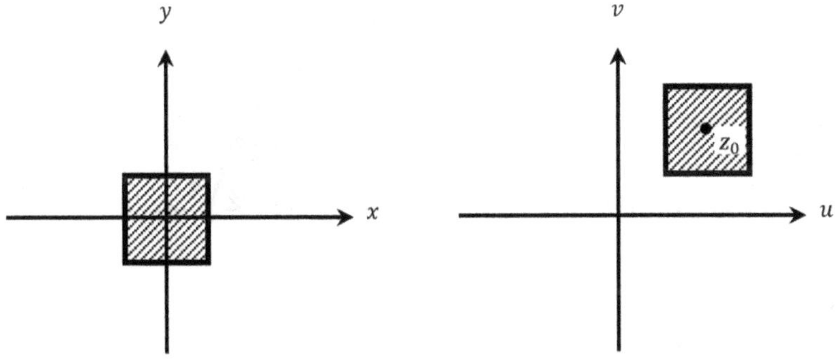

3. $x = z + z_0$

https://doi.org/10.1515/9783111351179-018

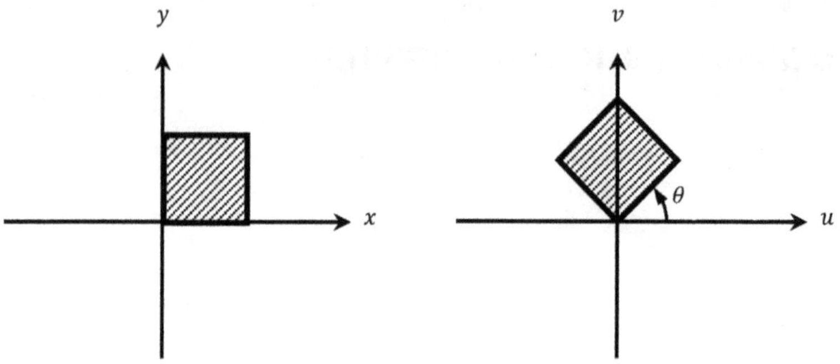

4. $w = e^{i\theta} z$

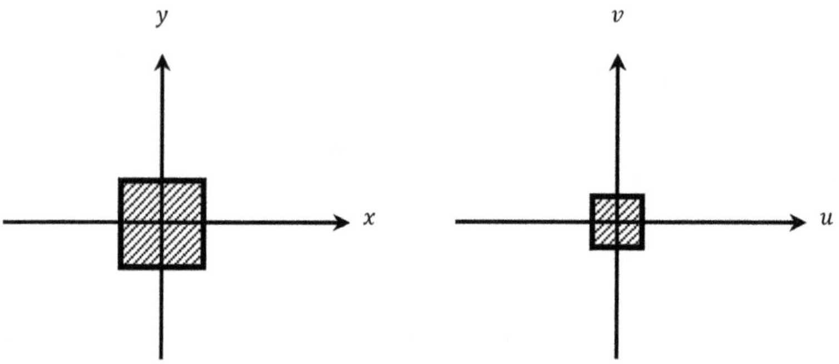

5. $w = \alpha z, \ \alpha > 0$

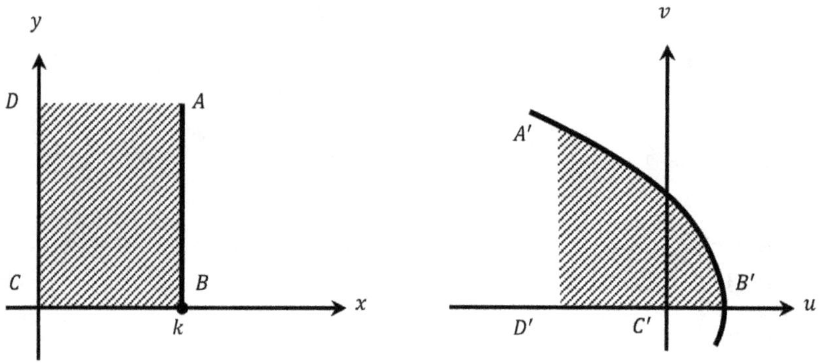

6. $w = z^2;\ A'B\colon\ \rho = \dfrac{2k^2}{1 + \cos\phi}$

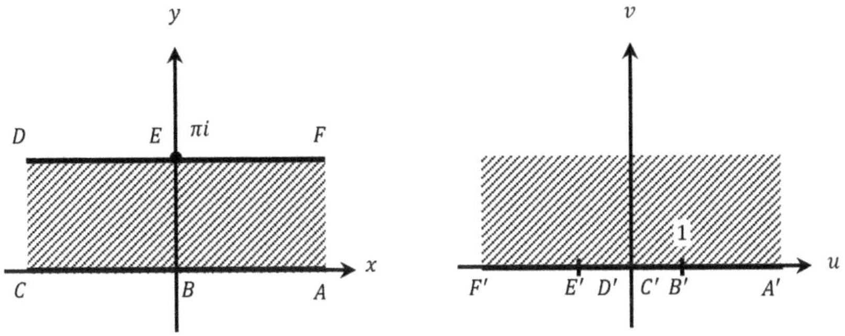

7. $w = e^z$

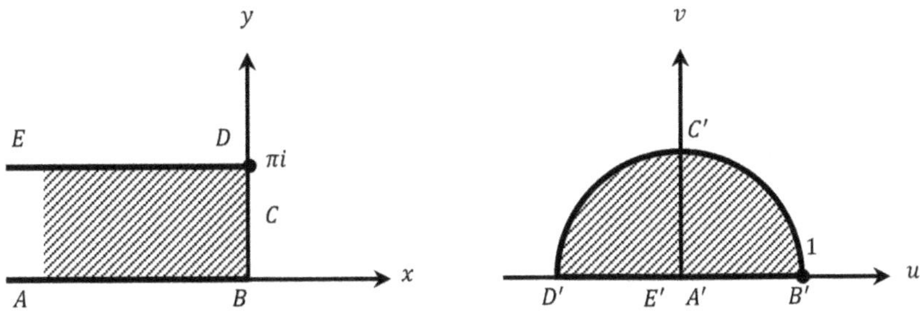

8. $w = e^z$

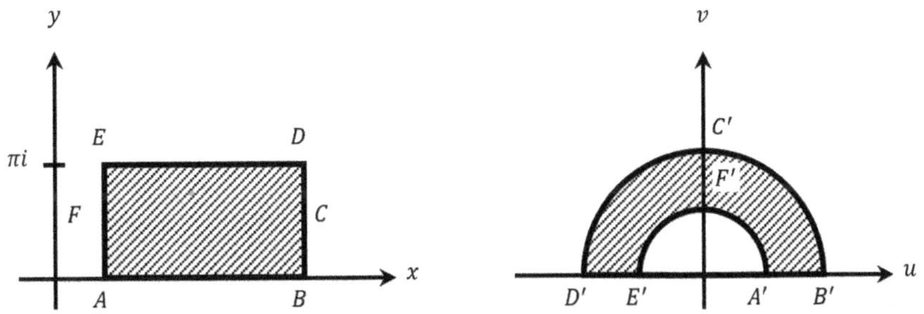

9. $w = e^z$

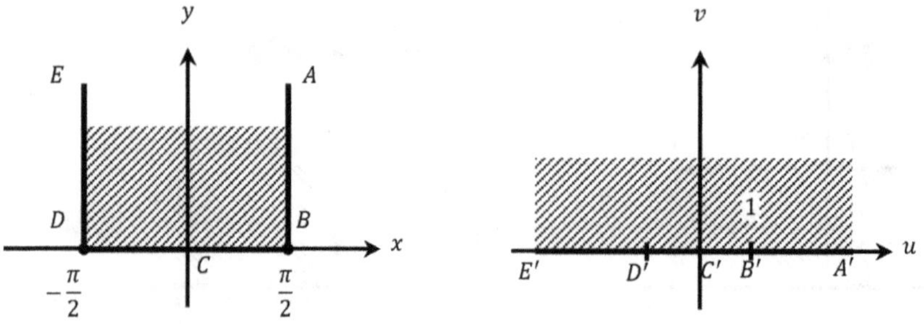

10. $w = \sin z$

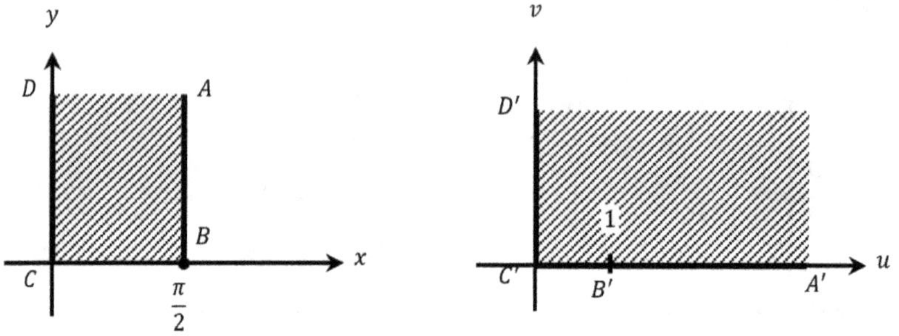

11. $w = \sin z$

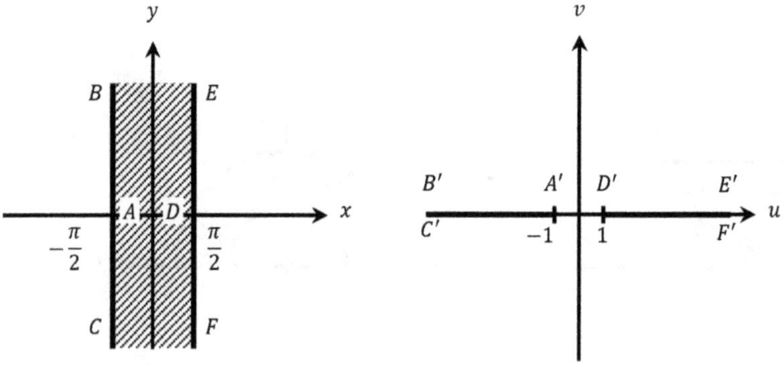

12. $w = \sin z,\ z = \sin^{-1} w$

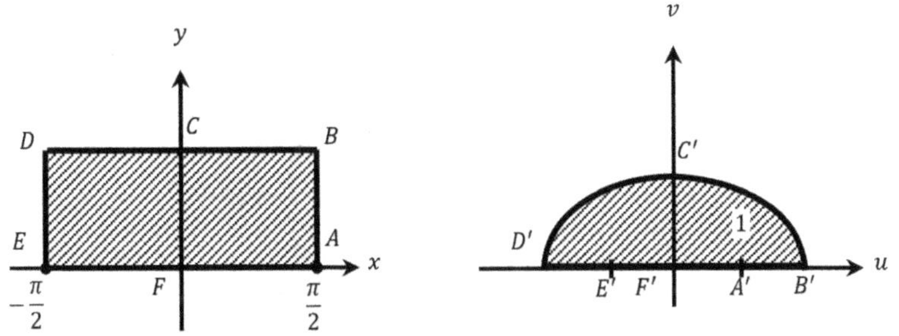

13. $w = \sin z$; BCD: $y = k$, $B'C'D'$: $\left(\frac{u}{\cosh k}\right)^2 + \left(\frac{v}{\sinh k}\right)^2 = 1$

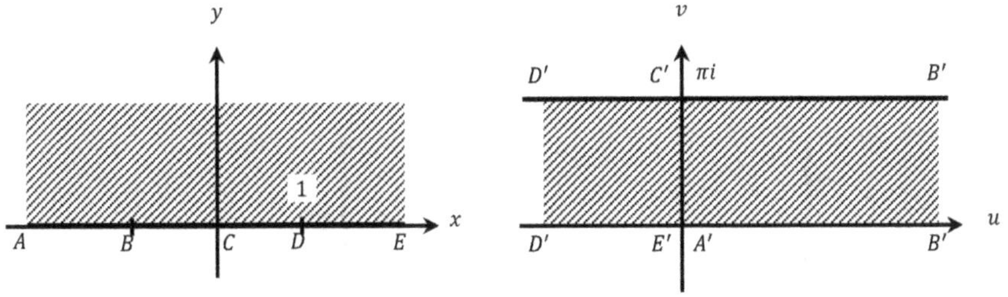

14. $w = \log \frac{z-1}{z+1} = \log \frac{r_1}{r_2} + i(\theta_1 - \theta_2)$; $z = -\coth \frac{w}{2}$

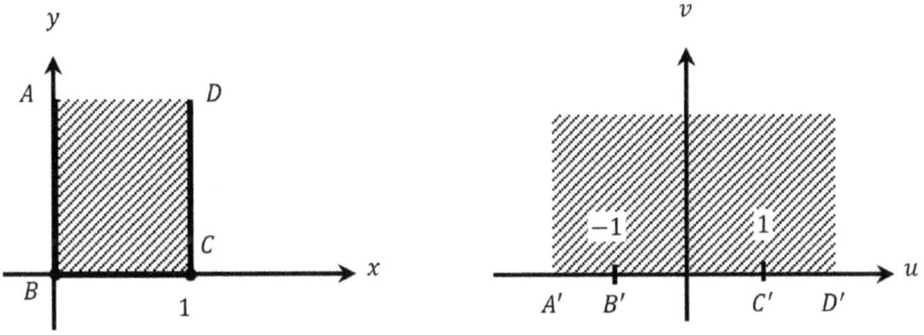

15. $w = -\cos \pi z$

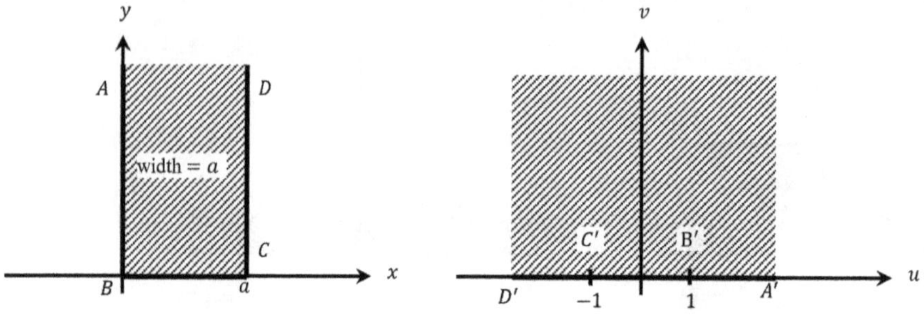

16. $w = \cos\left(\frac{\pi z}{a}\right)$

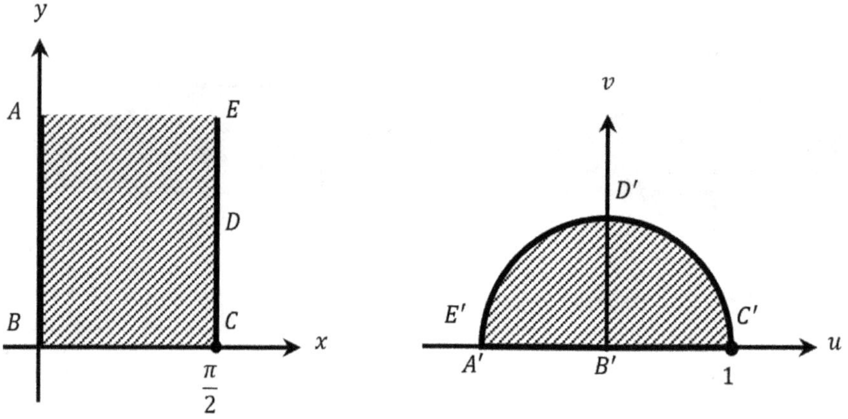

17. $w = \tan^2 \frac{z}{2} = \frac{1 - \cos z}{1 + \cos z}$

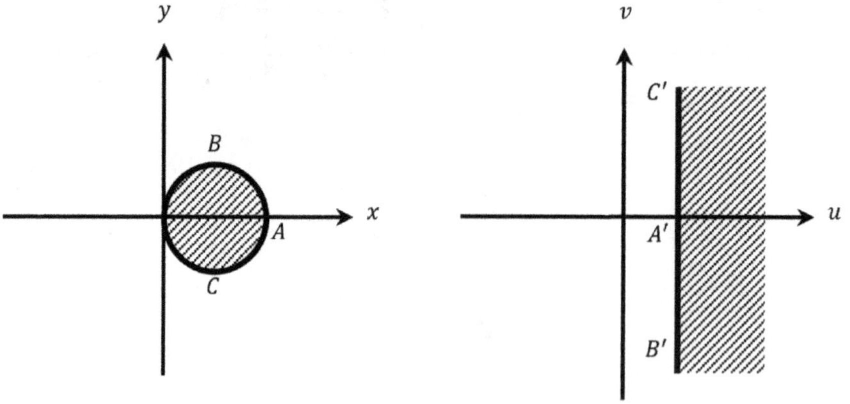

18. $w = 1/z$

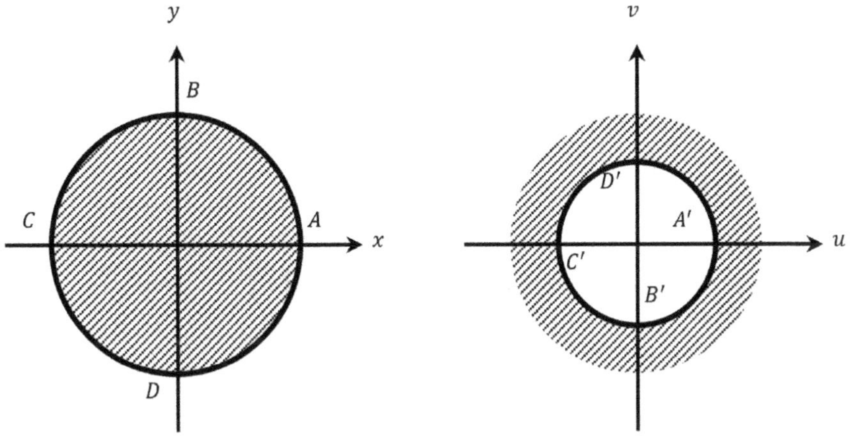

19. $w = 1/z$

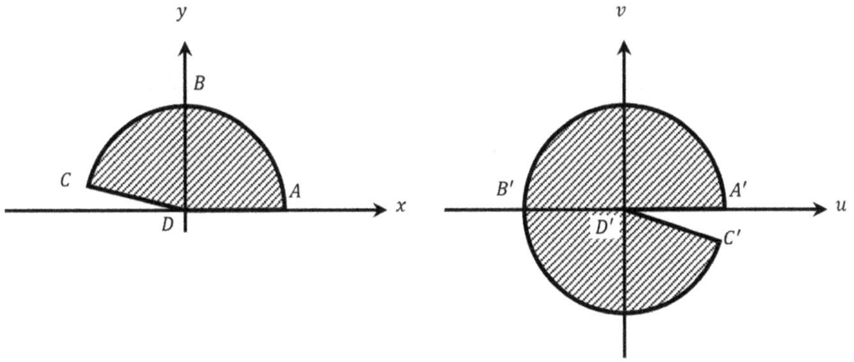

20. $w = z^2$

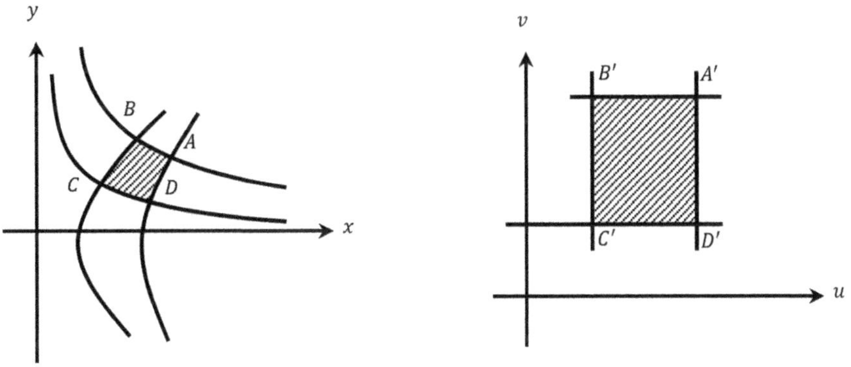

21. $w = z^2$

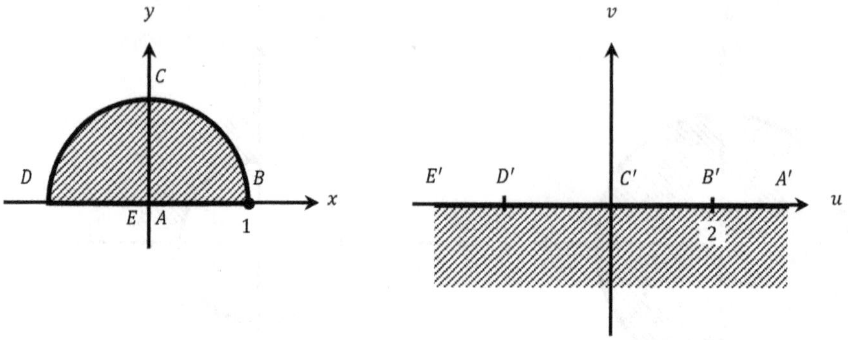

22. $w = z + 1/z$

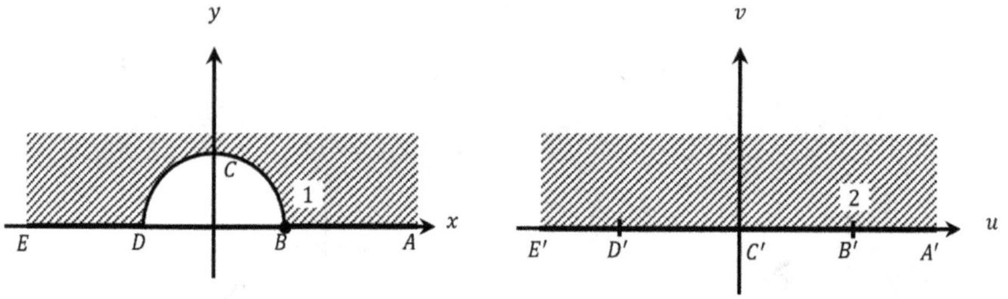

23. $w = z + 1/z$

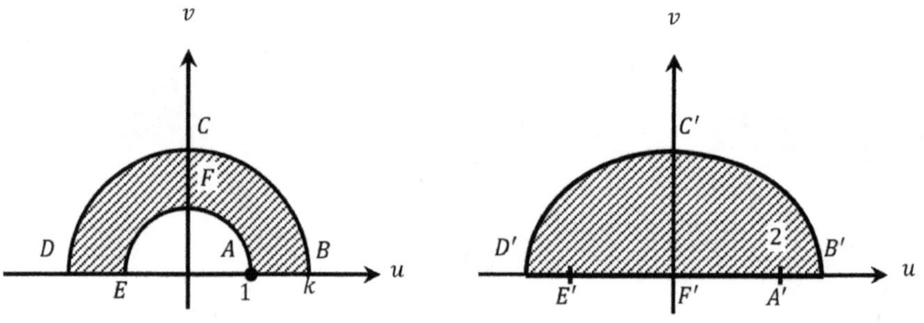

24. $w = z + \dfrac{1}{z}$; $B'C'D'$: $\left(\dfrac{ku}{k^2+1}\right)^2 + \left(\dfrac{kv}{k^2-1}\right)^2 - 1$

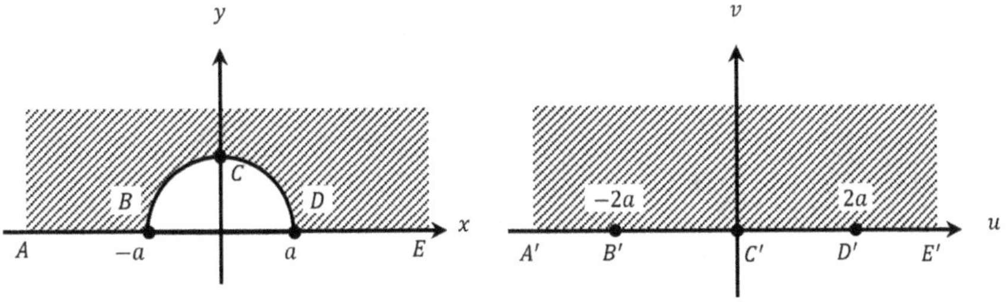

25. $w = z + \dfrac{a^2}{z}$

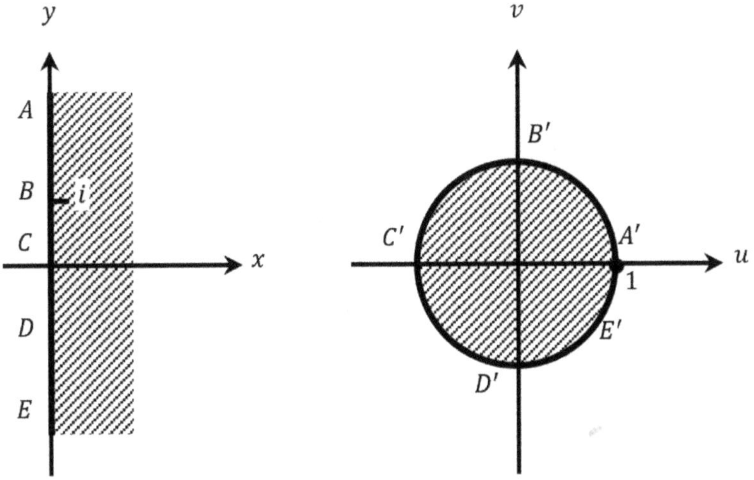

26. $w = \dfrac{z-1}{z+1}$

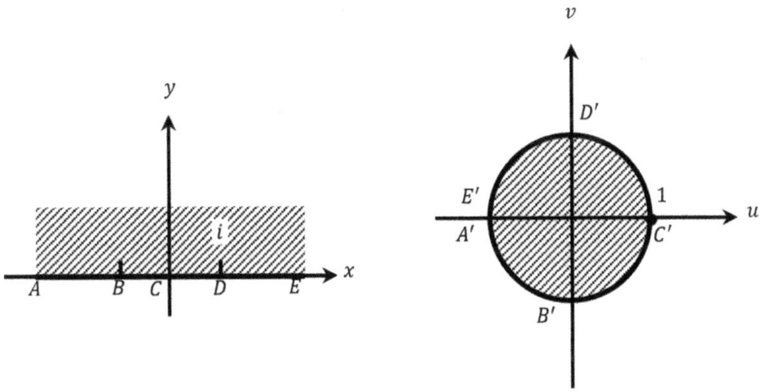

27. $w = \dfrac{i-z}{i+z}$

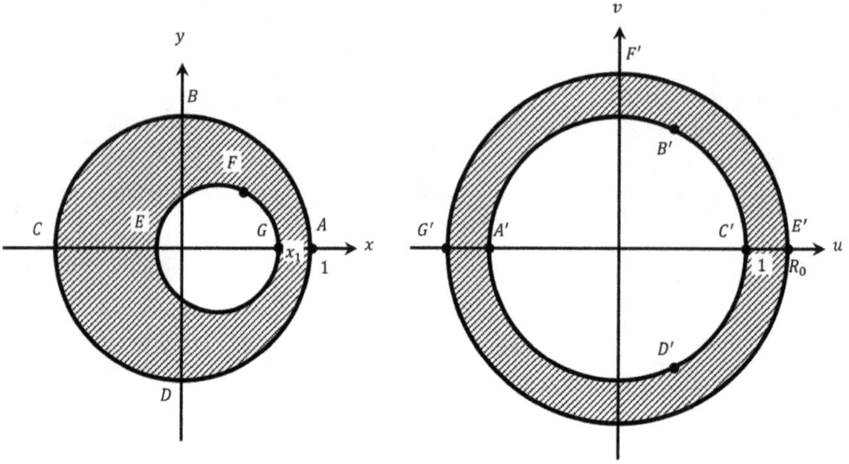

28. $w = \dfrac{z-a}{az-1}$; $a = \dfrac{1 + x_1x_2 + \sqrt{(1-x_1^2)(1-x_2^2)}}{x_1 + x_2}$; $R_0 = \dfrac{1 - x_1x_2 + \sqrt{(1-x_1^2)(1-x_2^2)}}{x_1 - x_2}$,

($a > 1$ and $R_0 > 1$ when $-1 < x_2 < x_1 < 1$)

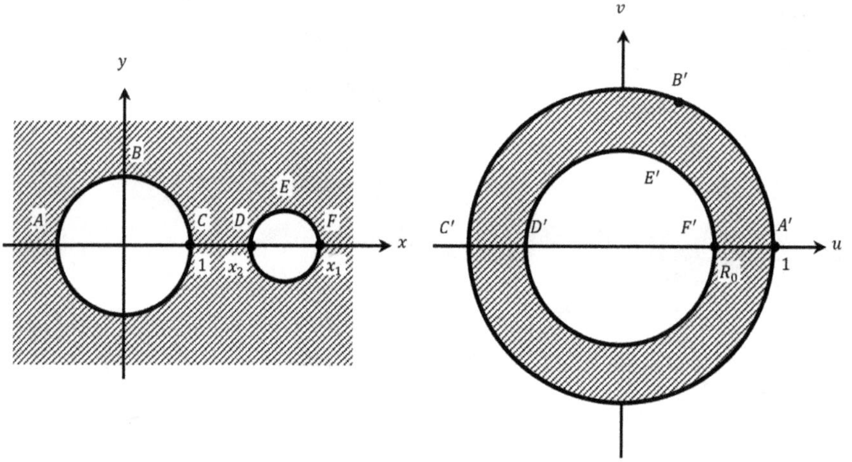

29. $w = \dfrac{z-a}{az-1}$; $a = \dfrac{1 + x_1x_2 + \sqrt{(x_1^2-1)(x_2^2-1)}}{x_1 + x_2}$; $R_0 = \dfrac{x_1x_2 - 1 - \sqrt{(x_1^2-1)(x_2^2-1)}}{x_1 - x_2}$,

($x_2 < a < x_1$ and $0 < R_0 < 1$ when $1 < x_2 < x_1$)

$$w = z + \frac{1}{z};\ B'C'D': \left(\frac{ku}{k^2+1}\right)^2 + \left(\frac{kv}{k^2-1}\right)^2 = 1$$

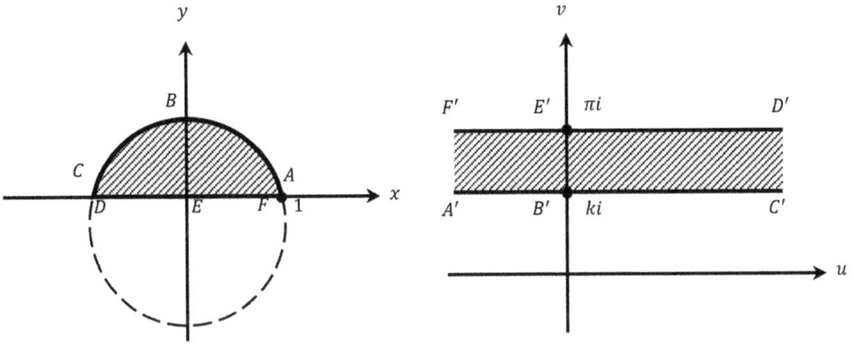

30. $w = \log\dfrac{z-1}{z+1}$; ABC: $x^2 + y^2 - 2y\cot k = 1$

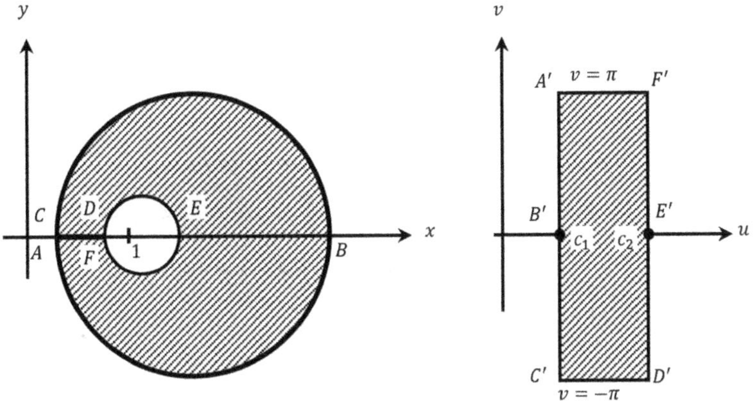

31. $w = \log\dfrac{z+1}{z-1}$; centers or circles at $z = \coth c_n$, radii: $\operatorname{csch} c_n$ $(n = 1,\ 2)$

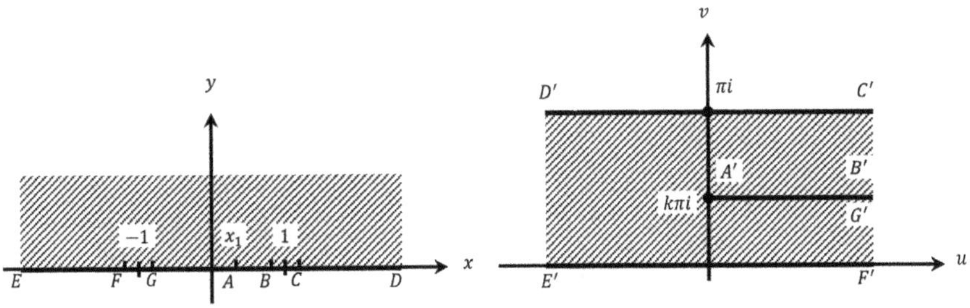

32. $w = k\log\dfrac{k}{1-k} + \log 2(1-k) + i\pi - k\log(z+1) - (1-k)\log(z-1)$; $x_1 = 2k - 1$

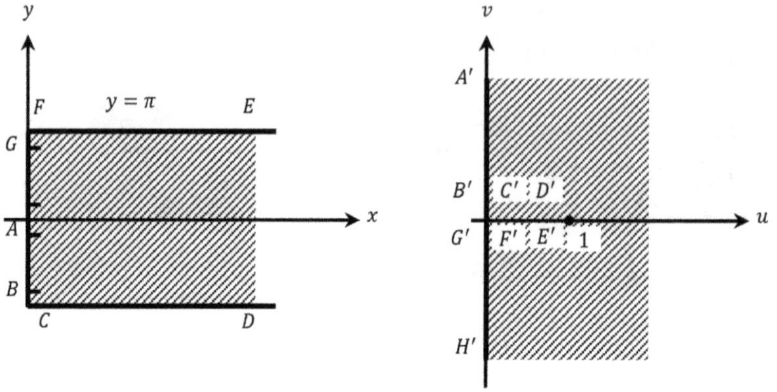

33. $w = \coth\frac{z}{2} = \frac{e^z + 1}{e^z - 1}$

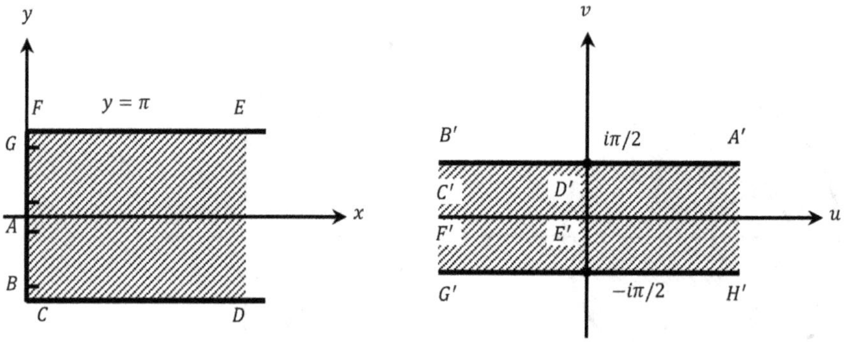

34. $w = \log\coth\frac{z}{2}$

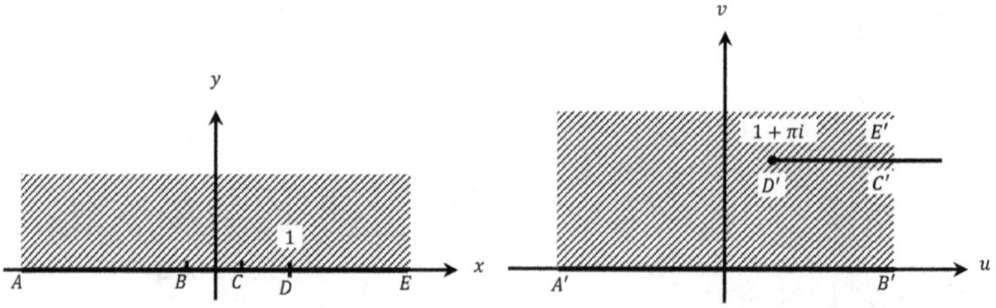

35. $w = \pi i + z - \log z$

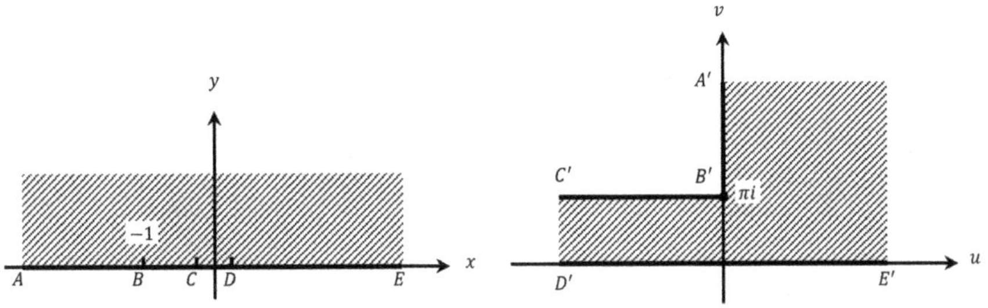

36. $w = 2(z+1)^{1/2} + \log\dfrac{(z+1)^{1/2}-1}{(z+1)^{1/2}+1}$

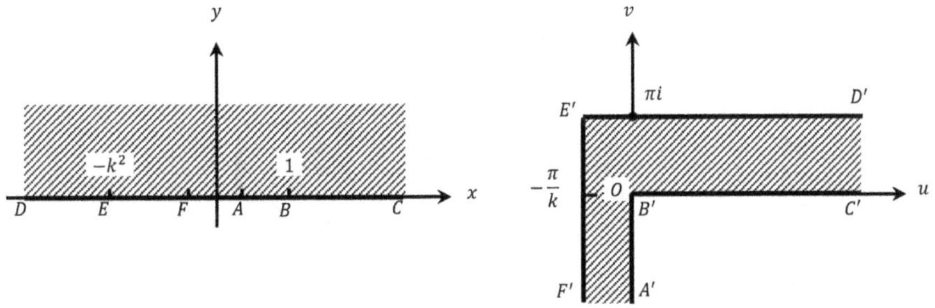

37. $w = \dfrac{i}{k}\log\dfrac{1+ikt}{1-ikt} + \log\dfrac{1+t}{1-t};\ t = \left(\dfrac{z-1}{z+k^2}\right)^{1/2}$

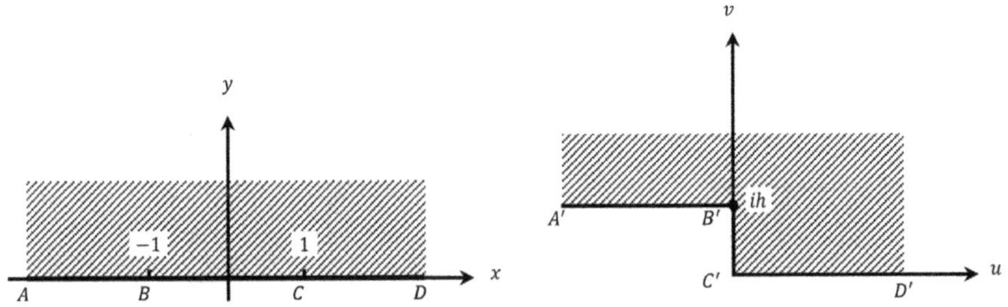

38. $w = \dfrac{h}{\pi}\left[(z^2-1)^{\frac{1}{2}} + \cosh^{-1} z\right]$

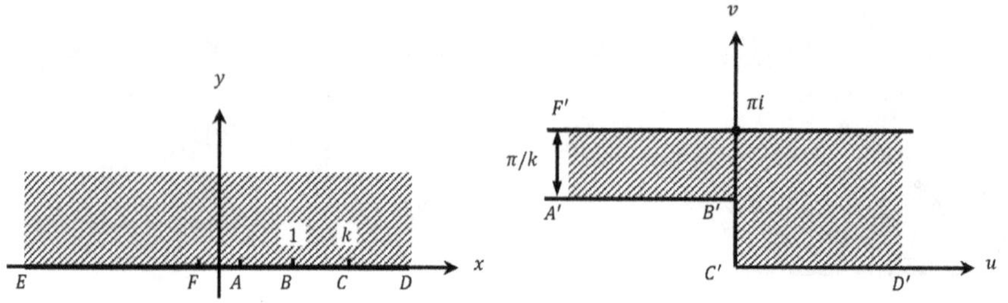

39. $w = \cosh^{-1}\left(\frac{2z-k-1}{k-1}\right) - \frac{1}{k}\cosh^{-1}\left[\frac{(k+1)z-2k}{(k-1)z}\right]$

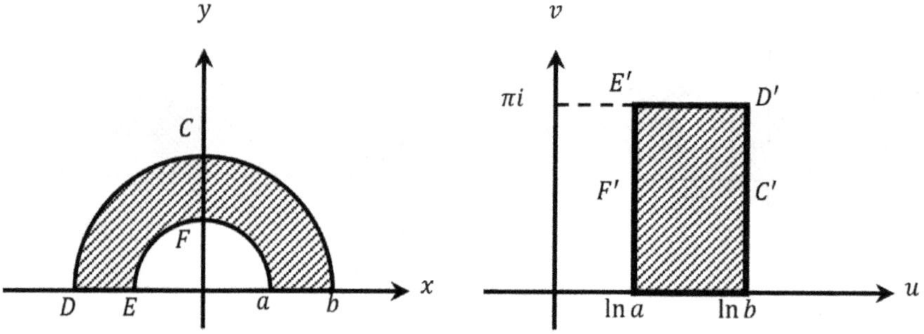

40. $w = \log_e|z| + i\arg z,\ a > 1$

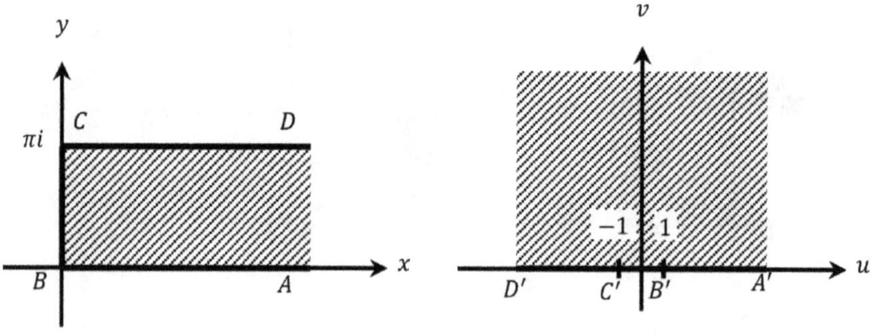

41. $w = \cosh z$

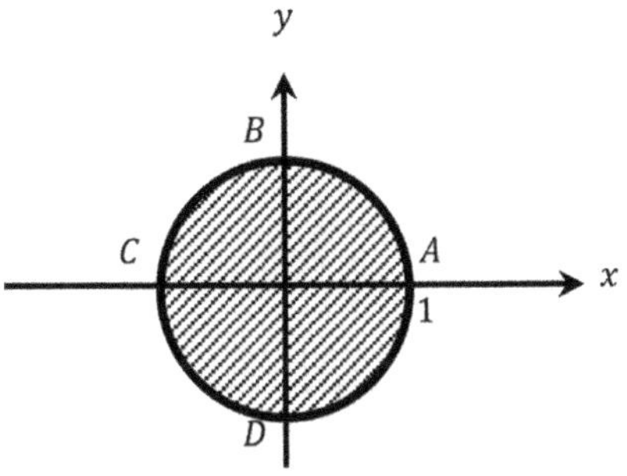

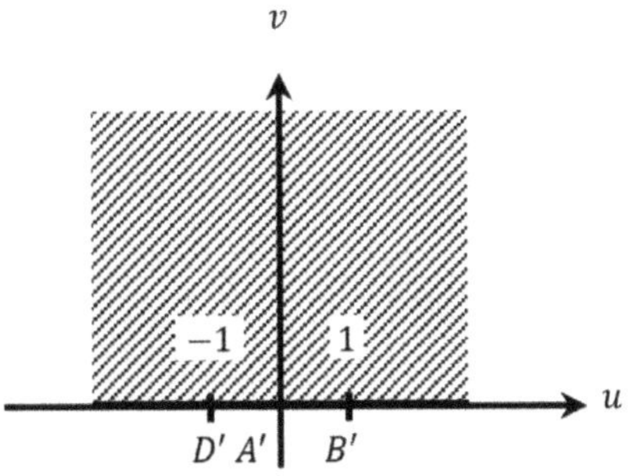

42. $w = i\dfrac{1-z}{1+z}$

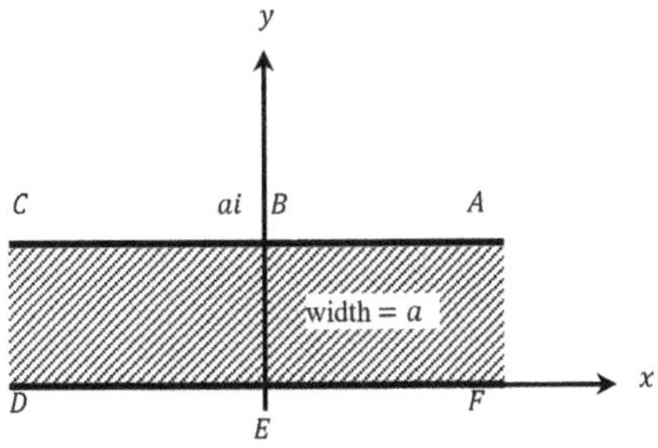

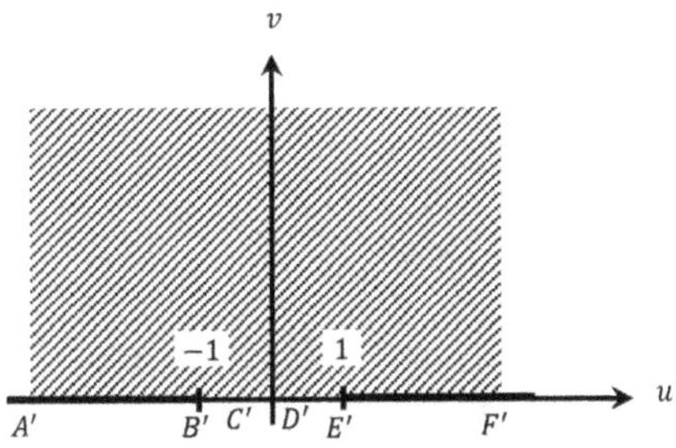

43. $w = e^{\pi z/a}$

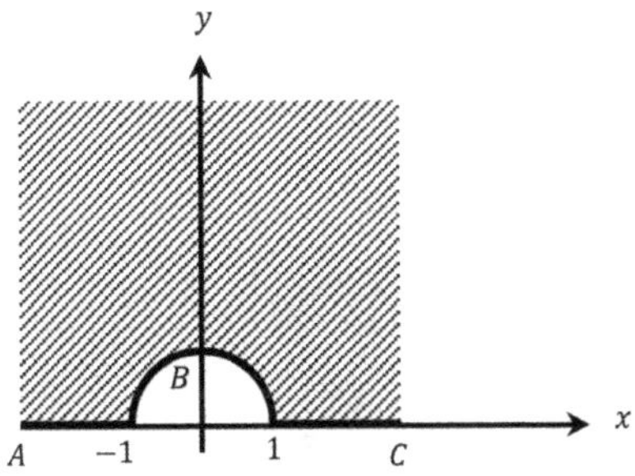

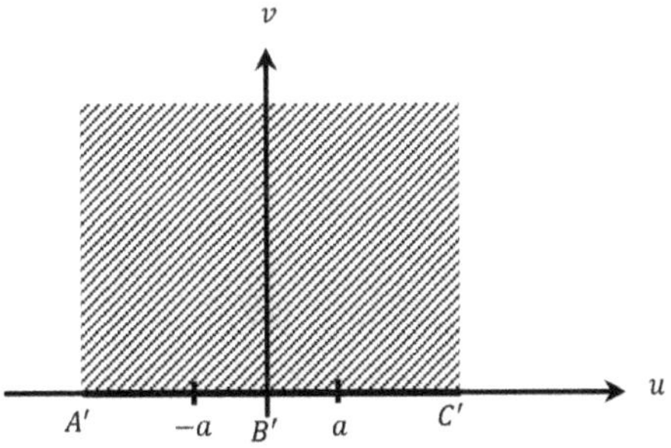

44. $w = \dfrac{a}{2}\left(z + \dfrac{1}{z}\right)$

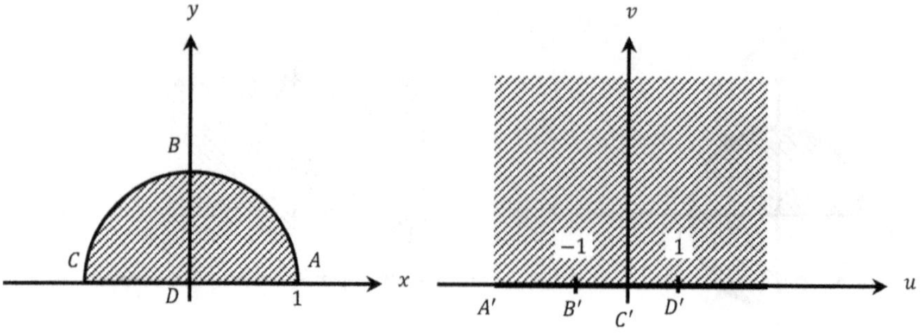

45. $w = \left(\dfrac{1+z}{1-z}\right)^2$

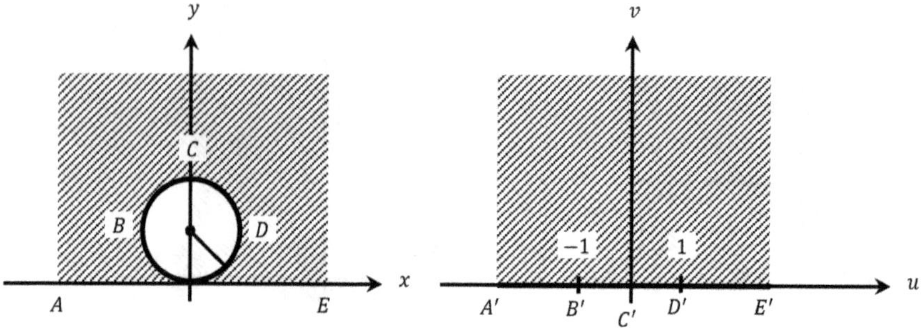

46. $w = \dfrac{e^{\pi/z} + e^{-\pi/z}}{e^{\pi/z} - e^{-\pi/z}}$

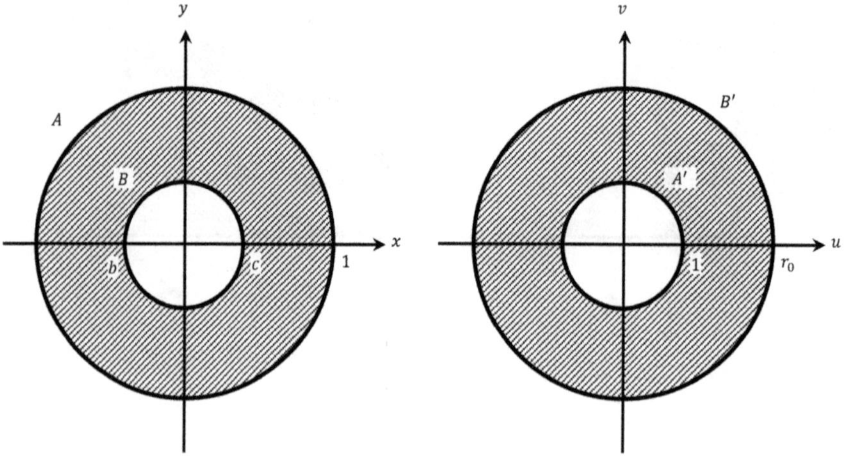

47. $a = \dfrac{1 + bc + \sqrt{(1-b^2)(1-c^2)}}{c+b}, \quad r_0 = \dfrac{1 - bc + \sqrt{(1-b^2)(1-c^2)}}{c-b}$

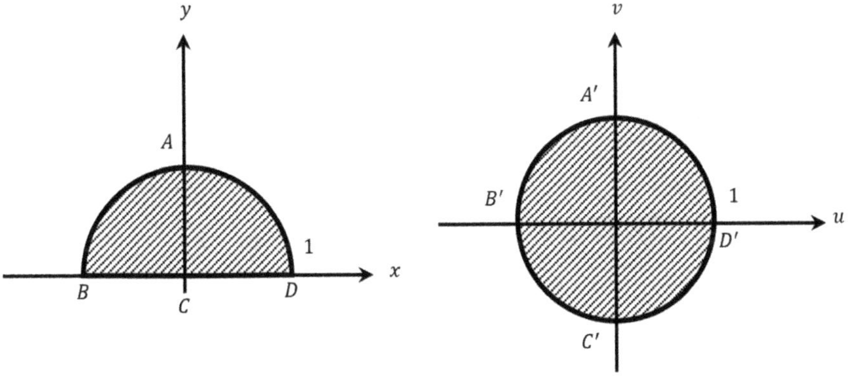

48. $w = i\dfrac{z^2 + 2iz + 1}{z^2 - 2iz + 1}$

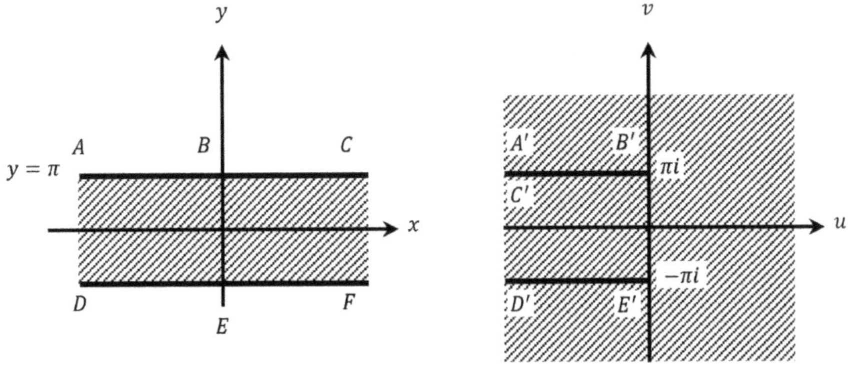

49. $w = z + e^z + 1$

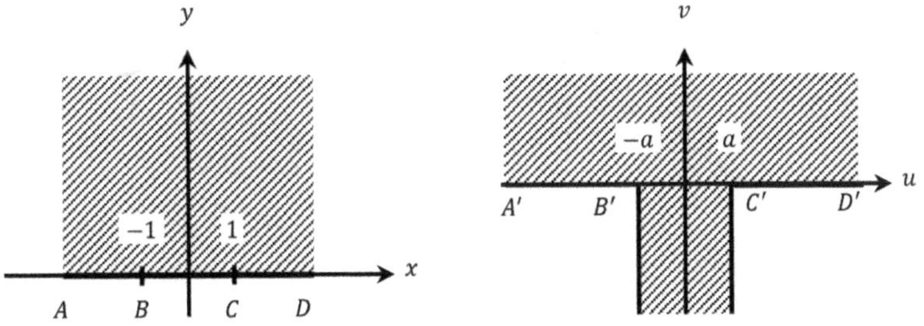

50. $w = \dfrac{2a}{\pi}\left[(z^2 - 1)^{1/2} + \sin^{-1}\left(\dfrac{1}{z}\right)\right]$

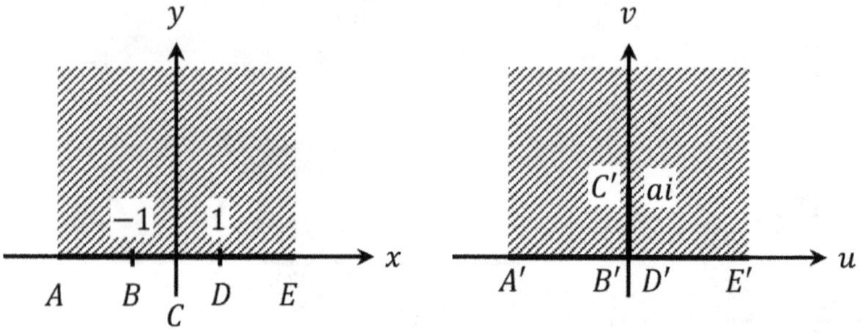

51. $w = a\left(z^2 - 1\right)^{1/2}$

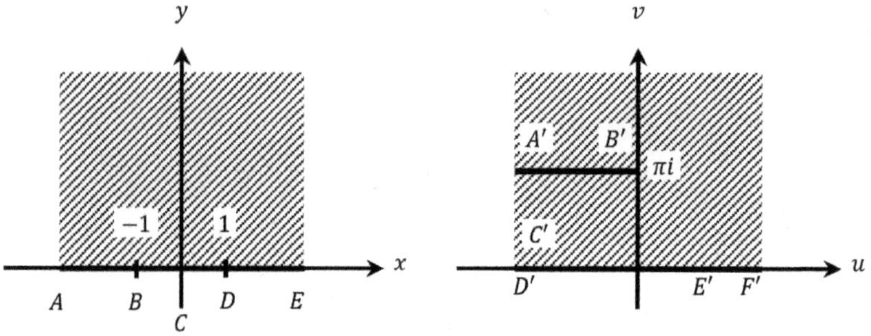

52. $w = z + \ln z + 1$

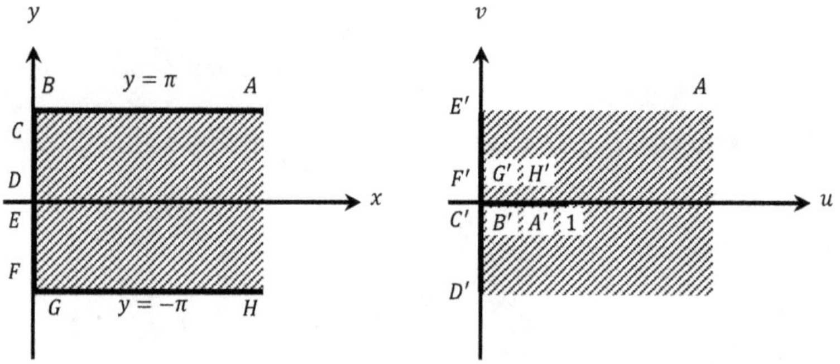

53. $w = \dfrac{e^z + 1}{e^z - 1}$

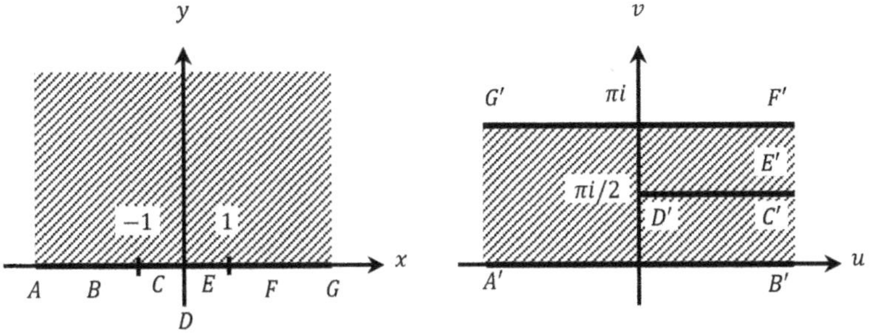

54. $w = \pi i - \dfrac{1}{2}[\ln(z+1) + \ln(z-1)]$

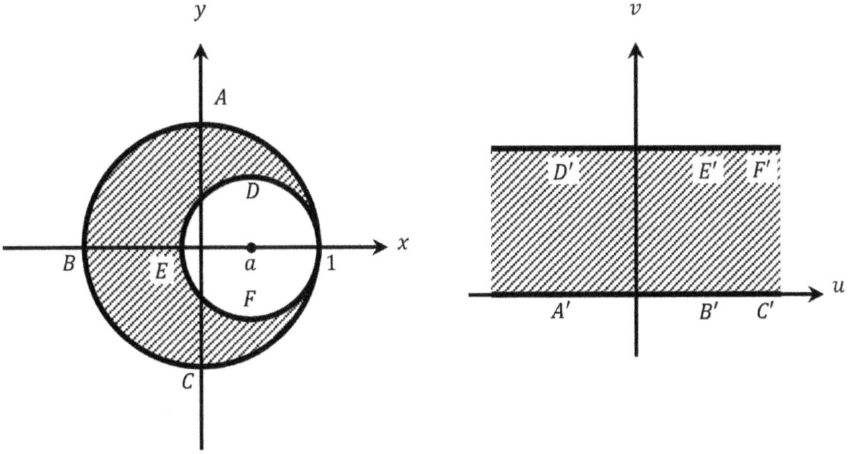

55. $w = (1-i)\dfrac{z-i}{z-1}$

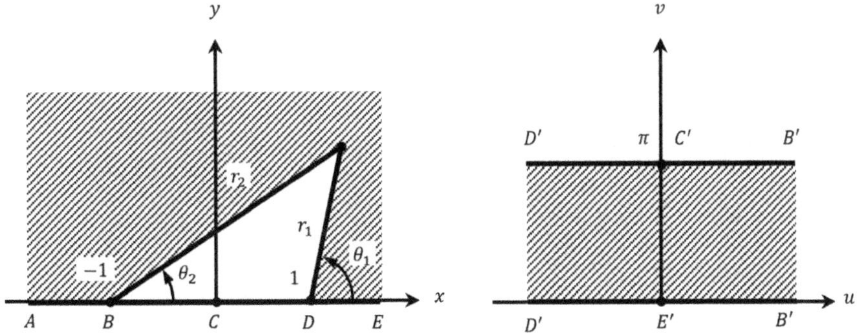

56. $w = \log\dfrac{z-1}{z+1} = \ln\dfrac{r_1}{r_2} + i(\theta_1 - \theta_2)$

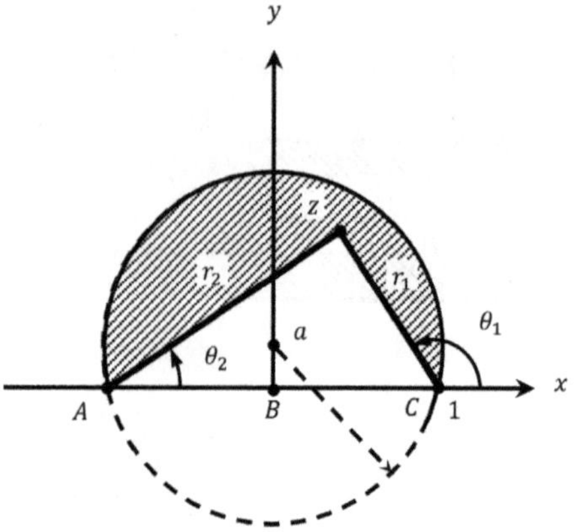

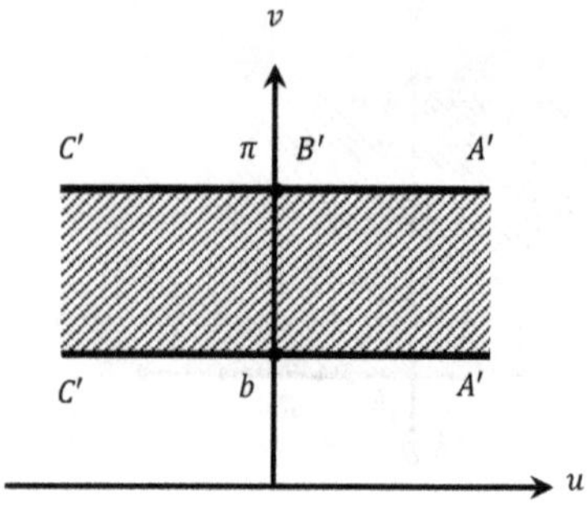

57. $w = \log\dfrac{z-1}{z+1} = \ln\dfrac{r_1}{r_2} + i(\theta_1 - \theta_2)$

List of figures

https://doi.org/10.1515/9783111351179-019

Index

Boldface page numbers indicate definitions.
Italic page numbers refer to tables or figures.

https://doi.org/10.1515/9783111351179-020

Erratum to Chapter 1

Foluso Ladeinde. *Applications of Complex Variables*. Berlin, Boston: De Gruyter, 2024.
https://doi.org/10.1515/9783111351179

Erratum

Despite careful production of our books, sometimes mistakes happen. We apologize that in the original version of this chapter, Figure 1.5 was unfortunately incorrect. This has been corrected.

The updated original chapter is available at DOI: https://doi.org/10.1515/9783111351179-002
https://doi.org/10.1515/9783111351179-021

www.ingramcontent.com/pod-product-compliance
Lightning Source LLC
Chambersburg PA
CBHW060941210326
41598CB00031B/4695
* 9 7 8 3 1 1 1 3 5 0 9 0 5 *